Achieving Proficiency in Mathematics

Achieving Proficiency in Mathematics

Revised

Mathew M. Mandery, Ed. D.
Adjunct Associate Professor
Hofstra University
Hempstead, NY

Executive Director, CARETECH
Brooklyn Tech Foundation
Brooklyn, NY

Marvin Schneider
Assistant Principal, Mathematics
Edward R. Murrow High School
New York City

AMSCO SCHOOL PUBLICATIONS, INC.
315 Hudson Street New York, N.Y. 10013

Please visit our Web site at:

www.amscopub.com

When ordering this book please specify: *either* **R 023 H** *or*
ACHIEVING PROFICIENCY IN MATHEMATICS, REVISED, HARDBOUND

ISBN 1-56765-537-8
NYC Item Number 56765-537-7

PRINTED IN THE UNITED STATES OF AMERICA

4 5 6 7 8 9 10 06 05 04

About this textbook

ACHIEVING PROFICIENCY IN MATHEMATICS reflects the following basic beliefs:

- Problem-solving approaches need to be made explicit, treating the concept as a philosophy and not merely as a topic.
- The calculator is a basic tool: eliminating drudgery, providing speed and accuracy, and encouraging discovery.
- A rudimentary understanding of statistical principles is essential for survival in our information-oriented society.

To implement the Standards of the National Council of Teachers of Mathematics and to assist students who have encountered difficulty in their study of mathematics, this AMSCO text stresses these axioms:

- Explanations of procedures and their underlying concepts must be mathematically sound, but presented in language that is understandable to the students.
- Sufficient time must be given to work with concepts, both in the classroom and at home.
- Applications should be presented in as realistic a framework as possible.
- Success with computational skills is founded on good quality practice.
- To achieve retention, students must be given the opportunity to use previously learned skills.

THE DESIGN OF THE TEXTBOOK

Theme, Module, Unit

The text is divided into 9 themes, each of which is subdivided into 2 or 3 modules. A module represents a broad topic within the theme, providing a convenient break for review and testing.

Within each module, a unit represents one (or more) day's instruction. Flexibility is deliberately built into this unit approach to allow for individual differences. The total number of units is 117, a comfortable number for a full school year (since many units will take more than one day and time for review and testing is required).

Calculator Strand

The first module of the introductory theme focuses on the basic mechanics of a 4-function calculator, including the memory and constant features.

The use of the calculator as a tool for exploration and implementation is then integrated with the work of the course. Specific uses for the calculator are highlighted in some illustrative examples under the heading "Keying In," and in some exercises that are flagged by a calculator symbol.

Problem-Solving Strand

The second module of the introductory theme treats problem solving as a skill in itself. Selected strategies and key vocabulary are presented, as well as exercises designed to sharpen decision-making skills.

After this initial presentation, a problem-solving approach is woven throughout the text. Attention is directed to the application of the strategies and the reinforcement of critical thinking by specifically treating appropriate illustrative examples under the heading "Thinking About the Problem."

Statistics Strand

The third module of the introductory theme introduces data reading as a basic skill. The initial focus on such statistical summaries as charts, tables, bar graphs, and line graphs is followed up by routinely presenting information in these modes.

Presentation of the circle graph is then integrated with later units on fractions and percents, and in a later theme, material on the traditional histogram is expanded to include such contemporary graphs as box-whisker plots and stem-leaf displays.

The Main Idea

At the outset of each unit, the essential concepts are explained in simple, but mathematically correct, language. Students should be encouraged to read "The Main Idea" aloud in class and again at home. Each Main Idea presents all the basic information necessary for the successful completion of that unit.

In some cases, to achieve maximum clarity and flexibility, the topic of a unit is separated into 2 or more subunits, each with its own Main Idea.

Illustrative Examples

Examples of every type covered in the unit are presented with detailed explanations and step-by-step solutions. These examples may be used as the basis for classroom illustration of the concepts. Students can refer to them as models while they work in class and at home.

Exercises

To provide a spiraled support system for the student, an abundance of exercise material is presented in 3 groups.

Class Exercises: These exercises immediately follow the illustrative examples and offer the first opportunity for students to work independently, still under the guidance of the teacher.

Homework Exercises: Armed with the detailed illustrative examples and the completed class exercises, students are equipped to handle the required homework exercises on their own.

Units that have 2 or more subunits, each with its own examples and class exercises, close with a composite set of homework exercises.

Spiral Review Exercises: These exercises provide a daily opportunity to stay in touch with previously presented material.

Included among the exercises are questions that aim to:

- build skills—help students acquire math power
- apply concepts—put mathematics in real-world contexts
- stimulate thought—require students to explain and justify

CONTENTS

INTRODUCTORY THEME Basic Tools

MODULE 2

THEME 2 Decimals

MODULE 3

MODULE 4

THEME 3 Rational Numbers

MODULE 5

MODULE 6

MODULE 7

THEME 4 Percents

MODULE 8

MODULE 9

THEME 5 Statistics and Measurement

MODULE 10

MODULE 16

THEME 8 Introduction to Geometry

MODULE 17

MODULE 18

MODULE 19

INTRODUCTORY THEME

Basic Tools

CALCULATOR SKILLS

PROBLEM-SOLVING SKILLS

READING DATA

OPERATIONS WITH WHOLE NUMBERS

CALCULATOR SKILLS

UNIT C-1 Beginning to Use Your Calculator

THE MAIN IDEA

1. There are many different types of calculators. Some are turned on by means of a switch. Others are turned on by pressing one of the following keys: [ON], [C], or [AC]. Still others are turned on simply by uncovering a light-sensitive panel.
2. A calculator has number keys and command keys. It also has a display that shows the numbers entered and the results of calculations. When a calculator is first turned on, the display should show [0.].
3. All calculators have one or more of the following keys that are used to clear (erase) the display and memory of numbers and/or operations previously entered:

   [C] (Clear) Use this key to clear both the display and any operations already entered. In some calculators, pressing this key once clears the last number entered and displays the result of the previous operation. Pressing the key twice clears all numbers and operations.

   [CE] (Clear Entry) Pressing this key allows a user to delete (take out) an incorrect entry, provided that a command key has not yet been pressed.

   [AC] (All Clear) Use this key to clear all numbers, operations, and memory.
4. Some calculators have combination keys for some of the functions listed above:

   [CE/C] [ON/C] [C/AC]

   Usually, the key is pressed once for the function on the left and twice for the function on the right.
5. To do a simple computation on a calculator:
   a. Clear both the memory and display.
   b. Use the number keys to enter the first number, one digit at a time from left to right.
   c. Press the command key for the operation to be performed.
   d. Enter the second number, and press the [=] key.

EXAMPLE 1 Enter the number 258 and then clear the display.

Key Sequence:	C	2	5	8	CE
Display:	0.	2.	25.	258.	0.

Observe that the display always shows a decimal point to the right of a whole number.

From now on, key sequences in this book will show numbers written in the usual way, not one key at a time for each digit as in Example 1. It is understood that numbers will be entered one digit at a time from left to right.

EXAMPLE 2 Write the calculator display for each step of the following key sequence:

Key Sequence:	C	15	+	23	=
Display:	0.	15.	15.	23.	38.

It is not until = is pressed that the result of the calculation, 38, is displayed.

EXAMPLE 3 Write the calculator display for each step of the following key sequences. Explain why the final displays are different.

a. Key Sequence:	C	5	+	9	CE	7	=	
Display:	0.	5.	5.	9.	0.	7.	12.	
b. Key Sequence:	C	5	+	9	C	7	=	
Display:	0.	5.	5.	9.	0.	7.	7.	

Answer: In the first sequence, CE cleared only the 9 but kept the 5 and the +. When 7 was entered and = pressed, the calculation 5 + 7 was completed. In the second sequence, when C was pressed the operation 5 + 9 was cleared. So, the final display showed only the 7.

EXAMPLE 4 Write the calculator display for each step of the following key sequences:

a. Key Sequence:	C	7	+	8	÷	3	=	
Display:	0.	7.	7.	8.	15.	3.	5.	
b. Key Sequence:	C	16	+	4	−	2	=	
Display:	0.	16.	16.	4.	20.	2.	18.	
c. Key Sequence:	C	123	−	45	CE	−	54	=
Display:	0.	123.	123.	45.	0.	123.	54.	69.
d. Key Sequence:	C	35	×	10	=	C	8	=
Display:	0.	35.	35.	10.	350.	0.	8.	8.

Observe that, when you enter a series of operations, each time an operation key is pressed, the calculator displays the result of the previous operation.

EXAMPLE 5 **a.** Write a key sequence that can be used to calculate $57 \times 9 \div 19$.
b. Write the final display.

a. Key Sequence:	C	57	×	9	÷	19	=
Display:	0.	57.	57.	9.	513.	19.	27.

b. Final Display: 27. .

CLASS EXERCISES

1. Write the calculator display for each step of the following key sequences:

   a. [C] 396 b. [C] 257 [C] c. [C] 37 [CE] 45 d. [C] 5 [+] 8
   e. [C] 5 [+] 8 [=] f. [C] 9 [×] 4 [+] 3 [=] g. [C] 8 [÷] 1 [=] h. [C] 6 [÷] 2 [−] 1

2. Write a key sequence that can be used to do each calculation. Then write the final display.

   a. $12 - 7$ b. 32×15 c. $50 \div 25$ d. $92 + 38$ e. $7 + 11 + 16$ f. $50 \div 2 \times 3$
   g. $9 + 8 - 6$ h. $11 \times 5 \times 7$

3. Without using the calculator, tell what would be the final display for each of the following key sequences.

   a. [C] 2 [C] b. [C] 358 c. [C] 84 [+] 16
   d. [C] 5 [−] 2 [=] e. [C] 8 [÷] 4 [C] 2 f. [C] 2 [+] 3 [+]

HOMEWORK EXERCISES

1. Write the calculator display for each step of the following key sequences:

   a. [C] 583 b. [C] 482 c. [C] 79 [C]
   d. [C] 4 [×] 3 e. [C] 9 [+] 5 [=] f. [C] 9 [+] 3 [CE] 2 [=]
   g. [C] 12 [÷] 4 [×] 5 h. [C] 3 [×] 5 [C] 4 [×] 2 [=] i. [C] 4 [+] 7 [+] 9 [=]
   j. [C] 3 [×] 4 [×] 2 k. [C] 15 [−] 9 [−] 3 [=] l. [C] 9 [CE] 6 [×] 7 [=]
   m. [C] 8 [×] 3 [CE] 4 [=] n. [C] 8 [×] 3 [C] 4 [=]

2. Write a key sequence that can be used to do each calculation. Then write the final display.

   a. 9×2 b. $6 - 1$ c. $18 + 7$ d. $8 \div 4$ e. $3 \times 2 \times 4$ f. $15 + 9 + 8$
   g. $25 - 18 + 2$ h. $12 \div 2 \times 3$

3. Without using the calculator, tell what would be the final display for each of the following key sequences.

   a. [C] 2 [+] 1 [C] b. [C] 123 [C] 456 c. [C] 2 [×] 5 [=]
   d. [C] 7 [−] 2 [−] e. [C] 2 [C] 4 [×] 3 [=] f. [C] 10 [×] 2 [−] 5 [=]
   g. [C] 12 [÷] 4 [=] h. [C] 3 [+] 5 [=] [C] i. [C] 3 [×] 3 [×]

UNIT C-2 Using the Memory Keys

THE MAIN IDEA

1. The memory of a calculator is like a separate note pad on which a number is jotted down while other calculations are performed.
2. The letter M on the display of a calculator indicates that a number other than zero is stored in memory.
3. The keys that control the memory are:
   [M+] Use this key to add the number on the display to the number in memory.
   [M−] Use this key to subtract the display number from the number in memory.
   [MRC] (Memory Recall/Memory Clear) Press this key once to display the number that is in memory. Press it twice to clear the number that is in memory, but not on the display.
   [AC] (All Clear) Use this key to clear the memory, the display, and any operation that has been entered. If a calculator does not have this key, use the following key sequence: [MRC] [MRC] [C]
4. Press [AC] before starting a calculation so that the memory, as well as the display, is cleared of any number that may remain from a previous problem.

EXAMPLE 1 Write and explain the display for each step of the following key sequence:
[AC] 2 [+] 7 [M+] [C] [MR]

Key Sequence	*Display*	*Explanation*
[AC]	0.	Clears the memory, display, and current operations.
2	2.	Enters the number 2.
[+]	2.	Enters the operation +.
7	7.	Enters the number 7.
[M+]	M 9.	The sum of 2 and 7 is stored in memory.
[C]	M 0.	Clears the display. Keeps the number stored in memory.
[MR]	M 9.	Displays the number stored in memory.

EXAMPLE 2 Write and explain the display for each step of the following key sequence:
[AC] 13 [×] 24 [M+] [AC] [MR]

Key Sequence	*Display*	*Explanation*
[AC]	0.	Clears the memory and display.
13	13.	Enters the number 13.
[×]	13.	Enters the operation ×.
24	24.	Enters the number 24.
[M+]	M 312.	Stores the result of 13 × 24 in memory.
[AC]	0.	Clears the memory, display, and current operations.
[MR]	0.	Shows that the memory was cleared by pressing [AC].

In place of [AC], memory can be cleared by pressing [MRC] twice, followed by [C] to clear the display. On some calculators, [MR] (Memory Recall) and [MC] are separate keys.

From now on, key sequences and displays will be simplified as follows:

- The clear key at the *beginning* of each computation will not be shown. However, it is still important to clear both the memory and display before each calculation.
- Often, the display will be shown at the end of a series of steps, not after every key is pressed.

EXAMPLE 3 For the expressions 8 × 5 and 12 – 2, write a key sequence to find:

a. the sum **b.** the difference **c.** the product **d.** the quotient

	Key Sequence	*Display*		*Key Sequence*	*Display*
a.	8 [×] 5 [M+]	0.	**b.**	8 [×] 5 [M+]	M 40.
	12 [–] 2 [M+]	M 10.		12 [–] 2 [M–]	M 10.
	[MR]	M 50.		[MR]	M 30.
c.	8 [×] 5 [M+]	M 40.	**d.**	12 [–] 2 [M+]	M 10.
	12 [–] 2 [=]	M 10.		8 [×] 5 [=]	M 40.
	[×] [MR] [=]	M 400.		[÷] [MR] [=]	M 4.

CLASS EXERCISES

1. Write and explain the display for each step of the key sequence.
   a. 6 [M+] [C] [MR]
   b. 9 [M+] 5 [M−] [C] [MR]
   c. 10 [÷] 2 [M+] [MR]
   d. 5 [×] 3 [M+] 8 [M−] [C] [MR]
   e. 9 [×] 2 [M+] 3 [×] 5 [M+] [MR]
   f. 6 [×] 7 [M+] 4 [+] 5 [M−] [C] [MR]
   g. 9 [+] 8 [M+] 12 [÷] 3 [M−] [MR]

2. For the expressions 6 × 4 and 12 ÷ 2, write a key sequence to find:
   a. the sum **b.** the difference **c.** the product **d.** the quotient

3. Write a key sequence that can be used to do the calculation and then write the result.
   a. Add 7 × 8 to 5 × 6.
   b. Subtract 5 + 4 from 18 − 2.
   c. Add 8 × 7 to 5 × 3.
   d. Subtract 4 × 7 from 132.
   e. Multiply 8 + 3 by 5 + 2.
   f. Divide 8 × 10 by 5 × 4.
   g. Subtract 12 ÷ 4 from 9 × 2.
   h. Divide 9 × 7 by 4 + 3.

HOMEWORK EXERCISES

1. Write and explain the display for each step of the key sequence.
   a. 9 [M+] [C] [MR]
   b. 5 [M+] [CE] [MR]
   c. 8 [M+] 3 [M−] [C] [MR]
   d. 9 [M+] 6 [M+] [MR]
   e. 15 [×] 3 [M+] 15 [M−] [MR]
   f. 4 [×] 5 [M+] 3 [×] 3 [M+] [C] [MR]
   g. 90 [÷] 6 [M+] 4 [+] 8 [M−] [MR]

In 2–5, for the given expressions, write a key sequence to find:
a. the sum **b.** the difference **c.** the product **d.** the quotient

2. 40 − 10 and 14 + 1
3. 18 × 2 and 12 − 3
4. 20 + 18 + 6 and 12 − 1
5. 12 × 5 and 8 + 3 − 1

6. Write a key sequence that can be used to do each calculation and write the result.
   a. Subtract 9 × 3 from 8 × 8.
   b. Multiply 4 + 7 by 8 × 2.
   c. Add 15 ÷ 3 to 16 × 2.
   d. Add 4 × 7 to 9 × 6.
   e. Subtract 3 × 5 from 9 + 8.
   f. Divide 6 × 8 by 9 − 6.
   g. Divide 9 + 6 by 3 × 5.
   h. Add 12 ÷ 3 to 27 − 9.
   i. Subtract 12 + 8 from 60 ÷ 3.
   j. Multiply 9 ÷ 3 by 5 − 5.

UNIT C-3 Using the Constant Features

THE MAIN IDEA

1. To solve some math problems, it may be necessary to repeat an operation several times using the same number. This number is called a ***constant***, because it does not change within that problem.
2. Special features for working with constants, called ***constant features***, are available on many calculators.
   - Use the ***constant feature for addition*** to add a constant to each of several numbers.
   - Use the ***constant feature for subtraction*** to subtract a constant from each of several numbers.
   - Use the ***constant feature for division*** to divide each of several numbers by a constant.
   - Use the ***constant feature for multiplication*** to multiply each of several numbers by a constant.
3. The constant features for addition, subtraction, and division, use the *second* number entered as the constant. The constant feature for multiplication is different; it uses the *first* number entered as the constant.
4. If you enter an operation and then press the [=] key repeatedly, the calculator continues to perform that same operation with the constant and each new result. That is, the calculator treats the previous result as the new number.

EXAMPLE 1 Write a key sequence that uses a constant feature to do each set of calculations.

a. Add 9 to each of the numbers 5, 8, and 7.

Key Sequence	*Display*	
5 [+] 9 [=]	14.	The constant is ⑨.
8 [=]	17.	← 8 + ⑨
7 [=]	16.	← 7 + ⑨

Answer: 5 [+] 9 [=] 8 [=] 7 [=]

b. Subtract 2 from each of the numbers 9, 6, and 7.

Key Sequence	*Display*	
9 [−] 2 [=]	7.	The constant is ②.
6 [=]	4.	← 6 − ②
7 [=]	5.	← 7 − ②

Answer: 9 [−] 2 [=] 6 [=] 7 [=]

c. Divide each of the numbers 9, 6, and 3 by 3.

Key Sequence	*Display*	
9 [÷] 3 [=]	3.	The constant is ③ .
6 [=]	2.	← 6 ÷ ③
3 [=]	1.	← 3 ÷ ③

Answer: 9 [÷] 3 [=] 6 [=] 3 [=]

d. Multiply each of the numbers 3, 5, and 2 by 4.

Key Sequence	*Display*	
4 [×] 3 [=]	12.	The constant is ④ .
5 [=]	20.	← ④ × 5
2 [=]	8.	← ④ × 2

Answer: 4 [×] 3 [=] 5 [=] 2 [=]

EXAMPLE 2 Use a constant feature to complete the table. The second number is 3 times the first number.

First Number	4	8	14
Second Number			

Key Sequence	*Display*
3 [×] 4 [=]	12.
8 [=]	24.
14 [=]	42.

Answer:

First Number	4	8	14
Second Number	12	24	42

EXAMPLE 3 Each number in a list is to be 6 less than the previous number. The first number is 100. Use a constant feature to complete the list.

100, ____, ____, ____, ____,

Key Sequence	*Display*
100 [−] 6 [=]	94.
[=]	88.
[=]	82.
[=]	76.

Answer: 100, 94 , 88 , 82 , 76

CLASS EXERCISES

1. Write a key sequence that uses a constant feature to:

a. Add 7 to each of the numbers 11, 15, and 22.

b. Subtract 9 from each of the numbers 17, 23, 34, and 50.

c. Multiply each of the numbers 5, 9, and 11 by 6.

d. Divide each of the numbers 25, 30, and 45 by 5.

e. Add 15 to each of the numbers 8, 12, and 20.

f. Multiply 9, 12 and 7 by 4.

2. Use a constant feature to complete each table.

a. The second number is 12 more than the first number.

First Number	3	6	9	12	30
Second Number					

b. The second number is the first number divided by 5.

First Number	25	35	45	85	95
Second Number					

c. Each price is reduced by $7.

Original Price	$20	$29	$46	$58	$99
New Price					

d. The number of inches is 12 times the number of feet.

Number of Feet	2	3	5	7	9
Number of Inches					

3. Use a constant feature to complete each list of numbers.

a. Each number is the previous number divided by 2.

144, _____, _____, _____, _____

b. Each number is 11 more than the previous number.

89, _____, _____, _____, _____

c. Each number is 4 times the previous number.

7, _____, _____, _____, _____

d. Each number is 20 less than the previous number.

208, _____, _____, _____, _____

HOMEWORK EXERCISES

1. Write a key sequence that uses a constant feature to:

a. Subtract 9 from each of the numbers 12, 19, and 32.

b. Multiply each of the numbers 4, 6, and 9 by 7.

c. Add 37 to each of the numbers 7, 19, 22, and 36.

d. Multiply 15, 19, and 42 by 6.

e. Divide each of the numbers 34, 68, 51, and 136 by 17.

f. Subtract 19 from each of the numbers 22, 47, and 68.

g. Add 56 to each of the numbers 47, 91, and 86.

h. Divide each of the numbers 36, 144, and 72 by 12.

2. Use a constant feature to complete each table.

a. The second number is 15 less than the first number.

First Number	37	59	77	115	211
Second Number					

b. The second number is the first number divided by 11.

First Number	77	99	187	242	308
Second Number					

c. Each salary is increased by $19.

Old Salary	$109	$152	$175	$229
New Salary				

d. To find the number of quarts, multiply the number of gallons by 4.

Number of Gallons	21	37	49	86
Number of Quarts				

3. Use a constant feature to complete each list of numbers.

a. Each number is 17 more than the previous number.

58, _____, _____, _____, _____

b. Each number is the previous number divided by 3.

729, _____, _____, _____, _____

c. Each number is 29 less than the previous number.

317, _____, _____, _____, _____

d. Each number is 5 times the previous number.

12, _____, _____, _____, _____

TEAMWORK

Jerry and Martha each invested $1,000. Jerry's investment earned $50 every month. Martha's investment multiplied the total amount by 1.02 every month. Use the constant features of your calculator to investigate the growth of these two investments. Whose investment earns more at the beginning? Does this pattern ever change? Who will reach $2,000 first? Who will reach $3,000 first? Who will be a millionaire first? Write a team report of your findings.

PROBLEM-SOLVING SKILLS

UNIT P-1 Reading Problems for Information

THE MAIN IDEA

1. Every problem *gives* you some information and asks you to *find* other information. The information that you have to find is called a ***solution*** or an ***answer***.
2. Often, the given information:
   - follows the word "if."
   - is described as something that someone *has* or *had*.
   - is mentioned as a fact at the beginning of a problem.
   - is given as numbers in a chart (table) or diagram.
   - can be found by counting.
3. Sometimes, some information is not stated, but you are expected to know it. For example, you should know that there are:
   - 12 months in a year.
   - 3 sides to a triangle.
   - 3 feet in a yard.
   - 4 aces in a standard deck of playing cards.
   - 12 items in a dozen.
   - 16 ounces in a pound.
   - 25 cents in a quarter.
4. Often, the information that you have to find:
   - follows the word "find."
   - follows words like "how many . . . ?" "how much . . . ?" "what is . . . ?"
5. Some information that is given may not be needed to find the solution. This is extra information that you ignore.

EXAMPLE 1 For each problem, state the given information and the information that you are asked to find.

	Problem	***Given Information***	***Information to Be Found***
a.	Mr. Harris earns $12.75 per hour. If he worked for 8 hours, how much did he earn?	The hourly rate is $12.75 (given as a fact at the beginning). The number of hours worked is 8 (following the word "if").	*How much* money was earned?

	Problem	Given Information	Information to Be Found
b.	Susan had a piece of ribbon 83 inches long. She wanted to cut it into 5-inch pieces. How many 5-inch pieces could she make?	There are 83 inches of ribbon. Each piece is to be 5 inches long. (Both are given as facts at the beginning.)	*How many* pieces can be made?
c.	What is the average height shown in the table? (see table below)	The respective heights are 60 in., 58 in., 62 in., 60 in., and 55 in. (given as numbers in the chart). There are 5 children. (Count names in the chart.)	*What is* the average height?
d.	Find the area of the rectangular room shown in the diagram. (Diagram: rectangle labeled 9 ft. and 20 ft.)	The dimensions are 9 ft. and 20 ft. (given as numbers in the diagram).	*Find* the area.

Name	Height
Joyce	60 in.
Barbie	58 in.
Mel	62 in.
Scott	60 in.
Jack	55 in.

9 ft.

20 ft.

EXAMPLE 2 For each problem, write the information that you are expected to know but that is not stated in the problem.

	Problem	Unstated Information
a.	Oranges cost $3.40 a dozen. What is the cost of one orange?	There are 12 oranges in a dozen.
b.	Mr. Ross earns $125 a day. He does not work on weekends. How much does he earn in a week?	There are 5 working days in a week, not counting the weekend.
c.	What is the total amount that Jenny can save in her Holiday Club at the bank in one year if she saves $5 a week?	There are 52 weeks in a year.
d.	Each side of a square measures 12.5 cm. What is the total distance around the square?	The square has 4 sides of equal length.

EXAMPLE 3 For each problem, state which information is extra, if any.

	Problem	***Extra Information***
a.	For the past 5 years, Mrs. Harris has been earning $37,000 a year. How much does she earn in 3 years?	"5 years" is extra. You only need to know that she earned $37,000 a year.
b.	Find the sales tax that Mrs. Santos paid on her 1995 used car if the ticket price of the car was $9,750 and the sales tax rate was 8%.	"1995" is extra. You only need to know the price and the sales tax rate.
c.	Find the distance around the rectangle.	"13 cm" is extra. You only need to know the lengths of the sides of the rectangle.

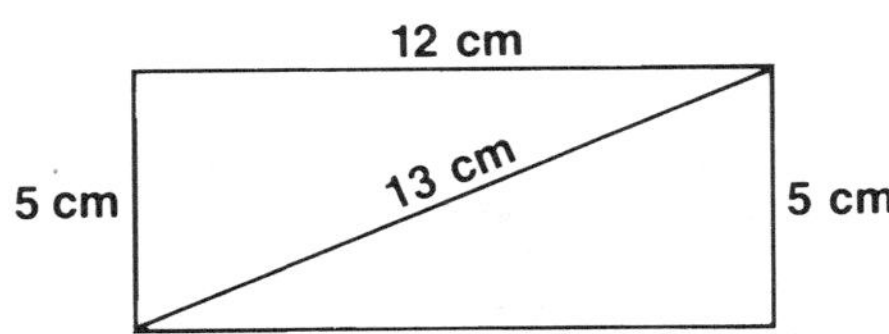

	Problem	***Extra Information***
d.	What is the total number of newspapers that Susan sold this week?	All the numbers except those in Susan's row are extra.

Number of Newspapers Sold Each Day

Salesperson	M	T	W	Th	F	S
Sally	35	37	34	30	29	39
James	16	27	20	18	22	25
Susan	23	17	29	35	18	19
Hank	14	18	20	13	17	22
Bill	20	15	17	19	21	18

	Problem	***Extra Information***
e.	Mr. Wilson made purchases of $12.50, $7.85, and $27.95. If the rate of sales tax was 5%, how much sales tax did he have to pay?	There is no extra information.

CLASS EXERCISES

For problems 1–8, state the given information and the information that you have to find.

1. If the area of a rectangle is 24 sq. yd. and its width measures 4 yd., what is the measure of its length?
2. Find the average of 11, 17, 18, 9, and 10.
3. The gas tank of a car had 15 gallons of gas in it. The car used one gallon of gas every 16 miles. If the car was driven 60 miles, how much gas was left?
4. What is 17% of 300?
5. Find $\frac{3}{4}$ of 128.
6. Find the largest value: 51%, $\frac{1}{2}$, 0.49
7. Find the total rainfall in Centerville for the five months shown in the table.

Rainfall in Centerville

Month	Rainfall
January	$1\frac{1}{4}$ in.
February	$\frac{1}{2}$ in.
March	$\frac{3}{4}$ in.
April	$2\frac{1}{2}$ in.
May	3 in.

8. Find the area of the triangular plot shown.

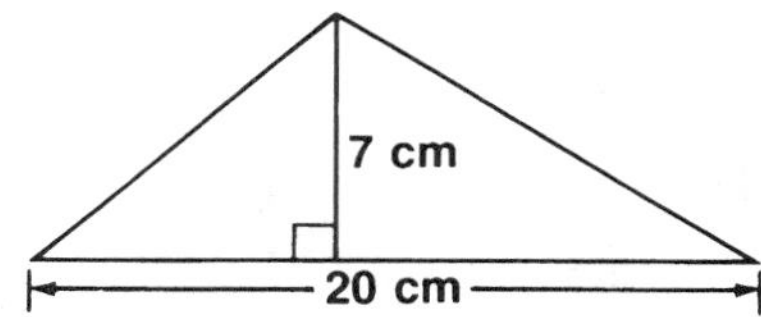

For problems 9–13, write the information that is not stated in the problem but that you are expected to know.

9. If a dozen pens cost \$2.40, what is the cost of one pen?
10. Mr. Dale earns \$3420 a month. How much does he earn in a year?
11. Each side of a triangle is 9 inches long. What is the total distance around the triangle?
12. What fractional part of the number of months of the year has names that begin with the letter "J"?
13. If an oil truck can pump 35 gallons of oil in a minute, how many gallons can it pump in a half-hour?

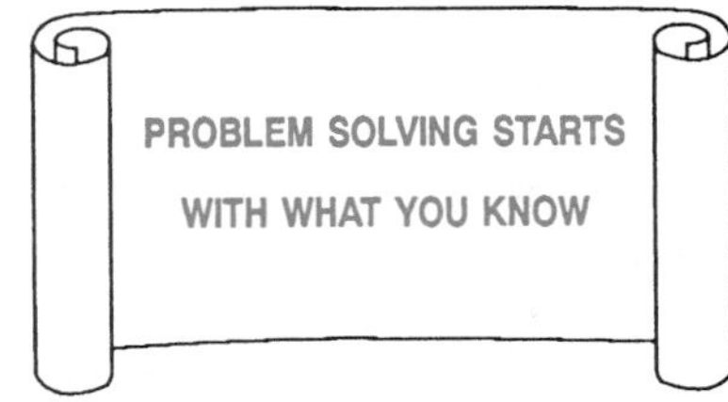

For problems 14–19, state which information is extra, if any.

14. A class of 28 students raised charity money by selling a total of 159 boxes of candy at $1.50 per box. How much money did they raise?

15. Jill bought 12 cans of soup that cost $1.19 per can and 8 pounds of tomatoes that cost 99¢ per pound. What was the total cost of her purchases?

16. Three children bought a 24-ounce bag of candy for $3.79. If they shared the candy equally, how many ounces did each child get?

17. Mrs. Farr drove 350 miles and used 30 gallons of gasoline. If gasoline costs $1.19 a gallon, how much did she spend on gasoline?

18. What is the average weight of these children?

Medical Information for Class 4–1

Name	Height	Weight	Age
Stan	62 in.	125 lb.	10
Phillip	56 in.	87 lb.	11
Marcia	59 in.	91 lb.	10
Nancy	60 in.	100 lb.	10
Barbara	58 in.	94 lb.	11

19. What is the total height of this house?

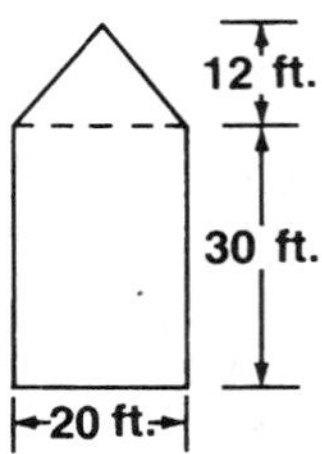

HOMEWORK EXERCISES

For problems 1–16, state the given information and the information that you have to find.

1. If the perimeter of a square is 48 cm, what is the measure of each side?

2. Find the sum of 28 and 57.

3. What is the product of 9 and 6?

4. Find the total amount that John earned if he was paid $8.40 an hour and he worked for 5 hours.

5. Robin bought 3 blouses that cost $19.50 each. How much did she spend?

6. Find the smallest value: $\frac{7}{8}$, 80%, 0.79

7. What is 25% of 400?

8. In Northern State High School, there are 29 classes with 30 students in each class. What is the total number of students in these classes?

9. William bought 3 pounds of potatoes at 52 cents per pound and 2 quarts of milk at $1.29 a quart. How much did he spend?

10. In a class, there are 10 boys and 20 girls. What fractional part of the total number of students in the class is the number of girls?

11. Last year, Dr. Jones paid $230,400 as rent for her office. What was the monthly rent?

12. Ms. Jacobs saves $\frac{7}{50}$ of her salary. What percent of her salary does she save?

13. Find the total number of newspapers that Jane delivered for the 5 days shown in the table.

Jane's Newspaper Record

Day	Newspapers Delivered
Monday	105
Tuesday	111
Wednesday	108
Thursday	115
Friday	102

14. Find the average number of points scored by the opponents of Westwood High School basketball team.

Westwood High School Basketball Team Record

Opponent	Points Scored
Clark H.S.	58
Franklin H.S.	72
Jackson H.S.	48
Randolph H.S.	52

15. Find the area of the right triangle shown.

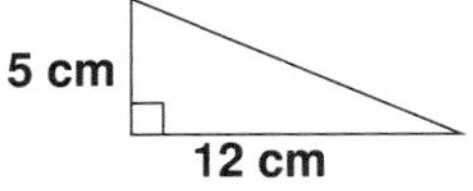

16. Find the amount of fencing needed to enclose the rectangular yard shown.

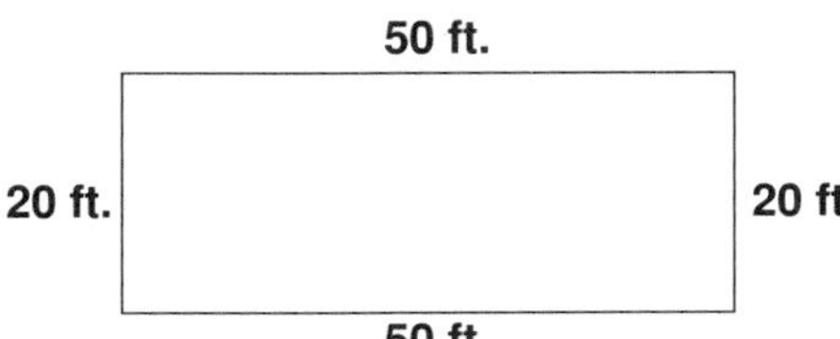

For problems 17–24, write the information that is not stated in the problem but that you are expected to know.

17. Sandra spends an average of $7.50 a day. How much does she spend in a week?

18. What fractional part of a pound is 10 ounces?

19. How many quart bottles can be filled from a 10-gallon can of milk?

20. Janet earns $6.50 per hour. How much did she earn if she worked 450 minutes?

21. At Kim's Market, the selling price for oranges is 3 for $1.08. What is the cost of one dozen oranges?

22. What is the total distance around a square whose sides measure 10 cm?

23. A machine can produce a toy every 3 minutes. How many toys can it produce in 5 hours?

24. A man walked 3 miles. How many yards did he walk?

For problems 25–30, state which information is extra, if any.

25. John bought 5 shirts that cost $21 each and 3 pairs of pants that cost $29 each. How much did he spend on the shirts?

26. Mr. Jackson paid 90 cents for a 3-mile bus ride to the train station. He then paid $4.50 for a 24-mile train ride to his job. What was the total cost of his trip?

27. James spent $4.99 for a 2-pound bag of peanuts and $3.45 for a gallon bottle of juice. How much change did he receive from a $10 bill?

28. Mr. Smith's car averages 25 miles per gallon of gasoline. How many gallons did he use to make a 250-mile trip?

29. How far is it from New Place to Saint Albane?

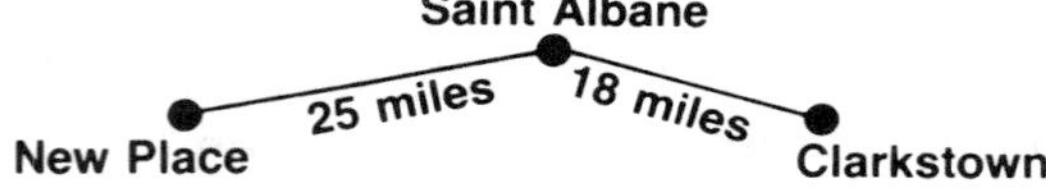

30. How much did Ms. Williams spend on food?

Ms. Williams' Expense Account

Date	Hotel	Food	Telephone
Monday 5/1	$70	$35	$8.50
Tuesday 5/2	$68	$42	$12.75
Wednesday 5/3	$68	$50	$23.45
Thursday 5/4	$80	$60	$9.80

UNIT P-2 Deciding Which Operation to Use

THE MAIN IDEA

1. The four arithmetic operations are ***addition*** (+), ***subtraction*** (–), ***multiplication*** (×), and ***division*** (÷).
2. Some problems contain key words that show which operation to use. For example:

Addition	***Subtraction***	***Multiplication***	***Division***
plus	minus	times	divide
added to	subtracted from	multiplied by	divided by
sum	difference	product	divided into
and	deduct	of (a fraction *of* a number; a percent *of* a number)	quotient
	remains		each
			per

3. Use certain operations in certain situations. For example:

   Use addition to find:
   - a total amount when each part is given.
   - a result when a number is increased.
   - an amount after tax is included.

   Use subtraction to find:
   - how much greater or less one number is than another.
   - a result when a number is decreased.
   - by how much a number has been increased.
   - an amount after a discount is made.
   - how much is left, or remains, after an amount is spent or used up.

   Use multiplication to find:
   - a total amount when given several equal amounts.
   - percents.

   Use division to find:
   - an amount for each item when given the total amount and the number of equal items.
   - the percent that is the equivalent of a given fraction.
   - a probability.
   - an average or mean.

EXAMPLE 1 For each problem, identify the key word(s) that help you to decide which operation to use.

	Problem	*Key Words*
a.	If a dozen eggs cost $1.32, what is the cost of each egg?	cost of each
b.	After a 20% discount, what is the final cost of a coat that is priced at $85?	discount, final cost
c.	In a class of 32 children, each child donated $1.50 to a toy drive. Find the total amount of money that was collected.	total amount

EXAMPLE 2 For each problem, tell which operation you would use to find the answer and write a number phrase to show this operation.

	Problem	*Operation*	*Number Phrase*
a.	Find the product of 85 and 153.	Multiplication, because of the word "product."	85×153
b.	If Mr. Harris earns $42,000 a year, how much does he earn each month?	Division, because a year is made up of 12 months with equal pay in each month.	$42,000 ÷ 12
c.	Find the total length of these line segments: 4 cm ______ 3.8 cm ______	Addition, because you are finding the total of the 2 parts.	4 cm + 3.8 cm
d.	In one day, the temperature outside Jim's room went from 75° to 98°. By how many degrees did the temperature increase?	Subtraction, because you are finding how much more one amount is than another.	$98° - 75°$

EXAMPLE 3 Which number phrase fits the following problem? Mrs. McGill's car gets 14.5 miles per gallon of gasoline. If her car used 7 gallons of gasoline, how far did she travel?

(a) $14.5 + 7$ (b) $14.5 - 7$ (c) 14.5×7 (d) $14.5 \div 7$

Since "per" means "for each," you have to find the total number of miles traveled when you are given the number of miles traveled for each gallon. Since the number of miles for each gallon is always the same, use multiplication.

Answer: (c)

CLASS EXERCISES

For problems 1–5, identify the key word(s) that help you to decide which operation to use.

1. A 50-foot length of wire is to be divided into 5-foot pieces. How many pieces can be cut?
2. David is 5 ft. 5 in. tall and Murray is 5 ft. 3 in. tall. How much taller is David than Murray?
3. Susan read 120 pages in 5 hours. How many pages did she read per hour?
4. The oil in 15 barrels is to be emptied into a storage tank. If each barrel contains 55 gallons of oil, how many gallons must the storage tank be able to hold?
5. How much change from a $20 bill will Joseph get if he buys a sweater that costs $17.98?

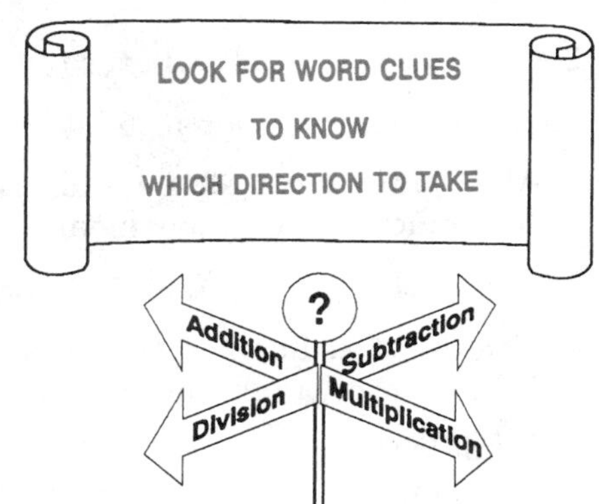

For problems 6–13, tell which operation you would use to find the answer and write a number phrase to show this operation.

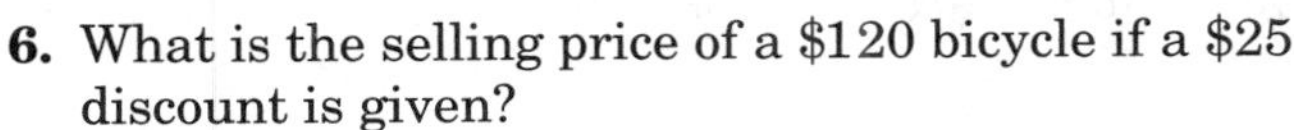

6. What is the selling price of a $120 bicycle if a $25 discount is given?

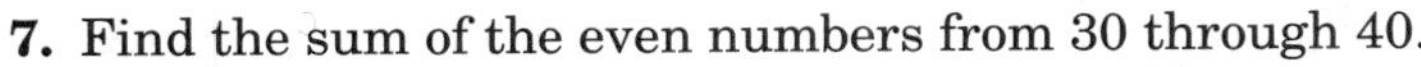

7. Find the sum of the even numbers from 30 through 40.

8. If each egg weighs 2.3 oz., find the weight of a dozen eggs.

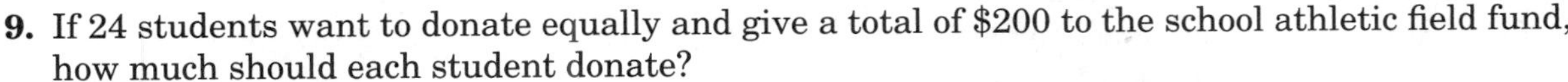

9. If 24 students want to donate equally and give a total of $200 to the school athletic field fund, how much should each student donate?
10. Mrs. Jameson used $3\frac{1}{2}$ oz. of butter from a full 1-pound container. How much was left?

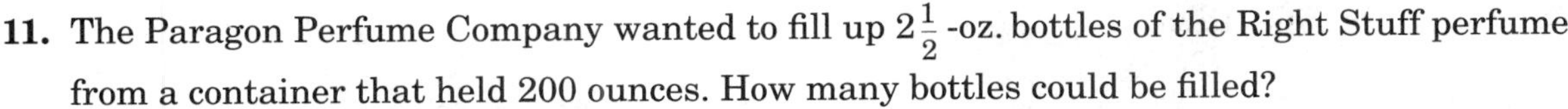

11. The Paragon Perfume Company wanted to fill up $2\frac{1}{2}$-oz. bottles of the Right Stuff perfume from a container that held 200 ounces. How many bottles could be filled?
12. The book collection in a library that owned 2,175 books was increased by 450 books. How many books did the library then have?
13. If an apartment requires 40 square yards of carpeting, how much carpeting do 25 identical apartments require?

For problems 14 and 15, tell which number phrase fits the problem.

14. Robert lost an average of $2\frac{1}{2}$ pounds per week when he was on a diet. He lost a total of 30 pounds. How many weeks did it take?

(a) $30 + 2\frac{1}{2}$ (b) $30 - 2\frac{1}{2}$ (c) $30 \times 2\frac{1}{2}$ (d) $30 \div 2\frac{1}{2}$

15. 240 students voted in the school election. If there were 410 students in the school, how many students did not vote?

(a) $410 + 240$ (b) $410 - 240$ (c) 410×240 (d) $410 \div 240$

HOMEWORK EXERCISES

For problems 1–8, identify the key word(s) that help you to decide which operation to use.

1. If 5 pens cost $1.60, what is the cost of each pen?

2. A box of candy weighs 1.5 kilograms. What is the total weight of 20 boxes?

3. Ms. Diamond bought a car for $10,800. If she paid for the car in 24 equal monthly payments, how much was each payment?

4. Mr. Warren borrowed $2,700 and has repaid $1,900 of his loan. How much does he still owe?

5. The Fast Car Company manufactured 5,348 two-door cars and 7,589 four-door cars last month. What is the total number of cars that it produced?

6. What is the total amount of money that Ms. James spent on her business trip? See chart at right.

Ms. James' Expense Account

Item	Cost
Meals	$48.50
Hotel	$75.80
Travel	$96.00

7. What is the total cost of 12 gallons of gasoline at $1.25 per gallon?

8. Mr. Arnold bought his house for $85,000 and sold it 10 years later for $123,000. What is the difference in the prices?

For problems 9–18, tell which operation you would use to find the answer and write a number phrase to show this operation.

9. Find the sum of 58 and 29.

10. Find the difference between 100 and 75.

11. Find the sum of the lengths of the sides of the triangle shown at right.

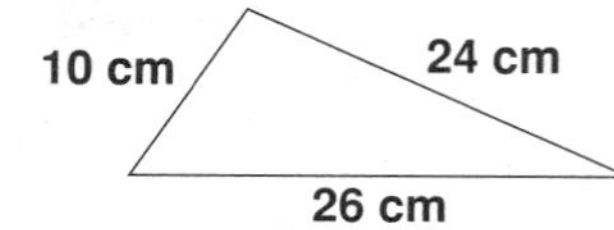

12. If the measure of the length of a rectangle is 9 inches and the measure of the width of the rectangle is 6 inches, how much longer is the length than the width?

13. A cake recipe calls for 2 cups of flour. How much flour should be used to make 8 such cakes?

14. Geysers are columns of boiling water that erupt from within the earth at fairly regular intervals. Old Faithful Geyser in Yellowstone National Park sends forth a 130-foot column about once every hour. Steamboat Geyser has been seen to erupt into a column as high as 380 feet. How much higher is the Steamboat Geyser than Old Faithful?

15. A man paid \$4 for a 20-pound bag of potatoes. How much did one pound of potatoes cost him?

16. Before he dieted, John weighed 189 pounds. If his present weight is 165 pounds, how many pounds did he lose?

17. Wilma ran 3.8 miles on Monday and 4.6 miles on Tuesday. Find the total number of miles that she ran.

18. Mr. Adams' son bought 3 compact discs that cost \$12.99 each. Find the total cost of the compact discs.

For problems 19–26, tell which number phrase fits the problem.

19. A crate of oranges contains 50 oranges. How many oranges are there in 25 crates?
(a) $50 + 25$ (b) $50 \div 25$ (c) 50×25 (d) $50 - 25$

20. A \$1,000 prize was shared equally by 5 people. How much was each person's share?
(a) $\$1{,}000 \times 5$ (b) $\$1{,}000 \div 5$ (c) $\$1{,}000 - 5$ (d) $\$1{,}000 + 5$

21. The population of Smithville was 3,485 in 1979. If the present population is 4,240, find the amount of increase.
(a) $3{,}485 + 1979$ (b) $3{,}485 - 1979$ (c) $4{,}240 + 3{,}485$ (d) $4{,}240 - 3{,}485$

22. At the beginning of a trip, Mr. Wilson's odometer (the mileage indicator) read 28,940 miles. If he traveled 289 miles, what did the odometer read at the end of the trip?
(a) $28{,}940 - 289$ (b) $28{,}940 + 289$ (c) $28{,}940 \times 289$ (d) $28{,}940 \div 289$

23. 384 cars are parked in rows containing 16 cars each. How many rows are there?
(a) $384 + 16$ (b) $384 - 16$ (c) 384×16 (d) $384 \div 16$

24. In Holtstown, 1,200 people voted in the election for mayor. If $\frac{2}{3}$ of the voters were women, how many women voted?
(a) $1{,}200 \div \frac{2}{3}$ (b) $1{,}200 \times \frac{2}{3}$ (c) $1{,}200 + \frac{2}{3}$ (d) $1{,}200 - \frac{2}{3}$

25. The Fantastic Boutique advertised a \$90 jacket at a \$15 discount. What is the sale price of the jacket?
(a) $\$90 + \$15$ (b) $\$90 - \$15$ (c) $\$90 \times \$15$ (d) $\$90 \div \$15$

26. Jenny's height is 162.5 cm and her younger sister, Beth, is 152 cm tall. What is the difference between their heights?
(a) 162.5 cm − 152 cm (b) 162.5 cm + 152 cm (c) 162.5 cm ÷ 152 cm (d) 162.5 cm × 152 cm

UNIT P-3 Estimation and Reasonable Answers

THE MAIN IDEA

1. ***Estimation*** is a way of getting a "rough" answer quickly. Reasons for using estimation include:
   - to check that an answer is reasonable.
   - to solve a problem when an exact answer is not needed.
   - to know if you've pressed a wrong key on the calculator.
   - to know if you've made a mistake in a computation with paper and pencil.
2. A ***reasonable answer*** is one that makes sense in light of the given facts. To check that an answer is reasonable, compare it to your estimate, to the facts of the problem, and to what you know from daily life.
3. An ***overestimate*** is an estimate that is greater than the exact answer. To be sure that you have enough time, space, or money, you may want to overestimate your needs.
4. An ***underestimate*** is an estimate that is less than the exact answer. You should underestimate supplies you have on hand, like available cash, time, or space.
5. To estimate an answer to a calculation, you can use ***approximate numbers*** that are close to the exact numbers and are easier to work with.

EXAMPLE 1 Explain why the number that is underlined is not reasonable.

a. The annual salary of a baker is $1,500.	*Answer:*	$1,500 is too low for the annual salary of a full-time worker. A person working 40 hours a week at minimum wage, about $6 an hour, earns about $240 each week. At $240 per week, the worker would earn $240 × 50, or about $12,000 in a year.
b. The length of a person's arm is 97 inches.	*Answer:*	97 inches is longer than 8 feet, which is too much for the length of an arm.
c. The driving distance between two cities in Ohio is 3,200 miles.	*Answer:*	Since the entire East-West distance of the United States is about 3,000 miles, 3,200 miles is too great for the distance between two cities in the same state.
d. The weight of a car is 150 pounds.	*Answer:*	Some adults weigh 150 pounds. The weight of a car is closer to 2,500 pounds.

EXAMPLE 2 For each situation, tell if an *exact* number is needed or if an *estimate* is good enough.

	Situation	Answer
a.	A clerk in a department store needs to give change to a customer.	*Answer:* A clerk is expected to give *exact* change.
b.	A sports announcer reports the attendance at a game.	*Answer:* An *estimate* is all the announcer needs.
c.	A ticket agent wants to determine if there are enough seats for an English class of 40 students at a performance of *Macbeth*.	*Answer:* If the performance is almost sold out, the ticket agent may have to find the *exact* number of seats available. If there are many unsold seats, an *estimate* may be good enough.

EXAMPLE 3 For each situation, tell if an *overestimate* or an *underestimate* is needed. If it doesn't matter, answer *either.*

	Situation	Think	Answer
a.	You want to know how much paint to buy to paint your room. Should you overestimate or underestimate the size of the room?	*Think:* If you underestimate, you might not buy enough paint to finish the job.	*Answer:* You're better off with an *overestimate.*
b.	When grocery shopping, you wonder if you have enough rice at home to make a recipe that calls for 2 cups of rice. Is it better to overestimate or underestimate the amount you have at home?	*Think:* Suppose you overestimate and you actually have less than 2 cups. You will have to go to the store again.	*Answer:* You should *underestimate* the amount on hand to be sure you will have enough.
c.	You wonder how many ducks there are in a flock as they pass overhead.	*Think:* It doesn't matter whether your estimate is high or low.	*Answer:* *Either* an overestimate or an underestimate.

EXAMPLE 4 Use approximate numbers to estimate each answer.

	Exact numbers		*Approximate numbers*		*Estimate*
a.	$98 + 63$	→	$100 + 60$	→	160
b.	$5{,}214 - 762$	→	$5{,}200 - 800$	→	4,400
c.	416×91	→	416×100 or 400×90	→	41,600 or 36,000
d.	$78 \div 21$	→	$75 \div 25$ or $80 \div 20$	→	3 or 4

Estimates may vary somewhat depending upon your choice of approximate numbers.

CLASS EXERCISES

1. Explain why the number that is underlined is not reasonable.
   a. A dozen pencils cost <u>$50</u>.
   b. The flying time between Los Angeles and San Francisco was <u>20</u> hours.
   c. A football player's waist measures <u>15</u> inches.
   d. The length of a giraffe's neck is <u>50</u> feet.
   e. The playing time for 5 compact discs is <u>90</u> minutes.

USE COMMON SENSE WHEN SOLVING PROBLEMS!

2. For each situation, tell if an *exact* number is needed or if an *estimate* is good enough.
   a. A customer is in a hurry but wants to know if the change received from the clerk is correct.
   b. A city planner needs to know how many people will use public transportation to commute into the city on weekdays.
   c. To fill a prescription, a pharmacist needs to know how many milligrams of a drug to give a patient.

3. For each situation, tell if an *overestimate* or an *underestimate* is needed. If it doesn't matter, answer *either*.
   a. You need to estimate the amount of food to serve at a party.
   b. Will you have enough time after school to meet with a friend before you go to work?
   c. At a fair, you are guessing the number of marbles in a jar.

4. Use estimation to tell if the calculator display shown is reasonable or not for the given question.

a. $1{,}142 + 2{,}365$	2507.	**b.** $67{,}492 - 11{,}016$	56479.
c. 145×981	142245.	**d.** $8{,}750 \div 125$	7.

5. Use approximate numbers to estimate each answer.
   a. $943 - 187$ **b.** 18×129 **c.** $5{,}781 \div 47$ **d.** $16{,}409 + 7{,}114$

6. Choose the most reasonable estimate.
   a. The total number of ballots cast for president of the student government if John received 112 votes, Melinda received 57 votes, Carmella received 295 votes, and Lin received 83 votes.
   (1) 550 (2) 600 (3) 650 (4) 700
   b. The total cost of 59 television sets if each costs $298.
   (1) $1,000 (2) $1,800 (3) $10,000 (4) $18,000
   c. The number of magazines a news delivery service must deliver to each newsstand if 510 magazines are to be distributed evenly to 18 newsstands.
   (1) 25 (2) 50 (3) 530 (4) 1,000

7. Describe a situation in which you would use an estimate rather than an exact answer.

HOMEWORK EXERCISES

1. Explain why each number that is underlined is *not* reasonable.
   a. A professional football player weighs 110 pounds.
   b. The distance from Miami, Florida to Seattle, Washington is 985 miles.
   c. The thickness of a dime is 3.5 in.
   d. There were 15,000 people at a high school play.
   e. There are 1,000 hours in a week.
   f. A stack of 100 one-dollar bills is 4 inches high.
   g. The distance traveled by a satellite in one complete orbit around the Earth is 10,000 miles.
   h. The temperature in Atlanta, Georgia is –10°F.

2. For each situation, tell if an *exact* number is needed or if an *estimate* is good enough.
   a. A zookeeper needs to know how much food to order for the animals for a week.
   b. A travel agent must tell a customer what time a flight departs.
   c. A rider on a commuter bus needs to know how much money to drop in the coin box.
   d. A pilot needs to know the cruising altitude for a flight.

3. For each situation, tell if an *overestimate* or an *underestimate* is needed. If it doesn't matter, answer *either*.
   a. To place tomorrow's order, the owner of a deli needs to estimate the amount of each item that will be left at the end of today and that can be sold tomorrow.
   b. The Parks Department needs to order grass seed.
   c. A family needs to estimate the number of juice boxes to take along on a car trip.

4. Use estimation to tell if the calculator answer is reasonable or not.
   a. $56,894 + 13,778$ 70672.
   b. $1,184 - 856$ 1099.
   c. 49×15 64.
   d. $1,411 \div 17$ 210.

5. Use approximate numbers to estimate each answer.
   a. $45 + 196$ **b.** $502 - 98$ **c.** $3,670 \times 11$ **d.** $8,852 \div 18$

6. Choose the most reasonable estimate.
   a. The total cost of 21 automobiles if each one costs $14,750.
   (1) $15,000 (2) $28,000 (3) $280,000 (4) $300,000
   b. The total length of 9 pieces of molding if each is 81 inches long.
   (1) 80 inches (2) 90 inches (3) 800 inches (4) 900 inches
   c. What is the best estimate of the total cost of the vacation with the following expenses?
   Airfare: $1,420, Hotel: $980, Food: $475, Sightseeing: $225
   (1) $2,600 (2) $3,100 (3) $4,500 (4) $5,000
   d. The total amount of money raised at a fund-raising ceremony if there were 27 tables each with 12 people, and if each person paid $75.
   (1) $120 (2) $14,000 (3) $20,000 (4) $24,000

7. Describe a situation in which you would overestimate a number or amount.

UNIT P-4 Using Problem-Solving Strategies

THE MAIN IDEA

There are procedures called ***strategies*** that can help you get the solution to a problem. For example:

1. ***Guessing and checking*** is a strategy to use when you are given a few possible solutions and want to find out which one is correct. Also, you can use your judgment to guess an answer and then check to see if the result obtained makes sense according to the given information. In this way, you narrow the given choices.
2. ***Drawing a diagram*** is a strategy to use to help you *see* how numbers in a problem are related to each other. By studying the diagram, you can get ideas for arriving at the solution.
3. ***Making a list*** is a strategy to use to display all the possible solutions to a problem. Once a list is made, you can examine each possible solution on the list, one at a time. Some possibilities can be eliminated easily. Then you can test those that remain for reasonableness.
4. ***Finding a pattern*** is a strategy to use when the solution is part of a sequence or a string of outcomes that repeats. By knowing the pattern, you can predict an outcome.
5. ***Breaking a problem into smaller problems*** is a strategy to use to simplify a problem that must have several steps. Instead of beginning by thinking about the final answer, you first work on each smaller problem and then use the results to get the answer.
6. ***Working backward*** is a strategy to use when you know a result and are trying to find an unknown piece of information that led to that result. In order to get back to the beginning of the problem, you perform the operation of arithmetic that is the *opposite* of the operation you would have performed if you had been looking for the result.

EXAMPLE 1 Tell which strategy *is being used* to solve the problem.

Problem: You are told the dimensions of a square and a rectangle, and you have to find how many times the square can fit into the rectangle.

You draw a diagram in which you place squares inside the rectangle. You count the number of squares that fit.

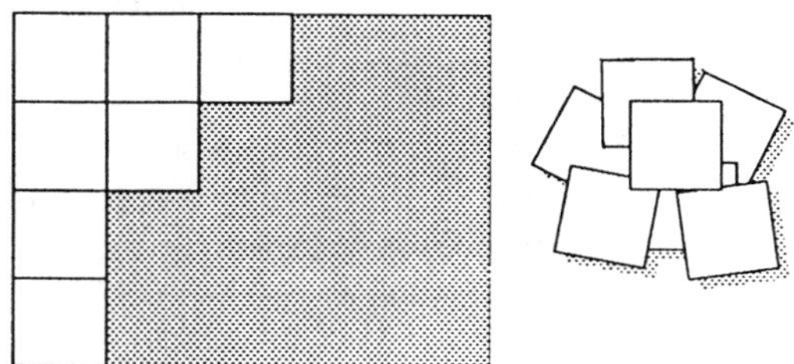

Think: While the problem can be solved by working only with the given numbers, drawing a diagram helped you to see the solution.

Answer: The strategy used is drawing a diagram.

EXAMPLE 2 Tell which strategy *is being used* to solve the problem.

Problem: Find how many multiples of 7, less than 100, end in the digit 1.

You write down all the multiples of 7 less than 100 and look for those that end in the digit 1.

Think: Writing down the multiples of 7 helps you keep track of possible answers, which would be difficult to do mentally since there are 14 of them under 100.

Answer: The strategy used is making a list.

EXAMPLE 3 Tell which strategy *is being used* to solve the problem.

Problem: You are told that May 2 was a Sunday and you are asked to find on which day of the week May 23 will fall.

You subtract 2 from 23 and find that the two dates are 21 days apart, exactly 3 sets of 7 days. Every seventh day after May 2 will be a Sunday. You conclude that May 23 will also be a Sunday.

Think: You used the idea that the days of the week repeat themselves in a pattern of repetition every 7 days to predict the outcome.

Answer: The strategy used is finding a pattern.

EXAMPLE 4 Tell which strategy *you could use* to solve the problem.

Problem: You are given three of Robert's test scores and the average of four of his test scores. You have to find his fourth test score.

Think: You could make a guess about Robert's fourth test score and use the guess to calculate the average score. If your guess leads to the same average that was given, you have found the solution. If not, guess again and check again. By comparing the wrong average with the known average, you would know if your next guess should be higher or lower.

Answer: The strategy to use is guessing and checking.

EXAMPLE 5 Tell which strategy *you could use* to solve the problem.

Problem: You are told a total price and the price of one of the two items that make up this total price. You have to find the price of the second item.

Think: You could work backward. Start with the total price and perform the arithmetic operation that is the opposite of the operation you would have used if you were trying to find the total price. Since you would have added, you subtract.

Answer: The strategy to use is working backward.

EXAMPLE 6 Tell which strategy *you could use* to solve the problem.

Problem: You have to find the total cost of gasoline used for Mr. Meyers' automobile trip. You are given three different amounts of gasoline that he bought at different times during the trip, and different prices per gallon for the three purchases.

Think: This problem has to be solved in four different parts. First, find the cost of each of the three purchases. Finally, use the three separate results to find the total cost of the gasoline.

Answer: The strategy to use is breaking the problem into smaller problems.

CLASS EXERCISES

In 1–7, tell which strategy *is being used* to solve the problem.

1. You are given the total weight of six students and the individual weights of five of them. You have to find the weight of the sixth student.

   You try a number for the sixth weight and see if this gives you the correct total.

2. You are asked to find the number of months in the year that have four-letter names.

You write the names of the months and next to each name the number of letters in it.

3. Two different rectangles are described verbally, and the measures of their sides are given. You have to find the total distance around both rectangles.

You sketch the rectangles and write all the measures on the sketch.

4. You are told your aunt's age now. You are asked to find her age a certain number of years ago.

You begin with the present age.

5. You are asked to find how much carpeting is needed to cover the floor of a hotel lobby that has a very complicated shape.

You sketch the given shape so that you can see the individual geometric figures that make up the shape.

6. You are asked to find what month it will be 27 months after September.

You use the idea that the months repeat in groups of 12.

7. You are asked to compute $11{,}111 \times 11{,}111$. However, your calculator cannot display a number that large. On the calculator, you find the results shown at the right.

$$1 \times 1 = 1$$
$$11 \times 11 = 121$$
$$111 \times 111 = 12321$$
$$1111 \times 1111 = 1234321$$

By examining this list of results, you predict that $11{,}111 \times 11{,}111 = 123{,}454{,}321$

In 8–13, tell which strategy *you could use* to solve the problem.

8. The front of a house is described verbally as a square topped by a triangle, and the dimensions of the square and triangle are given. You have to find the height of the house.

9. You have to find the final price of a purchase of two differently priced items after a discount.

10. You are given the final balance in a savings account. You are also given the amounts of deposits and withdrawals that have been made before this final balance was calculated. You have to find the original balance.

11. A high school student's age has been doubled and that result has been subtracted from 100. You are given the final answer. You have to find the student's age.

12. A traffic light changes from red to green to yellow to red. If it is green now, what color will it be 17 changes from now?

13. You are asked to find how many different four-digit identification numbers can be formed by rearranging the digits 5, 6, 7, and 8.

In 1–10, tell which strategy *is being used* to solve the problem.

1. You have to find a pair of whole numbers that give a result of 48 when they are multiplied, and 14 when they are added.

You write all the pairs of whole numbers that give the result 48 when they are multiplied. Next to each pair, you write the result when they are added. Then you select those pairs that give the result 14 when added.

2. You are given the present price of a stock and the amounts by which the price has increased and decreased over the past several days. You have to find the original price of the stock.

You begin with the present price and reduce it by the amounts of increase and raise it by the amounts of decrease.

3. You are told how much water two different containers hold. After water has been poured back and forth from one container to another, you have to find how much water is left in one of them.

You make a picture and draw lines to show the increase and decrease in water levels.

4. When a number is multiplied by 7, and 5 is added to the product, the result is 54. You have to find the original number.

You keep trying numbers, multiplying by 7 and adding 5, until you find one that works.

5. Two students each took five exams. You are asked to find the difference between their exam score averages.

You first find each student's average, and then you find the difference of the averages.

6. You are given Mr. Manning's hourly rate of pay, his overtime rate, and the number of hours he worked during a week (regular and overtime). You have to find his total pay for the week.

You separately calculate his regular pay and his overtime pay, then find the sum.

7. A cafeteria lunch consists of one sandwich, one beverage, and one dessert. You are asked to find the number of different lunches that can be selected from the choices shown on the menu.

Sandwiches	Beverages	Desserts
tuna egg salad ham	milk orange juice	apple pie ice cream

You assign a code letter to each choice and write all possible selections, such as EMI (egg salad, milk, ice cream). Then you count the different selections.

8. You are given the results: $12 \times 11 = 132$, $13 \times 11 = 143$, and $14 \times 11 = 154$

You have to find 18×11.

You notice that you can find each result in the following way:

Write the two digits of the number that is multiplying 11 as the first and third digits of the answer. Add these digits to find the second digit of the answer.

9. You are told how much change Mark received from a ten-dollar bill after buying a certain number of pens, all at the same price.

   In order to find the price of one pen, you subtract the amount of change from \$10 and divide that result by the number of pens.

10. You are asked to compute $37{,}037 \times 18$ after first finding the results shown at the right.

   $37037 \times 3 = 111111$
   $37037 \times 6 = 222222$
   $37037 \times 9 = 333333$

   You predict that $37{,}037 \times 18$ must equal 666,666.

In 11–20, tell which strategy *you could use* to solve the problem.

11. You are told the total price of a certain number of pounds of ham and the total price of a different number of pounds of turkey. You are asked to compare the price of one pound of ham to the price of one pound of turkey.
12. How many whole numbers between 10 and 50 are divisible by 4?
13. After being given Joseph's present weight and his changes in weight over the past eight weeks, you want to find what his weight was eight weeks ago.
14. You are asked to find how the area of a large rectangle compares to the area of a small rectangle, given a relationship between the dimensions of the rectangles.
15. What is the greatest number of pieces that can be made when you cut a circular piece of paper with three straight cuts?
16. A digital clock now reads 7:00. What will it read 37 hours from now?
17. On a certain exam, each correct answer on part I is worth a given number of points, and each correct answer on part II is worth a different given number of points. You want to find the number of correct answers a student had on each part if her total score is known.
18. Mrs. Smith buys a candy mixture. You are given the different weights and prices of three different kinds of candy that make up the mixture. You want to find the price of the mixture.
19. How many times a day does a digital clock display the same numeral in every position?
20. On Day 1 of your new exercise program, you jog around the block, going one block in each direction. On Day 2, you plan to jog 2 blocks in each direction. On Day 3, you will jog 3 blocks in each direction, and so on. You want to know how far you will jog on Day 10 if you continue to increase your distance like this.

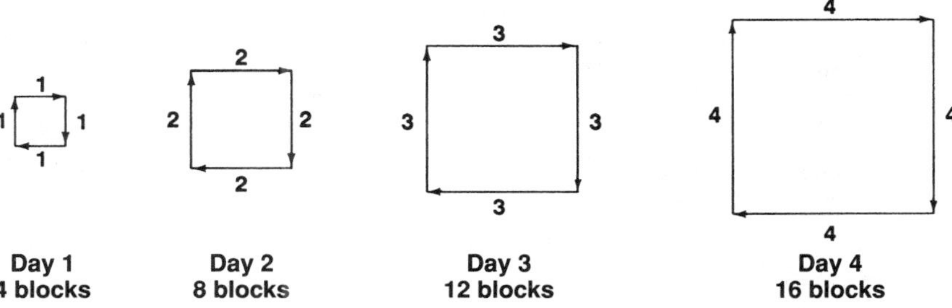

You predict that on Day 10 you will jog 40 blocks.

READING DATA

UNIT D-1 Reading Charts and Tables

THE MAIN IDEA

1. In newspapers, magazines, books, etc., information is often displayed in the form of charts and tables that present information, or ***data***, in an organized way that is easy to use.
2. Charts and tables are often organized in vertical columns (reading up and down) and horizontal rows (reading left and right). Each column and each row is labeled to show how the information given in that row or column is related to the other data.

   In the table below, the arrows point to the row labeled TECHNO 900 and the column labeled MILES PER GALLON. All the information in the row relates to one car, the Techno 900. All the information in the column relates to one measurement, miles per gallon.

FUEL USE OF CARS

Car	Price	Cruising Range in Miles	Miles per Gallon
Apex 6000	$17,598	365	17
Techno 900	$10,245	406	16
Taft Mark II	$24,837	500	21

3. Specific items of data can be found by locating a column and a row and looking at the box where they meet. (The Techno 900 travels 16 miles per gallon.)
4. You can also find which car has a price of $24,837 by finding which row crosses the price column at $24,837. (The Taft Mark II)

FUEL USE OF CARS

Car	Price	Cruising Range in Miles	Miles per Gallon
Apex 6000	$17,598	365	17
Techno 900	$10,245	406	16
Taft Mark II	$24,837	500	21

5. Examining the information in a table often reveals general patterns in the numbers, called ***trends***.

EXAMPLE The following chart shows the sizes and widths in which Bigfoot Shoes are available. Use this chart to answer the questions.

Sizes of Bigfoot Shoes

Widths	6	$6\frac{1}{2}$	7	$7\frac{1}{2}$	8	$8\frac{1}{2}$	9	$9\frac{1}{2}$	10	$10\frac{1}{2}$	11	$11\frac{1}{2}$	12	13	14	15
B					■	■	■	■	■	■	■	■	■	■	■	■
C			■	■	■	■	■	■	■	■	■	■	■	■	■	■
D	■	■	■	■	■	■	■	■	■	■	■	■	■	■	■	■
E	■	■	■	■	■	■	■	■	■	■	■	■	■	■		
EEE	■	■	■	■	■	■	■	■	■	■	■	■	■	■		

Key: ■ indicates that the size is available.

a. In which widths is size 9-1/2 available?

In the column labeled 9-1/2, there are shaded squares for widths B, C, D, E, and EEE.

Answer: Size 9-1/2 is available in widths B, C, D, E, EEE.

b. What is the smallest size that is available in C width?

Look at the row labeled C. The first shaded square is at size 7.

Answer: The smallest size available in C width is size 7.

c. Which width is available in every size?

In the row labeled D, there is a shaded square under every size.

Answer: D width is available in every size.

d. Which width is available in the fewest sizes?

In the row labeled B, there are 12 shaded squares. Every other width has more than 12 shaded squares.

Answer: B width is available in the fewest sizes.

e. What is the total number of available combinations of sizes and widths?

There are 70 shaded squares.

Answer: There are 70 combinations of sizes and widths available.

f. Explain how the answer to part **e** could have been found without counting all the shaded squares.

Answer: There are 16 columns and 5 rows, giving a total number of boxes of 16×5, or 80. Since there are 10 empty boxes, the number of shaded squares is $80 - 10$, or 70.

CLASS EXERCISES

1. The table shows the minimum age required for a person to obtain a driver's license in each state of the United States in 2000. Where two ages appear, a driving license was issued at the lower age to a person who had completed an approved driver education course.

State	Min. Age	State	Min. Age	State	Min. Age
Alabama	16	Louisiana	15/17	Ohio	16/18
Alaska	16	Maine	15/17	Oklahoma	16
Arizona	16	Maryland	16/18	Oregon	16
Arkansas	16	Massachusetts	17/18	Pennsylvania	17/18
California	16/18	Michigan	16/18	Rhode Island	16/18
Colorado	18	Minnesota	16/18	South Carolina	16
Connecticut	16/18	Mississippi	15	South Dakota	16
Delaware	16/18	Missouri	16	Tennessee	16
Florida	16	Montana	15/18	Texas	16/18
Georgia	16	Nebraska	16	Utah	16/18
Hawaii	15	Nevada	16	Vermont	18
Idaho	16	New Hampshire	16/18	Virginia	16/19
Illinois	16/18	New Jersey	17	Washington	16/18
Indiana	16/18	New Mexico	15/16	West Virginia	16/18
Iowa	16/18	New York	17/18	Wisconsin	16/18
Kansas	16	North Carolina	16/18	Wyoming	16
Kentucky	16	North Dakota	16		

   a. What was the minimum age for obtaining a license in Kansas in 2000?

   b. What was the age requirement for obtaining a license in most states?

   c. Which state or states had the lowest age for obtaining a driver's license?

   d. How many states issued a license at a lower age upon completion of an approved driver education course?

2. Use the given table of state populations to answer each question.

State	Census Year			
	1960	1970	1980	1990
Alaska	226,167	302,583	401,851	550,043
California	15,717,204	19,971,069	23,667,764	29,785,857
Florida	4,951,560	6,791,418	9,746,961	12,938,071
Kansas	2,178,611	2,249,071	2,364,236	2,477,588
New York	16,782,304	18,241,391	17,558,165	17,990,778
Texas	9,579,677	11,198,655	14,225,513	16,986,335

   a. What was the population of Kansas in 1970?

   b. What was the population of Texas in 1960?

   c. In what year was the population of Florida 9,746,961?

   d. Which state had a population of 16,782,304 in 1960?

   e. For which state did the population decrease from 1970 to 1980?

   f. For how many states listed in the table was the population in 1970 greater than 10,000,000?

   g. In which of the given years did California have its greatest population?

   h. Estimate: Between which two given years did Alaska have its greatest population increase?

   i. Describe the general trend of state populations from 1960 to 1990.

1. The chart shows the features of five models of Turbojet dishwashers. Use this chart to answer the questions.

Turbojet Dishwasher Models

Features	Classic I	Classic II	Turbostar	Turbo King	Hydro Queen
Energy Monitor				■	■
Light Wash Cycle	■	■	■	■	■
China/Crystal Cycle		■	■	■	■
Electronic Control			■	■	■
Delay Start					■
Self-Clean Filter	■			■	■
Price	$310	$340	$430	$490	$570

Key: ■ Model has feature

a. Which models have a china/crystal cycle?

b. What is the least expensive model with the electronic control feature?

c. What feature does every model have?

d. What is the most expensive model without the self-clean filter?

e. To choose between the Turbo King and Hydro Queen, what decision must you make?

2. The map shows the concentration of sulfates in the air in the United States for two recent years. Sulfates are chemicals produced by factories and power plants that burn coal. They are a major cause of acid rain, which kills forests and wildlife.

Where the Acid Rain Falls

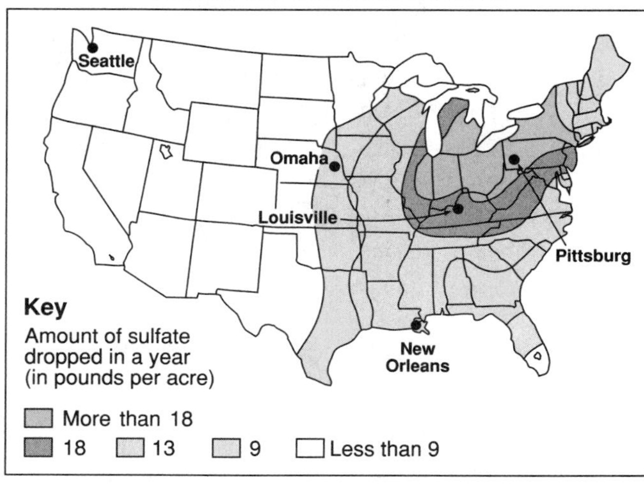

a. Where are the states that have the highest concentrations of sulfates? Where are the states that have the lowest?

b. Sulfates are carried long distances by the winds before they come down in an area. Knowing this and using the map, what can you tell about the main direction of the winds in the United States?

c. Which area along the East Coast is most free of sulfates? Why do you think this is so?

d. Using the map, make a table showing the concentrations of sulfates that fell on Louisville, New Orleans, Omaha, Seattle, and Pittsburgh. Rank the data in decreasing order.

3. The table shows worldwide motor vehicle production for each of the given years. Use the table to answer the questions.

World Motor Vehicle Production

Year	United States	Europe	Japan	Other
1965	11,138,000	9,549,000	1,876,000	1,705,000
1975	8,987,000	13,590,000	6,942,000	3,480,000
1985	11,653,000	15,988,000	12,271,000	4,871,000
1995	11,985,000	17,045,000	10,196,000	8,349,000

a. How many motor vehicles were produced in the United States in 1965?

b. In what year did Japan produce 6,942,000 motor vehicles?

c. About how many more motor vehicles did Europe produce in 1985 than in 1965?
(a) 6,000,000 (b) 8,000,000 (c) 10,000,000 (d) 12,000,000

d. The 1995 production in Europe was about how many times as great as the 1965 production?
(a) 2 (b) 4 (c) 6 (d) 8

e. Which area shows the greatest increase in production between 1965 and 1995?

f. When and where was the number of motor vehicles produced closest to 5 million?

g. Estimate to the nearest million the total number of motor vehicles produced in 1975.

h. Explain how the number of motor vehicles produced in the United States generally compares with the number of motor vehicles produced in Japan.

i. Describe a trend in the number of motor vehicles produced in Europe for the given years.

4. The table gives data about the careers of the given members of the Basketball Hall of Fame. Use the table to answer the questions.

Player	Games	Points	Rebounds	Assists
Wilt Chamberlain	1,045	31,419	23,924	4,643
Bob Cousy	924	16,960	4,786	6,955
Elvin Hayes	1,303	27,313	16,279	2,398
Oscar Robertson	1,040	26,710	7,804	9,887
Jerry West	932	25,192	5,376	6,238

a. How many assists did Bob Cousy make in his career?

b. Which player was credited with 9,887 assists in his career?

c. Which player scored the greatest number of points?

d. Which player played the greatest number of games?

e. Which number phrase would tell how many more assists Oscar Robertson made than Jerry West?
(a) 7,804 – 5,376 (b) 9,887 + 6,238 (c) 9,887 – 7,804 (d) 9,887 – 6,238

f. How many of the players listed played in more than 1,000 games?

g. Which players made more than 15,000 rebounds?

h. How many players have more rebounds than assists?

UNIT D-2 Reading and Making Bar Graphs

READING BAR GRAPHS

THE MAIN IDEA

1. A ***graph*** is a visual way to organize and present data.
2. A ***bar graph*** is a way to display data used to compare differences among several unconnected items that are measured in the same unit.
3. In a bar graph, the data are shown as bars of different lengths, which correspond to the numbers that they represent, shown on the ***scale***.

EXAMPLE 1 This bar graph shows the average number of calories consumed at breakfast by 8 school children.

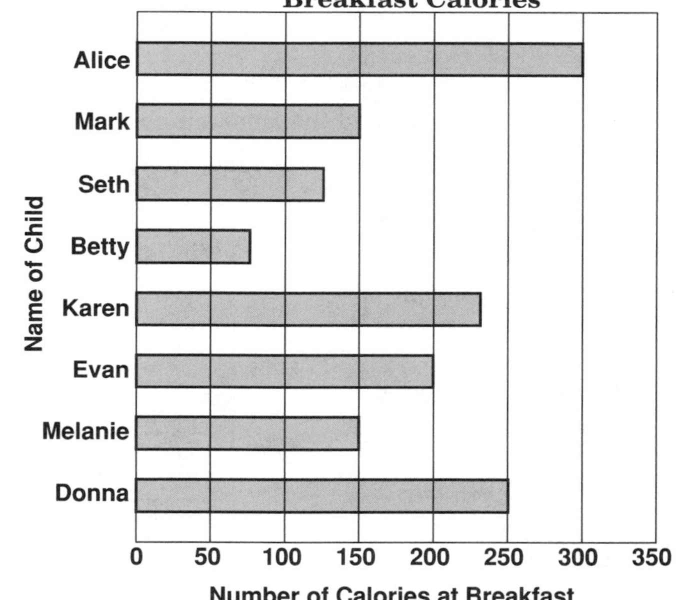

a. Which child consumes the greatest number of calories at breakfast?

Look for the longest bar.

Answer: Alice

b. Who consumes the fewest calories at breakfast?

Look for the shortest bar.

Answer: Betty

c. Which children consume the same number of calories at breakfast?

Look for bars that are equal in length.

Answer: Mark and Melanie

d. About how many calories does Seth consume at breakfast?

The bar for Seth stops about halfway between 100 and 150.

Answer: 125 calories

e. Who consumes twice as many calories as Melanie?

Melanie consumes 150 calories. The child who consumes 300 calories is Alice.

Answer: Alice

CLASS EXERCISES

1. This bar graph shows the number of books read by some students during the month of October.

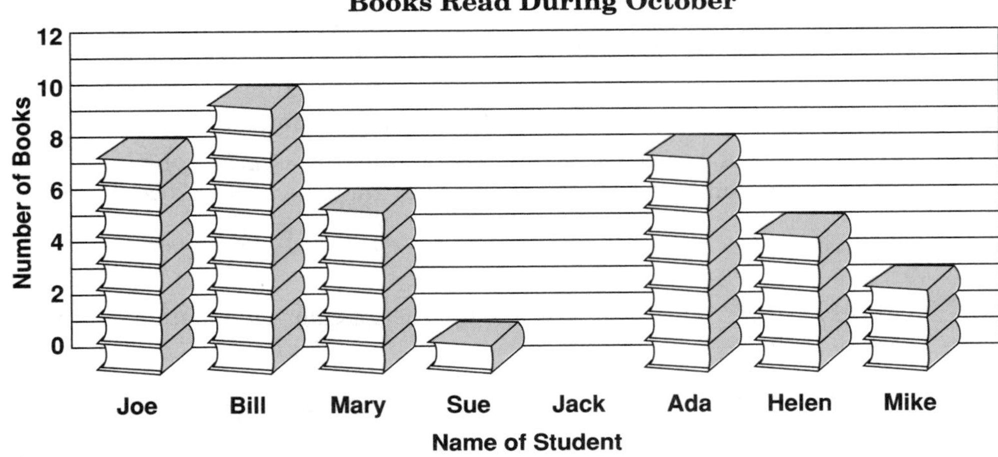

a. Who read the greatest number of books in October?
b. Who read no books in October?
c. Which two students read the same number of books?
d. Who read twice as many as Mike?
e. How many more books did Joe read than Helen?
f. Who read half as many books as Bill?

2. This graph shows the expected enrollment for Plainview School from 1995 to 2001. (Observe that the horizontal scale is "in hundreds." The number 3, for example, means 300.)

a. During which year is enrollment expected to be the lowest?
b. When is enrollment expected to reach its highest level?
c. During which two years is it expected that enrollment will stay the same?
d. How many more students are expected to enroll in 1996 than in 2001?

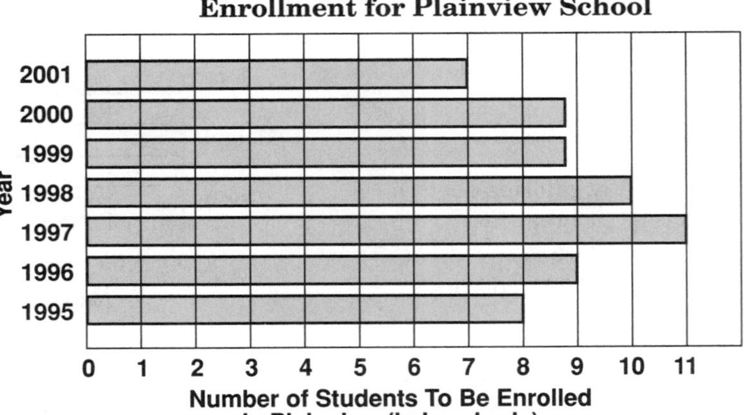

e. In which year are 300 fewer students expected than in 1997?

MAKING BAR GRAPHS

THE MAIN IDEA

To make a bar graph:

1. Decide whether to use a horizontal scale or a vertical scale.
2. Draw the scale with numbers from 0 to a little more than the greatest number to be represented. Use equal lengths to represent equal quantities. Write a label to describe the numbers on the scale.
3. Draw a base line (or zero line) at which each bar will start. Write a label for each bar and a general description of all the bars.
4. Use the scale to determine the length of each bar and draw a bar for each quantity to be represented, using the same width for all the bars.
5. Write a title for the graph.

EXAMPLE 2 Make a bar graph to show the given information.

Number of Pairs of Sneakers Sold at the Sneaker Den					
Mon.	Tues.	Wed.	Thurs.	Fri.	Sat.
15	20	10	18	24	37

Make either a horizontal or vertical graph.

Choose a scale to fit the numbers. Starting at 0, mark off and number the scale.

Label the scale.

Draw the proper lengths for the bars. Make all bars the same width.

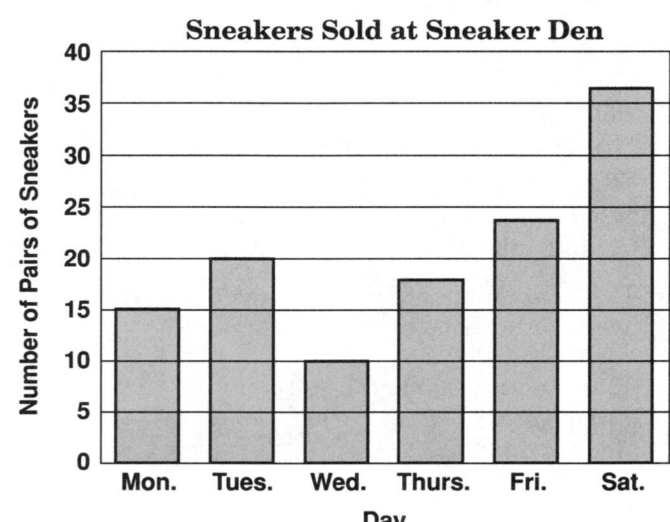

CLASS EXERCISES

Make a bar graph for the information in each table.

1.

Salaries of Executives at Topper Company			
Ms. Jones	$60,000	Mr. Smith	$57,000
Mr. Brown	$35,000	Mr. Green	$45,000
Miss White	$37,000	Mrs. Roy	$65,000

2.

Number of Trees Sold by Green Nursery			
March	150	June	280
April	170	July	190
May	220	August	100

HOMEWORK EXERCISES

1. This bar graph shows Mary's baby-sitting earnings for 6 months.

   a. How much did Mary earn in August?
   (a) $20 (b) $30
   (c) $40 (d) $60

   b. In which month did Mary earn the least?

   c. In which two months did Mary earn the same amount?

   d. How much more did Mary earn in November than in October?

   e. For which two months combined did Mary earn a total of $200?

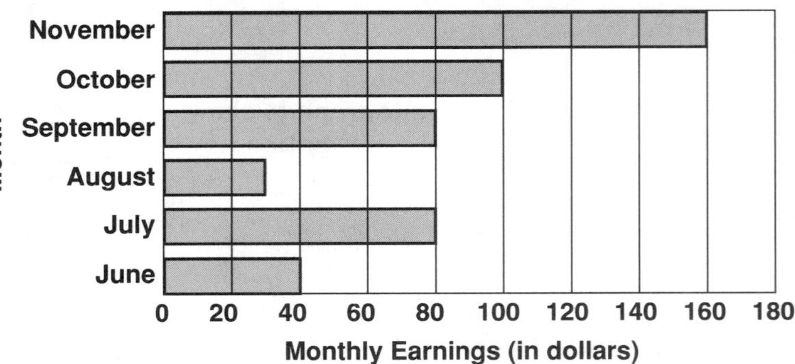

2. This bar graph shows the number of miles traveled by a salesperson in each of five months.

   a. What is the best estimate of the number of miles traveled in January?
   (a) 400 (b) 375 (c) 350 (d) 300

   b. The total number of miles traveled over the five-month period is
   (a) under 1,000
   (b) between 1,000 and 1,500
   (c) between 1,500 and 2,000
   (d) over 2,000

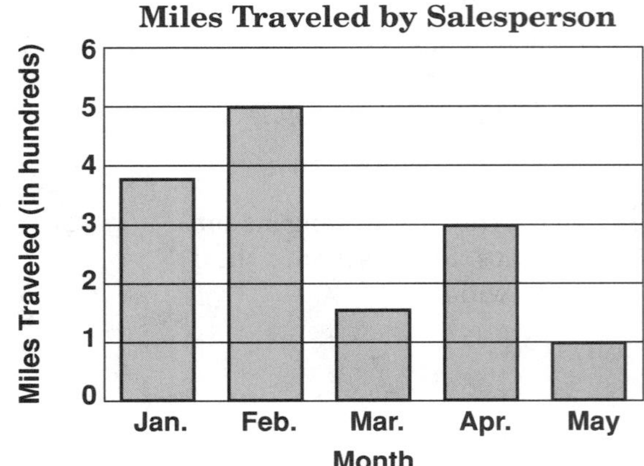

3. This bar graph shows the number of milligrams of cholesterol in an average loaf of 5 different brands of bread.

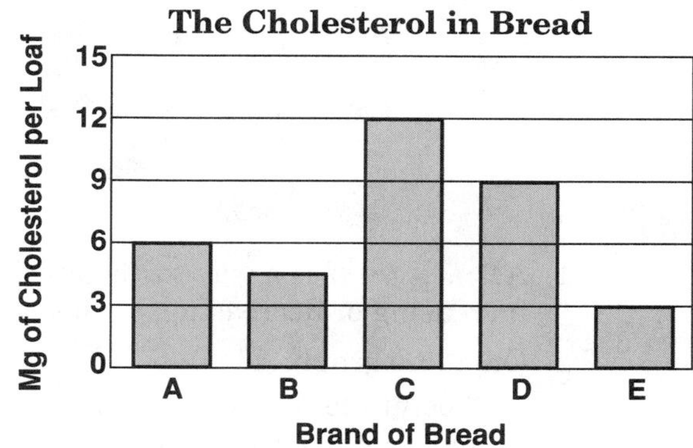

   a. Which brand has about half the amount of cholesterol as Brand D?
      (a) Brand A (b) Brand B
      (c) Brand C (d) Brand E

   b. Which two brands have as their sum the same amount of cholesterol as Brand D?

4. This bar graph shows the number of lunches served each month at Beefburg High School.

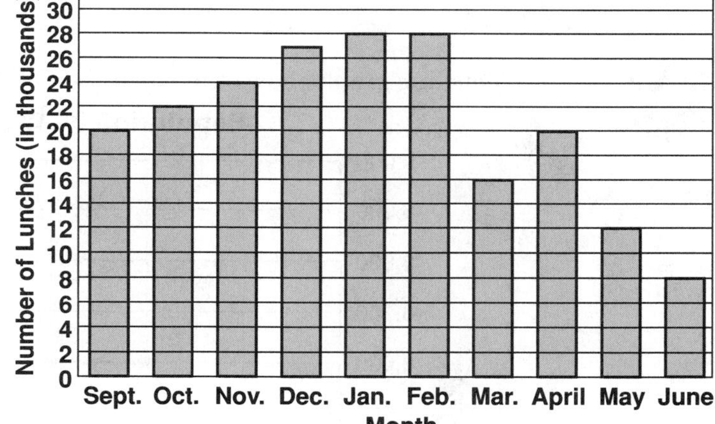

   a. What was the greatest number of lunches served in one month?
   b. In which months was the greatest number of lunches served?
   c. What was the smallest number of lunches served in one month?
   d. Estimate the number of lunches served in December.
   e. In which month was the same number of lunches served as in September?
   f. How many more lunches were served in April than in May?
   g. Between which two months was the greatest drop in the number of lunches?
   h. In which month was twice the number of lunches served as in May?
   i. In which month was one-third the number of lunches served as in November?
   j. During how many months were fewer than 20,000 lunches served?

In 5 and 6, make a bar graph for the information in the table.

5.

Number of Students Taking Languages at Endwood High School			
French	550	German	300
Latin	200	Italian	250
Spanish	700	Chinese	100

6.

Number of Shoe Sales for May at the Branch Stores of Comf Co.			
East	460	North	520
West	180	South	380
City	340	Town	400

UNIT D-3 Reading and Making Line Graphs

READING LINE GRAPHS

THE MAIN IDEA

1. A *line graph* is a way to display data used to call attention to how a quantity is increasing or decreasing, or to show a *trend*.
2. On a line graph:
   a. A point identifies a relation between a number on a vertical scale, or *axis*, and a number on a horizontal scale.
   b. Pairs of points are connected to form a *broken line*.
   c. The rise or fall of the line shows the increase or decrease between numbers.

EXAMPLE 1 This line graph shows the population of Apex City for a period of 7 years.

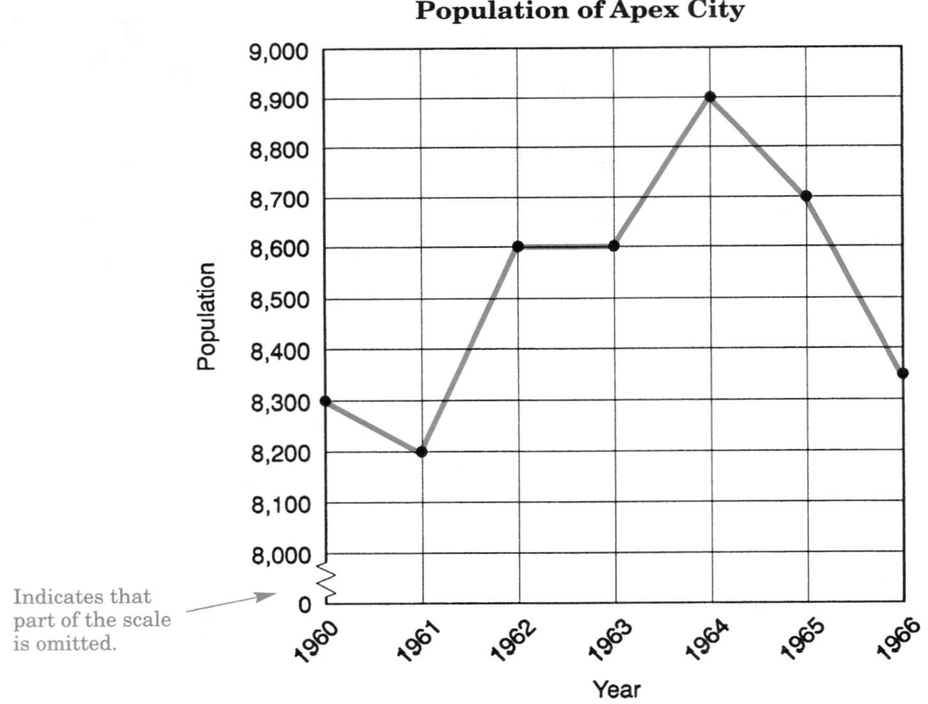

a. What was the population during the year when the population was the greatest?

Answer: 8,900 people

b. Estimate the population for 1966.

The point above 1966 lies halfway between 8,300 and 8,400 on the vertical scale. The number halfway between 8,300 and 8,400 is 8,350.

Answer: 8,350 people

c. For which two consecutive years did the population remain the same?

Look for a piece of the line *(line segment)* that is neither rising nor falling.

Answer: 1962 and 1963

d. Describe the trend in population growth in Apex City from 1961 to 1964.

Look at the portion of the line graph from 1961 to 1964.

Answer: Generally, the population increased from 1961 to 1964 but stayed the same for one year in that period from 1962 to 1963.

CLASS EXERCISES

1. This graph shows the average price of one share of stock of the Bionic Dog Food Co. over the period of 8 months.

a. When was the stock price lowest?

b. What was the lowest average price of the stock?

c. When was the average price highest?

d. What was the highest average price of the stock?

e. What was the average price of one share of stock in February?

f. Between which two months did the average price fall the most?

g. During which month was the average price of the stock the same as it was in June?

h. How much did the average price of the stock drop from January to February?

i. Describe the trend in prices of this stock from January to August.

Price of Bionic Dog Food Co. (One Share)

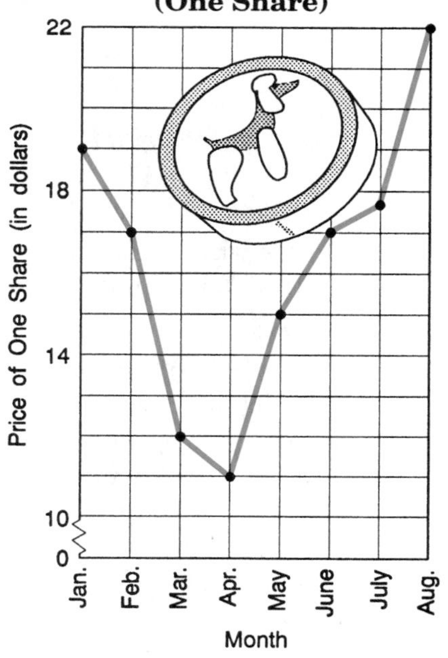

2. This line graph shows the attendance figures at Beefburg High School football games for a period of 7 weeks.

   a. Which game had the lowest attendance?

   b. What was the lowest attendance?

   c. What was the highest attendance?

   d. Which game had 550 in attendance?

   e. How many fewer people attended the Sept. 27 game than the Sept. 20 game?

   f. Estimate the attendance at the Oct. 4 game.

   g. How many games had fewer than 500 present?

   h. What was the total attendance for the 4-week period beginning Oct. 11?

   i. By about how much did the attendance for the first two October games exceed that for the two September games?

   j. Describe the trend from Oct. 18 to Nov. 1.

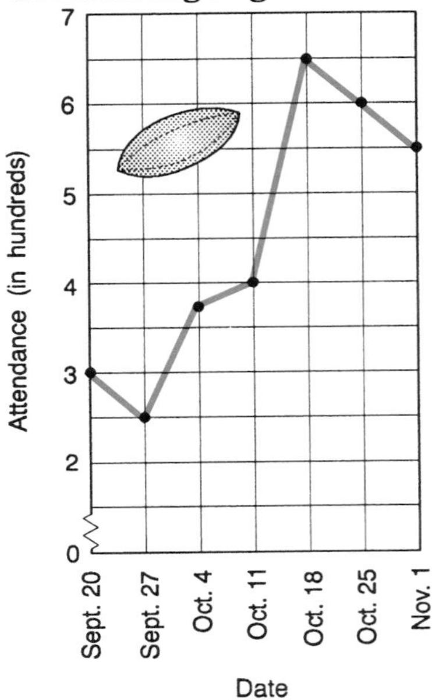

MAKING LINE GRAPHS

THE MAIN IDEA

To make a line graph:

1. Decide which information to represent on the horizontal axis and which on the vertical axis. (Time is usually shown on the horizontal axis.)
2. Draw a scale on each axis. Number and label the scales, and write a general description of the quantities represented.
3. For each item of data:
   a. Imagine a horizontal line through the corresponding point on the vertical axis, and a vertical line through the corresponding point on the horizontal axis.
   b. Make a dot where the horizontal and vertical lines would intersect if they were drawn.
4. Connect the dots in sequence, and write a title to describe the graph.

EXAMPLE 2 Make a line graph to show the information in the table.

Mr. Harris' Systolic Blood Pressure	
July 4	140
July 5	138
July 6	138
July 7	135
July 8	139
July 9	140
July 10	137
July 11	135
July 12	132
July 13	132

Represent time on the horizontal axis. For the vertical axis, choose a scale to fit the numbers between 130 and 140. Break the axis between 0 and 130. Mark off equal lengths, number and label the scale.

Place a point on the graph for each piece of information.

Draw line segments between consecutive points, and write a title for the graph.

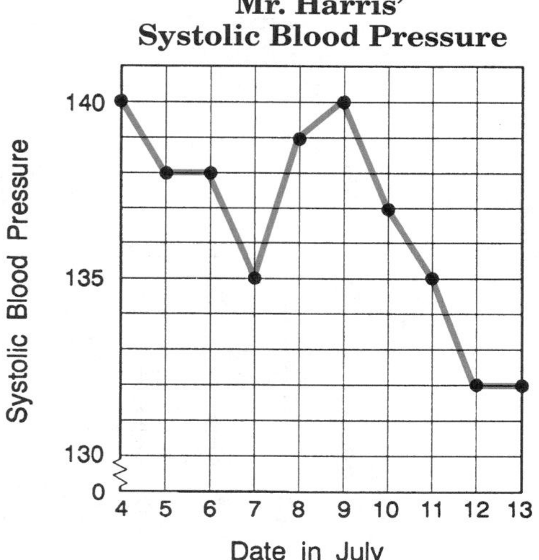

CLASS EXERCISES

Make a line graph to show the information in each table.

1.

Ms. William's Weight	
August 1	137
September 1	135
October 1	138
November 1	135
December 1	135
January 1	133

2.

Number of Points Scored by Brookville H.S. Football Team			
October 3	10	October 31	17
October 10	10	November 7	14
October 17	10	November 14	14
October 24	17	November 21	17

HOMEWORK EXERCISES

1. The line graph shows how the number of books borrowed from a school library changed over a five-week period.

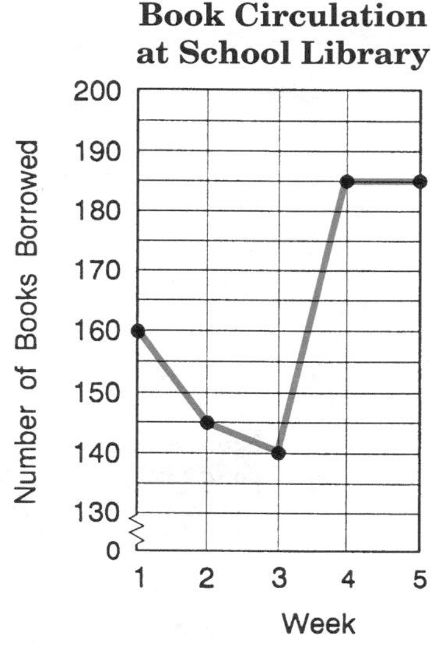

a. Find the difference between the number of books borrowed in weeks 2 and 4.

b. In which week did the number of books begin to increase?

c. During which weeks did the number of books decrease?

d. For which weeks did the number of books remain the same?

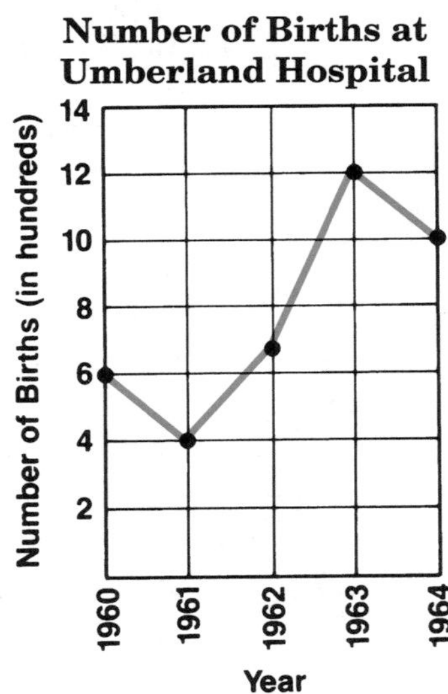

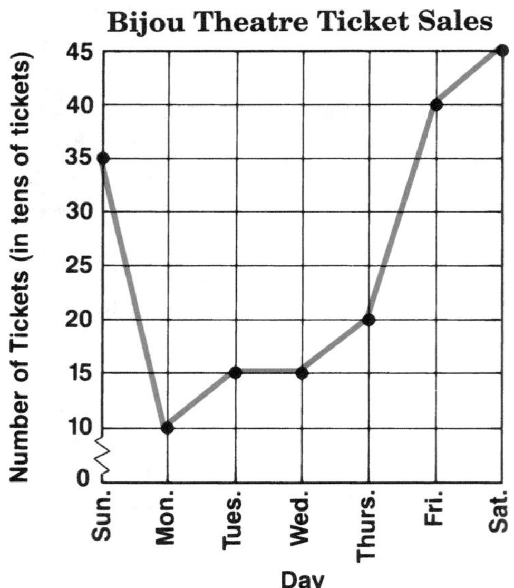

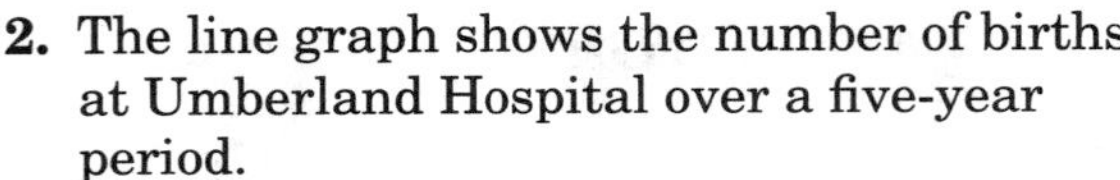

2. The line graph shows the number of births at Umberland Hospital over a five-year period.

a. Between which two years did the number of births increase the most?
(a) 1960–1961 (b) 1961–1962
(c) 1962–1963 (d) 1963–1964

b. During which interval did the number of births double?
(a) 1960–1962 (b) 1960–1963
(c) 1961–1963 (d) 1961–1964

c. Identify a period of time when the trend in the number of births was increasing.

3. The line graph shows the number of tickets sold each night at the Bijou Theatre during one week.

a. How many tickets were sold on Wednesday?

b. How many more tickets were sold on Saturday than on Wednesday?
(a) 45 (b) 450 (c) 30 (d) 300

c. What was the total number of tickets sold that week?

d. Name the day that marks the end of a decreasing trend and the beginning of an increasing trend in ticket sales.

4. Make a line graph to show the given information.

	Mon.	Tues.	Wed.	Thurs.	Fri.	Sat.	Sun.
Number of Miles Cindy Bicycled	5	3	7	1	0	5	4

5. The line graph shows Manny's scores on weekly reading tests.

   a. What was the difference between Manny's highest score and his lowest score?

   b. Estimate Manny's score in week 4.

   c. What was the increase from week 6 to 7?

   d. What was the greatest decrease in Manny's reading score from one week to the next?

   e. In which week did Manny score 30 points higher than he did in week 1?

   f. Which week showed a drop of about 2 points?

   g. Describe the trend in Manny's scores.

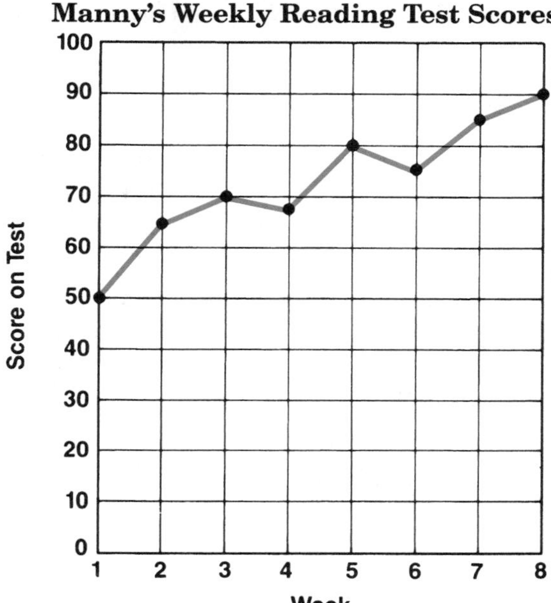

6. This graph shows the number of cars sold by Happy Harry's Car Lot during an eight-week period. Explain what you would do to improve the graph. Be specific.

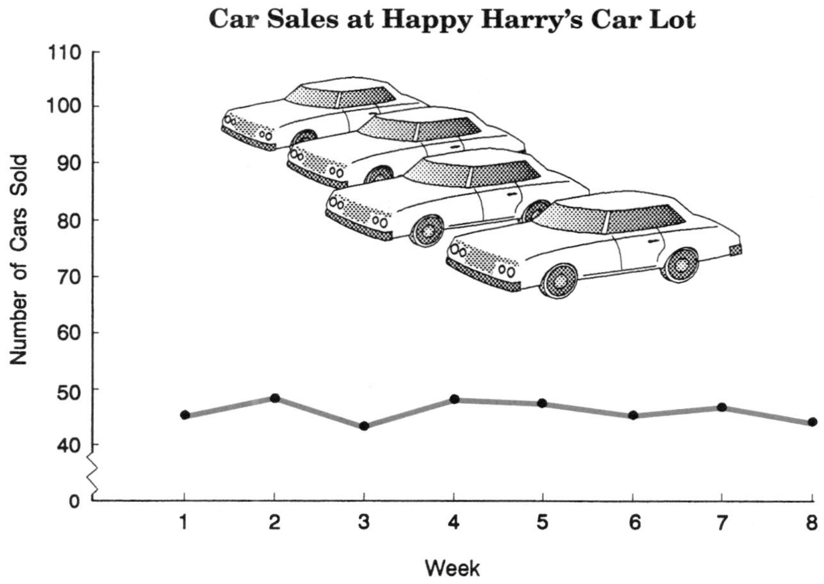

OPERATIONS WITH WHOLE NUMBERS

UNIT W-1 Place Value; Writing, Reading Whole Numbers

MEANING OF PLACE VALUE

THE MAIN IDEA

1. A ***whole number*** is a member of the set {0, 1, 2, 3, 4,...}.
2. A ***numeral*** is a symbol for a number. Each of the whole numbers can be written as a numeral by using the ***digits*** 0, 1, 2, 3, 4, 5, 6, 7, 8, and 9.
3. The ***position*** or ***place*** of each digit in a numeral is named as follows:

Billions			,	*Millions*			,	*Thousands*			,	*Ones*		
hundred billions	ten billions	billions	,	hundred millions	ten millions	millions	,	hundred thousands	ten thousands	thousands	,	hundreds	tens	ones

The positions are grouped into sets of three, separated by commas as shown above. Each group is called a ***period***, and each period has a name such as billions, millions, thousands, and ones.

4. The ***place value of a position*** is the numerical value taken from its name. *As you move to the left, each position has a place value of ten times the previous position.*
5. To find the ***place value of a digit*** in a numeral, multiply the digit by the value of the place in which the digit appears.

EXAMPLE 1 Name the position and tell the value of each of the underlined digits.

Numeral	*Answer*
a. $7\underline{8}4$	8 is in the tens position and has the value of 8×10 or 80.
b. $2\underline{3},491$	3 is in the thousands position and has the value of $3 \times 1,000$ or 3,000.
c. $6\underline{5}2,418$	5 is in the ten thousands position and has the value of $5 \times 10,000$ or 50,000.

CLASS EXERCISES

1. Name the position of 7 in each numeral.
a. 972 **b.** 67,431 **c.** 8,517 **d.** 375,913 **e.** 1,762

2. Name the position and tell the value of each of the underlined digits.
a. 5,$\underline{4}$73 **b.** $\underline{3}$,528,916 **c.** 8$\underline{3}$,419 **d.** 5,$\underline{9}$61,482 **e.** $\underline{3}$9,615

3. Complete each statement to make it true.

a. In 375, the digit 3 is in the __________ position.

b. In 28,972, the digit 8 is in the __________ position and has the value of __________.

c. The value of 2 in 7,241 is __________ times greater than the value of 2 in 623.

WRITING WHOLE NUMBERS AS NUMERALS

THE MAIN IDEA

To write as a numeral a whole number that is expressed in words:

1. Determine the number of places needed. (Use the name of the position farthest to the left.)
2. Moving from right to left, separate the number of places into periods of three by commas. The period farthest to the left may have fewer than three places.
3. Fill each period with numerals from left to right.
4. If there is no digit for a place in the whole number, use 0 as a placeholder.

EXAMPLE 2 Write three million, fifty-two thousand as a numeral.

___, ___ ___ ___, ___ ___ ___	"Million" tells us that we need seven places.
3, ___ ___ ___, ___ ___ ___	Fill periods from left to right.
3, 0 5 2, ___ ___ ___	Since there is no digit for the hundred thousands place, use 0 as a placeholder there.
3, 0 5 2, 0 0 0	Since there are no digits for the hundreds, tens, or ones places, use three 0's as placeholders there.

CLASS EXERCISES

Write as a numeral:

1. four hundred twenty-three
2. seven thousand, eight hundred sixty-two
3. fifteen thousand, three hundred twelve
4. one million, four hundred sixty-two
5. four million, three hundred sixty-two thousand, one hundred twenty
6. nine hundred twenty-one thousand, two hundred seventeen
7. six thousand, fifty-two
8. two hundred eight thousand
9. two hundred thousand, five hundred
10. seven million, seventeen thousand

READING WHOLE NUMBERS THAT ARE WRITTEN AS NUMERALS

THE MAIN IDEA

To read a whole number that is written as a numeral:

1. Begin with the largest period, the farthest to the left. Read the 1-, 2-, or 3-digit number followed by the name of the period, dropping the final "s."
2. Repeat this process for each of the remaining periods. When reading the ones period, do not say the name "ones."

EXAMPLE 3 Read: 3,000,809

3,	000,	809
millions	thousands	ones

Begin with the largest period, the farthest to the left. Read the number in this period followed by the word "million," that is: three million

Since the next period, thousands, has only zeros that are used as placeholders, skip it and read the following period.

Read the number in the ones period, without the period name: eight hundred nine

Answer: 3,000,809 is read: three million, eight hundred nine

EXAMPLE 4 Read the calculator display: 14572328.

KEYING IN

Since a calculator display does not separate groups of digits with commas, to read large numbers on a calculator display, put the digits into groups of three, from right to left.

14572328. ⟶ 14,572,328

Answer: fourteen million, five hundred seventy-two thousand, three hundred twenty-eight

CLASS EXERCISES

Read each whole number.

1. 397 **2.** 5,146 **3.** 23,412 **4.** 352,000 **5.** 4,370,000

Read the calculator display and write its word name.

6. 5642. **7.** 836. **8.** 379812. **9.** 89652. **10.** 21826519.

WRITING A WHOLE NUMBER AS A WORD NAME

THE MAIN IDEA

To write a whole number as a word name:

1. Write each period as you read it, separating period names with commas.
2. Use a hyphen to connect tens and ones in each period.

EXAMPLE 5 Write the word name for 541,096.

$$\underbrace{541,}_{\text{thousands}} \quad \underbrace{096}_{\text{ones}}$$

Begin with the largest period, thousands. Read the number in this period (541) and write its name followed by the word "thousand": five hundred forty-one thousand

Read the number in the next period (96). Since this is the ones period, omit the period name: ninety-six

Notice that the 0 is used as a placeholder, but it is not written in the word name.

Notice also that the tens and ones in each period are connected by a hyphen.

Answer: The word name for 541,096 is five hundred forty-one thousand, ninety-six.

CLASS EXERCISES

1. Write a word name for each whole number.

a. 48 **b.** 783 **c.** 2,475 **d.** 257,402 **e.** 3,000,400

2. Write a word name for the number on each calculator display.

a. 21. **b.** 104. **c.** 706000. **d.** 1234. **e.** 9470082.

HOMEWORK EXERCISES

1. Name the position of 9 in each number.

a. 93 **b.** 198 **c.** 4,009 **d.** 907 **e.** 97,832 **f.** 9,637,728
g. 1,903,275 **h.** 9,134,000,000 **i.** 9,752,186,824

2. Determine the value of each of the underlined digits.

a. $4\underline{9}2$ **b.** $4,\underline{5}82$ **c.** $27\underline{6}$ **d.** $8\underline{1},079$ **e.** $\underline{3},457,892$ **f.** $2\underline{7}5,342$
g. $\underline{4},132,789,076$ **h.** $871,64\underline{2},185$ **i.** $659,2\underline{8}1$

3. Write as a numeral:
   a. two hundred six b. one thousand, ninety-two
   c. fifteen thousand, seventy d. one hundred thirty thousand, nine
   e. two million, seventy-three thousand, one hundred eleven
   f. seventy thousand, six hundred eighty-six g. five billion, three hundred thousand

4. Write a word name for each whole number.
   a. 251 b. 45,075 c. 3,010 d. 4,150,000 e. 505,028
   f. 7,002 g. 205,918 h. 1,005,870 i. 209 j. 970,091

5. Read each calculator display.
   a. 7941. b. 84002. c. 3700965. d. 401. e. 6023007.

6. In which place is the digit 6 in the numeral 986,452?

7. What is the value of 7 in 1,732,584?

8. In which numeral does the digit 4 have the value of forty?
   (a) 4,580 (b) 17,410 (c) 40,000 (d) 92,040

9. In the numeral 782,639, the digit 2 has the value
   (a) 20 (b) 200 (c) 2,000 (d) 20,000

10. By how much will the number 26,493 be increased if you change the digit 6 to an 8?

11. In which numeral is the digit 1 read as "one"?
    (a) 410 (b) 411 (c) 401 (d) 10,000

In 12–14, the first three numerals follow a pattern. Write the next numeral in the pattern.

12. 5, 8, 11, ? 13. 111, 222, 333, ? 14. 7–1, 8–2, 9–3, ?

15. One way to describe the pattern followed by the numerals 19, 28, 37, . . . , 91 is that 9 is added to get the next numeral. Describe the pattern in another way.

16. Determine a constant that can multiply 10, 11, 12, . . . , 20 to give the following sequence, and use the constant feature to fill in the missing terms.

    1010., 1111., 1212., . . . , 2020.,

UNIT W-2 **Rounding Whole Numbers**

THE MAIN IDEA

1. A ***rounded number*** is an *approximation* for an exact whole number.
2. To *round* a whole number to a desired place:
   a. Circle the digit in the place to be rounded.
   b. Look at the digit to the immediate right of the circled digit.
      (1) If the digit to the immediate right is 5 or more, add 1 to the circled digit *(round up)*.
      (2) If the digit to the immediate right is less than 5, keep the circled digit as is *(round down)*.
   c. Replace all the digits to the right of the rounded digit by zeros.
3. Rounded numbers are often used in estimation because they are easier to work with than the exact numbers.

EXAMPLE 1 Round 8,273 to the nearest ten.

8,2⑦3	"The nearest ten" tells you that the digit to be rounded is in the tens place. Circle the 7. The digit to the immediate right of the circled digit is 3.
8,2⑦_	Since the digit to the right is less than 5, *round down* (leave the circled digit as is).
8,2⑦0	Replace all the digits to the right of the rounded digit by zeros.

Answer: 8,273 rounded to the nearest ten is 8,270.

EXAMPLE 2 Round 21,895 to the nearest thousand.

2①,895	"The nearest thousand" tells you that the digit to be rounded is in the thousands place. Circle the 1. The digit to the right of the circled digit is 8.
2②,_ _ _	As the digit to the right is greater than 5, *round up* (add 1 to the circled digit).
2②,000	Replace all the digits to the right of the rounded digit by zeros.

Answer: 21,895 rounded to the nearest thousand is 22,000.

EXAMPLE 3 In one year, 3,951,386 people were born in the United States. Round the number of births to the nearest hundred thousand.

3,(9) 51,386	"The nearest hundred thousand" tells you that the digit to be rounded is in the hundred thousands place. Circle the 9. The digit to the immediate right of the circled digit is 5.
4,(0) _ _, _ _ _	Since the digit to the right is 5, *round up* (in this case, 39 + 1 = 40).
4,(0)00,000	Replace all the digits to the right of the rounded digit by zeros.

Answer: 3,951,386 rounded to the nearest hundred thousand is 4,000,000.

EXAMPLE 4 1,111,111 rounded to the nearest million is:
(a) 1,000,000 (b) 1,100,000 (c) 1,110,000 (d) 1,111,000

Answer: (a)

CLASS EXERCISES

1. Round 1,465 to the nearest hundred.
2. Round 293,124 to the nearest ten thousand.
3. Round 82,563 to the nearest ten.
4. Round 4,986,416 to the nearest hundred thousand.
5. In 1982, Star Company had total sales of $51,418,000. Round the total sales to the nearest million.
6. The population of Belgium is 10,182,034. Round the population to the nearest ten thousand.
7. 55,555,555 rounded to the nearest ten million is
   (a) 50,000,000 (b) 56,000,000 (c) 55,600,000 (d) 60,000,000
8. The area of Japan is approximately 370,325 square kilometers. Round the area to the nearest thousand square kilometers.
9. In 1998, Japan produced 11,200,000 metric tons of rice. Round this amount of rice to the nearest:
   a. million metric tons **b.** hundred thousand metric tons **c.** thousand metric tons

HOMEWORK EXERCISES

1. Round to the nearest ten:

   a. 187 **b.** 47,095 **c.** 3,063 **d.** 583,498

2. Round to the nearest hundred:

   a. 592 **b.** 15,813 **c.** 819,457 **d.** 1,596,962

3. Round to the nearest thousand:

   a. 4,527 **b.** 153,697 **c.** 73,918 **d.** 1,249,897

4. Write in your own words how you would round 592,483 to the nearest thousand.

5. 372,516 rounded to the nearest hundred thousand is

   (a) 300,000 (b) 370,000 (c) 400,000 (d) 470,000

6. The Empire State Building is 1,250 feet tall. Round its height to the nearest hundred feet.

7. In 1988, George Bush received 48,881,278 votes in the presidential election. Round the number of votes to the nearest ten thousand.

8. The Nile River is 4,145 miles long. Round the length of the Nile River to the nearest ten miles.

9. In one season, about 34,540,000 households watched the Super Bowl. Round this number to the nearest hundred thousand.

10. One year, the world production of cattle was about 1,263,584,000 head. Round this number of cattle to the nearest:

    a. billion head **b.** million head **c.** hundred thousand head

11. In 1998, the production of wheat in Afghanistan was about 2,834,612 metric tons. Round this quantity of wheat to the nearest:

    a. hundred thousand metric tons **b.** ten thousand metric tons **c.** thousand metric tons

12. Two different radio stations broadcasted the same baseball game. One announcer said that the attendance that day was 6,489 people. The other announcer said that 6,500 people attended. Could they both be right? Explain your answer.

UNIT W–3 Adding Whole Numbers

THE MAIN IDEA

1. The result of an addition is called the ***sum***.
2. The numbers in an addition are named in the following way:

$$\begin{array}{r} 2 \leftarrow \text{addend} \\ +5 \leftarrow \text{addend} \\ \hline 7 \leftarrow \text{sum} \end{array} \qquad 2 + 5 = 7$$

(In $2 + 5 = 7$: 2 is an addend, 5 is an addend, 7 is the sum.)

3. To find the sum of whole numbers:
   a. Write the numerals in vertical columns, lining up the ones digits.
   b. Add the digits in the columns, beginning with the ones column.
   c. Remember to carry when necessary.

EXAMPLE 1 Add 2,765; 842; and 98.

Write the numerals in vertical columns, lining up the ones digits.

Add in the ones column (5 + 2 + 8 = 15). Write 5 and carry 1.

$$\begin{array}{r} {}^{1} \\ 2{,}765 \\ 842 \\ +98 \\ \hline 5 \end{array}$$

Add in the tens column (1 + 6 + 4 + 9 = 20). Write 0 and carry 2.

$$\begin{array}{r} {}^{21} \\ 2{,}765 \\ 842 \\ +98 \\ \hline 05 \end{array}$$

Add in the hundreds column (2 + 7 + 8 = 17). Write 7 and carry 1.

$$\begin{array}{r} {}^{1\ 21} \\ 2{,}765 \\ 842 \\ +98 \\ \hline 705 \end{array}$$

Add in the thousands column (1 + 2 = 3). Write 3.

$$\begin{array}{r} {}^{1\ 21} \\ 2{,}765 \\ 842 \\ +98 \\ \hline 3{,}705 \end{array}$$

Answer: The sum is 3,705.

EXAMPLE 2 Given the populations of the five boroughs of New York City, as listed by the 1990 census, find the total.

Borough	***Population***
Bronx	1,203,789
Brooklyn	2,300,664
Manhattan	1,487,536
Queens	1,951,598
Staten Island	378,977

KEYING IN

To find the sum with a calculator, press [+] after entering each number. Then press the [=].

Key Sequence: 1168972 [+] 2230936 [+] 1428285 [+] 1891325 [+] 352121 [=]

Final Display: 7071639.

Answer: The total population of New York City in 1980 was 7,071,639.

EXAMPLE 3 Abe wants to buy a stereo system. The receiver costs $479, the speakers $396, and the CD player $82. Find the total cost.

THINKING ABOUT THE PROBLEM

The given information are the three amounts, $479, $396, and $82. The key words "total cost" tell you to use addition.

Find the total cost by adding the price of each of the separate parts.

$$\begin{array}{r} 479 \\ 396 \\ +82 \\ \hline 957 \end{array}$$

Answer: The system costs $957.

EXAMPLE 4 At a football game, 15,283 fans bought general admission tickets, 2,815 box seats, and 11,017 bleacher seats. About how many fans attended the game?

(a) 28,000 (b) 39,000
(c) 19,000 (d) 29,000

THINKING ABOUT THE PROBLEM

You can use rounded numbers to estimate the sum. Replace 15,283 by 15,000, 2,815 by 3,000, and 11,017 by 11,000.

$$\begin{array}{r} 15{,}000 \\ 3{,}000 \\ +11{,}000 \\ \hline 29{,}000 \end{array}$$

Add the rounded numbers.

Answer: (d)

EXAMPLE 5 The Changs are driving from their home in Titus to attend a reunion in Laren.

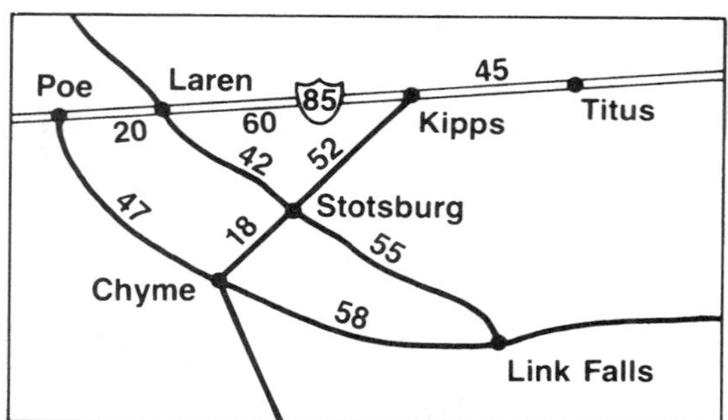

a. If they travel by way of Stotsburg, how many miles must they drive from Titus to Laren?

The drive is from Titus to Kipps to Stotsburg to Laren: 45 + 52 + 42 = 139 mi.

b. If they go back to Titus from Laren directly along Route 85, how many miles will they drive?

Add the two distances along Route 85: 60 + 45 = 105 miles

CLASS EXERCISES

1. Find the sum of each pair of whole numbers.

a. 215 and 381 **b.** 472 and 749 **c.** 1,896 and 4,528 **d.** 805 and 96
e. 5,947 and 859 **f.** 40,816 and 92,458 **g.** 429 and 1,571

2. Add:

a. $\begin{array}{r} 57 \\ 962 \\ \underline{439} \end{array}$ **b.** $\begin{array}{r} 1{,}092 \\ 4{,}842 \\ \underline{658} \end{array}$ **c.** $\begin{array}{r} 579 \\ 6{,}842 \\ \underline{39{,}275} \end{array}$ **d.** $\begin{array}{r} 26{,}468 \\ 5{,}904 \\ \underline{17{,}629} \end{array}$ **e.** $\begin{array}{r} 8 \\ 392 \\ 486 \\ \underline{114} \end{array}$

3. Add: 2,581; 698; and 57 **4.** Add: 5,476; 2,398; and 11,465

5. At a football game between Beefburg and Fillmore High Schools, there were 378 students from Beefburg and 589 students from Fillmore. What was the total number of students at the game?
(a) 857 (b) 1,967 (c) 967 (d) 1,000

6. The base price of a car is $13,752. The cost of a stereo CD system is $286, and the cost of air conditioning is $837. Find the total cost of the car.

7. In March, Mr. Harris drove 3,729 miles, in April, 4,162 miles, and in May, 5,236 miles.

a. Round each number to the nearest thousand.

b. Add the rounded numbers to estimate his total mileage in this 3-month period.

c. Use the estimate to choose the exact answer: (a) 30,127 (b) 23,127 (c) 13,127 (d) 10,127

In 8 and 9, use the road distances shown on the map on page 60.

8. How far is it by the shortest route from Chyme to Kipps?

9. What is the shortest distance between Poe and Stotsburg?

10. Government employees may work for the United States (federal) government, the state, or the local city or town.

What is the total number of government employees shown in the list?

Level	*Number of Employees*
Federal	3,091,000
State	4,115,000
Local	10,076,000

11. Which group of rounded numbers gives the best estimate of the sum of 58 and 81?
(a) 50 + 80 (b) 60 + 90 (c) 60 + 80 (d) 55 + 90

HOMEWORK EXERCISES

1. Add:

a. $638 + 275$ **b.** $1,240 + 760$ **c.** $3,087 + 13$ **d.** $739 + 31 + 251$ **e.** $401 + 97 + 3,967$ **f.** $2,003 + 4,884 + 3,127$

2. Find each sum.

a. 22 + 49 **b.** 113 + 99 **c.** 279 + 88 **d.** 109 + 202 + 0 **e.** 196 + 28 + 42
f. 97 + 246 + 524 **g.** 28 + 563 + 72 **h.** 398 + 47 + 53 **i.** 2,756 + 39 + 842

3. John's weekly salary is $730. Last week he earned $180 more by working overtime. How much did he earn last week?

4. A used car costs $7,980 and the sales tax is $634. What is the car's cost, including the sales tax?

5. Mary delivered 132 newspapers on Monday, 143 newspapers on Tuesday, and 118 newspapers on Wednesday. Find the total number of papers delivered on these three days.

In 6 and 7, use the road distances shown on the map on page 60.

6. How far is it from Poe to Titus? **7.** What is the shortest distance between Laren and Chyme?

8. Losses caused by three tornadoes were $93,230,840; $39,500,000; and $11,747,500.

   a. Round each number to the nearest million.

   b. Add the rounded numbers and estimate the total losses for the three tornadoes.

   c. Use the estimate to choose the exact answer:
   (a) $136,482,360 (b) $139,584,490 (c) $142,386,720 (d) $144,478,340

9. The number of births in the United States for five consecutive years are given in the table. Find the total number of births in the United States during this five-year period.

Year	Number of Births
1	3,680,537
2	3,638,936
3	3,669,141
4	3,760,561
5	3,735,000

10. Each of the letters A, B, and C represents a different digit in this addition.

    Find the value of each of the three letters.

$$\begin{array}{r} 4\ 9 \\ +\ 5\ A \\ \hline C\ B\ 7 \end{array}$$

11. The graph shows the number of fares sold from Capital City to Rolling Hills for each of five months. Find the total number of tickets sold.

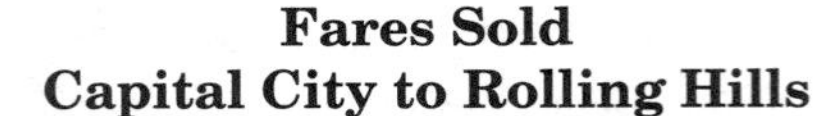

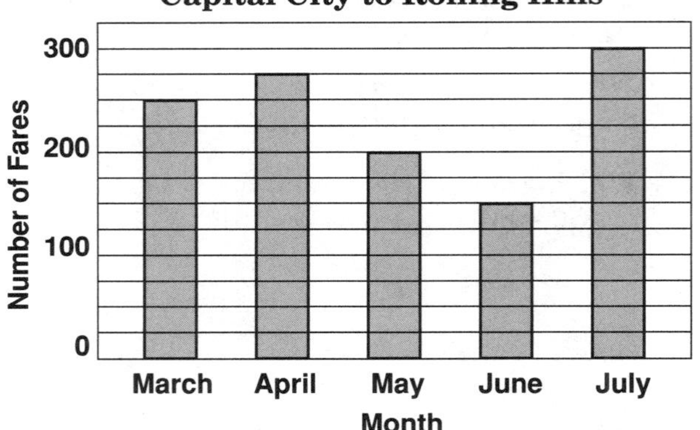

12. Which group of rounded numbers gives the best estimate of the sum 19 + 21 + 62?

    (a) 10 + 20 + 60 (b) 20 + 20 + 60
    (c) 20 + 30 + 60 (d) 20 + 30 + 70

13. Which numbers give the best estimate for 838 + 113 + 42?

    (a) 800 + 100 + 40 (b) 830 + 110 + 40
    (c) 900 + 200 + 50 (d) 840 + 110 + 40

14. The Abrams family wants to display their house number, 1492, as a written word name. Ace Hardware sells brass words that can be used as house numbers. The price for each word is given in the table.

    a. What will be the total cost of the brass words for the Abrams' house number (1492)?

    b. Study the prices and explain how the price for each word was determined.

Word	Price	Word	Price
One	$3	Twenty	$6
Two	3	Thirty	6
Three	5	Forty	5
Four	4	Fifty	5
Five	4	Sixty	5
Six	3	Seventy	7
Seven	5	Eighty	6
Eight	5	Ninety	6
Nine	4	- (hyphen)	1
Ten	3	Hundred	7
		Thousand	8

UNIT W-4 Subtracting Whole Numbers; Addition and Subtraction as Inverse Operations

SUBTRACTING WHOLE NUMBERS

THE MAIN IDEA

1. The result of a subtraction is called the ***difference***.
2. The numbers in a subtraction are named in the following way:

$$\begin{array}{rl} 7 & \leftarrow \text{minuend} \\ -5 & \leftarrow \text{subtrahend} \\ \hline 2 & \leftarrow \text{difference} \end{array}$$

3. To subtract two whole numbers:
   a. Write the numerals in vertical columns, lining up the ones digits. (The "from" number goes on top.)
   b. Subtract the digits in the columns, beginning with the ones column.
   c. If a digit in the subtrahend is greater than the digit above it in the minuend, rename the minuend.

EXAMPLE 1 Subtract 97 from 238.

The "from" number is the minuend. Write the numerals in vertical columns.

$$\begin{array}{r} 238 \\ -97 \\ \hline \end{array}$$

Subtract in the ones column (8 – 7 = 1).

$$\begin{array}{r} 238 \\ -97 \\ \hline 1 \end{array}$$

In the tens column, the digit in the subtrahend is greater than the digit in the minuend (9 is greater than 3).

You must rename the minuend: Write one of the hundreds as 10 tens. Thus, 238 is renamed as 1 hundred + 13 tens + 8 ones.

$$\begin{array}{rrr} {\scriptstyle 1} & {\scriptstyle 13} & \\ \not{2} & \not{3} & 8 \\ - & 9 & 7 \\ \hline & 4 & 1 \end{array}$$

Subtract in the tens column (13 – 9 = 4).

Subtract in the hundreds column (1 – 0 = 1).

$$\begin{array}{rrr} {\scriptstyle 1} & {\scriptstyle 13} & \\ \not{2} & \not{3} & 8 \\ - & 9 & 7 \\ \hline 1 & 4 & 1 \end{array}$$

Answer: The difference is 141.

EXAMPLE 2 Mr. Smith borrowed $7,542 to buy a car. He has repaid $4,789. How much does he still owe?

THINKING ABOUT THE PROBLEM

Since Mr. Smith has repaid some of the money he borrowed, the original amount has been decreased. This tells you to subtract.

You must find the difference between 7,542 and 4,789.

$$\begin{array}{rrrr} {\scriptstyle 6} & {\scriptstyle 14} & {\scriptstyle 13} & {\scriptstyle 12} \\ \not{7}, & \not{5} & \not{4} & \not{2} \\ -4, & 7 & 8 & 9 \\ \hline 2, & 7 & 5 & 3 \end{array}$$

Answer: Mr. Smith still owes $2,753.

EXAMPLE 3 The population of California in 1990 was 29,785,857. In 1970, the population was 19,971,069. By how much did the population increase from 1970 to 1990?

KEYING IN

Key Sequence: 29785857 [−] 19971069 [=] Display: [9814788.]

Answer: The population increased by 9,814,788.

EXAMPLE 4 Of 150 students in a school, 7 were absent from school because of illness, and 25 were away on a school trip. How many students remained in school?

THINKING ABOUT THE PROBLEM

Break the problem into simpler problems. First find the *total* number of students who were not in school. Then, subtract to find the number of students left.

To find the total number of students not in school: $7 + 25 = 32$

To find the number of students left: $150 - 32 = 118$

Answer: 118 students remained in school.

CLASS EXERCISES

1. Subtract: **a.** $\begin{array}{r} 89 \\ -54 \\ \hline \end{array}$ **b.** $\begin{array}{r} 586 \\ -271 \\ \hline \end{array}$ **c.** $\begin{array}{r} 9{,}873 \\ -7{,}452 \\ \hline \end{array}$ **d.** $\begin{array}{r} 803 \\ -486 \\ \hline \end{array}$ **e.** $\begin{array}{r} 961 \\ -875 \\ \hline \end{array}$ **f.** $\begin{array}{r} 1{,}009 \\ -546 \\ \hline \end{array}$
2. Subtract: **a.** 96 – 43 **b.** 881 – 472 **c.** 709 – 546 **d.** 5,002 – 3,895
3. From 4,971 subtract 875.
4. Subtract 279 from 352.
5. From the sum of 461 and 29, subtract 113.
6. A hospital needs $425,600 to buy a piece of new equipment. It has received $283,700 in donations. How much money is still needed?
7. Teresa had $45. She spent $21 for a blouse and $12 for a belt. How much money did she have left?
8. The number of marriages in the United States in 1976 was 2,154,807. In 1968 the number of marriages was 2,069,258. How many more marriages took place in 1976 than in 1968?
9. Which expression provides the best estimate of the difference 62 – 19?

   (a) 60 – 10 (b) 60 – 20 (c) 70 – 10 (d) 70 – 20

ADDITION AND SUBTRACTION AS INVERSE OPERATIONS

THE MAIN IDEA

1. Addition and subtraction are called ***inverse operations*** because each can be used to undo the other.
2. Use the inverse of an operation to check the operation.

EXAMPLE 5 Use $7 + 9 = 16$ to show that subtraction is the inverse of addition.

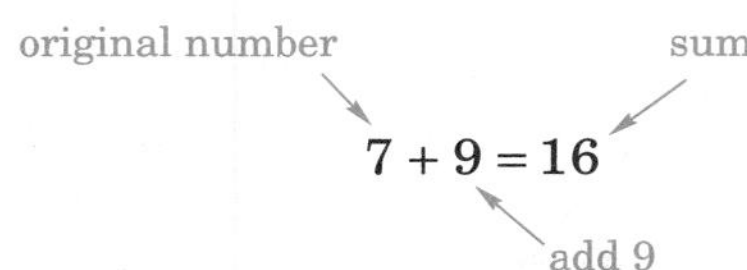

To undo this addition, subtract 9 from the sum.

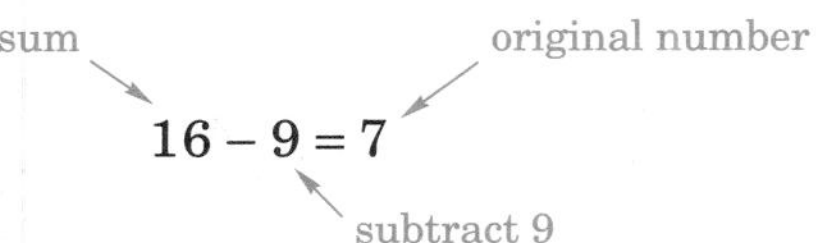

EXAMPLE 6 Find the difference between 327 and 712. Check by using the inverse operation.

712 ←	original number
−327 ←	subtract
385 ←	The difference is 385.

Check by using addition, the inverse of subtraction.

385 ←	difference
+327 ←	add
712 ←	Since 712 is the original number, the subtraction is correct.

EXAMPLE 7 Use a calculator to subtract 92 from 126.

Then check by using the inverse operation.

KEYING IN

To check an answer on the calculator, do not clear the answer from the calculator. Instead, use the answer and the inverse operation.

	Key Sequence	*Display*
Subtract:	126 [−] 92 [=]	34.
Check:	[+] 92 [=]	126.

Answer: The difference is 34.

CLASS EXERCISES

1. Use 11 + 7 = 18 to show that subtraction is the inverse of addition.
2. Use 57 – 23 = 34 to show that addition is the inverse of subtraction.
3. Perform each operation. Check by using the inverse operation.

   a. 197 + 56 **b.** 2,006 – 492 **c.** 83 + 126 **d.** 549 – 387
4. Show that 4,197 + 384 = 4,581 is correct by using the inverse operation as a check.
5. Show that 851 – 675 = 176 is correct by using the inverse operation as a check.
6. Write in your own words why 53 + 15 – 15 = 53.
7. Write a calculator key sequence to do each operation and to check the answer.

   a. 147 + 62 **b.** 315 – 137

HOMEWORK EXERCISES

1. Subtract:

	a.	**b.**	**c.**	**d.**	**e.**	**f.**
	96 –43	487 –236	2,948 –528	5,689 –2,627	706 –253	3,002 –993

	g.	**h.**	**i.**	**j.**	**k.**	**l.**
	6,219 –1,474	3,223 –1,911	2,016 –49	9,000 –111	5,000 –2,784	2,975 –1,897

2. Subtract: **a.** 32 – 9 **b.** 164 – 38 **c.** 400 – 76 **d.** 222 – 123

   e. 2,841 – 726 **f.** 7,049 – 3,237 **g.** 5,004 – 2,475 **h.** 1,000 – 555
3. Subtract 89 from 275.
4. From the sum of 306 and 64, subtract 101.
5. Find the difference between 593 and 479.
6. How much larger is 976 than 842?
7. How much smaller is 246 than 783?
8. Perform each operation. Check by using the inverse operation.

   a. 549 – 474 **b.** 74 + 89 **c.** 187 – 96 **d.** 231 – 0 **e.** 2,007 + 3,049

   f. 502 – 496 **g.** 1,807 – 1,698 **h.** 321–123 **i.** 3,250 + 3,250
9. Using the inverse operation as a check, determine if each statement is correct.

   a. 893 – 486 = 407 **b.** 125 + 236 = 351 **c.** 1,209 – 675 = 434

10. Write a calculator key sequence to do each operation and to check the answer.
   a. 515 – 236   b. 1,047 + 2,414

11. There were 13 states in the United States in 1790. Now there are 50 states. How many more states are there now than there were in 1790?

12. Christopher Columbus sailed to America in 1492. The Declaration of Independence was signed in 1776. How many years apart were the two events?

13. Alaska is our largest state. It has 586,400 square miles of land. Our smallest state, Rhode Island, has 1,200 square miles of land.
   a. How much bigger is our largest state than our smallest state?
   b. What is the total number of square miles in Alaska and Rhode Island?

14. John, Betty, and Sam ran for president of their class. John received 40 votes, and Betty received 53 votes. If 200 votes were cast, how many votes did Sam receive?

15. Joseph's car weighs 3,175 pounds and Marcia's car weighs 2,950 pounds.
   a. Round each number to the nearest hundred.
   b. By subtracting the rounded numbers, estimate how many pounds heavier Joseph's car is.
   c. Use the estimate to choose the exact answer: (a) 3,205 (b) 325 (c) 2,255 (d) 225

16. In 1986, the number of female babies born in the United States was 1,831,679. The number of male babies was 1,924,868. How many more males than females were born?

17. When Maria's age is subtracted from Tama's age, the result is the same as when Tama's age is subtracted from Maria's age. Explain how this can be true.

18. Peggy sells souvenirs at a baseball stadium. The chart shows her sales for three weeks.
   a. For which type of souvenir was the total number of sales the greatest? the least?
   b. How many more buttons did she sell than pennants?

Item	Week 1	Week 2	Week 3
Buttons	133	137	152
Posters	148	112	128
Pennants	123	144	139

19. Which expression gives the closest estimate of the difference between 158 and 81?

   (a) 200 – 100 (b) 100 – 100 (c) 150 – 80 (d) 160 – 80

20. Choose the closest estimate for 2,605 – 1,897.

   (a) 2,600 – 1,900 (b) 3,000 – 2,000 (c) 2,000 – 1,000 (d) 2,600 – 2,000

UNIT W-5 Multiplying Whole Numbers

MULTIPLYING BY A ONE-DIGIT MULTIPLIER

THE MAIN IDEA

1. Multiplication is a way to accomplish repeated additions of the same number.

   $5 \times 9 = \underbrace{9 + 9 + 9 + 9 + 9}_{\text{9 is written 5 times}} = 45$

2. The result of a multiplication is called the ***product***.

3. Each number in a multiplication is named in the following way:

   $3 \times 2 = 6$ — multiplier: 3; multiplicand: 2; product: 6

   $$\begin{array}{r} 2 \\ \times 3 \\ \hline 6 \end{array}$$
   2 ← multiplicand, ×3 ← multiplier, 6 ← product

4. To multiply a whole number by a one-digit multiplier, multiply *each digit* of the whole number by the multiplier. Start with the ones digit, carrying as needed.

5. To perform multiplications, you must know the multiplication facts.

×	0	1	2	3	4	5	6	7	8	9	10
0	0	0	0	0	0	0	0	0	0	0	0
1	0	1	2	3	4	5	6	7	8	9	10
2	0	2	4	6	8	10	12	14	16	18	20
3	0	3	6	9	12	15	18	21	24	27	30
4	0	4	8	12	16	20	24	28	32	36	40
5	0	5	10	15	20	25	30	35	40	45	50
6	0	6	12	18	24	30	36	42	48	54	60
7	0	7	14	21	28	35	42	49	56	63	70
8	0	8	16	24	32	40	48	56	64	72	80
9	0	9	18	27	36	45	54	63	72	81	90
10	0	10	20	30	40	50	60	70	80	90	100

EXAMPLE 1 Jessie earns $8 for each lawn that she mows. How much will Jessie earn if she mows 4 lawns?

KEYING IN

On the calculator, you can add $8 + 8 + 8 + 8 = 32$ by using the constant feature for addition:

Key Sequence: 8 [+] [=] [=] [=] [=]   Display: [32.]

A quicker solution is to multiply 4×8.

Answer: Jessie will earn $32.

EXAMPLE 2 Find the product of 42 and 3.

Multiply each digit of the multiplicand by the multiplier.

Multiply the ones digit of the multiplicand by the multiplier ($3 \times 2 = 6$).

$$\begin{array}{r} 42 \\ \times 3 \\ \hline 6 \end{array}$$

Multiply the tens digit of the multiplicand by the multiplier ($3 \times 4 = 12$).

$$\begin{array}{r} 42 \\ \times 3 \\ \hline 126 \end{array}$$

Answer: The product is 126.

EXAMPLE 3 Multiply 59 by 8.

Multiply the ones digit of the multiplicand by the multiplier. Since $8 \times 9 = 72$, write 2 in the ones place and carry 7.

$$\begin{array}{r} {}^{7} \\ 59 \\ \times 8 \\ \hline 2 \end{array}$$

Multiply the tens digit of the multiplicand by the multiplier ($8 \times 5 = 40$). Add the number carried ($40 + 7 = 47$).

$$\begin{array}{r} {}^{7} \\ 59 \\ \times 8 \\ \hline 472 \end{array}$$

Answer: The product is 472.

CLASS EXERCISES

1. Write each addition as a multiplication: **a.** $7 + 7 + 7$ **b.** $3 + 3 + 3 + 3 + 3 + 3$
2. Find each product mentally: **a.** 5×3 **b.** 8×2 **c.** 7×9 **d.** 9×4 **e.** 6×6
3. Find each product: **a.** 28×7 **b.** 96×9 **c.** 135×4 **d.** 476×7
4. Write a calculator key sequence that uses the constant feature for addition to multiply 9×5.

5. Use the constant for multiplication to create the "12 times table."

MULTIPLYING BY A MULTIPLIER THAT HAS MORE THAN ONE DIGIT

THE MAIN IDEA

1. To multiply a whole number by a two-digit multiplier:
   a. Start with the ones digit of the multiplier and multiply each digit of the multiplicand by that digit. The result is called a ***partial product***.
   b. Next use the tens digit of the multiplier to multiply each digit of the multiplicand. Write a zero in the ones place of this partial product to maintain place value.
2. When the multiplier has more than two digits, continue the procedure.

EXAMPLE 4 Find the product of 358 and 46.

To obtain the first partial product, multiply each digit of the multiplicand by the ones digit of the multiplier, carrying as necessary.

```
  358
  ×46
2,148  ← first partial product
```

Begin the second partial product by writing a zero in the ones place. Then multiply each digit of the multiplicand by the tens digit of the multiplier, carrying as necessary.

```
   358
   ×46
 2,148
14,320  ← second partial product
```

Add the two partial products, carrying as necessary. (In this case, there is no carrying.)

```
   358
   ×46
 2,148
14,320
16,468  ← product
```

Answer: The product is 16,468.

EXAMPLE 5 Multiply 519 by 806.

```
    519
   ×806
  3,114  ← partial product of 519 × 6 ← ones digit
         ← partial product of 519 × 0 ← tens digit
           (does not have to be written because it equals 0)
415,200  ← partial product of 519 × 8 ← hundreds digit
418,314    (two zeros are needed as placeholders)
```

Answer: The product is 418,314.

EXAMPLE 6 A theater had 100 rows of seats, with 30 seats in each row. If 2,500 seats were occupied, how many seats were not occupied?

THINKING ABOUT THE PROBLEM

Use the strategy of breaking the problem into smaller problems. First, find the total number of seats in the theater. Then, from this total, subtract the number of occupied seats.

Find the total number of seats: $100 \times 30 = 3{,}000$

Subtract 2,500 occupied seats from the total: $3{,}000 - 2{,}500 = 500$

Answer: 500 seats were not occupied.

EXAMPLE 7 There are 144 items in a gross. Use the constant feature for multiplication on a calculator to complete the table.

Number of Gross	2	5	7	9	15
Number of Items					

KEYING IN

Key Sequence	*Display*
144 × 2 =	288.
5 =	720.
7 =	1008.
9 =	1296.
15 =	2160.

Answer:

Number of Gross	2	5	7	9	15
Number of Items	288	720	1,008	1,296	2,160

CLASS EXERCISES

1. Multiply:
   a. 29×58 b. 76×32 c. 932×65 d. 90×38 e. 504×87 f. 375×402
   g. 562×212 h. 967×38 i. 897×564 j. 989×879 k. 359×231 l. 455×822
2. Find the product of 567 and 54.
3. Find the product of 487 and 608.
4. If each of 36 boxes of shirts contains 24 shirts, how many shirts are there?
5. Channel 79 received 189 calls from listeners, each promising to donate $75 to the station. How much money was promised?
6. A sports stadium has 275 sections. Each section contains 157 seats. How many seats are there?
7. A carton contains 2,715 pieces of candy. Find the number of pieces of candy in 58 cartons.
8. For a school trip, 5 buses each took 48 students to the museum. Thirty-eight students took the train to get there. How many students from the school were at the museum?
9. Mr. Roth's automobile uses, on the average, one gallon of gasoline to travel 27 miles. Use the constant feature for multiplication on a calculator to complete the table.

Number of Gallons Used	2	5	8	11	19
Number of Miles Traveled					

10. Choose the best estimate for 19×91: (a) 10×90 (b) 20×100 (c) 10×100 (d) 20×90

HOMEWORK EXERCISES

1. Find each product: **a.** 9×7 **b.** 13×5 **c.** 82×9 **d.** 47×3 **e.** 125×6 **f.** 99×7 **g.** 211×5 **h.** 49×8 **i.** 36×7 **j.** 28×4

2. Multiply: **a.** 52×37 **b.** 46×54 **c.** 83×71 **d.** 57×48 **e.** 106×42 **f.** 412×55 **g.** 652×92 **h.** 829×88 **i.** 315×95 **j.** 227×109 **k.** 415×223 **l.** 946×385

3. Write a calculator key sequence using the constant feature for addition to multiply 15×6.
4. Find the product of 712 and 75.
5. What is the product of 437 and 52?
6. Find the product of 39 and 516.
7. What is the product of 5 and 5?
8. Ramon earns $58 a week. How much will he earn in 32 weeks?
9. A dealer has 58 cases of soda. If each case contains 24 bottles, how many bottles are there?
10. The manager of a store bought 290 shirts. Each shirt cost $12. What was the total cost?
11. The library in Midtown High School has 418 shelves. Each shelf can hold 36 books. What is the greatest number of books that the library can have on all its shelves?
12. Jennifer bought 3 CD's at $13 each. If she started with $50, how much money remained?

13. In 1927, Charles Lindbergh made the first solo nonstop transatlantic airplane flight. His average speed was approximately 107 miles per hour. Use the constant feature for multiplication on a calculator to complete the table.

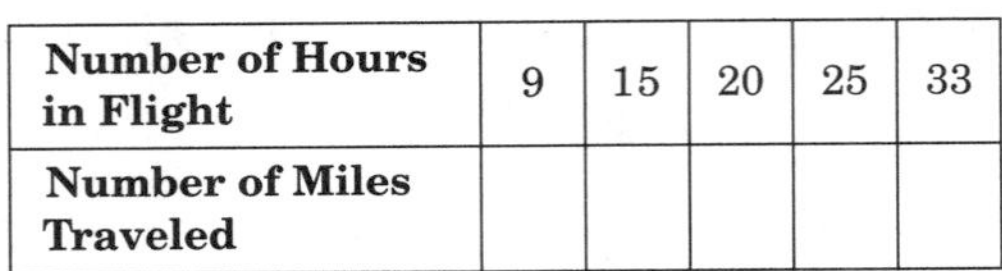

Number of Hours in Flight	9	15	20	25	33
Number of Miles Traveled					

14. The total attendance at a baseball game was 3,245. In the stadium, 48 sections of seats each had 65 fans who bought tickets. The rest of the fans received free passes. How many fans received free passes?
15. The graph shows the number of lunches sold by a school cafeteria for one week. If the cost of a lunch is $2, what is the difference between the greatest amount of money collected and the least amount collected in one day?

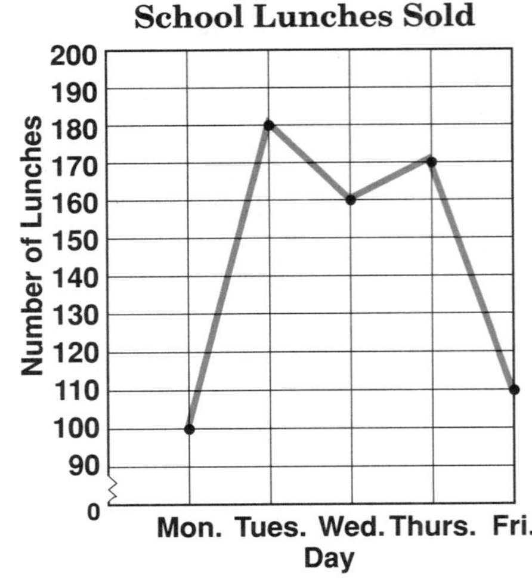

16. Which expression gives the closest estimate for the product of 119 and 182?
    (a) 100×100 (b) 100×200
    (c) 200×200 (d) 200×300
17. Which is the closest estimate for $1{,}989 \times 2{,}206$?
    (a) $2{,}000 \times 2{,}200$ (b) $1{,}000 \times 2{,}000$
    (c) $2{,}000 \times 2{,}000$ (d) $1{,}900 \times 2{,}200$

UNIT W-6 Dividing Whole Numbers; Division Involving Zero

DIVIDING WHOLE NUMBERS

THE MAIN IDEA

1. Division is the inverse of multiplication.

   $7\overline{)63}$ with quotient 9 (divisor 7, dividend 63) means $7 \times 9 = 63$ (divisor 7, quotient 9, dividend 63)

2. The result of a division is called the ***quotient***.
3. Division can be shown in other ways.

   $35 \div 7 = 5$ (dividend 35, divisor 7, quotient 5)

   $\frac{35}{7} = 5$ (dividend 35, divisor 7, quotient 5)

4. If the divisor does not divide the dividend exactly, the "extra" is called the ***remainder***.
5. To check a division, first find the product of the divisor and the quotient. Then add the remainder (if there is one). The resulting sum should equal the dividend.

   Quotient × Divisor + Remainder = Dividend

EXAMPLE 1 Find each quotient and check.

	Divide	***Answer***	***Check***
a.	$4\overline{)28}$	$4\overline{)28}$ = 7; 28 − 28 = 0	$7 \times 4 = 28$
b.	$72 \div 9$	$72 \div 9 = 8$	$8 \times 9 = 72$
c.	$\frac{45}{9}$	$\frac{45}{9} = 5$	$5 \times 9 = 45$

EXAMPLE 2 Divide 96 by 8 and check.

Solution:

(1) *Divide* $8\overline{)9}$. 1 is the largest whole-number quotient.

(2) *Multiply* $1 \times 8 = 8$

(3) *Subtract* $9 - 8 = 1$

$$\begin{array}{r} 1 \\ 8\overline{)96} \\ 8 \\ \hline 1 \end{array}$$

(4) *Bring down* the next digit from the dividend.

$$\begin{array}{r} 1 \\ 8\overline{)96} \\ 8\downarrow \\ \hline 16 \end{array}$$

This four-step process is repeated.

(1) *Divide* $8\overline{)16} = 2$

(2) *Multiply* $2 \times 8 = 16$

(3) *Subtract* $16 - 16 = 0$

$$\begin{array}{r} 12 \\ 8\overline{)96} \\ 8\downarrow \\ \hline 16 \\ 16 \\ \hline 0 \end{array}$$

(4) *Bring down.* There is no digit to bring down.

Check: Since division and multiplication are inverse operations, check a division by using multiplication.

Quotient × Divisor + Remainder = Dividend

$$12 \times 8 + 0 = 96$$

Another way to check a division is to estimate. Replace the exact numbers either with rounded numbers or approximate numbers.

Exact Numbers	***Rounded Numbers***
$8\overline{)96}$	$\begin{array}{r} 10 \\ 10\overline{)100} \end{array}$ ← estimate
	The estimate of 10 is close to 12. So, the answer 12 is reasonable.

Answer: The quotient is 12.

EXAMPLE 3 Divide 2,819 by 7 and check.

Solution:

(1) *Divide* $7\overline{)2}$. This division is not possible. Include the next digit.

$$7\overline{)28} = 4$$

(2) *Multiply* $4 \times 7 = 28$

(3) *Subtract* $28 - 28 = 0$

$$\begin{array}{r} 4 \\ 7\overline{)2,819} \\ 28 \\ \hline 0 \end{array}$$

(4) *Bring down* the next digit from the dividend.

$$\begin{array}{r} 4 \\ 7\overline{)2,819} \\ 28\downarrow \\ \hline 01 \end{array}$$

Repeat this four-step process.

(1) *Divide* $7\overline{)1}$. This division is not possible. Use 0 as a placeholder in the quotient, and bring down the next digit from the dividend.

$$7\overline{)19} = 2$$

$$\begin{array}{r} 40 \\ 7\overline{)2,819} \\ 28\downarrow\downarrow \\ \hline 019 \end{array}$$

(2) *Multiply* $2 \times 7 = 14$

(3) *Subtract* $19 - 14 = 5$

(4) *Bring down.* There is no digit to bring down.

$$\begin{array}{r} 402 \\ 7\overline{)2,819} \\ 28 \\ \hline 019 \\ 14 \\ \hline 5 \end{array}$$ ← remainder

Check:

$$\begin{array}{r} 402 \\ \times 7 \\ \hline 2,814 \\ +5 \\ \hline 2,819 \end{array}$$

402 ← quotient
×7 ← divisor
+5 ← Add the remainder.
2,819 ← dividend

Answer: The quotient is 402, and the remainder is 5.

EXAMPLE 4 A group of 210 students is to be divided into 15 teams with an equal number of students on each team. Find how many students should be on each team.

THINKING ABOUT THE PROBLEM

Since each team will be smaller than the whole group, the answer must be a number smaller than 210. The problem uses the key words "divided into" and "each."

Divide 210 by 15. *Check:*

$$\begin{array}{r} 14 \\ 15\overline{)210} \\ \underline{15} \\ 60 \\ \underline{60} \\ 0 \end{array} \qquad \begin{array}{rl} 14 & \leftarrow \text{quotient} \\ \underline{\times 15} & \leftarrow \text{divisor} \\ 70 & \\ \underline{140} & \\ 210 & \leftarrow \text{dividend} \end{array}$$

Answer: 14 students on each team

EXAMPLE 5 Treewick High School has 153 seniors scheduled to go on a trip. Each bus can hold 45 students. How many buses should be hired?

KEYING IN

The 153 students are to be divided into groups of not more than 45 each.

Key Sequence: 153 [÷] 45 [=]

Display: [3.4]

The digit after the decimal point shows that the quotient is 3 and that there is also a remainder. Thus, more than 3 buses are needed.

Answer: 4 buses are needed.

CLASS EXERCISES

1. Find each quotient: **a.** $9\overline{)54}$ **b.** $3\overline{)18}$ **c.** $64 \div 8$ **d.** $32 \div 8$ **e.** $\frac{42}{7}$ **f.** $\frac{18}{9}$
2. Divide and check: **a.** $8\overline{)400}$ **b.** $9\overline{)207}$ **c.** $6\overline{)1,830}$ **d.** $4\overline{)2,024}$ **e.** $7\overline{)42,007}$
3. Divide and check: **a.** $16\overline{)192}$ **b.** $48\overline{)672}$ **c.** $76\overline{)856}$ **d.** $83\overline{)6,308}$ **e.** $76\overline{)18,336}$
4. Mr. Coles made 300 cookies. He wanted to make packages that would each contain 25 cookies. Find how many packages he could make, and check.
5. Show that $(36 \div 6) \div 2 = 36 \div (6 \div 2)$ is a false statement.
6. A distributor needs to ship 587 compact disc players. The largest carton available can hold 18 players. What is the least number of cartons that must be shipped?
7. Which rounded numbers give the best estimate for $38 \div 21$?

   (a) $30 \div 20$ (b) $30 \div 30$ (c) $40 \div 20$ (d) $40 \div 30$

DIVISION INVOLVING ZERO

THE MAIN IDEA

1. When zero is divided by any nonzero number, the quotient is zero.

$$0 \div 714 = 0$$

2. Dividing a number by zero is not possible.

EXAMPLE 6 Explain why $5 \div 0$ is not possible.

Since division is the inverse of multiplication, $5 \div 0 = ?$ means that $? \times 0 = 5$. There is no number to replace ? since any number $\times\, 0 = 0$, not 5.

KEYING IN

If you try to divide by zero on a calculator, an error message will appear.

Key Sequence: 5 [÷] 0 [=] Display: [ERROR] or [E 0.]

CLASS EXERCISES

1. Find the quotient if it can be found, or write "the division is not possible."

a. $0 \div 15$ **b.** $18 \div 0$ **c.** $37\overline{)0}$ **d.** $0\overline{)37}$ **e.** $0 \div 0$

2. Replace ? with a number to make a true statement.

a. $\frac{?}{5} = 0$ **b.** $\frac{0}{9} = ?$ **c.** $\frac{17}{17} = ?$ **d.** $\frac{?}{2,752} = 1$

HOMEWORK EXERCISES

1. Find each quotient: **a.** $19\overline{)38}$ **b.** $15\overline{)60}$ **c.** $50 \div 10$ **d.** $\frac{39}{13}$ **e.** $\frac{15}{3}$

2. Using mathematical symbols, write "60 divided by 15" in three different ways.

3. Divide and check: **a.** $7\overline{)742}$ **b.** $9\overline{)252}$ **c.** $6\overline{)1,530}$ **d.** $8\overline{)24,040}$ **e.** $7\overline{)325}$ **f.** $8\overline{)1,557}$

4. Divide and check:

a. $12\overline{)144}$	**b.** $17\overline{)119}$	**c.** $28\overline{)336}$	**d.** $12\overline{)1{,}080}$	**e.** $53\overline{)10{,}865}$
f. $27\overline{)875}$	**g.** $43\overline{)2{,}559}$	**h.** $81\overline{)5{,}450}$	**i.** $37\overline{)555}$	**j.** $35\overline{)490}$

5. Replace ? by a whole number to make a true statement.

   a. $15 \div 5 = ?$ means $5 \times ? = 15$. **b.** $? \div 9 = 5$ means $9 \times 5 = ?$. **c.** $48 \div ? = 4$ means $? \times 4 = 48$.

6. Tell whether each statement is *true* or *false*.

   a. $\frac{0}{357} = 0$ **b.** $\frac{211}{211} = 1$ **c.** $\frac{137}{0} = 0$ **d.** $\frac{5{,}283}{5{,}283} = 1$ **e.** $\frac{864}{27} = 32$

7. Divide 1,564 by 17.

8. What is the quotient of 9,546 and 43?

9. If 52 cards are dealt to 4 players, how many cards does each player receive?

10. If there are 8 maps on each page of an atlas and 256 maps are shown, how many pages are there in the atlas?

11. The Manville Public Library has 6,045 books on its shelves. Each shelf holds 65 books. How many shelves are there in the library?

12. Forty-eight pounds of food were distributed equally among 16 campers. Each camper carried an additional 17 pounds of equipment. What was the total weight carried by each camper?

13. Quick-Calc calculators are sold to schools only in cases of 36. What is the least number of cases that should be purchased so that 520 students will each have a calculator?

14. Cheryl subtracted the number of games lost by her volleyball team from the total number of games played, and then divided the total number of games played by this result. Her calculator displayed an error message. What does this tell you about the number of games her team won?

15. Ms. Edmunds, the supply secretary of Northeast High School, forgot to enter all the information in her book records. Use the book record card shown to find:

    a. the number of Mathematics books.

    b. the total cost of the English books.

    c. the cost per book of the Science books.

Type of Book	Number of Books	$ Cost per Book	Total Cost
Mathematics	?	15	630
English	65	8	?
Science	91	?	910
			$2,060

16. Which is the best estimate of the quotient of 294 and 31?

    (a) $300 \div 30$ (b) $200 \div 30$ (c) $300 \div 40$ (d) $250 \div 50$

17. Choose the best estimate for $4{,}895 \div 99$.

    (a) $5{,}000 \div 100$ (b) $4{,}000 \div 100$ (c) $5{,}000 \div 90$ (d) $4{,}000 \div 90$

THEME 1

Whole Numbers and Integers

MODULE 1

MODULE 2

UNIT 1-1 Using a Number Line to Compare Whole Numbers

THE MAIN IDEA

1. A ***number line*** is used to draw the ***graph*** of the whole numbers. Each whole number is represented by a point on the number line. The next whole number on a number line can be found by adding 1 to the previous whole number.

2. The symbol > means ***is greater than***. If one number is *to the right of* a second number on a number line, then the first number is greater than the second.
3. The symbol < means ***is less than***. If one number is *to the left of* a second number on a number line, then the first number is less than the second.
4. The symbols > and < always point to the smaller number.
5. If a number is ***between*** two other numbers, then we can write the three numbers in *increasing order* of size separated by < symbols, or we can write the three numbers in *decreasing order* of size separated by > symbols.

EXAMPLE 1 Use the constant feature for addition on your calculator to find the next three whole numbers after 9.

Key Sequence	*Display*
9 [+] 1 [=]	10.
[=]	11.
[=]	12.

Answer: The next three whole numbers after 9 are 10, 11, and 12.

EXAMPLE 2 Use the symbol > or < to rewrite each statement based on the number line shown.

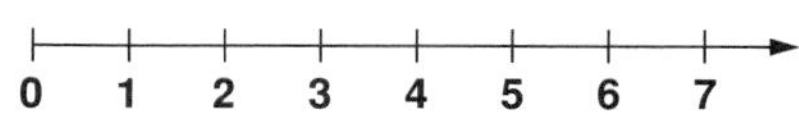

	Statement	*Answer*
a.	5 is greater than 2	$5 > 2$
b.	2 is less than 5	$2 < 5$
c.	2 is between 0 and 5	$0 < 2 < 5$ or $5 > 2 > 0$

When you use more than one inequality symbol in a number statement, the symbols must point in the same direction.

EXAMPLE 3 Tell which of the numbers in each set is *between* the others. Then write the three numbers in order using > or <.

Number Set	*Answer*
a. 2, 3, 5	3 is between 2 and 5 2 < 3 < 5 or 5 > 3 > 2
b. 4, 0, 7	4 is between 0 and 7 0 < 4 < 7 or 7 > 4 > 0

EXAMPLE 4 Tell whether each statement is *true* or *false*. If the statement is false, correct it to make a true statement.

Statement	*Answer*
a. 26 > 32	false; 26 < 32
b. 19 < 20 < 21	true
c. 20 < 19 < 18	false; 20 > 19 > 18
d. 32 > 16 > 3	true
e. 7 + 8 > 12	true
f. 21 − 6 < 6 × 2	false; 21 − 6 > 6 × 2

EXAMPLE 5 Use the data in the table to list the Great Lakes in order from least area to greatest area.

Lake	Area (sq. mi.)	Depth (ft.)
Erie	9,940	210
Huron	23,010	750
Michigan	22,400	923
Ontario	7,540	802
Superior	31,820	1,333

First, compare the areas and list the areas in order from least to greatest. You can compare these numbers by looking at the thousands.

7,540; 9,940; 22,400; 23,010; 31,820

Then write the name of each lake in the same order as its area.

Answer: Ontario, Erie, Michigan, Huron, Superior

CLASS EXERCISES

1. Use the constant feature for addition on your calculator to find the next three whole numbers after 998. Write the calculator key sequence.

2. Use the symbol > or < to rewrite each statement based on the number line shown.

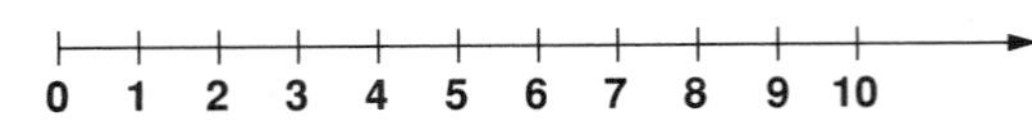

a. 1 is greater than 0 **b.** 0 is less than 1
c. 1 is less than 6 **d.** 6 is greater than 1
e. 1 is between 0 and 6 **f.** 5 is between 2 and 7

3. Tell which of the numbers in each pair is farther to the *right* on a number line.

a. 2; 3 **b.** 27; 19 **c.** 706; 607 **d.** 454; 445 **e.** 1,000; 999

4. Tell which of the numbers in each pair is farther to the *left* on a number line.

a. 5; 4 **b.** 121; 123 **c.** 423; 342 **d.** 7,003; 3,007 **e.** 9,999; 10,103

5. Tell which of the numbers in each set is farthest to the *right* on a number line.

a. 6; 8; 1 **b.** 13; 17; 19 **c.** 441; 404; 440

6. Tell whether each comparison is *true* or *false*. If the statement is false, correct it to make a true statement.

a. $1 > 0$ **b.** $100 < 97$ **c.** $263 > 189$ **d.** $2{,}007 < 2{,}008 < 2{,}009$
e. $332 < 471 < 512$ **f.** $84 > 85 > 86$ **g.** $14 - 9 > 16 - 7$ **h.** $23 - 4 < 17 + 2$

7. Each letter on the number line shown represents a number. Replace each ? by < or > to write a true statement.

a. E ? O **b.** I ? A **c.** A ? U **d.** U ? O ? E **e.** A ? E ? U **f.** I ? O ? U

8. Use the table in Example 5 to list the Great Lakes in order by depth from least to greatest.

HOMEWORK EXERCISES

1. Use the constant feature for addition on your calculator to find the next five whole numbers after 108. Write the calculator key sequence.

2. For each whole number, give the next whole number on a number line.

a. 9 **b.** 57 **c.** 139 **d.** 999 **e.** 1,063

3. Write all the whole numbers that are between each pair of whole numbers.

a. 1; 5 **b.** 8; 11 **c.** 98; 101 **d.** 482; 483 **e.** 999; 1,011

4. Use the symbol > or < to rewrite each statement based on the number line shown.

a. 8 is greater than 4 **b.** 4 is less than 8
c. 0 is less than 3 **d.** 4 is between 0 and 8

5. Tell which of the numbers in each pair is farther to the *right* on a number line.

a. 10; 0 **b.** 17; 27 **c.** 92; 89 **d.** 304; 403 **e.** 1,001; 1,000

6. Tell which of the numbers in each pair is farther to the *left* on a number line.

a. 0; 1 **b.** 20; 19 **c.** 222; 333 **d.** 996; 699 **e.** 2,999; 3,000

7. Tell which of the numbers in each set is farthest to the *left* on a number line.

a. 5; 1; 9 **b.** 21; 19; 7 **c.** 101; 111; 100

8. Tell whether each comparison is *true* or *false*. If the statement is false, correct it to make a true statement.

   a. $19 > 21$ **b.** $189 < 263$ **c.** $2{,}741 < 2{,}742$ **d.** $8{,}999 > 9{,}000$

   e. $701 > 711 > 747$ **f.** $101 < 110 < 111$ **g.** $98 + 3 > 100 - 1$ **h.** $204 - 5 < 199 + 1$

9. Referring to the number line shown, replace each ? by $<$ or $>$ to write a true statement about each pair of numbers.

   a. A ? C **b.** C ? A **c.** D ? E **d.** B ? D **e.** D ? C **f.** A ? C ? B

   g. D ? B ? E **h.** E ? B ? D

10. Complete each statement with $>$ or $<$ to make it true.

    a. If A is to the left of B on a number line, then A __ B.

    b. If A is to the right of B on a number line, the A __ B.

    c. If B is between A and C on a number line, then A $<$ B __ C or C __ B $>$ A.

11. The five largest football stadiums are shown in the table. List them in order from largest seating capacity to smallest.

Stadium	**Seating Capacity**
LA Coliseum	92,516
Michigan Stadium	101,701
Neyland Stadium	91,110
Ohio Stadium	86,071
Rose Bowl	104,091

12. On the number line shown, the letters P, L, A, C, E, and S are to stand for numbers. Given the following information about these numbers, determine their correct order on the number line.

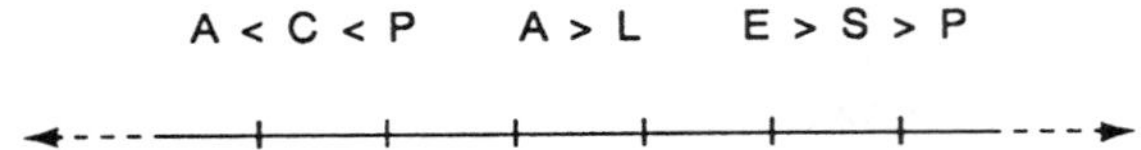

13. A sports announcer reported that 82,000 people attended a football game.

    a. If the number of people attending was rounded to the nearest thousand, what was the least exact number of people who could have attended? What was the greatest exact number?

    b. If the number of people attending was rounded to the nearest hundred, what was the least exact number of people who could have attended? What was the greatest exact number?

UNIT 1–2 Properties of Addition; Properties of Multiplication

PROPERTIES OF ADDITION

THE MAIN IDEA

1. The sum of two whole numbers is always a whole number. This is called the property of ***closure*** for addition of whole numbers.
2. The order in which you add two whole numbers does not change the sum.

   $$5 + 9 = 9 + 5$$

   This is called the ***commutative property for addition***.
3. The way you group numbers in order to perform a series of additions does not change the sum.

   $$(6 + 3) + 7 = 6 + (3 + 7)$$

   This is called the ***associative property for addition***.
4. The sum of 0 and a whole number is that same whole number.

   $$9 + 0 = 9$$

   This is the ***identity property for addition***.
   0 is called the ***additive identity***.

EXAMPLE 1 Tell whether each statement is *true* or *false*.

	Statement	*Answer*
a.	$11 + 5 = 5 + 11$	*True*. The order in which you add numbers does not change the sum.
b.	$(28 + 10) + 2 > 28 + (10 + 2)$	*False*. The way you group numbers does not change the sum.
c.	$0 + 37 < 37$	*False*. The sum of 0 and a number is that number.
d.	If $\triangle$ and $\square$ are whole numbers, then $\triangle + \square$ is a whole number.	*True*. The sum of two whole numbers is a whole number.

EXAMPLE 2 Use the calculator to show that $5 + 7 = 7 + 5$. Then name the property illustrated.

KEYING IN

Add 5 + 7.		Add 7 + 5.	
Key Sequence:	5 [+] 7 [=]	Key Sequence:	7 [+] 5 [=]
Display:	12.	Display:	12.

Answer: Both $5 + 7$ and $7 + 5$ give the same result.
This is the commutative property for addition.

EXAMPLE 3 Show that $(7 + 2) + 8 = 7 + (2 + 8)$ is a true statement.

$$(7 + 2) + 8 \stackrel{?}{=} 7 + (2 + 8)$$

Add first. ↗ $$9 + 8 \stackrel{?}{=} 7 + 10$$ ↖ Add first.

$$17 = 17 \quad \textit{True}$$

EXAMPLE 4 Replace each ? by a number to make a true statement.

	Statement	***Answer***
a.	$(17 + 6) + 8 = 17 + (? + 8)$	$(17 + 6) + 8 = 17 + (\underline{6} + 8)$
b.	$352 + 178 = ? + 352$	$352 + 178 = \underline{178} + 352$
c.	$2{,}511 + ? = 2{,}511$	$2{,}511 + \underline{0} = 2{,}511$
d.	$? + 47 = 47 + ?$	any number (the same number in both places)

CLASS EXERCISES

1. Show that $(8 + 11) + 17 = 8 + (11 + 17)$ is a true statement.

2. Tell whether each statement is *true* or *false*.
a. $(12 + 11) + 9 = 12 + (11 + 9)$ **b.** $26 + 5 < 5 + 26$ **c.** $79 + 1 = 79$ **d.** $73 + 0 = 73$

3. Replace □ by a number to form a true statement.
a. $\square + 296 = 296 + 85$ **b.** $(32 + 40) + \square = 32 + (40 + 19)$ **c.** $\square + 512 = 512$
d. $(11 + 17) + 19 = \square + (17 + 19)$ **e.** $2{,}575 + 137 = 137 + \square$ **f.** $1{,}720{,}000 + 0 = \square$

4. Use the calculator to show that $7 + 0 = 7$. Then name the property illustrated.

PROPERTIES OF MULTIPLICATION

THE MAIN IDEA

1. The product of two whole numbers is a whole number. This is called the property of ***closure*** for multiplication of whole numbers.
2. The order in which you multiply two whole numbers does not change the product.

   $$2 \times 9 = 9 \times 2$$

   This is called the ***commutative property for multiplication***.
3. The way you group numbers in a series of multiplications does not change the final product.

   $$(2 \times 5) \times 8 = 2 \times (5 \times 8)$$

   This is called the ***associative property for multiplication***.
4. The product of 1 and a whole number is that same whole number.

   $$9 \times 1 = 9$$

   This is the ***identity property for multiplication***.
   1 is called the ***multiplicative identity***.
5. The product of 0 and a whole number is 0.

   $$7 \times 0 = 0$$

   This is called the ***zero property for multiplication***.
6. The operation of multiplication can be *distributed* over the operations of addition or subtraction.

   The ***distributive property of multiplication over addition***: $6 \times (5 + 7) = 6 \times 5 + 6 \times 7$

   The ***distributive property of multiplication over subtraction***: $9 \times (8 - 2) = 9 \times 8 - 9 \times 2$

EXAMPLE 5 Tell whether each statement is *true* or *false*.

	Statement	***Answer***
a.	$11 \times 89 < 89 \times 11$	*False*. The order in which you multiply does not change the product.
b.	$(5 \times 7) \times 6 = 5 \times (7 \times 6)$	*True*. The way you group numbers does not change the product.
c.	$193 \times 1 = 193$	*True*. The product of a whole number and 1 is that same whole number.
d.	$17 \times 0 = 0$	*True*. The product of 0 and any whole number is 0.

EXAMPLE 6 Use the calculator to show that $8 \times 5 = 5 \times 8$. Then name the property illustrated.

KEYING IN

Multiply 8×5.	Multiply 5×8.
Key Sequence: 8 [×] 5 [=]	Key Sequence: 5 [×] 8 [=]
Display: [40.]	Display: [40.]

Answer: Both 8×5 and 5×8 give the same result.
This is the commutative property for multiplication.

EXAMPLE 7 Show that $(5 \times 3) \times 7 = 5 \times (3 \times 7)$ is a true statement.

$$(5 \times 3) \times 7 \stackrel{?}{=} 5 \times (3 \times 7)$$

Multiply first. ↗ $15 \times 7 \stackrel{?}{=} 5 \times 21$ ↖ Multiply first.

$$105 = 105 \quad \textit{True}$$

EXAMPLE 8 Replace each ? by a number to make a true statement.

	Statement	*Answer*
a.	$15 \times 8 = ? \times 15$	$15 \times 8 = 8 \times 15$
b.	$2{,}351 \times ? = 2{,}351$	$2{,}351 \times 1 = 2{,}351$
c.	$? \times 0 = 0$	any number $\times 0 = 0$
d.	$(17 \times 9) \times 6 = ? \times (9 \times 6)$	$(17 \times 9) \times 6 = 17 \times (9 \times 6)$

EXAMPLE 9 A dealer sold 37 boats at $4,921 each. What is the total amount of the sale?

THINKING ABOUT THE PROBLEM

Since this problem gives one of many equal amounts, the word "total" tells you to multiply.

You can use estimation to get an approximate answer: 37 is close to 40; $4,921 is close to $5,000. A reasonable solution would be about 40 × $5,000 = $200,000.

You must multiply 37 by $4,921 to find the total amount of the sale. Since the order in which you multiply does not change the product, you can write:

$$\begin{array}{r} 37 \\ \times 4{,}921 \\ \hline \end{array} \quad \text{or} \quad \begin{array}{r} 4{,}921 \\ \times 37 \\ \hline \end{array}$$

It is easier to use the smaller number as the multiplier.

$$\begin{array}{r} 4{,}921 \\ \times 37 \\ \hline 34{,}447 \\ 147{,}630 \\ \hline 182{,}077 \end{array}$$

This answer rounded to the nearest hundred thousand dollars is $200,000. It is a reasonable answer because it agrees with the estimate.

Answer: The total amount of the sale is $182,077.

EXAMPLE 10 Replace ? by a number so that each statement illustrates distributing multiplication.

	Statement	*Answer*
a.	$6 \times (11 + 8) = 6 \times 11 + 6 \times ?$	$6 \times (11 + 8) = 6 \times 11 + 6 \times \underline{8}$
b.	$7 \times (14 + 12) = ? \times 14 + ? \times 12$	$7 \times (14 + 12) = \underline{7} \times 14 + \underline{7} \times 12$
c.	$9 \times (6 - ?) = 9 \times 6 - 9 \times 2$	$9 \times (6 - \underline{2}) = 9 \times 6 - 9 \times 2$
d.	$14 \times (? - 7) = 14 \times 20 - 14 \times 7$	$14 \times (\underline{20} - 7) = 14 \times 20 - 14 \times 7$

EXAMPLE 11 Find the product 5×23 by distributing multiplication.

$$\begin{aligned} 5 \times 23 = 5 \times (20 + 3) &= 5 \times 20 + 5 \times 3 \\ &= 100 \quad + \quad 15 \\ &= 115 \quad \textit{Ans.} \end{aligned}$$

EXAMPLE 12 The seats in a theater are divided into two sections. Section A has 10 rows of seats, and section B has 8 rows of seats. If each row has 9 seats, find the number of seats in the theater.

Method I

Multiply to find the number of seats in section A. 9×10

Multiply to find the number of seats in section B. 9×8

Add to find the total number of seats. $9 \times 10 + 9 \times 8$

$= 90 + 72$

$= 162$ *Ans.*

Method II

Add to find the number of rows. $10 + 8$

Multiply to find the total number of seats. $9 \times (10 + 8)$

$= 9 \times 18$

$= 162$ *Ans.*

Notice that both methods give the same result and, therefore, they show that multiplication distributes over addition: $9 \times 10 + 9 \times 8 = 9 \times (10 + 8)$.

CLASS EXERCISES

1. Show that $(7 \times 9) \times 4 = 7 \times (9 \times 4)$ is a true statement.
2. Tell whether each statement is *true* or *false*.

a. $837 \times 592 = 592 \times 837$ **b.** $1 \times 474 = 474$ **c.** $(9 \times 8) \times 7 > 9 \times (8 \times 7)$

d. $0 \times 13 = 13$ **e.** $57 \times 0 < 57 \times 1$ **f.** $385 \times 1 > 385$

3. Replace each ? by a number to make a true statement.

   a. $42 \times 847 = ? \times 42$ **b.** $129 \times 1 = ?$ **c.** $(15 \times ?) \times 9 = 15 \times (12 \times 9)$
   d. $? \times 1 = 1{,}300{,}000$ **e.** $5{,}200 \times ? = 4{,}500 \times 5{,}200$ **f.** $1 \times 0 = 2 \times ?$

4. A theater has 128 rows. Each row has 40 seats. How many seats are there in the theater?

5. What is the total number of pieces of candy in 105 boxes if each box contains 36 pieces?

6. A computer is on sale for $995. What is the cost of 18 such computers?

7. Show that $11 \times (8 + 5) = 11 \times 8 + 11 \times 5$ is a true statement.

8. Replace ? by a number so that each statement illustrates distributing multiplication.

   a. $5 \times (9 + 2) = 5 \times ? + 5 \times 2$ **b.** $5 \times (? + 8) = 5 \times 7 + 5 \times 8$
   c. $12 \times (4 - ?) = 12 \times 4 - 12 \times 2$ **d.** $? \times (7 + 5) = ? \times 7 + 3 \times 5$
   e. $14 \times (12 - 3) = ? \times 12 - ? \times 3$

9. Write a key sequence to show that any number times 0 equals 0.

10. Find each product by distributing multiplication.

    a. 4×52 **b.** 7×81 **c.** 5×38 **d.** 8×46

HOMEWORK EXERCISES

1. Tell whether each statement is *true* or *false*.

   a. $(18 + 9) + 6 = 18 + (9 + 6)$ **b.** $262 + 0 = 262$ **c.** $48 + 0 = 0 + 48$

2. Show that $(29 + 42) + 56 = 29 + (42 + 56)$ is a true statement.

3. Replace □ by a number that will make a true statement.

   a. $9 + 3 = \square + 9$ **b.** $27 + \square = 17 + 27$ **c.** $(8 + \square) + 11 = 8 + (7 + 11)$
   d. $(3 + 15) + 6 = \square + (15 + 6)$ **e.** $(9 + 22) + 83 = 9 + (\square + 83)$

4. Tell whether each statement is *true* or *false*.

   a. $(17 \times 2) \times 3 = 17 \times (2 \times 3)$ **b.** $87 \times 3 > 3 \times 87$ **c.** $0 = 18 \times 0$
   d. $1 \times 253 > 253$ **e.** $1 \times 54 < 0 \times 54$ **f.** $95 \times 1 > 95$

5. Replace each ? by a number that will make a true statement.

   a. $4 \times 3 = 3 \times ?$ **b.** $(8 \times ?) \times 7 = 8 \times (3 \times 7)$ **c.** $7 \times ? = 1 \times 7$ **d.** $312 \times ? = 0$

6. Replace ? so that each expression is an illustration of distributing multiplication.

 a. $8 \times (5 + 9) = 8 \times 5 + 8 \times ?$ b. $6 \times (3 + 4) = ? \times 3 + ? \times 4$

 c. $2 \times (? + 9) = 2 \times 7 + 2 \times 9$ d. $5 \times (2 + ?) = 5 \times 2 + 5 \times 11$

7. Mr. Harris said he would contribute three times the total amount that his two children donated to the charity drive. If one child donated $12 and the other donated $17, which number phrase can be used to find Mr. Harris' donation?

 (a) $3 \times (12 + 17)$ (b) $(3 \times 12) + 17$ (c) $3 + (12 \times 17)$ (d) $(3 + 12) \times 17$

8. Write a key sequence to show that any number times 1 equals that number.

SPIRAL REVIEW EXERCISES

1. Replace ? by <, >, or = to make a true comparison.
 a. $8 + 9 \ ? \ 17 + 5$
 b. $15 + 32 \ ? \ 42 + 18$
 c. $96 + 73 \ ? \ 84 + 68$
 d. $230 + 0 \ ? \ 0 + 197$
 e. $2{,}248 + 3{,}741 \ ? \ 3{,}741 + 2{,}248$

2. Write as a numeral:
 a. one million, sixty thousand
 b. forty-seven thousand, thirty-six
 c. five thousand, seven

3. In the election for city council, Mr. Adams received 123,798 votes, Ms. Berkeley received 204,078 votes, and Mr. Jackson received 112,218 votes.
 a. Round each number to the nearest thousand.
 b. By adding the rounded numbers, estimate the total number of votes cast for the three candidates.
 c. Use the estimate to choose the exact answer:
 (a) 440,094 (b) 340,094
 (c) 400,094 (d) 300,094

4. Which key should you press on your calculator to find out what number is in memory?
 (a) [M+] (b) [MR] (c) [C] (d) [M−]

5. The population of Greenfalls is 12,762 and the population of Brainboro is 42,089. Find the total population of the two towns, rounded to the nearest ten thousand.

6. In 1998, the population of Colorado was 3,970,971. Round the population to the nearest hundred thousand.

7. Replace ? by < or > to make a true statement.
 a. $123 \ ? \ 132$ b. $86{,}340 \ ? \ 9{,}999$
 c. $24 + 23 \ ? \ 48$ d. $60 + 3 \ ? \ 63 + 1$
 e. the sum of ninety-nine and one ? 101
 f. the value of 2 in 29 ? the value of 9 in 39

UNIT 1-3 Using the Order of Operations to Evaluate Numerical Expressions

THE MAIN IDEA

1. A *numerical expression* contains numerals and operational symbols.
2. We *evaluate* a numerical expression by performing the operations shown to obtain a single numerical value.
3. To evaluate a numerical expression that contains two or more operations, we use the following *order of operations*:
   a. First, do all the multiplications and divisions, working from left to right.
   b. Next, do all the additions and subtractions, working from left to right.

EXAMPLE 1 Evaluate: $2 + 8 \times 5$

$2 + 40$	According to the order of operations, multiply first.
42 *Ans.*	Next, add.

KEYING IN

Some calculators automatically use the correct mathematical order (MO) of operations. Other calculators use the order in which operations are entered. These are called entry-order (EO) calculators. Compare the results of Example 1 for the same key sequence:

Key Sequence:	2 [+] 8 [×] 5 [=]
MO Display:	42.
EO Display:	50.

The key sequence above gives an incorrect result on an entry-order calculator.

If you are using an entry-order calculator, you will have to remember to enter the multiplication first.

EO Key Sequence:	8 [×] 5 [+] 2 [=]
EO Display:	42.

The key sequences in this book will be for entry-order (EO) calculators.

EXAMPLE 2 Evaluate $22 - 56 \div 7$.

Divide first.	$22 - 8$
Next, subtract.	14 *Ans.*

EXAMPLE 3 Evaluate $100 \div 50 \times 2$.

Work from left to right.	2×2
	4 *Ans.*

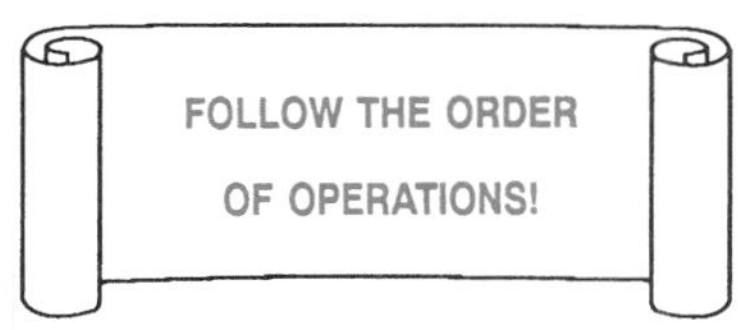

EXAMPLE 4 Evaluate $9 \times 5 - 3 \times 7$.

The given numerical expression contains two multiplications and a subtraction.	$9 \times 5 - 3 \times 7$
Do all the multiplications first.	$45 - 21$
Next, subtract.	24 *Ans.*

KEYING IN

To do this example on the calculator, use the memory feature to subtract the *result* of the second multiplication, 3×7, from the *result* of the first, 9×5.

Key Sequence	***Display***
9 [×] 5 [M+]	M 45.
3 [×] 7 [M−]	M 21.
[MR]	M 24.

Answer: 24

EXAMPLE 5 Evaluate $50 - 20 \div 5 + 3 \times 2$.

The given numerical expression contains a subtraction, a division, an addition, and a multiplication.	$50 - 20 \div 5 + 3 \times 2$
Do all the multiplications and divisions first.	$50 - 4 + 6$
Next, do the additions and subtractions. Remember to work from left to right.	$46 + 6$ 52 *Ans.*

EXAMPLE 6 Choose the expression that can be used to solve this problem.

A scenic tour ride costs \$5 plus \$2 for each mile traveled. What is the cost of an 8-mile ride?

(a) $5 + 8 \times 2$ (b) $8 \times (2 + 5)$
(c) $8 + 2 \times 5$ (d) $8 \times 5 + 2$

The phrase "\$2 for each mile" tells us that 2 must multiply 8, the number of miles. This rules out choices (c) and (d), which show an addition sign before the 2.

The phrase "\$5 plus" tells us that 5 is to be added to the product of 8 and 2. This rules out choice (b), in which the parentheses tell us to add 5 only to the number 2.

Choice (a) adds 5 to the product of 2 and 8.

Answer: (a)

EXAMPLE 7 The toll charged for a car to cross a bridge is \$2. The toll for a trailer truck is \$8. Write an expression for the amount of money that will be collected in tolls from the drivers of 9 cars and 12 trailer trucks. Then evaluate the expression.

First, write an expression for the car tolls:	2×9
The truck tolls must be *added* to this product:	$2 \times 9 +$
Write an expression for the truck tolls:	$2 \times 9 + 8 \times 12$

Evaluate the expression, using the order of operations:

$$2 \times 9 + 8 \times 12 = 18 + 96 \quad \leftarrow \text{Multiply first.}$$
$$= 114 \quad \leftarrow \text{Then, add.}$$

Answer: \$114 in tolls will be collected.

CLASS EXERCISES

1. Tell which operation you would perform first.

   a. $9 + 8 \times 15$ **b.** $12 \div 2 + 10 \div 5$ **c.** $30 - 15 + 8$ **d.** $81 - 12 \times 3$

2. Evaluate each numerical expression.

   a. $12 + 5 \times 2$ **b.** $30 \div 5 + 8$ **c.** $100 - 80 \div 4$ **d.** $8 \times 9 + 7$

   e. $75 - \frac{20}{4}$ **f.** $52 - 12 \times 4$ **g.** $12 \times 5 + 2$ **h.** $52 \times 12 - 4$

3. Evaluate each numerical expression.

   a. $72 \div 9 - 15 \div 3$ **b.** $20 + 5 \times 8 - 16$ **c.** $12 \times 3 + 3 \times 9$ **d.** $42 - 28 \div 7 + 30$

   e. $30 \div 10 + 5 \times 7$ **f.** $15 + 3 \times 7 - 8 \times 4$ **g.** $15 \times 3 \div 3 + 5$ **h.** $30 \div 30 \times 30 + 30$

4. Choose the expression that can be used to solve each problem.

   a. Each issue of a magazine contains 32 pages of text and 16 pages of advertisements. What is the total number of pages in 12 issues?

   (1) $12 + 32 \times 16$ (2) $12 \times 32 + 16$ (3) $12 \times (32 + 16)$ (4) $16 + 32 \times 12$

   b. Admission to an amusement park is $6, and each ride costs an additional $2. How much would you be charged for admission to the park and 5 rides?

   (1) $5 \times 2 + 6$ (2) $5 + 6 \times 2$ (3) $5 + 2 \times 6$ (4) $5 \times (6 + 2)$

   c. Each camper is allowed to catch a maximum of 6 fish per day at Crystal Lake. First, a group of 8 campers and then, another group of 5 campers fished at the lake. What is the maximum number of fish the two groups of campers are allowed to catch?

   (1) $6 \times 8 + 6 \times 5$ (2) $8 \times (6 + 5)$ (3) $6 \times 8 + 5$ (4) $6 + 8 \times 6 + 5$

5. Write an expression for the number of calories used in each set of activities. Then, evaluate the expression.

   a. 1 hour of running and 3 hours of slow walking

   b. 3 hours of tennis and 2 hours of brisk walking

   c. 2 hours of running, 1 hour of swimming, and 2 hours of slow walking

   d. 1 hour of tennis, 1 hour of swimming, 2 hours of bicycling, and 2 hours of running

Calories Used

Activity	Number of Calories per hour
Running	900
Bicycling (13 mph)	660
Swimming	450
Tennis	450
Brisk walking	300
Slow walking	200

6. A photographer has 3 rolls of 24-exposure film and 4 rolls of 36-exposure film. What is the maximum number of frames she can expose?

7. Write a calculator key sequence that can be used to evaluate each expression on: **(1)** a mathematical-order calculator **(2)** an entry-order calculator

   a. $8 + 9 \times 6$ **b.** $\frac{80}{2} - 3$ **c.** $7 \times 9 - 30 - 6$ **d.** $\frac{21}{3} + 9 \times 8$

1. Evaluate each numerical expression.
   a. $36 - 14 \times 2$ b. $18 + 12 \div 6$ c. $20 \times 3 - 40$ d. $18 \div 6 + 12$
   e. $28 - 14 \div 7$ f. $3 \times 0 + 7$ g. $8 \times 1 - 1$ h. $35 \div 7 + 7$

2. Explain in your own words why we must all follow the same order of operations.

3. Evaluate each numerical expression.
   a. $18 + 9 \times 2$ b. $\frac{15}{3} - 3$ c. $9 \times 7 - 40$ d. $36 + 8 \times 5$
   e. $50 - \frac{100}{5}$ f. $6 \times 1 - 6$ g. $\frac{50}{2} \times 2$ h. $\frac{70}{10} + 7 + 10$

4. Evaluate each numerical expression.
   a. $5 \times 8 + 3 \times 5$ b. $47 - 4 \times 7 + 6$ c. $8 + 5 \times 7 + 12$ d. $50 \div 5 + 10 \times 4$
   e. $5 \times 9 + 7 \times 6$ f. $\frac{20}{4} + \frac{18}{6}$ g. $90 \div 6 - 15 \div 3$ h. $100 - 90 \div 10 \times 9$
   i. $35 - 18 \div 9 + 2$ j. $90 - 14 \times 2 \times 3$ k. $7 - 12 \times 0 + 8 \div 1$ l. $72 \div 3 - 8 \times 3$
   m. $100 \div 2 + 4 - 3$ n. $13 \times 1 + 2 + 0$

5. Write a calculator key sequence that can be used to evaluate each expression. Tell whether your calculator is an entry-order or a math-order calculator.
   a. $4 + 9 \times 2$ b. $15 - 6 - 3$ c. $9 \times 4 - \frac{16}{8}$ d. $40 - 8 + 4 \times 9$

6. Evaluate each numerical expression.
   a. $18 \times 2 + 17 \times 3$ b. $49 - 7 \times 8 \div 4$ c. $37 - 18 \times 2 + 12$ d. $29 - 3 \times 8$
   e. $15 \times 4 + 8 \times 3$ f. $32 + 25 \div 5 \times 2$ g. $18 \div 6 - \frac{90}{45}$ h. $42 \div 7 + 9$
   i. $96 \div 16 - 3 \times 2$ j. $18 + 40 \div 8 - 9$ k. $15 + 15 \div 15$ l. $54 \div 9 - 6$

7. Choose the expression that can be used to solve each problem.
   a. To rent a motor bike from the Speedaway Bike Shop, a customer pays \$5 plus an additional \$4 for every hour. How much will it cost to rent a motor bike for 3 hours?
      (1) $3 \times 4 + 5$ (2) $3 + 4 \times 5$ (3) $3 \times (4 + 5)$
   b. The Big Apple Fruit Company puts 4 citrus fruits and 5 other fruits in every fruit basket it sells. If a customer buys 3 baskets, how many fruits is that person buying?
      (1) $3 \times 4 + 5$ (2) $3 + 4 \times 5$ (3) $3 \times (4 + 5)$
   c. A customer at a hardware store is buying a package of nails for \$3 and 4 electric switches that cost \$5 each. How much should that person expect to pay?
      (1) $3 \times 4 + 5$ (2) $3 + 4 \times 5$ (3) $3 \times (4 + 5)$

8. A theater charges \$8 for adults and \$4 for children under 12. What is the cost of admission for 2 adults and 3 children?

9. Write an expression for the shipping weight of each mail order. Then, evaluate the expression.
   a. 1 tape cassette and 3 adaptor plug sets
   b. 2 sets of headphones, 3 sets of ear pads, and a roll of speaker wire

Shipping Weights (Ounces)	
Set of headphones	4
Tape cassette	4
Roll of speaker wire	8
Cleaning kit	6
Set of ear pads	2
Adaptor plug set	2

10. Replace $\square$ by $<$, $>$, or $=$ to make each statement a true comparison.

a. $30 - 7 \times 4 - 1 \square 1 \times 0$ **b.** $176 - 95 \square 6 \times 12 + 8$ **c.** $28 - 4 \times 7 \square 17 \div 17$

d. $36 - 6 \times 6 + 11 \square 121 \div 11$ **e.** $21 \div 3 \times 11 \square 616 \div 8 + 0$ **f.** $1{,}001 - 1 \times 763 \square 27 - 13 \times 2$

SPIRAL REVIEW EXERCISES

1. Find the sum: 12, 848 + 3,497 + 23,575

2. Find the product of 157 and 29.

3. Divide 4,320 by 48 and check.

4. Subtract 7,909 from 11,823.

5. 18,965 rounded to the nearest ten is
   (a) 19,000 (b) 18,900
   (c) 18,960 (d) 18,970

6. Fifty thousand three hundred is
   (a) 5,300 (b) 50,300
   (c) 500,300 (d) 503,000

7. Adding 1 to which digit of the number 9,476,523 will increase the number by one hundred thousand?
   (a) 9 (b) 4 (c) 7 (d) 6

8. Which statement is not true?
   (a) $5 + (9 + 6) = (5 + 9) + 6$
   (b) $18 \times 72 = 72 \times 18$
   (c) $(40 - 18) - 7 = 40 - (18 - 7)$
   (d) $320 + 89 = 89 + 320$

9. The [M−] key
   (a) subtracts the number displayed from whatever is in memory.
   (b) displays the number in memory.
   (c) subtracts the number in memory from the displayed number.
   (d) displays the opposite of the number in memory.

10. A section of a stadium has 125 rows with 20 seats in each row. If 2,140 seats were occupied, how many seats were not occupied?

11. One tablespoon of butter contains 32 milligrams of cholesterol, and one tablespoon of salad dressing contains 8 milligrams of cholesterol. Find the total amount of cholesterol contained in 3 tablespoons of butter and 5 tablespoons of salad dressing.

UNIT 1-4 Evaluating Numerical Expressions Containing Parentheses

THE MAIN IDEA

1. ***Parentheses*** ( ) are used as a grouping symbol. The numerical expression inside them names a single number.
2. To evaluate numerical expressions that contain parentheses:
   a. First, evaluate within the parentheses. Be sure to follow the order of operations.
   b. Perform whatever operations remain, following the order of operations.

EXAMPLE 1 Evaluate:

a. $4 + (5 \times 3)$
$4 + (15)$
19 *Ans.*

b. $(4 + 5) \times 3$
$(9) \times 3$
27 *Ans.*

Notice how the parentheses give different results in **a** and **b** even though the numbers and operations are the same.

EXAMPLE 2 Evaluate: $12 - (40 \div 4 - 6)$

Work in parentheses.	$12 - (40 \div 4 - 6)$
Divide, then subtract.	$12 - (\ 10 \ - 6)$
Subtract again.	$12 - (4)$
	8 *Ans.*

KEYING IN

Key Sequence	*Display*
40 [÷] 4 [−] 6 [M+]	M 4.
12 [−] [MR] [=]	M 8.

EXAMPLE 3 Evaluate:

$11 \times (15 - 12) + 5 \times (12 - 2 \times 5)$
$11 \times (3) + 5 \times (12 - 10)$
$11 \times (3) + 5 \times (2)$
$33 + 10$
43 *Ans.*

EXAMPLE 4 Bruno sells T-shirts at \$7. Before lunch he sold 13, and after lunch he sold 28. How much money did he earn?

THINKING ABOUT THE PROBLEM

Break down the problem. First, find the number sold, by addition. Then, multiply by \$7 to find the amount earned.

$$(13 + 28) \times 7 = 41 \times 7 = 287$$

Answer: Bruno earned \$287.

EXAMPLE 5 Joseph had 20 pieces of candy. After eating 4 of them, he divided the remaining candy equally between 2 friends. Write a number phrase to find the number of pieces that each friend got.

THINKING ABOUT THE PROBLEM

This is a two-step problem in which the order of operations is important.

First, use subtraction to find how much candy was left after Joseph ate some: $20 - 4$
Then, use division to find how much each friend got: $(20 - 4) \div 2$ *Ans.*

EXAMPLE 6 Replace the boxes by operation symbols to produce the given result. Use parentheses where necessary.

$$152 \square 147 \square 17 \square 13 = 150$$

Use a calculator to help you try different possibilities. To begin, note that 152 has to be reduced. Try subtraction. Now $152 - 147 = 5$. If you multiply by 17, getting 85, you cannot then attach 13. Work with 17 and 13.

Answer: $(152 - 147) \times (17 + 13) = 150$

CLASS EXERCISES

In1–6, evaluate the numerical expressions.

1. **a.** $5 + 3 \times 8 + 7$ **b.** $(5 + 3) \times 8 + 7$ **2.** **a.** $(47 - 11) \times (3 + 5)$ **b.** $47 - 11 \times 3 + 5$

3. **a.** $(5 + 9) \times 7$ **b.** $9 \times (12 - 8)$ **4.** **a.** $5 + (3 \times 8 \div 2)$ **b.** $17 + 2 \times (4 + 3 \times 5)$

5. **a.** $(17 - 11) \times (3 + 5)$ **b.** $50 \div (8 + 2) \times 6$ **6.** $(10 + 5 \times 8) - (20 + 42 \div 7)$

7. Mary earns $45 a day from her employer. She earns an additional $5 a day in tips. Which number phrase can be used to find the amount of money that Mary earns in 6 days?

(a) $45 + 5 + 6$ (b) $(45 + 5) \times 6$ (c) $45 + (5 \times 6)$ (d) $(45 + 5) \div 6$

8. Dan travels 15 miles by bus and 12 miles by train to work. How far does he go:

a. in 5 trips to work? **b.** in 5 round trips?

9. Use a calculator to help you decide which operation symbols should be placed in the boxes to produce the given results. Use parentheses where necessary.

a. $256 \square 8 \square 13 = 19$ **b.** $256 \square 8 \square 13 = 261$ **c.** $256 \square 8 \square 13 = 360$

10. Which calculation does *not* represent 3,000?

(a) $(10 \times 100) + (40 \times 50)$ (b) $(15 + 100) \times (15 + 50)$

(c) $(15 + 135) \times (10 + 10)$ (d) $(20 \times 100) + (20 \times 50)$

HOMEWORK EXERCISES

1. Evaluate each numerical expression.

a. $(8 + 5) \times 7$ **b.** $8 + 5 \times 7$ **c.** $36 \div 9 + 3$ **d.** $36 \div (9 + 3)$

e. $(24 - 8) \div 2$ **f.** $24 - 8 \div 2$ **g.** $15 \times 2 + 3 \times 5$ **h.** $15 \times (2 + 3) \times 5$

i. $30 \div (5 - 2) \times 8$ **j.** $30 \div 5 + 2 \times 8$ **k.** $12 + 8 \times 3 + 5$

l. $(12 + 8) \times (3 + 5)$ **m.** $(90 - 60) \div (20 + 10)$ **n.** $90 - 60 \div 20 + 10$

o. $(32 + 8) \div 4 - 2$ **p.** $32 + 8 \div (4 - 2)$ **q.** $90 - 60 \div (20 + 10)$

2. A company produced 3,000 toys. It sold 22 boxes, each containing 100 toys. Which number phrase can be used to find the number of toys that were not sold?
(a) $(3{,}000 - 100) \times 22$ (b) $(3{,}000 \times 22) - 100$ (c) $(3{,}000 - 22) \times 100$ (d) $3{,}000 - (22 \times 100)$

3. Isaac needed $188 to buy a new bicycle. He saved $15 a week for 8 weeks. How much more money did he need?

4. Each of six children donated $3 toward a gift. Their parents gave them $25 toward the gift. How much money was collected in all?

5. Use a calculator to help you decide which operation symbols should be placed in the boxes to produce the given results. Use parentheses where necessary.

a. $147 \square 15 \square 12 \square 90 = 139$ **b.** $144 \square 4 \square 2 \square 1 = 19$

c. $680 \square 142 \square 142 = 0$ **d.** $150 \square 12 \square 15 \square 95 = 375$

6. Angela and Mr. Collins are checking the inventory of holiday ornaments at the Greet 'n Treat card store. There are 15 turkey ornaments and 17 pumpkin ornaments, each with a selling price of $4. Mr. Collins wrote the expression $15 + 17 \times 4$ and recorded the value of the ornaments as $128. Angela said that the value should be $83. Whose answer was correct? Explain why they got two different answers.

7. Explain why the calculation $(5 + 8) \div (12 - 4 \times 3)$ will result in an error message on an MO calculator.

8. The graph shows the number of regular members and new members who attended classes at a health club during a special membership drive.

If regular members pay \$10 per class and new members pay \$5 per class, how much money was paid for Saturday's classes?

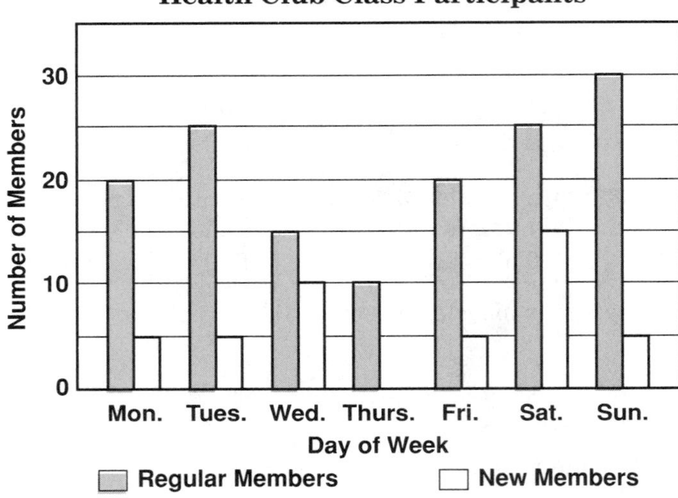

SPIRAL REVIEW EXERCISES

1. Find the difference: 4,846 – 2,797

2. Find the quotient: 41,654 ÷ 59

3. Find the product: 742 × 26

4. Written as a numeral, seventy thousand, one hundred two is

(a) 7,102 (b) 70,102
(c) 700,102 (d) 70,002

5. The value of "9" in the number 795,418 is

(a) 900 (b) 9,000
(c) 90,000 (d) 900,000

6. 875,483 rounded to the nearest ten thousand is

(a) 875,000 (b) 876,000
(c) 870,000 (d) 880,000

7. John wanted to add 2,476 and 3,458 on his calculator. He mistakenly entered 3,548. Which key should he press to clear the incorrect entry without changing the previous entries?

(a) [M+] (b) [AC] (c) [CE] (d) [MC]

8. At a Mets baseball game, 22,481 people paid for admission. Another 9,576 persons attended on free passes. What was the total attendance for that game?

9. Each ticket to a baseball game costs \$16. A total of \$239,168 was collected. How many tickets were sold?

UNIT 1-5 Factors; Prime and Composite Numbers; Greatest Common Factor

MEANING OF FACTOR

THE MAIN IDEA

1. When two nonzero whole numbers are multiplied, each is called a ***factor*** of the product.

   $2 \times 5 = 10$ (2 and 5: factors; 10: product)

2. If a number is divided by one of its factors, the remainder is 0 and the quotient is the other factor. Thus, a factor of a number is an ***exact divisor***.

EXAMPLE 1 Determine if 42 is a factor of 2,394.

```
       57
42 )2,394
    2 10↓
      294
      294
        0  ←— remainder
```

Since the remainder is 0, then 42 is an exact divisor, or factor, of 2,394.

EXAMPLE 2 Find all the factors of 40.

The exact divisors of 40 are 1, 2, 4, 5, 8, 10, 20, and 40. To be sure that you are getting *all* the factors, think of the factors in *pairs:*

CLASS EXERCISES

1. Tell whether the first number is a factor of the second.

a. 5; 75 **b.** 7; 56 **c.** 9; 73 **d.** 30; 90 **e.** 11; 121 **f.** 21; 84 **g.** 17; 33

2. Tell whether each statement is *true* or *false*.

a. 19 is a factor of 608. **b.** 27 is an exact divisor of 1,161.

c. 7 is not a factor of 1,428. **d.** 42 and 37 are factors of 1,554.

3. Use a calculator to determine if the first number is a factor of the second.

a. 37; 555 **b.** 49; 2,455 **c.** 196; 18,700

4. Write all the pairs of factors of each number. **a.** 15 **b.** 36 **c.** 81 **d.** 100

MEANING OF PRIME AND COMPOSITE NUMBERS

THE MAIN IDEA

1. If a whole number is greater than 1 and has 1 and itself as its only factors, then that number is called a ***prime number***. 2, 3, 5, 7, 11, 13, 17, 19, 23, 29, 31, 37 are examples of primes.
2. If a whole number also has factors other than 1 and itself, then that whole number is called a ***composite number***. 8, 9, 14, 15, 25, 26 are examples of composite numbers.
3. The number 1 is not called prime or composite.

EXAMPLE 3 Tell whether each number is prime or composite.

	Number	*Answer*
a.	2	prime
b.	12	composite
c.	17	prime
d.	28	composite
e.	35	composite

EXAMPLE 4 Show that 123 is not prime.

$$\begin{array}{r} 41 \\ 3\overline{)123} \\ \underline{12}\downarrow \\ 3 \\ \underline{3} \\ 0 \end{array}$$

Answer: Since 3 and 41 are factors of 123, then 123 is not prime.

CLASS EXERCISES

1. Tell whether each number is prime or composite.

a. 4 **b.** 7 **c.** 22 **d.** 37 **e.** 55 **f.** 81 **g.** 99 **h.** 29 **i.** 87 **j.** 91

2. Show that each number is not prime.

a. 40 **b.** 328 **c.** 125 **d.** 500 **e.** 729

3. What is the next prime number greater than 97?

GREATEST COMMON FACTOR

THE MAIN IDEA

1. A ***common factor*** of two numbers is a number that is a factor of both of them.
2. The ***greatest common factor*** (GCF) of two numbers is the largest number that is a common factor of both numbers.
3. To find the greatest common factor (GCF):
   a. Find all the factors of each of the numbers.
   b. Determine the factors common to both numbers.
   c. Select the largest common factor.

EXAMPLE 5 Find the greatest common factor of 8 and 12.

Factors of 8: 1, 2, 4, 8

Factors of 12: 1, 2, 3, 4, 6, 12

Common factors of 8 and 12: 1, 2, 4

Answer: The greatest common factor of 8 and 12 is 4.

EXAMPLE 6 Find the greatest common factor of 21 and 42.

Factors of 21: 1, 3, 7, 21

Factors of 42: 1, 2, 3, 6, 7, 14, 21, 42

Common factors of 21 and 42: 1, 3, 7, 21

Answer: The greatest common factor of 21 and 42 is 21.

EXAMPLE 7 Factor 100 until all the factors are primes.

You may begin with several different pairs of factors for 100.

2×50
$2 \times 25 \times 2$
$2 \times 5 \times 5 \times 2$

4×25
$2 \times 2 \times 5 \times 5$

10×10
$5 \times 2 \times 5 \times 2$

Notice that no matter how you begin to factor 100, if you continue factoring until you reach prime numbers, the result has the same prime factors. That is, $100 = 2 \times 2 \times 5 \times 5$.

CLASS EXERCISES

1. Find the common factors of each pair of numbers: **a.** 10 and 20 **b.** 24 and 36 **c.** 12 and 30
2. Find the greatest common factor of each pair of numbers: **a.** 22 and 33 **b.** 12 and 21 **c.** 25 and 100
3. Factor each number until all the factors are primes: **a.** 30 **b.** 27 **c.** 48 **d.** 500

HOMEWORK EXERCISES

1. Tell whether the first number is a factor of the second number.

   a. 12; 36 **b.** 5; 30 **c.** 18; 72 **d.** 14; 29 **e.** 24; 144 **f.** 50; 250 **g.** 19; 82
2. Write in your own words how you can tell if 6 is a factor of 24.
3. Find all the factors of each number: **a.** 12 **b.** 20 **c.** 48 **d.** 72 **e.** 144
4. Tell whether each statement is *true* or *false*.

   a. 23 is a factor of 1,403. **b.** 37 is a factor of 221. **c.** 13 is not a factor of 3,913.
   d. 19 and 42 are factors of 798. **e.** 56 is an exact divisor of 225.

5. Use a calculator to determine if the first number is a factor of the second.

   a. 37; 2,370 **b.** 65; 3,770 **c.** 98; 4,606
6. Tell whether each number is prime or composite: **a.** 19 **b.** 30 **c.** 31 **d.** 51 **e.** 47
7. Find the common factors of each pair of numbers.

   a. 20 and 40 **b.** 18 and 32 **c.** 12 and 36 **d.** 15 and 18 **e.** 10 and 25 **f.** 24 and 30
8. Find the greatest common factor of each pair of numbers.

   a. 24 and 42 **b.** 45 and 50 **c.** 9 and 15 **d.** 16 and 32 **e.** 56 and 72 **f.** 14 and 42
9. Factor each number until all the factors are primes: **a.** 18 **b.** 36 **c.** 60 **d.** 121 **e.** 50
10. Explain when the greatest common factor of two numbers will be the smaller of the two.
11. A marching band of 56 members wants to march in rows that have the same number of members in each row. How many ways can they do this without leaving any members out? Explain.
12. Jane and Paul want to sell their muffins in local shops. Jane suggests packing 25 muffins in a box. Paul thinks that 24 in a box is better. Whose plan do you think is best? Explain.
13. A number is evenly divisible by 7, but 14 is not a factor of the number. Which of the following statements is true of the number?

    (a) It is odd. (b) 28 is a factor. (c) 4 is a factor. (d) 2 is a factor.

SPIRAL REVIEW EXERCISES

1. Perform each operation.

a. Add: 2,752 + 1,496 + 847

b. Subtract: 7,809 − 2,489

c. Multiply: 53 × 29

d. Divide: 72)14,472

2. Subtract 358 from 4,027 and check.

3. Find the product of 435 and 36 and check.

4. Show that "1,305 ÷ 29 = 45" is a true statement by using the inverse operation.

5. Write as a numeral:

a. thirty-five thousand, sixty-two
b. thirty thousand, six
c. three hundred thousand, six hundred twenty
d. three million, three hundred sixty-two

6. Round 47,496 to the nearest thousand.

7. Round 42,552 to the nearest hundred.

8. Replace ? by <, >, or = to make a true comparison.

a. 1 ? the product of 475 and 0
b. the quotient of 0 and 295 ? 295
c. the quotient of 357 and 357 ? 0
d. the difference of 139 and 129 ? the product of 20 and 1

9. Nayda earns $5 per hour. She works 35 hours every week. How much does she earn in a week?

10. a. Sixteen identical pieces of lumber weigh 240 pounds. How much does one piece of lumber weigh?

b. Five pieces of a different type of lumber weigh 100 pounds. How much weight must Luis carry to the car if he buys one piece of each type of lumber?

11. A key sequence that can be used to produce all the positive multiples of 5 is

(a) 5 [×] 5 [=] [=] [=] ...
(b) 5 [×] 1 [=] [=] [=] ...
(c) 5 [+] [=] [=] [=] ...
(d) 5 [−] 1 [=] [=] [=] ...

12. If a bus can carry 35 children and costs $210 to hire, how much would it cost to transport 294 children?

13. Which expression can be used to represent the total cost of seven books that cost $5 each and eight more books that cost $5 each?

(a) $5 \times (7 + 8)$ (b) $5 + 7 \times 8$
(c) $8 + 7 \times 5$ (d) $5 \times 7 \times 8$

UNIT 2-1 The Meaning and Use of Signed Numbers; The Set of Integers

THE MAIN IDEA

1. Numbers that have the signs + (positive) and – (negative) are called ***signed numbers***. Zero (0) is neither positive nor negative. If a nonzero number is written without a sign, it is understood to be positive.

2. To show the set of signed numbers on a graph, we use a number line that contains 0 and that extends without end to the right of 0 for the positive numbers and to the left of 0 for the negative numbers.

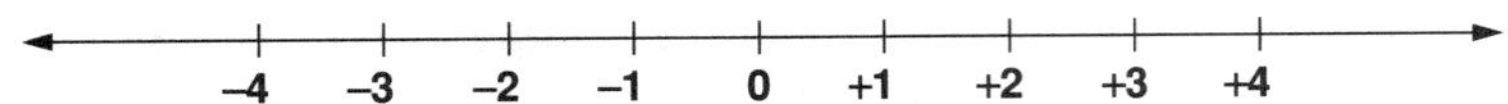

3. Every signed number has an opposite. ***Opposites*** are two signed numbers that are the *same distance from 0 on a number line but are in different directions,* right or left. Zero is its own opposite.

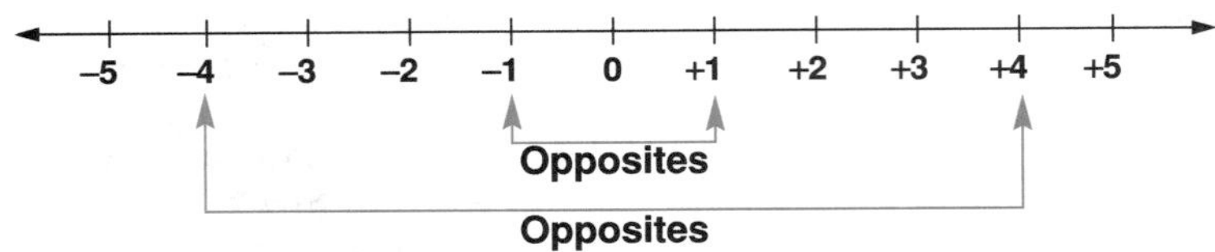

4. The distance of a signed number from zero on the number line is called the ***absolute value*** of the signed number. The absolute value of –6 is written $|-6|$.

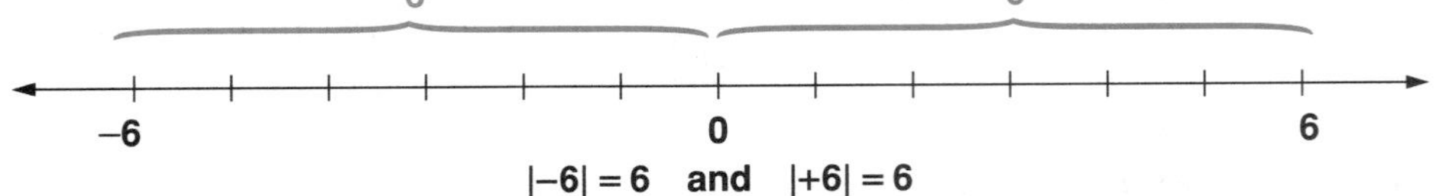

$|-6| = 6$ and $|+6| = 6$

5. The whole numbers and their opposites make up the set of ***integers***.

{Integers} = {. . . , –3, –2, –1, 0, +1, +2, +3, . . .}

6. Signed numbers are also called ***directed numbers***. These numbers allow us to work with situations in which there are two opposite directions, such as:

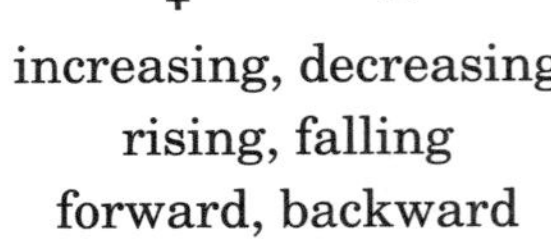

+ –	+ –
increasing, decreasing	deposit, withdrawal
rising, falling	above, below
forward, backward	future, past

EXAMPLE 1 Name the integer that is represented by each of the letters shown on the number line.

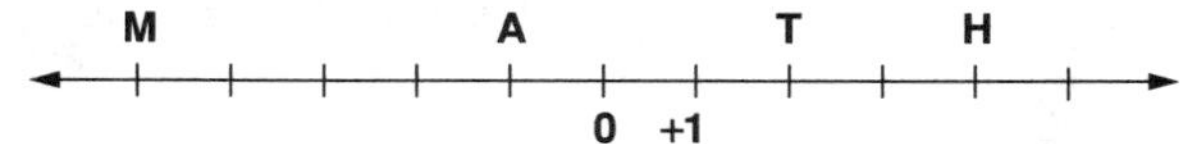

	Letter		*Answer*
a.	M	M is 5 units to the left of 0.	–5
b.	A	A is 1 unit to the left of 0.	–1
c.	T	T is 2 units to the right of 0.	2
d.	H	H is 4 units to the right of 0.	4

EXAMPLE 2 Graph each integer on a number line.

	Integer
a.	+3
b.	–4
c.	0
d.	2

Answer

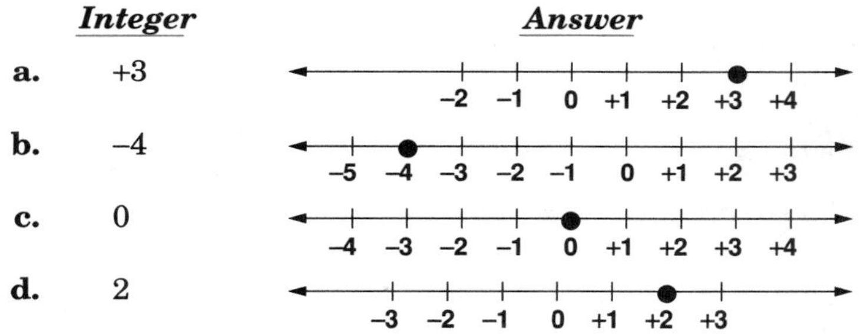

EXAMPLE 3 Find the absolute value of each integer.

	Integer	*Distance from Zero*	*Answer*
a.	+12	+12 is 12 units from zero.	$\lvert +12 \rvert = 12$
b.	–8	–8 is 8 units from zero.	$\lvert -8 \rvert = 8$
c.	0	0 is 0 units from zero.	$\lvert 0 \rvert = 0$

EXAMPLE 4 Tell the opposite of each integer.

	Integer	*Answer*
a.	+7	–7
b.	–50	+50
c.	–1,700	+1,700
d.	0	0
e.	12	–12

EXAMPLE 5 Use a signed number to show each situation.

	Situation	*Answer*
a.	a loss of 9 pounds	–9
b.	a profit of $15	+15
c.	5 years ago	–5
d.	a $20 increase in price	+20
e.	17° above zero	+17

EXAMPLE 6 Use a calculator to find the opposite of −3.

KEYING IN

The [+/−] key on a calculator is called the "change sign" key. Pressing this key will display the opposite of a number that is entered.

Key Sequence	***Display***
3 [+/−]	− 3.
[+/−]	3.

Notice that the calculator does not display the + sign for positive numbers.

Answer: The opposite of −3 is 3.

EXAMPLE 7 Name the integers described.

a. the integer that is 6 places to the left of +4 on a number line

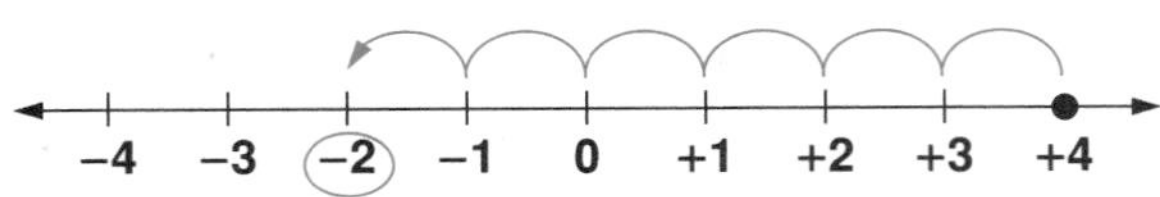

Count 6 places to the left of +4 and read the number below that point.

Answer: −2

b. the integer that is 4 places to the right of −7 on a number line

Count 4 places to the right of −7.

Answer: −3

c. the integer that has the same absolute value as −8

Opposites have the same absolute value. The opposite of −8 is +8.

Answer: +8

d. the integer that is midway between a pair of opposites on a number line

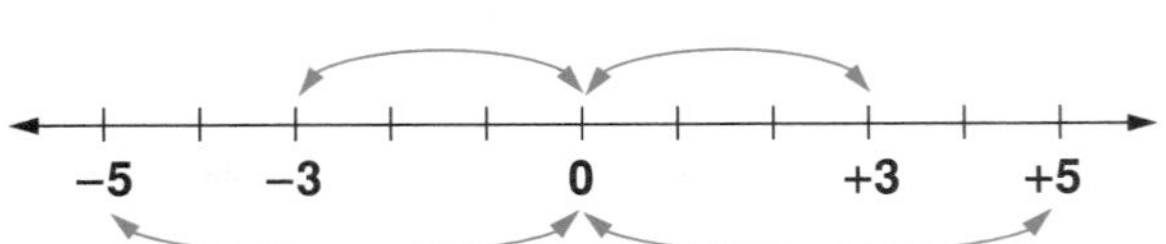

Opposites are always equal distances from 0, on either side of 0.

Answer: 0

EXAMPLE 8 The graph shows the depth, in inches, of a tidal pool over an 8-hour period.

a. Represent as an integer the change in depth for each hour.

Read the height of each point from the vertical scale. Determine the sign of the change between each pair of consecutive values by observing the direction of the line as you move from left to right. If the line moves upward, the change is an increase and is positive (+). If the line moves downward, the change is a decrease and is negative (–). The changes can be displayed in a table.

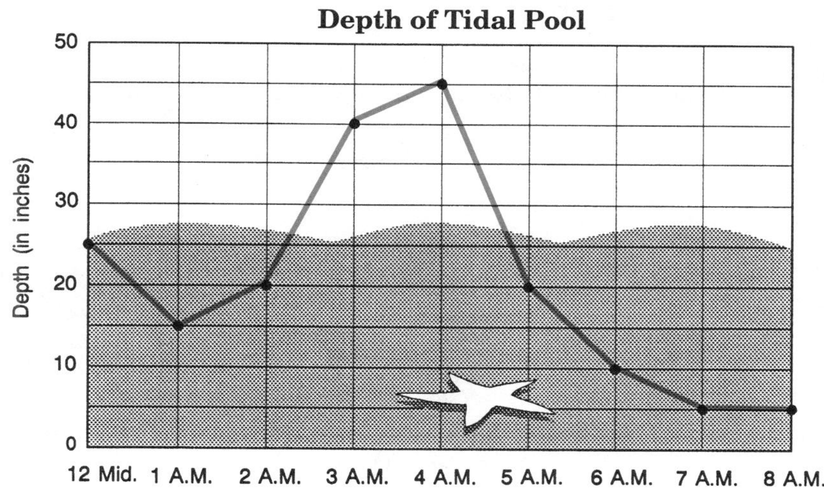

Answer:

Hour	12 A.M. – 1 A.M.	1 A.M. – 2 A.M.	2 A.M. – 3 A.M.	3 A.M. – 4 A.M.	4 A.M. – 5 A.M.	5 A.M. – 6 A.M.	6 A.M. – 7 A.M.	7 A.M. – 8 A.M.
Change in Depth	–10	+5	+20	+5	–25	–10	–5	0

b. Between which hours did the depth have the greatest change?

The greatest change is represented by the integer with the greatest absolute value, which is –25.

Answer: The greatest change took place between 4 A.M. and 5 A.M.

c. Between which hours did the depth have the least change?

The least change is represented by 0, which indicates no change.

Answer: The least change took place between 7 A.M. and 8 A.M.

d. Between which hours did the depth have the greatest increase?

The greatest increase is represented by the greatest positive integer, +20.

Answer: The greatest increase took place between 2 A.M. and 3 A.M.

e. Use an integer to represent the overall change in depth from 12 midnight to 8 A.M.

From 12 midnight to 8 A.M. the depth changed from 25 inches to 5 inches. This is a decrease of 20 inches.

Answer: –20

CLASS EXERCISES

1. Name the integer that is represented by each of the letters shown on the number line.

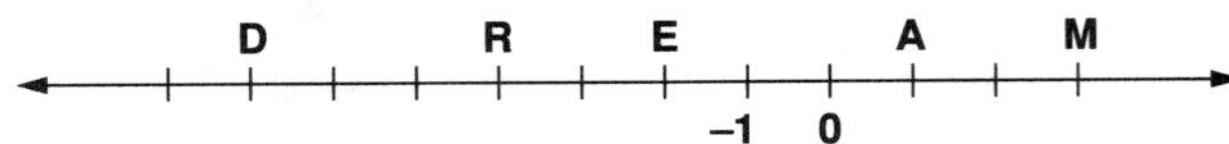

2. Graph each integer on a number line.

a. +6 **b.** –2 **c.** 0 **d.** 4 **e.** –1

3. Find the absolute value of each integer.

a. +15 **b.** –9 **c.** –110 **d.** +36 **e.** –36

4. Tell the opposite of each integer.

a. 2 **b.** –8 **c.** –12 **d.** +15 **e.** –30

5. Use a calculator to find:

a. the opposite of 4 **b.** the opposite of –5 **c.** the opposite of the opposite of 2

6. Tell if the number described is positive or negative.

a. the opposite of 7 **b.** the opposite of –6 **c.** the opposite of a positive number

7. Use a signed number to show each situation.

a. 7 years from now **b.** 10° below zero **c.** a gain of 14 pounds
d. a $50 deposit **e.** a fall of 25 feet **f.** a $10 decrease in price
g. 5 seconds before launch time **h.** 1,000 meters above sea level

8. Describe the opposite of each situation in Exercise 7, and write a signed number to represent it.

9. Name the integers described.

a. the least positive integer
b. the integer that is 5 places to the left of –2 on a number line
c. the integer that is 2 places to the right of –1 on a number line
d. the integer that is its own opposite

10. The number of hours that Sharlene worked as a baby-sitter each week for 6 weeks is shown on the line graph.
    a. Make a table in which each weekly change in the number of hours is represented by an integer.
    b. Between which weeks did the number of hours change the most?
    c. Between which weeks did the number of hours change the least?
    d. Between which weeks did the number of hours increase the most?
    e. From week 2 to week 5, how did the total decrease compare with the total increase?

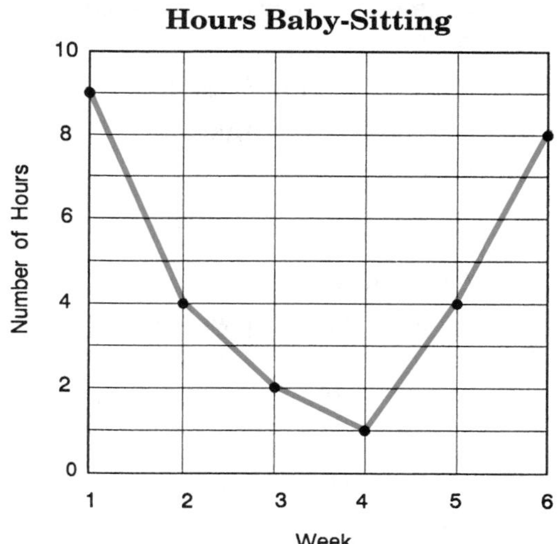

HOMEWORK EXERCISES

1. Name the integer that is represented by each of the letters shown on the number line.

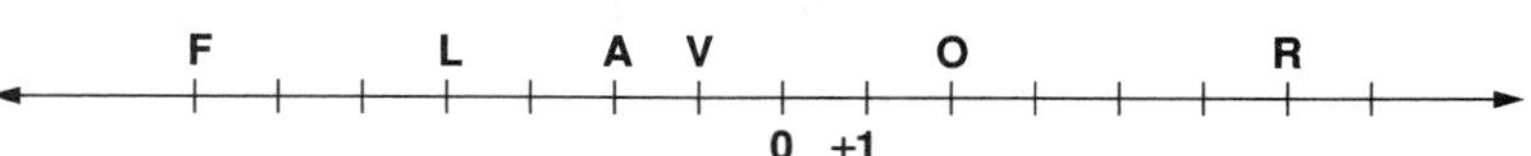

2. Graph each integer on a number line.
   a. –1 **b.** –5 **c.** +7 **d.** 2 **e.** –6 **f.** +6 **g.** –2 **h.** –11

3. Find the absolute value of each integer.
   a. –7 **b.** +16 **c.** +41 **d.** +1,001 **e.** –137

4. Find the value of $|-10|$.
   (a) 10 (b) 1 (c) –10 (d) 0

5. Tell the opposite of each integer.
   a. –17 **b.** +8 **c.** –9 **d.** 9 **e.** 3 **f.** 0

6. Use a calculator to find:
   a. the opposite of –7 **b.** the opposite of the opposite of –1

7. Tell if the number described is positive or negative.
   a. the absolute value of –8 **b.** the opposite of a negative number
   c. the opposite of the absolute value of a number

8. Use a signed number to show each situation.
   - **a.** a loss of 5 pounds
   - **b.** 50 feet below sea level
   - **c.** a 10-yard gain in football
   - **d.** a \$30 decrease in price
   - **e.** 10 seconds before takeoff
   - **f.** 20° above zero
   - **g.** 12 years ago
   - **h.** a \$10 withdrawal

9. Describe the opposite of each situation in Exercise 8, and write a signed number to represent it.

10. Name the integers described.
    - **a.** the positive integer whose absolute value is 12
    - **b.** the integer that is 5 places to the right of +5 on a number line
    - **c.** the integer that is 6 places to the left of +5 on a number line
    - **d.** the two integers that are opposites of each other and are 8 units apart on a number line

11. What is the distance between –17 and –23 on a number line?

12. After the number –17 was entered on a calculator, the [+/–] key was pressed 67 times. What was the number on the display? Explain how you got your answer.

SPIRAL REVIEW EXERCISES

1. Name the place and tell the value of each of the underlined digits.
   a. 3$\underline{7}$1 **b.** 5,$\underline{9}$42 **c.** $\underline{2}$96,417

2. Write twenty-three thousand, four hundred seven as a numeral.

3. Round each of the following to the nearest hundred.
   a. 2,791 **b.** 42,547 **c.** 568

4. Add: 3,760 + 946 + 47,869

5. Subtract 4,782 from 6,891.

6. From 7,813 subtract 908.

7. Find the product of 702 and 83.

8. Divide 391 by 17.

9. Use a calculator to help you decide which operation symbols should replace [?] to produce the given result. Use parentheses where necessary.
   7 [?] 47 [?] 23 [?] 16 = 2,303

10. Find the total number of ice-cream bars in 15 cartons if each carton contains 36 bars.

11. How many newspapers will each newsstand receive if 1,200 newspapers are to be equally divided among 8 newsstands?

12. Frank sold 73 magazines on Monday and 35 magazines on each of the next four days. What was the total number of magazines that he sold during these five days?

13. Find the value:
    a. $8 + 5 \times 4$ **b.** $(8 + 5) \times 4$

UNIT 2-2 Using a Number Line to Compare Integers

THE MAIN IDEA

To compare two integers:

1. Think of their locations on a number line.

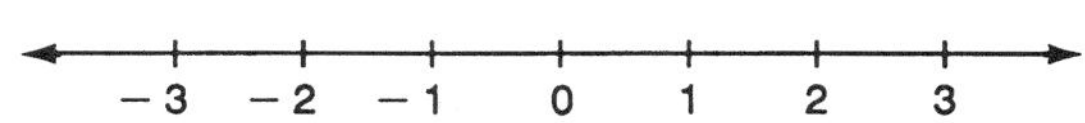

2. The number to the right is the greater number. Because 1 is to the right of –2 on the number line, 1 > –2.
3. A negative number is always less than a positive number.

EXAMPLE 1 Tell which of the integers in each pair is farther to the right on a number line.

	Number Pair	*Answer*
a.	–7 and 5	5
b.	–4 and –5	–4
c.	–50 and –51	–50
d.	0 and 3	3
e.	–4 and 0	0

EXAMPLE 2 Replace each ? with < or > to make a true comparison.

	Number Pair	*Placement on a Number Line*	*Answer*
a.	3 ? –12	3 is to the right of –12	3 > –12
b.	–4 ? –2	–4 is to the left of –2	–4 < –2
c.	–5 ? –8	–5 is to the right of –8	–5 > –8

EXAMPLE 3 Tell which of the numbers in each set is between the others. Next, write the three numbers in order, using <. Then, write the three numbers in order, using >.

	Number Set	*Answer*
a.	–4, 3, 0	0 is between –4 and 3. 4 < 0 < 3; 3 > 0 > –4
b.	5, –6, –2	–2 is between –6 and 5. –6 < –2 < 5; 5 > –2 > –6

EXAMPLE 4 Tell whether the given statement is *true* or *false*.

	Statement	***Answer***
a.	$-6 > -11$	true
b.	$2 < -9$	false
c.	$-40 > 1$	false
d.	$0 > -2$	true
e.	$-5 < 2 < 5$	true
f.	$5 > -2 > -1$	false

EXAMPLE 5 The table shows the temperature in Detroit over an eight-hour period on January 15.

Time	Temperature
6 A.M.	–11°
7 A.M.	–5°
8 A.M.	0°
9 A.M.	5°
10 A.M.	3°
11 A.M.	6°
12 Noon	10°
1 P.M.	2°
2 P.M.	–8°

a. What was the lowest temperature during this period?

b. What was the highest temperature during this period?

c. Was the temperature higher at 7 A.M. or at 9 A.M.?

d. Was the temperature higher at 6 A.M. or at 2 P.M.?

e. Did the temperature increase or decrease from 7 A.M. to 8 A.M.?

THINKING ABOUT THE PROBLEM

Draw a diagram in the form of a number line. This strategy will help you to compare the given numbers.

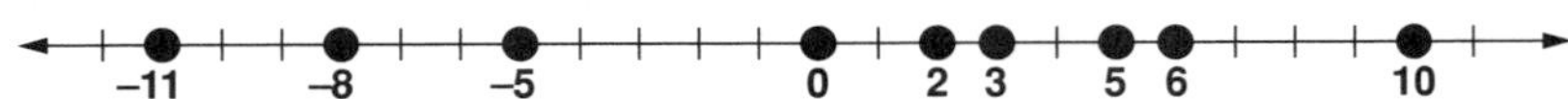

	Reasoning From the Number Line	***Conclusion***
a.	Since –11 is farthest to the left,	then the lowest temperature was –11°.
b.	Since 10 is farthest to the right,	then the highest temperature was 10°.
c.	Since 5 is to the right of –5, 5 is greater than –5.	Thus, the 9 A.M. temperature, 5°, was higher than the 7 A.M. temperature, –5°.
d.	Since –8 is to the right of –11, –8 is greater than –11.	Thus, the 2 P.M. temperature, –8°, was higher than the 6 A.M. temperature, –11°.
e.	Since 0 is to the right of –5, 0 is greater than –5, and the 8 A.M. temperature, 0°, was higher than the 7 A.M. temperature, –5°.	Thus, from 7 A.M. to 8 A.M., the temperature increased.

CLASS EXERCISES

1. Tell which of the integers in each pair is farther to the right on a number line.

a. 2 and –5 **b.** –10 and 4 **c.** –7 and –8 **d.** –24 and –30 **e.** –8 and 0

2. Replace ? with < or > to make a true comparison.

a. –3 ? 1 **b.** 0 ? –2 **c.** 5 ? –10 **d.** –7 ? –11 **e.** –10 ? –9 **f.** –5 ? 0

3. Tell which of the numbers in each set is between the others. Next, write the three numbers in order, using <. Then, write the three numbers in order, using >.

a. –2, 3, 1 **b.** –6, –9, –2 **c.** –5, 6, –8 **d.** 3, –6, 0 **e.** –10, –30, –20

4. Tell whether the given statement is *true* or *false*.

a. –10 < –12 **b.** –6 > 0 **c.** 5 > –15 **d.** –30 > –20 **e.** 28 < –50

f. –2 < 0 < 1 **g.** –6 < –4 < –10 **h.** 8 > –2 > –3 **i.** –10 > –8 > 0

5. For ten weeks, Robert made a table by recording his weight change at the end of each week.

a. At the end of which week did Robert's weight increase the most?

b. At the end of which week did Robert's weight decrease the most?

c. Did Robert's weight increase more at the end of week 8 or at the end of week 1?

d. Did Robert's weight decrease more at the end of week 3 or at the end of week 1?

e. At the end of which week(s) did Robert's weight change the least?

Week	Weight Change (in lb.)
1	+2
2	+3
3	–1
4	–4
5	0
6	–1
7	+2
8	+1
9	0
10	–3

HOMEWORK EXERCISES

1. Tell which of the integers in each pair is farther to the right on a number line.

a. –7 and 6 **b.** –12 and –8 **c.** –10 and 12 **d.** –14 and –18

e. –100 and –101 **f.** –9 and 0 **g.** –6 and 6 **h.** 0 and –9

2. Replace ? with < or > to make a true comparison.

a. –5 ? –4 **b.** –10 ? 0 **c.** 9 ? –10 **d.** –12 ? –11 **e.** –17 ? –20

f. –26 ? 18 **g.** –19 ? –18 **h.** –25 ? –35 **i.** –98 ? –99 **j.** 5 ? –4

3. Tell which of the numbers in each set is between the others. Next, write the three numbers in order, using <. Then, write the three numbers in order, using >.

a. 3, –1, 0 **b.** –8, 4, –2 **c.** –11, –7, –15 **d.** 4, 7, –5 **e.** –6, –4, –1

f. –18, –100, –25 **g.** 85, –72, –71 **h.** –33, –30, –35 **i.** 0, –11, –7

4. Tell whether the given statement is *true* or *false*.

a. 5 > –5 **b.** –10 > –6 **c.** –50 < –60 **d.** –100 > 10 **e.** 0 > –12

f. –18 > –38 **g.** –21 < –22 **h.** –10 < –12 < –14 **i.** –5 < –4 < 1

5. During each quarter of the last 2 games, Coach Sherman kept a record of his football team's total yardage.

a. During which quarter did the team gain the greatest yardage?

b. During which quarter did the team lose the greatest yardage?

c. Was the team more successful in the 4th quarter of the October 5 game or in the 1st quarter of the October 12 game?

d. During which quarter of the October 5 game did the team do the worst?

e. Did the team do better in the 1st quarter or in the 4th quarter of the October 12 game?

Game	Quarter	Yardage
October 5	1	45
	2	0
	3	30
	4	–15
October 12	1	–25
	2	50
	3	25
	4	–30

6. Mrs. Price kept a record of deposits and withdrawals in her savings account.

a. When did Mrs. Price withdraw the greatest amount of money?

b. When did she deposit the greatest amount?

c. Did she take a greater amount of money from her account on 1/19 or on 1/25?

d. Did she decrease her balance more on 2/2 or on 2/25?

Date	Deposit / Withdrawal (in dollars)
1/12	+150
1/19	–75
1/25	–50
1/30	+200
2/2	–100
2/12	+100
2/20	+75
2/25	–200

7. In Anchorage, Alaska, the afternoon temperature rose 15° from the morning temperature, which was –12°. What was the afternoon temperature?

SPIRAL REVIEW EXERCISES

1. Use a signed number to show each of the following situations.

a. a \$50 deposit **b.** 10° below zero

c. a 20° decrease in temperature

d. a loss of \$10

2. Name the opposite of each signed number.

a. –15 **b.** 8 **c.** 23 **d.** 0

3. 5,864,293 rounded to the nearest ten thousand is

(a) 5,860,000 (b) 5,870,000
(c) 5,900,000 (d) 5,800,000

4. Which comparison is true?

(a) $0 > 7$ (b) $9 < 8$
(c) $8 < 5 < 3$ (d) $15 > 9 > 3$

5. Perform each operation. Check your answer by using the inverse operation.

a. $384 - 295$ **b.** 906×53

c. $5{,}384 + 12{,}957$ **d.** $646 \div 19$

6. Evaluate each expression.

a. $5 \times 7 - 10$ **b.** $30 + 6 \times 9$

c. $5 \times 8 + 3 \times 11$

d. $144 \div 36 + 24 \div 8$

7. The value of the expression $(8 + 5) \times (11 - 7)$ is

(a) 234 (b) 56 (c) 52 (d) 42

8. The greatest common factor of 36 and 54 is

(a) 9 (b) 12 (c) 15 (d) 18

9. Which number is composite?

(a) 3 (b) 5 (c) 7 (d) 9

10. The product of four whole numbers is 72,352. If three of the numbers are 14, 16, and 17, what is the fourth number?

11. A prize of \$609 will be shared equally among Fran, Bill, and Karen. How much money will each person receive?

12. The table shows the number of meters Jane jogged last week.

a. What is the total number of meters that she jogged?

b. On which day did she jog the longest distance?

c. On which day did she jog the shortest distance?

d. What is the difference between the longest and shortest distances that she jogged?

Day	Meters Jogged
Monday	3,550
Tuesday	2,980
Wednesday	1,840
Thursday	3,050
Friday	4,700

13. The organizers of a contest planned to divide \$1,260 equally among three award winners. Because of a tie, four people actually qualified for the award, and the prize money was divided equally among them. How much less did each award winner receive than was originally planned?

UNIT 2-3 Adding Integers

THE MAIN IDEA

1. To add two integers on a number line:
   a. Locate the first number.
   b. From the location of the first number, move the number of units indicated by the second number.
      (1) If the second number is +, move to the right.
      (2) If the second number is –, move to the left.
   c. The sum is the number at the last location.
2. To add two integers without using a number line:
   a. If both numbers are +, the sum is +. Add the absolute values.
   b. If both numbers are –, the sum is –. Add the absolute values.
   c. If one number is + and the other is –:
      (1) Compare the absolute values. The sum has the same sign as the number with the greater absolute value.
      (2) Subtract the absolute values.

EXAMPLE 1 Use a number line to add –5 and 8.

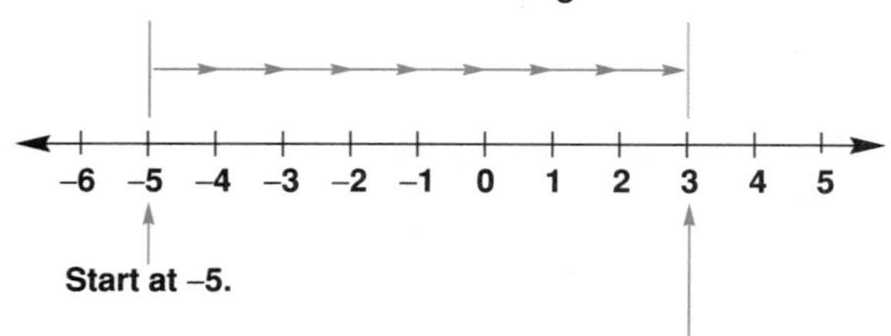

EXAMPLE 2 Using a number line, add: –2 + (–3)

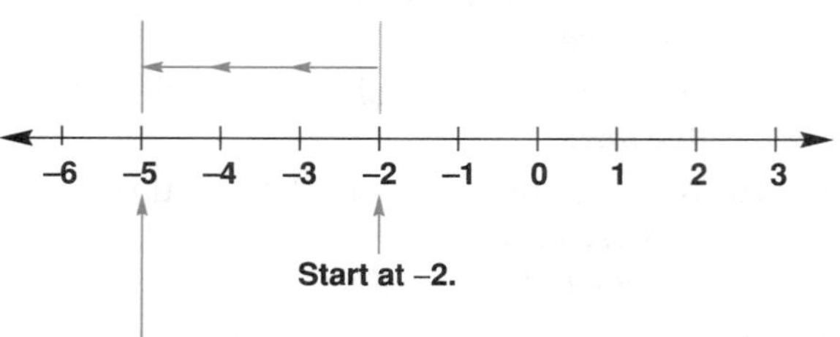

EXAMPLE 3 Add: –6 + (–3)

The signs are the same, both –. So, the sum is –.

Add the absolute values: 6 + 3 = 9.

–6 + (–3) = –?

–6 + (–3) = –9 *Ans.*

EXAMPLE 4 Add: +12 + (–8)

The signs are different, one + and one –. So, compare the absolute values: 12 > 8. The number with the greater absolute value is positive.

+12 + (–8) = +?

Subtract the absolute values: 12 – 8 = 4

+12 + (–8) = +4 *Ans.*

EXAMPLE 5 Write a key sequence that can be used to find the sum of –15 and –9.

KEYING IN

Key Sequence	*Display*
15 [+/–]	– 15.
[+] 9 [+/–]	– 9.
[=]	– 14.

Answer: –15 + (–9) = –14

EXAMPLE 6 Add: –5 + (+1) + (–3)

THINKING ABOUT THE PROBLEM

Use the strategy of breaking the problem into two smaller problems.

Add the first two signed numbers. –5 + (+1) = –4

To this sum, add the third signed number. –4 + (–3) = –7

Answer: –5 + (+1) + (–3) = –7

EXAMPLE 7 The Mudville Tigers football team gained 35 yards during the first quarter and lost 40 yards during the second quarter of their game. Write a number phrase that can be used to find the total number of yards gained or lost during the two quarters.

THINKING ABOUT THE PROBLEM

The key words in this problem tell us to:
(1) Use signed numbers: "gain" (+), "loss" (–).
(2) Use addition: "total."

Show a gain of 35 yards by a positive number. +35

Show a loss of 40 yards by a negative number. –40

Show the total by an addition sign, +. +35 + (–40) *Ans.*

EXAMPLE 8 In January, a Cub Scout troop gained 11 members, and in February, it lost 13 members. For the two months, find the total gain or loss of members.

THINKING ABOUT THE PROBLEM

Write a number phrase that can be used to find the sum of two signed numbers, and find the sum.

A gain of 11 members.	$+11$
A loss of 13 members.	-13
Write a number phrase that shows the total gain or loss.	$+11 + (-13)$
Find the sum.	$+11 + (-13) = -2$

Answer: In the two months, the troop lost 2 members.

CLASS EXERCISES

1. Use a number line to add:

a. $3 + (-1)$ **b.** $5 + (-7)$ **c.** $-2 + (-5)$ **d.** $-4 + (-6)$ **e.** $-5 + (8)$
f. $8 + (-5)$ **g.** $10 + (-10)$ **h.** $-10 + (10)$ **i.** $-3 + (+2)$ **j.** $6 + (3)$

2. Add: **a.** $-7 + (6)$ **b.** $-7 + (-2)$ **c.** $-11 + (-15)$ **d.** $-7 + (-6)$ **e.** $-5 + (9)$
f. $-18 + (7)$ **g.** $-3 + (9)$ **h.** $11 + (-6)$ **i.** $-22 + (0)$ **j.** $-12 + (5)$
k. $-8 + (-4)$ **l.** $28 + (-30)$ **m.** $-2 + (5)$ **n.** $-1 + 1$ **o.** $-3 + (-3)$

3. Add:

a.	**b.**	**c.**	**d.**	**e.**	**f.**	**g.**
-10	-12	-19	-50	-32	-91	-14
$+8$	-17	$+24$	-27	$+32$	$+89$	-14

4. Add: **a.** $50 + (-17)$ **b.** $-150 + (70)$ **c.** $-120 + (55)$ **d.** $-225 + (-145)$
e. $-29 + (101)$ **f.** $-175 + (+175)$ **g.** $-47 + (-133)$ **h.** $250 + (-250)$
i. $-8 + (+11) + (+9)$ **j.** $+4 + (-9) + (-6)$ **k.** $-12 + (-8) + (-7)$

In 5–8, write a number phrase that can be used to find the total gain or loss.

5. The temperature dropped 12 degrees and then dropped another 7 degrees.

6. Mr. Borman made a withdrawal of $87 and then made a deposit of $150 in his savings account.

7. Marie gained 5 pounds and then lost 6 pounds.

8. Rex Clothing Store reduced the price of a suit by $8 and then reduced it further by $3.

9. Mr. Briscoll, a mathematics teacher, uses pennies to represent signed numbers. A penny showing heads stands for +1, and a penny showing tails stands for –1. The number –5 is represented with 5 pennies showing tails.

To add two numbers, Mr. Briscoll represents each number with the proper number of pennies turned to the appropriate side, and then "cancels" each head with a tail until there are no more pairs left. The pennies that remain show the sum.

a. For the pennies shown below, write a number phrase that represents the addition problem illustrated.

b. Explain how the pairing of pennies leads to the correct answer.

10. Write a key sequence that can be used to find each sum.

a. 12 + (–9) **b.** –11 + (–16) **c.** 19 + (–46) **d.** –20 + (36)

11. Explain why the sum of two negative numbers is a negative number.

HOMEWORK EXERCISES

1. Use a number line to add:

a. 8 + (–6) **b.** 2 + (–7) **c.** 3 + (+4) **d.** –5 + (–2) **e.** –6 + (9)
f. –7 + (0) **g.** –1 + (–7) **h.** 4 + (–4) **i.** –7 + (–5) **j.** 3 + (–4)
k. 5 + (–1) + (3) **l.** –2 + (0) + (4) **m.** –3 + (1) + (–1) **n.** 4 + (2) + (–4)

2. Add: **a.** –4 + (9) **b.** –6 + (–3) **c.** –8 + (3) **d.** 7 + (–10) **e.** 0 + (–7)
f. –15 + (15) **g.** 12 + (–7) **h.** –16 + (8) **i.** –20 + (–5) **j.** –7 + (7)
k. –22 + (10) **l.** 24 + (–30) **m.** –4 + (–2) **n.** –8 + (6) **o.** –8 + (0)

3. Add:

a.	**b.**	**c.**	**d.**	**e.**	**f.**	**g.**
–10	–15	+36	–40	+80	–15	32
+20	–18	–21	–35	–90	+19	32

h.	**i.**	**j.**	**k.**	**l.**	**m.**	**n.**
+42	–21	+48	–100	0	0	–5
–42	–21	–50	+100	+14	–21	–10

4. Add:

a. –35 + (–50) **b.** –20 + (42) **c.** –81 + (81) **d.** 46 + (–25) **e.** 32 + (–50)
f. –75 + (–25) **g.** –19 + (43) **h.** –17 + (–82) **i.** –29 + (–80)
j. –100 + (0) **k.** +200 + (–200) **l.** –96 + (–24) **m.** 51 + (0) + (–27)
n. –32 + (25) + (–61) **o.** –70 + (41) + (–28) **p.** –25 + (25) + (25)

5. Find the sum of –48 and –24.

6. The sum of –50 and 35 is (a) –15 (b) 15 (c) –85 (d) 85

In 7–9, write a number phrase that can be used to find the total gain or loss.

7. A kite was flying at a height of 250 m. It was raised 50 m and then lowered 125 m.

8. Dennis had $18 and spent $10.

9. On a winter morning, the temperature was 5°*F*. In the afternoon, the temperature rose 17°, and in the evening fell 9°.

10. Read the explanation in Class Exercise 9. Then, write a number phrase to represent the addition problem shown below, and explain how you can pair pennies to find the correct sum.

11. Write a key sequence that can be used to find each sum.
a. –12 + (29) **b.** +53 + (7) **c.** 17 + (–19) **d.** –12 + (–16)

In 12–14, use signed numbers to find the final result.

12. A football team:

a. gained 20 yards and lost 6 yards. **b.** lost 5 yards and lost 9 yards.

c. lost 11 yards, gained 8 yards, and lost 3 yards.

d. gained 15 yards, lost 12 yards, and gained 8 yards.

13. In a game, Ben started with 35 points. Then he won 9 points and lost 12 points.

14. The temperature one day was $-4°F$, and it dropped 9°.

15. a. A person is \$12 "in the red" (owes \$12). Write a signed number to represent the situation.

b. Write a signed number to represent the least number of whole dollars that person needs to earn in order to be "in the black" (have money).

16. a. What is the least positive integer that, when added to –17, gives a positive sum?

b. Explain how you would find the least positive integer that, when added to –685, gives a positive sum.

Michael Dwyer/Stock Boston, Inc.

17. Roller coasters are designed as loops so that riders return to the starting point. The design for a roller coaster in a new amusement park calls for changes in altitude (number of feet up or down) that are represented by the following sum:

$$50 + (-10) + 30 + (-40) + 15 + ?$$

What must the last signed number be in order for the roller coaster to complete a loop?

18. Which of the following properties of addition of whole numbers are also properties of addition of integers? Use examples to explain your answer.

a. Closure **b.** Commutative property **c.** Associative property

19. One morning, the temperature in Detroit was $8°F$, and the temperature in Cleveland was $-5°F$. How much warmer was it in Detroit?

20. In a boutique, a shirt sold for \$8 above the suggested retail price. At a discount store, the same shirt sold for \$5 below the suggested retail price. What is the difference between the prices?

21. Explain why we subtract absolute values when adding a positive number to a negative.

SPIRAL REVIEW EXERCISES

1. Write the opposite of each signed number.

a. -4 **b.** -9 **c.** 7 **d.** 23

e. -68 **f.** $+11$ **g.** -2 **h.** -30

2. Tell whether each given statement is *true* or *false*.

a. $-10 > -8$ **b.** $0 > -4$

c. $-13 < 13$ **d.** $-9 > -5 > -2$

3. Evaluate each expression.

a. $100 - 22 \times 3$

b. $(100 - 22) \times 3$

c. $4 \times 9 + 8 \times 3$

d. $4 \times (9 + 8) \times 3$

4. Find the product of 240 and 52.

5. Find the sum of 472 and 5,948.

6. Replace each ? by a whole number to make a true statement.

a. $79 + 62 = ? + 79$

b. $71 \times ? = 71$

c. $93 - ? = 93$

d. $(27 + 46) + 85 = ? + (46 + 85)$

7. Which is a true statement?

(a) $-17 + (+17) = 0$
(b) $-17 + (-17) = 0$
(c) $+17 + (+17) = 0$
(d) $-17 + (0) = 0$

8. A calculation that will cause an error symbol to be displayed is

(a) multiplying a large number by zero.
(b) dividing zero by a large number.
(c) dividing a number by zero.
(d) subtracting a large number from zero.

9. Replace each ? by an integer to make a true statement.

a. $9 + ? = 0$

b. $-12 + (-17) = ? + (-12)$

c. $+3 + ? = -5 + (+3)$

d. $(-7 + 2) + ? = -7 + (2 + 9)$

10. Mr. Conte had $12,500 in his bank account. He took $8,560 from his account to pay for a car. How much money was left in his account?

11. The table shows the total number of career homeruns hit by some famous baseball players.

Name of Player	Number of Home Runs
Hank Aaron	755
Reggie Jackson	563
Mickey Mantle	536
Willie Mays	660
Babe Ruth	714

Make a table in which these players are ranked by the number of career homeruns in decreasing order.

12. A personnel director tested five typists. Scores for four of the typists were 50, 35, 40, and 60 words per minute. The difference between the highest and lowest of the five scores was 30 words per minute. What scores are possible for the fifth typist? Explain your reasoning.

UNIT 2-4 Subtracting Integers

THE MAIN IDEA

1. Subtracting a number means adding its opposite.
2. To subtract:
   a. Replace the number being subtracted by its opposite, and change the subtraction symbol to addition.
   b. Then add, remembering the rules for addition of integers.
3. As with subtraction of whole numbers, you can check an answer to a subtraction of integers by addition.

EXAMPLE 1 Using a number line, subtract: $2 - (5)$

Subtracting a number means adding its opposite.

$$2 - (5) = 2 + (-5)$$

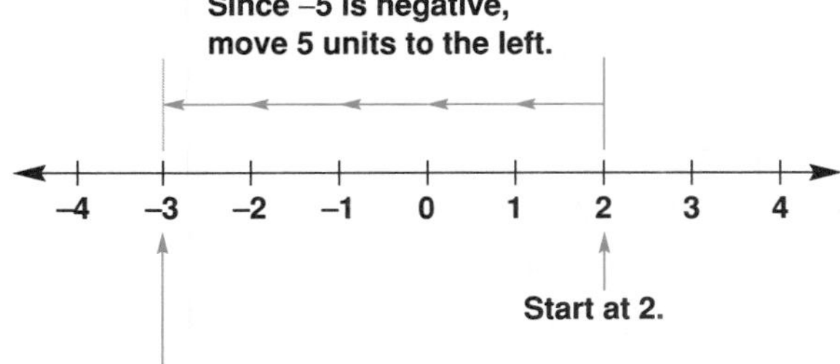

EXAMPLE 2 Subtract: $-9 - (8)$

Replace 8, the number being subtracted, by its opposite, and change to addition.	$-9 - (8)$ $= -9 + (-8)$
Add, remembering the rules for addition of integers.	$= -17$ *Ans.*

EXAMPLE 3 Subtract -6 from -7. Then check the answer.

Replace -6, the number being subtracted, by its opposite, and change to addition.	$-7 - (-6)$ $= -7 + (+6)$
Add, remembering the rules for addition of integers.	$= -1$ *Ans.*

Check: Is $-7 - (-6)$ equal to -1?

Use the inverse operation.

Add -6 to the answer: Does $-1 + (-6) = -7$?

Yes! The answer checks. ✔

EXAMPLE 4 From the sum of $+5$ and -7, subtract -1.

THINKING ABOUT THE PROBLEM

Break this problem into two smaller problems.

Add $+5$ and -7.	$+5 + (-7) = -2$
To subtract -1, add its opposite, $+1$.	$-2 + (+1) = -1$

Answer: $+5 + (-7) - (-1) = -1$

EXAMPLE 5 Find the change in temperature between a morning temperature of $-12°F$ and an afternoon temperature of $+15°F$.

THINKING ABOUT THE PROBLEM

The word "change" tells you to subtract to find the difference. Begin with the most recent temperature, the afternoon temperature, and subtract the earlier temperature, the morning temperature.

Write a number phrase. $+15 - (-12)$

Find the difference by adding the opposite. $+15 + (+12) = +27$

Answer: The temperature increased 27°.

EXAMPLE 6 Write two different key sequences you can use to subtract 5 – (–8) on the calculator.

KEYING IN

Key Sequence	*Display*
5 [–]	5.
8 [+/–]	– 8.
[=]	13.

Key Sequence	*Display*
5 [+]	5.
8 [=]	13.

Both results are the same. This shows that subtracting –8 is the same as adding +8.

CLASS EXERCISES

1. Use a number line to do each subtraction.
   a. 2 – (5) **b.** 6 – (–3) **c.** –7 – (–2) **d.** –9 – (+3) **e.** –10 – (–5)
   f. –3 – (–3) **g.** 8 – (–4) **h.** –4 – (8) **i.** 8 – (10) **j.** 6 – (–4)

2. Subtract and check.
   a. –9 – (5) **b.** 4 – (+7) **c.** –2 – (–18) **d.** 7 – (–3) **e.** –6 – (2)
   f. –15 – (9) **g.** –9 – (–7) **h.** –10 – (5) **i.** 25 – (–25) **j.** 5 – (–8)
   k. –10 – (–5) **l.** –25 – (25) **m.** –13 – (13) **n.** 18 – (2) **o.** 0 – (–2)

3. Subtract and check.

a.	**b.**	**c.**	**d.**	**e.**	**f.**	**g.**	**h.**
–14	+76	16	–38	+10	–51	–42	–82
9	–12	–12	–38	+25	+39	–19	–47

4. From the sum of –7 and 3, subtract 2.

5. Subtract –3 from the sum of 1 and –4.

6. Subtract the sum of 5 and –1 from the sum of –2 and 6.

7. **a.** Find the change in temperature from $+22°C$ to $-12°C$.
   b. Find the change in altitude from 1,200 feet above sea level to 200 feet below sea level.

8. Write a key sequence that can be used to find each difference.
   a. –12 – (5) **b.** 12 – (7) **c.** 19 – (21) **d.** –17 – (–31)

9. **a.** Is the difference of two integers always an integer? Use examples to explain your answer.
   b. What is this property called?

HOMEWORK EXERCISES

1. Using a number line, subtract:
   a. 3 – (5) **b.** 2 – (–4) **c.** –6 – (–3) **d.** –7 – (4) **e.** –8 – (–9)
   f. –4 – (–2) **g.** 6 – (–3) **h.** –3 – (–3) **i.** 5 – (2) **j.** –7 – (2)

2. Subtract:
   a. –10 – (6) **b.** –15 – (–9) **c.** –22 – (–30) **d.** 17 – (23) **e.** 26 – (–18)
   f. –9 – (–5) **g.** –4 – (4) **h.** –4 – (–4) **i.** 20 – (–10) **j.** –10 – (20)
   k. –25 – (–15) **l.** –15 – (–25) **m.** 0 – (–2) **n.** 0 – (+5) **o.** 1 – (–4)

3. Subtract and check.

a.	**b.**	**c.**	**d.**	**e.**	**f.**	**g.**	**h.**
–19	42	–40	58	–36	–58	50	–18
–20	50	+18	–32	–15	+32	–30	–18

4. Subtract –17 from 26. 5. From –26 subtract –9. 6. Subtract 51 from –80.

7. From –75 subtract 25. 8. Subtract 9 from the sum of –2 and –8.

9. From the sum of 6 and –8, subtract –15. 10. Subtract –12 from the sum of –9 and 14.

11. Subtract the sum of –2 and –5 from the sum of –4 and 8.

12. Find the difference between an altitude of 250 feet below sea level and an altitude of 3,000 feet above sea level.

13. Find the change in time between 4 minutes before a rocket blast-off and 1 minute before the blast-off.

14. Write a key sequence that can be used to find each difference.
    a. –17 – (–12) **b.** –9 – (9) **c.** 26 – (–15) **d.** 81 – (92)

15. John used the key sequence 15 [+/–] [–] 8 [=] to subtract –8 from 15. Explain his error.

16. Add: –50 + (+17) + (+42) + (–21) + (+31) + (–29) + (+10)

17. Use a number line and the fact that addition and subtraction of the same number are inverse operations to explain why 5 – (–3) is the same as 5 + (+3).

18. Does the subtraction of integers have a commutative property? Use examples on a number line to explain your answer.

SPIRAL REVIEW EXERCISES

1. The temperature at 8 A.M. was $-6°F$ and it rose 19 degrees. Find the new temperature.

2. The temperature at noon was $8°C$ and it dropped 12 degrees by midnight. Find the temperature at midnight.

3. Barry's science class weighed 10 packages of food to see how close the actual weight of the food came to the advertised weight on the packages. The table shows how the weights compared.

Package	A	B	C	D	E	F	G	H	I	J
Comparison (ounces)	+1	+1	+2	0	–1	+2	0	0	–3	–4

a. Which package(s) had the most extra weight?

b. Which package was closer to the advertised weight, package C or package E?

c. Which package was further below its advertised weight, package I or J?

4. Add:

a. $-9 + (-16)$ **b.** $-19 + (12)$

c. $32 + (-18)$ **d.** $-27 + (30)$

e. $-14 + 0 + (-12)$

f. $-32 + (-1) + (16)$

5. Find the sum of –22 and –45.

6. Which number is prime?
(a) 81 (b) 21 (c) 31 (d) 121

7. The value of 9 in the numeral 79,523 is
(a) 90 (b) 900
(c) 9,000 (d) 90,000

8. Which number rounded to the nearest hundred has the value 2,700?
(a) 2,740 (b) 2,790
(c) 2,640 (d) 2,649

9. When 23 is divided by 5, the remainder is
(a) 1 (b) 2 (c) 3 (d) 4

10. When the key sequence
[CM] 9 [×] 3 [M+] 4 [×] 6 [M–] [MR]
is entered, the calculator displays
(a) 3 (b) 51 (c) 486 (d) 648

11. John bowled five games, scoring 159, 175, 146, 188, and 122. The range of his scores was:
(a) 66 (b) 42 (c) 24 (d) 16

12. How many packages of 16 crackers can be made from a batch of 412 crackers?

13. Philip ate three slices of pizza, and Maria ate one. If each slice of pizza contains 145 calories, how many more calories did Philip consume than Maria?

TEAMWORK

Explore how pennies might be used to show the subtraction of integers. Begin with simple examples such as +10 – (+3). Think of subtraction as removing pennies. Show how removing pennies can be a model for the example –8 – (–3). Move on to more difficult examples such as +9 – (–3). Hint: if you don't have enough heads or tails showing, add "zero" pairs — one head with one tail. These pairs do not change the value, but they give you enough pennies to work with. Write an explanation of how pennies can be used to show the subtraction of integers.

UNIT 2-5 Multiplying Integers

THE MAIN IDEA

To multiply two integers:

1. Determine whether the product will be positive or negative.

Positive × Positive = Positive	Positive × Negative = Negative
Negative × Negative = Positive	Negative × Positive = Negative

2. Then multiply as you would whole numbers.

EXAMPLE 1 Multiply: $-3 \times (-5)$

Negative × Negative = Positive $\qquad -3 \times (-5) = +15$ or 15 *Ans.*

EXAMPLE 2 Find the product of 9 and −4.

"Find the product" means multiply.

Positive × Negative = Negative $\qquad +9 \times (-4) = -36$ *Ans.*

EXAMPLE 3 Use a calculator to multiply −27 by +3.

KEYING IN

Use the change sign key to enter each number with the correct sign. The calculator will automatically display a positive product with no sign and a negative product with a − sign.

Key Sequence	*Display*
27 [+/−]	− 27.
[×] 3 [=]	− 81.

Answer: $-27 \times (+3) = -81$

EXAMPLE 4 Multiply: $-4 \times (-6) \times (-3)$

THINKING ABOUT THE PROBLEM

Use the strategy of breaking the problem into smaller problems.

Find the product of the first two integers. $-4 \times (-6) = +24$

Multiply this product by the third integer. $+24 \times (-3) = -72$

Answer: $-4 \times (-6) \times (-3) = -72$

CLASS EXERCISES

1. Represent each situation as the product of two signed numbers. Then multiply.

a. a 4-degree drop in temperature every day for 3 days
b. paying \$2 a day for bus fare for 5 days
c. earning \$40 a day for 5 days **d.** paying \$12 for each of 10 installments
e. the number of extra pounds a person had 3 weeks ago if he has been losing 2 pounds a week

2. Multiply: **a.** $9 \times (-3)$ **b.** $-9 \times (-3)$ **c.** $-5 \times (7)$ **d.** $8 \times (-4)$ **e.** $-7 \times (-8)$
f. $-6 \times (9)$ **g.** $-4 \times (3)$ **h.** $3 \times (-4)$ **i.** $-2 \times (-5)$ **j.** $6 \times (-4)$
k. $-6 \times (-5)$ **l.** $-9 \times (7)$ **m.** $-2 \times (-2)$ **n.** $8 \times (3)$ **o.** $-5 \times (1)$

3. Find the product of each of the given pairs of integers.
a. 3 and –5 **b.** –2 and 7 **c.** –8 and –4 **d.** –6 and +3 **e.** –9 and –7
f. 5 and –2 **g.** –7 and –4 **h.** –4 and 7 **i.** –2 and +2 **j.** –5 and –5
k. 6 and –4 **l.** –7 and 0 **m.** –1 and 6 **n.** –1 and –5 **o.** 8 and –1

4. Multiply: **a.** $-10 \times (+2) \times (-4)$ **b.** $+5 \times (-6) \times (7)$ **c.** $+1{,}467 \times (-7{,}631) \times (0)$

5. Write a key sequence that can be used to find each product. Then find each product on your calculator.
a. $9 \times (-6)$ **b.** $-17 \times (3)$ **c.** $12 \times (5)$ **d.** $-11 \times (-7)$

6. Use the situations in Exercise 1 to explain why the product of a positive number and a negative number is a negative number.

7. Which of the following properties of multiplication of whole numbers are also properties of multiplication of integers? Use examples to explain your answer.
(a) Closure (b) Commutative property (c) Associative property

8. Describe the result obtained when an integer is multiplied: **a.** by –1 **b.** by 0

HOMEWORK EXERCISES

1. Represent each situation as the product of two signed numbers.

a. a shop sells 4 sweaters that cost \$20 each
b. 4 sweaters that were bought at \$20 each are returned to the shop
c. the greater amount of money you had 5 days ago after paying \$2 each day for carfare
d. the total donation to a charity made by 30 people, each of whom gave \$3
e. the total amount lost by a store owner who sells 12 radios below cost and loses \$4 on each

2. Multiply: **a.** $-9 \times (5)$ **b.** $-10 \times (-2)$ **c.** $4 \times (-7)$ **d.** $-8 \times (-3)$ **e.** $-12 \times (4)$
f. $0 \times (-2)$ **g.** $-6 \times (-6)$ **h.** $-7 \times (9)$ **i.** $3 \times (-9)$ **j.** $-10 \times (5)$
k. $5 \times (-10)$ **l.** $-9 \times (0)$ **m.** $9 \times (6)$ **n.** $8 \times (-2)$ **o.** $-3 \times (-5)$

3. Find the product of each of the given pairs of integers.
a. 7 and –2 **b.** –10 and 6 **c.** –9 and –8 **d.** –3 and –3 **e.** –14 and 0
f. –8 and 7 **g.** –4 and –5 **h.** –9 and 2 **i.** –11 and –5 **j.** –2 and –7
k. –8 and 4 **l.** 4 and –8 **m.** 12 and –3 **n.** –8 and +6 **o.** 1 and 1

4. Multiply: **a.** $-8 \times (+3) \times (+4)$ **b.** $-8 \times (-2) \times (-3)$ **c.** $5 \times (-6) \times (-4)$

5. Write a key sequence that can be used to find each product. Then find each product.
a. $-9 \times (3)$ **b.** $-15 \times (-6)$ **c.** $12 \times (-9)$ **d.** $51 \times (6)$

6. The key sequence 5 [+/–] [×] 3 [+/–] [=] gives the same result as the key sequence 5 [×] 3 [=]. However, the key sequence 5 [+/–] [×] 3 [+/–] [×] 2 [+/–] [=] does not give the same result as the key sequence 5 [×] 3 [×] 2 [=]. Explain why.

7. Use the situations in Homework Exercise 1 to explain why the product of two negative numbers is positive.

8. The order of operations used for whole numbers is also used for integers. Evaluate each expression.
a. $-5 + 3 \times (-8)$ **b.** $-40 - 7 \times (-6)$ **c.** $16 \times (-2) + 14$ **d.** $-6 \times 2 - 2$

9. As with whole numbers, multiplication of integers is distributive over addition and subtraction. Complete the following:
a. $-6 \times (-4 + 2) = -6 \times (-4) + \boxed{?} \times 2$ **b.** $4 \times (7 - (-3)) = 4 \times 7 - 4 \times \boxed{?}$

10. Replace each □ by an integer to make a true statement.
a. $+5 \times \square = +20$ **b.** $+7 \times \square = -35$ **c.** $-9 \times \square = 36$ **d.** $-8 \times \square = -64$

SPIRAL REVIEW EXERCISES

1. The table shows how membership increased and decreased at the Wonderbody Health Club each month.

Month	Jan.	Feb.	Mar.	Apr.	May	June	July	Aug.	Sept.	Oct.	Nov.	Dec.
Change in Membership	+4	+18	+21	+18	+11	+6	0	0	−4	−10	−18	−6

a. In which month did membership decrease the most?

b. Did the membership decrease more in July or in December?

2. Add:
   a. $-9+(-12)$ **b.** $-17+(3)$
   c. $24+(-11)$ **d.** $-63+(-42)+(8)$
   e. $101+(-11)+(-100)$

3. Subtract:
   a. $-7-(-8)$ **b.** $-11-(4)$
   c. $19-(-4)$ **d.** $-20-(-10)$

4. Find the sum of −15 and 10.

5. Subtract −9 from −16.

6. From the sum of +7 and −2, subtract −23.

7. Replace each ? by < or > to make a true statement.
   a. $-14 \; ? -16$ **b.** $-14+(+14) \; ? -1$
   c. $20+(-5) \; ? \; 20-(-5)$

8. Replace each ? by an integer to make a true statement.
   a. $+2\times(-8)= \; ? \times(+2)$
   b. $-10\times \; ? =0$
   c. $-7\times \; ? =-7$
   d. $(? \times -2)\times 4=6\times(-2\times 4)$
   e. $-15\times \; ? =+15$

9. Nine thousand, twenty-three is
   (a) 92,300 (b) 90,230
   (c) 9,230 (d) 9,023

10. 6,542,183 rounded to the nearest million is
   (a) 6,000,000 (b) 7,000,000
   (c) 6,500,000 (d) 6,600,000

11. The value of 7 in 57,382 is
   (a) 70,000 (b) 7,000
   (c) 700 (d) 70

12. The greatest common factor of 24 and 48 is
   (a) 2 (b) 6 (c) 12 (d) 24

13. The value of $6+4\times 3$ is
   (a) 30 (b) 13 (c) 18 (d) 72

14. Which of the following key sequences will display an error symbol?
   (a) 492 [−] 586 [=]
   (b) 492 [×] 586 [=]
   (c) 492 [÷] 0 [=]
   (d) 492 [×] 0 [=]

15. An oil tank that holds 5,000 gallons of oil has a leak and loses 75 gallons an hour. After 6 hours, how many gallons of oil are left in the tank?

16. The price of a ticket for a school concert was \$4. If \$508 were collected and 15 free passes were given out, how many seats were accounted for?

UNIT 2-6 Dividing Integers

THE MAIN IDEA

1. The rules for dividing integers are similar to the rules for multiplying integers.
   a. First, determine the sign of the quotient:

      Positive ÷ Positive = Positive   Positive ÷ Negative = Negative

      Negative ÷ Negative = Positive   Negative ÷ Positive = Negative

   b. Then, divide as you would whole numbers.
2. You can check the answer to a division by using multiplication, the inverse operation.

EXAMPLE 1 Divide: $-18 \div (-9)$ Check.

First, determine whether the quotient will be positive or negative.

Negative ÷ Negative = Positive

Then, divide as you would whole numbers.

$-18 \div (-9) = +2$ *Ans.*

Use the inverse operation to check. $2 \times (-9) = -18$

EXAMPLE 2 Divide the sum of -6 and 30 by -3.

THINKING ABOUT THE PROBLEM

Use the strategy of breaking the problem into smaller problems.

Find the sum of -6 and 30. $-6 + (+30) = +24$

Divide this sum by -3. $+24 \div (-3) = -8$ *Ans.*

CLASS EXERCISES

1. Divide: **a.** $-10 \div (2)$ **b.** $-15 \div (-3)$ **c.** $20 \div (-5)$ **d.** $-12 \div (-4)$ **e.** $8 \div (-2)$
f. $-16 \div (8)$ **g.** $-50 \div (10)$ **h.** $-48 \div (6)$ **i.** $-28 \div (-7)$ **j.** $-40 \div (10)$
k. $-100 \div (-20)$ **l.** $-81 \div (-9)$ **m.** $-40 \div (5)$ **n.** $-16 \div (-16)$ **o.** $4 \div (-4)$

2. Divide: **a.** $120 \div (-6)$ **b.** $56 \div (-7)$ **c.** $-32 \div (-8)$ **d.** $-96 \div (-12)$ **e.** $100 \div (-25)$
f. $-150 \div (75)$ **g.** $-108 \div (9)$ **h.** $225 \div (-15)$ **i.** $-144 \div (-8)$ **j.** $14 \div (-1)$

3. Divide the sum of –12 and +20 by –2.

4. Divide the difference of +15 and –15 by +3.

5. Divide the product of –12 and –2 by –3.

6. Write a key sequence that can be used to find each quotient.

a. –16 ÷ (4) **b.** 27 ÷ (3) **c.** –15 ÷ (–5) **d.** 24 ÷ (–8)

7. Explain why the rules of sign for dividing and multiplying integers are similar.

HOMEWORK EXERCISES

1. Divide:
a. 20 ÷ (–5) **b.** –14 ÷ (7) **c.** –24 ÷ (–4) **d.** –12 ÷ (6) **e.** –36 ÷ (–9)
f. –90 ÷ (10) **g.** 0 ÷ (–8) **h.** –45 ÷ (9) **i.** –63 ÷ (–7) **j.** –80 ÷ (8)
k. 56 ÷ (–7) **l.** –49 ÷ (7) **m.** –14 ÷ (–2) **n.** –14 ÷ (2) **o.** 14 ÷ 2

2. Divide:
a. –88 ÷ (11) **b.** –100 ÷ (–10) **c.** –64 ÷ (16) **d.** 48 ÷ (–12)
e. –96 ÷ (–24) **f.** 200 ÷ (–50) **g.** –99 ÷ (33) **h.** –120 ÷ (–40)
i. –54 ÷ (–18) **j.** –78 ÷ (26) **k.** 99 ÷ (–11) **l.** –100 ÷ (–25)

3. Divide the sum of –14 and +18 by +4.

4. Divide the difference of –28 and –21 by 7.

5. Divide the product of 0 and –14 by –7.

6. Divide the sum of –9 and –5 by the product of –2 and +1.

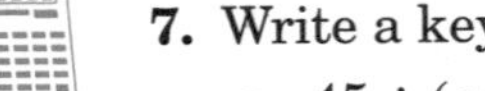

7. Write a key sequence that can be used to find each quotient.

a. 45 ÷ (+5) **b.** –100 ÷ (–20) **c.** 28 ÷ (–7) **d.** –36 ÷ (12)

8. Replace each □ by a number to make a true statement.

a. □ × 24 = –24 **b.** □ + 24 = –24 **c.** □ ÷ 24 = –24 **d.** □ – 24 = –24

9. Explain why the division of any nonzero integer by its opposite always gives the same result.

10. Which of the following are properties of division of integers? Use examples to explain.

a. Closure **b.** Commutative property **c.** Associative property
d. Distributive property over addition **e.** Distributive property over subtraction

SPIRAL REVIEW EXERCISES

1. Multiply:

a. $-6 \times (-3) \times (-4)$

b. $12 \times (+10) \times (-2)$

2. Find the sum of 29 and -17.

3. Subtract -18 from 18.

4. Perform the indicated operation.

a. $-6 + (-10)$ **b.** $-6 - (-10)$

c. $-6 \times (-10)$ **d.** $-16 - (-4)$

e. $28 + (-7)$ **f.** $28 - (-7)$

5. Find the product of -18 and -5.

6. From -27 subtract -43.

7. Which is a true statement?
(a) $8 \times (-1) = 8$ (b) $5 - 7 = 7 - 5$
(c) $-9 + 1 < -8$ (d) $-7 \times (-1) = 7$

8. Replace ? by < or > to make a true statement: $-2 \times (-3) \times (-1)$? 0

9. When 494 is divided by 16, the remainder is (a) 14 (b) 12 (c) 10 (d) 0

10. At the beginning of a trip, the odometer of a car read 23,760 miles, and at the end of the trip, 24,095 miles. How long was the trip?

11. Mr. Harris travels 37 miles round-trip each day to work. How many miles will he travel in 20 workdays?

12. At 8 P.M., the temperature was 9°*C*. The temperature dropped 5° between 8 P.M. and 9 P.M., and 6° during the next hour. Find the temperature at 10 P.M.

13. A football team gained 15 yards, lost 12 yards, and then gained 8 yards. Find the total change.

14. It is estimated that in every 2 seconds, 9 babies are born and 3 people die in the world. At this rate, what is the change in the world population in a 365-day year?

15. The table shows how the Chinese calendar matches each year with an animal.

Chinese Year 1972–2007

Rat	Ox	Tiger	Rabbit	Dragon	Snake	Horse	Sheep	Monkey	Rooster	Dog	Pig
1972	1973	1974	1975	1976	1977	1978	1979	1980	1981	1982	1983
1984	1985	1986	1987	1988	1989	1990	1991	1992	1993	1994	1995
1996	1997	1998	1999	2000	2001	2002	2003	2004	2005	2006	2007

a. What animal is associated with 1976? With the year of your birth?

b. What animal is associated with the year 2000? With the year 2050?

c. What year in the 21st century will be the first year of the rat? The last year of the rat?

d. What animal was associated with 1776, the year of American Independence?

THEME 2

Decimals

MODULE 3

MODULE 4

UNIT 3–1 The Meaning of Decimals; Reading and Writing Decimals

THE MEANING OF DECIMALS

THE MAIN IDEA

1. The number 1 represents a whole quantity. A part of a whole quantity is represented by a fraction.
2. A fraction is written with two numerals separated by a division line:

$$\frac{\textit{numerator}}{\textit{denominator}}\text{ (not zero)}$$

3. The denominator of a fraction tells the number of equal parts that make up the whole quantity. The numerator shows the number of equal parts represented by the fraction.

The whole quantity

The whole quantity divided into 10 equal parts

$\frac{1}{10}$ means 1 out of 10 equal parts

4. A ***decimal*** is another way of writing a fraction.
5. The place names and place values for decimals are:

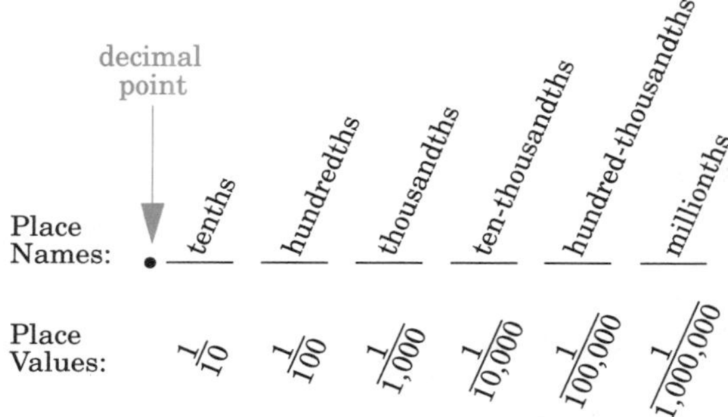

6. As we move to the right from the decimal point, each place stands for a fraction that is $\frac{1}{10}$ (or 0.1) of the place before it. To find the value of a digit in a decimal, multiply the digit by its place value.

EXAMPLE 1 For each of the underlined digits, write the place name as a word and as a fraction.

Digit	***Place Name as a Word***	***Place Name as a Fraction***
a. $0.82\underline{3}$	thousandths	$\frac{1}{1,000}$
b. $0.\underline{9}41$	tenths	$\frac{1}{10}$
c. $0.6\underline{8}52$	hundredths	$\frac{1}{100}$
d. $0.739\underline{6}$	ten-thousandths	$\frac{1}{10,000}$

EXAMPLE 2 For each digit of the decimal number 0.9387, write the value of the digit as a fraction.

Digit	***Value***
a. $0.\underline{9}387$	$9 \times \frac{1}{10}$ or $\frac{9}{10}$
b. $0.9\underline{3}87$	$3 \times \frac{1}{100}$ or $\frac{3}{100}$
c. $0.93\underline{8}7$	$8 \times \frac{1}{1,000}$ or $\frac{8}{1,000}$
d. $0.938\underline{7}$	$7 \times \frac{1}{10,000}$ or $\frac{7}{10,000}$

CLASS EXERCISES

1. For each of the underlined digits, write the name of the place as a word and as a fraction.
a. $0.7\underline{2}6$ **b.** $0.\underline{8}17$ **c.** $0.23\underline{8}9$ **d.** $0.465\underline{7}$ **e.** $0.3\underline{4}68$ **f.** $0.\underline{3}18$ **g.** $0.42\underline{1}6$

2. Write as a fraction the value of 7 in each decimal.
a. $0.2\underline{7}$ **b.** $0.\underline{7}39$ **c.** $0.52\underline{7}8$ **d.** $0.8426\underline{7}$ **e.** $0.493\underline{7}2$ **f.** $0.\underline{7}$ **g.** $0.0\underline{7}$

READING DECIMALS

THE MAIN IDEA

To read a decimal:

1. Ignore the decimal point and any leading zeros (zeros immediately following the decimal point).
2. Read the digits as if they were naming a whole number.
3. Say the name of the place value of the last digit.
4. If the decimal has an integer part:
   a. Read the integer part.
   b. Say the word "and."
   c. Read the decimal part.

EXAMPLE 3 Read each decimal and write its word name.

a. 0.8 Read the digit 8 as "eight" and say the place value of the 8, which is tenths.

Answer: eight tenths

b. 0.0075 Ignore the leading zeros. Read 75 as "seventy-five" and say the place value of the 5, which is ten-thousandths.

Answer: seventy-five ten-thousandths

EXAMPLE 4 Read each decimal and write its word name.

	Decimal	*Word Name*
a.	0.27	twenty-seven hundredths
b.	0.0027	twenty-seven ten-thousandths
c.	0.00056	fifty-six hundred-thousandths
d.	8.07	eight and seven hundredths
e.	72.0005	seventy-two and five ten-thousandths

EXAMPLE 5 Read the decimal displayed on a calculator for the computation 5 ÷ 8.

Key Sequence	*Display*
5 [÷] 8 [=]	0.625

Answer: six hundred twenty-five thousandths

CLASS EXERCISES

1. Read each decimal and write its word name.

a. 0.14 **b.** 0.5 **c.** 0.07 **d.** 0.27 **e.** 0.037 **f.** 0.05 **g.** 0.7 **h.** 0.083
i. 0.374 **j.** 0.19 **k.** 0.007 **l.** 0.0023 **m.** 0.310 **n.** 0.204 **o.** 0.14 **p.** 0.592

2. Read each decimal and write its word name.

a. 21.07 **b.** 124.023 **c.** 132.4 **d.** 91.091 **e.** 5.129 **f.** 17.19 **g.** 702.05

3. Read the decimal displayed on a calculator after each key sequence.

a. 4 [÷] 5 [=] **b.** 3 [÷] 8 [=] **c.** 11 [÷] 16 [=] **d.** 22 [÷] 20 [=]

WRITING DECIMALS GIVEN AS WORD NAMES

THE MAIN IDEA

To write a decimal that is given as a word name:

Think of the decimal as a fraction with a denominator whose place name is mentioned. Write the fraction as a decimal.

EXAMPLE 6 Write twenty-three hundredths as a decimal.

Think of twenty-three *hundredths* as a fraction with a denominator whose place name is mentioned.

twenty-three — The numerator is 23.
hundredths — The denominator is 100.

Write $\frac{23}{100}$ as a decimal. $\frac{23}{100} = 0.23$ *Ans.*

EXAMPLE 7 Write each word name as a decimal.

Word Name	***Decimal***
a. nine hundred thirty-seven thousandths	0.937
b. nine and thirty-seven thousandths	9.037
c. four hundred seven and nine tenths	407.9

CLASS EXERCISES

Write each word name as a decimal.

1. forty-seven hundredths
2. fifteen thousandths
3. two hundred and three hundredths
4. nine tenths
5. eleven and seven tenths
6. two and twelve hundredths
7. one hundred forty-five and twenty thousandths
8. one hundred forty-five thousandths
9. two hundred twenty-two and two thousandths
10. four and eighty-one ten-thousandths

HOMEWORK EXERCISES

1. For each of the underlined digits, write the name of the place as a word and as a fraction.

a. 0.935 **b.** 0.429 **c.** 0.264 **d.** 0.8243 **e.** 0.07 **f.** 0.826 **g.** 0.3758
h. 0.3492 **i.** 0.4759 **j.** 0.0007 **k.** 0.18 **l.** 0.3254 **m.** 0.892 **n.** 0.6178

2. In each decimal, write the value of 3 as a fraction.

a. 0.135 **b.** 0.2538 **c.** 0.317 **d.** 0.5143 **e.** 0.9365 **f.** 0.03 **g.** 0.003

3. Read each decimal and write its word name.

a. 0.09 **b.** 0.017 **c.** 0.213 **d.** 0.79 **e.** 0.2 **f.** 0.9 **g.** 0.6
h. 0.0213 **i.** 0.0003 **j.** 0.002 **k.** 7.21 **l.** 8.09 **m.** 12.015 **n.** 115.003
o. 207.702 **p.** 9.3 **q.** 16.57 **r.** 96.08 **s.** 89.03 **t.** 73.5 **u.** 0.111

4. Read the decimal displayed on a calculator after each key sequence.

a. 9 [÷] 12 [=] **b.** 15 [÷] 8 [=] **c.** 22 [÷] 25 [=] **d.** 67 [÷] 1000 [=]

5. Write each word name as a decimal.

a. seven tenths **b.** eighty-three hundredths **c.** two hundred thirty-two thousandths
d. seventy-five thousandths **e.** five hundredths **f.** six and nine tenths
g. twenty-one and fifteen hundredths **h.** two hundred nine thousandths
i. two hundred and nine thousandths **j.** two and one hundred nine thousandths

6. Without actually calculating, complete the following table of divisions and their resulting decimals.

Division	1 ÷ 20	2 ÷ 20	3 ÷ 20	7 ÷ 20	10 ÷ 20	?
Decimal	0.05	0.1	0.15	?	?	0.95

7. Write the sum of $\frac{5}{10,000}$, $\frac{3}{1,000}$, $\frac{7}{100}$, and $\frac{1}{10}$ as a decimal.

8. The graph gives the average annual snowfall for five selected cities, in tenths of an inch.

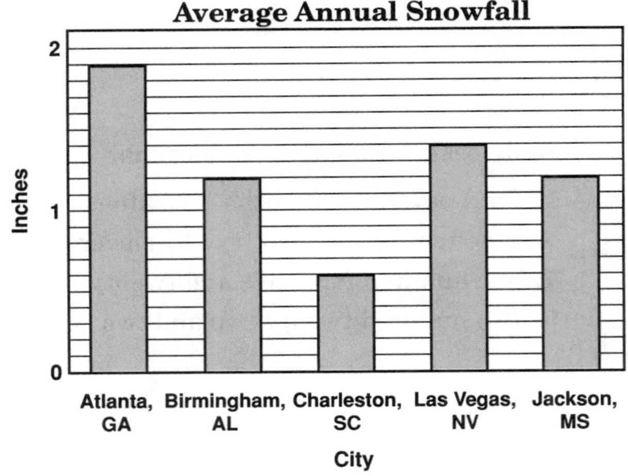

a. Read the average snowfall of each city on the graph. Then copy and complete the following table, using decimals.

City	Average Snowfall
Atlanta	
Birmingham	
Charleston	
Las Vegas	
Jackson	

b. If the average snowfall of a city is halfway between the average snowfall for Birmingham and Las Vegas, what is it?

SPIRAL REVIEW EXERCISES

1. A prime number between 32 and 40 is
(a) 33 (b) 34 (c) 37 (d) 39

2. The [+/-] key changes a negative number displayed to a positive number. It also
(a) determines if you should add or subtract.
(b) displays a number from memory.
(c) changes a positive number to a negative number.
(d) clears the calculator.

3. What is the remainder when 414 is divided by 18?

4. Divide –140 by –7.

5. Rearrange the numbers 100, –100, 10, and –10 in order, from least to greatest.

6. If 456 eggs are to be packed into boxes holding 12 eggs per box, how many boxes will be needed?

7. Mr. James bought a $399 VCR on sale for $280. With the money that he saved, he bought video tapes of films, each of which cost $17. How many tapes did he buy?

8. Ryan's father lives 22 miles from work. How many miles does he commute in a five-day work week?

9. Find the value of $9 \times 4 + 10 \div 2$.

10. Henry is making a poster to show the number of students who attended the Senior Class Show. On his poster, a small picture of a person represents 25 students. A part of the picture represents fewer than 25. If 212 students attended, how many copies of the picture should he make?

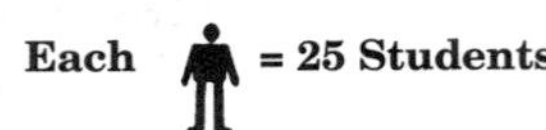

UNIT 3–2 Comparing Decimals; Rounding Decimals

COMPARING DECIMALS

THE MAIN IDEA

1. Just as we represent integers on a number line, we also represent decimals on a number line.

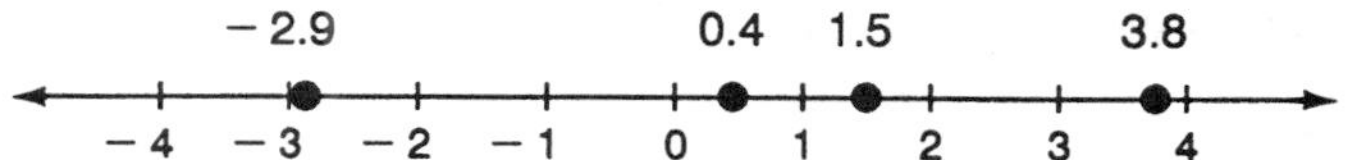

2. You can use a number line to compare two decimals. On the number line, the greater number is on the right .

   3.8 > 0.4 –2.9 < 1.5

3. To compare two or more decimals without using a number line:
   a. Look at the decimals to see what the greatest number of decimal places is.
   b. Write each decimal as an equivalent decimal having that same number of decimal places. Do this by writing zeros on the right when necessary.
   c. Ignore the decimal point and compare the numbers as if they were integers.

EXAMPLE 1 Replace ? by < or > to make a true comparison: 0.7 ? 0.63

The greatest number of decimal places is 2.	0.7 ? 0.63
Write 0.7 as 0.70 so that both decimals have 2 places.	0.70 ? 0.63
Ignore the decimal point and compare as with integers, starting with the first nonzero number.	70 > 63

Answer: 0.7 > 0.63

EXAMPLE 2 Write the numbers 0.25, –0.052, 0.52, and 0.025 in order from least to greatest.

The greatest number of decimal places is 3.	0.25	–0.052	0.52	0.025
Insert zeros as needed.	0.250	–0.052	0.520	0.025
Think of the integers.	250	–52	520	25
Arrange the integers in order from least to greatest.	3rd	1st	4th	2nd

Answer: –0.052, 0.025, 0.25, 0.52

EXAMPLE 3 Jeff ran the mile in 5.3 minutes. Jerome's time was 5.26 minutes, and Cindi's time was halfway between. What was Cindi's time? Who ran fastest?

Jeff	*Cindi*	*Jerome*	
5.30		5.26	Write each with 2 decimal places.
530		526	Think of the integers. Find the one halfway between
	528		

Answer: Cindi ran in 5.28 minutes. Jerome was fastest.

CLASS EXERCISES

1. Replace ? by $<$ or $>$ to make a true statement.

   a. 0.3 ? 0.4 **b.** 0.3 ? 0.04 **c.** 0.72 ? 0.702 **d.** 57.38 ? –58.38 **e.** 23.7 ? 2.37 **f.** –0.37 ? 0.5

2. The decimal that has the same value as 0.25 is

   (a) 0.205 (b) 0.025 (c) 0.2500 (d) 0.2005

3. The decimal that has the greatest value is

   (a) 0.7 (b) 0.707 (c) 0.077 (d) 0.77

4. The decimal that has the least value is

   (a) 0.003 (b) 0.029 (c) –0.01 (d) 0.3

5. Write the numbers 7.8, 0.078, –0.78, and 0.807 in order from least to greatest.

6. At the end of a day of fishing, Carol had caught 4.8 kilograms of fish and Maria had caught 4.68 kilograms of fish. Which girl caught the greater weight of fish?

7. Which number is between 4.72 and 4.73?

   (a) 4.7 (b) 4.725 (c) 4.736 (d) 4.8

8. The Dewey Decimal System is a way of classifying library books according to topic. A whole number is assigned to each general category, and specific topics within that general category are assigned decimals starting with that same whole number. Books about geography and travel have numbers in the 900's, with books about Europe assigned the whole number 914, and books about Asia, 915.

   a. The number for Italy is halfway between 914 and 915. What is the Dewey Decimal number for books about Italy?

   b. If 914.8 is the number for books about Scandinavia, what number, halfway between the numbers for Italy and Scandinavia, is assigned to a book on the Spanish Pyrenees?

ROUNDING DECIMALS

THE MAIN IDEA

Rounding a decimal is similar to rounding a whole number.

1. Circle the digit in the place to be rounded.
2. Look at the digit immediately to the right of the circled digit.
   a. If the digit to the right of the circled digit is 5 or more, increase the circled digit by one (round up).
   b. If the digit to the right of the circled digit is less than 5, keep the circled digit (round down).
3. Drop all digits to the right of the circled digit.

EXAMPLE 4 Round 0.793 to the nearest tenth.

Circle 7, the digit in the tenths place. 9 is the digit in the next position to the right. 0.(7)93

Since 9 > 5, round up. Drop the digits to the right of the tenths place. 0.8 *Ans.*

EXAMPLE 5 Round 1.2475 to the nearest

a. tenth **b.** hundredth **c.** thousandth

	Place	***Circle Digit***	***Next Digit Compared to 5***	***Answer***
a.	tenth	1.(2)4̲75	4̲ < 5	1.2
b.	hundredth	1.2(4)7̲5	7̲ > 5	1.25
c.	thousandth	1.24(7)5̲	5̲ = 5	1.248

EXAMPLE 6 Round 0.3965 inch to the nearest thousandth of an inch.

Circle 6, the digit in the thousandths place. 5 is in the position to the right. 0.39(6)5

Since the next digit is 5, round up. Drop the digit to the right of the thousandths place. 0.397 inch *Ans.*

EXAMPLE 7 Round 3.981 gallons to the nearest tenth of a gallon.

Since 8 > 5, round up. 3.(9)81

4.0 or 4 gallons *Ans.*

CLASS EXERCISES

1. Which digit do we circle in 0.592 to round correct to the nearest tenth?
2. When we are rounding, to what number do we compare the digit to the right of the circled digit?
3. Round to the nearest tenth: **a.** 7.25 **b.** 19.375 **c.** 3.94
4. Round to the nearest hundredth: **a.** 0.596 **b.** 6.245 **c.** 42.6842
5. Round to the nearest thousandth: **a.** 0.2468 **b.** 20.1935 **c.** 32.7124
6. Round 5.395624 to the nearest: **a.** ten-thousandth **b.** hundredth **c.** whole number
7. Barbara spent $29.52 for a dress. Round the cost of the dress to the nearest dollar.
8. Mr. Jones needs 125.86 meters of fencing for his yard. Round the amount of fencing needed to the nearest tenth of a meter.
9. A cook wants to divide 32 ounces of ground beef into 7 equal portions. Find the size of each portion to the nearest tenth of an ounce.

HOMEWORK EXERCISES

1. Replace ? by < or > to make a true statement.
   a. 0.302 ? 0.23 **b.** 5.9 ? –6.9 **c.** 10.517 ? 10.175 **d.** 8.275 ? 7.975 **e.** –0.7 ? –0.52
   f. 6.77 ? 6.707 **g.** –0.52 ? 0.045 **h.** 0.483 ? 0.438 **i.** –0.6982 ? –0.7 **j.** 0 ? –0.06
2. A decimal with the same value as 0.37 is (a) 0.370 (b) 0.307 (c) 0.037 (d) 0.3007
3. The decimal with the greatest value is (a) 0.404 (b) 0.0444 (c) 0.44 (d) 0.04444
4. Write the numbers 0.59, 5.9, 9.5, 0.95, and 0.509 in order from greatest to least.
5. Yvonne's practice times for the 100-yard dash on Monday and Tuesday were 11 seconds and 12 seconds. On Wednesday, her time was halfway between her times for Monday and Tuesday. Her time for Thursday was halfway between her times for Tuesday and Wednesday. What was her time for Thursday?
6. Using the number line, write the letter that shows the position of each given decimal.
   a. 2.087 **b.** 0.087 **c.** –1.5 **d.** 0.6 **e.** –0.7 **f.** 0.87

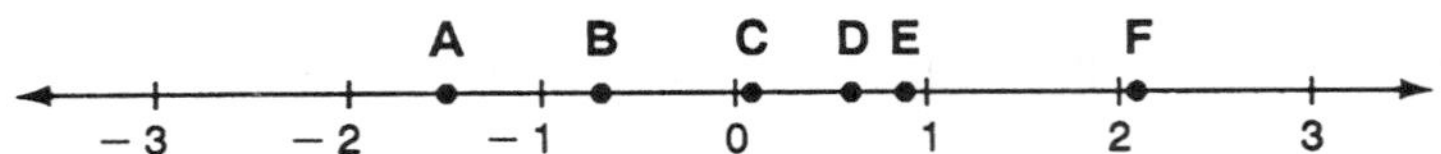

7. When we round 2.943 to the nearest hundredth, which digit do we circle?

8. Round to the nearest tenth: **a.** 0.72 **b.** 0.35 **c.** 0.891 **d.** 8.92 **e.** 27.96

9. Round to the nearest hundredth: **a.** 0.283 **b.** 0.517 **c.** 0.123 **d.** 2.861 **e.** 6.998

10. Round to the nearest thousandth: **a.** 0.8176 **b.** 0.1462 **c.** 0.4739 **d.** 12.8754

11. Round 0.549283 to the nearest: **a.** hundredth **b.** thousandth **c.** ten-thousandth

12. Bennett spent $42.85 on a pair of shoes. Round the amount of money he paid to the nearest dollar.

13. Rita swam 47.29 meters. Round the distance she swam to the nearest tenth of a meter.

14. A machinist needs to cut metal rods that are 3.956 inches long from a 20-inch piece of metal. What is the maximum number of rods that can be cut?

15. John was asked to round 0.748 to the nearest tenth. He said that the 8 rounds the 4 up to 5, and 5 rounds the 7 up to 8. Thus, John's answer is that 0.748 rounded to the nearest tenth is 0.8. Is John correct? Explain.

SPIRAL REVIEW EXERCISES

1. Write the word name of each decimal.

   a. 0.95 **b.** 0.04 **c.** 0.2 **d.** 0.125

2. For the decimal 10.0625, the value of 2 as a fraction is

   (a) $\frac{2}{10}$ (b) $\frac{2}{100}$ (c) $\frac{2}{1,000}$ (d) $\frac{2}{10,000}$

3. James uses about 90 calories for every mile he jogs. If he jogs 12 miles, about how many calories does he use?

4. 502 × 98 is closest to

   (a) 5,000,000 (b) 500,000
   (c) 50,000 (d) 5,000

5. The dot on the number line represents the number

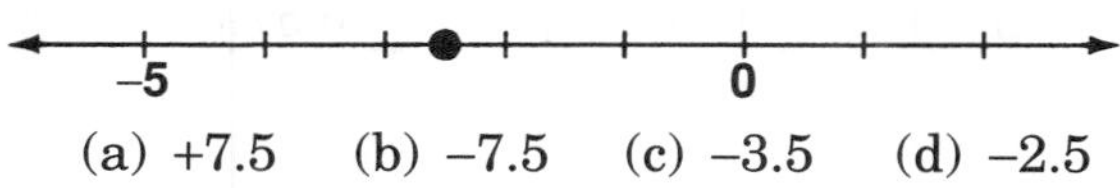

   (a) +7.5 (b) –7.5 (c) –3.5 (d) –2.5

6. Find the least common multiple of 16 and 12.

7. Find the sum of –23 and –43.

8. Use a calculator to help you decide which operation symbols should replace [?] to produce the given result.

   1210 [?] 11 [?] 280 [?] 400 = 77

9. Mr. Damsky's truck can safely carry 25,000 pounds. If he loads the truck with 18,750 pounds of bricks, and 2,500 pounds of sand, how much more weight can be safely put on the truck?

10. Mrs. Klein planted tomatoes and corn in her garden. There are 12 plants in each row of tomatoes and 10 plants in each row of corn. If there are 10 rows altogether, what is the least number of plants in Mrs. Klein's garden?

11. Add: –4 + (–12)

12. Divide 3,914 by 19.

UNIT 3-3 Adding and Subtracting Decimals; Using Decimals to Make Change

ADDING AND SUBTRACTING DECIMALS

THE MAIN IDEA

To add or subtract decimals:

1. Line up the decimals vertically so that the decimal points are beneath one another. Doing this aligns digits that have the same place value. In a whole number, a decimal point is understood to be at the end.
2. If necessary, insert zeros so that the numbers have the same number of places after the decimal point.
3. Add or subtract the numbers in the same way that you add or subtract whole numbers.
4. Place the decimal point in the sum or difference directly beneath the points above.

EXAMPLE 1 Joann spent \$27.80 on a skirt, \$15.99 on a blouse, and \$32.45 on shoes. Find the total amount she spent.

THINKING ABOUT THE PROBLEM

The key word "total" tells you to add the three costs.

Since dollars and cents are written as decimals, line up the decimal points. When you add, carry numbers over the decimal point.

```
 $27.80
  15.99
 +32.45
 $76.24
```

Answer: \$76.24 is the total.

EXAMPLE 2 Add 0.300 + 0.27 + 0.009 on a calculator.

KEYING IN

By entering the decimal points in their proper positions, you are indicating which digits are to be lined up with each other. Whole-number parts that are 0, or final zeros in the decimal part, need not be entered.

Key Sequence	***Display***
.3 [+] .27 [+] .009 [=]	0.579

EXAMPLE 3 Jacqueline ran 12 kilometers and Francine ran 9.28 kilometers. How many more kilometers did Jacqueline run than Francine?

THINKING ABOUT THE PROBLEM

To find how much more one number is than another, subtract.

Line up the decimals vertically. In the whole number 12, the decimal point is at the end.

$$\begin{array}{r} 12. \\ -9.28 \\ \hline \end{array}$$

Insert zeros. Then subtract, and insert the decimal point.

$$\begin{array}{r} 12.00 \\ -9.28 \\ \hline 2.72 \end{array}$$

Answer: Jacqueline ran 2.72 kilometers more.

EXAMPLE 4 At the start of an experiment, a flask contained 25 cubic centimeters (cc) of fluid. During the course of the experiment, the following changes were recorded:

an increase of 0.5 cc　a decrease of 0.23 cc

a decrease of 2.7 cc　an increase of 1.08 cc

How much fluid was in the flask at the end?

THINKING ABOUT THE PROBLEM

You can work with these changes as a series of additions and subtractions:

$$25 + 0.5 - 2.7 - 0.23 + 1.08 = 23.65$$

Another way is to think of the changes as a sum of signed numbers:

$$(25) + (+0.5) + (-2.7) + (-0.23) + (+1.08)$$

Answer: The flask has 23.65 cc of fluid.

CLASS EXERCISES

1. Add: **a.** 37.6 + 0.015 **b.** 0.9 + 27.05 + 10 **c.** 0.29 + 0.0037 + 0.378
2. Janette's auto insurance costs $351.52 for bodily injury liability, $162.48 for property damage, and $237.14 for collision coverage. Find the total cost of Janette's auto insurance.
3. Jason mixed 0.46 kilogram of cashews, 0.68 kilogram of walnuts, and 1.2 kilograms of peanuts together. What is the total weight of the mixture?
4. Subtract: **a.** 0.75 – 0.32 **b.** 59 – 46.8 **c.** 11.65 – 8.9 **d.** 22.7 – 17.68
5. Subtract 0.97 from 2.5.
6. From 37.2 subtract 18.35.
7. Find the difference between 235.8 and 194.25.
8. How much greater is 49.7 than 37.9?
9. Charlene ran 100 meters in 10.9 seconds. Suso ran the same distance in 12.8 seconds. By how many seconds was Charlene faster than Suso?
10. Robin saved $9.70 for a blouse that costs $21. How much more money does she need?
11. Ray had $118.83 in the cash register of his grocery. Customers paid him $72.53, he paid $38.50 for a delivery, and he paid $15 for window signs. How much money remained?
12. On two consecutive days, the maximum temperatures of Mr. Leung, a hospital patient, were 102.3 and 99.7. The corresponding maximum temperatures of Mr. Katz, another patient, were 98.6 and 101.9. Which patient had the greater change in maximum temperature?

USING DECIMALS TO MAKE CHANGE

THE MAIN IDEA

1. You receive ***change*** when the amount of money given is greater than the price of the purchase.
2. To find the amount of change, subtract the price of the purchase from the amount of money given.

Amount of Change = Amount Given – Purchase Price

EXAMPLE 5 If Andy pays for a $7.29 shopping order with a $20 bill, how much change will he receive?

To find the change, subtract the amount of the purchase from the amount of money given to the cashier.

$$\begin{array}{r} \$20.00 \\ -7.29 \\ \hline \$12.71 \end{array}$$

Answer: Andy will receive $12.71 as change.

EXAMPLE 6 Suki bought school supplies. She gave a $5 bill to the cashier and got $1.73 change. What was the charge?

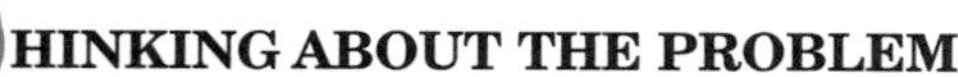

THINKING ABOUT THE PROBLEM

Knowing the amount of change, you can find the amount spent by subtraction.

Subtract the change from the amount given to the cashier.

$$\begin{array}{r} \$5.00 \\ -1.73 \\ \hline \$3.27 \end{array}$$

Answer: Suki paid $3.27 for school supplies.

EXAMPLE 7 Joan bought a hamburger for $3.25, milk for 79¢, and a salad for $1.65. If Joan gave the cashier a $10 bill, how much change did she receive?

THINKING ABOUT THE PROBLEM

To find the amount of the purchase, add the three costs. Subtract the total from the amount given to the cashier.

$$\begin{array}{r} \$3.25 \\ 0.79 \\ +1.65 \\ \hline \$5.69 \end{array} \qquad \begin{array}{r} \$10.00 \\ -5.69 \\ \hline \$4.31 \end{array}$$

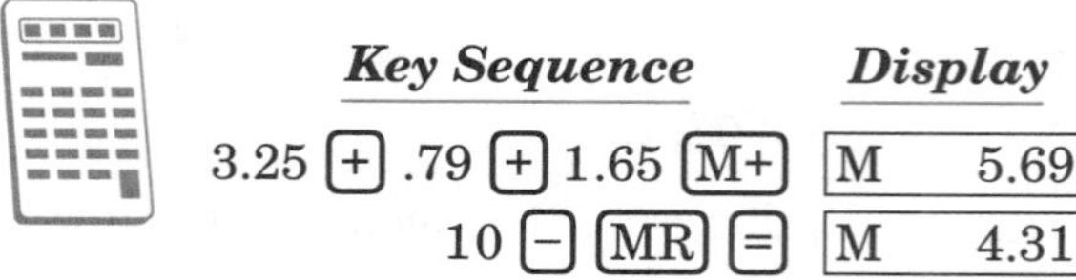

The M+ key stores 5.69 in memory. Then 10 is entered and MR recalls 5.69.

Answer: Joan received $4.31 in change.

CLASS EXERCISES

1. Find the amount of change from a $10 bill for each of the following purchases:
   a. $2.75 **b.** $5.39 **c.** $.86 **d.** $9.47
2. Henri bought a shirt for $12.90. How much change did he receive from a $20 bill?
3. The cost of Nancy's breakfast is $2.70. If Nancy gave the cashier a $5 bill, how much change did she receive?
4. Marilyn bought a scarf for $8.99 and a bow for $3.70. How much change did she receive from a $20 bill?
5. Mr. Ramirez bought $9.25 worth of gasoline and one quart of oil costing $1.80. How much change did he receive from a $20 bill?
6. Frank bought a cake mix costing $2.30, a quart of milk for $1.25, and a dozen eggs for $1.29. How much change did he receive from a $5 bill?
7. If Julio received $5.90 change from a $10 bill, how much did he spend?
8. Ms. Coles had a grocery order delivered to her home. She received $6.23 in change from a $20 bill. If the total value for the groceries was $12.52, how much did the store charge for delivery?
9. Luis paid for a purchase of groceries with a $20 bill. The money shown is what he received as change. What was the amount of Luis' purchase?

HOMEWORK EXERCISES

1. Add.
a. $\begin{array}{r} 0.6 \\ +0.9 \\ \hline \end{array}$
b. $\begin{array}{r} 0.47 \\ +2.89 \\ \hline \end{array}$
c. $\begin{array}{r} 5.93 \\ 6.02 \\ +0.917 \\ \hline \end{array}$
d. $\begin{array}{r} 8.026 \\ 9.1 \\ +23.25 \\ \hline \end{array}$
e. $\begin{array}{r} 2.6 \\ +0.37 \\ \hline \end{array}$
f. $\begin{array}{r} 42.7 \\ 11.43 \\ +240.084 \\ \hline \end{array}$

2. Add. **a.** 0.92 + 0.7 **b.** 2.8 + 0.89 **c.** 10.75 + 118.073 **d.** 9.3 + 0.29 + 4.07
e. 842.7 + 24.78 + 208.07 **f.** 97.45 + 3.719 + 24.9 **g.** 47 + 406.215 + 21.59

3. Subtract.
a. $\begin{array}{r} 0.9 \\ -0.6 \\ \hline \end{array}$
b. $\begin{array}{r} 0.7 \\ -0.29 \\ \hline \end{array}$
c. $\begin{array}{r} 14.87 \\ -9.79 \\ \hline \end{array}$
d. $\begin{array}{r} 346.8 \\ -156.264 \\ \hline \end{array}$
e. $\begin{array}{r} 16.0 \\ -8.7 \\ \hline \end{array}$
f. $\begin{array}{r} 31 \\ -19.28 \\ \hline \end{array}$

4. Subtract. **a.** 14.87 − 9.79 **b.** 9.3 − 7.4 **c.** 10.6 − 8.47 **d.** 0.3 − 0.28 **e.** 33.05 − 18.6

5. Subtract 8.35 from 11 and check.

6. Add +18.9 to −5.8.

7. How much greater is −6.28 than −9.7?

8. Mr. Joseph's $28.40 gas bill was reduced by 5¢. He thought he would have to pay $23.40. Explain Mr. Joseph's mistake.

9. The estimate closest to 16.53 + 8.29 is (a) 25.0 (b) 24.8 (c) 24.7 (d) 24.0

10. Better breeding techniques have made hens more productive. The graph below shows the number of eggs per hen per week for the years 1925 to 1990.

Number of Eggs per Hen per Week (1925–1990)

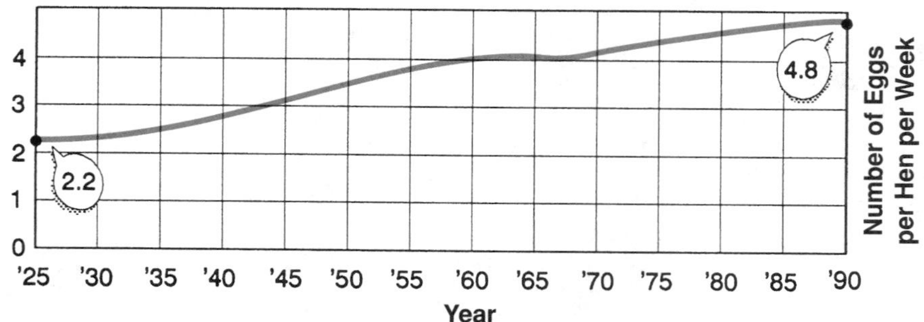

a. What is the difference in the number of eggs per hen per week between 1925 and 1990?

b. If the same trend continues, which of the following would be a reasonable estimate for the year 2000? (a) 4.8 (b) 4.9 (c) 5.0 (d) 5.2

11. Mabel walked 3.1 miles on Monday, 2.8 miles on Tuesday, and 3.7 miles on Wednesday. Find the total distance that she walked.

12. Ramon bought a shirt for $15. Joe bought the same shirt on sale for $11.89. How much less did Joe pay?

13. Jane weighs 101.3 pounds. If Mike weighs 6.8 pounds less than Jane, how much does he weigh?

14. A man gives a $10 bill to the cashier when he buys a hat that costs $6.36. How much change will he receive?

15. Rosario bought two toys. One cost $1.63 and the other cost $0.92. How much change did he receive if he paid with a $5 bill?

16. Fran received $1.51 in change when she used a $20 bill to pay for a blouse. How much was she charged for the blouse?

17. Frank bought a baseball, a bat that cost $19.75, and a baseball glove that cost $55.25. He received $19.35 in change from a $100 bill. What was the price of the baseball?

18. Mrs. Chung began the day with $20. She spent $14.30 at the supermarket, found a $5 bill, lost $1.50 in change, and received a $10 refund at the cleaners. How much money did she have at the end of the day?

19. Katisha paid a grocery bill of $28.32 from her paycheck of $97.85. The best estimate of the change that she should receive is (a) $126 (b) $125 (c) $70 (d) $69

SPIRAL REVIEW EXERCISES

1. Write 0.98 as a fraction.

2. Use one of the symbols <, >, or = to describe the relationship between 0.004 and 0.00047.

3. Round 12.397 to the nearest hundredth.

4. 640×0.49 is closest to
(a) 30 (b) 320 (c) 700 (d) 3,000

5. Which digit is in the hundredths place in 487.132?

6. Evaluate: $100 - 8 \times 3 \div 12$

7. Find the greatest common factor of 75 and 60.

8. Which is a composite number?
(a) 31 (b) 51 (c) 61 (d) 7

9. The numbers 653 and 749 are multiplied. The hundreds digit and units digit of each number are then interchanged and the new numbers are multiplied. What is the difference between the original product and the new product?

10. An airplane flying at an altitude of 35,000 feet experienced the following changes in altitude: +1,200 feet; +900 feet; –2,000 feet; +300 feet; –300 feet. What was its altitude after these changes?

11. What is the least number that can be represented by a number phrase that uses each of the numbers –3 and +5 once and one of the symbols +, –, ×, ÷?

UNIT 3-4 Using Decimals in Checking Accounts

THE MAIN IDEA

1. You can use ***checks*** in place of cash to make payments. In order to write checks, you maintain a ***checking account*** in a bank. The amount of money in your account is called the ***balance***.
2. When you write a check, the amount of the check is *deducted* from the original balance to find the new balance.
3. When you make a ***deposit***, the amount of the deposit is *added* to the original balance to find the new balance. A ***deposit slip*** is a record of the date and amount of the deposit to an account.
4. A ***checkbook register*** is used to record all deposits and checks and to keep the balance up to date.

EXAMPLE 1 Mr. Darby had a balance of $492.53 in his checking account. Then he wrote the check pictured below. What was his new balance after he wrote the check?

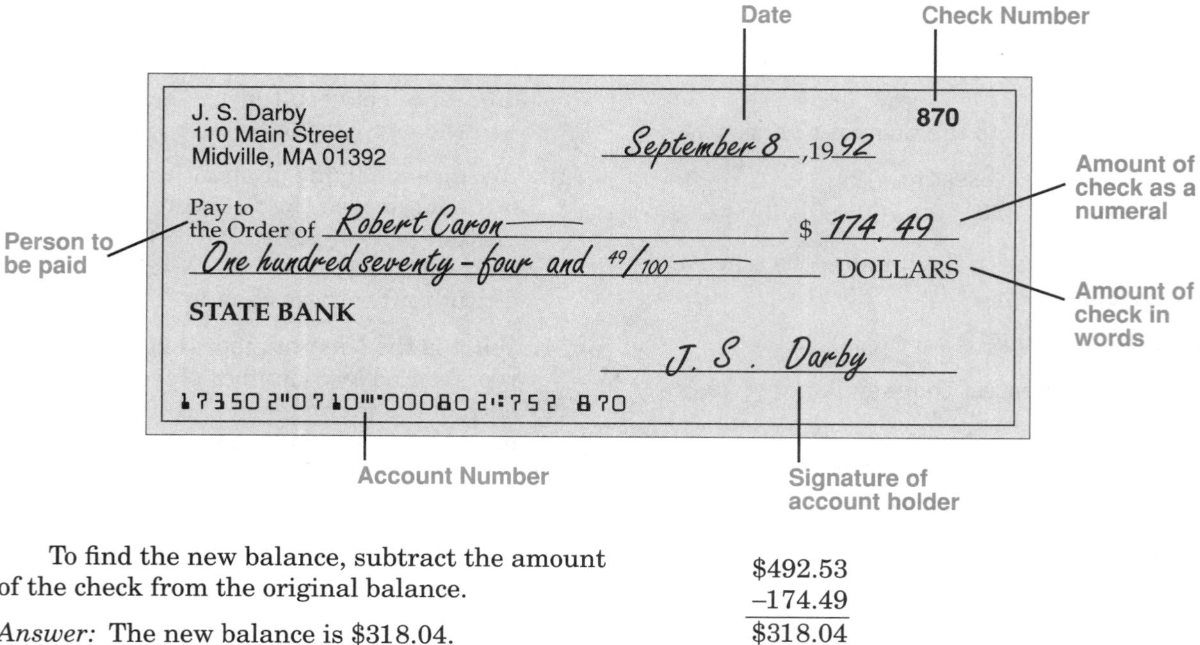

To find the new balance, subtract the amount of the check from the original balance.

$$
\begin{array}{r}
\$492.53 \\
-174.49 \\
\hline
\$318.04
\end{array}
$$

Answer: The new balance is $318.04.

EXAMPLE 2 Ms. Carlson had a balance of $265.87 in her checking account. She used the deposit slip shown below to deposit some cash and two checks. How much money did she then have in her account?

STATE BANK

Lila Carlson
123 Fuller St.
Kokomo, IN 46902

Date *January 7*, 19*93*

	Dollars	Cents
Cash	*120*	*00*
Checks	*258*	*40*
	71	*60*
Total		

2"0710"'17350"00080 2 ":752

Add the cash amount to the amount of the two checks to find the total deposit.

$$
\begin{array}{r}
\$120.00 \\
258.40 \\
+71.60 \\
\hline
\$450.00
\end{array}
$$

The total deposit is $450.00.

To find the new balance, add the deposit to the original balance.

$$
\begin{array}{r}
\$265.87 \\
+450.00 \\
\hline
\$715.87
\end{array}
$$

Answer: The new balance is $715.87.

EXAMPLE 3 Mary had a balance of $357.89 in her checking account. She deposited $182.43 and then wrote a check for $98.37. How much money does she have left in her account?

THINKING ABOUT THE PROBLEM

Use the strategy of breaking the problem into smaller, separate problems.

To find the first new balance, add the deposit to the original balance.

$$
\begin{array}{r}
\$357.89 \\
+182.43 \\
\hline
\$540.32
\end{array}
$$

To find the final balance, subtract the amount of the check from the previous balance.

$$
\begin{array}{r}
\$540.32 \\
-98.37 \\
\hline
\$441.95
\end{array}
$$

Answer: Mary has $441.95 left in her checking account.

EXAMPLE 4 After writing a check for $159.87, John had a balance of $207.33 in his checking account. What was the balance in John's account before writing this check?

THINKING ABOUT THE PROBLEM

Use the strategy of working backward. Before John wrote the check, he had more money in his account.

To find John's balance before writing the check, add the amount of the check to his present balance.

$$\begin{array}{r} \$207.33 \\ +159.87 \\ \hline \$367.20 \end{array}$$

Answer: The original balance was $367.20.

EXAMPLE 5 The balance in Frank's checking account was $1,452.09. He wrote checks for $378.71 and $532.96, and made a deposit of $125. How much money is there now in his account?

KEYING IN

To find the new balance, subtract the amount of each check from the previous balance and then add the deposit.

Key Sequence

1452.09 [−] 378.71 [−] 532.96 [+] 125 [=]

Display

665.42

Answer: The new balance is $665.42.

CLASS EXERCISES

1. The balance in Mrs. Wilson's checking account was $709.12. Find the new balance after she wrote the check shown at the right.

J. M. Wilson
124 Edge wood St.
Evansville, IL 60102
328
December 3, 1992
Pay to the Order of Able's Clothing Store $ 246.75
Two hundred forty-six and 75/100 DOLLARS
National Bank
J. M. Wilson
2"0710"'17350'00080 2"752

2. Ms. Chan had $1,042.82 in her checking account. If she wrote a check for $798.99, did she have enough money left in her account to pay for a $250 television set by check?

3. Dr. Malonski deposited $842.00 into his checking account, which had a previous balance of $59.67. What is the new balance in Dr. Malonski's account?

4. The balance in Mrs. Plummer's checking account was $406.28 before she made a deposit of three checks in the amounts of $127.53, $98.50, and $12.75. Does she now have enough money in her account to pay bills that total $650? If not, how much more must she deposit?

5. Before writing a check for $295.07, Mr. Lemmon had $327.08 in his checking account. How much money remains in Mr. Lemmon's account?

6. Ms. Redd deposited $45.29 and then wrote a check for $27.63. If Ms. Redd's checking account originally had a balance of $56.82, what is the present balance in her account?

HOMEWORK EXERCISES

1. Mr. Hoppe had a balance of $636.13 in his checking account. Find the new balance after he wrote a check for $129.17.

2. To-buy a new refrigerator, Mrs. Bouvier wrote a check for $599.78. If her original balance was $2,363.49, what is her new balance after writing the check?

3. George paid his electricity bill of $63.17 and his telephone bill of $22.11 by check. Find his new balance if his old balance was $279.38.

4. Susan has a balance of $93.16 in her checking account. Find her new balance after she made a deposit of $167.34.

5. How much money does Mr. Adams have in his account if he deposited $273.09 and he had an old balance of $564.18?

6. Mrs. Wilson receives a check for $97.61 each week. If she deposited her check directly into her account for two consecutive weeks and she had an old balance of $31.11, what was her balance at the end of two weeks?

7. Mrs. Phuong had a balance of $243.97 in her account. If she made a deposit of $191.73 and then wrote a check for $362.38, what was her new balance?

8. Mr. Franklin had a balance of $636.87 in his checking account. Then he wrote checks for $187.29 and $212.37, and made a deposit as shown. What was his new balance?

Michael Franklin
148 Bright Road
San Marino, California 91108

Date *November 12*, 19*92*

Orange County Bank

1735O2"O71O"'OOO8O2"7 O48

Checking Account Deposit

	Dollars	Cents
Cash	85	00
Checks	187	55
	132	83
Total	405	38
Less Cash Received	—	—
Net Deposit	405	38

9. Mr. Rowe's bank charges a service fee of $0.20 for each check. If he had a balance of $1,783.56 in his account and then wrote checks in the amounts of $75.83, $196.42, $46.90, and $317.80, what was his new balance?

10. Mr. Rabinowitz was charged for the following work on his car: oil change, $27.80; tune-up, $119.50; muffler replacement, $89.75. If he writes a check for the total, how should he write the amount of the check as a word phrase?

11. A check "bounces," or a checking account is "overdrawn," if a check is written for an amount greater than the balance in the account. Four persons made deposits, then wrote checks. Whose account is overdrawn?

Name	Old Balance	Deposit	Check
Mr. Arthur	$222.17	$63.92	$281.11
Mrs.Brooks	$122.64	$387.19	$593.71
Mr. Czerny	$273.81	$15.68	$287.49
Ms. Hwang	$1,369.00	$2,463.00	$3,821.00

SPIRAL REVIEW EXERCISES

1. Ms. Foster buys a handbag that costs \$18.95. How much change will she receive from a \$20 bill?

2. Find the sum of 0.295, 3.7, and 18.19.

3. From 3.47 subtract 2.49.

4. Write 0.127 as a fraction.

5. Round 29.4783 to the nearest thousandth.

6. By how much is 2 greater than –5?

7. Add: –12 + (–17)

8. Which key sequence would result in the display of a negative number?
   (a) 4 ÷ 20 × 1 − .1 =
   (b) 10 − 5 − 3 − 2 =
   (c) 100 − 99 ÷ 2 =
   (d) 57 − 58 × 3 + 2 =

9. If a bus holds 48 people, estimate to the nearest hundred the number of people that can be carried in 11 buses.

10. The temperature of a solution is 10°C. If the temperature drops 2° each hour, what will be the temperature of the solution 8 hours from now?

11. The graph below shows the height and year of construction of each of the six tallest building in the world.
    a. How much taller is One World Trade Center than Two World Trade Center?
    b. How much older is the Sears Tower than the Jin Mao Building?
    c. What is the difference in the heights of the tallest and the shortest of these buildings?

The World's Tallest Buildings

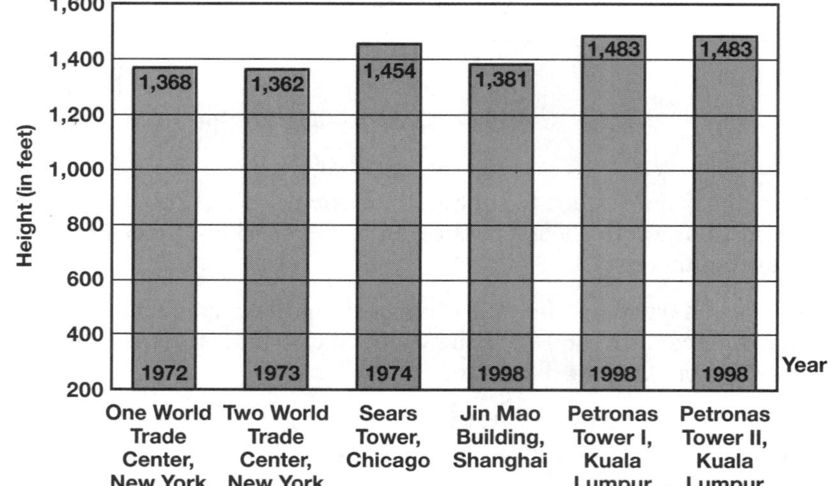

MODULE 4

UNIT 4-1 Multiplying Decimals

THE MAIN IDEA

To multiply a decimal by a whole number or another decimal:

1. Multiply as if both numbers were whole numbers.
2. Count the total number of places after the decimal points in both numbers. (A whole number has 0 places after the decimal point.)
3. Place a decimal point in the product so that the number of places after the decimal point in the product equals the total number of places after the decimal points in both numbers.
4. If necessary, write zeros to the left of the digits in the product to allow for the correct number of decimal places.

EXAMPLE 1 Multiply: 0.047×0.08

Multiply as with whole numbers and count the number of places after the decimal point.

$$\begin{array}{r} 0.047 \\ \times\, 0.08 \\ \hline 376 \end{array}$$

0.047 ← 3 decimal places
× 0.08 ← 2 decimal places

The product has 3 + 2, or 5 decimal places.

0 • 0 0 3 7 6

Begin at the right and count off 5 decimal places. To do this, insert two zeros to the right of the decimal point.

Answer: 0.00376

EXAMPLE 2 Josephine is buying a barbecued chicken that weighs 2.45 pounds. If the price of the chicken is $2.99 per pound, what should Josephine pay?

KEYING IN

Key Sequence	***Display***
2.45 [×] 2.99 [=]	7.3255

Since amounts of money in dollars and cents contain only two decimal places, round the result to the nearest hundredth.

Answer: She should pay $7.33.

EXAMPLE 3 Marcia's company will pay her 33¢ per mile for car expenses. At the beginning of the week, her odometer read 13452.9 miles, and at the end of the week, 13721.4 miles. How much are Marcia's car expenses for the week?

THINKING ABOUT THE PROBLEM

Use the strategy of breaking the problem into two separate problems.

To find the distance traveled, subtract the two odometer readings.

13721.4	miles
−13452.9	miles
268.5	miles traveled

Multiply the number of miles traveled by the allowance per mile, writing 33¢ as 0.33.

268.5	miles traveled
× 0.33	allowance per mile
88.605	

Round to the nearest hundredth.

Answer: The car expenses are $88.61.

EXAMPLE 4 The table shows the cost of CD's at Ace Record. Mai bought 4 CD's of Type B, 3 of C, and 2 of E. Find the total cost.

Compact Discs

Type	Price
A	$7.99
B	$9.99
C	$10.99
D	$12.99
E	$13.99

KEYING IN

Store the cost of the Type B purchases. Add the cost of the Type C purchases to the number in memory. Add on the cost of E.

Key Sequence	*Display*	
4 [×] 9.99 [M+]	M 39.96	cost of Type B
3 [×] 10.99 [M+]	M 32.97	cost of Type C
2 [×] 13.99 [M+]	M 27.98	cost of Type E
[MR]	M 100.91	total cost

The partial results $39.96, $32.97, and $27.98, are ***subtotals***. Pressing [MR] displays the ***grand total***.

Answer: The total cost was $100.91.

EXAMPLE 5 Cinema 66 charges $6.50 for adults and $3.75 for children. How many adults and how many children were in a camp group that paid $222.75 for 55 people?

THINKING ABOUT THE PROBLEM

Use the calculator and a guess-and-check strategy.

Suppose there are 4 adults. Then, there are 55 – 4, or 51 children. Find the cost for 4 adults and 51 children. This is less than the desired result.

Key Sequence	*Display*	
4 [×] 6.5 [M+]	M 26.	
51 [×] 3.75 [M+]	M 191.25	
[MR]	M 217.25	← Too small.

Increase the number of adults to 5, then 6, then 7, and so on, until the sum is $222.75. Arrange the results in a table.

Number of Adults	Number of Children	Total Admission	
4	51	$217.25	
5	50	$220.00	
6	49	$222.75	← Stop!

Answer: There were 6 adults and 49 children.

CLASS EXERCISES

1. Multiply.

a.	b.	c.	d.	e.
0.75×3	71.08×25	0.6×0.3	0.28×0.09	17.01×0.32

f. 0.045×0.5 **g.** 25.5×0.08 **h.** 0.2×0.4 **i.** 0.12×2.10

2. Find the total cost of 5 shirts if each shirt costs $15.85.
3. Mr. O'Keefe's car averages 18.3 miles per gallon. How many miles can he travel using 12.8 gallons of gasoline?
4. A skier is traveling at 29.3 feet per second. How many feet can she travel in 10.7 seconds?
5. The weight of a gallon of water is 8.336 pounds. Choose the best estimate of the weight of the water in a twenty-gallon aquarium.
   (a) 165 pounds (b) 1,660 pounds (c) 16.6 pounds (d) 833 pounds
6. A jeweler bought 8.4 ounces of gold at $366.90 per ounce. The following day, the price of gold increased to $370.25. What was the increase in the value of his purchase?

7. Roberto took lunch orders from his coworkers, as written at the left below.

   Use the prices listed on the take-out menu for the Feel Fine Health Bar to find the total amount of money that Roberto should collect from the people who gave him these orders.

2 large health salads
1 pasta salad
3 vegetables and rice
2 fruit salads

Feel Fine Health Bar Take-Out Menu

Item	Price
Health Salad (small)	$2.75
Health Salad (large)	$4.25
Pasta Salad	$3.75
Vegetables & Rice	$3.50
Fruit Salad	$2.50

MULTIPLYING A DECIMAL BY A POWER OF 10

THE MAIN IDEA

To multiply a decimal by a ***power of 10*** (numbers such as 10, 100, 1,000, 10,000, etc.), there are two methods:

1. Use the method you have learned to multiply a decimal by a whole number, or
2. a. Count the number of zeros in the power of 10.
   b. Move the decimal point to the *right* the same number of places as there are zeros in the power of 10.
   c. Write extra zeros to the right if there are not enough decimal places for moving the decimal point.

EXAMPLE 6 Multiply: 4.35 × 100

Method 1:

Multiply as with whole numbers.

Since 4.35 has 2 decimal places, the product must have 2 decimal places.

Begin at the right and count off 2 decimal places.

```
    4.35 ←— 2 decimal places
   ×100 ←— 0 decimal places
    000
   0000
  43500
  43500

4 3 5 • 0 0
      ←—←—
```

Answer: 435.00 or 435

Method 2:

Count the number of zeros in 100. There are 2 zeros in 100.

Move the decimal point in 4.35 2 places to the right.

```
4 • 3 5 •
  —→—→
```

Answer: 435

EXAMPLE 7 In one state, the gasoline tax is $0.17 per gallon. How much tax is collected on 1 million gallons of gasoline? (One million = 1,000,000 [6 zeros])

To multiply 0.17 by 1,000,000, move the decimal point 6 places to the right. Insert as many zeros as needed after the last digit.

```
0.17 × 1,000,000 = • 1 7 0 0 0 0 •
                   —→—→—→—→—→—→
```

Answer: $170,000 is collected.

CLASS EXERCISES

1. Multiply.

   a. 25.34×10 **b.** $10{,}000 \times 5.11756$ **c.** $7.006 \times 1{,}000$ **d.** 100.71×100 **e.** $4.08 \times 1{,}000$

   f. 8.4×100 **g.** 0.156×10 **h.** $17.6 \times 1{,}000$ **i.** $0.7 \times 10{,}000$ **j.** 100×0.100

2. There are 2.54 centimeters in an inch. Find the number of centimeters in 100 inches.

3. There are 1.6 kilometers in a mile. Find the number of kilometers in 1,000 miles.

HOMEWORK EXERCISES

1. Multiply.

   a. 23×0.7 **b.** 18×3.14 **c.** 5.42×6

   d. $\begin{array}{r} 14.62 \\ \underline{\times 45} \end{array}$ **e.** $\begin{array}{r} 90.07 \\ \underline{\times 21} \end{array}$ **f.** $\begin{array}{r} 47 \\ \underline{\times 0.026} \end{array}$

2. Multiply.

   a. $\begin{array}{r} 4.8 \\ \underline{\times 3.2} \end{array}$ **b.** $\begin{array}{r} 5.19 \\ \underline{\times 4.6} \end{array}$ **c.** $\begin{array}{r} 9.2 \\ \underline{\times 0.09} \end{array}$

3. Multiply. **a.** 3.75×10 **b.** 19.84×100 **c.** $8.9 \times 1{,}000$ **d.** 0.4×100 **e.** 0.3×100

   f. $94.25 \times 1{,}000$ **g.** 8.4×10 **h.** $0.91 \times 10{,}000$ **i.** $3.14 \times 1{,}000$ **j.** 10.10×10

4. Find the product of 209 and 0.08.

5. What is the total cost of 6 pounds of ham if the price is $3.55 per pound?

6. If one yard of material costs $3.98, which is the best estimate of the cost of 4.25 yards?
   (a) $12 (b) $16 (c) $39 (d) $42

7. Estimate each product correct to the nearest whole number.

   a. 5.9×0.96 **b.** 12.1×3.98 **c.** 999.9×4.028

8. There is about 0.914 meter in a yard. The number of meters in 10,000 yards is
   (a) 9.14 (b) 91.4 (c) 914 (d) 9,140

9. We use the same order of operations with decimals as with whole numbers. Use the order of operations to find the value of: **a.** $6 + 0.5 \times 8 - 3.2$ **b.** $2.7 - 1 + 8 \times 1.5$

10. Mr. Alcalde's car averages 25.7 miles per gallon. How many miles can he drive using 11.8 gallons of gasoline?

11. A crate weighs 37.8 pounds. What is the weight of 1,000 crates?

12. Marcia was offered a job in a clothing shop where she would work 17.5 hours per week and earn \$5.85 per hour. She was also offered a job in a health food store where she would work 21.5 hours per week for \$5.15 an hour. At which job would she earn more per week?

13. On a science exam, each correct answer in Part I receives 2.5 points and each correct answer on Part II receives 3.5 points. George answered 26 questions correctly in Part I and 8 in Part II. Write a key sequence that can be used on the calculator to find his total score on the exam.

14. The table shows film prices advertised by a photo supply house.

The school camera club places the following order: 8 rolls of 135, 8 rolls of 120, and 4 packages of 4 × 5. What is the total cost of the order?

Size	Price
135	\$7.60
120	\$4.30
220	\$8.60
4 × 5	\$17.50
8 × 10	\$62.90

15. Mrs. Ascher spent \$13.90 for 60 stamps. Some of the stamps were 29¢ stamps and some were 19¢ stamps. How many stamps of each type did she buy?

16. Bernice wanted to mark off lengths of 4.5 cm on a roll of ribbon that was 450 cm long. Explain how she determined, without dividing, that there should be 100 pieces.

17. The best estimate of the value of 2.89 × 1.13 is (a) 1 (b) 2 (c) 3 (d) 4

18. The best estimate for the product 583 × 0.85 is (a) 300 (b) 500 (c) 700 (d) 900

19. A reasonable estimate for 5.6 × 4.2 is (a) 0.23 (b) 2.3 (c) 23 (d) 230

SPIRAL REVIEW EXERCISES

1. Find the sum of 12.75, 8.02, and 3.98.

2. Mary had \$125.83 in her checking account and then wrote a check for \$87.90. What was her new balance?

3. Leo bought 2 cases of bottled water that cost \$6.95 per case. How much change should he have received from a \$20 bill?

4. Replace ? with < or > to make a true comparison. **a.** 0.8 ? 0.08 **b.** 0.903 ? 0.93

5. Round 25.976 to the nearest hundredth.

6. How much smaller is 19.87 than 21.6?

7. Dr. Volpe flew 2,853 miles to New City, drove another 389 miles, and then took a train for 450 miles. What was her total?

8. What number in place of [?] would give an error message?

   24 [−] [?] [M+] 100 [÷] [MR] [=]

9. The best estimate for the total of \$4.25, \$10.98, \$0.57, \$0.57, \$6.14, \$3.99, \$2.99, \$0.37, \$3.20, \$16.29, \$5.49, and \$3.78 is between

   (a) \$20 and \$30 (b) \$30 and \$40
   (c) \$40 and \$50 (d) \$50 and \$60

10. Mary has \$50 to spend on gifts. She has already bought 3 ties at \$12 each. What is the maximum number of posters she can buy if the price of each poster is \$4?

11. Replace each ? with +, −, ×, or ÷ to make each a true statement.

   a. −8 ? −8 = −16 **b.** 0 ? 5 = 0

UNIT 4-2 Dividing Decimals

DIVIDING A DECIMAL BY A WHOLE NUMBER

THE MAIN IDEA

1. To divide a decimal by a whole number:
   a. Place a decimal point in the quotient just above the decimal point of the dividend.
   b. Ignore the decimal points and divide as with whole numbers. If necessary, use zeros as placeholders in the quotient.
2. As with division of whole numbers, a division with decimals may be checked using the inverse operation of multiplication by finding the product of the quotient and the divisor.

EXAMPLE 1 Divide and check: 7.25 ÷ 5

Set up the division.

divisor → $5\overline{)7.25}$ ← dividend; quotient goes above

Place a decimal point in the quotient just above the decimal point of the dividend.

$5\overline{)7.25}$ (decimal point placed above)

Ignore the decimal points and divide as with whole numbers.

$$\begin{array}{r} 1.45 \\ 5\overline{)7.25} \\ \underline{5} \\ 2\,2 \\ \underline{2\,0} \\ 25 \\ \underline{25} \\ 0 \end{array}$$

Check

$$\begin{array}{rl} 1.45 & \leftarrow \text{quotient} \\ \underline{\times 5} & \leftarrow \text{divisor} \\ 7.25 & \leftarrow \text{dividend} \end{array}$$

The answer checks. ✔

Answer: 1.45

EXAMPLE 2 Divide: $0.144 \div 12$

Place a decimal point in the quotient just above the decimal point of the dividend.

```
    0.
12)0.144
```

Ignore the decimal point and divide as with whole numbers. Since 12 is too large to divide into 1, place a zero above the 1.

```
    0.0
12)0.144
```

Then divide 12 into 14.

```
    0.012
12)0.144
     0
     14
     12
      24
      24
       0
```

Answer: 0.012

EXAMPLE 3 Divide: $72.3 \div 5$

Sometimes, to complete the division of a decimal, we must write zeros to the right in the dividend.

```
  14.4
5)72.3
  5
  22
  20
   2 3
   2 0
     3
```

There is still a remainder of 3.

Write a zero to the right, and continue dividing. Since the remainder is now 0, the division is complete.

```
  14.46
5)72.30
  5
  22
  20
   2 3
   2 0
     30
     30
      0
```

Answer: 14.46

EXAMPLE 4 If 4 bars of soap sell for \$0.97, the amount charged for one bar is

(a) \$0.20 (b) \$0.23 (c) \$0.24 (d) \$0.25

THINKING ABOUT THE PROBLEM

To find the price of one bar when the total cost of several bars is given, divide the total cost by the number of bars.

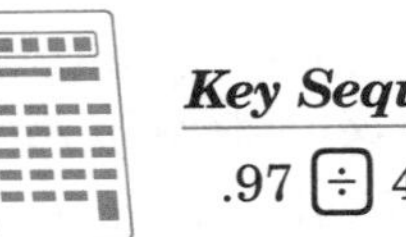

Key Sequence	*Display*
.97 ÷ 4 =	0.2425

The result, rounded to the nearest cent, is 0.24 or 24¢. However, stores usually increase the price to the next whole cent if there is any remainder, even if the next decimal place is less than 5.

Answer: (d)

EXAMPLE 5 If 5 oranges cost \$2.25, what is the cost of 7 oranges?

THINKING ABOUT THE PROBLEM

First, find the cost of one orange.

To find the cost of one orange, divide the total cost by the number of oranges.

```
  $0.45
5)$2.25
```

To find the cost of 7 oranges, multiply the cost of one orange by 7.

```
 $0.45
   × 7
 $3.15
```

Answer: The cost of 7 oranges is \$3.15.

CLASS EXERCISES

1. Divide and check.
   a. $3.125 \div 5$ **b.** $28.8 \div 12$ **c.** $12.58 \div 5$ **d.** $2.87 \div 14$ **e.** $227.7 \div 9$
   f. $0.54 \div 12$ **g.** $21.48 \div 24$ **h.** $4{,}847.55 \div 85$ **i.** $6.25 \div 1$ **j.** $14.50 \div 2$
2. If 12 boxes of cheese weigh 5.4 kilograms, find the weight of each box.
3. The height of a stack of 8 cans is 99.2 cm. Find the height of each can.
4. If 5 rolls of paper towels cost $2.45, what is the cost of one roll of paper towels?
5. If 4 bagels cost $1.60, what is the cost of 9 bagels?
6. If 3 slices of pizza cost $3.75, what is the cost of 5 slices?
7. If 2 quarts of orange juice cost $2.70, then 5 quarts cost (a) $13.50 (b) $7.75 (c) $6.75 (d) $5.40

DIVIDING BY A DECIMAL

THE MAIN IDEA

To divide by a decimal:

1. Move the decimal point of the divisor all the way to the right, after the last digit.
2. Count the number of places that the decimal point was moved.
3. Move the decimal point of the dividend the same number of places to the right. If necessary, write extra zeros as placeholders after the last digit. (Moving the decimal point of both numbers the same number of places is the same as multiplying the divisor and the dividend by the same power of ten.)
4. Now divide as if by a whole number.

EXAMPLE 6 Divide: $4.824 \div 0.12$

Since the decimal point of the divisor must be moved 2 places right, the decimal point of the dividend must also be moved 2 places right.

0.12.)4.82.4

40.2 *Ans.*
12)482.4
48↓↓
02 4
2 4
0

EXAMPLE 7 Divide: $175.5 \div 0.25$

Since the decimal point of the divisor must be moved 2 places right, that of the dividend must also be moved 2 places right. Insert 1 zero in the dividend.

0.25.)175.50.

702. *Ans.*
25)17550.
175↓↓
50
50
0

EXAMPLE 8 A bottle contains 124 ounces of perfume. How many sample bottles can be filled if each holds 0.248 ounce?

Divide: 124 ÷ 0.248

The decimal point in the dividend is at the end of the whole number 124.

$$0.248\overline{)124.}$$

Move the decimal points of the divisor and dividend 3 places to the right. Insert 3 zeros in the dividend.

$$0.248.\overline{)124.000.}$$

Complete the division. 500 bottles can be filled. *Ans.*

$$\begin{array}{r} 500. \\ 248\overline{)124000.} \\ \underline{1240} \\ 000 \end{array}$$

CLASS EXERCISES

1. Divide and check.

a. 0.525 ÷ 0.05 **b.** 86.48 ÷ 0.2 **c.** 244 ÷ 0.08 **d.** 3.6048 ÷ 0.012 **e.** 0.968 ÷ 0.008

f. 1,575 ÷ 2.5 **g.** 0.0297 ÷ 0.11 **h.** 0.01512 ÷ 0.0006 **i.** 2.5 ÷ 0.1 **j.** 3.25 ÷ 0.25

2. If 9.5 gallons of a liquid weigh 79.23 pounds, what is the weight of one gallon?

3. In 2.25 hours, a printing press uses 348.75 pounds of paper. How many pounds does it use each hour?

4. Write a key sequence that can be used to divide and check 2.37 ÷ 0.25.

DIVIDING A DECIMAL BY A POWER OF 10

THE MAIN IDEA

To divide a decimal by a power of 10, there are two methods:

1. Use the method you have learned to divide a decimal by a whole number, or
2. a. Count the number of zeros in the power of 10.
   b. Move the decimal point to the *left* the same number of places as there are zeros in the power of 10.
   c. Write extra zeros to the left if there are not enough decimal places for moving the decimal point.
3. Since division and multiplication are inverse operations, the movements of the decimal places in the two operations are in opposite directions. For example:

   0.15 ÷ 100 = 0.0 0.15 but 0.15 × 100 = .1 5 .

EXAMPLE 9 Divide: 0.046 ÷ 100

Method 1:

Set up in the usual way.

$$100\overline{)0.046}$$

Insert 3 zeros in the quotient because 100 is too large to divide into 0, 4, or 46. Add a zero in the dividend in order to begin the division.

$$\begin{array}{r} 0.000 \\ 100\overline{)0.0460} \end{array}$$

Begin the division. Add another zero in the dividend to complete the division.

$$\begin{array}{r} 0.00046 \\ 100\overline{)0.04600} \\ 400 \\ \hline 600 \\ 600 \\ \hline 0 \end{array}$$

Answer: 0.00046

Method 2:

Count the number of zeros in the divisor, 100. — There are 2 zeros in 100.

To move the decimal point 2 places to the left in 0.046, insert 2 zeros. — . 0 0 . 0 4 6

Answer: 0.00046

EXAMPLE 10 One thousand bolts cost $97. At the same rate, what do 450 bolts cost?

THINKING ABOUT THE PROBLEM

First find the item price, the cost of one bolt. Do not round costs until all calculations are completed.

To find the cost of one bolt, divide the total cost by the number of bolts.

$97 ÷ 1,000 = 0 . 0 9 7 .

To find the cost of 450 bolts, multiply the cost of one bolt by 450.

$$\begin{array}{r} \$0.097 \\ \times 450 \\ \hline 000 \\ 4\ 85 \\ 38\ 8 \\ \hline \$43.650 \end{array}$$

Answer: The cost of 450 bolts is $43.65.

CLASS EXERCISES

1. Divide.

a. 14.732 ÷ 10 **b.** 9.683 ÷ 100 **c.** 148.9 ÷ 1,000 **d.** 87.1 ÷ 1,000

e. 42.6 ÷ 10,000 **f.** 0.017 ÷ 100 **g.** 0.917 ÷ 10 **h.** 0.005 ÷ 100

2. If 100 paper containers cost $24.75, what is the cost of one container?

3. If 10,000 key chains cost $850, what is the cost of one key chain?

4. Ten signs cost $465. At the same rate, what is the cost of 6 signs?

5. One thousand envelopes cost $120. At the same rate, what is the cost of 750 envelopes?

HOMEWORK EXERCISES

1. Divide.

 a. $42.85 \div 5$ b. $3.79 \div 2$ c. $0.081 \div 3$ d. $9.18 \div 9$ e. $13.28 \div 16$
 f. $69.72 \div 12$ g. $159.12 \div 17$ h. $644.8 \div 26$ i. $25.5 \div 5$ j. $20.50 \div 2$

2. Divide and check.

 a. $37.5 \div 0.5$ b. $806 \div 0.2$ c. $24.12 \div 0.04$ d. $8{,}127 \div 0.009$ e. $3.278 \div 0.11$
 f. $10.03 \div 1.7$ g. $155 \div 0.25$ h. $0.576 \div 0.018$ i. $13.5 \div 13.5$ j. $400 \div 0.50$

3. Divide.

 a. $25.9 \div 10$ b. $37.1 \div 100$ c. $936 \div 1{,}000$ d. $4.8 \div 100$ e. $92 \div 10$
 f. $103 \div 100$ g. $15.7 \div 10$ h. $97.09 \div 1{,}000$ i. $100 \div 10$ j. $10 \div 100$

4. How many pieces measuring 0.6 meter each can be cut from a metal rod 7.2 meters long?

5. Fifteen boxes of nails weigh 6 pounds. How much does one box weigh?

6. If 5 cans of peas cost $2.25, what is the cost of one can?

7. If 3 packages of looseleaf paper cost $3.59, what is the cost of one package?

8. If 4 cans of juice cost $0.92, then the cost of one can is
 (a) $0.20 (b) $0.23 (c) $0.30 (d) $0.31

9. If 3 lemons cost $0.84, what is the cost of 5 lemons?

10. If 6 boxes of raisins cost $1.20, what is the cost of 9 boxes?

11. A printer's catalog shows the prices of envelopes of various sizes in the following table:

Price per Box of 100

Size	Number of Boxes Ordered			
	1–4	5–19	20–99	100+
4×5	3.95	3.55	3.15	2.95
5×7	5.00	4.50	4.00	3.75
8×10	10.30	9.25	8.25	7.75

 a. If you order 6 boxes of 8×10 envelopes, what is the price of each box?
 b. If you order 1 box of 5×7 envelopes, what is the price of each envelope?
 c. How much money do you save per envelope if you order 20 boxes of 4×5 envelopes instead of 19 boxes?
 d. What is the total cost of an order of 25 boxes of 4×5 envelopes and 100 boxes of 8×10 envelopes?
 e. What is the maximum number of boxes of 8×10 envelopes that can be bought for $100? Explain the reasoning that led to your answer.

SPIRAL REVIEW EXERCISES

1. Find each of the products.
   a. 9.35×12 b. 0.8×2.65
   c. 12.45×0.08 d. 2.9×12.5
2. How much greater is 13.65 than 9.2?
3. Mr. Lobell bought 2 shirts that were on sale for $8.75 each. How much change did he receive from a $20 bill?
4. Choose the best estimate for 0.097×0.82.
   (a) 8.0 (b) 0.8 (c) 0.08 (d) 0.008
5. Mrs. Marker had $248.52 in her checking account. She deposited $49.89 and wrote a check for $95.80. What was her new balance?

6. Write a key sequence that uses the constant feature of addition to calculate $12 + 5, 17 + 5, 23 + 5$, and $25 + 5$.
7. The largest decimal of the group 0.8, 0.82, 0.28, and 0.802 is
   (a) 0.8 (b) 0.82 (c) 0.28 (d) 0.802
8. Choose the correct decimal for the number "seven hundred seven thousandths."
   (a) 707 (b) 700.007 (c) 0.707 (d) 0.07007
9. What is the greatest prime number less than 100?
10. Divide 2,688 by 48.
11. Peter works in the shipping department of a grocery distribution company. If he packs 3,749 quarts of oil into boxes that hold 24 quarts each, how many quarts will be left over?
12. A toll bridge token costs one dollar. Jessica's coins are shown in the photograph. Does she have enough money in coins to buy a token?

UNIT 4-3 Using Decimals in Computing Wages

THE MAIN IDEA

1. The amount of money earned by a person is called ***salary*** or ***wage***. Wages are often paid by the hour.
2. To find a person's total wages when the hourly wage is known, multiply the number of hours worked by the hourly wage.

   Total Wages = Number of Hours Worked × Hourly Wage

3. To find a person's hourly wage when the total wages are known, divide the total wages by the number of hours worked.

   Hourly Wage = Total Wages ÷ Number of Hours Worked

4. To find the number of hours worked when the hourly wage and the total wages are known, divide the total wages by the hourly wage.

   Number of Hours Worked = Total Wages ÷ Hourly Wage

5. Sometimes, for working extra hours, a person is paid ***overtime***. Generally, the number of extra hours worked is counted as either 1.5 times or 2 times the actual number of overtime hours.
6. From a person's total wages, or ***gross earnings***, money is deducted for various reasons. ***Deductions*** may include Federal Income Tax, State Income Tax, Social Security Tax, and Health Insurance fees. The amount that remains after deductions are subtracted is called the ***net earnings***, or ***take-home pay***.

EXAMPLE 1 Janice earns $6.80 per hour. If she works for 7 hours, how much money will she earn?

Multiply the hourly wage by the number of hours worked.

```
  $6.80  <-- hourly wage
     ×7  <-- number of hours
 $47.60  <-- total wages
```

Answer: Janice will earn $47.60 in 7 hours.

EXAMPLE 2 Mr. Harris worked 40 hours last week and earned $530. What is Mr. Harris' hourly wage?

Divide the total wages by the number of hours worked.

```
       13.25
 40)530.00
    40
    130
    120
     100
      80
      200
      200
        0
```

Answer: Mr. Harris earns $13.25 per hour.

EXAMPLE 3 Rebecca earns $5.20 per hour and time and one-half for each hour that she works overtime. Last week, Rebecca worked 35 hours and 6 hours overtime. How much money did she earn?

```
 1.5 ← time and one-half        35 ← regular hours
 ×6  ← overtime hours           +9 ← extra hours
  9  ← extra hours              44 ← total hours

   $7.20 ← hourly wage
    ×44  ← total hours
   28 80
  288 0
 $316.80 ← total wages
```

Answer: Rebecca earned $316.80.

EXAMPLE 4 Mrs. Franklin is paid double time for any hours worked above 40 hours per week. Last week, she worked 49 hours. How much money did she earn if her hourly wage is $10.40?

```
 49 ← total hours               9 ← overtime hours
−40 ← regular hours            ×2 ← double time
  9 ← overtime hours           18 ← extra hours

 40 ← regular hours         $10.40 ← hourly wage
+18 ← extra hours             ×58  ← sum of hours
 58 ← sum of hours           83 20
                            520 0
                           $603.20 ← total wages
```

Answer: Mrs. Franklin earned $603.20.

EXAMPLE 5 Teresa earns $18.75 an hour and works 35 hours a week. Each week, the following deductions are made from her paycheck:

Federal Income Tax $131.25
Social Security Tax. $50.20
Health Insurance $17.50

What are her gross earnings, total deductions, and take-home pay?

KEYING IN

First, find the gross earnings by multiplying 35 hours by $18.75 and store this result in memory. Then, find the total deductions by adding $131.25, $50.20, and $17.50. Finally, subtract this sum from the total earnings already in memory to find the take-home pay.

Key Sequence	***Display***	
35 [×] 18.75 [M+]	M 656.25	← gross earnings
131.25 [+] 50.2 [+] 17.5 [M−]	M 198.95	← total deductions
[MR]	M 457.3	← take-home pay

Answer: Teresa's gross earnings are $656.25, her total deductions are $198.95, and her take-home pay is $457.30.

CLASS EXERCISES

1. Jane earns $8.25 per hour. If she works for 12 hours, how much money will she earn?
2. Mr. Roberts earns $15.70 per hour. How much will he earn if he works 35 hours?
3. Mr. Smith earns $7.50 an hour as a typist. What are his wages if he works 40 hours?
4. The Perfect Crating Company pays its workers $10.90 per hour for unloading its trucks. Find the amount of money earned for each of the following numbers of hours worked:

   a. 20 hours **b.** 18 hours **c.** 30 hours **d.** 35 hours
5. Ms. Hope worked 35 hours last week and earned $332.50. What is Ms. Hope's hourly wage?
6. Bob worked 25 hours packing groceries and earned $170.00. What is Bob's hourly wage?
7. Jessica worked for 16 hours, and her take-home pay was $99.60. If her total deductions were $24.40, what was her hourly wage?
8. An accountant billed a client $350 for 10 hours work. What is the accountant's hourly wage?
9. Bill worked 7 hours on Monday, 4 hours on Tuesday, 8 hours on Wednesday, 5 hours on Thursday, and 6 hours on Friday. If Bill earns $6.10 per hour, how much did he earn?
10. Julie earns $6.90 per hour working as a carpenter's assistant. Last week, she worked 3 hours on Monday, 5 hours on Wednesday, and 2 hours on Friday. How much did Julie earn last week?
11. Mr. James earns $16.80 per hour and time and one-half for each hour that he works overtime. Last week, Mr. James worked 35 hours plus 10 hours overtime. How much did Mr. James earn?
12. One week, Kim had net earnings of $155 and total deductions of $35.50. After subtracting, she thought that her gross earnings were $119.50. Explain her mistake.
13. Working at the rate of $8.40 per hour, Mr. Jones earned $63. How many hours did he work?
14. Mr. Cousins is paid time and one-half for any hours worked above 35 hours per week. Last week, he worked 41 hours. How much money did he earn if his hourly wage is $7.20?
15. Mrs. Robins is paid double time for any hours worked above 40 hours per week. Last week, she worked 50 hours. If her hourly wage is $8.50, how much did she earn?
16. One corner of Mr. Prenner's check stub was accidently torn off. Use the information on the check stub shown to find the amount that was deducted for union dues.

Deductions:

Federal Income Tax: $518.20 Union Dues:
Social Security Tax: $136.59

Gross: $1,785.50 Net: $1,050.76

1. Phil earns $7.40 an hour. If he worked 15 hours this week, how much did he earn?
2. How much did Mrs. Osawa earn last week if she worked 40 hours at the rate of $18.50 an hour?
3. Mr. Ayala worked 30 hours last week. If he earned a total of $450, what was his hourly wage?
4. What is Chuck's hourly wage if he earned $157.50 for 25 hours of work?
5. Helen worked 10 hours on Monday, 12 hours on Wednesday, and 8 hours on Friday. If she earns $6.80 an hour, how much money did she earn?
6. Alvin earns $6.20 an hour and time and a half for any hours above 40. If he worked 43 hours, how much did he earn?
7. Fatima kept a record, as shown at the right, of the hours she worked last week. If she makes $6.80 an hour, how much did she earn last week?

Day	Hours Worked
Monday	3
Tuesday	2.5
Wednesday	2
Thursday	3
Friday	3.5

8. Mrs. Chan earned $90.40 working at a rate of $11.30 an hour. How many hours did she work?
9. Mr. Torre earns $10.50 an hour. One day, he earned a total of $57.75. How many hours did he work?
10. Sue earns double time for any hours that she works over 35. Her regular hourly wage is $5.60. Last week, she worked 39 hours. How much did she earn?
11. The bar graph shows the number of regular hours and overtime hours worked in 4 departments at the Zapp Electric Company for one week. The regular hours are represented by unshaded bars and the overtime hours by shaded bars. For all the workers, the hourly wage is $10.00 and all the workers are paid double for overtime.
    a. What was the greatest number of hours worked in any department?
    b. In which department did the workers work the greatest number of overtime hours?
    c. How much money was earned by all the workers in the Transformers Department?
    d. In which department did the workers earn the greatest amount of money?

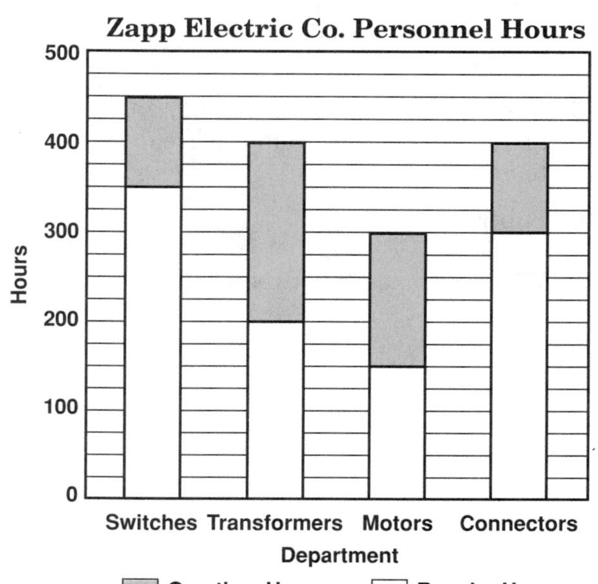

12. Mr. Allison earned \$9.70 an hour and worked 37.5 hours. His payroll deductions were \$80.03 for Federal Income Tax, \$27.83 for Social Security Tax, and \$12.85 for State Income Tax. What was his take-home pay?

13. A computer programmer earned \$19.75 an hour and worked a 40-hour week. Her total deductions came to \$318.90. Estimate her take-home pay to the nearest hundred dollars.

14. Debbie worked a 35-hour week at the rate of \$6.75 an hour. The next week she received a raise of 50 cents an hour. If she worked 35 hours on her new salary, how much money did she earn for the two weeks?

SPIRAL REVIEW EXERCISES

1. Write 0.14 as a fraction.

2. Add:

   $27.02 + 93.8 + 18.75$

3. Divide:

   $27.063 \div 0.03$

4. The manager of an electronics warehouse ordered 1,000 VCRs for \$184.72 each. What was the total cost of the order?

5. Rewrite the numbers 1.8, 0.81, 0.18, and 0.801 in order from the greatest to least.

6. Subtract: $-12 - (+7)$

7. Which number is between 5.2 and 5.3?
   - (a) 5.023
   - (b) 5.032
   - (c) 5.230
   - (d) 5.320

8. Choose the correct decimal for the number "five hundred and five hundredths."
   - (a) 505
   - (b) 500.5
   - (c) 500.05
   - (d) 0.505

9. A utility crew had a spool containing 20 meters of electrical cable. The job at Northport required the use of 6.45 meters of this cable. How much was left on the spool?

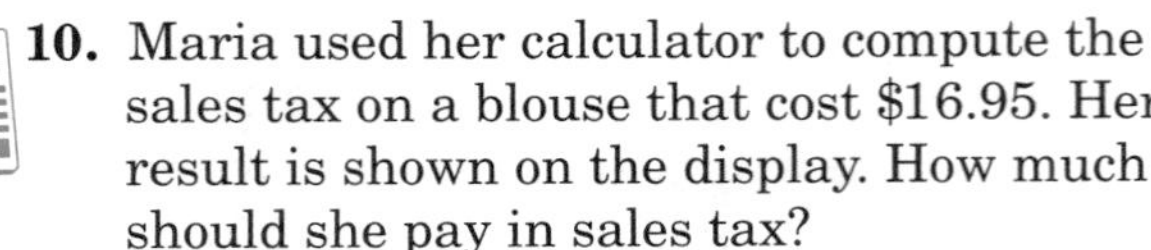

10. Maria used her calculator to compute the sales tax on a blouse that cost \$16.95. Her result is shown on the display. How much should she pay in sales tax?

1.1865

UNIT 4-4 Using Decimals in Stepped Rates

THE MAIN IDEA

1. Some services are priced so that the rate charged changes for different quantities purchased.
2. The rates are stepped so that quantities in different ranges are priced at different levels. Often, as the quantity increases the cost per unit decreases.
3. To find the total cost of a rate that has two steps:
   a. Find the cost of the first step.
   b. If more than the number of units allowed in the first step have been used, subtract the number of units allowed in the first step from the total number of units. Multiply the difference by the second step rate.
   c. Add the costs of the two steps to find the total cost.

EXAMPLE The rates for electricity are:

$5.50 usage charge
$0.14 per kilowatt-hour for the first 700 kw-hr.
$0.13 per kilowatt-hour after 700 kw-hr.

a. What is the cost of 200 kilowatt-hours?

Since the usage is less than 700 kilowatt-hours, only the first step rate is charged, in addition to the usage charge: 5.50 + 200 (0.14) = 5.50 + 28.00 = $33.50 *Ans.*

b. What is the cost of 1,000 kilowatt-hours?

Since the usage is more than the amount allowed in the first step, find the cost for each step.

(1) To find the separate costs, find the number of kw-hr. at each step.

1,000	← total number of kw-hr.
−700	← number of kw-hr. in first step
300	← number of kw-hr. over first step

(2) To find the first-step charge, multiply the first-step allowance by its cost.

$0.14	← charge per kw-hr.
× 700	← number of kw-hr.
$98.00	← cost of first step

(3) To find the additional charge, multiply the remaining 300 kw-hr. by their cost.

$0.13	← charge per kw-hr.
× 300	← number of kw-hr.
$39.00	← cost of second step

(4) To find the total cost, add the cost of both steps and the usage charge.

$98.00	← cost of first step
39.00	← cost of second step
5.50	← usage charge
$142.50	← total cost *Ans.*

CLASS EXERCISES

1. The rates for electricity are:
   $5.25 usage charge
   $0.14 per kilowatt-hour for the first 500 kw-hr.
   Find the cost of 500 kilowatt-hours.

2. The rates charged by South Shore Electric are:
   $5.00 usage charge
   $0.13 per kilowatt-hour for the first 300 kw-hr.
   $0.14 per kilowatt-hour after 300 kw-hr.
   Find the cost of 650 kilowatt-hours.

3. Montgomery Power and Light charges the following rates:
   $5.20 usage charge
   $0.12 per kilowatt-hour for the first 250 kw-hr.
   $0.13 per kilowatt-hour after 250 kw-hr.
   Find the cost of 750 kilowatt-hours.

4. A mailing service charges:
   $2.20 for the first pound
   $0.90 for each additional pound up to 5 pounds
   Find the cost of mailing a package that weighs:
   a. 1 pound **b.** 3 pounds **c.** 5 pounds

5. The cost for a classified ad in a newspaper is:
   $2.50 for the first 15 words
   $0.25 for each additional word
   How much does a classified ad of 21 words cost?

6. The rates at the Speedy Duplicating Shop are:
   $0.10 for each of the first 10 copies
   $0.05 for each additional copy
   How much will 25 copies cost?

7. The Write-Right Pen Company sells pens at the following rate:
   $0.50 for each of the first 20 pens bought
   $0.30 for each additional pen
   Find the cost of: **a.** 15 pens **b.** 20 pens **c.** 30 pens **d.** 50 pens

HOMEWORK EXERCISES

1. The rates for electricity are:
   $5.52 usage charge
   $0.14 per kilowatt-hour for the first 250 kw-hr.
   $0.13 per kilowatt-hour after the first 250 kw-hr.
   Find the electric cost for the following:
   a. 200 kw-hr. **b.** 300 kw-hr. **c.** 600 kw-hr.

2. Island Electric Company charges the following rates:
   $5.70 usage charge
   $0.14 per kilowatt-hour for the first 300 kw-hr.
   $0.15 per kilowatt-hour after the first 300 kw-hr.
   What is the cost of 750 kilowatt-hours?

3. The rates that Midwest Power charges are:
   $5.00 usage charge
   $0.12 per kilowatt-hour for the first 250 kw-hr.
   $0.16 per kilowatt-hour after 250 kw-hr.
   Find the electric cost for the following:
   a. 250 kw-hr. **b.** 500 kw-hr. **c.** 650 kw-hr.

4. The Speedy Cab Co. charges $1.50 for the first $\frac{1}{10}$ of a mile and $0.12 for each additional $\frac{1}{10}$ of a mile.
   What is the charge for a 2-mile ride?

5. Graystone Bus Co. charges $12 for the first 50 miles and $0.15 for each additional mile. What is the price of a ticket for a 75-mile ride?

6. An ad in the school newspaper costs $5 for the first line and $3 for each additional line. Find the cost of each ad: **a.** 1 line **b.** 2 lines **c.** 5 lines **d.** 10 lines

7. The chart at the right appeared in an advertisement for a messenger service. Explain how you would reword it to avoid disputes over pricing.

8. Mr. Li mailed a 9-ounce letter first class and paid $2.13 postage. If the postal service charges $0.29 for the first ounce and a fixed charge for each additional ounce, how much was the fixed charge per ounce?

$5 for First Mile, Plus:

Number of Additional Miles	Price
1 – 5	$2
5 – 10	$4
10 – 15	$7
15 or more	$11

SPIRAL REVIEW EXERCISES

1. Sandra worked a 35-hour week at an hourly wage of $6.20, with time and a half for overtime. One week she worked 47 hours. What were her total earnings?

2. How much less than 13 is 12.96?

3. If 12 candles cost $18.60, what is the price of one candle? of 5 candles?

4. Sara opened a checking account with $500. After writing checks for $117.85, $36.50, and $210, what was the balance?

5. The value of the digit 7 in 238.175 is

   (a) 700 (b) 70 (c) $\frac{7}{10}$ (d) $\frac{7}{100}$

6. Round 52,489 to the nearest thousand.

7. Which calculator key will place a sum in memory?

   (a) [CE] (b) [M−] (c) [M+] (d) [A/C]

8. Which group of decimals is in order from least to greatest, reading from left to right?

   (a) 0.575, 0.4, 0.625, 0.37, 0.07
   (b) 0.4, 0.37, 0.07, 0.625, 0.575
   (c) 0.07, 0.4, 0.37, 0.575, 0.62
   (d) 0.07, 0.37, 0.4, 0.575, 0.62

9. A house painter charges $250 to paint two rooms and $95 for each additional room. How much will this painter charge to paint 7 rooms?

THEME 3

Rational Numbers

MODULE 5

MODULE 6

MODULE 7

UNIT 5-1 Meaning of Rational Numbers; Representing Rational Numbers on a Number Line

MEANING OF RATIONAL NUMBERS

THE MAIN IDEA

1. A *rational number* is a number that can be written as a fraction in which the numerator and denominator are integers and the denominator is not zero.
2. The set of rational numbers includes the whole numbers and integers as well as fractions and some decimals. The set of rational numbers is a subset of a larger set of numbers called the *real numbers*.

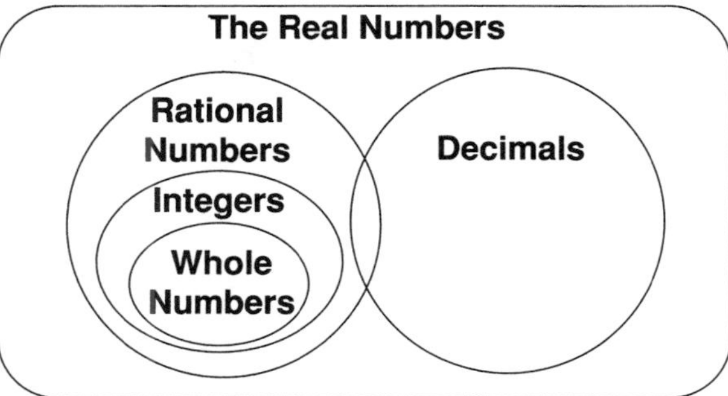

3. a. If a positive fraction is less than 1, with its numerator less than its denominator, it is called a *proper fraction*.

   b. If a positive fraction is greater than or equal to 1, with its numerator greater than or equal to its denominator, it is called an *improper fraction*.

EXAMPLE 1 Name each of the shaded regions as a fraction.

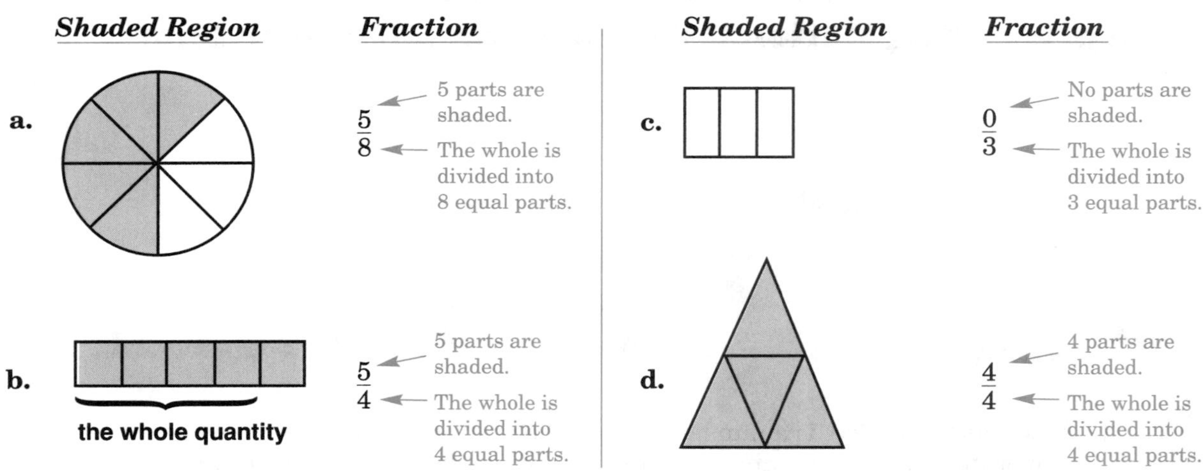

EXAMPLE 2 Tell whether each fraction is proper or improper. Then use <, >, or = to compare each fraction with 1.

	Fraction	***Type of Fraction***	***Comparison With 1***
a.	$\frac{5}{7}$ ← numerator / ← denominator	proper (5 < 7)	$\frac{5}{7} < 1$
b.	$\frac{9}{9}$	improper (9 = 9)	$\frac{9}{9} = 1$
c.	$\frac{5}{3}$	improper (5 > 3)	$\frac{5}{3} > 1$
d.	$\frac{3}{5}$	proper (3 < 5)	$\frac{3}{5} < 1$

EXAMPLE 3 Show that each integer is a rational number by writing it as a fraction in which the numerator and denominator are both integers.

	Integer	***Fraction***
a.	6	$6 = \frac{6}{1}$
b.	0	$0 = \frac{0}{1}$
c.	–3	$-3 = \frac{-3}{1}$

EXAMPLE 4 A Study Hall has 60 seats. The shaded portion of each circle below shows the fraction of the seats that are used for each period.

a. What fraction of the seats are used for each period?

Use of 60 Study Hall Seats

Period 1 *Period 2* *Period 3*

Answer: $\frac{1}{2}$ of seats used in Period 1 $\frac{1}{3}$ of seats used in Period 2 $\frac{1}{4}$ of seats used in Period 3

b. How many students are there in Study Hall for each period?

Period 1: $60 \div 2 = 30$ $\frac{1}{2}$ of 60 is 30

Period 2: $60 \div 3 = 20$ $\frac{1}{3}$ of 60 is 20

Period 3: $60 \div 4 = 15$ $\frac{1}{4}$ of 60 is 15

Answer: There are 30 students in Study Hall during Period 1, 20 during Period 2, and 15 during Period 3.

CLASS EXERCISES

1. Name each of the *shaded* regions as a fraction.

a. 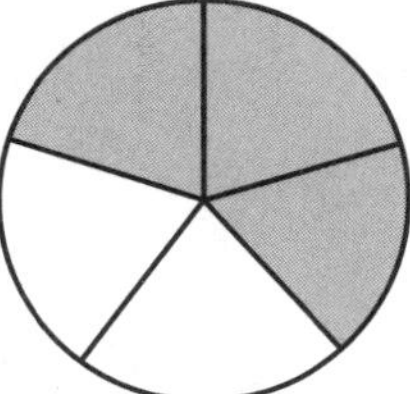

b. 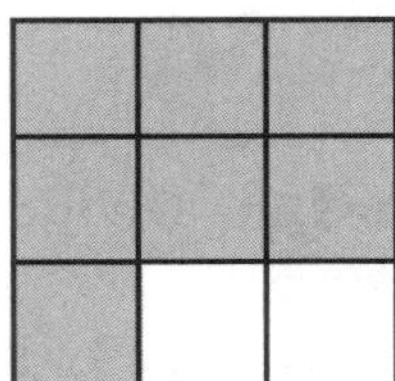

c. 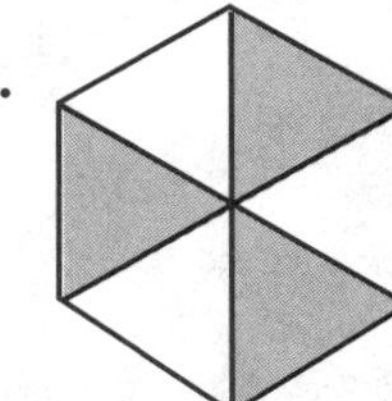

d. 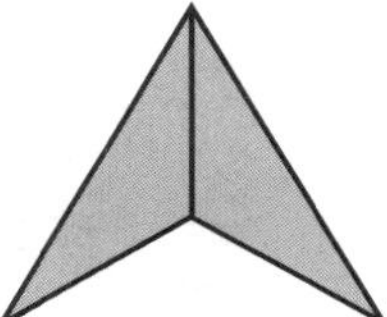

e. 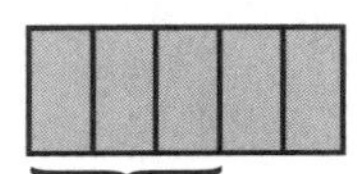

f. 

2. Show that each integer is a rational number by writing it as a fraction in which the numerator and denominator are both integers.

a. 9 **b.** –8 **c.** –1 **d.** +2

3. Tell whether each fraction is proper or improper.

a. $\frac{9}{11}$ **b.** $\frac{15}{17}$ **c.** $\frac{13}{11}$ **d.** $\frac{81}{81}$ **e.** $\frac{50}{75}$ **f.** $\frac{0}{17}$ **g.** $\frac{112}{113}$ **h.** $\frac{113}{112}$

4. Replace ? by $<$, $>$, or $=$ to make a true comparison with 1.

a. $\frac{5}{7}$? 1 **b.** $\frac{7}{5}$? 1 **c.** $\frac{7}{7}$? 1 **d.** $\frac{0}{9}$? 1 **e.** $\frac{121}{120}$? 1 **f.** $\frac{397}{397}$? 1 **g.** $\frac{1{,}000}{1{,}001}$? 1

5. A survey asked 720 students whether they usually bring their own lunch to school or buy lunch at school. Use the diagram to answer each question.

a. How many students usually bring lunch to school?

b. How many students usually buy lunch at school?

Lunch Survey Results
(720 students)

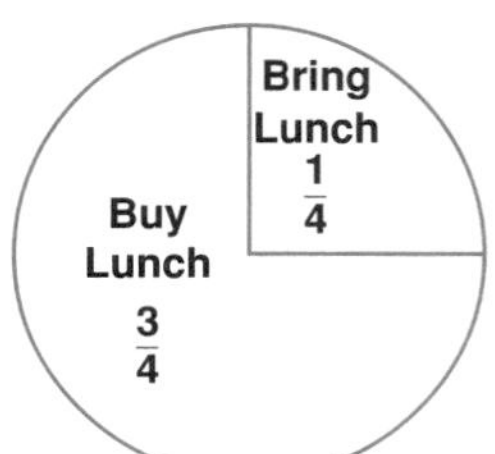

REPRESENTING RATIONAL NUMBERS ON A NUMBER LINE

THE MAIN IDEA

Just as we represent integers as points on a number line, we also represent rational numbers as points on a number line.

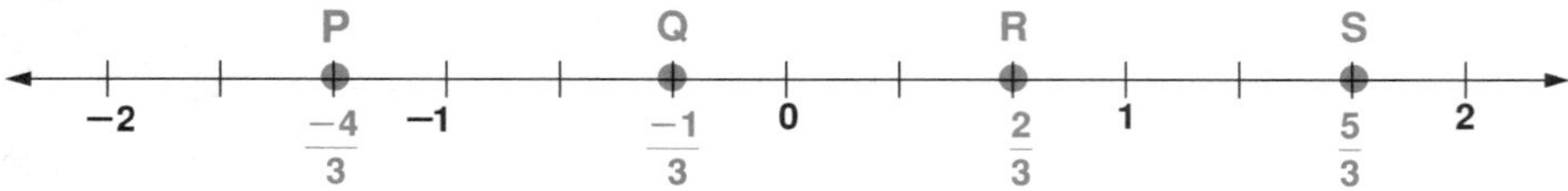

Number Line Divided Into Thirds

EXAMPLE 5 Use a rational number to name each point on the number lines shown.

	Point	*Rational Number*
a.	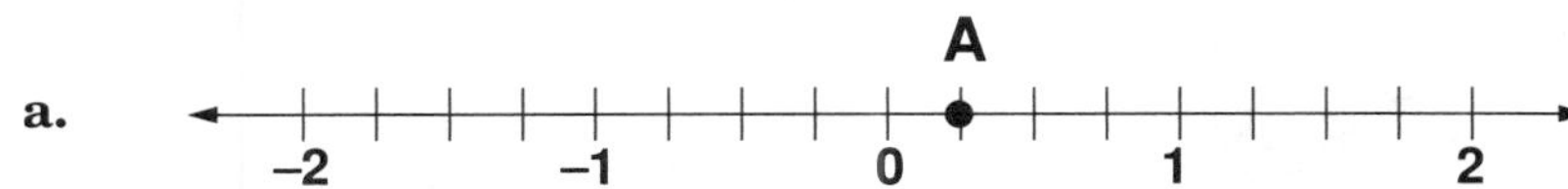	$\frac{1}{4}$

Think: A whole quantity has been divided into 4 equal parts. Thus, 4 is the *denominator* of the rational number. Since point A is on the first division, 1 is the *numerator* of the rational number.

	Point	*Rational Number*
b.	 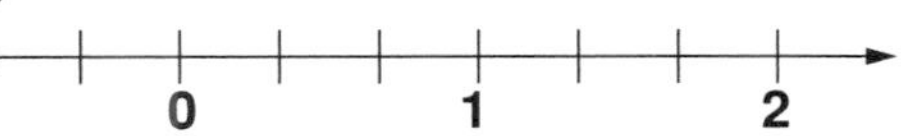	$\frac{-2}{3}$
c.		$\frac{7}{8}$
d.		$\frac{-5}{4}$
e.		$\frac{12}{6}$ or 2
f.		$\frac{0}{3}$ or 0

CLASS EXERCISES

Use a rational number to name each point on the number lines shown.

1. 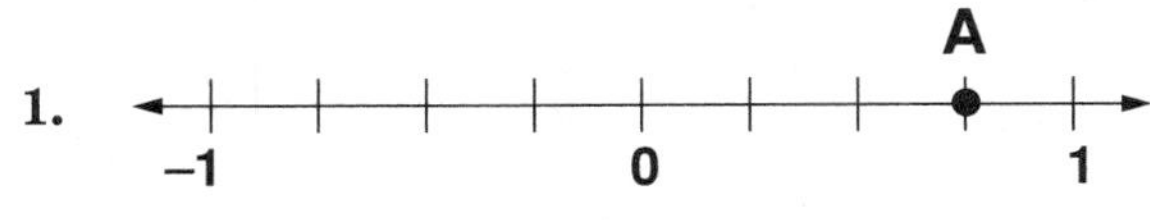

2. 

3. 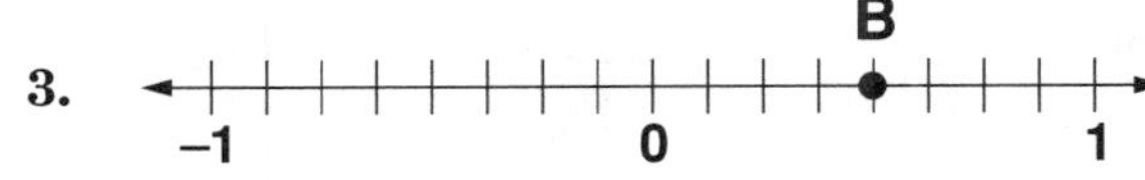

4. 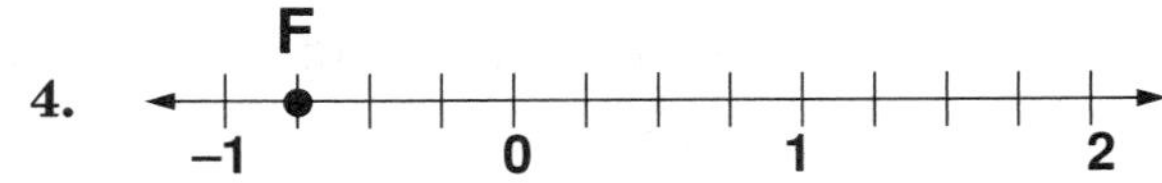

5. 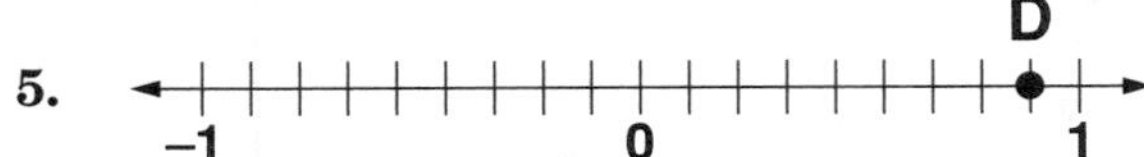

6. 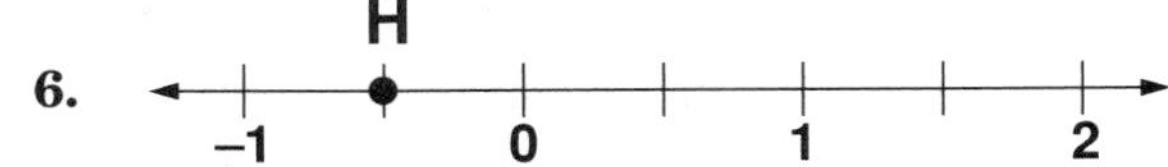

1. Name each of the *shaded* regions as a fraction.

a. 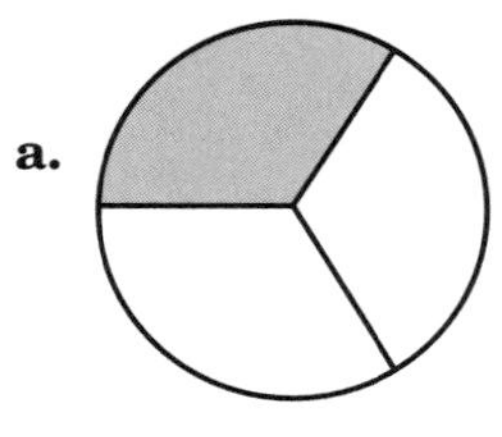 **b.** 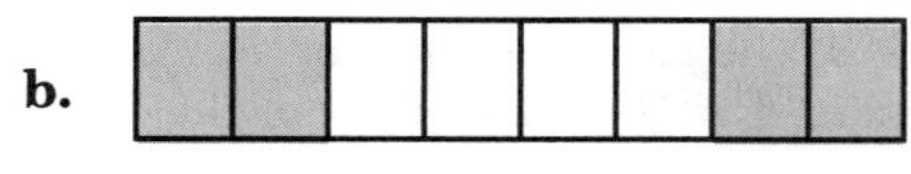 **c.** 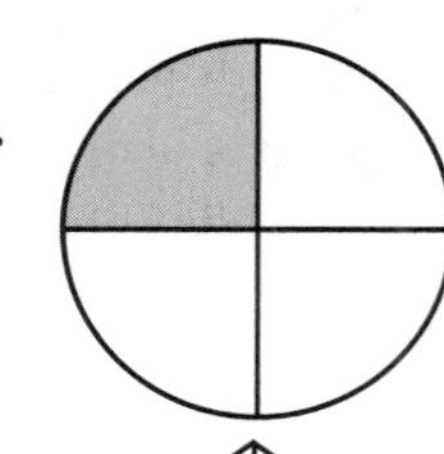

d. 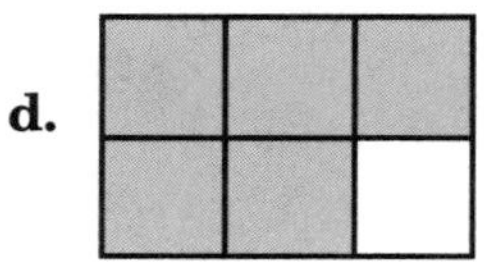 **e.** 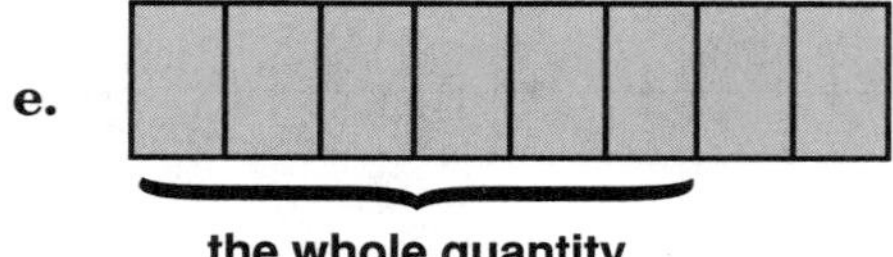

f. 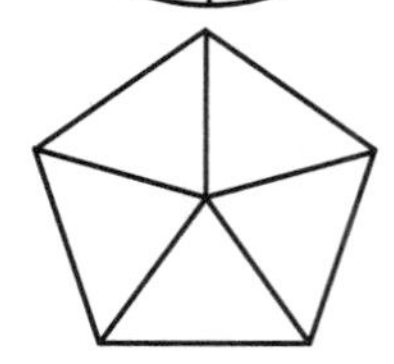

2. Show that each integer is a rational number by writing it as a fraction in which the numerator and denominator are both integers.

   a. 12 **b.** +8 **c.** −5 **d.** −17 **e.** 199

3. Tell whether each fraction is proper or improper.

   a. $\frac{5}{7}$ **b.** $\frac{3}{4}$ **c.** $\frac{0}{16}$ **d.** $\frac{17}{15}$ **e.** $\frac{12}{21}$ **f.** $\frac{21}{12}$ **g.** $\frac{75}{75}$ **h.** $\frac{75}{74}$

4. Replace ? by <, >, or = to make a true comparison with 1.

   a. $\frac{9}{11}$? 1 **b.** $\frac{0}{6}$? 1 **c.** $\frac{7}{8}$? 1 **d.** $\frac{8}{7}$? 1 **e.** $\frac{20}{17}$? 1 **f.** $\frac{17}{20}$? 1 **g.** $\frac{52}{52}$? 1

5. Explain why the absolute value of an improper fraction must always be greater than the absolute value of a proper fraction.

6. A survey asked 540 consumers to compare the taste of different brands of breath mints. Use the diagram to answer each question.

   a. How many people in the survey prefer Brand A?

   b. How many prefer Brand B?

   c. How many prefer a brand other than Brand A or Brand B?

Breath Mint Preferences
(540 consumers)

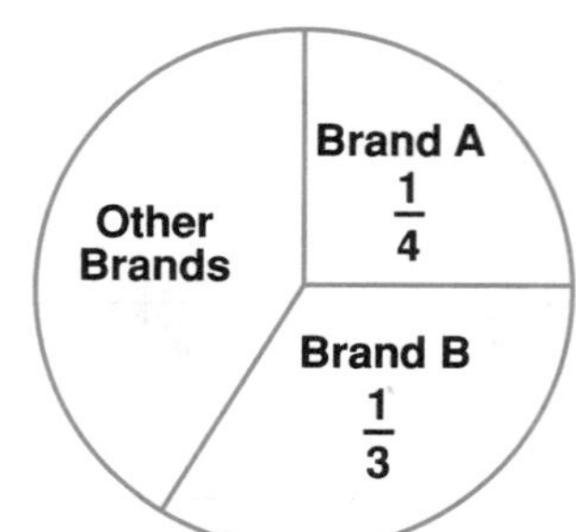

7. Use a rational number to name each point on the number lines shown.

a. 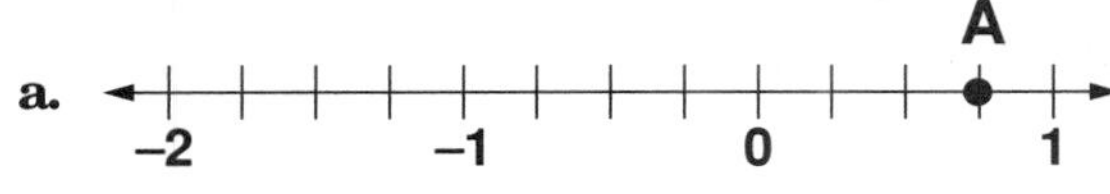

b. 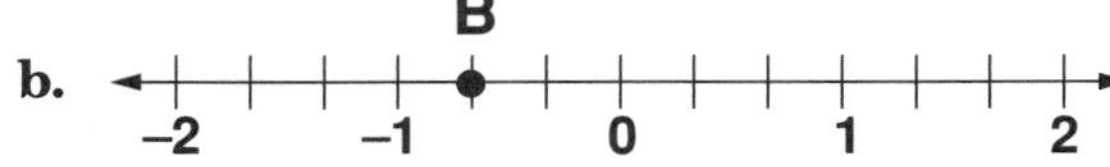

c. 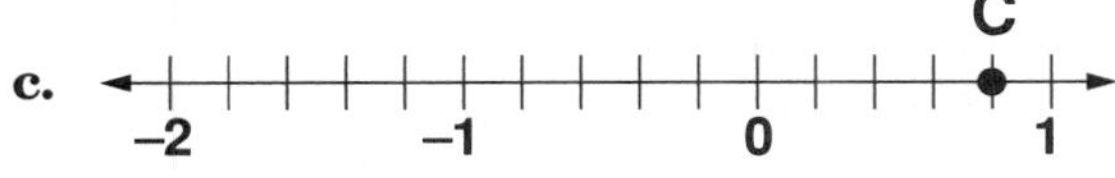

d. 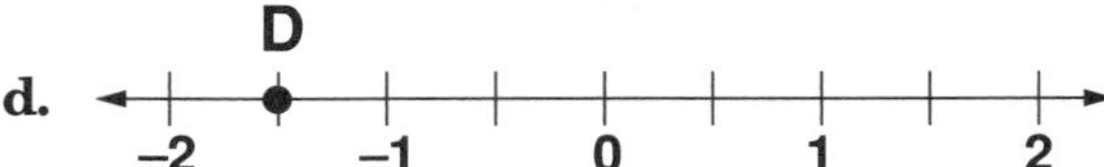

e. 

f. 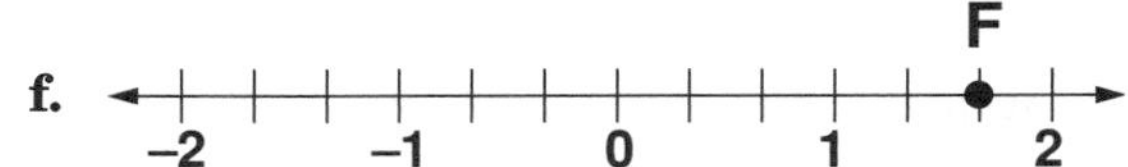

8. Use the number line to answer each question.

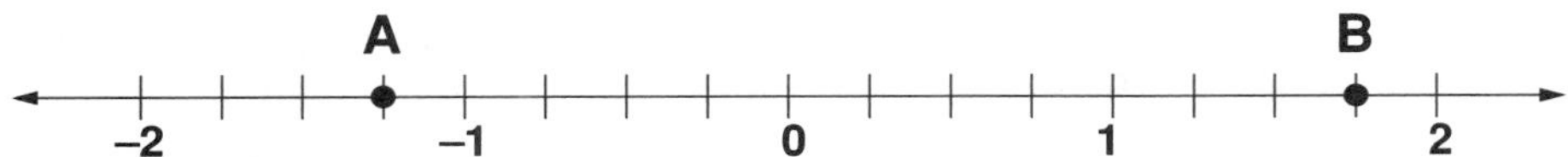

a. What number is represented by Point A?
b. What number is represented by Point B?
c. How many units apart are Point A and Point B?

9. An event at a track meet requires runners to go around a track 8 times. What fraction of the distance has a runner completed after having run around the track:

a. 1 time **b.** 3 times **c.** 4 times **d.** 7 times **e.** 8 times

10. A store advertised $\frac{1}{3}$ off the price of all sweaters and $\frac{1}{4}$ off the price of all equipment for winter sports. What fraction of the original price will a customer pay for:

a. a sweater **b.** a pair of skis

11. Answer *true* or *false*. Use examples to explain your answers.
a. Whole numbers are rational numbers.
b. Rational numbers are whole numbers.

12. The table shows rainfall in Winburg for a recent year.
a. The year had 366 days. Express as a fraction the part of the year shown on the table.
b. In the same year, there was a total of 13 inches of rain. Express as a fraction the part of the yearly rain that fell during March and May together.

Rainfall in Winburg

Month	Days in Month	Inches of Rain
February	29	1
March	31	2
April	30	3
May	31	1

SPIRAL REVIEW EXERCISES

1. Round 27,592 to the nearest thousand.

2. The product of −8 and −7 is
(a) −15 (b) +15
(c) +56 (d) −56

3. Which comparison is true?
(a) $-11 < -3$ (b) $-25 > -19$
(c) $-5 > 0 > 2$ (d) $0 < 5 < -10$

4. The value of $9 \times 4 - 5 \times 3$ is
(a) 93 (b) 51 (c) 41 (d) 21

5. Find the greatest common factor of 24 and 60.

6. In the month of April, Ms. Johnson earned \$2,746 and Mr. Frank earned \$2,498. How much more did Ms. Johnson earn?

7. Which whole number is prime?
(a) 1 (b) 15 (c) 17 (d) 21

8. Fifty thousand, three hundred seven written as a numeral is
(a) 5,307 (b) 53,007
(c) 50,307 (d) 53,070

9. Write a key sequence that uses the constant feature of multiplication to calculate 15×3, 27×3, 35×3, and 41×3.

10. Mr. Gorham bought 6 shirts at \$23 a shirt and 4 ties at \$9 a tie. What is the total amount of money that he spent?

11. Francine uses 18 strips of balsa wood to make a model. She bought 2 boxes of balsa wood that contain 36 strips each. How many models can she make?

12. The value of $|+8| + |-6| - |-2|$ is
(a) 16 (b) 12 (c) 4 (d) 0

TEAMWORK

Choose five stocks and find these stocks in the daily newspaper report of stock prices. Record the daily change in price of each of the five stocks for one day. Convert each number to an improper fraction. Draw a number line and locate these five numbers on it. Identify which stock shows the greatest increase and which shows the greatest decrease. On the number line, determine the number of dollars between the smallest number and the largest number. Prepare one number line with your team results.

UNIT 5-2 Mixed Numbers

THE MEANING OF A MIXED NUMBER

THE MAIN IDEA

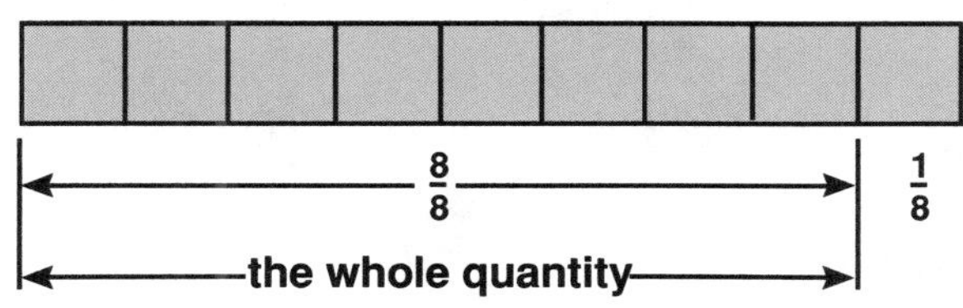

1. The diagram shows that the improper fraction $\frac{9}{8}$ is equal to $\frac{8}{8} + \frac{1}{8}$ or $1 + \frac{1}{8}$. Write $1 + \frac{1}{8}$ as the mixed number $1\frac{1}{8}$ (read "one and one-eighth").

2. A ***positive mixed number*** represents the sum of a positive integer and a proper fraction. A ***negative mixed number*** represents the sum of a negative integer and a negative fraction.

$$9\frac{1}{8} = 9 + \frac{1}{8} \qquad -9\frac{1}{8} = -9 + \left(\frac{-1}{8}\right)$$

integer part — fraction part

EXAMPLE 1 Write each sum as a mixed number.

	Sum	*Mixed Number*
a.	$3 + \frac{7}{8}$	$3\frac{7}{8}$
b.	$7 + \frac{5}{16}$	$7\frac{5}{16}$
c.	$-12 + \left(\frac{-9}{11}\right)$	$-12\frac{9}{11}$

EXAMPLE 2 Name the integer part and the fraction part of each mixed number.

	Mixed Number	*Integer Part*	*Fraction Part*
a.	$7\frac{2}{9}$	7	$\frac{2}{9}$
b.	$12\frac{3}{5}$	12	$\frac{3}{5}$
c.	$-19\frac{11}{17}$	−19	$\frac{-11}{17}$
d.	$65\frac{110}{111}$	65	$\frac{110}{111}$

EXAMPLE 3 Write the display obtained when the mixed number $6\frac{13}{20}$ is entered on a calculator.

KEYING IN

To enter a mixed number on a calculator, enter the fraction part first; then add the integer part.

Key Sequence	***Display***
13 [÷] 20 [+] 6 [=]	6.65

Answer: 6.65

CLASS EXERCISES

1. Write each sum as a mixed number.

a. $6 + \frac{5}{9}$ **b.** $3 + \frac{6}{13}$ **c.** $8 + \frac{3}{7}$ **d.** $10 + \frac{1}{6}$ **e.** $-15 + \left(\frac{-8}{11}\right)$ **f.** $-19 + \left(\frac{-5}{12}\right)$

2. Name the integer part and the fraction part of each mixed number.

a. $5\frac{3}{4}$ **b.** $6\frac{7}{12}$ **c.** $7\frac{9}{11}$ **d.** $10\frac{5}{8}$ **e.** $4\frac{2}{5}$ **f.** $17\frac{8}{19}$ **g.** $-11\frac{12}{19}$ **h.** $-23\frac{7}{8}$

3. Write the display that results when each mixed number is entered on a calculator.

a. $1\frac{3}{5}$ **b.** $2\frac{3}{8}$ **c.** $2\frac{3}{10}$ **d.** $2\frac{2}{3}$ **e.** $-2\frac{1}{5}$

WRITING AN IMPROPER FRACTION AS A MIXED NUMBER

THE MAIN IDEA

To write an improper fraction as an equivalent mixed number, divide the denominator of the improper fraction into the numerator.

The quotient is the integer part of the mixed number.

The $\frac{\text{remainder}}{\text{divisor}}$ is the fraction part of the mixed number.

EXAMPLE 4 Write $\frac{19}{8}$ as a mixed number.

Divide the denominator 8 into the numerator 19. $8\overline{)19}$

Obtain a quotient and a remainder.

$$\begin{array}{r} 2 \\ 8\overline{)19} \\ \underline{16} \\ 3 \end{array}$$

(quotient: 2; divisor: 8; remainder: 3)

The quotient is the integer part of the mixed number, and the $\frac{\text{remainder}}{\text{divisor}}$ is the fraction part of the mixed number.

$\frac{19}{8} = 2\frac{3}{8}$ *Ans.*

EXAMPLE 5 Use a calculator to write the fraction $\frac{261}{20}$ as a mixed number.

KEYING IN

Divide 261 by 20 and subtract the integer part of the result on the display. Then multiply by the divisor (denominator) to obtain the remainder (numerator).

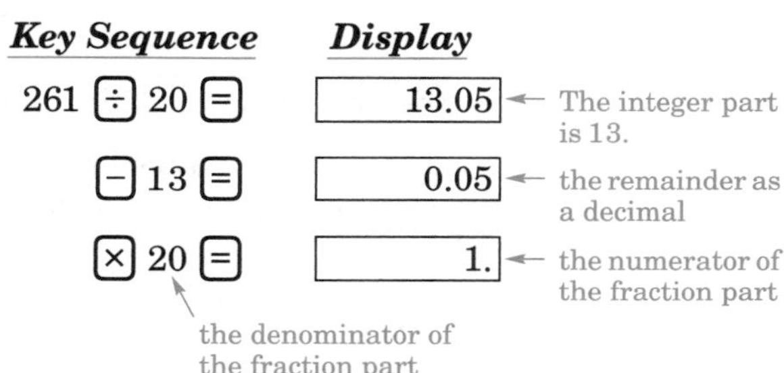

Key Sequence	***Display***	
261 [÷] 20 [=]	13.05	The integer part is 13.
[−] 13 [=]	0.05	the remainder as a decimal
[×] 20 [=]	1.	the numerator of the fraction part

Answer: $\frac{261}{20} = 13\frac{1}{20}$

CLASS EXERCISES

Write each improper fraction as an equivalent mixed number.

1. $\frac{15}{2}$ **2.** $\frac{17}{5}$ **3.** $\frac{23}{4}$ **4.** $\frac{5}{3}$ **5.** $\frac{81}{7}$ **6.** $\frac{109}{9}$ **7.** $\frac{-58}{5}$ **8.** $\frac{143}{10}$ **9.** $\frac{-99}{8}$ **10.** $\frac{120}{11}$

11. Use a calculator to write each improper fraction as a mixed number.

a. $\frac{187}{8}$ **b.** $\frac{253}{16}$ **c.** $\frac{941}{20}$ **d.** $\frac{1,427}{50}$ **e.** $\frac{-151}{12}$

WRITING A MIXED NUMBER AS AN IMPROPER FRACTION

THE MAIN IDEA

To write a mixed number as an equivalent improper fraction:

1. Multiply the integer part by the denominator of the fraction part.
2. Add this product to the numerator of the fraction part.
3. The improper fraction is:

$$\frac{\text{the sum obtained in Step 2}}{\text{the denominator of the fraction part of the mixed number}}$$

A good way to remember this procedure is to start at the denominator and work in a clockwise direction as follows:

$3 \; \frac{2}{5}$ ← start (× then +)

$$3\frac{2}{5} = \frac{5 \times 3 + 2}{5} = \frac{17}{5}$$

EXAMPLE 6 Write $5\frac{3}{4}$ as an equivalent improper fraction.

$$5\frac{3}{4} = \frac{4 \times 5 + 3}{4}$$

(4 in the numerator: the denominator of the fraction part; 5: the integer part; 3: the numerator of the fraction part; 4 in the denominator: the denominator of the fraction part)

$$= \frac{20 + 3}{4} = \frac{23}{4} \quad \textit{Ans.}$$

CLASS EXERCISES

Write each mixed number as an equivalent improper fraction.

1. $4\frac{1}{9}$ **2.** $3\frac{4}{7}$ **3.** $1\frac{7}{13}$ **4.** $8\frac{4}{5}$ **5.** $6\frac{7}{10}$ **6.** $9\frac{5}{7}$ **7.** $6\frac{9}{11}$ **8.** $12\frac{3}{4}$

9. $10\frac{1}{3}$ **10.** $20\frac{2}{3}$ **11.** $18\frac{2}{5}$ **12.** $40\frac{1}{3}$ **13.** $25\frac{3}{4}$ **14.** $30\frac{3}{8}$ **15.** $50\frac{1}{3}$ **16.** $101\frac{1}{2}$

HOMEWORK EXERCISES

1. Write each sum as a mixed number.

   a. $3 + \frac{2}{7}$ **b.** $5 + \frac{2}{3}$ **c.** $7 + \frac{3}{5}$ **d.** $2 + \frac{15}{16}$ **e.** $11 + \frac{4}{9}$ **f.** $-9 + \left(\frac{-3}{11}\right)$ **g.** $-10 + \left(\frac{-11}{12}\right)$

2. Name the integer part and the fraction part of each mixed number.

   a. $1\frac{5}{8}$ **b.** $3\frac{4}{5}$ **c.** $9\frac{2}{7}$ **d.** $7\frac{8}{11}$ **e.** $10\frac{1}{2}$ **f.** $-12\frac{7}{8}$ **g.** $20\frac{15}{17}$ **h.** $-42\frac{17}{25}$

3. Write each improper fraction as an equivalent mixed number.

   a. $\frac{5}{4}$ **b.** $\frac{9}{5}$ **c.** $\frac{11}{2}$ **d.** $\frac{20}{3}$ **e.** $\frac{15}{7}$ **f.** $\frac{16}{3}$ **g.** $\frac{25}{8}$ **h.** $\frac{-17}{2}$

   i. $\frac{26}{5}$ **j.** $\frac{35}{6}$ **k.** $\frac{-42}{9}$ **l.** $\frac{56}{11}$ **m.** $\frac{107}{20}$ **n.** $\frac{-91}{9}$ **o.** $\frac{135}{11}$ **p.** $\frac{65}{7}$

4. Give an example of an improper fraction that can*not* be written as a mixed number.

5. Write each mixed number as an equivalent improper fraction.

   a. $1\frac{3}{4}$ **b.** $1\frac{7}{16}$ **c.** $2\frac{5}{8}$ **d.** $3\frac{4}{7}$ **e.** $5\frac{9}{10}$ **f.** $8\frac{2}{3}$ **g.** $10\frac{1}{2}$ **h.** $14\frac{3}{5}$

   i. $20\frac{5}{6}$ **j.** $9\frac{7}{16}$ **k.** $-6\frac{10}{11}$ **l.** $11\frac{5}{9}$ **m.** $24\frac{1}{3}$ **n.** $30\frac{5}{7}$ **o.** $10\frac{10}{13}$ **p.** $-2\frac{5}{6}$

6. $\frac{19}{3}$ is equivalent to (a) $4\frac{2}{3}$ (b) $5\frac{1}{3}$ (c) $6\frac{1}{3}$ (d) $6\frac{2}{3}$

7. $6\frac{3}{4}$ is equivalent to (a) $\frac{27}{4}$ (b) $\frac{25}{4}$ (c) $\frac{25}{3}$ (d) $\frac{22}{3}$

8. Using each of the three numbers 2, 7, and 9 exactly once in each answer, write:
   a. three proper fractions **b.** three improper fractions **c.** three mixed numbers

9. Runners must go around a track 4 times to run one mile. Use a fraction or mixed number to express the distance, in miles, that a runner has completed after having run around the track:

   a. 1 time **b.** 3 times **c.** 5 times **d.** 7 times **e.** 9 times

10. Marcy bought 15 new baseball cards at a collectors' show. This was $\frac{1}{3}$ the number of cards she already owned. How many cards did Marcy have before the show?

11. Julio added 22 buttons to his collection and gave 5 to a friend. Then he told his friend that the increase in his collection was $\frac{1}{4}$ the number of buttons he originally had. How many buttons did Julio have after the increase?

12. Use a calculator to write each improper fraction as a mixed number.

    a. $\frac{99}{16}$ **b.** $\frac{117}{32}$ **c.** $\frac{1,058}{25}$ **d.** $\frac{684}{625}$

SPIRAL REVIEW EXERCISES

1. Find the product of 209 and 96.

2. Divide: $28\overline{)11{,}396}$

3. What is the fraction that represents the shaded region in the diagram?

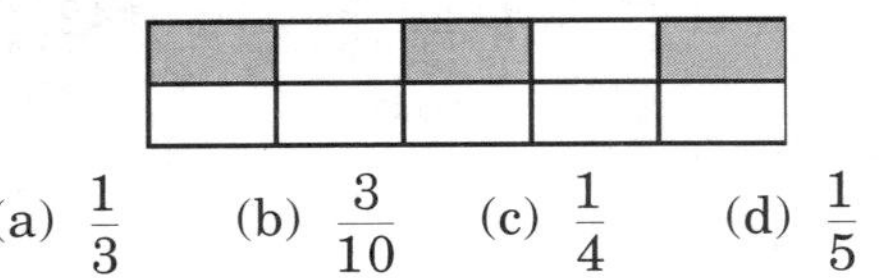

(a) $\frac{1}{3}$ (b) $\frac{3}{10}$ (c) $\frac{1}{4}$ (d) $\frac{1}{5}$

4. Subtract 738 from 926.

5. Felix earned $29 on Monday, $42 on Tuesday, $52 on Wednesday, $18 on Thursday, and $37 on Friday. What is the total amount of money he earned that week?

6. What is the rational number that is represented on the number line shown?

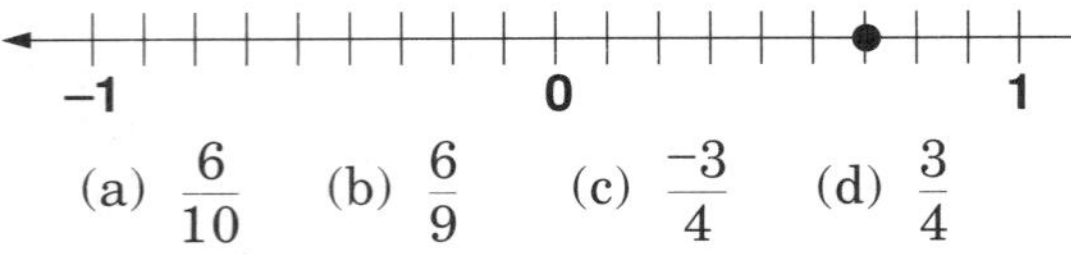

(a) $\frac{6}{10}$ (b) $\frac{6}{9}$ (c) $\frac{-3}{4}$ (d) $\frac{3}{4}$

7. Perform each of the indicated operations.

   a. $-18 + (+9)$ b. $-18 \times (+9)$

   c. $-18 - (+9)$ d. $-18 \div (+9)$

8. In a nine-inning baseball game, Jim pitched four innings. What fractional part of the game did he pitch?

9. Last year, Ms. Pepper earned $44,512. If she worked 52 weeks at the same wage, how much did she earn each week?

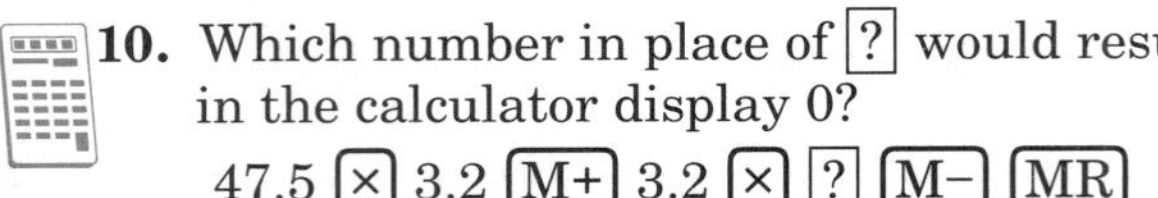

10. Which number in place of [?] would result in the calculator display 0?

    47.5 [×] 3.2 [M+] 3.2 [×] [?] [M–] [MR]

11. A baker had 126 pounds of flour. He made 50 cakes and used 2 pounds of flour for each cake. How much flour did he have left?

12. Evaluate: $58 - 4 \times 9$

13. At the beginning of a 2,200-mile auto trip, the odometer on Mrs. Wilson's car read 32,897 miles. At the end of two days, it read 33,124 miles. How many miles did Mrs. Wilson still have to drive to complete the trip?

UNIT 5-3 Equivalent Fractions

THE MEANING OF EQUIVALENT FRACTIONS

THE MAIN IDEA

1. Fractions that represent the same quantity are called ***equivalent fractions***.

   $\frac{3}{6}, \frac{4}{8}, \frac{5}{10}$, etc. are equivalent fractions since each represents the same amount, $\frac{1}{2}$ of a whole quantity:

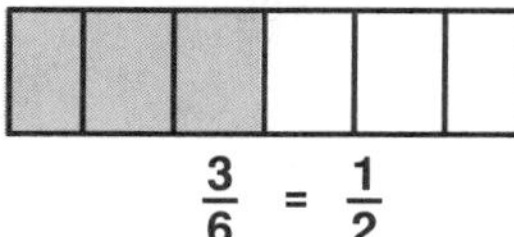

$\frac{3}{6} = \frac{1}{2}$

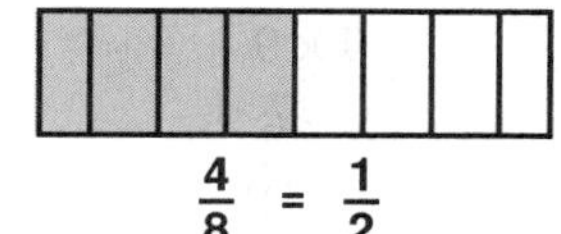

$\frac{4}{8} = \frac{1}{2}$

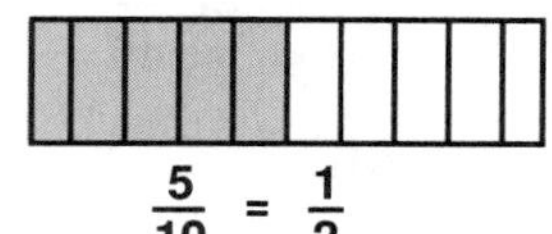

$\frac{5}{10} = \frac{1}{2}$

2. If we multiply or divide both the numerator and the denominator of a fraction by the same nonzero number, an equivalent fraction will result:

$$\frac{1 \times 3}{2 \times 3} = \frac{3}{6} \qquad \frac{3}{6} \text{ is equivalent to } \frac{1}{2}$$

$$\frac{9 \div 3}{12 \div 3} = \frac{3}{4} \qquad \frac{3}{4} \text{ is equivalent to } \frac{9}{12}$$

EXAMPLE 1 Find the fraction with a denominator of 21 that is equivalent to $\frac{2}{3}$.

$$\frac{2}{3} = \frac{?}{21}$$

Notice that we must multiply 3 by 7 in order to obtain 21. Therefore, we must multiply 2 by 7 to obtain the numerator of the equivalent fraction.

$$\frac{2 \times 7}{3 \times 7} = \frac{14}{21} \quad \textit{Ans.}$$

EXAMPLE 2 Use a calculator to complete the equivalent fraction:

$$\frac{9}{17} = \frac{?}{442}$$

KEYING IN

First , divide 442 by 17 to find by what number the denominator, 17, was multiplied. Then, multiply the numerator, 9, by that multiplier.

Key Sequence	***Display***
442 [÷] 17 [×] 9 [=]	234.

Answer: $\frac{9}{17} = \frac{234}{442}$

EXAMPLE 3 Which fraction is *not* equivalent to $\frac{3}{5}$?

	Fraction	***Has the Same Multiplier Been Used?***	***Is the Fraction Equivalent to*** $\frac{3}{5}$***?***
(a)	$\frac{6}{10}$	$\frac{3\times 2}{5\times 2}$	yes
(b)	$\frac{12}{20}$	$\frac{3\times 4}{5\times 4}$	yes
(c)	$\frac{27}{45}$	$\frac{3\times 9}{5\times 9}$	yes
(d)	$\frac{18}{60}$	$\frac{3\times 6}{5\times 12}$	no

Answer: (d)

CLASS EXERCISES

1. Write each fraction as an equivalent fraction that has the denominator shown in parentheses.

a. $\frac{1}{2}$, (40) **b.** $\frac{3}{8}$, (32) **c.** $\frac{2}{5}$, (35) **d.** $\frac{5}{16}$, (96) **e.** $\frac{7}{12}$, (36)

f. $\frac{7}{9}$, (72) **g.** $\frac{4}{11}$, (66) **h.** $\frac{9}{10}$, (90) **i.** $\frac{3}{4}$, (100) **j.** $\frac{1}{4}$, (360)

2. Replace ? to make a true statement.

a. $\frac{3}{5}=\frac{?}{20}$ **b.** $\frac{7}{11}=\frac{?}{55}$ **c.** $\frac{5}{8}=\frac{45}{?}$ **d.** $\frac{5}{12}=\frac{?}{144}$ **e.** $\frac{11}{20}=\frac{44}{?}$ **f.** $\frac{13}{24}=\frac{?}{432}$

3. Which fraction is *not* equivalent to $\frac{4}{9}$? (a) $\frac{12}{27}$ (b) $\frac{28}{63}$ (c) $\frac{32}{81}$ (d) $\frac{400}{900}$

4. Which fraction is *not* equivalent to $\frac{8}{15}$? (a) $\frac{16}{30}$ (b) $\frac{32}{75}$ (c) $\frac{24}{45}$ (d) $\frac{80}{150}$

5. Explain why $\frac{2}{3}$ and $\frac{12}{18}$ are names for the same number.

DETERMINING IF TWO FRACTIONS ARE EQUIVALENT

THE MAIN IDEA

A short way to tell if two fractions are equivalent is to find the ***cross products***.

1. If the cross products are equal, then the fractions are equivalent.
2. If the cross products are not equal, then the fractions are not equivalent.

EXAMPLE 4 Are $\frac{3}{10}$ and $\frac{15}{50}$ equivalent?

$\frac{3}{10} \times \frac{15}{50}$

first cross product: $3 \times 50 = 150$

second cross product: $10 \times 15 = 150$

Answer: Since the cross products are equal, the fractions are equivalent.

EXAMPLE 5 Are $\frac{7}{14}$ and $\frac{5}{8}$ equivalent?

$\frac{7}{14} \times \frac{5}{8}$

first cross product: $7 \times 8 = 56$

second cross product: $14 \times 5 = 70$

Answer: Since the cross products are *not* equal, the fractions are *not* equivalent.

EXAMPLE 6 Are $\frac{-6}{8}$ and $\frac{-39}{52}$ equivalent?

$\frac{-6}{8} \times \frac{-39}{52}$

first cross product: $-6 \times 52 = -312$

second cross product: $8 \times (-39) = -312$

Answer: Since the cross products are equal, the fractions are equivalent.

CLASS EXERCISES

In each pair, tell if the fractions are equivalent.

1. $\frac{2}{3}$ and $\frac{10}{15}$ **2.** $\frac{7}{11}$ and $\frac{5}{8}$ **3.** $\frac{30}{40}$ and $\frac{9}{12}$ **4.** $\frac{8}{9}$ and $\frac{24}{25}$ **5.** $\frac{3}{7}$ and $\frac{15}{28}$

6. $\frac{-5}{10}$ and $\frac{-15}{30}$ **7.** $\frac{9}{10}$ and $\frac{27}{30}$ **8.** $\frac{-7}{11}$ and $\frac{-20}{33}$ **9.** $\frac{8}{30}$ and $\frac{4}{15}$ **10.** $\frac{-3}{8}$ and $\frac{15}{40}$

HOMEWORK EXERCISES

1. Write each fraction as an equivalent fraction that has the denominator shown in parentheses.

a. $\frac{2}{3}$, (12) **b.** $\frac{7}{8}$, (16) **c.** $\frac{4}{5}$, (25) **d.** $\frac{5}{9}$, (72) **e.** $\frac{9}{16}$, (48) **f.** $\frac{11}{12}$, (72)

g. $\frac{3}{10}$, (80) **h.** $\frac{6}{11}$, (99) **i.** $\frac{4}{13}$, (52) **j.** $\frac{12}{25}$, (75) **k.** $\frac{17}{20}$, (100) **l.** $\frac{11}{30}$, (120)

2. Replace ? to make a true statement.

a. $\frac{4}{7} = \frac{?}{28}$ **b.** $\frac{3}{11} = \frac{?}{44}$ **c.** $\frac{7}{9} = \frac{42}{?}$ **d.** $\frac{3}{4} = \frac{?}{40}$ **e.** $\frac{11}{16} = \frac{55}{?}$

f. $\frac{5}{12} = \frac{?}{96}$ **g.** $\frac{9}{25} = \frac{45}{?}$ **h.** $\frac{13}{20} = \frac{?}{160}$ **i.** $\frac{17}{50} = \frac{68}{?}$ **j.** $\frac{3}{4} = \frac{?}{100}$

3. Which fraction is *not* equivalent to $\frac{1}{2}$? (a) $\frac{7}{14}$ (b) $\frac{15}{30}$ (c) $\frac{20}{40}$ (d) $\frac{30}{50}$

4. Which fraction is *not* equivalent to $\frac{7}{12}$? (a) $\frac{21}{36}$ (b) $\frac{30}{60}$ (c) $\frac{56}{96}$ (d) $\frac{63}{108}$

5. Explain why $\frac{3}{7}$ and $\frac{24}{56}$ are names for the same number.

6. Explain why $\frac{7}{10}$ and $\frac{42}{50}$ are *not* names for the same number.

7. In each pair, tell if the fractions are equivalent.

a. $\frac{1}{2}$ and $\frac{15}{30}$ **b.** $\frac{3}{4}$ and $\frac{27}{36}$ **c.** $\frac{5}{12}$ and $\frac{20}{46}$ **d.** $\frac{-7}{11}$ and $\frac{-70}{110}$ **e.** $\frac{7}{16}$ and $\frac{42}{92}$

f. $\frac{8}{15}$ and $\frac{-40}{75}$ **g.** $\frac{7}{10}$ and $\frac{35}{60}$ **h.** $\frac{9}{13}$ and $\frac{18}{25}$ **i.** $\frac{7}{24}$ and $\frac{13}{48}$ **j.** $\frac{-9}{40}$ and $\frac{-6}{30}$

k. $\frac{20}{25}$ and $\frac{12}{15}$ **l.** $\frac{-14}{20}$ and $\frac{21}{30}$ **m.** $\frac{12}{18}$ and $\frac{6}{8}$ **n.** $\frac{27}{36}$ and $\frac{12}{16}$ **o.** $\frac{25}{40}$ and $\frac{20}{36}$

8. Which of the following statements are equivalent? Explain.
   (a) Of 680 people surveyed, 510 preferred Brand X.
   (b) Three-fourths of the people surveyed preferred Brand X.
   (c) Only 170 out of 680 people surveyed did not prefer Brand X.
   (d) In a survey of about 600 people, over 500 preferred Brand X.

9. Mike walked $\frac{7}{8}$ of a mile, while Marilyn walked $\frac{23}{24}$ of a mile. Did they walk the same distance? If not, who walked farther?

10. Write a key sequence that can be used on the calculator to determine whether or not $\frac{13}{17}$ is equivalent to $\frac{247}{323}$.

SPIRAL REVIEW EXERCISES

1. Write $\frac{47}{3}$ as an equivalent mixed number.

2. Write $5\frac{7}{15}$ as an equivalent improper fraction.

3. Perform the indicated operation.
   a. $-8 + (-6)$ b. $-12 - (-9)$
   c. $-8 \times (-4)$ d. $-8 \div (-4)$

4. Which key sequence on an entry-order calculator would display the result of $25(17 + 3)$?
   (a) 25 [×] 17 [+] 3 [=]
   (b) 17 [+] 3 [×] 25 [=]
   (c) 25 [×] 17 [+] 25 [×] 3 [=]
   (d) 17 [+] 3 [M+] 25 [×] [=]

5. Round 298,472 to the nearest ten thousand.

6. John bought $\frac{99}{10}$ gallons of gasoline. Express the number of gallons as a mixed number.

7. At a fund-raising dinner, each person paid $25. If the total amount collected was $4,050, how many people attended the dinner?

8. A dealer bought 14 cars that cost $11,500 each. If he sold the cars at $12,200 each, what was the total amount of his profit?

9. Replace ? by <, >, or = to make a true statement.
   a. $5 + 10 \times 2 \ ? \ (5 + 10) \times 2$
   b. $8 + (3 \times 2) \ ? \ 8 + 3 \times 2$
   c. $(15 + 6) \times 5 + 8 \ ? \ 15 + 6 \times 5 + 8$

10. At the Dress Smart Clothing Shop, each salesperson receives a $20 gift certificate for every 3 new charge accounts that she brings in. Copy and complete the table below.

Dress Smart Gift Certificates

Salesperson	Number of Charge Accounts	Value of Gift Certificates
Melanie	15	
Ann Marie	13	
Kristen	2	
Sonya		$60

UNIT 5-4 Comparing Fractions; Least Common Multiple and Least Common Denominator

COMPARING LIKE FRACTIONS

THE MAIN IDEA

1. ***Like fractions*** are fractions that have a *common* denominator:

$$\frac{2}{7} \text{ and } \frac{5}{7} \text{ are like fractions}$$

2. To compare two like fractions, look at the numerators. The fraction with the greater numerator is the greater fraction:

$$\frac{5}{7} > \frac{2}{7}$$

EXAMPLE 1 Select the greatest fraction and the least fraction from $\frac{9}{31}$, $\frac{21}{31}$, $\frac{7}{31}$, and $\frac{4}{31}$.

Since the fractions have a common denominator, the fractions are like fractions.

Since 21 is the greatest numerator, $\frac{21}{31}$ is the greatest fraction.

Since 4 is the least numerator of the like fractions, $\frac{4}{31}$ is the least fraction.

EXAMPLE 2 Replace ? by < or > to make a true comparison.

	Like Fractions	***True Comparison***
a.	$\frac{3}{8}$? $\frac{-3}{8}$	$\frac{3}{8} > \frac{-3}{8}$
b.	$\frac{1}{4}$? $\frac{3}{4}$	$\frac{1}{4} < \frac{3}{4}$
c.	$\frac{-1}{4}$? $\frac{-3}{4}$	$\frac{-1}{4} > \frac{-3}{4}$
d.	$\frac{5}{12}$? $\frac{-7}{12}$	$\frac{5}{12} > \frac{-7}{12}$

CLASS EXERCISES

1. Select the greater fraction from each pair of fractions.

a. $\frac{7}{11}$ and $\frac{5}{11}$ **b.** $\frac{5}{8}$ and $\frac{7}{8}$ **c.** $\frac{9}{57}$ and $\frac{12}{57}$ **d.** $\frac{-11}{15}$ and $\frac{9}{15}$ **e.** $\frac{34}{91}$ and $\frac{42}{91}$

2. Select the least fraction from each group of fractions.

a. $\frac{5}{7}$ and $\frac{6}{7}$ **b.** $\frac{-9}{11}$ and $\frac{7}{11}$ **c.** $\frac{14}{17}$ and $\frac{16}{17}$ **d.** $\frac{5}{27}$, $\frac{8}{27}$ and $\frac{7}{27}$ **e.** $\frac{17}{35}$, $\frac{-21}{35}$ and $\frac{32}{35}$

3. Replace ? by < or > to make a true comparison.

a. $\frac{3}{11}$? $\frac{7}{11}$ **b.** $\frac{9}{31}$? $\frac{6}{31}$ **c.** $\frac{-14}{81}$? $\frac{26}{81}$ **d.** $\frac{43}{44}$? $\frac{37}{44}$ **e.** $\frac{-59}{100}$? $\frac{-61}{100}$ **f.** $\frac{77}{112}$? $\frac{59}{112}$

THE MEANING OF LEAST COMMON MULTIPLE AND LEAST COMMON DENOMINATOR

THE MAIN IDEA

1. The ***least common multiple*** (LCM) of two or more whole numbers is the *least* whole number that is a multiple of each.

   Some multiples of 4 are: 4, 8, 12, 16, 20, 24, 28, 32, 36, . . .

   Some multiples of 6 are: 6, 12, 18, 24, 30, 36, . . .

   Of those shown, the common multiples of 4 and 6 are 12, 24, and 36. The *least* common multiple, LCM, of 4 and 6 is 12.

2. To find the LCM of two or more numbers, work with the greatest of the numbers.
   a. If the other numbers are exact divisors of the greatest number, then the greatest number is the LCM.
   b. If the greatest number is not the LCM, then try, in order, multiples of the greatest number, until you find a multiple that all the other numbers divide exactly.

3. The ***least common denominator*** (LCD) of two or more fractions is the LCM of the denominators.

   The LCD of $\frac{1}{4}$ and $\frac{5}{6}$ is 12, the LCM of 4 and 6.

EXAMPLE 3 Find the LCM of 14 and 21.

Work with the greater number, 21, and see if 14 is an exact divisor of 21.

$$14\overline{)21}\quad \text{quotient } 1,\ 21 - 14 = 7 \leftarrow \text{remainder not } 0$$

Since 14 is not an exact divisor of 21, try 21×2, or 42, the next multiple of 21, and see if 14 is an exact divisor of 42.

$$14\overline{)42}\quad \text{quotient } 3,\ 42 - 42 = 0\ ✔$$

Answer: 42 is the LCM of 14 and 21.

EXAMPLE 4 Find the LCD of $\frac{4}{9}$ and $\frac{5}{12}$.

To find the LCD, find the LCM of 9 and 12.

Work with the greater number, 12.

Try multiples of 12:

9 is not an exact divisor of 12.
9 is not an exact divisor of 24.
9 is an exact divisor of 36.

Answer: 36 is the LCD of $\frac{4}{9}$ and $\frac{5}{12}$.

EXAMPLE 5 Find the LCD of $\frac{1}{4}$, $\frac{2}{5}$, and $\frac{-7}{20}$.

Find the LCM of 4, 5, and 20.

Work with the greatest number, 20. Since 4 is an exact divisor of 20 and 5 is an exact divisor of 20, then 20 is the LCD.

Answer: 20 is the LCD of $\frac{1}{4}$, $\frac{2}{5}$, and $\frac{-7}{20}$.

CLASS EXERCISES

1. Find the least common multiple (LCM) for each group of whole numbers.

a. 4 and 8 **b.** 6 and 15 **c.** 7 and 5 **d.** 27 and 54 **e.** 10 and 12
f. 6, 9, and 12 **g.** 8, 10, and 20 **h.** 3, 4, and 12 **i.** 7, 8, and 112

2. Find the least common denominator (LCD) for each group of fractions.

a. $\frac{3}{5}$ and $\frac{9}{35}$ **b.** $\frac{1}{3}$ and $\frac{5}{12}$ **c.** $\frac{1}{6}$ and $\frac{7}{8}$ **d.** $\frac{7}{10}$ and $\frac{3}{4}$ **e.** $\frac{3}{8}$ and $\frac{-3}{10}$
f. $\frac{9}{24}$, $\frac{11}{36}$, and $\frac{7}{12}$ **g.** $\frac{8}{15}$, $\frac{11}{30}$, and $\frac{3}{5}$ **h.** $\frac{4}{5}$, $\frac{3}{20}$, and $\frac{-1}{10}$ **i.** $\frac{1}{2}$, $\frac{1}{3}$, and $\frac{1}{4}$

COMPARING UNLIKE FRACTIONS

THE MAIN IDEA

1. ***Unlike fractions*** are fractions that do *not* have a common denominator.
2. To compare unlike fractions:
   a. Find the LCD for the fractions.
   b. Change each fraction into an equivalent fraction that has the LCD as the denominator.
   c. Compare the resulting like fractions.
3. A positive mixed number can be rounded to the nearest integer by comparing its fraction part to $\frac{1}{2}$. If the fraction part is greater than or equal to $\frac{1}{2}$, round up to the next integer. If the fraction part is less than $\frac{1}{2}$, round the mixed number down to the integer part.

EXAMPLE 6 Which is greater, $\frac{3}{7}$ or $\frac{11}{28}$?

Find the LCD. Since 7 is an exact divisor of 28, the LCD is 28.

Rewrite $\frac{3}{7}$ as an equivalent fraction with denominator 28. $\frac{3 \times 4}{7 \times 4} = \frac{12}{28}$

Since $\frac{12}{28}$ and $\frac{11}{28}$ are like fractions, the fraction with the greater numerator is the greater fraction.

Answer: $\frac{12}{28}$ or $\frac{3}{7}$ is the greater fraction.

EXAMPLE 7 Roger walked $\frac{7}{10}$ of a mile and Grace walked $\frac{5}{8}$ of a mile. Who walked the shorter distance?

To find the LCD, try multiples of 10. 8 is not an exact divisor of 10, 20, or 30. The LCD is 40.

Rewrite $\frac{7}{10}$ and $\frac{5}{8}$ with denominator 40. $\frac{7 \times 4}{10 \times 4} = \frac{28}{40}$ $\frac{5 \times 5}{8 \times 5} = \frac{25}{40}$

Compare $\frac{28}{40}$ and $\frac{25}{40}$. The smaller fraction is $\frac{25}{40}$ or $\frac{5}{8}$.

Answer: Grace walked the shorter distance.

EXAMPLE 8 Round each mixed number to the nearest whole number.

A fraction is less than $\frac{1}{2}$ if the numerator is less than half of the denominator.

	Mixed Number	*Compare Fraction Part with $\frac{1}{2}$*	*Answer*
a.	$3\frac{1}{4}$	$\frac{1}{4} < \frac{1}{2}$ (round down)	3
b.	$4\frac{7}{8}$	$\frac{7}{8} > \frac{1}{2}$ (round up)	5
c.	$8\frac{1}{2}$	$\frac{1}{2} = \frac{1}{2}$ (round up)	9

CLASS EXERCISES

1. Select the greatest fraction from each group of fractions.

a. $\frac{2}{5}$ and $\frac{12}{35}$ **b.** $\frac{7}{12}$ and $\frac{5}{8}$ **c.** $\frac{15}{36}$ and $\frac{21}{24}$ **d.** $\frac{-3}{5}$ and $\frac{1}{2}$ **e.** $\frac{-1}{2}$ and 0

f. $\frac{3}{5}, \frac{7}{15}, \frac{28}{45}$ **g.** $\frac{5}{7}, \frac{13}{14}, \frac{10}{21}$ **h.** $\frac{1}{2}, \frac{1}{3}, \frac{1}{4}$ **i.** $\frac{-1}{2}, \frac{-1}{3}, \frac{-1}{4}$

2. Replace ? by < or > to make a true comparison.

a. $\frac{2}{3}$? $\frac{3}{4}$ **b.** $\frac{3}{8}$? $\frac{7}{16}$ **c.** $\frac{23}{25}$? $\frac{7}{10}$ **d.** $\frac{13}{20}$? $\frac{5}{8}$ **e.** $\frac{17}{36}$? $\frac{5}{12}$ **f.** $\frac{3}{5}$? $\frac{5}{9}$ **g.** $\frac{-1}{3}$? $\frac{-7}{8}$

3. James ran $\frac{8}{10}$ of a mile, and Mario ran $\frac{7}{8}$ of a mile. Who ran farther?

4. One bottle contained $\frac{3}{4}$ of a quart of juice, and another bottle contained $\frac{6}{10}$ of a quart of juice. Which bottle contained less juice?

5. Round each mixed number to the nearest whole number.

a. $6\frac{1}{3}$ **b.** $9\frac{5}{12}$ **c.** $12\frac{15}{16}$ **d.** $15\frac{3}{4}$ **e.** $20\frac{1}{2}$

6. To compare two unlike fractions, Jamil does not always use the LCD. To compare $\frac{5}{6}$ and $\frac{7}{10}$, he writes both fractions with a denominator of 6×10, or 60. Will his method always work? Explain.

HOMEWORK EXERCISES

1. Select the greatest fraction from each group of fractions.

 a. $\frac{3}{17}$ and $\frac{9}{17}$ **b.** $\frac{5}{16}$ and $\frac{3}{16}$ **c.** $\frac{4}{9}$ and $\frac{7}{9}$ **d.** $\frac{21}{51}, \frac{13}{51}, \frac{-24}{51}$ **e.** $\frac{47}{101}, \frac{39}{101}, \frac{42}{101}$

2. Select the least fraction from each group of fractions.

 a. $\frac{9}{11}$ and $\frac{6}{11}$ **b.** $\frac{7}{19}$ and $\frac{10}{19}$ **c.** $\frac{14}{27}$ and $\frac{12}{27}$ **d.** $\frac{18}{67}, \frac{22}{67}, \frac{20}{67}$ **e.** $\frac{-81}{91}, \frac{-85}{91}, \frac{-87}{91}$

3. Replace ? by < or > to make a true comparison.

 a. $\frac{3}{5}\ ?\ \frac{2}{5}$ **b.** $\frac{5}{7}\ ?\ \frac{6}{7}$ **c.** $\frac{9}{17}\ ?\ \frac{12}{17}$ **d.** $\frac{18}{25}\ ?\ \frac{-21}{25}$ **e.** $\frac{75}{91}\ ?\ \frac{72}{91}$ **f.** $\frac{50}{99}\ ?\ \frac{53}{99}$

 g. $\frac{81}{111}\ ?\ \frac{71}{111}$ **h.** $\frac{92}{101}\ ?\ \frac{97}{101}$ **i.** $\frac{17}{116}\ ?\ \frac{15}{116}$ **j.** $\frac{25}{83}\ ?\ \frac{-26}{83}$ **k.** $\frac{-19}{72}\ ?\ \frac{-21}{72}$

4. Find the least common multiple (LCM) for each group of integers.

 a. 3 and 6 **b.** 7 and 21 **c.** 8 and 12 **d.** 20 and 15 **e.** 18 and 27

 f. 24 and 16 **g.** 10, 20, and 25 **h.** 6, 15, and 18 **i.** 4, 6, and 24

5. Find the least common denominator (LCD) for each group of fractions.

 a. $\frac{5}{12}$ and $\frac{3}{4}$ **b.** $\frac{7}{20}$ and $\frac{4}{5}$ **c.** $\frac{11}{18}$ and $\frac{25}{36}$ **d.** $\frac{-2}{3}$ and $\frac{1}{2}$ **e.** $\frac{5}{8}$ and $\frac{5}{6}$

 f. $\frac{7}{10}$ and $\frac{8}{15}$ **g.** $\frac{5}{6}, \frac{2}{9}, \frac{7}{18}$ **h.** $\frac{-7}{12}, \frac{1}{4}, \frac{3}{8}$ **i.** $\frac{5}{6}, \frac{11}{12}, \frac{1}{2}$

6. Select the greatest fraction from each group of fractions.

 a. $\frac{2}{3}$ and $\frac{7}{12}$ **b.** $\frac{-3}{8}$ and $\frac{1}{4}$ **c.** $\frac{5}{18}$ and $\frac{8}{27}$ **d.** $\frac{11}{12}$ and $\frac{4}{5}$ **e.** $\frac{2}{3}$ and $\frac{5}{6}$

 f. $\frac{2}{5}$ and $\frac{3}{8}$ **g.** $\frac{7}{8}$ and $\frac{9}{10}$ **h.** $\frac{3}{10}$ and $\frac{1}{4}$ **i.** $\frac{3}{4}, \frac{11}{16}, \frac{5}{8}$ **j.** $\frac{3}{5}, \frac{2}{3}, \frac{7}{10}$

7. Select the least fraction from each group of fractions.

 a. $\frac{5}{6}$ and $\frac{13}{18}$ **b.** $\frac{7}{9}$ and $\frac{29}{36}$ **c.** $\frac{2}{3}$ and $\frac{3}{5}$ **d.** $\frac{5}{8}$ and $\frac{-7}{10}$ **e.** $\frac{-3}{4}$ and $\frac{1}{2}$

 f. $\frac{8}{27}, \frac{7}{18}, \frac{1}{3}$ **g.** $\frac{5}{6}, \frac{2}{3}, \frac{7}{8}$ **h.** $\frac{3}{50}, \frac{2}{25}, \frac{4}{100}$ **i.** $\frac{2}{7}, \frac{2}{8}, \frac{2}{9}$

8. Replace ? by < or > to make a true comparison.

a. $\frac{3}{5}$? $\frac{7}{10}$ **b.** $\frac{1}{2}$? $\frac{3}{8}$ **c.** $\frac{11}{20}$? $\frac{3}{4}$ **d.** $\frac{9}{16}$? $\frac{2}{3}$ **e.** $\frac{-2}{5}$? $\frac{-3}{8}$ **f.** $\frac{7}{9}$? $\frac{5}{7}$

g. $\frac{7}{12}$? $\frac{3}{5}$ **h.** $\frac{5}{18}$? $\frac{1}{4}$ **i.** $\frac{7}{15}$? $\frac{5}{12}$ **j.** $\frac{-1}{2}$? $\frac{3}{4}$ **k.** $\frac{-1}{2}$? $\frac{-3}{4}$ **l.** $\frac{-1}{2}$? $\frac{-7}{8}$

9. Helen memorized $\frac{4}{5}$ of a vocabulary list, and Bob memorized $\frac{5}{8}$ of the list. Who memorized more of the list?

10. Which oil tank is less full, one that is $\frac{5}{16}$ full or one that is $\frac{1}{4}$ full?

11. Between any two rational numbers, there are many other rational numbers.

For example, to name a rational number between $\frac{3}{5}$ and $\frac{9}{10}$, write the fractions with a common denominator, $\frac{6}{10}$ and $\frac{9}{10}$. Then note that $\frac{7}{10}$ or $\frac{8}{10}$ can be named as a rational number between $\frac{3}{5}$ and $\frac{9}{10}$.

To name a rational number between $\frac{5}{8}$ and $\frac{3}{4}$, changing $\frac{3}{4}$ to $\frac{6}{8}$ does not allow you to fit a number between $\frac{5}{8}$ and $\frac{6}{8}$. However, you can see a fit if you use a greater denominator. That is, $\frac{5}{8} = \frac{10}{16}$ and $\frac{6}{8} = \frac{12}{16}$, showing $\frac{11}{16}$ as a rational number between $\frac{5}{8}$ and $\frac{3}{4}$.

Name one rational number between each pair of rational numbers.

a. $\frac{3}{8}$ and $\frac{3}{4}$ **b.** $\frac{1}{2}$ and 1 **c.** $\frac{1}{4}$ and $\frac{3}{8}$ **d.** $\frac{-3}{4}$ and $\frac{-1}{2}$

12. When you add 1 to an integer, the result is the next integer. Is this true for rational numbers? Explain.

13. Round each mixed number to the nearest integer.

a. $7\frac{3}{4}$ **b.** $2\frac{1}{3}$ **c.** $6\frac{3}{8}$ **d.** $13\frac{1}{2}$ **e.** $25\frac{9}{10}$ **f.** $16\frac{2}{5}$ **g.** $5\frac{4}{9}$

14. Round each mixed number to the nearest integer, and use the rounded numbers to estimate an answer for each expression.

a. $15\frac{1}{4} + 7\frac{1}{2} - 3\frac{3}{4}$ **b.** $32 \div 7\frac{5}{8}$

c. $1\frac{1}{2} \times 11\frac{7}{8} \div 2\frac{2}{3}$ **d.** $3\frac{1}{2} \times \left(8\frac{5}{6} - 3\frac{1}{8}\right)$

15. If 1 is added to the numerator and to the denominator of the fraction $\frac{7}{9}$, is the resulting fraction less than, equal to, or greater than the original fraction?

16. Describe how you can use a calculator to compare two unlike fractions. Use examples to explain.

SPIRAL REVIEW EXERCISES

1. Change $\frac{29}{7}$ to an equivalent mixed number.

2. Perform each of the indicated operations and check.

   a. $2{,}752 + 947$ b. $837 - 592$

   c. 58×43 d. $507 \div 3$

3. Evaluate: $15 + 9 \times 3 - 8$

4. Write five thousand, twenty-one as a numeral.

5. The value of 9 in 792,465 is
   (a) 900,000 (b) 90,000
   (c) 9,000 (d) 900

6. Replace ? by $<$ or $>$ to make a true comparison.

   a. -9 ? -10 b. -115 ? 2

   c. 0 ? -75 d. $-1{,}001$? $-1{,}010$

7. Subtract -50 from 92.

8. The greatest common factor of 56 and 28 is
   (a) 7 (b) 14 (c) 28 (d) 56

9. Write a key sequence to display the new balance in a checking account that had an opening balance of \$500, after the amount of \$121.50 was deducted to cover a check.

10. Juan pitched $\frac{22}{3}$ innings without allowing a hit. Express the number of hitless innings as a mixed number.

11. Mr. Jones, a traveling salesman, is paid \$4 a mile traveling expenses by his company. If he traveled 178 miles, how much was he paid?

12. A baseball bat costs \$11 and a baseball glove costs \$14. What is the total cost of supplying bats and gloves for the 15 members of the Bantam little league team?

13. In the sentence

$$\frac{\square}{17} = \frac{\square}{23}$$

replace each $\square$ with the same number so that the resulting fractions are equivalent.

UNIT 5-5 Simplifying Fractions

THE MAIN IDEA

1. To write a fraction in *simplest form*:
   a. Find a number that is a common factor (exact divisor) of the numerator and the denominator.
   b. Divide the numerator and the denominator by this common factor.
   c. Repeat this process until 1 is the only common factor of the numerator and the denominator.
2. If the numerator and the denominator of a fraction are each divided by the greatest common factor (GCF), the resulting fraction will be in simplest form.

EXAMPLE 1 Simplify: $\frac{24}{36}$

Divide the numerator and the denominator by the greatest common factor, 12.

$$\frac{24 \div 12}{36 \div 12} = \frac{2}{3}$$

EXAMPLE 2 For each fraction, find the GCF of the numerator and the denominator, and write the fraction in simplest form.

	Fraction	*GCF*	*Simplified*
a.	$\frac{8}{20}$	4	$\frac{2}{5}$
b.	$\frac{36}{48}$	12	$\frac{3}{4}$
c.	$\frac{-70}{210}$	70	$\frac{-1}{3}$

EXAMPLE 3 A baseball team won 18 out of 27 games. Express in simplest form the fractional part of the games won.

THINKING ABOUT THE PROBLEM

The key words "out of" tell you to compare the numbers 18 and 27.

Translate the words into a fraction.

$\frac{18}{27}$ ← number of games won / total number of games

Simplify the fraction by dividing the numerator and the denominator by the GCF.

$$\frac{18 \div 9}{27 \div 9} = \frac{2}{3}$$

Answer: The team won $\frac{2}{3}$ of the games.

CLASS EXERCISES

1. Name the GCF of the numerator and the denominator of each fraction. Then, write each fraction in simplest form.

   a. $\frac{4}{8}$ **b.** $\frac{42}{63}$ **c.** $\frac{24}{36}$ **d.** $\frac{56}{72}$ **e.** $\frac{77}{99}$ **f.** $\frac{-120}{130}$ **g.** $\frac{60}{72}$ **h.** $\frac{-16}{64}$ **i.** $\frac{-80}{200}$ **j.** $\frac{70}{105}$

2. Simplify:

   a. $\frac{3}{6}$ **b.** $\frac{11}{33}$ **c.** $\frac{8}{12}$ **d.** $\frac{20}{25}$ **e.** $\frac{9}{18}$ **f.** $\frac{-6}{10}$ **g.** $\frac{-12}{48}$ **h.** $\frac{21}{28}$ **i.** $\frac{-75}{100}$ **j.** $\frac{88}{121}$

3. Jane delivers newspapers to 72 homes on her paper route. If she has already delivered papers to 60 homes, express as a fraction in simplest form the part of the route that she has completed.

4. Roberto has $100 and plans to spend $55 on a sports jacket. Express as a fraction in simplest form the part of his money that he plans to spend on the sports jacket.

HOMEWORK EXERCISES

1. Write the fraction in simplest form.

   a. $\frac{20}{25}$ **b.** $\frac{18}{36}$ **c.** $\frac{36}{48}$ **d.** $\frac{40}{60}$ **e.** $\frac{42}{54}$ **f.** $\frac{40}{64}$ **g.** $\frac{75}{100}$ **h.** $\frac{-24}{32}$

   i. $\frac{72}{81}$ **j.** $\frac{48}{72}$ **k.** $\frac{-75}{120}$ **l.** $\frac{50}{200}$ **m.** $\frac{90}{120}$ **n.** $\frac{-54}{90}$ **o.** $\frac{-80}{400}$ **p.** $\frac{15}{25}$

2. Simplify:

   a. $\frac{12}{14}$ **b.** $\frac{15}{20}$ **c.** $\frac{30}{48}$ **d.** $\frac{25}{50}$ **e.** $\frac{20}{24}$ **f.** $\frac{21}{28}$ **g.** $\frac{-40}{100}$ **h.** $\frac{30}{32}$

   i. $\frac{48}{96}$ **j.** $\frac{50}{75}$ **k.** $\frac{-12}{18}$ **l.** $\frac{77}{88}$ **m.** $\frac{42}{63}$ **n.** $\frac{-70}{90}$ **o.** $\frac{-40}{45}$ **p.** $\frac{75}{100}$

3. Franklin answered 42 of 50 questions correctly. Express as a fraction in simplest form the part of the test that he got correct.

4. Sheila is saving money to buy a portable CD player that costs $200. If she has saved $150, express as a fraction in simplest form the part of the money that she has already saved.

5. Match each fraction in Column I with the letter of the number line in Column II that represents the graph of the rational number in simplest form.

	Column I		***Column II***
a.	$\frac{28}{16}$	(A)	 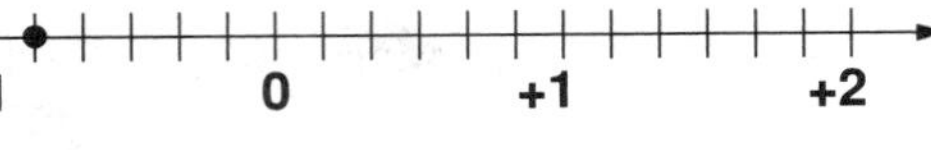
b.	$\frac{-35}{42}$	(B)	
c.	$\frac{33}{77}$	(C)	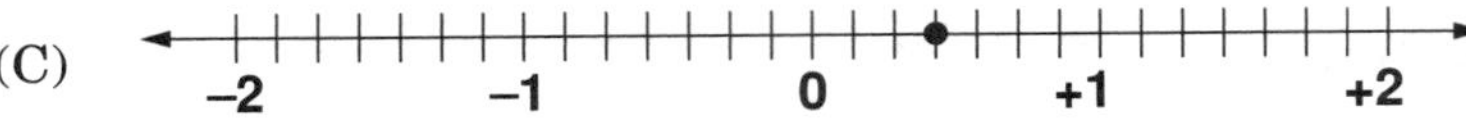

SPIRAL REVIEW EXERCISES

1. Round each of the decimals to the nearest hundredth: **a.** 3.612 **b.** 52.535

2. Replace ? by < or > to make a true statement: **a.** 0.8 ? 6.1 **b.** 3.52 ? 2.68

3. Add: 312.84 + 29.13 + 5.194

4. Explain why $\frac{3}{7}$ cannot be written as an equivalent fraction having a denominator of 24.

5. Find the least common multiple of 14 and 21.

6. Which calculation, if done with an entry-order calculator, would require the use of the memory?

(a) $53 + 19 \times 12$ (b) $18 + 17 \div 15$
(c) $73\ (14 + 85)$ (d) $48 \div (14 + 16)$

7. $40 \div (-8)$ is (a) -320 (b) -5 (c) $+5$ (d) -6

8. Which fraction is *not* equivalent to $\frac{5}{8}$?

(a) $\frac{15}{24}$ (b) $\frac{35}{56}$ (c) $\frac{20}{32}$ (d) $\frac{45}{81}$

9. 27,563,418 rounded to the nearest hundred thousand is

(a) 28,000,000 (b) 27,000,000
(c) 27,600,000 (d) 27,500,000

10. Mr. Goodstone makes $8 profit for each tire sold. What is the amount of profit if 86 tires are sold?

11. A dealer plans to pack 2,080 Ping-Pong balls in 40 cases so that each case will contain the same number of Ping-Pong balls. How many balls will be in each case?

12. Write eight hundred thousand, two hundred five as a numeral.

13. The value of $5 + 8 \times 3 + 2$ is

(a) 65 (b) 31 (c) 41 (d) 45

14. Determine the relationship between the First and Second Numbers in the table, and find the missing numbers.

First Number	2	3	10	12	?	−4
Second Number	7	9	23	?	33	?

UNIT 5-6 Writing Fractions in Decimal Form; Writing Decimals in Fraction Form

WRITING FRACTIONS IN DECIMAL FORM

THE MAIN IDEA

To write a fraction in decimal form, divide the numerator by the denominator.

1. If the division ends, the decimal is called a ***terminating decimal.***
2. When the division does not end, we have a ***repeating decimal.*** A bar is sometimes used to show the digits that repeat.

EXAMPLE 1 Write $\frac{5}{8}$ in decimal form.

Divide the numerator by the denominator. $8\overline{)5}$

Insert a decimal point in the dividend and quotient. $8\overline{)5.}$

Insert as many zeros as necessary in the dividend to complete the division.

$$\begin{array}{r} 0.625 \\ 8\overline{)5.000} \\ \underline{4\,8} \\ 20 \\ \underline{16} \\ 40 \\ \underline{40} \\ 0 \end{array}$$

Answer: $\frac{5}{8} = 0.625$

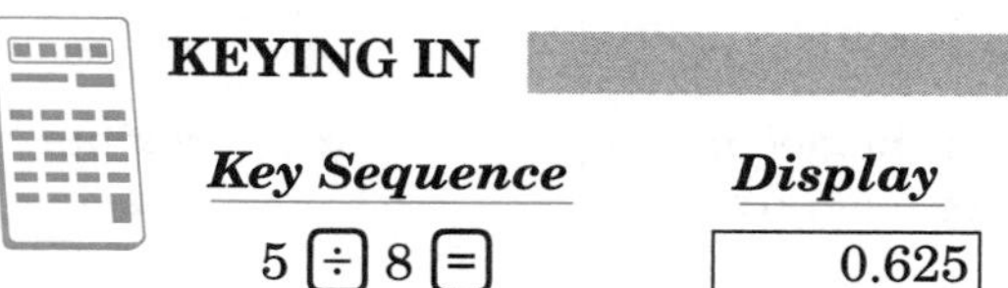

KEYING IN

Key Sequence	*Display*
5 ÷ 8 =	0.625

Most calculators will display up to 8 digits. Because the decimal equivalent of $\frac{5}{8}$ is a terminating decimal with only 3 decimal places, the number on the display is exactly the decimal form of $\frac{5}{8}$.

EXAMPLE 2 Write $\frac{13}{3}$ as a repeating decimal.

Since $\frac{13}{3}$ is an improper fraction, the equivalent decimal has an integer part and a fraction part.

The digit 3 repeats. The remainder is never 0.

$$\begin{array}{r} 4.333 \\ 3\overline{)13.000} \\ \underline{12} \\ 1\,0 \\ \underline{9} \\ 10 \\ \underline{9} \\ 10 \end{array}$$

Answer: $\frac{13}{3} = 4.\overline{3}$

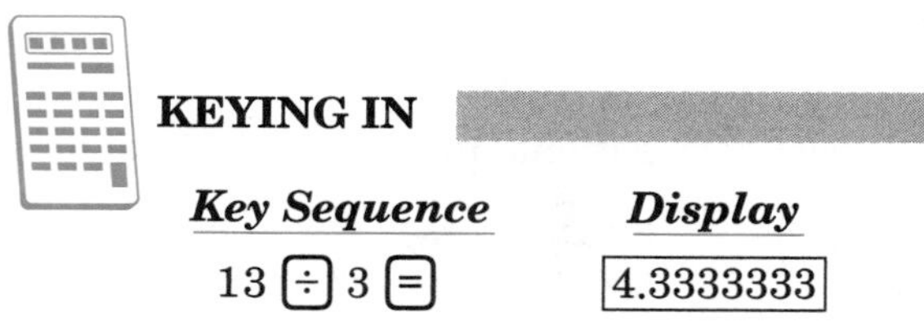

KEYING IN

Key Sequence	*Display*
13 ÷ 3 =	4.3333333

Because $\frac{13}{3}$ is equivalent to a repeating decimal and the calculator can display only 7 decimal places, the number on the display is an *approximation* of $\frac{13}{3}$.

EXAMPLE 3 Use a calculator to write $\frac{21}{47}$ in decimal form correct to the nearest hundredth.

Key Sequence	*Display*
21 [÷] 47 [=]	0.4468085
The symbol ≈ means "is about equal to."	$0.446805 \approx 0.45$

Answer: $\frac{21}{47} \approx 0.45$

EXAMPLE 4 Use a calculator to write the mixed number $6\frac{13}{20}$ in decimal form.

First, divide the numerator of the fraction part by the denominator, and then add the integer part.

Key Sequence	*Display*
13 [÷] 20 [+] 6 [=]	6.65

Answer: $6\frac{13}{20} = 6.65$

EXAMPLE 5 Use a calculator to write $\frac{23}{25}$, $\frac{15}{17}$, and $\frac{18}{19}$ in increasing order.

Find the decimal equivalent or approximation for each fraction.

	Key Sequence	*Display*
$\frac{23}{25}$	23 [÷] 25 [=]	0.92
$\frac{15}{17}$	15 [÷] 17 [=]	0.8823529
$\frac{18}{19}$	18 [÷] 19 [=]	0.9473684

Compare the decimals and rank them in increasing order.

0.8823529, 0.92, 0.9473684

Replace the decimals with the original fractions.

$\frac{15}{17}, \frac{23}{25}, \frac{18}{19}$ *Ans.*

CLASS EXERCISES

1. Write each rational number in decimal form.

a. $\frac{3}{4}$ **b.** $\frac{7}{16}$ **c.** $\frac{14}{25}$ **d.** $\frac{41}{8}$ **e.** $\frac{2}{5}$ **f.** $9\frac{1}{2}$ **g.** $\frac{-7}{40}$ **h.** $\frac{51}{20}$ **i.** $-5\frac{1}{4}$

2. Write each fraction as a repeating decimal.

a. $\frac{1}{3}$ **b.** $\frac{5}{6}$ **c.** $\frac{9}{11}$ **d.** $\frac{2}{9}$ **e.** $\frac{11}{3}$ **f.** $\frac{-2}{3}$ **g.** $\frac{10}{11}$ **h.** $\frac{-4}{9}$ **i.** $\frac{11}{9}$

3. Write $\frac{2}{3}$ in decimal form correct to the nearest hundredth.

4. Write $\frac{9}{17}$ in decimal form correct to the nearest thousandth.

5. Use a calculator to write $\frac{37}{59}$ in decimal form correct to the nearest thousandth.

6. Use a calculator to write $12\frac{13}{16}$ in decimal form.

7. Use a calculator to write $\frac{38}{79}$, $\frac{21}{43}$, and $\frac{54}{111}$ in decreasing order.

8. It is helpful to memorize some fraction-decimal equivalents.

Fraction Form	$\frac{1}{2}$	$\frac{1}{3}$	$\frac{1}{4}$	$\frac{1}{5}$	$\frac{1}{10}$	$\frac{1}{20}$
Decimal Form	0.5	0.333. . .	0.25	0.2	0.1	0.05

You can use the values in the table above to find other equivalents mentally.

$$\frac{3}{4} = 3 \times \frac{1}{4} = 3 \times 0.25 = 0.75$$

Use this method to find the decimal form for: **a.** $\frac{5}{2}$ **b.** $\frac{7}{20}$ **c.** $\frac{2}{3}$ **d.** $\frac{3}{5}$

WRITING DECIMALS IN FRACTION FORM

THE MAIN IDEA

1. Terminating decimals and repeating decimals are rational numbers and may be written in fraction form.
2. To write a decimal in fraction form, read the decimal using its word name.
   a. The denominator of the fraction is the place name that you read.
   b. The numerator of the fraction is the whole number that your read.
   c. Simplify the fraction by dividing the numerator and the denominator by common factors.

EXAMPLE 6 Write 0.059 in fraction form.

Read 0.059 as "fifty-nine thousandths."

"Thousandths" is the denominator of the fraction. $\frac{}{1,000}$

"Fifty-nine" is the numerator of the fraction. $\frac{59}{1,000}$

Answer: $0.059 = \frac{59}{1,000}$

EXAMPLE 7 Write 0.45 as a fraction in simplest form.

Read 0.45 as "forty-five hundredths" and write it as a fraction. $\frac{45}{100}$

Simplify $\frac{45}{100}$ by dividing the numerator and denominator by 5. $\frac{45 \div 5}{100 \div 5} = \frac{9}{20}$

Answer: $0.45 = \frac{9}{20}$

EXAMPLE 8 Write 1.75 as a mixed number.

Write the integer part. 1

Write the decimal part as a fraction. $0.75 = \frac{75}{100}$

Simplify the fraction. $\frac{75 \div 25}{100 \div 25} = \frac{3}{4}$

Answer: $1.75 = 1\frac{3}{4}$

EXAMPLE 9 Write –5.4 as a mixed number.

Write the integer part. –5

Write the decimal part as a fraction. $0.4 = \frac{4}{10}$

Simplify the fraction. $\frac{4 \div 2}{10 \div 2} = \frac{2}{5}$

Answer: $-5.4 = -5\frac{2}{5}$

CLASS EXERCISES

1. Write each decimal in fraction form.

a. –0.3 **b.** 0.051 **c.** 0.417 **d.** 0.019 **e.** –0.27 **f.** 0.0793 **g.** 0.00027 **h.** 0.3007

2. Write each decimal as a fraction in simplest form.

a. 0.6 **b.** 0.25 **c.** 0.125 **d.** 0.550 **e.** 0.18 **f.** –0.95 **g.** 0.875 **h.** 0.048

3. Write each decimal as a mixed number.

a. 3.4 **b.** 5.27 **c.** 2.5 **d.** –8.25 **e.** 10.03 **f.** –9.2 **g.** 8.42 **h.** 5.8

4. Which of the following mixed numbers is equivalent to –4.6?

(a) $-4\frac{1}{6}$ (b) $-6\frac{4}{10}$ (c) $-5\frac{6}{10}$ (d) $-4\frac{3}{5}$

5. Raymond ran 3.85 miles. This distance is the same as

(a) $3\frac{4}{5}$ miles (b) $3\frac{8}{10}$ miles (c) $2\frac{17}{20}$ miles (d) $3\frac{17}{20}$ miles

6. Which of the following is *not* equivalent to $\frac{-36}{5}$?

(a) $-7\frac{1}{5}$ (b) -7.2 (c) $-7 + \left(\frac{-1}{5}\right)$ (d) $-7 - \left(\frac{-1}{5}\right)$

7. a. Use a calculator to find the decimal equivalents for $\frac{1}{9}$, $\frac{2}{9}$, and $\frac{3}{9}$. Look for a pattern.

b. Make a *conjecture* (reasonable prediction) about the decimal equivalent for $\frac{4}{9}$.

c. Test your conjecture on the calculator.

d. Still using the same pattern, make a conjecture about a fraction that is equivalent to 0.77777…. Test your conjecture on the calculator.

8. A calculator displays answers in decimal form only. If you need to write an answer in fraction form, it will help if you know the equivalents shown in the table on page 211, Exercise 8.

For example, because 0.8 is 4×0.2 and $0.2 = \frac{1}{5}$, it follows that 0.8 is $4 \times \frac{1}{5}$, or $\frac{4}{5}$.

Write the number on each display as a fraction or mixed number.

a. [0.6] b. [1.3333333] c. [4.5] d. [– 7.35]

HOMEWORK EXERCISES

1. Write each rational number in decimal form.

a. $\frac{5}{8}$ b. $\frac{11}{25}$ c. $\frac{23}{50}$ d. $\frac{22}{5}$ e. $\frac{-3}{5}$ f. $\frac{1}{4}$ g. $5\frac{1}{2}$ h. $\frac{15}{200}$ i. $\frac{-5}{10}$

j. $\frac{-13}{20}$ k. $\frac{3}{16}$ l. $3\frac{7}{8}$ m. $\frac{40}{16}$ n. $\frac{3}{4}$ o. $7\frac{1}{8}$ p. $3\frac{2}{5}$ q. $\frac{34}{10}$ r. $\frac{-1}{8}$

2. Write each fraction as a repeating decimal.

a. $\frac{2}{3}$ b. $\frac{4}{9}$ c. $4\frac{10}{11}$ d. $\frac{28}{9}$ e. $\frac{-5}{11}$ f. $\frac{3}{9}$ g. $14\frac{1}{11}$ h. $\frac{-40}{3}$

3. Write $\frac{5}{17}$ in decimal form correct to the nearest hundredth.

4. Write $\frac{6}{11}$ in decimal form correct to the nearest thousandth.

5. Use a calculator to write $\frac{43}{91}$ in decimal form correct to the nearest hundredth.

6. Use a calculator to write $21\frac{7}{32}$ in decimal form.

7. Use a calculator to write $\frac{59}{176}$, $\frac{46}{139}$, and $\frac{37}{110}$ in decreasing order.

8. Write each decimal in fraction form.

a. 0.29 **b.** 0.127 **c.** 0.509 **d.** −0.31 **e.** 0.0293 **f.** −0.7 **g.** 0.09 **h.** 0.059 **i.** 0.003

9. Write each decimal as a fraction in simplest form.

a. 0.4 **b.** 0.08 **c.** 0.375 **d.** −0.50 **e.** 0.8750 **f.** 0.24 **g.** −0.8 **h.** 0.58
i. 0.064 **j.** −0.05 **k.** 0.75 **l.** 0.35 **m.** 0.625 **n.** 0.006 **o.** 0.025 **p.** 0.15

10. Write each decimal as a mixed number.

a. 8.2 **b.** 6.45 **c.** 5.24 **d.** −3.5 **e.** 9.37 **f.** −4.8 **g.** 3.1 **h.** 5.04

11. 0.005 written in fraction form is (a) $\frac{1}{2}$ (b) $\frac{1}{20}$ (c) $\frac{1}{200}$ (d) $\frac{1}{2,000}$

12. 0.375 names the same point on the number line as (a) $\frac{1}{8}$ (b) $\frac{1}{4}$ (c) $\frac{3}{8}$ (d) $\frac{5}{8}$

13. 9.55 has the same value as (a) $9\frac{1}{2}$ (b) $9\frac{5}{11}$ (c) $9\frac{11}{20}$ (d) $9\frac{3}{4}$

14. Rosa traveled 23.15 miles. This distance is the same as

(a) $23\frac{3}{20}$ (b) $23\frac{3}{10}$ (c) $23\frac{3}{5}$ (d) $23\frac{3}{4}$

15. Which of the following is *not* equivalent to $-8\frac{1}{4}$?

(a) −8.25 (b) $\frac{-33}{4}$ (c) $-8+\left(\frac{-1}{4}\right)$ (d) $-8-\left(\frac{-1}{4}\right)$

16. a. Use a calculator to find the decimal equivalents for $\frac{1}{99}$, $\frac{7}{99}$, $\frac{14}{99}$, and $\frac{98}{99}$. Look for a pattern.

b. Make a conjecture about the decimal equivalents for $\frac{25}{99}$ and $\frac{74}{99}$.

c. Test your conjecture on the calculator.

d. Using the same pattern, make a conjecture about a fraction that is equivalent to 0.62626262 .... Test your conjecture on the calculator.

17. Use the guess-and-check strategy and a calculator to write an equivalent fraction for each repeating decimal.

a. 0.5555555 ... **b.** $0.\overline{23}$ **c.** 0.3453453 ...

18. Decimals that neither terminate nor repeat the identical group of digits cannot be written as a quotient of two integers. Such decimals are called ***irrational numbers***.

Look for a pattern in the digits of each of the following irrational numbers. Then, use the pattern to write the next five digits of each number.

a. 0.10110011100011110000 ... **b.** −1.23456789101112131415 ...

19. Make a large copy of this diagram.

a. Write the name of each of the following kinds of real numbers on the appropriate blank (a – e) of your diagram:
(1) Integers
(2) Whole Numbers
(3) Irrational Numbers
(4) Terminating Decimals
(5) Repeating Decimals

b. For each section of your diagram, write two examples of each kind of real number.

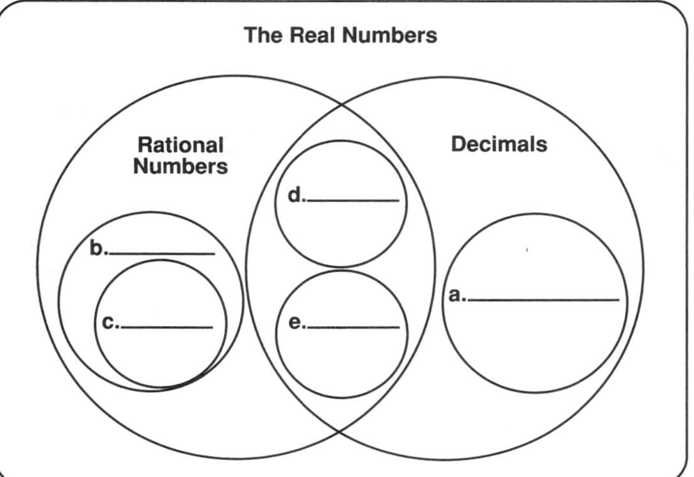

SPIRAL REVIEW EXERCISES

1. Write each word name as a decimal.
a. seventeen thousandths
b. three tenths
c. one hundred and twenty-three hundredths
d. five and five hundredths

2. 0.003 is read as
(a) three tenths
(b) three hundredths
(c) three thousandths
(d) three ten-thousandths

3. Round 0.481 to the nearest tenth.

4. The sum of −14 and +14 is
(a) 0 (b) −28 (c) 28 (d) −196

5. Find the product of −12 and −4.

6. When 5 is subtracted from −8, the result is
(a) −13 (b) −3 (c) 13 (d) 3

7. In 6,417,235, what is the value of the digit 4?
(a) 400,000 (b) 4,000
(c) 40,000 (d) 400

8. Use your calculator to find the total cost of 3 notebooks at a cost of $1.39 each, 5 rolls of tape at $1.79 each, and 4 markers at $2.49 each.

9. Perform the indicated operations: $(47 + 3) \times (24 - 9)$

10. Add: 173 + 21,460 + 8

11. Fast Burgers sells 1,250 burgers each day. How many burgers are sold in a seven-day week?

12. The population of Yorkville is 22,430 and the population of Carlsbury is 18,195. How many more people live in Yorkville than live in Carlsbury?

13. Is 0.00734 closer to $\frac{73}{10,000}$ or $\frac{74}{10,000}$?

UNIT 6–1 Multiplying Fractions; Multiplying a Fraction and an Integer

MULTIPLYING FRACTIONS

THE MAIN IDEA

To multiply two fractions:

1. Try to simplify. If possible, divide a numerator and a denominator by a common factor.
2. Multiply the resulting numerators.
3. Multiply the resulting denominators.
4. If possible, simplify the resulting fraction.

EXAMPLE 1 Multiply: $\frac{3}{5} \times \frac{2}{7}$

No simplification is possible.

Multiply the numerators and multiply the denominators.

$$\frac{3}{5} \times \frac{2}{7} = \frac{3 \times 2}{5 \times 7} = \frac{6}{35} \quad \textit{Ans.}$$

EXAMPLE 2 Multiply: $\frac{-5}{8} \times \frac{4}{15}$

There is a common factor of 4. Divide 8 and 4 by 4.

$$\frac{-5}{\overset{}{\underset{2}{\cancel{8}}}} \times \frac{\overset{1}{\cancel{4}}}{15}$$

There is a common factor of 5. Divide 5 and 15 by 5.

$$\frac{\overset{-1}{\cancel{-5}}}{\underset{2}{\cancel{8}}} \times \frac{\overset{1}{\cancel{4}}}{\underset{3}{\cancel{15}}}$$

Multiply the results.

$$\frac{-1 \times 1}{2 \times 3} = \frac{-1}{6} \quad \textit{Ans.}$$

EXAMPLE 3 Find the product of $\frac{42}{75}$ and $\frac{50}{63}$.

"Product" tells you to multiply.

$$\frac{42}{75} \times \frac{50}{63}$$

To simplify, divide 42 and 63 by 7. Divide 75 and 50 by 25.

$$\frac{\overset{6}{\cancel{42}}}{\underset{3}{\cancel{75}}} \times \frac{\overset{2}{\cancel{50}}}{\underset{9}{\cancel{63}}}$$

Before multiplying, you can simplify further. Divide 6 and 3 by 3.

$$\frac{\overset{2}{\cancel{\overset{6}{\cancel{42}}}}}{\underset{1}{\cancel{\underset{3}{\cancel{75}}}}} \times \frac{\overset{2}{\cancel{50}}}{\underset{9}{\cancel{63}}}$$

Multiply the results.

$$\frac{2 \times 2}{1 \times 9} = \frac{4}{9} \quad \textit{Ans.}$$

EXAMPLE 4 Sherry has $\frac{6}{8}$ of a pound of fruit and nut mix. She wants to give half to Mia. Find the amount that Sherry should give to Mia.

THINKING ABOUT THE PROBLEM

A key word in the problem is "of." To find half of $\frac{6}{8}$, multiply $\frac{6}{8}$ by $\frac{1}{2}$.

Write a number phrase. $\frac{1}{2} \times \frac{6}{8}$

To simplify, divide 2 and 6 by 2.
Multiply the results. $\frac{1}{\cancel{2}_{1}} \times \frac{\cancel{6}^{3}}{8} = \frac{3}{8}$

Answer: Sherry should give $\frac{3}{8}$ of a pound of the mix to Mia.

CLASS EXERCISES

1. Multiply. Write answers in simplest form.

a. $\frac{1}{4} \times \frac{2}{5}$ **b.** $\frac{3}{8} \times \frac{2}{3}$ **c.** $\frac{5}{6} \times \frac{3}{7}$ **d.** $\frac{7}{8} \times \frac{16}{21}$ **e.** $\frac{3}{5} \times \frac{5}{3}$ **f.** $\frac{8}{3} \times \frac{27}{16}$

2. a. Find $\frac{3}{7}$ of $\frac{35}{48}$. **b.** Find $\frac{7}{8}$ of $\frac{32}{49}$. **c.** Find $\frac{9}{10}$ of $\frac{45}{72}$. **d.** Find $\frac{1}{2}$ of $\frac{18}{30}$.

3. Jane's house is $\frac{3}{4}$ of a mile from Marcia's house. They agree to meet at a point that is $\frac{1}{2}$ the distance between the two houses. How far is the meeting point from Jane's house?

4. A metal part is $\frac{24}{32}$ of an inch long. A mechanic wants to replace it with a part that is $\frac{2}{3}$ the length. How long is the replacement part?

5. Mr. Norris bought $\frac{8}{10}$ of a pound of meat and wanted to prepare $\frac{1}{4}$ of this amount for his lunch.

a. How much meat should he prepare for lunch?

b. Is the amount of meat to be prepared for lunch more or less than $\frac{1}{2}$ pound?

MULTIPLYING A FRACTION AND AN INTEGER

THE MAIN IDEA

To multiply a fraction and an integer:

1. Write the integer as an improper fraction with a denominator of 1.
2. Use the method for multiplying two fractions.

EXAMPLE 5 Multiply: $20 \times \frac{3}{5}$

Write 20 as a fraction. $\frac{20}{1} \times \frac{3}{5}$

To simplify, divide 20 and 5 by 5. $\frac{\overset{4}{\cancel{20}}}{1} \times \frac{3}{\underset{1}{\cancel{5}}}$

Multiply. $\frac{4 \times 3}{1 \times 1} = \frac{12}{1} = 12$ *Ans.*

EXAMPLE 6 Two-thirds of the people who took a taste test preferred Zappo Orange Juice. If there were 120 people in the test, how many preferred Zappo?

Write a number phrase. $\frac{2}{3} \times 120$

Write 120 as a fraction. $\frac{2}{3} \times \frac{120}{1}$

To simplify, divide 3 and 120 by 3. Then multiply. $\frac{2}{\underset{1}{\cancel{3}}} \times \frac{\overset{40}{\cancel{120}}}{1} = \frac{80}{1} = 80$

Answer: 80 people preferred Zappo.

CLASS EXERCISES

1. Multiply. Write answers in simplest form.

a. $40 \times \frac{1}{5}$ **b.** $\frac{1}{6} \times 72$ **c.** $\frac{2}{3} \times 33$ **d.** $48 \times \frac{3}{8}$ **e.** $100 \times \frac{2}{5}$ **f.** $\frac{5}{6} \times 81$

g. $98 \times \frac{4}{7}$ **h.** $144 \times \frac{7}{12}$ **i.** $\frac{9}{10} \times 100$ **j.** $55 \times \frac{7}{10}$ **k.** $64 \times \frac{5}{24}$ **l.** $108 \times \frac{7}{72}$

2. **a.** Find $\frac{3}{4}$ of 48. **b.** Find $\frac{2}{3}$ of 63. **c.** Find $\frac{7}{8}$ of 40. **d.** Find $\frac{4}{5}$ of 120.

e. Find $\frac{3}{10}$ of 25. **f.** Find $\frac{7}{18}$ of 81. **g.** Find $\frac{3}{4}$ of -40. **h.** Find $\frac{2}{3}$ of -150.

3. How many pounds of ground beef would be needed to make 60 hamburgers if each hamburger weighs $\frac{1}{4}$ of a pound before cooking?

4. Four hundred fifty students voted for the president of the student government. If $\frac{3}{5}$ of these students voted for Carlton Besk, find the number of students that voted for Carlton.

5. In a shipment of 96 automobiles, $\frac{2}{3}$ of them had air conditioning installed. In another shipment of 124 automobiles, $\frac{3}{4}$ of them had air conditioning. What is the total number of automobiles that had air conditioning?

6. Describe the result obtained when a rational number is multiplied

a. by 1 **b.** by –1 **c.** by 0

7. Students at Wedgewood High School raised $4,800 to help fix up a community recreation area. The graph below shows how the different grades contributed to the fund-raising effort.

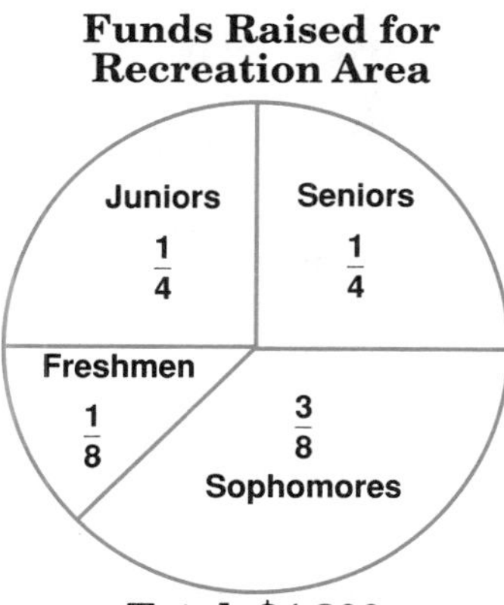

How much money was raised by:

a. the Freshmen **b.** the Sophomores **c.** the Juniors **d.** the Seniors

HOMEWORK EXERCISES

1. Multiply. Write answers in simplest form.

a. $\frac{1}{2} \times \frac{4}{5}$ **b.** $\frac{2}{3} \times \frac{5}{8}$ **c.** $\frac{3}{5} \times \frac{5}{7}$ **d.** $\frac{2}{3} \times \frac{1}{5}$ **e.** $\frac{7}{24} \times \frac{8}{14}$ **f.** $\frac{5}{12} \times \frac{8}{15}$

g. $\frac{12}{24} \times \frac{24}{60}$ **h.** $\frac{9}{10} \times \frac{20}{45}$ **i.** $\frac{-1}{3} \times \frac{9}{10}$ **j.** $\frac{-3}{4} \times \frac{-4}{5}$ **k.** $\frac{3}{8} \times \frac{-9}{10}$ **l.** $\frac{-8}{9} \times \frac{-15}{16}$

2. Which is greater, $\frac{2}{3}$ of $\frac{1}{5}$ or $\frac{2}{5}$ of $\frac{1}{3}$? Explain your answer.

3. Find each product. Write answers in simplest form.

a. $5 \times \frac{1}{8}$ **b.** $\frac{2}{3} \times 4$ **c.** $12 \times \frac{1}{5}$ **d.** $\frac{3}{4} \times 11$ **e.** $\frac{1}{3} \times 24$ **f.** $50 \times \frac{2}{5}$

g. $33 \times \frac{3}{22}$ **h.** $\frac{5}{18} \times 63$ **i.** $\frac{1}{2} \times -40$ **j.** $6 \times \frac{-2}{3}$ **k.** $\frac{3}{4} \times 0$ **l.** $\frac{1}{8} \times -8$

4. **a.** Find $\frac{5}{8}$ of 40. **b.** Find $\frac{3}{5}$ of 45. **c.** Find $\frac{9}{10}$ of 80. **d.** Find $\frac{3}{7}$ of 42. **e.** Find $\frac{2}{3}$ of $\frac{4}{5}$.

f. Find $\frac{1}{8}$ of $\frac{4}{5}$. **g.** Find $\frac{2}{5}$ of -50. **h.** Find $\frac{5}{28}$ of 42. **i.** Find $\frac{2}{75}$ of 100. **j.** Find $\frac{1}{3}$ of -3.

5. As with multiplication of integers, multiplication of rational numbers is also distributive over both addition and subtraction. Use the distributive properties to complete each expression.

a. $\frac{1}{2} \times \left(8 + \frac{3}{4}\right) = \left(\frac{1}{2} \times 8\right) + \left(\frac{1}{2} \times ?\right)$

b. $\frac{2}{3} \times \left(9 - \frac{3}{4}\right) = \left(\frac{2}{3} \times 9\right) - (? \times ?)$

c. $\left(\frac{3}{4} \times 8\right) + \left(\frac{3}{4} \times \frac{2}{3}\right) = \frac{3}{4} \times (?)$

6. **a.** Multiply: $\frac{1}{2} \times \frac{1}{4}$

b. Use a calculator to write $\frac{1}{2}$ and $\frac{1}{4}$ in equivalent decimal form.

c. Multiply the decimals, using the calculator.

d. How is the answer in part **a** related to the answer in part **c**?

7. Write a calculator key sequence that can be used to multiply:

a. $\frac{3}{4} \times 140$ **b.** $\frac{5}{8} \times \frac{3}{10}$ **c.** $3\frac{1}{8} \times 5$

8. Two-thirds of a road that is $\frac{9}{10}$ of a mile long is to be repaved. Is the section to be repaved greater than or less than $\frac{1}{2}$ mile?

9. Sixty beads, each $\frac{3}{8}$ of an inch long, are strung together on a wire. What is the total length of the 60 beads?

10. Of the 640 students at Oakdale School, $\frac{2}{5}$ are sophomores. How many students are not sophomores?

11. Which of the following properties of multiplication of integers are also properties of multiplication of rational numbers?

a. closure **b.** commutative property **c.** associative property

12. Carlos has 380 stamps in his collection. One-fifth of his collection consists of stamps he bought last week.

a. What fraction of his collection did Carlos have before last week?

b. How many stamps did Carlos have before last week?

13. Elisa Santos is driving to an important business meeting, but she left her house a little late. The fuel gauge of her car tells her that she has almost $\frac{1}{8}$ of a tank of gas. When full, the tank holds 17 gallons. If Elisa stops for gas, she will certainly be late for the meeting. She wonders whether she needs to stop for gas.

a. When estimating the amount of fuel she has left, should Elisa overestimate or underestimate?

b. When estimating the distance she still must drive, should she overestimate or underestimate?

c. Elisa estimates that she must still drive 30 to 40 miles. She knows that her car can go 22 to 24 miles on one gallon of gas. Does she have enough fuel left to drive to the meeting on time? Explain your answer.

14. a. Copy and complete the following table to show how a 12-in. length of wire can be divided into equal segments of various sizes.

Number of Equal Segments	2	8	12	?	?	?
Length of Each Segment (inches)	?	?	?	$\frac{1}{2}$	$\frac{1}{4}$	$\frac{1}{8}$

b. Explain what would happen to the number in the second row if the number in the first row were doubled.

c. If one half of the 12-in. length of wire is divided into pieces that are $\frac{1}{4}$-in. long and the other half is divided into pieces $\frac{1}{2}$-in. long, how many pieces will result?

SPIRAL REVIEW EXERCISES

1. From –12, subtract –3.

2. When 25 is divided by 7, the remainder is
(a) 3 (b) 4 (c) 5 (d) 7

3. The least common denominator of the fractions $\frac{1}{3}$ and $\frac{2}{5}$ is
(a) 3 (b) 5 (c) 8 (d) 15

4. The dot on the number line represents the number
(a) –5 (b) –4 (c) +4 (d) $\frac{4}{5}$

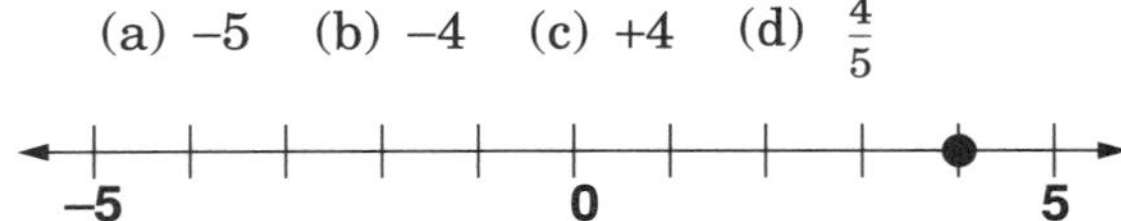

5. In the following key sequence, what number in place of [?] would have the effect of moving the decimal point of 57.43 two places to the left?
57.43 [×] [?] [=]

6. If each mathematics book contains 273 pages, how many pages are there in 35 such books?

7. Mr. and Mrs. Jonas began their trip with \$1,000. First, they spent \$250 on train fare, and then they spent \$325 on hotel rooms. How much money was left?

8. The value of –9 – (–9) is
(a) –18 (b) +18 (c) +9 (d) 0

9. What digit will be in the 100th decimal place of each repeating decimal? (The bar indicates which digit(s) repeat.)
a. $0.\overline{2}$ **b.** $0.\overline{36}$ **c.** $0.3\overline{6}$
d. $0.\overline{125}$ **e.** $0.1\overline{25}$

10. The chart below summarizes the ratings of five restaurants by different restaurant critics.

Critic	Restaurant				
	Blue Sea	**Burger Haven**	**Ribs 'n' 'Taters**	**Health Carousel**	**Grease Pit**
A	****	***	***	**	****
B	***	**	*	**	**
C	****	**	***	***	***

* = poor ** = acceptable *** = good **** = superior

Which restaurants were rated good or superior by at least $\frac{2}{3}$ of the critics?

UNIT 6–2 The Meaning of Reciprocal; Using Reciprocals in Dividing Fractions

THE MEANING OF RECIPROCAL

THE MAIN IDEA

1. The *reciprocal* of a fraction is formed by exchanging the numerator and the denominator of the original fraction. This is the same as *inverting* a fraction, that is, turning it upside down. — The reciprocal of $\frac{2}{3}$ is $\frac{3}{2}$.

2. To form the reciprocal of an integer, think of the integer as a fraction with a denominator of 1; then invert. — The reciprocal of 5 is $\frac{1}{5}$.

3. 0 is the only number that does not have a reciprocal.

4. The product of any number and its reciprocal is 1. — $\frac{2}{3} \times \frac{3}{2} = 1$

5. In a negative fraction, the minus sign can be in the numerator, in the denominator, or before the fraction. — $\frac{-2}{3} = \frac{2}{-3} = -\frac{2}{3}$

EXAMPLE 1 Find the reciprocal.

	Number	*Reciprocal*
a.	$\frac{5}{9}$	$\frac{9}{5}$
b.	$\frac{11}{16}$	$\frac{16}{11}$
c.	$\frac{-13}{2}$	$\frac{2}{-13}$ or $-\frac{2}{13}$
d.	-21	$\frac{1}{-21}$ or $-\frac{1}{21}$
e.	0	0 does not have a reciprocal.

EXAMPLE 2 It took Jill eight days to paint her boat. Assuming a steady rate of work, what part of the job did she do each day?

THINKING ABOUT THE PROBLEM

Represent the whole job by 1 (one boat painted). Dividing the job into 8 equal parts gives the reciprocal of 8:

$$1 \div 8 = \frac{1}{8}$$

Answer: She painted $\frac{1}{8}$ of the boat each day.

CLASS EXERCISES

Write the reciprocal of each number.

1. $\frac{7}{9}$ **2.** $\frac{1}{6}$ **3.** $\frac{3}{2}$ **4.** 7 **5.** 32 **6.** $-\frac{2}{5}$ **7.** 1 **8.** $\frac{22}{15}$ **9.** $\frac{9}{10}$ **10.** $-\frac{10}{3}$

11. It took Mario 5 hours to wax his car. Assuming a steady rate of work, what part of the job did he do in one hour?

USING RECIPROCALS IN DIVIDING FRACTIONS

THE MAIN IDEA

1. Division by a number gives the same result as multiplication by the reciprocal of that number:

$$20 \div 5 \text{ gives a result of } 4$$

$$20 \times \frac{1}{5} \text{ gives a result of } 4$$

2. A division can be written as an equivalent multiplication by using the reciprocal of the divisor:

$$\frac{1}{5} \div \frac{2}{3} \text{ is the same as } \frac{1}{5} \times \frac{3}{2}$$

3. To divide by a fraction, use the reciprocal of the divisor to write an equivalent multiplication. Then multiply.

EXAMPLE 3 Use the reciprocal of the divisor to rewrite each division as an equivalent multiplication.

	Division	***Multiplication***
a.	$\frac{1}{12} \div \frac{2}{5}$	$\frac{1}{12} \times \frac{5}{2}$
b.	$\frac{4}{9} \div 7$	$\frac{4}{9} \times \frac{1}{7}$

EXAMPLE 4 Divide: $\frac{3}{5} \div \frac{2}{3}$

Invert the divisor and write an equivalent multiplication. $\frac{3}{5} \times \frac{3}{2}$

Since there are no common factors, multiply. $\frac{3 \times 3}{5 \times 2} = \frac{9}{10}$ *Ans.*

EXAMPLE 5 Divide: $\frac{18}{25} \div \frac{27}{20}$

Use the reciprocal of the divisor and write an equivalent multiplication. $\frac{18}{25} \times \frac{20}{27}$

To simplify, divide 18 and 27 by 9. Divide 20 and 25 by 5. $\frac{\overset{2}{\cancel{18}}}{\underset{5}{\cancel{25}}} \times \frac{\overset{4}{\cancel{20}}}{\underset{3}{\cancel{27}}}$

Multiply the results. $\frac{2 \times 4}{5 \times 3} = \frac{8}{15}$ *Ans.*

EXAMPLE 6 Divide: $-6 \div \frac{3}{5}$

Write the integer −6 as a fraction. $\frac{-6}{1} \div \frac{3}{5}$

Write an equivalent multiplication. $\frac{-6}{1} \times \frac{5}{3}$

To simplify, divide −6 and 3 by 3. $\frac{\overset{-2}{\cancel{-6}}}{1} \times \frac{5}{\underset{1}{\cancel{3}}}$

Multiply the results. $\frac{-2 \times 5}{1 \times 1} = \frac{-10}{1} = -10$ *Ans.*

EXAMPLE 7 A piece of ribbon $\frac{3}{4}$ of a meter long is to be cut into 6 equal pieces. The length of each piece is

(a) 2 meters (b) $4\frac{1}{2}$ meters

(c) $\frac{1}{6}$ meter (d) $\frac{1}{8}$ meter

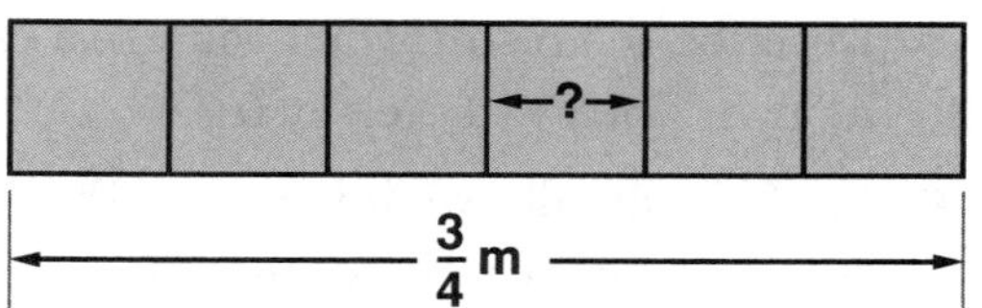

THINKING ABOUT THE PROBLEM

To separate a given amount into a number of equal parts, you divide.

Write a number phrase. $\frac{3}{4} \div 6$

Write the integer 6 as a fraction. $\frac{3}{4} \div \frac{6}{1}$

Use the reciprocal of the divisor and write an equivalent multiplication. $\frac{3}{4} \times \frac{1}{6}$

Simplify and multiply. $\frac{\overset{1}{\cancel{3}} \times 1}{4 \times \underset{2}{\cancel{6}}} = \frac{1}{8}$

Answer: (d)

CLASS EXERCISES

1. Rewrite each division as an equivalent multiplication.

   a. $12 \div 3$ b. $8 \div \frac{1}{12}$ c. $\frac{18}{5} \div \frac{3}{2}$ d. $\frac{1}{3} \div 2$ e. $4 \div \frac{-1}{2}$ f. $\frac{1}{3} \div \frac{1}{3}$

2. Divide. Write answers in simplest form.

   a. $\frac{3}{4} \div \frac{1}{8}$ b. $\frac{6}{5} \div \frac{2}{3}$ c. $\frac{5}{3} \div \frac{4}{3}$ d. $\frac{4}{3} \div \frac{5}{3}$ e. $\frac{9}{2} \div \frac{-1}{6}$ f. $\frac{2}{5} \div \frac{2}{5}$

3. Divide. Write answers in simplest form.

   a. $12 \div \frac{2}{3}$ b. $\frac{3}{5} \div 6$ c. $9 \div \frac{3}{4}$ d. $3 \div \frac{9}{5}$ e. $-8 \div \frac{3}{7}$ f. $\frac{7}{15} \div 14$

4. Why does 0 not have a reciprocal?

5. Describe the result when you use a calculator to divide a rational number

   a. by 1 b. by –1 c. by 0

6. a. If you divide two integers, is the result always an integer?

   b. If you divide two rational numbers, is the result always a rational number?

7. How many pieces of plastic tubing $\frac{3}{4}$ of an inch long can be cut from a 6-inch piece of tubing?

8. Mr. Adams wants to put a "No Trespassing" sign on every $\frac{1}{4}$-mile section of his 3-mile-long fence. How many signs will he need?

9. The distance around a racetrack is $\frac{3}{8}$ of a mile. How many times around the track must Cynthia run each day in order to run 27 miles in one 6-day week?

HOMEWORK EXERCISES

1. Write the reciprocal of each number.

   a. $\frac{2}{5}$ b. 3 c. $\frac{1}{2}$ d. $\frac{1}{25}$ e. $\frac{5}{9}$ f. 1 g. 100 h. $\frac{2}{4}$ i. 48 j. $\frac{4}{3}$

2. Samantha can mow the lawn in 3 hours.

   a. What fractional part of the lawn can she mow in one hour?

   b. What fractional part can she mow in 2 hours?

3. Rewrite each division as an equivalent multiplication.

a. $12 \div \frac{1}{3}$ **b.** $\frac{2}{5} \div \frac{3}{10}$ **c.** $\frac{5}{8} \div 2$ **d.** $\frac{11}{3} \div \frac{3}{4}$ **e.** $\frac{1}{2} \div -1$ **f.** $1 \div \frac{-1}{2}$

4. Divide. Write answers in simplest form.

a. $\frac{3}{4} \div \frac{1}{8}$ **b.** $\frac{3}{10} \div \frac{4}{5}$ **c.** $\frac{1}{2} \div \frac{5}{8}$ **d.** $\frac{5}{12} \div \frac{5}{9}$ **e.** $\frac{14}{35} \div \frac{2}{7}$ **f.** $\frac{7}{10} \div \frac{3}{10}$

g. $\frac{25}{48} \div \frac{16}{30}$ **h.** $\frac{49}{24} \div \frac{7}{8}$ **i.** $\frac{11}{17} \div \frac{11}{17}$ **j.** $\frac{3}{4} \div \frac{-3}{1}$ **k.** $\frac{-7}{8} \div \frac{7}{8}$ **l.** $\frac{-3}{5} \div \frac{-3}{5}$

5. Divide. Write answers in simplest form.

a. $2 \div \frac{1}{2}$ **b.** $\frac{2}{3} \div 3$ **c.** $24 \div \frac{1}{6}$ **d.** $6 \div \frac{4}{5}$ **e.** $\frac{9}{10} \div -3$ **f.** $12 \div \frac{9}{4}$

g. $\frac{44}{50} \div 20$ **h.** $\frac{45}{60} \div 18$ **i.** $10 \div \frac{18}{3}$ **j.** $4 \div \frac{-1}{4}$ **k.** $\frac{1}{3} \div 6$ **l.** $\frac{-2}{3} \div \frac{-2}{3}$

6. In her homework, Lily had to divide 8 by $\frac{1}{2}$. Her result was 4. Explain her mistake.

7. **a.** What is the decimal equivalent of $\frac{2}{5}$?

b. What is the decimal equivalent of its reciprocal, $\frac{5}{2}$?

c. Multiply the decimals obtained in part **a** and part **b**. What is the product?

d. Try this with other fractions and their reciprocals. Is the result always the same? Explain.

8. Which of the following are properties of division of rational numbers? Use examples to explain your answer.

a. closure **b.** commutative property **c.** associative property

9. How many postage stamps, each $\frac{11}{16}$ of an inch wide, are in a row of stamps 22 inches wide?

10. Mrs. Cody needed 15 pounds of cheese to make sandwiches for her company picnic. The cheese comes in packages containing $\frac{5}{8}$ of a pound. How many packages did she need?

11. Wakena has a rope that is 3 ft. long. She wants to cut it into pieces that are $\frac{3}{4}$ ft. long.

a. Draw a diagram showing where she should cut the rope.

b. How many pieces $\frac{3}{4}$ ft. long will she have?

c. Use your diagram to explain how to divide 3 by $\frac{3}{4}$.

12. Mario challenged Edward to stump him with a fraction problem. Edward gave him the following problem to compute. He was amazed when Mario quickly calculated it mentally. Explain how Mario was able to do this.

$$\left(\frac{132}{512} \times \frac{715}{273} \times 25 \times \frac{273}{715} \times \frac{512}{132}\right) \div 5$$

13. After Chen exchanged $\frac{1}{4}$ of his collection of baseball cards for 4 rare cards, he then had 286 cards in all. How many cards did Chen start with?

14. Explain why the reciprocal of an improper fraction is not always a proper fraction. Give an example.

SPIRAL REVIEW EXERCISES

1. Round 25,364 to the nearest hundred.

2. Find $\frac{3}{4}$ of 48.

3. Add:

   a. $3.25 + 10.5$ **b.** $312.5 + 31.25 + 3.125$

4. Use estimation to tell which of the following key sequences produced an incorrect result because the clear key was not pressed first.

Key Sequence	*Display*
(a) 1.987 [×] 14.98 [=]	29.76526
(b) 119.5 [÷] 4.02 [=]	278.48258
(c) 101.1 [−] 58.9 [=]	42.2
(d) 1987 [+] 6001 [=]	7988.

5. Subtract:

   a. $39.2 - 6.15$ **b.** $150.5 - 0.5$

6. Replace ? with < or > to make a true comparison.

   a. 3.2 ? 0.64 **b.** 13.5 ? 13.6
   c. 0.215 ? 1.00 **d.** 2.15 ? 1

7. Bob bought 3 shirts and 4 pairs of slacks. The price of a shirt was $14, and the price of a pair of slacks was $24. The difference between the cost of the slacks and the cost of the shirts can be represented by the number phrase

   (a) $4 \times 24 - 3 \times 14$ (b) $3 \times 24 - 4 \times 14$
   (c) $(3 + 4) \times (14 + 24)$ (d) $(14 + 24) - (3 + 4)$

8. $\frac{2}{3}$ of 300 is

   (a) 100 (b) 200 (c) 300 (d) 400

9. Which is a longer run, 8 laps around a 300-yard track or 6 laps around a 360-yard track?

10. Mary earned $19 on each of 5 days and then earned $25 on the sixth day. What was the total of her earnings?

11. Evaluate:

    a. $2 + 3 \times 4$
    b. $-3 - (-5)$
    c. $6 \times 8 \div 4 + 12$
    d. $6 \times 8 \div (4 + 12)$

12. Replace ? by < or > to make a true comparison.

    a. $\frac{7}{10}$? $\frac{3}{5}$ **b.** $\frac{4}{7}$? $\frac{5}{9}$ **c.** $\frac{7}{18}$? $\frac{5}{12}$

13. There are three thousand, seventeen students at Westwood High School, and two thousand, eight hundred nine students at McKinley High School. How many more students are at Westwood?

UNIT 6–3 Adding and Subtracting Like Fractions

THE MAIN IDEA

To add or subtract like fractions (fractions that have the same denominator):

1. Add or subtract the numerators.
2. Keep the original denominator.
3. If necessary, simplify the resulting fraction.

EXAMPLE 1 Add: $\frac{5}{17} + \frac{7}{17}$

Since the fractions are alike (have the same denominator), add the numerators and keep the original denominator:

$$\frac{5}{17} + \frac{7}{17} = \frac{5+7}{17} = \frac{12}{17} \quad \textit{Ans.}$$

EXAMPLE 2 $\frac{13}{18} + \frac{5}{18} + \frac{7}{18}$ equals

(a) $2\frac{5}{18}$ (b) $3\frac{7}{18}$ (c) $2\frac{7}{18}$ (d) $1\frac{7}{18}$

Since the fractions are alike, add the numerators and keep the original denominator:

$$\frac{13}{18} + \frac{5}{18} + \frac{7}{18} = \frac{13+5+7}{18} = \frac{25}{18}$$

Write the improper fraction $\frac{25}{18}$ as a mixed number:

$$\begin{array}{r} 1 \\ 18\overline{)25} \\ \underline{18} \\ 7 \end{array}$$

divisor → 18; 1 ← quotient; 7 ← remainder

$$\frac{25}{18} = 1\frac{7}{18}$$

Answer: (d)

EXAMPLE 3 John ran $\frac{29}{32}$ of a mile in the morning and $\frac{19}{32}$ of a mile in the evening. How much farther did he run in the morning?

THINKING ABOUT THE PROBLEM

The key words "how much farther" tell you to subtract to find how much greater one number is than another.

Write a number phrase. $\frac{29}{32} - \frac{19}{32}$

Since the fractions are alike, subtract the numerators and keep the original denominator.

$$\frac{29}{32} - \frac{19}{32} = \frac{29-19}{32} = \frac{10}{32}$$

Simplify.

$$\frac{10 \div 2}{32 \div 2} = \frac{5}{16}$$

Answer: John ran $\frac{5}{16}$ of a mile farther in the morning.

EXAMPLE 4 Andy trimmed $\frac{3}{16}$ of an inch from one side of a photograph and $\frac{9}{16}$ of an inch from the opposite side. What was the total amount that he trimmed from the photograph?

THINKING ABOUT THE PROBLEM

The key word "total" tells you to add the two fractions.

Write a number phrase. $\frac{3}{16} + \frac{9}{16}$

Add the like fractions. $\frac{3}{16} + \frac{9}{16} = \frac{3+9}{16} = \frac{12}{16}$

Simplify. $\frac{12 \div 4}{16 \div 4} = \frac{3}{4}$

Answer: The total amount trimmed was $\frac{3}{4}$ of an inch.

CLASS EXERCISES

1. Add. Write answers in simplest form.

a. $\frac{5}{9} + \frac{2}{9}$ **b.** $\frac{6}{17} + \frac{9}{17}$ **c.** $\frac{13}{41} + \frac{27}{41}$ **d.** $\frac{24}{69} + \frac{12}{69} + \frac{15}{69}$ **e.** $\frac{1}{3} + \frac{1}{3}$

f. $\frac{69}{100} + \frac{6}{100}$

g. $\frac{5}{23} + \frac{7}{23} + \frac{9}{23}$

h. $\frac{8}{25} + \frac{4}{25} + \frac{3}{25}$

i. $\frac{7}{24} + \frac{11}{24} + \frac{5}{24}$

j. $\frac{3}{10} + \frac{2}{10} + \frac{1}{10}$

k. $\frac{1}{5} + \frac{1}{5} + \frac{2}{5}$

2. Subtract. Write answers in simplest form.

a. $\frac{5}{9} - \frac{2}{9}$ b. $\frac{6}{7} - \frac{4}{7}$ c. $\frac{7}{8} - \frac{1}{8}$ d. $\frac{11}{19} - \frac{7}{19}$ e. $\frac{11}{12} - \frac{5}{12}$ f. $\frac{29}{37} - \frac{22}{37}$

g. $\frac{51}{77} - \frac{31}{77}$ h. $\frac{12}{25} - \frac{7}{25}$ i. $\frac{4}{5} - \frac{2}{5}$ j. $\frac{7}{8} - \frac{3}{8}$ k. $\frac{13}{20} - \frac{5}{20}$ l. $\frac{14}{15} - \frac{6}{15}$

3. On his bike, James rode $\frac{6}{8}$ of a mile to his friend's house, $\frac{3}{8}$ of a mile to the cleaners, and $\frac{7}{8}$ of a mile to his home. The total mileage that James traveled is

(a) $1\frac{5}{8}$ (b) $\frac{5}{8}$ (c) 2 (d) $\frac{1}{2}$

4. Vera bought $\frac{5}{6}$ of a pizza. Donna bought $\frac{3}{6}$ of a pizza. How much more did Vera buy than Donna?

5. Mrs. Ramos ordered $\frac{3}{4}$ of a pound of ham, $\frac{1}{4}$ of a pound of cheese, and $\frac{3}{4}$ of a pound of roast beef from the deli.

a. What was the total weight of her purchases?
b. Her purchases were weighed on a digital scale at the deli. What was the weight of each purchase as a decimal?
c. What was the total weight as a decimal?

6. Joseph ran $\frac{9}{10}$ of a mile, and Michael ran $\frac{4}{10}$ of a mile. How much more did Joseph run than Michael?

HOMEWORK EXERCISES

1. Add. Write answers in simplest form.

a. $\frac{3}{7} + \frac{2}{7}$ b. $\frac{5}{8} + \frac{2}{8}$ c. $\frac{2}{9} + \frac{6}{9}$ d. $\frac{11}{48} + \frac{14}{48}$ e. $\frac{5}{24} + \frac{11}{24}$ f. $\frac{6}{21} + \frac{10}{21}$

g. $\frac{8}{31} + \frac{11}{31} + \frac{7}{31}$ h. $\frac{23}{72} + \frac{10}{72} + \frac{15}{72}$ i. $\frac{21}{100} + \frac{37}{100} + \frac{32}{100}$ j. $\frac{7}{15} + \frac{2}{15}$ k. $\frac{1}{6} + \frac{1}{6} + \frac{1}{6}$

2. Betty assembled 3 electronic parts. The first was $\frac{5}{16}$ of an inch long, the second was $\frac{3}{16}$ of an inch long, and the third was $\frac{7}{16}$ of an inch long. What was the total length of the assembly?

3. Michael bought $\frac{1}{4}$ of a yard of cotton, $\frac{3}{4}$ of a yard of rayon, and $\frac{1}{4}$ of a yard of orlon. What was the total number of yards of cloth that he bought?

4. Subtract. Write answers in simplest form.

a. $\frac{5}{6} - \frac{4}{6}$ **b.** $\frac{6}{8} - \frac{3}{8}$ **c.** $\frac{5}{9} - \frac{2}{9}$ **d.** $\frac{9}{10} - \frac{6}{10}$ **e.** $\frac{9}{12} - \frac{4}{12}$ **f.** $\frac{9}{22} - \frac{5}{22}$

g. $\frac{9}{16} - \frac{4}{16}$ **h.** $\frac{11}{26} - \frac{5}{26}$ **i.** $\frac{9}{20} - \frac{1}{20}$ **j.** $\frac{3}{5} - \frac{1}{5}$ **k.** $\frac{7}{8} - \frac{1}{8}$ **l.** $\frac{3}{4} - \frac{2}{4}$

5. The diameter of one coin was $\frac{9}{16}$ of an inch long, and of another was $\frac{15}{16}$ of an inch long. How much longer was the diameter of the larger coin than the smaller one?

6. The diagram shows how a baseball team distributed souvenir bats to fans at the first game of the season. Some bats were saved to be given away at another game.
   a. What fractional part of the total number of bats was given away at the first game?
   b. How many bats were given away?
   c. What fractional part of the total was saved for another game?

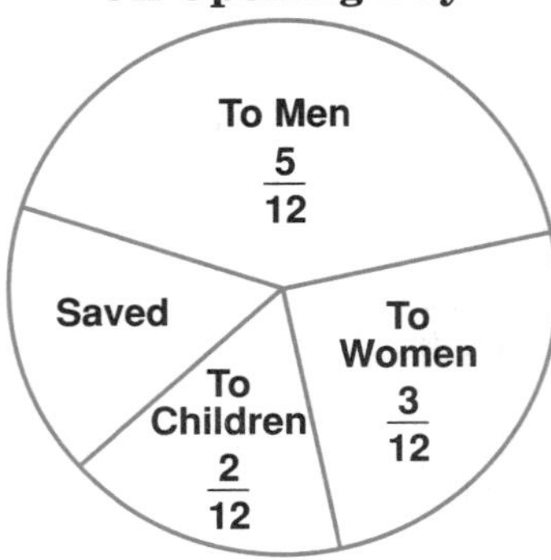

SPIRAL REVIEW EXERCISES

1. The value of 7 in 17,428 is
(a) 70,000 (b) 7,000 (c) 700 (d) 7

2. The product of –7 and –3 is
(a) 21 (b) –21 (c) –10 (d) 10

3. The least common multiple (LCM) of 8 and 12 is
(a) 8 (b) 12 (c) 24 (d) 96

4. Of the fractions $\frac{1}{2}$, $\frac{1}{3}$, $\frac{2}{3}$, and $\frac{2}{5}$, the one with the greatest value is
(a) $\frac{1}{2}$ (b) $\frac{1}{3}$ (c) $\frac{2}{3}$ (d) $\frac{2}{5}$

5. Which key sequence and display show that multiplication and division are inverse operations?

	Key Sequence	***Display***
(a)	24 ÷ 24 × 72 =	7.2
(b)	24 ÷ 72 × 24 =	7.9999992
(c)	72 × 24 ÷ 24 =	7.2
(d)	72 ÷ 24 × 72 =	216.2

6. Simplify the fraction $\frac{12}{75}$.

7. Write $8\frac{2}{9}$ as an improper fraction.

8. On Saturday, the attendance at the ballpark was 9,075 and on Sunday, it was 10,162. How many more people attended the game on Sunday than on Saturday?

9. A distributor must ship 300 compact discs using boxes that can hold, at most, 16 discs each. How many boxes will be needed?

10. The price of a shirt is \$15, and the price of a tie is \$10. Which costs more, 12 shirts or 19 ties?

11. The heights of six students, in inches, are 72, 67, 59, 64, 73, and 67. What is the greatest difference in height between any two of these students?

UNIT 6–4 Adding and Subtracting Unlike Fractions

THE MAIN IDEA

To add or subtract fractions that have unlike denominators:

1. Find the LCD.
2. Write the given fractions as equivalent fractions that have the LCD as their denominators.
3. Use the method for adding or subtracting like fractions.

EXAMPLE 1 Add: $\frac{3}{8} + \frac{1}{6}$

Find the LCD of 8 and 6. The LCD is 24.

Write the fractions as equivalent fractions that have 24 as the denominator.

$$\frac{3 \times 3}{8 \times 3} = \frac{9}{24}$$

$$\frac{1 \times 4}{6 \times 4} = \frac{4}{24}$$

Add the equivalent like fractions.

$$\frac{9}{24} + \frac{4}{24} = \frac{13}{24} \quad \textit{Ans.}$$

EXAMPLE 2 Subtract: $\frac{17}{36} - \frac{1}{18}$

Find the LCD of 36 and 18. The LCD is 36.

Write $\frac{1}{18}$ as an equivalent fraction that has 36 as the denominator.

$$\frac{1 \times 2}{18 \times 2} = \frac{2}{36}$$

Subtract the like fractions.

$$\frac{17}{36} - \frac{2}{36} = \frac{15}{36}$$

Simplify.

$$\frac{15 \div 3}{36 \div 3} = \frac{5}{12} \quad \textit{Ans.}$$

EXAMPLE 3 Express as a mixed number: $\frac{3}{4} + \frac{5}{6} - \frac{1}{3}$

The LCD of 4, 6, and 3 is 12.

(1) Rewrite each fraction.

$$\frac{3}{4} + \frac{5}{6} - \frac{1}{3}$$

$$\frac{3 \times 3}{4 \times 3} + \frac{5 \times 2}{6 \times 2} - \frac{1 \times 4}{3 \times 4}$$

$$\frac{9}{12} + \frac{10}{12} - \frac{4}{12}$$

(2) Following the order of operations, work from left to right. Add, then subtract.

$$\frac{19}{12} - \frac{4}{12} = \frac{15}{12}$$

$$\frac{15}{12} = 1\frac{3}{12} = 1\frac{1}{4} \quad \textit{Ans.}$$

CLASS EXERCISES

1. Add. Write answers in simplest form.

 a. $\frac{1}{4}+\frac{2}{3}$ b. $\frac{3}{5}+\frac{3}{10}$ c. $\frac{15}{27}+\frac{5}{9}$ d. $\frac{5}{6}$ $+\frac{3}{11}$ e. $\frac{3}{4}+\frac{1}{5}+\frac{5}{8}$ f. $\frac{4}{27}+\frac{1}{3}+\frac{5}{9}$

2. Subtract. Write answers in simplest form.

 a. $\frac{3}{4}-\frac{5}{16}$ b. $\frac{5}{6}-\frac{1}{2}$ c. $\frac{4}{5}-\frac{1}{3}$ d. $\frac{12}{7}$ $-\frac{2}{3}$ e. $\frac{9}{10}$ $-\frac{3}{4}$ f. $\frac{8}{12}$ $-\frac{2}{3}$

3. Perform the indicated operations. Write answers in simplest form.

 a. $\frac{1}{3}+\frac{3}{5}-\frac{1}{2}$ b. $\frac{3}{4}-\frac{3}{7}+\frac{1}{4}$ c. $\frac{4}{5}-\frac{1}{2}-\frac{1}{4}$ d. $\frac{2}{3}+\frac{1}{3}-\frac{1}{3}$ e. $\frac{8}{15}-\frac{1}{3}+\frac{2}{5}$

4. Carol bought $\frac{1}{2}$ of a pound of cashews, $\frac{2}{3}$ of a pound of walnuts, and $\frac{3}{4}$ of a pound of peanuts. Find the total weight of the nuts that Carol bought.

5. Fred had to lose $\frac{7}{8}$ of a pound to qualify for his wrestling match. He dieted and lost $\frac{7}{10}$ of a pound. How much does he still have to lose?

6. Which of the following properties of addition of integers are also properties of addition of rational numbers? (a) closure (b) commutative property (c) associative property Use examples to explain your answer.

HOMEWORK EXERCISES

1. Add. Write answers in simplest form.

 a. $\frac{2}{3}+\frac{1}{4}$ b. $\frac{3}{5}+\frac{1}{2}$ c. $\frac{3}{4}+\frac{2}{5}$ d. $\frac{3}{10}+\frac{1}{5}$ e. $\frac{1}{3}+\frac{1}{12}$ f. $\frac{1}{4}+\frac{3}{6}$

 g. $\frac{1}{8}$ $+\frac{1}{2}$ h. $\frac{1}{6}$ $+\frac{2}{3}$ i. $\frac{4}{5}$ $+\frac{3}{10}$ j. $\frac{3}{8}+\frac{1}{12}+\frac{5}{24}$ k. $\frac{4}{15}+\frac{2}{5}+\frac{2}{3}$

2. Subtract. Write answers in simplest form.

a. $\frac{5}{8} - \frac{1}{4}$ **b.** $\frac{7}{16} - \frac{1}{8}$ **c.** $\frac{3}{4} - \frac{1}{8}$ **d.** $\frac{3}{4} - \frac{1}{3}$ **e.** $\frac{3}{4} - \frac{1}{6}$ **f.** $\frac{1}{2} - \frac{3}{10}$

g. $\begin{array}{r} \frac{3}{5} \\ -\frac{3}{10} \\ \hline \end{array}$ **h.** $\begin{array}{r} \frac{15}{16} \\ -\frac{3}{4} \\ \hline \end{array}$ **i.** $\begin{array}{r} \frac{9}{10} \\ -\frac{4}{5} \\ \hline \end{array}$ **j.** $\frac{11}{12} - \frac{2}{5}$ **k.** $\frac{5}{6} - \frac{2}{5}$ **l.** $\frac{18}{20} - \frac{9}{10}$

3. Perform the indicated operations. Write answers in simplest form.

a. $\frac{3}{4} - \frac{1}{2} + \frac{1}{6}$ **b.** $\frac{2}{5} + \frac{7}{10} - \frac{3}{10}$ **c.** $\frac{6}{7} - \frac{1}{2} - \frac{1}{4}$ **d.** $\frac{2}{3} - \frac{1}{2} + \frac{1}{6}$

4. Answer *true* or *false*. Then explain your answer.

a. $0.3333\ldots + 0.4444\ldots = 0.7777\ldots$ **b.** $\frac{1}{3} + \frac{1}{4} = \frac{1}{7}$ **c.** $\frac{3}{1} + \frac{4}{1} = \frac{7}{1}$

5. A highway crew built a fence $\frac{1}{2}$ of a mile long and then added two more sections, one that was $\frac{1}{5}$ of a mile long and another $\frac{1}{4}$ of a mile long. What was the total length of the fence?

6. A jeweler removed $\frac{3}{16}$ of an inch from the length of a gold wire that was originally $\frac{5}{8}$ of an inch long. What was the final length of the wire?

7. a. Is the difference of two rational numbers always a rational number? Use examples to explain your answer.

b. What is this property called?

c. Give examples to show whether or not subtraction of integers has this property.

8. Is the commutative property true for subtraction of rational numbers? Use examples to explain your answer.

9. The order of operations that you used with integers is also used with rational numbers. Use the order of operations to evaluate:

a. $\frac{3}{4} + 5 \times \left(\frac{1}{2}\right)$ **b.** $\frac{-1}{2} - \frac{3}{4} \times \frac{2}{3}$ **c.** $\frac{4}{5} \times (-10) - 3$ **d.** $\frac{2}{5} \times \left(\frac{-3}{8} + \frac{1}{4}\right) + 1$

10. The Clarkville Library Association wants to post a large circular chart to let the community know how much various groups gave to the building fund for the new library.

a. The parts of the circle have been marked and the labels have been printed. Match each label with the appropriate part of the circle.

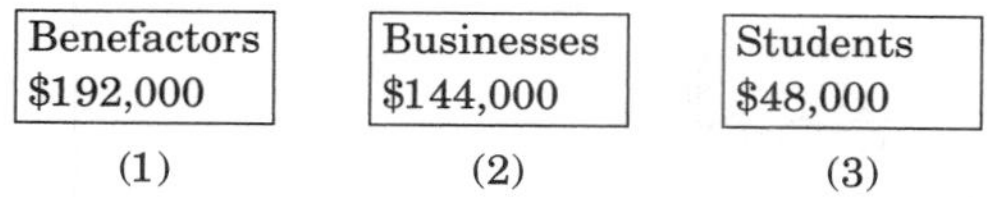

Library Building Fund
Community Gifts: $480,000

$\frac{2}{5}$ | $\frac{3}{10}$ | $\frac{1}{10}$ | Other

b. What fraction of the circle is "Other?"

c. How much money does "Other" represent?

11. Sasha subtracted $\frac{1}{5}$ from $\frac{1}{4}$. Khalil didn't want to find the LCD. He wrote the problem in decimal form as $0.25 - 0.20$ and subtracted the decimals.

a. What answer did Sasha get? Khalil? **b.** Are their answers the same? Explain.

SPIRAL REVIEW EXERCISES

1. A prime number between 62 and 70 is
(a) 63 (b) 67 (c) 68 (d) 69

2. The greatest common factor of 75 and 21 is
(a) 1 (b) 3 (c) 5 (d) 21

3. The sum of $\frac{1}{12}$ and $\frac{7}{12}$ is
(a) $\frac{2}{3}$ (b) $\frac{8}{24}$ (c) $\frac{1}{3}$ (d) $\frac{3}{4}$

4. From $\frac{12}{21}$, subtract $\frac{5}{21}$.

5. Simplify the fraction $\frac{42}{70}$.

6. Divide: $-20 \div (-5)$

7. Round 275,892 to the nearest thousand.

8. Which key on the calculator allows you to correct the number that was just entered without changing anything else?
(a) MR (b) CE (c) C (d) +/−

9. Which fraction is equivalent to $\frac{2}{3}$?
(a) $\frac{3}{4}$ (b) $\frac{12}{20}$ (c) $\frac{10}{21}$ (d) $\frac{16}{24}$

10. A stadium contains 64 sections with 128 seats in each section. What is the total number of seats in the stadium?

11. How many 12-inch lengths of rope can be cut from a piece of rope that is 335 inches long?

12. The value of $24 + 16 \div 8 - 4$ is
(a) 1 (b) 28 (c) 10 (d) 22

13. What fractional part of the English alphabet is made up of vowels? (The vowels are *a, e, i, o,* and *u*.)

14. The distance between +10 and −4 on a number line is represented by
(a) $|+10|$ (b) $|-4|$
(c) $|+10| - |-4|$ (d) $|+10| + |-4|$

MODULE 7

UNIT 7-1 Adding Mixed Numbers

ADDING MIXED NUMBERS WITH LIKE-FRACTION PARTS

THE MAIN IDEA

To add mixed numbers with like-fraction parts:

1. Add the like-fraction parts. If possible, simplify the resulting fraction.
2. Add the integer parts.
3. If the fraction is an improper fraction, write it as an equivalent mixed number and add it to the integer part.

EXAMPLE 1 Add: $5\frac{7}{16} + 3\frac{2}{16}$

First add the like-fraction parts. Then add the integer parts. No simplification is possible since $\frac{9}{16}$ is in lowest terms.

$$\begin{array}{r} 5\frac{7}{16} \\ +3\frac{2}{16} \\ \hline 8\frac{9}{16} \end{array} \quad \textit{Ans.}$$

like fractions

EXAMPLE 2 Add: $2\frac{3}{8} + 5\frac{1}{8}$

Add the like-fraction parts and add the integer parts.

$$\begin{array}{r} 2\frac{3}{8} \\ +5\frac{1}{8} \\ \hline 7\frac{4}{8} \end{array}$$

Simplify the fraction part.

$$7\frac{4}{8} = 7\frac{4 \div 4}{8 \div 4} = 7\frac{1}{2} \quad \textit{Ans.}$$

EXAMPLE 3 Add: $15\frac{7}{13} + 12\frac{9}{13}$

Add the like-fraction parts and add the integer parts.

$$\begin{array}{r} 15\frac{7}{13} \\ +12\frac{9}{13} \\ \hline 27\frac{16}{13} \end{array}$$

Since the resulting fraction part is an improper fraction, write it as an equivalent mixed number.

$$\frac{16}{13} = 1\frac{3}{13}$$

Add the resulting mixed number to the integer part.

$$27 + 1\frac{3}{13} = 28\frac{3}{13} \quad \textit{Ans.}$$

CLASS EXERCISES

1. Add. Write answers in simplest form.

   a. $2\frac{3}{5} + 4\frac{1}{5}$ b. $9\frac{3}{8} + 7\frac{1}{8}$ c. $11\frac{13}{32} + 20\frac{5}{32}$ d. $3\frac{1}{4} + 2\frac{1}{4}$ e. $1\frac{1}{2} + 2\frac{1}{2}$

   f. $12\frac{3}{4} + 5\frac{1}{4}$ g. $22\frac{5}{17} + 14\frac{13}{17}$ h. $9\frac{5}{7} + 2\frac{3}{7} + 1\frac{6}{7}$ i. $5\frac{5}{12} + 8\frac{7}{12}$ j. $13\frac{4}{15} + 12\frac{11}{15} + 26\frac{3}{15}$

2. X-Ray Industries' stock was priced at $19\frac{7}{8}$. The stock gained $1\frac{5}{8}$ points. What was the new price of the stock?

3. June jogged $3\frac{3}{4}$ miles one morning, and she jogged another $4\frac{3}{4}$ miles that afternoon. What was the total distance she jogged that day?

4. At a deli counter, all cold cuts are on sale for $3 a pound. What is the total cost of $1\frac{1}{4}$ pounds of turkey, $1\frac{3}{4}$ pounds of roast beef, $\frac{3}{4}$ pound of ham, and $\frac{1}{4}$ pound of bologna?

ADDING MIXED NUMBERS WITH UNLIKE-FRACTION PARTS

THE MAIN IDEA

To add mixed numbers with unlike-fraction parts:

1. Find the LCD of the fraction parts.
2. Write the fraction parts as equivalent fractions with the LCD as the denominator.
3. Follow the procedure for adding mixed numbers with like-fraction parts.

EXAMPLE 4 Add: $3\frac{1}{2} + 7\frac{2}{5}$

The LCD of the unlike fractions $\frac{1}{2}$ and $\frac{2}{5}$ is 10.

Rewrite the unlike-fraction parts as equivalent like fractions.

$$3\frac{1 \times 5}{2 \times 5} = 3\frac{5}{10}$$

$$7\frac{2 \times 2}{5 \times 2} = 7\frac{4}{10}$$

Add the resulting like-fraction parts and the integer parts. No simplification is possible.

$$3\frac{5}{10} + 7\frac{4}{10} = 10\frac{9}{10} \quad \textit{Ans.}$$

EXAMPLE 5 Add: $6\frac{1}{3} + 5\frac{3}{4} + 2\frac{5}{6}$

The LCD of the unlike fractions is 12.

Rewrite the unlike fractions as equivalent like fractions.

$$6\frac{1 \times 4}{3 \times 4} = 6\frac{4}{12}$$

$$5\frac{3 \times 3}{4 \times 3} = 5\frac{9}{12}$$

Add the resulting like-fraction parts and the integer parts.

$$2\frac{5 \times 2}{6 \times 2} = 2\frac{10}{12}$$

$$13\frac{23}{12}$$

Write the improper fraction as a mixed number.

$$\frac{23}{12} = 1\frac{11}{12}$$

Add on the mixed number.

$$13 + 1\frac{11}{12} = 14\frac{11}{12} \quad \textit{Ans.}$$

CLASS EXERCISES

1. Add. Write answers in simplest form.

   a. $5\frac{2}{3} + 3\frac{1}{6}$ b. $6\frac{2}{5} + 3\frac{2}{3}$ c. $5\frac{3}{4} + 1\frac{3}{8}$ d. $4\frac{1}{3} + 2\frac{3}{5}$ e. $8\frac{1}{2} + 2\frac{1}{3} + 4\frac{1}{6}$

   f. $7\frac{5}{6} + 2\frac{3}{4}$

   g. $3\frac{1}{5} + 2\frac{2}{3} + 5\frac{3}{10}$

   h. $6\frac{1}{4} + 3\frac{5}{10} + 8\frac{1}{2}$

   i. $4\frac{5}{12} + 5\frac{1}{6} + 9\frac{5}{8}$

   j. $\frac{7}{10} + \frac{15}{20} + \frac{75}{100}$

2. Mr. McCarthy bought $2\frac{1}{4}$ pounds of ham, $3\frac{3}{8}$ pounds of cheese, and $1\frac{7}{16}$ pounds of roast beef. What was the total weight of his purchases?

3. An electric cable is made from a connector on each end and a piece of wire between. If one connector is $\frac{5}{8}$ of an inch long, the wire is $18\frac{3}{4}$ inches long, and the second connector is $\frac{1}{2}$ inch long, find the total length of the cable.

4. Answer *true* or *false*. Give examples to support your answer.

   a. The sum of two mixed numbers is always a mixed number.

   b. The sum of two mixed numbers is always a rational number.

5. Estimate each sum by first rounding each mixed number to the nearest whole number.

a. $5\frac{1}{4}$ $7\frac{2}{3}$ $+6\frac{5}{8}$

b. $13\frac{1}{3}$ $6\frac{1}{2}$ $+23\frac{3}{8}$

c. $9\frac{3}{5}$ $7\frac{1}{5}$ $+12\frac{9}{10}$

d. $42\frac{1}{2}$ $16\frac{7}{8}$ $+9\frac{3}{5}$

e. $32\frac{7}{10}$ $16\frac{4}{9}$ $+8\frac{7}{8}$

HOMEWORK EXERCISES

1. Add. Write answers in simplest form.

a. $5\frac{1}{6}+3\frac{4}{6}$

b. $9\frac{11}{32}+5\frac{13}{32}$

c. $48\frac{9}{15}+52\frac{8}{15}$

d. $5\frac{1}{8}+3\frac{3}{8}+2\frac{5}{8}$

e. $7\frac{3}{8}$ $+8\frac{2}{8}$

f. $11\frac{7}{12}$ $+13\frac{1}{12}$

g. $24\frac{3}{10}$ $32\frac{7}{10}$ $+16\frac{9}{10}$

h. $22\frac{11}{40}+11\frac{11}{40}$

2. Add. Write answers in simplest form.

a. $7\frac{1}{2}+4\frac{1}{3}$

b. $23\frac{1}{12}+18\frac{2}{3}$

c. $6\frac{2}{3}+11\frac{3}{4}$

d. $6\frac{1}{8}+4\frac{5}{12}+1\frac{9}{24}$

e. $9\frac{7}{8}$ $+17\frac{5}{12}$

f. $6\frac{1}{2}$ $5\frac{1}{3}$ $+4\frac{1}{4}$

g. $35\frac{7}{15}$ $48\frac{7}{12}$ $+52\frac{7}{20}$

h. $3\frac{1}{3}$ $2\frac{5}{6}$ $+4\frac{7}{12}$

3. Joe added $1\frac{1}{3}$ cups of chlorine to his swimming pool. After testing the water, he added another $\frac{1}{2}$ cup of chlorine. What was the total number of cups of chlorine that he added?

4. It costs about $24\frac{1}{2}$¢ per mile to operate a car. How much would it cost to drive it for 2 miles?

5. Bob caught fish weighing $4\frac{1}{4}$ lb. and $3\frac{5}{8}$ lb. How much did the fish he caught weigh altogether?

6. Mr. Leung and his two daughters, Jennifer and Mary, worked together to paint the trim on the windows of their home. Each started with a new one-quart can of paint. After they finished, Mr. Leung had $\frac{3}{8}$ of a quart left, Jennifer had $\frac{1}{4}$ of a quart left, and Mary had $\frac{5}{8}$ of a quart left.
   a. How much paint did they have left over?
   b. How could the painters have used the paint more efficiently?

7. One week Rebecca worked 3 days. The first day she worked $5\frac{2}{3}$ hours, the next day she worked $6\frac{3}{4}$ hours, and the third day she worked $4\frac{1}{2}$ hours. How many hours did she work in all?

8. Verna needs 5 yards of fabric to make a quilt. She has scraps of fabric with lengths of $1\frac{1}{4}$ yd., $\frac{7}{8}$ yd., $\frac{1}{2}$ yd., $1\frac{1}{2}$ yd., and $2\frac{1}{8}$ yd.
   a. Round each fraction or mixed number to the nearest whole number.
   b. Estimate the sum of all the lengths.
   c. Does Verna have enough fabric to make the quilt? Explain.

9. Copy and complete the chart to obtain the final answer.

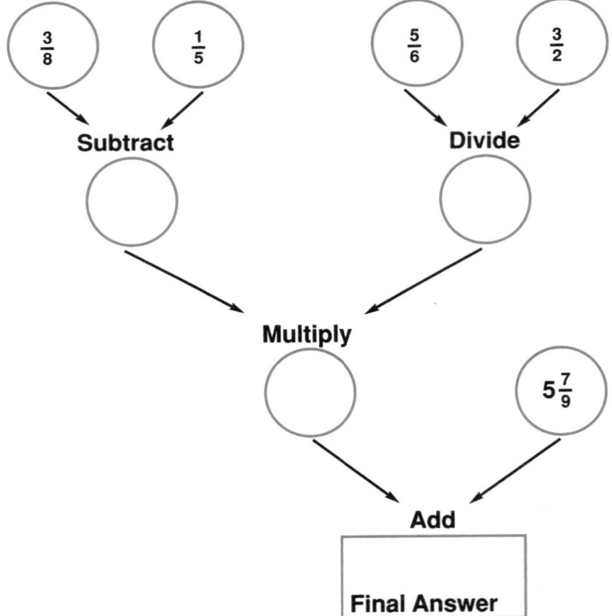

SPIRAL REVIEW EXERCISES

1. Which fraction has the largest value?
   (a) $\frac{2}{3}$ (b) $\frac{3}{4}$ (c) $\frac{3}{5}$ (d) $\frac{4}{5}$

2. Find the sum of $\frac{2}{5}$ and $\frac{3}{10}$.

3. Find the product of $\frac{3}{5}$ and $\frac{10}{27}$.

4. $\frac{22}{3}$ has the same value as
   (a) $8\frac{1}{3}$ (b) $7\frac{1}{3}$ (c) $6\frac{2}{3}$ (d) $5\frac{2}{3}$

5. Subtract –10 from –20.

6. What is the value of 2 in 524,689?

7. Which key sequence can be used to make a table of all positive multiples of 7?
   (a) 1 [×] 7 [×] 7 [×] 7 [×] 7 [×] 7 [×] 7 [=]
   (b) 1 [×] 2 [×] 3 [×] 4 [×] 5 [×] 6 [×] 7 [=]
   (c) 7 [×] 7 [=] [=] [=] [=] [=] [=]
   (d) 7 [+] 7 [=] [=] [=] [=] [=] [=]

8. Which is a prime number?
   (a) 21 (b) 31 (c) 51 (d) 81

9. Find the least common multiple (LCM) of 15 and 25.

10. How many 6-ounce portions can be made from 304 ounces of ground beef?

11. A number phrase showing that the total price of 12 books, each costing $8, has been reduced by $5 is
    (a) $(12 - 8) \times 5$ (b) $(12 - 5) \times 8$
    (c) $12 \times 8 - 5$ (d) $12 \times (8 - 5)$

12. Hector Blair ordered 14 tires, at $47 each, for his auto service center. He received credit for 6 tires, costing $43 each, that he returned to the supplier. How much does he still owe?

UNIT 7-2 Subtracting Mixed Numbers

SUBTRACTING MIXED NUMBERS WITH LIKE-FRACTION PARTS

THE MAIN IDEA

To subtract mixed numbers with like-fraction parts:

1. Subtract the like-fraction parts.
   a. If necessary, rename the minuend (the first, or top, number).
   b. If possible, simplify the resulting fraction.
2. Subtract the integer parts.

EXAMPLE 1 Subtract: $7\frac{7}{11} - 2\frac{5}{11}$

No renaming is necessary, since $\frac{7}{11}$ is greater than $\frac{5}{11}$.

First, subtract the like-fraction parts. No simplification is possible. Then, subtract the integer parts.

$$\begin{array}{r} 7\frac{7}{11} \\ -2\frac{5}{11} \\ \hline 5\frac{2}{11} \end{array} \quad \textit{Ans.}$$

EXAMPLE 2 Subtract: $7\frac{1}{12} - 5\frac{5}{12}$

Since $\frac{5}{12} > \frac{1}{12}$, rename the minuend. $7\frac{1}{12} = 6\frac{13}{12}$

Subtract the like-fraction parts and subtract the integer parts.

$$\begin{array}{r} 6\frac{13}{12} \\ -5\frac{5}{12} \\ \hline 1\frac{8}{12} \end{array}$$

Simplify the resulting fraction part.

$$1\frac{8 \div 4}{12 \div 4} = 1\frac{2}{3} \quad \textit{Ans.}$$

CLASS EXERCISES

1. Subtract. Write answers in simplest form.

a. $\begin{array}{r} 12\frac{9}{17} \\ -8\frac{3}{17} \\ \hline \end{array}$ **b.** $\begin{array}{r} 22\frac{9}{10} \\ -18\frac{3}{10} \\ \hline \end{array}$ **c.** $\begin{array}{r} 7\frac{11}{15} \\ -3\frac{8}{15} \\ \hline \end{array}$ **d.** $\begin{array}{r} 9\frac{3}{8} \\ -5\frac{1}{8} \\ \hline \end{array}$

e. $2\frac{3}{10} - 1\frac{7}{10}$ **f.** $8\frac{1}{6} - 2\frac{5}{6}$

g. $9 - 6\frac{3}{4}$ **h.** $12 - 10\frac{11}{16}$

2. Jenny promised her mother she would practice piano for 7 hours this week. If she has practiced for $4\frac{3}{4}$ hours so far, how many hours does she have left to practice?

3. The average employee in Switzerland works 36 hours per week, while the average employee in France works $33\frac{4}{5}$ hours. How many hours more per week does the average Swiss employee work than the average French employee?

SUBTRACTING MIXED NUMBERS WITH UNLIKE-FRACTION PARTS

THE MAIN IDEA

To subtract mixed numbers with unlike-fraction parts:

1. Subtract the unlike-fraction parts.
   a. Use the LCD to write equivalent like fractions.
   b. If necessary, rename the minuend.
   c. If possible, simplify the resulting fraction.
2. Subtract the integer parts.

EXAMPLE 3 Subtract: $8\frac{2}{3} - 4\frac{1}{5}$

The LCD of the unlike fractions $\frac{2}{3}$ and $\frac{1}{5}$ is 15.

(1) Rewrite the unlike-fraction parts as equivalent like fractions.

$8\frac{2\times5}{3\times5} = 8\frac{10}{15}$ $\quad$ $4\frac{1\times3}{5\times3} = 4\frac{3}{15}$

(2) Subtract the resulting like-fraction parts and the integer parts. No simplification is possible.

$$\begin{array}{r} 8\frac{10}{15} \\ -4\frac{3}{15} \\ \hline 4\frac{7}{15} \end{array}$$ *Ans.*

CLASS EXERCISES

1. Subtract. Write answers in simplest form.

a. $\begin{array}{r} 7\frac{3}{10} \\ -2\frac{2}{5} \\ \hline \end{array}$ **b.** $\begin{array}{r} 6\frac{9}{10} \\ -4\frac{3}{4} \\ \hline \end{array}$ **c.** $\begin{array}{r} 4\frac{7}{12} \\ -1\frac{3}{8} \\ \hline \end{array}$ **d.** $\begin{array}{r} 10\frac{1}{2} \\ -4\frac{7}{8} \\ \hline \end{array}$ **e.** $\begin{array}{r} 9\frac{5}{16} \\ -6\frac{5}{8} \\ \hline \end{array}$ **f.** $\begin{array}{r} 12\frac{2}{5} \\ -3\frac{3}{4} \\ \hline \end{array}$

g. $\begin{array}{r} 14\frac{3}{8} \\ -5\frac{2}{3} \\ \hline \end{array}$ **h.** $\begin{array}{r} 20\frac{7}{10} \\ -9\frac{4}{5} \\ \hline \end{array}$ **i.** $\begin{array}{r} 8\frac{1}{4} \\ -4\frac{3}{8} \\ \hline \end{array}$ **j.** $\begin{array}{r} 4\frac{1}{6} \\ -2\frac{2}{3} \\ \hline \end{array}$ **k.** $\begin{array}{r} 6\frac{1}{3} \\ -2\frac{1}{8} \\ \hline \end{array}$ **l.** $\begin{array}{r} 3\frac{1}{6} \\ -1\frac{3}{4} \\ \hline \end{array}$

2. Mary's home is $2\frac{3}{4}$ miles from school. If Mary walked $1\frac{1}{10}$ miles toward school, how much farther must she walk?

3. Alan had $12\frac{1}{8}$ pounds of peanuts. After he sold $6\frac{3}{4}$ pounds, how much does he have?

4. A recipe for fish chowder calls for $2\frac{1}{2}$ pounds of fish. Marika has pieces of fish that weigh $\frac{3}{4}$ lb. and $1\frac{5}{16}$ lb. Does she need to buy more fish to make the recipe? If so, how much more?

HOMEWORK EXERCISES

1. Subtract. Write answers in simplest form.

a. $24\frac{11}{12} - 18\frac{10}{12}$ b. $16\frac{3}{4} - 11\frac{1}{4}$ c. $20\frac{7}{8} - 17\frac{4}{8}$ d. $42\frac{9}{10} - 24\frac{7}{10}$ e. $19\frac{5}{6} - 9\frac{1}{6}$ f. $53 - 47\frac{17}{30}$ g. $16\frac{3}{4} - 2\frac{1}{4}$

h. $21\frac{9}{36} - 18\frac{7}{36}$ i. $29\frac{16}{17} - 19\frac{15}{17}$ j. $37 - 26\frac{1}{4}$ k. $12\frac{7}{9} - 9$ l. $21 - 11\frac{3}{8}$ m. $15\frac{1}{2} - 2\frac{1}{2}$

2. Subtract. Write answers in simplest form.

a. $18\frac{3}{4} - 6\frac{1}{2}$ b. $21\frac{5}{8} - 17\frac{1}{4}$ c. $38\frac{4}{5} - 22\frac{3}{10}$ d. $29\frac{5}{12} - 15\frac{1}{6}$ e. $5\frac{2}{3} - 3\frac{1}{2}$ f. $24\frac{3}{8} - 17\frac{1}{5}$ g. $21\frac{1}{6} - 12\frac{1}{4}$

h. $40\frac{5}{9} - 36\frac{2}{5}$ i. $36\frac{1}{8} - 24\frac{3}{5}$ j. $52\frac{3}{7} - 40\frac{3}{5}$ k. $29\frac{7}{10} - 12\frac{7}{8}$ l. $19\frac{1}{20} - 5\frac{1}{15}$ m. $8\frac{1}{8} - 2\frac{1}{4}$

3. From $13\frac{7}{8}$, subtract $9\frac{1}{5}$. 4. Subtract $32\frac{4}{5}$ from $41\frac{1}{7}$. 5. Subtract 15 from $30\frac{1}{2}$.

6. Steve ran $6\frac{3}{4}$ miles Sunday. After he had run $3\frac{1}{4}$ miles, Joy joined him for the rest of the run. How far did Joy run with Steve?

7. It is $4\frac{4}{5}$ miles from Rebecca's house to school. She walked $\frac{2}{5}$ mile and then took the bus. How far did she ride the bus?

8. The largest brook trout caught so far weighed $14\frac{1}{2}$ pounds. The largest rainbow trout caught weighed $42\frac{1}{8}$ pounds. How much more did the largest rainbow trout weigh than the largest brook trout?

9. A package of hamburger weighing $2\frac{1}{2}$ pounds was cooked. After cooking, the hamburger weighed $1\frac{7}{8}$ pounds. How much less did the cooked hamburger weigh than the raw hamburger?

10. From an 8-foot board, a carpenter sawed off pieces that were $3\frac{3}{4}$ feet and $2\frac{1}{2}$ feet long. How much of the board is left?

11. After school, Juan has 7 hours of time available. The chart shows Juan's plan to use this time. How much time does this leave Juan to spend with his friends during a five-day school week?

Activity	Time Budgeted (in Hours)
Homework	$2\frac{3}{4}$
Track Practice	$2\frac{1}{2}$
Dinner	$\frac{3}{4}$
Friends	?

SPIRAL REVIEW EXERCISES

1. Find the sum of $9\frac{7}{8}$ and $2\frac{2}{3}$.

2. Jennifer needs $4\frac{1}{3}$ cups of milk for one recipe and $6\frac{5}{6}$ cups of milk for a second recipe. How many cups of milk does she need for both recipes?

3. The sum of −12 and −8 is
(a) −4 (b) 4 (c) −20 (d) 20

4. 47 ÷ 8 is closest to
(a) 5 (b) 6 (c) 7 (d) 8

5. The greatest common factor of 27 and 75 is
(a) 3 (b) 5 (c) 9 (d) 15

6. Divide 40 by $\frac{1}{4}$.

7. Find $\frac{1}{4}$ of 40.

8. Which key sequence and display show that addition and subtraction are inverse operations?

Key Sequence	Display
(a) 128 [−] 59 [+] 128 [=]	197.
(b) 128 [+] 128 [+] 59 [=]	315.
(c) 128 [−] 0 [+] 128 [=]	256.
(d) 128 [+] 59 [−] 59 [=]	128.

9. Zena spends $\frac{1}{2}$ hour delivering newspapers each day. How many hours does she spend in one 7-day week?

10. Each line of a computer printout contains 75 characters.
    a. How many characters are there on 56 such lines?
    b. A printer can print 150 characters each second. How long will it take to print these lines?

11. What is the quotient when 12,060 is divided by 60?
(a) 21 (b) 210 (c) 200 (d) 201

12. Find all the prime numbers between 70 and 80.

13. Write the word name for 23,802,010.

14. Write each group of numbers in increasing order.
    a. 0, −16, −5 b. $\frac{11}{4}$, $2\frac{3}{5}$, $\frac{13}{6}$

UNIT 7-3 Multiplying Mixed Numbers

THE MAIN IDEA

To multiply mixed numbers:

1. Write each mixed number as an improper fraction.
2. Follow the procedure for multiplying two fractions.

EXAMPLE 1 Multiply: $9 \times 4\frac{2}{3}$

Write the integer as a fraction. $\frac{9}{1} \times 4\frac{2}{3}$

Write the mixed number as an improper fraction. $\frac{9}{1} \times \frac{14}{3}$

To simplify, divide 9 and 3 by 3. $\frac{\overset{3}{\cancel{9}}}{1} \times \frac{14}{\underset{1}{\cancel{3}}}$

Multiply the results. $\frac{3 \times 14}{1 \times 1} = \frac{42}{1} = 42$ *Ans.*

EXAMPLE 2 A recipe calls for $2\frac{1}{4}$ cups of flour. How much flour is needed to make $\frac{2}{3}$ of the recipe?

THINKING ABOUT THE PROBLEM

The key phrase "$\frac{2}{3}$ of" tells you to multiply.

Write a number phrase. $\frac{2}{3} \times 2\frac{1}{4}$

Write the mixed number as an improper fraction. $\frac{2}{3} \times \frac{9}{4}$

Simplify and multiply. $\frac{\overset{1}{\cancel{2}}}{\underset{1}{\cancel{3}}} \times \frac{\overset{3}{\cancel{9}}}{\underset{2}{\cancel{4}}} = \frac{3}{2}$

Answer: $\frac{3}{2}$ cups or $1\frac{1}{2}$ cups of flour are needed.

EXAMPLE 3 Ann bought $1\frac{3}{4}$ pounds of beef at $5.20 a pound. How much did she spend?

Write a number phrase. $1\frac{3}{4} \times 5.20$

Write $1\frac{3}{4}$ as an improper fraction and write 5.20 as a fraction. $\frac{7}{4} \times \frac{5.20}{1}$

Multiply. To simplify, divide 4 and 5.20 by 4. $\frac{7}{\underset{1}{\cancel{4}}} \times \frac{\overset{1.30}{\cancel{5.20}}}{1} = \$9.10$ *Ans.*

CLASS EXERCISES

1. Multiply. Write answers in simplest form.

a. $3\frac{1}{2} \times \frac{1}{4}$ **b.** $2\frac{1}{3} \times 2\frac{3}{4}$ **c.** $3\frac{2}{5} \times 5\frac{1}{3}$ **d.** $1\frac{3}{4} \times \frac{1}{5}$ **e.** $5\frac{3}{8} \times 3\frac{1}{2}$ **f.** $8 \times 2\frac{1}{6}$

g. $4\frac{2}{3} \times 9$ **h.** $7\frac{3}{4} \times 1\frac{1}{3}$ **i.** $\frac{5}{8} \times 1\frac{1}{4}$ **j.** $1\frac{1}{5} \times \frac{5}{12}$ **k.** $\frac{2}{3} \times 5\frac{1}{2}$ **l.** $2 \times 3\frac{1}{4}$

2. The product of $4\frac{4}{5}$ and $3\frac{5}{8}$ is (a) $15\frac{4}{5}$ (b) $16\frac{3}{5}$ (c) $17\frac{2}{5}$ (d) $18\frac{7}{40}$

3. The distance from Michelle's house to school is $\frac{3}{5}$ of a mile. How far did Michelle walk if she made this trip 20 times?

4. A bug landed on a phonograph record that was making $33\frac{1}{3}$ revolutions per minute. The bug stayed on the record for 6 minutes before flying off. How many complete revolutions did the bug make on the record?

5. Phil Rizzo worked $12\frac{1}{2}$ hours last week. If his hourly wage is \$6.50, how much did he earn?

6. Alex spent $\frac{2}{3}$ of his allowance. If his allowance is \$7.50, how much money does he have left?

HOMEWORK EXERCISES

1. Multiply. Write answers in simplest form.

a. $5\frac{1}{3} \times \frac{1}{2}$ **b.** $3\frac{3}{4} \times \frac{1}{5}$ **c.** $10\frac{1}{2} \times 9$ **d.** $4\frac{1}{2} \times 3\frac{1}{3}$ **e.** $5\frac{5}{8} \times 1\frac{2}{3}$

f. $12 \times 3\frac{3}{8}$ **g.** $\frac{7}{9} \times 3\frac{3}{7}$ **h.** $5\frac{3}{5} \times 2\frac{1}{5}$ **i.** $16\frac{3}{10} \times 4\frac{3}{5}$ **j.** $8\frac{5}{8} \times 6\frac{2}{3}$

2. When $3\frac{2}{3}$ is multiplied by $2\frac{1}{2}$, the product is (a) $\frac{37}{6}$ (b) 11 (c) $\frac{22}{15}$ (d) $9\frac{1}{6}$

3. Find $\frac{7}{16}$ of $2\frac{2}{7}$.

4. A pound of flour fills $2\frac{1}{3}$ jars. How many jars would be filled by $1\frac{1}{2}$ pounds of flour?

5. A bag of grapefruit weighs $8\frac{1}{2}$ pounds. How much would 3 bags of grapefruit weigh?

6. A melon weighed $1\frac{7}{16}$ pounds. If you ate $2\frac{1}{2}$ melons in a week, how many pounds of melon would you have eaten?

7. On a package of Sweet Stuff sugar substitute, the table shows what amounts of the powdered and liquid forms are equivalent to given amounts of granulated sugar.

Granulated Sugar	$\frac{1}{4}$ Cup	$\frac{1}{3}$ Cup	$\frac{1}{2}$ Cup	1 Cup
Sweet Stuff Powder	1 tsp.	$1\frac{1}{4}$ tsp.	2 tsp.	4 tsp.
Sweet Stuff Liquid	$1\frac{1}{2}$ tsp.	2 tsp.	3 tsp.	6 tsp.

If $\frac{2}{3}$ of a cup of sugar is to be replaced by powdered Sweet Stuff and $\frac{3}{4}$ of a cup of sugar is to be replaced by liquid Sweet Stuff, what is the total number of teaspoons (tsp.) of Sweet Stuff that has to be measured out?

8. Mei Ling needs $3\frac{1}{4}$ yards of fabric for a pattern she is sewing. If she buys fabric that costs $7.50 per yard, how much will she pay?

9. Kathy had a very profitable year with her investments. Each of her four stocks showed a profit, as indicated on her "stock profit sheet."

Stock Profit Sheet

Name of Stock	Number of Shares	Profit per Share	Total Profit for Stock
Dynamic Dolls	400	$2\frac{3}{8}$	
Striking Oil	1,000	$10\frac{2}{5}$	
Blue Coal	800	$12\frac{3}{16}$	
Fast Bikes	1,200	$11\frac{1}{12}$	

a. Copy the table and complete each of the entries in the last column.
b. What was Kathy's total profit from the four stocks?

SPIRAL REVIEW EXERCISES

1. John's father baked 3 loaves of bread. John and his friends ate $1\frac{1}{6}$ loaves. How many loaves were left?

2. Add: $12\frac{3}{5} + 8\frac{2}{3}$

3. From 5, subtract $3\frac{3}{8}$.

4. Olga practiced playing the piano $\frac{3}{4}$ hour Monday, $\frac{1}{2}$ hour Tuesday, $\frac{1}{2}$ hour Wednesday, $1\frac{1}{3}$ hours Thursday, and $\frac{2}{3}$ hour Friday. How much time did she spend practicing the piano in all?

5. Find the remainder when 189 is divided by 13.

6. Subtract 13 from −20.

7. Simplify: $\frac{48}{80}$

8. Use your calculator to find the number to replace $\boxed{?}$ so that the result is a true statement.

$$\frac{27}{19} = \frac{621}{\boxed{?}}$$

9. Joy's diet allowed her to eat $\frac{1}{8}$ of a pound of cheese in a sandwich. How many sandwiches could she make from $2\frac{3}{4}$ pounds of cheese?

10. On Martin's block, there are 13 garages that can hold 2 cars each and 8 garages that can hold 1 car each. What is the total number of cars that can be garaged on the block?

11. How many blood specimens can be held on 8 trays if each tray has 12 rows of test tubes with 10 test tubes in each row?

12. Which is not a factor of 144?
(a) 8 (b) 9 (c) 12 (d) 14

UNIT 7-4 Dividing Mixed Numbers

THE MAIN IDEA

To divide mixed numbers:

1. Write each mixed number as an improper fraction.
2. Follow the procedure for dividing two fractions.

EXAMPLE 1 Divide: $2\frac{3}{5} \div 5\frac{1}{2}$

Write each mixed number as an improper fraction.

$$2\frac{3}{5} = \frac{5 \times 2 + 3}{5} = \frac{13}{5}$$

$$5\frac{1}{2} = \frac{2 \times 5 + 1}{2} = \frac{11}{2}$$

Divide the fractions. $\frac{13}{5} \div \frac{11}{2}$

Use the reciprocal of the divisor and change the division to multiplication. $\frac{13}{5} \times \frac{2}{11}$

There are no common factors. Multiply. $\frac{13 \times 2}{5 \times 11} = \frac{26}{55}$ *Ans.*

EXAMPLE 2 Divide: $2\frac{5}{8} \div 3$

Write the mixed number as an improper fraction. $2\frac{5}{8} = \frac{8 \times 2 + 5}{8} = \frac{21}{8}$

Express the integer as an improper fraction. $3 = \frac{3}{1}$

Divide the fractions. $\frac{21}{8} \div \frac{3}{1}$

Invert the divisor and multiply. $\frac{\overset{7}{\cancel{21}}}{8} \times \frac{1}{\underset{1}{\cancel{3}}} = \frac{7 \times 1}{8 \times 1} = \frac{7}{8}$ *Ans.*

EXAMPLE 3 June makes pasta salad and sells it in containers that hold $1\frac{1}{4}$ pounds. How many containers does she need to hold 660 pounds of pasta salad?

THINKING ABOUT THE PROBLEM

To separate a given amount into a number of smaller equal portions, you divide.

Write a number phrase. $660 \div 1\frac{1}{4}$

Change the mixed number to an improper fraction. $1\frac{1}{4} = \frac{5}{4}$

Express the whole number as an improper fraction. $660 = \frac{660}{1}$

Divide the fractions. $\frac{660}{1} \div \frac{5}{4}$

Invert the divisor and multiply. $\frac{\overset{132}{\cancel{660}}}{1} \times \frac{4}{\underset{1}{\cancel{5}}}$

$$\frac{132 \times 4}{1 \times 1} = \frac{528}{1} = 528$$

Answer: June needs 528 containers.

CLASS EXERCISES

1. Divide. Write answers in simplest form.

   a. $1\frac{3}{5} \div 1\frac{1}{3}$ b. $3\frac{1}{3} \div 5\frac{1}{2}$ c. $\frac{5}{6} \div 6\frac{2}{3}$ d. $2\frac{2}{3} \div 1\frac{3}{8}$ e. $2\frac{1}{4} \div 2\frac{2}{5}$ f. $3\frac{3}{8} \div 9$

   g. $1\frac{4}{5} \div \frac{3}{5}$ h. $6\frac{7}{8} \div 1\frac{2}{3}$ i. $42 \div 4\frac{1}{5}$ j. $5\frac{1}{2} \div \frac{1}{2}$ k. $1\frac{1}{3} \div \frac{1}{3}$ l. $1\frac{1}{3} \div \frac{2}{3}$

2. The quotient of $4\frac{2}{3}$ and $1\frac{1}{6}$ is (a) $3\frac{20}{21}$ (b) 4 (c) $4\frac{5}{21}$ (d) $5\frac{1}{21}$

3. Jack has $6\frac{2}{5}$ pounds of raisins that he wants to share equally among himself and three friends. How many pounds should he give to each person?

4. How many pieces, each $4\frac{3}{4}$ inches long, can be cut from a string that is $85\frac{1}{2}$ inches long?

5. It took $4\frac{1}{2}$ hours for Bo and Shelly to unload a truckload of bricks. What fractional part of the job did they complete in one hour?

6. Paco earned $110.70 last week. If he worked $13\frac{1}{2}$ hours, what was his hourly wage?

HOMEWORK EXERCISES

1. Divide. Answers should be in simplest form.

   a. $2\frac{1}{3} \div \frac{1}{2}$ b. $\frac{3}{5} \div 1\frac{1}{2}$ c. $4\frac{1}{3} \div 2\frac{1}{4}$ d. $5\frac{2}{3} \div 1\frac{3}{5}$ e. $5\frac{1}{4} \div 7$ f. $1\frac{3}{8} \div \frac{22}{32}$

   g. $2\frac{2}{3} \div 3\frac{1}{3}$ h. $7\frac{2}{3} \div 46$ i. $6\frac{1}{2} \div 8\frac{2}{3}$ j. $2\frac{1}{4} \div \frac{1}{4}$ k. $2\frac{1}{2} \div 1\frac{1}{4}$ l. $1\frac{1}{4} \div 2\frac{1}{2}$

2. When 48 is divided by $\frac{1}{2}$, the quotient is (a) $48\frac{1}{2}$ (b) 96 (c) 24 (d) $\frac{24}{48}$

3. A bag of 12 oranges weighed $7\frac{1}{2}$ pounds. About how much did each orange weigh?

4. There were $5\frac{1}{2}$ sandwiches for 4 children. How much should each child get?

5. Steve cut a wire that was $20\frac{1}{4}$ feet long into pieces $3\frac{3}{8}$ feet long. How many pieces did he have?

6. It takes $6\frac{3}{4}$ days to photograph the entire senior class for the yearbook. At a steady rate, what part of the class can be photographed in one day?

7. Karl has a board $4\frac{1}{2}$ ft. long. He wants to cut it into pieces that are $1\frac{1}{2}$ ft. long.

   a. Draw a diagram showing where he should cut the board.

   b. How many pieces $1\frac{1}{2}$ ft. long will he have?

   c. Use your diagram to explain how to divide $4\frac{1}{2}$ by $1\frac{1}{2}$.

8. What number must replace ? to make a true statement? $3\frac{1}{2} \times ? = 17\frac{1}{2}$

9. Suso bought 2 packages of hamburger meat. Each package weighed $2\frac{3}{4}$ pounds. When he got home, he divided the meat into $\frac{1}{4}$-pound patties. How many patties did he make?

10. If $3\frac{1}{2}$ pounds of tomatoes cost \$6.30, what is the cost of one pound of tomatoes?

SPIRAL REVIEW EXERCISES

1. Jane took a trip to Cape Cod. She averaged 50 miles per hour for $4\frac{1}{2}$ hours. How far is Cape Cod from Jane's home?

2. Multiply: $5\frac{1}{2}$ by $3\frac{2}{3}$

3. Find the sum of $8\frac{1}{4}$ and $4\frac{1}{8}$.

4. Subtract $2\frac{3}{7}$ from 5.

5. The opposite of –11 is (a) 11 (b) $\frac{1}{11}$ (c) $\frac{-1}{11}$

6. What is the reciprocal of 20?

7. Is 2,349 a multiple of 87?

8. Find $\frac{4}{5}$ of $\frac{5}{16}$.

9. Replace ? by <, >, or = to make a true statement:

   a. $\frac{7}{16}$? $\frac{21}{48}$ b. $\frac{1}{5}$? $\frac{1}{4}$ c. $\frac{2}{3}$? $\frac{1}{3}$

10. On a $5\frac{1}{2}$-hour trip, Rich drove $\frac{2}{3}$ of the time. How long did he drive?

11. If a printer can produce 450 copies in one hour, how long would be required for 9,900 copies?

12. Write a number phrase that represents the total cost of 8 paperback books costing \$5 each and 2 calculators costing \$8 each.

13. If Earl can construct $\frac{2}{5}$ of a wall each day, how long will it take him to do the wall?

THEME 4

Percents

MODULE 8

MODULE 9

MODULE 8

UNIT 8-1 The Meaning of Percent: Writing Percents as Decimals and Decimals as Percents

THE MEANING OF PERCENT

THE MAIN IDEA

1. *Percent* means hundredths. (The symbol for percent is %.)
2. 100% means 100 hundredths or $\frac{100}{100}$ or 1, which represents a whole quantity.
3. Percents less than 100% are numbers less than 1, and represent less than a whole quantity.
4. Percents greater than 100% are numbers greater than 1, and represent more than a whole quantity.

EXAMPLE 1 For each percent given in numbers, write the percent in words. Tell, in words and in numbers, how many hundredths there are in each percent.

	Percent in Numbers	*Percent in Words*	*Hundredths in Words*	*Hundredths in Numbers*
a.	1%	one percent	one hundredth	$\frac{1}{100}$
b.	95%	ninety-five percent	ninety-five hundredths	$\frac{95}{100}$
c.	100%	one hundred percent	one hundred hundredths	$\frac{100}{100}$
d.	250%	two hundred fifty percent	two hundred fifty hundredths	$\frac{250}{100}$

EXAMPLE 2 Write 57 hundredths as a percent.

Since "percent" means "hundredths," 57 hundredths is the same as 57 percent.

Answer: 57%

EXAMPLE 3 Replace ? by $<$, $=$, or $>$ to make each comparison a true statement.

	Comparison	*True Statement*
a.	$12\%\ ?\ 1$	$12\% < 1$
b.	$99\frac{1}{2}\%\ ?\ 1$	$99\frac{1}{2}\% < 1$
c.	$100\%\ ?\ 1$	$100\% = 1$
d.	$101\%\ ?\ 1$	$101\% > 1$
e.	$1{,}000\%\ ?\ 1$	$1{,}000\% > 1$

EXAMPLE 4 Jonathan has finished 68% of his homework. What percent of his homework remains to be done?

THINKING ABOUT THE PROBLEM

The key word "remains" tells you to subtract.

$$\begin{array}{r l} 100\% & \leftarrow \text{the whole quantity} \\ -68\% & \leftarrow \text{the percent already done} \\ \hline 32\% & \leftarrow \text{the percent remaining to be done} \end{array}$$

Answer: Jonathan must do the remaining 32% of his homework.

CLASS EXERCISES

1. For each percent given in numbers, write the percent in words. Tell, in words and in numbers, how many hundredths there are in each percent.

a. 35% **b.** 42% **c.** 10% **d.** 100% **e.** 87% **f.** 110% **g.** 2% **h.** 200%

2. Write as a percent.

a. $\frac{37}{100}$ **b.** 42 hundredths **c.** 3 hundredths **d.** $\frac{175}{100}$ **e.** 5 hundred hundredths

3. Replace ? by $<$, $=$, or $>$ to make each comparison a true statement.

a. $300\%\ ?\ 1$ **b.** $30\%\ ?\ 1$ **c.** $90\%\ ?\ 1$ **d.** $4\%\ ?\ 1$ **e.** $120\%\ ?\ 1$

f. $100\%\ ?\ 1$ **g.** $50\%\ ?\ 1$ **h.** $150\%\ ?\ 1$ **i.** $5\frac{1}{2}\%\ ?\ 1$ **j.** $31.5\%\ ?\ 1$

4. Every morning, Samantha does her exercise routine.

a. If she has completed 26% of the routine, what percent remains to be done?

b. If she has completed 111% of the routine, by what percent has she exceeded her routine?

WRITING PERCENTS AS DECIMALS AND DECIMALS AS PERCENTS

THE MAIN IDEA

1. To write a percent as a decimal:
   a. Drop the % sign.
   b. Divide by 100 by moving the decimal point two places to the *left*, adding zeros if necessary.
2. To write a decimal as a percent:
   a. Multiply by 100 by moving the decimal point two places to the *right*, adding zeros if necessary.
   b. Write the resulting number followed by a % sign.

EXAMPLE 5 Write each percent as a decimal.

	Percent	*Moving the Decimal Point*	*Decimal*
a.	52%	.52.	0.52
b.	8%	.08.	0.08
c.	4.3%	.04.3	0.043
d.	31.5%	.31.5	0.315
e.	0.6%	.00.6	0.006
f.	200%	2.00.	2

EXAMPLE 6 Write each decimal as a percent.

	Decimal	*Moving the Decimal Point*	*Percent*
a.	0.34	0.34.	34%
b.	0.253	0.25.3	25.3%
c.	0.047	0.04.7	4.7%
d.	7	7.00.	700%
e.	8.5	8.50.	850%
f.	29.3	29.30.	2,930%

CLASS EXERCISES

1. Write each percent as a decimal.

a. 17% **b.** 9% **c.** 50% **d.** 6.2% **e.** 48.6% **f.** 0.8% **g.** 300%
h. 120% **i.** 250% **j.** 115% **k.** 410% **l.** 500% **m.** 33.3% **n.** 1%

2. A certain fruit drink is 32.5% fruit juice. What is this percent written as a decimal?

3. The inflation rate in a certain country was recorded at 160%. What is this percent written as a decimal?

4. Of the eggs packed by Red Hen Dairy, 1.5% break before they reach the market. Write this percent as a decimal.

5. Stuart is 120.5% of his normal weight. Write this percent as a decimal.

6. Write each decimal as a percent.

a. 0.90 **b.** 0.06 **c.** 2 **d.** 0.2 **e.** 0.02 **f.** 1.5 **g.** 0.003
h. 0.486 **i.** 0.038 **j.** 1.08 **k.** 0.0043 **l.** 0.184 **m.** 0.75 **n.** 1.75

7. The sales tax rate in a certain city is 0.075, written as a decimal. What is this tax rate written as a percent?

8. Carmen's family spends 0.33 of their income for rent. What percent of their income is spent for rent?

HOMEWORK EXERCISES

1. For each percent given in numbers, write the percent in words. Tell, in words and in numbers, how many hundredths there are in each percent.

a. 1% **b.** 100% **c.** 24% **d.** 400% **e.** 95% **f.** 33% **g.** 50%

2. Write as a percent.

a. $\frac{5}{100}$ **b.** 12 hundredths **c.** one hundred hundredths **d.** $\frac{43}{100}$ **e.** 82 hundredths

3. Replace ? by $<$, $=$, or $>$ to make each comparison a true statement.

a. 24% ? 1 **b.** 100% ? 1 **c.** 8.5% ? 1 **d.** 125% ? 1 **e.** 48% ? 1 **f.** 1% ? 1

4. On his diet, Mr. Watson is allowed no more than a certain number of calories daily.

a. If he has consumed 82% of the allowed number of daily calories, what percent remains to be consumed?

b. If he has consumed 136% of the allowed number of daily calories, by what percent has he exceeded his allowed number of calories?

5. Write each percent as a decimal.

a. 98% **b.** 2% **c.** 0.7% **d.** 31.4% **e.** 8.5% **f.** 65% **g.** 99.9%
h. 1.1% **i.** 0.08% **j.** 140% **k.** 500% **l.** 250% **m.** 101% **n.** 1,000%

6. A certain credit card company charges 11.5% as a service fee. Write this percent as a decimal.

7. Five percent of every dollar spent in Townville goes for sales tax. Write this percent as a decimal.

8. 1% written as a decimal is (a) 0.001 (b) 0.01 (c) 0.1 (d) 1

9. A soap company claims that its soap is 99.44% pure. Write this percent as a decimal.

10. Write each decimal as a percent.

a. 0.6 **b.** 0.48 **c.** 0.09 **d.** 3.2 **e.** 3 **f.** 0.89 **g.** 1.1
h. 0.70 **i.** 0.025 **j.** 0.512 **k.** 0.008 **l.** 3.04 **m.** 1.0 **n.** 10.0

11. In a certain high school, 0.258 of the senior class was accepted by Central University. What percent of the class was accepted by Central?

12. Thirteen thousandths of a certain yogurt is fat. Write this decimal as a percent.

13. Which is equal to 200%? (a) 200 (b) 0.200 (c) 2% (d) 2

14. Rearrange in order, from least to greatest:

13% $\frac{2}{100}$ 0.08 0.8 3% 5 0.5

SPIRAL REVIEW EXERCISES

1. Multiply: 63.1×4.6

2. Mrs. Farland's checking account had a balance of $117.50. Then she wrote checks for $17.86, $25, and $35.98. What was her new balance?

3. B & G Clothing Store reduced the price of a $35 bathing suit by $7.50. What was the new price?

4. Three-eighths of the cars in a dealer's lot were blue and one-fourth were red. If there were 160 cars, how many were neither blue nor red?

5. To make the statement $\frac{7}{8} = \frac{35}{?}$ true, the symbol ? should be replaced by
(a) 48 (b) 8 (c) 40 (d) 56

6. What is the cost of 7 oranges if 12 cost $4.68?

7. $\frac{5}{6}$ is equivalent to
(a) $0.8\overline{3}$ (b) $0.\overline{83}$ (c) $0.\overline{38}$ (d) $0.\overline{8}$

8. The sum of +7 and −7 is
(a) −49 (b) −14 (c) +14 (d) 0

9. On an entry-order calculator, which key sequence and display show that 1 is the identity element for multiplication?

Key Sequence	***Display***
(a) 47 [+] 1 [×] 5 [=]	240.
(b) 58 [−] 57 [×] 39 [=]	39.
(c) 139 [+] 1 [−] 1 [=]	139.
(d) 1 [+] 6 [×] 6 [−] 1 [=]	41.

10. A jet airplane flew at an altitude of 28,000 feet and averaged 565 miles per hour for exactly 4.3 hours. How many miles did the jet airplane travel during this time?

UNIT 8–2 Writing Percents as Fractions and Fractions as Percents; Fraction, Decimal, and Percent Equivalents

WRITING PERCENTS AS FRACTIONS

THE MAIN IDEA

1. To write a percent as a fraction:
   a. Write a fraction whose numerator is the amount of the percent and whose denominator is 100.
   b. If possible, simplify the fraction.
2. To write a percent that contains a decimal (such as 3.8%) as a fraction:
   a. Write the percent as a decimal.
   b. Write the decimal as a fraction.
   c. If possible, simplify the fraction.

EXAMPLE 1 Write 24% as a fraction.

Write a fraction with numerator 24 and denominator 100. Simplify the fraction.

$$24\% = \frac{24}{100}$$

$$\frac{24 \div 4}{100 \div 4} = \frac{6}{25} \quad \textit{Ans.}$$

EXAMPLE 2 Write 175% as a fraction.

Write a fraction with numerator 175 and denominator 100. Simplify the fraction.

$$175\% = \frac{175}{100}$$

$$\frac{175 \div 25}{100 \div 25} = \frac{7}{4} \text{ or } 1\frac{3}{4} \quad \textit{Ans.}$$

EXAMPLE 3 Write $66\frac{2}{3}\%$ as a fraction.

Write a fraction whose numerator is $66\frac{2}{3}$ and whose denominator is 100.

$$66\frac{2}{3}\% = \frac{66\frac{2}{3}}{100}$$

Since dividing by 100 is the same as multiplying by $\frac{1}{100}$, rewrite the fraction as a product.

$$66\frac{2}{3} \times \frac{1}{100}$$

Write $66\frac{2}{3}$ as a fraction. Multiply.

$$\frac{200}{3} \times \frac{1}{100} = \frac{200}{300} = \frac{2}{3} \quad \textit{Ans.}$$

EXAMPLE 4 Write 6.5% as a fraction.

Write the percent as a decimal. $6.5\% = 0.065$

Write the decimal as a fraction. $0.065 = \frac{65}{1{,}000}$

Simplify. $\frac{65 \div 5}{1{,}000 \div 5} = \frac{13}{200}$ *Ans.*

CLASS EXERCISES

1. Write each percent as a fraction in lowest terms.

a. 9% **b.** 30% **c.** $33\frac{1}{3}\%$ **d.** 7.5% **e.** 47% **f.** 80% **g.** $12\frac{1}{2}\%$ **h.** 24.8%

i. 75% **j.** 495% **k.** $8\frac{1}{4}\%$ **l.** 0.04% **m.** 60% **n.** 1,000% **o.** $\frac{3}{4}\%$ **p.** 10%

2. A nut mixture contains 15% cashews. Write as a fraction in simplest form the part of the mixture that is cashews.

3. A quarterback completed 72% of the passes that he attempted. Write as a fraction in simplest form the part of the passes that he completed.

4. Marvelous Bond Trust Company offers a $6\frac{1}{4}\%$ rate of interest on all investments. Write this percent as a fraction in simplest form.

WRITING FRACTIONS AS PERCENTS

THE MAIN IDEA

To write a fraction as a percent:

1. Write the fraction as a decimal by dividing the denominator into the numerator.
2. Write the resulting decimal as a percent by moving the decimal point two places to the right and writing a % sign.

EXAMPLE 5 Write $\frac{3}{4}$ as a percent.

Write $\frac{3}{4}$ as a decimal by dividing the denominator 4 into the numerator 3.

$$\begin{array}{r} 0.75 \\ 4\overline{)3.00} \\ \underline{2\,8} \\ 20 \\ \underline{20} \\ 0 \end{array}$$

Write 0.75 as a percent by moving the decimal point two places to the right and writing a % sign.

0.75. = 75% *Ans.*

EXAMPLE 6 Write $\frac{1}{3}$ as a percent.

Write $\frac{1}{3}$ as a decimal by dividing the denominator 3 into the numerator 1. Since the decimal repeats, stop the division at the hundredths place.

$$\begin{array}{r} 0.33... \\ 3\overline{)1.00} \\ \underline{9} \\ 10 \\ \underline{9} \\ 1 \end{array}$$

Write 0.33 as a percent.

0.33. = 33%

Use the remainder 1 and the divisor 3 to write the fraction $\frac{1}{3}$. Add this fraction to the percent.

$33\frac{1}{3}\%$ *Ans.*

EXAMPLE 7 Use a calculator to write $\frac{3}{8}$ as a percent.

KEYING IN

Use this key sequence to write $\frac{3}{8}$ as a decimal:

Key Sequence	***Display***
3 [÷] 8 [=]	0.375

Then, write the resulting decimal as a percent: 0.375 = 37.5%

If your calculator has a [%] key, the process can be shortened by using this key sequence and then adding the % sign:

Key Sequence	***Display***
3 [÷] 8 [%]	37.5

Answer: $\frac{3}{8}$ = 37.5%

EXAMPLE 8 Write $3\frac{1}{2}$ as a percent.

Write $3\frac{1}{2}$ as a fraction. $3\frac{1}{2} = \frac{7}{2}$

Write $\frac{7}{2}$ as a decimal by dividing the denominator 2 into the numerator 7.

$$\begin{array}{r} 3.5 \\ 2\overline{)7.0} \\ \underline{6} \\ 1\,0 \\ \underline{1\,0} \\ 0 \end{array}$$

Write 3.5 as a percent. 3.50. = 350% *Ans.*

CLASS EXERCISES

1. Write as a percent:

a. $\frac{1}{2}$ **b.** $\frac{2}{3}$ **c.** $3\frac{1}{4}$ **d.** $\frac{9}{10}$ **e.** $\frac{3}{5}$ **f.** $\frac{1}{6}$ **g.** $5\frac{2}{5}$ **h.** $\frac{17}{20}$

i. $1\frac{5}{8}$ **j.** $\frac{4}{7}$ **k.** $7\frac{1}{12}$ **l.** $4\frac{3}{25}$ **m.** $\frac{5}{16}$ **n.** $\frac{7}{9}$ **o.** 9 **p.** $5\frac{1}{2}$

2. Mr. Wise saves $\frac{13}{50}$ of his salary. What percent of his salary does he save?

3. A basketball team lost $\frac{7}{20}$ of the games that it played. What percent of the games did it lose?

4. Seven-eighths of the students in Ms. Jones's class passed the mathematics test. What percent of the class passed the test?

5. Of 110 juniors, 28 were elected to the Honor Society. Of 89 seniors, 23 were elected to the Honor Society. Which class has a greater percent of honor students?

FRACTION, DECIMAL, AND PERCENT EQUIVALENTS

THE MAIN IDEA

Some often-used fractions and their decimal and percent equivalents are:

Fraction	*Decimal*	*Percent*	*Fraction*	*Decimal*	*Percent*
$\frac{1}{2}$	0.5	50%	$\frac{1}{8}$	0.125	$12\frac{1}{2}\%$
$\frac{1}{3}$	$0.3\overline{3}$	$33\frac{1}{3}\%$	$\frac{3}{8}$	0.375	$37\frac{1}{2}\%$
$\frac{2}{3}$	$0.6\overline{6}$	$66\frac{2}{3}\%$	$\frac{5}{8}$	0.625	$62\frac{1}{2}\%$
$\frac{1}{4}$	0.25	25%	$\frac{7}{8}$	0.875	$87\frac{1}{2}\%$
$\frac{3}{4}$	0.75	75%	$\frac{1}{10}$	0.1	10%
$\frac{1}{5}$	0.2	20%	$\frac{3}{10}$	0.3	30%
$\frac{2}{5}$	0.4	40%	$\frac{7}{10}$	0.7	70%
$\frac{3}{5}$	0.6	60%	$\frac{9}{10}$	0.9	90%
$\frac{4}{5}$	0.8	80%			

EXAMPLE 9 Half of Marie's paycheck went for food this week. What percent of her money did she spend on food?

Without calculating, write $\frac{1}{2}$ as a percent. $\frac{1}{2} = 50\%$

Answer: 50% was spent on food.

EXAMPLE 10 75% of the class voted for Barry for class president. What fractional part of the class voted for Barry?

Without calculating, write 75% as a fraction. $75\% = \frac{3}{4}$

Answer: $\frac{3}{4}$ of the class voted for Barry.

CLASS EXERCISES

1. Without calculating, write each fraction as a decimal and as a percent.

   a. $\frac{1}{10}$ **b.** $\frac{1}{4}$ **c.** $\frac{3}{8}$ **d.** $\frac{2}{5}$ **e.** $\frac{3}{4}$ **f.** $\frac{1}{3}$ **g.** $\frac{7}{8}$ **h.** $\frac{7}{10}$

2. Without calculating, write each decimal as a fraction and as a percent.

   a. 0.125 **b.** 0.75 **c.** $0.\overline{6}$ **d.** 0.375 **e.** 0.1 **f.** 0.25 **g.** 0.5 **h.** 0.9

3. Without calculating, write each percent as a fraction and as a decimal.

   a. 25% **b.** 60% **c.** $87\frac{1}{2}\%$ **d.** 50% **e.** 10% **f.** $33\frac{1}{3}\%$ **g.** 90% **h.** $12\frac{1}{2}\%$

4. Sam saves 10% of his allowance. What fractional part of his money does Sam save?

5. A fourth of the puppies in a litter were tan. What percent of the litter was tan?

6. Copy and complete the following table of fraction, decimal, and percent equivalents.

Fraction	Decimal	Percent	Fraction	Decimal	Percent
$\frac{1}{16}$			$\frac{9}{16}$		
$\frac{3}{16}$			$\frac{11}{16}$		
$\frac{5}{16}$			$\frac{13}{16}$		
$\frac{7}{16}$			$\frac{15}{16}$		

HOMEWORK EXERCISES

1. Write each percent as a fraction in simplest form.

   a. 7% **b.** 5% **c.** 51% **d.** 73% **e.** 97% **f.** 40% **g.** 70%

   h. 65% **i.** 25% **j.** 50% **k.** 130% **l.** 300% **m.** $66\frac{2}{3}\%$ **n.** $87\frac{1}{2}\%$

   o. $\frac{1}{4}\%$ **p.** $83\frac{1}{3}\%$ **q.** 12.5% **r.** 8.25% **s.** 0.8% **t.** 0.06% **u.** 75%

2. A football team won 85% of the games it played. Write as a fraction in simplest form the part of the games that the team won.

3. Mr. Baker sold 63% of the shirts that he had on sale. Write as a fraction in simplest form the part of the shirts that he sold.

4. Money Savings Bank offers a $5\frac{1}{2}\%$ rate of interest. Write the interest rate as a fraction in simplest form.

5. Write as a percent:

a. $\frac{3}{4}$ **b.** $\frac{7}{10}$ **c.** $\frac{5}{8}$ **d.** $\frac{7}{20}$ **e.** $\frac{11}{50}$ **f.** $\frac{8}{9}$ **g.** $\frac{13}{25}$ **h.** $\frac{2}{3}$

i. $\frac{3}{16}$ **j.** $\frac{5}{12}$ **k.** 4 **l.** $2\frac{1}{8}$ **m.** $3\frac{4}{5}$ **n.** $\frac{9}{32}$ **o.** $2\frac{5}{6}$ **p.** $\frac{83}{90}$

6. Ms. Donland's children ate $\frac{6}{8}$ of a pizza pie. What percent of the pie did they eat?

7. Steve's height is $\frac{7}{8}$ of his father's height. What percent of his father's height is this?

8. Without calculating, write each fraction as a decimal and as a percent.

a. $\frac{3}{10}$ **b.** $\frac{7}{10}$ **c.** $\frac{1}{2}$ **d.** $\frac{5}{8}$ **e.** $\frac{2}{3}$ **f.** $\frac{4}{5}$ **g.** $\frac{3}{4}$ **h.** $\frac{7}{8}$

9. Without calculating, write each decimal as a fraction and as a percent.

a. 0.875 **b.** 0.8 **c.** 0.25 **d.** 0.375 **e.** 0.5 **f.** 0.75 **g.** 0.9 **h.** 0.4

10. Without calculating, write each percent as a fraction and as a decimal.

a. $66\frac{2}{3}\%$ **b.** $12\frac{1}{2}\%$ **c.** 80% **d.** 62.5% **e.** 30% **f.** 75% **g.** 50% **h.** 40%

11. A carpenter completed 40% of a job. Without calculating, tell what fractional part of the job he completed.

12. A bolt is $\frac{3}{8}$ of an inch long. Without calculating, write this length as a decimal.

13. Copy and complete the following table.

Fraction	Decimal	Percent
$\frac{3}{32}$		
$\frac{5}{64}$		
$\frac{31}{32}$		

14. Which of the following is not equal to the others?

$$0.5 \quad 50\% \quad \frac{1}{2} \quad \frac{50}{100} \quad \frac{5}{10} \quad 0.05$$

15. Hotstuff High School boasted that its students were doing better than the students at O'Kaye High School because 57 Hotstuff students and only 23 O'Kaye students received state scholarships. What additional information do you need to know in order to determine whether this boast is justified?

16. Carlos is doing a science project. His research reveals the following information:

- 65% of the human body is oxygen.
- The body of a 100-pound person contains 3 pounds of nitrogen.
- $\frac{9}{50}$ of the human body is carbon.
- Phosphorus makes up 0.012 of the human body.
- The body of a 150-pound individual contains 3 pounds of calcium.

Use this information to answer the following questions.

a. What percent of the human body is made up of elements other than oxygen, nitrogen, carbon, phosphorus, and calcium?

b. Arrange the listed elements in order by percent from most abundant to least abundant in the human body.

SPIRAL REVIEW EXERCISES

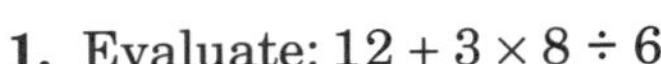

1. Evaluate: $12 + 3 \times 8 \div 6$

2. The greatest common factor of 36, 54, 72, and 108 is

(a) 6 (b) 9 (c) 16 (d) 18

3. The key sequence
5 [÷] 12 [M+] 35 [÷] 84 [−] MR [=] displays 0.

This shows that:

(a) $\frac{5}{12} = \frac{35}{84}$ (b) $\frac{5}{12} < \frac{35}{84}$

(c) $\frac{5}{12} > \frac{35}{84}$ (d) $\frac{5 \times 35}{112 \times 84} = 1$

4. What is the total thickness of 1,000 sheets of paper if each sheet is 0.006 inch thick?

5. Divide –360 by 8.

6. Find $\frac{5}{9}$ of 81.

7. Bill opened a checking account with a balance of $750. He wrote two checks, for $12.78 and $550, and deposited $200. What was his new balance?

8. The first rental of a videotape in the TV-Mania Video Club costs $5.50. Each rental after that costs $2. If the Carson family rented 10 tapes, what was the total cost?

UNIT 8-3 Finding a Percent of a Number; Using Percents to Find Commissions

FINDING A PERCENT OF A NUMBER

THE MAIN IDEA

To find a percent of a number:

1. Write the percent as a decimal.
2. Multiply the number by the decimal.

EXAMPLE 1 Find 25% of 84.

Write 25% as a decimal. 0.25

Multiply 84 by this decimal.

$$\begin{array}{r} 84 \\ \times 0.25 \\ \hline 4\ 20 \\ 16\ 8 \\ \hline 21.00 \end{array}$$

21.00 or 21 *Ans.*

EXAMPLE 2 This year the Russo family spent 150% of what they spent last year for clothes. If they spent $1,000 last year for clothes, what did they spend this year?

Write 150% as a decimal. 1.50

Multiply 1.50 by 1,000 by moving the decimal point three places to the right. 1 . 5 0 0.

Answer: The Russos spent $1,500 for clothes.

EXAMPLE 3 Of the approximately 1,500,000 wood bats sold in 1990, 12.5% of them were bought by major and minor league baseball teams. How many bats does this represent?

Write 12.5% as a decimal. 12.5% = 0.125

Multiply.

$$\begin{array}{r} 1{,}500{,}000 \\ \times \quad 0.125 \\ \hline 7\ 500\ 000 \\ 30\ 000\ 00 \\ 150\ 000\ 0 \\ \hline 187{,}500.000 \end{array}$$

KEYING IN

Key Sequence	*Display*
1500000 [×] 12.5 [%]	187500.

Answer: 187,500 bats

EXAMPLE 4 Estimate 49% of $148.

49% is very close to 50%, which is $\frac{1}{2}$. $\frac{1}{2}$ of $148 = $74 *Ans.*

CLASS EXERCISES

1. Find the value of each expression.

a. 6% of 120 **b.** 42% of 80 **c.** 1% of 445 **d.** 15.4% of 200 **e.** 4.2% of 0.5

2. The Johnson family spent 34% of its weekly income for food. If the Johnsons' weekly income was $300, how much did they spend for food?

3. Carl saved 25% of the money that he earned. If Carl earned $140, how much did he save?

4. The basketball team won 60% of the games that it played. If the team played 40 games, how many games did it win?

5. Membership this year in the Student Organization is 220% of last year's membership. There were 300 members last year. How many members are there this year?

6. Estimate each of the following:

a. 51% of 2,000 **b.** 99% of 77 **c.** 24.5% of 480 **d.** 0.9% of 5,500

USING PERCENTS TO FIND COMMISSIONS

THE MAIN IDEA

1. A ***commission*** is the amount of money that a salesperson earns on a sale.
2. A percent is used to show the rate of commission.
3. To find the amount of commission:
   a. Change the rate of commission to a decimal.
   b. Multiply the dollar amount of the sales by this decimal.

   Amount of Commission = Amount of Sales × Rate of Commission

EXAMPLE 5 Ms. Holt is a salesperson for the Zippy Used Car Company. She earns a 9% commission on each sale. How much did she earn on the sale of a $5,400 car?

To find the commission, find 9% of $5,400.

$5,400	← amount of sale
× 0.09	← rate of commission
$486.00	← amount of commission

KEYING IN

This key sequence gives the amount of commission directly:

Key Sequence	***Display***
5400 [×] 9 [%]	486.

Answer: Ms. Holt earned $486 in commission.

EXAMPLE 6 Mr. James earns $250 a week plus a 10% commission on his sales. Last week his sales totaled $1,490. Find his total earnings for last week.

Find the amount of commission by finding 10% of $1,490.

$1490	← amount of sales
×0.10	← rate of commission
$149.00	← amount of commission

Add the commission to the weekly salary.

$250	← weekly salary
+149	← commission
$399	← total earnings

Answer: Mr. James' total earnings for last week were $399.

EXAMPLE 7 Mrs. Baxter earns a 7% commission on her sales. She sold $240 worth of merchandise on Monday, $300 on Tuesday, and $190 on Wednesday. Find the commission for her total sales on these three days.

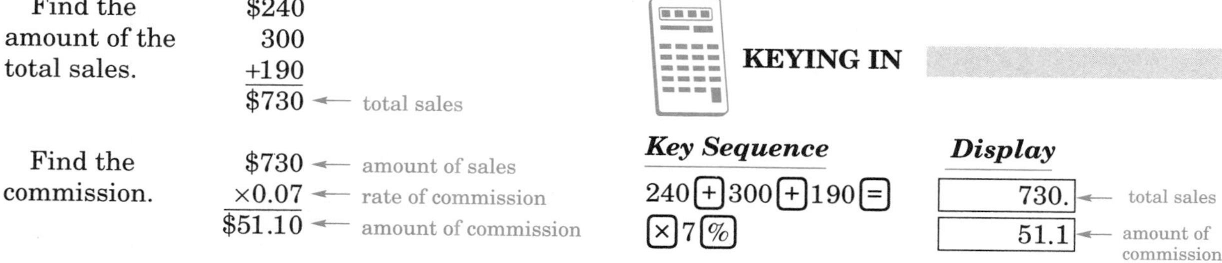

Find the amount of the total sales.

$240	
300	
+190	
$730	← total sales

Find the commission.

$730	← amount of sales
×0.07	← rate of commission
$51.10	← amount of commission

KEYING IN

Key Sequence	***Display***	
240 [+] 300 [+] 190 [=]	730.	← total sales
[×] 7 [%]	51.1	← amount of commission

Answer: Mrs. Baxter's commission was $51.10.

CLASS EXERCISES

1. For each of the given sales and rates of commission, find the amount of commission.

   a. \$180, 10% b. \$235, 5% c. \$12,500, 3% d. \$6,480, $4\frac{1}{2}$% e. \$3,000, 4.8%

2. Bill is paid a 5% commission on his magazine sales. He sold \$180 worth of magazines. Find his commission.
3. Wanda earns an 8% commission. Find Wanda's commission on total sales of \$450.
4. Betty's weekly salary is \$520, plus a 9% commission on her sales. Betty's sales for the week totaled \$800. Find Betty's total earnings for that week.
5. Jim's sales for four months were \$1,500, \$2,900, \$960, and \$3,400. Jim is paid a 10% commission. Find his commission for the four months.

HOMEWORK EXERCISES

1. Find the value of each expression.

   a. 15% of 182 b. 4% of 98 c. 1% of 1,000 d. 98% of 200 e. 200% of 138
   f. 110% of 42.5 g. 52.4% of 800 h. 1.1% of 57 i. 0.3% of 125 j. 10% of 125

2. Of Mr. Franklin's income, 32% is spent for rent. Mr. Franklin's monthly income is \$3,250. How much does he spend for rent?
3. Of Mike's books, 58% are paperbacks. Mike has 50 books. How many are paperbacks?
4. Thirty-six percent of the members of the French Club voted to have their party at Chez Luis Restaurant. The club has 25 members. How many voted for Chez Luis?
5. Mary's weight is 82% of Jim's weight. Jim weighs 150 pounds. How much does Mary weigh?
6. This year, the price of a subscription to Sports Fair Magazine is 115% of what it was last year. Last year, a subscription cost \$18. What is the price this year?
7. If there are 240 cubic centimeters of a solution and 3.2% of the solution is acid, how many cubic centimeters of acid are there?
8. Estimate each of the following:

   a. 101% of 75 b. 9.5% of 60 c. 50.2% of 98 d. 33.1% of 300 e. 76% of 20 f. 76.4% of 64

9. Mr. Jacobs is paid 6% commission on his carpet sales. If he sold $450 worth of carpet, what was his commission?

10. How much commission did Mrs. Florio make on the sale of a $150,000 house if her rate of commission is 3%?

11. Valerie earns a salary of $175 a week, plus a 25% commission on her sales of vacuum cleaners. One week, she sold $950 worth of vacuums. What were Valerie's total earnings for the week?

12. Mr. Wolf is paid 11% commission on his sales of appliances. One day, he sold appliances for $348, $175.90, $244.50, and $79.60. What was his commission for the day?

13. Replace ? by $<$, $=$, or $>$ to make a true statement.

a. 48% of 72 ? 72% of 48 **b.** 0.5% of 20 ? 5% of 20

SPIRAL REVIEW EXERCISES

1. Which is equal to 20%?

(a) $\frac{1}{5}$ (b) $\frac{1}{20}$ (c) $\frac{2}{100}$ (d) $\frac{1}{2}$

2. In a group of 9 children, 6 are wearing blue. The percent of children wearing blue is

(a) 6% (b) 60% (c) $33\frac{1}{3}\%$ (d) $66\frac{2}{3}\%$

3. Multiply: 185×0.12

4. Find the least common denominator for the fractions $\frac{3}{20}$, $\frac{12}{25}$, and $\frac{38}{100}$.

5. Find $\frac{32}{100}$ of 485.

6. Change $\frac{13}{25}$ to a decimal.

7. Round 15.645 to the nearest hundredth.

8. Divide $\frac{20}{39}$ by $\frac{2}{3}$.

9. Which digit in 748,325 is in the ten-thousands place?

10. What percent of the figure is shaded?

(a) 3% (b) 30%
(c) 50% (d) 75%

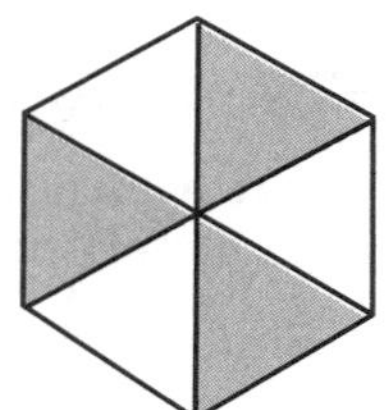

11. Which key sequence would display a number equivalent to $\frac{7}{50}$?

(a) 14 [÷] 100 [=]
(b) 7 [÷] 25 [×] 2 [=]
(c) 14 [÷] 2 [×] 1 [÷] 25 [=]
(d) 7 [÷] 100 [÷] 2 [=]

12. One year, amateur photographers in the United States took more than 16 billion photographs and spent $11.4 billion on film, photofinishing, cameras, and related products. If $4.6 billion was spent on photofinishing, $2.7 billion on film, and $2.6 billion on related products, how much was spent on cameras?

UNIT 8-4 Finding What Percent One Number Is of Another; Finding a Number When a Percent of It Is Known

FINDING WHAT PERCENT ONE NUMBER IS OF ANOTHER

THE MAIN IDEA

To find what percent one number is of another:

1. Write a fraction with the "is" number in the numerator and the "of" number in the denominator: $\frac{\text{is}}{\text{of}}$ ("is-over-of"). The "of" number is the whole quantity to which the "is" number is being compared.
2. Simplify the fraction.
3. Write the fraction as a percent.

EXAMPLE 1 What percent is 8 of 40?

Write a fraction with the "is" number in the numerator and the "of" number in the denominator. $\frac{8}{40}$

Simplify the fraction.

$$\frac{8 \div 8}{40 \div 8} = \frac{1}{5}$$

Write $\frac{1}{5}$ as a decimal.

$$\frac{1}{5} = 0.20$$

Write 0.20 as a percent.

$$0.20. = 20\%$$

Answer: 8 is 20% of 40.

EXAMPLE 2 What percent of 1,500 is 120?

Use $\frac{\text{is}}{\text{of}}$. $\frac{120}{1{,}500}$

Simplify the fraction. $\frac{120 \div 60}{1{,}500 \div 60} = \frac{2}{25}$

Write $\frac{2}{25}$ as a decimal. $25\overline{)2.00}$ = 0.08

Write 0.08 as a percent. $0.08. = 8\%$

Answer: 120 is 8% of 1,500.

KEYING IN

There is no need to simplify the fraction $\frac{120}{1{,}500}$. This key sequence gives the answer directly.

Key Sequence	***Display***
120 [÷] 1500 [%]	8.

Write the number on the display followed by a % sign: 8%

EXAMPLE 3 Mr. Rosen saved $300 and spent $42 of it on some new clothes. What percent of his savings did he spend?

THINKING ABOUT THE PROBLEM

"What percent of his savings did he spend?" can be read as "What percent *of* the $300 he saved *is* the $42 he spent?"

Use $\frac{\text{is}}{\text{of}}$. $\frac{42}{300}$

Simplify the fraction. $\frac{42 \div 6}{300 \div 6} = \frac{7}{50}$

Write $\frac{7}{50}$ as a decimal. $50\overline{)7.00}$ = 0.14

Write 0.14 as a percent. 0.14. = 14%

Answer: Mr. Rosen spent 14% of his savings.

EXAMPLE 4 A television network uses the time for the evening news as follows:

Type of Report	Number of Minutes
Special reports	10
Commercials	8
National news	6.6
International news	4.4
Entertainment, etc.	0.7
Greetings, etc.	0.3

What percent of the time for evening news is spent on national news?

Find the total length of the broadcast:

$10 + 8 + 6.6 + 4.4 + 0.7 + 0.3 = 30$

Find what percent the time for national news is of the total time:

$\frac{6.6}{30} = 0.22$ or 22%

Answer: 22% of the time is on national news.

CLASS EXERCISES

1. What percent is 36 of 60? **2.** What percent of 140 is 28? **3.** What percent of 2 is 5?

4. What percent is 120 of 300? **5.** What percent of 40 is 100? **6.** What percent of 5 is 2?

7. In each case, find what percent the first number is of the second number.
a. 5; 75 **b.** 100; 200 **c.** 3; 120 **d.** 12; 48 **e.** 15; 60 **f.** 22.8; 200

8. Alyse spent $18 of her $150 paycheck on a gift. What percent of her paycheck did she spend?

9. Of 600 lenses inspected, 9 were defective. What percent of the lenses were defective?

10. One hundred seventy-five of the 1,400 students at Greenbay College have registered to vote. What percent of the students have registered?

11. The table shows the results of a telephone poll about the eating habits of American families. What percent of the families polled took about 60 minutes for dinner?

Amount of Time Spent for Dinner

Number of Families	Length of a Dinner
111	less than 30 minutes
266	30 minutes
117	45 minutes
44	60 minutes
12	more than 60 minutes

12. Phil has 40 CDs, of which 24 are by jazz artists. What percent of Phil's CDs are jazz?

13. In an NBA playoff game with the Boston Celtics, a player for the New York Knicks made 17 of 25 attempted shots from the floor. On what percent of his shots from the floor did he score?

FINDING A NUMBER WHEN A PERCENT OF IT IS KNOWN

THE MAIN IDEA

To find a number (the whole) when a percent of it is known:

1. Write a fraction with the known part in the numerator and the percent in the denominator.
2. Write the percent that is in the denominator as a decimal.
3. Divide this decimal into the numerator to find the answer.

EXAMPLE 5 16 is 8% of what number?

Use $\frac{\text{known part}}{\text{percent}}$. $\frac{16}{8\%}$

Write 8% as a decimal. $\frac{16}{0.08}$

Divide the denominator 0.08 into the numerator 16.

$0.08\overline{)16.00}$

$$\begin{array}{r} 200 \\ 8\overline{)1600} \\ \underline{16} \\ 000 \end{array}$$

KEYING IN

Use the following key sequence on a calculator with a percent key:

Key Sequence	*Display*
16 ÷ 8 %	200.

Answer: 16 is 8% of 200.

EXAMPLE 6 Thirty-five percent of Mrs. Chin's grocery bill was spent for meat. If she spent \$14.70 for meat, how much was her grocery bill?

THINKING ABOUT THE PROBLEM

You must answer the question:

\$14.70 is 35% of what number?

Use $\frac{\text{known part}}{\text{percent}}$. $\frac{14.70}{35\%}$

Write 35% as a decimal. $\frac{14.70}{0.35}$

Divide 0.35 into 14.70.

$$\begin{array}{r} 42 \\ 0.35\overline{)14.70} \\ \underline{14\ 0} \\ 70 \\ \underline{70} \\ 0 \end{array}$$

Answer: The total grocery bill was \$42.

CLASS EXERCISES

1. In each case, find the missing number.
   a. 30 is 25% of what number?
   b. 10.2 is 17% of what number?
   c. 57 is 100% of what number?
   d. 34 is 200% of what number?
   e. 58.016 is 11.2% of what number?
   f. 100 is 50% of what number?
2. Eight percent of the price of a camera is $20. Find the price of the camera.
3. Forty-five percent of the senior class went on the senior-class trip. If 135 seniors went on the senior-class trip, how many students are there in the senior class?
4. The 420 freshmen at North Central High School are 28% of the entire student body. How many students are there at North Central?
5. The 8% service charge on Mrs. Spyros' bill came to $2.60. What was the amount of the bill?
6. Four of Mr. White's mathematics students got 100% on their exam. This was 16% of the class. How many students were in the class?

HOMEWORK EXERCISES

1. What percent is 15 of 75?
2. What percent of 640 is 32?
3. What percent is 7 of 140?
4. What percent of 110 is 11?
5. What percent is 21 of 420?
6. What percent of 92 is 4.6?
7. What percent of 11.5 is 4.6?
8. What percent is 32 of 48?
9. What percent is 48 of 32?
10. Johnny saved $24 of the $96 he received as birthday gifts. What percent did he save?
11. Four of the 12 members of the Paulson family have red hair. What percent of the family has red hair?
12. What percent of Mr. Rawlings' bicycles are red if 18 of his 60 bicycles are red?
13. In a recent year, approximately 22 million men and 43 million women were dieters. Five years later, approximately 17 million men and 31 million women were dieters. In which year were men a greater percentage of the total number of dieters for that year?
14. Use the table to determine the percent of tractors sold that were medium-sized tractors.

Tractors Sold in the U.S. in One Year

Size	Number of Tractors Sold
Small (under 40 horsepower)	50,642
Medium (40–99 horsepower)	42,917
Large (100 or more horsepower)	29,061

15. In each case, find the missing number.

a. 50 is 75% of what number? **b.** 26 is 25% of what number?

c. 15.8 is 5% of what number? **d.** 36.4 is 150% of what number?

16. Twelve of the city's buses were taken out of service for repairs. This represented 5% of all the city's buses. How many buses did the city have?

17. Fifteen percent of the pizzas made by J & W Pizzeria were made with mushrooms. If 48 pizzas were made with mushrooms, what was the total number of pizzas made?

18. Mr. Holt saves 20% of his total earnings. Last week, he earned a salary of $280, plus a 6% commission on sales totaling $6,000. How much did Mr. Holt save?

SPIRAL REVIEW EXERCISES

1. Change 48% to a fraction in simplest form.

2. Find 12% of $6.50.

3. What is the commission on a $450 sale if the rate of commission is 4%?

4. The price of a computer is $948.98, and the East Townsend School District needs to buy 18 computers. What will be the total cost of the computers?

5. What is the total length of 12 wood blocks if each block is $4\frac{3}{4}$ inches long?

6. From $\frac{5}{12}$ subtract $\frac{1}{10}$.

7. Find the product of –8 and –12.

8. From the product of –4 and 3, subtract the sum of –7 and –1.

9. Use a calculator to write $\frac{2,759}{255}$ as a mixed number.

10. Evaluate: $10 \div 2 + 3$

11. Which of the following is not equal to the others?

(a) 0.01 (b) 1% (c) $\frac{1}{100}$ (d) 0.1%

12. John Dalton earned $9.80 an hour. His time card for one week is shown below. How much did he earn that week?

JOHN DALTON	
Day	**Hours**
Monday	$3\frac{1}{2}$
Tuesday	4
Wednesday	$2\frac{1}{2}$
Thursday	3
Friday	$2\frac{1}{2}$

UNIT 8–5 Using Percents With Circle Graphs

READING CIRCLE GRAPHS

THE MAIN IDEA

1. A ***circle graph*** is a display of data in which:
   a. a circle represents the whole of a quantity, 100%, and
   b. pie-shaped wedges of the circle, called ***sectors***, represent percents of the whole quantity.

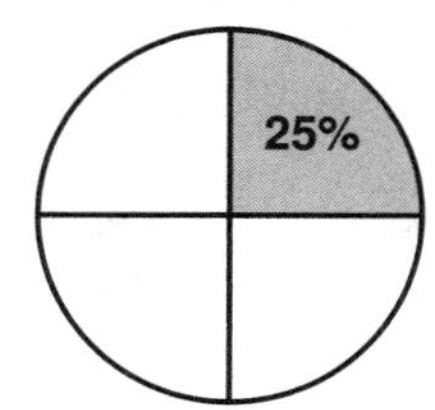

25% of the circle represents 25% of the quantity

3. Each sector is labeled to identify the part and the percent of the whole that it represents.
4. A circle graph is used to emphasize the size of each part in relation to the whole quantity and to the other parts.

EXAMPLE 1 This circle graph shows how the Patel family plans to spend its money for one month. What percent of the money does the Patel family plan to spend for transportation?

Patel Family Monthly Plan

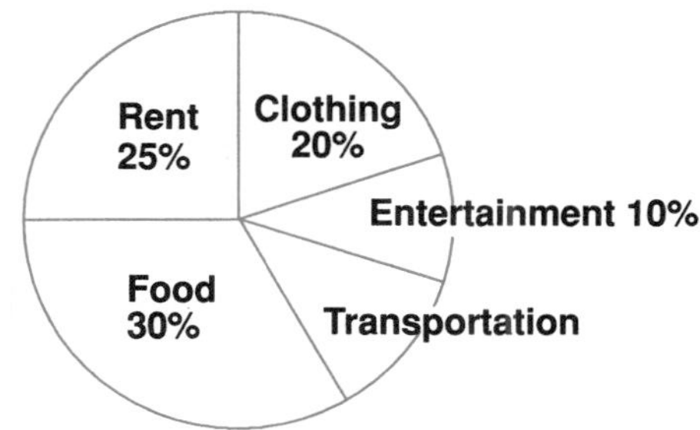

THINKING ABOUT THE PROBLEM

Since the sum of all the percents must be 100%, add to find the sum of the four known percents, then subtract from 100%.

$30\% + 25\% + 20\% + 10\% = 85\%$

$100\% - 85\% = 15\%$

Answer: The Patel family plans to spend 15% of its money for transportation.

EXAMPLE 2 An international pizza chain has 7,147 restaurants. The circle graph shows the percent of these restaurants in different parts of the world. How many are in Europe?

Europe 6.4%
Mideast/Africa 1%
Other 0.6%
United States 92%

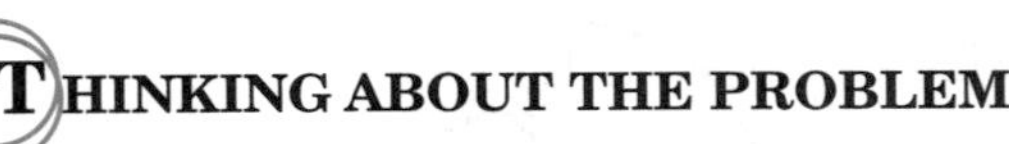

THINKING ABOUT THE PROBLEM

Since the whole circle represents the total number of restaurants, and 6.4% of all the restaurants are in Europe, find 6.4% of 7,147.

$$7{,}147 \times 0.064 = 457.408$$

Round to the nearest whole number, since a fraction of a restaurant has no meaning.

Answer: This chain has 457 restaurants in Europe.

EXAMPLE 3 The total number of tickets printed for a school dance performance was 250. The circle graph shows ticket sales to various groups. How many more tickets were sold to students than to faculty members?

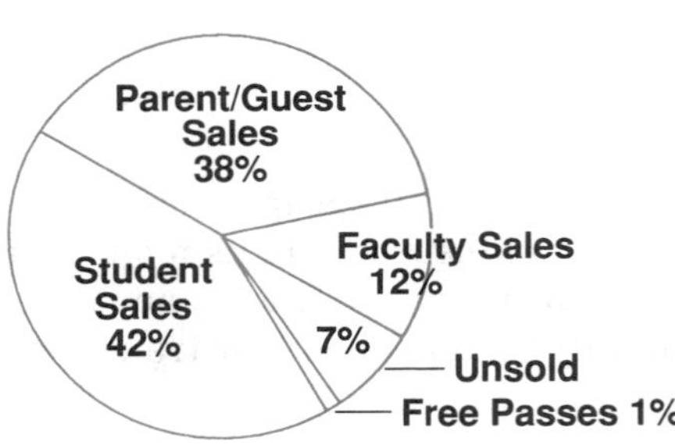

THINKING ABOUT THE PROBLEM

Method 1: Work with the number of tickets.

$250 \times 0.42 = 105$ ← tickets sold to students
$250 \times 0.12 = \underline{-30}$ ← tickets sold to faculty
75 ← difference in number

Answer: There were 75 more tickets sold to students than to faculty.

Method 2: Work with the percent of tickets.

42% ← student sales
$\underline{-12\%}$ ← faculty sales
30% ← % difference

$250 \times 0.30 = 75$ ← number for student sales

CLASS EXERCISES

1. In each circle graph below, find the percent for the sector that is not labeled.

a.

17%
62%
12%
?

b.

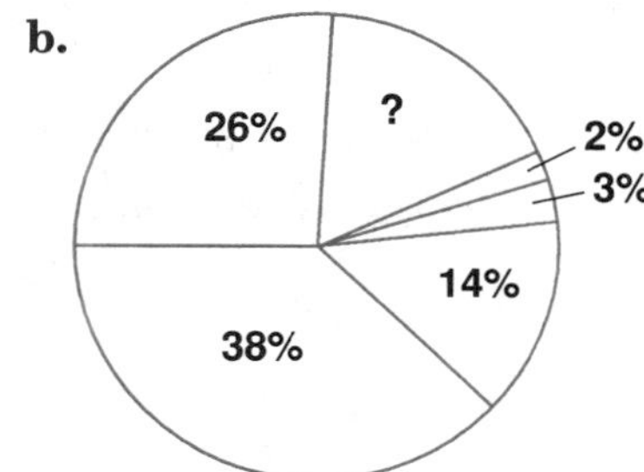

c.

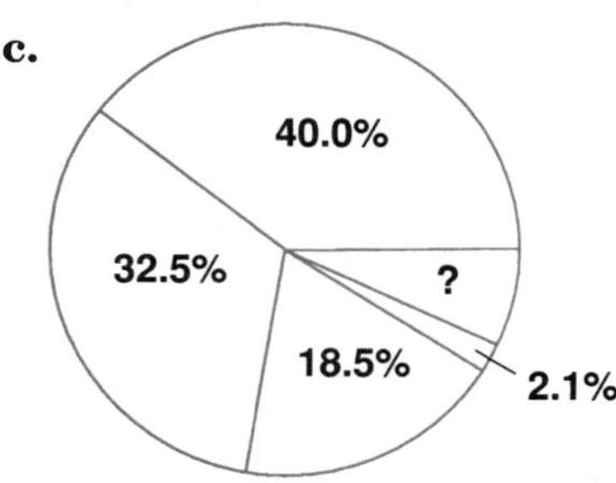

2. The motorcycle industry announced that it expects to sell 461,000 motorcycles this year. The graph shows expected sales for each type of motorcycle. How many dirt bikes does the industry expect to sell this year?

Projected Motorcycle Sales

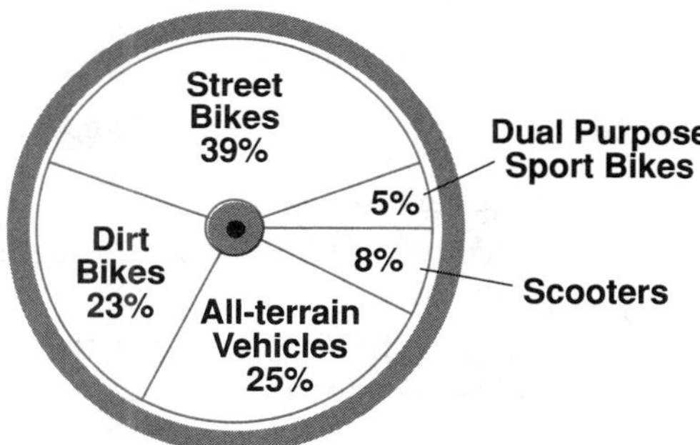

3. A survey of 1,235 adults asked people to name their favorite spectator sport. Results were reported in the circle graph shown. How many of the people surveyed indicated that their favorite spectator sport was either basketball or ice hockey?

Favorite Spectator Sports

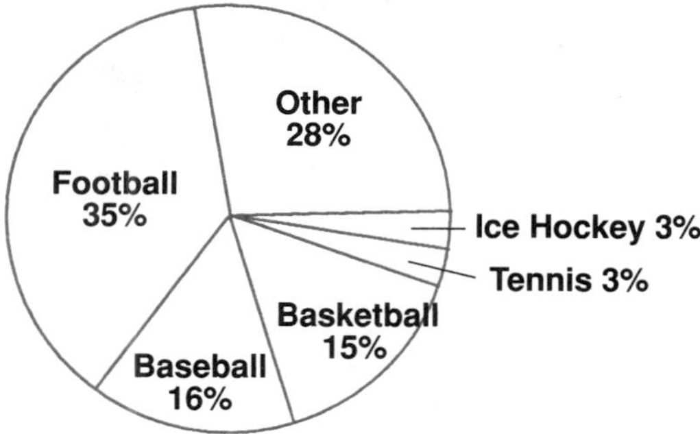

MAKING CIRCLE GRAPHS

THE MAIN IDEA

1. A *percent circle* is helpful for drawing the sectors of a circle graph. A percent circle is divided into 100 equal spaces, each representing $\frac{1}{100}$, or 1%, of the whole.

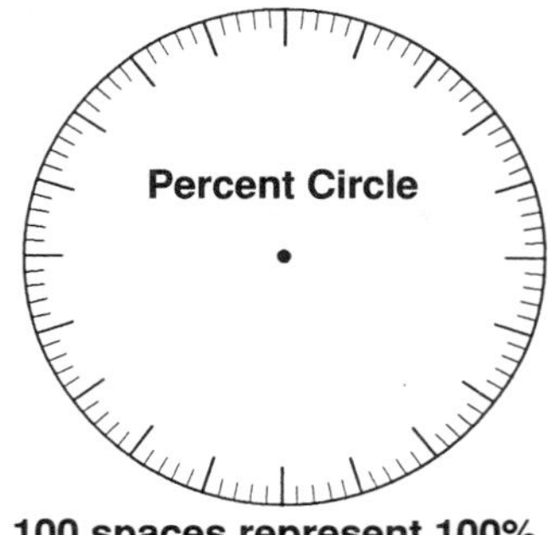

100 spaces represent 100%

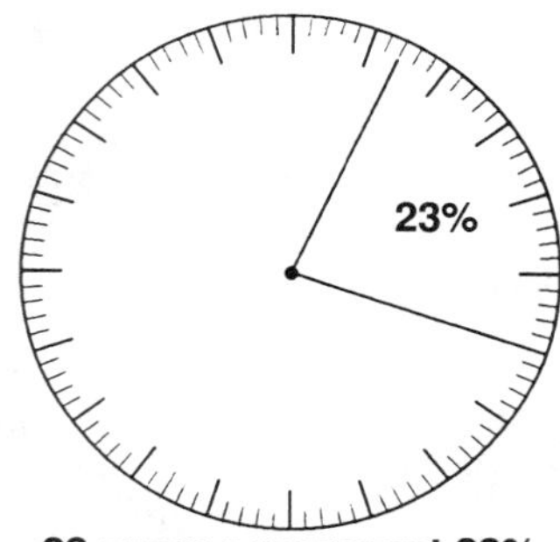

23 spaces represent 23%

2. To make a circle graph using a percent circle:
   a. Pick a starting point on the circle and draw a line segment, called a *radius*, from the center to that point.
   b. Count to locate another point on the circle that is the same number of spaces from the starting point as the percent to be represented. Draw a radius to this point.
   c. Repeat Step b for each sector to be drawn, using the last radius drawn as the first radius for the next sector.
   d. Continue in the same direction around the circle, counting spaces and drawing a radius for each sector until each part is represented.
   e. Label each sector with the title and percent of the whole that it represents and write a title to describe the whole graph.

EXAMPLE 4 An automobile sales agency kept a record of customer complaints on new car sales. Make a circle graph to display this data.

Customer Complaints on New Car Sales

Kind of Complaint	% of Total Complaints
Car not ready on time	57%
Car scratched	26%
Car dirty	8%
Missing papers	3%
Other	6%

(1) Start by drawing a radius; then count 57 spaces on the percent circle to represent 57%. Draw a second radius to a point that is 57 spaces from the endpoint of the first radius. Label the sector "Not Ready 57%."

(2) To represent 26% for the second sector, start on the circle at the endpoint of one of the radii for the first sector, count 26 spaces, and draw a third radius. Label the new sector "Scratched 26%."

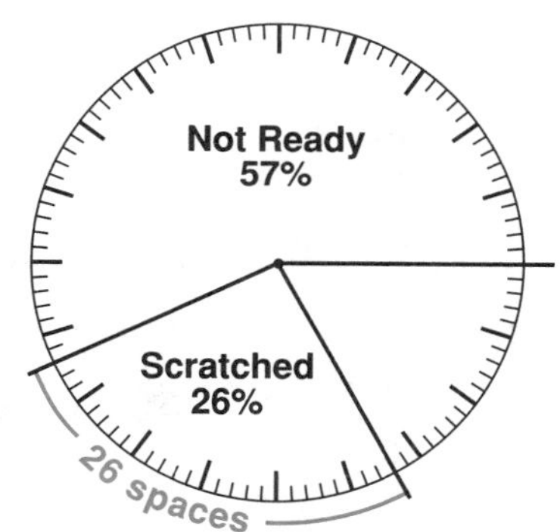

(3) Keep going around the circle, sector by sector, until each percent is represented. Label each sector and give a title to the graph.

Customer Complaints on New Car Sales

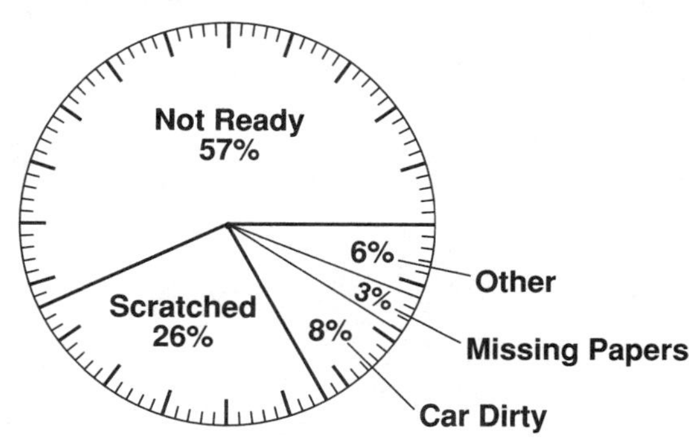

EXAMPLE 5 The manager of a chain of department stores kept a record of the number of letters of commendation received by salespersons in one year. Make a circle graph using this data.

Salesperson Commendations

Number of Commendations	Number of Salespersons
0	12
1	66
2	58
3	42
4	18
More than 4	4

THINKING ABOUT THE PROBLEM

Since the table shows *numbers* of commendations and not percents of the total, first find the percent that each category is of the total.
Start by finding the total of all the categories: $12 + 66 + 58 + 42 + 18 + 4 = 200$

Then, compute the percent that each category is of the total, 200:

Number of Commendations	*Number Salespersons*	*Key Sequences*	*Displays (% of Total)*
0	12	[C] 12 [÷] 200 [%]	6.
1	66	[C] 66 [÷] 200 [%]	33.
2	58	[C] 58 [÷] 200 [%]	29.
3	42	[C] 42 [÷] 200 [%]	21.
4	18	[C] 18 [÷] 200 [%]	9.
More than 4	4	[C] 4 [÷] 200 [%]	2.

Mark off the sectors for these percents on a percent circle. Add labels and a title.

Salesperson Commendations

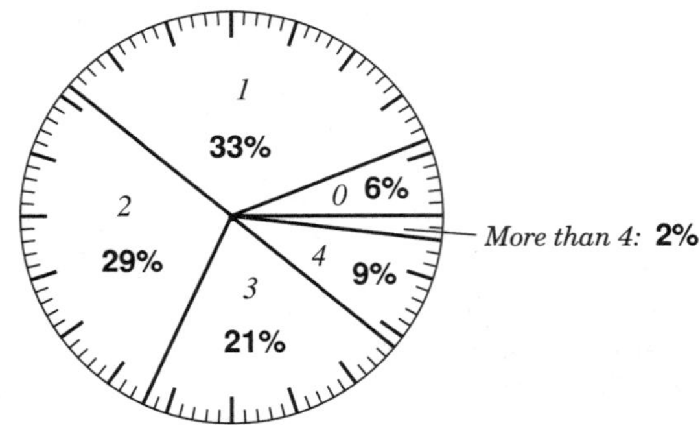

CLASS EXERCISES

1. The table shows the results of a survey of 400 students about their method of transportation to school. Using a percent circle, make a circle graph to display this data.

Method of Transportation to School

Type of Transportation	% of Students
Walking	31%
Bicycle	8%
School Bus	43%
Drive Own Car	6%
Driven by Other Driver	10%
Other	2%

2. The table shows the sales of detergent in one week at Acme Market. Make a circle graph showing the portion of the total sales represented by each brand.

Detergent Sold in 1 Week

Brand	Number of Containers
Wham	85
Zap	53
Blot	28
Glare	72
Prize	102

3. Marine biologists recorded the number of each kind of specimen in a sample. Both the number of each kind and the percent of the total are shown in the table.
   a. Use the data to make a circle graph. Round each percent to the nearest tenth of a percent and approximate, by eye, on the percent circle.
   b. Explain the effect on the graph if you rounded to the nearest whole percent instead of tenth.

Sea Organisms in Sample

Kind of Organism	Number	% of Total
Fish (12 in. or more)	17	10.17964
Fish (less than 12 in.)	49	29.341317
Crustaceans	86	51.497005
Mammals	3	1.7964071
Coral Colonies	12	7.1856287
Total	167	

HOMEWORK EXERCISES

1. Find the percent for the sector that is not labeled in each circle graph.

a.

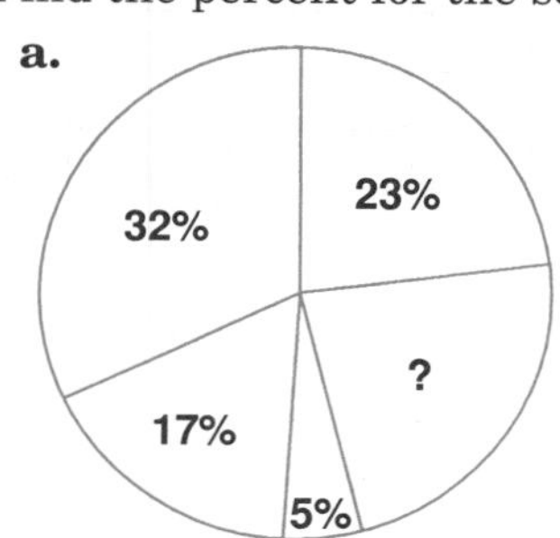

b.

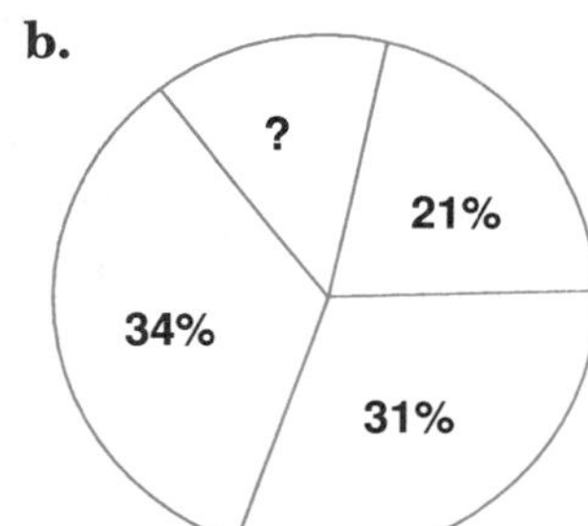

c.

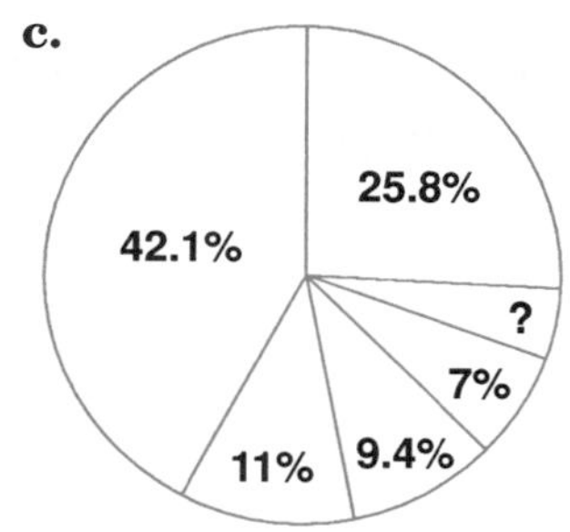

2. There are 405 endangered species in the United States. The circle graph shows how these species are classified.

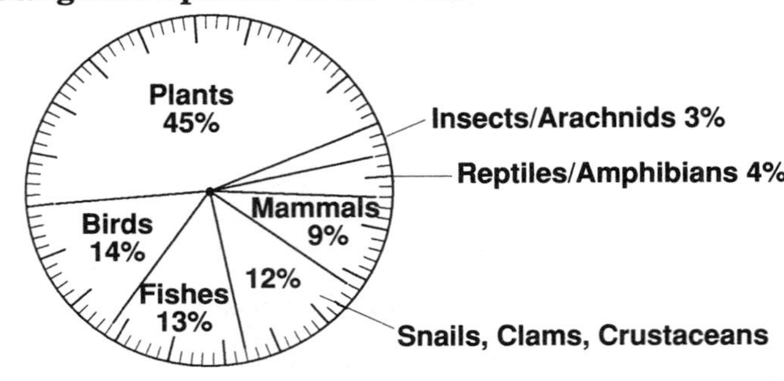

   a. How many species of birds are endangered?
   b. What percent of the endangered species are either mammals or birds?
   c. How many more species of fish are endangered than species of reptiles/amphibians?
   d. Are there more species of plants or of animals (non-plants) that are endangered? Explain how you arrived at your conclusion.
   e. From this graph, can you conclude anything about the numbers of individual specimens of plants or animals that are endangered?

3. In a recent year, 2.5 billion pounds of cereal were sold in the United States. The circle graph shows these sales by manufacturer.

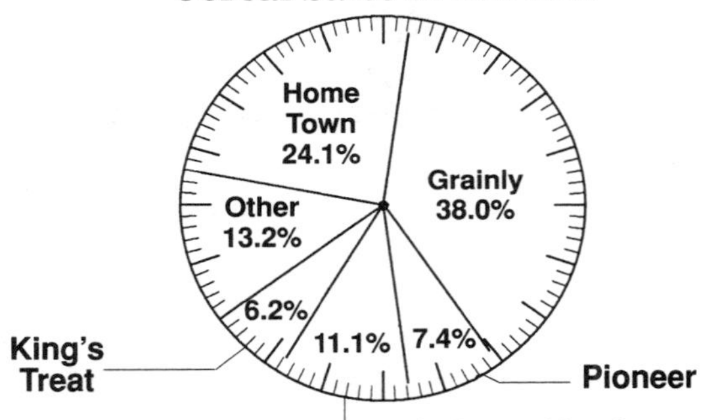

   a. What percent of the total sales were made either by Industry Foods or Pioneer?
   b. By what percent did the sales of Grainly cereals exceed the sales of Home Town cereals?
   c. How many pounds of cereal were sold by Grainly?
   d. How many more pounds of cereal were sold by Pioneer than King's Treat?
   e. Many circle graphs have a sector labeled "Other." Why do you think this is so?

4. Expenses for a five-mile road race are shown in the table.
   a. Find the percent of the total for each category.
   b. Make a circle graph to show how the money was spent.

Road Race Expenses

Type of Expense	Amount Spent
Free tee shirts	$4,200
Printing & mailing	2,200
Police/Medical/Parking	1,000
Timing equipment	800
Tents & signs	600
Prizes & trophies	400
Runners' numbers	400
Post-race snacks	400
Total	**$10,000**

5. The table shows the sales of an electronics company by type of product in billions of dollars.
   a. Find the percent of total sales for each product category.
   b. Make a circle graph to show how the sales in each category contributed to the total.

Apex Electronics

Product	Sales (billions)
Video equipment	$10
Audio equipment	4
Home appliances	6
Communications/ Industrial Equip.	9
Electronic components	5
Batteries/Kitchen-related	2
Other	4

SPIRAL REVIEW EXERCISES

1. What is 0.873 expressed as a percent?
   (a) 0.873% (b) 8.73% (c) 87.3% (d) 873%

2. Which fraction is equal to $\frac{5}{6}$?
   (a) $\frac{10}{18}$ (b) $\frac{20}{24}$ (c) $\frac{15}{30}$ (d) $\frac{20}{30}$

3. Which number is represented by point A on the number line?

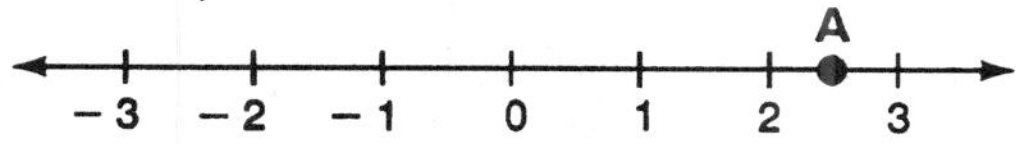

   (a) $\frac{12}{30}$ (b) $\frac{22}{18}$ (c) $\frac{30}{9}$ (d) $\frac{29}{12}$

4. Last year, approximately 13,000 new consumer products were introduced. If 2,340 of these products were health and beauty aids, what percent of the total number of new products were health and beauty aids?

5. Use a calculator to help you write $\frac{135}{349}$ in decimal form correct to the nearest thousandth.

6. Subtract –6 from –1.

7. The price of a monthly commuter train ticket increased 25 percent over a three-year period. To find the amount of increase in the price of a ticket originally priced at $89.20, you would multiply $89.20 by
   (a) $\frac{1}{25}$ (b) $\frac{3}{25}$ (c) $\frac{1}{2}$ (d) $\frac{1}{4}$

8. Jerry used $3\frac{2}{3}$ cups of flour to make a cake and $4\frac{1}{2}$ cups of flour to make a loaf of bread. What is the total number of cups of flour that he used?

9. Which set of decimals is in order from least to greatest?
   (a) 0.7, 0.407, 0.704, 0.47
   (b) 0.47, 0.407, 0.704, 0.7
   (c) 0.704, 0.7, 0.47, 0.407
   (d) 0.407, 0.47, 0.7, 0.704

10. The product of 29.8 and 0.49 is about
    (a) 35 (b) 30 (c) 20 (d) 15

11. Jaclyn brought her bicycle for repairs. She was charged $8.50 to fix a flat, $6.75 to adjust the gears, $9.80 to true a bent wheel, and $5.95 to replace a gear cable. What was the total cost?

UNIT 8-6 Finding the Percent of Increase or Decrease

THE MAIN IDEA

To find the percent of increase or decrease:

1. Find the amount of increase or decrease by subtracting the lesser number from the greater number.
2. Write a fraction in which the numerator is the amount of increase or decrease and the denominator is the original amount.
3. If possible, simplify the fraction.
4. Write the fraction as a percent.

EXAMPLE 1 The temperature increased from 60°*F* to 75°*F*. Find the percent of increase.

Find the amount of increase by subtraction.	75 ← −60 ← 15 ←	new temperature original temperature amount of increase
Write a fraction in which 15, the amount of increase, is the numerator and 60, the original amount, is the denominator.	$\frac{15}{60}$ ←	amount of increase original amount
Simplify $\frac{15}{60}$.	$\frac{15 \div 15}{60 \div 15} = \frac{1}{4}$	
Write $\frac{1}{4}$ as a percent.	$4\overline{)1.00}$ = 0.25 = 25%	

Answer: The percent of increase is 25%.

EXAMPLE 2 Before dieting, Walter weighed 180 pounds. He now weighs 153 pounds. Use a calculator to find the percent of decrease in his weight.

Key Sequence	***Display***	
180 [M+]	M 180.	← Store original weight in memory.
[−] 153 [=]	M 27.	← Subtract to find the amount of decrease.
[÷] [MR]	M 180.	← Divide by the number in memory.
[%]	M 15.	← Find the quotient as a percent.

Answer: The percent of decrease is 15%.

EXAMPLE 3 A student organization had 400 members. The membership was increased to 1,000 students. Find the percent of increase.

Find the amount of increase.

$$\begin{array}{r} 1{,}000 \\ -400 \\ \hline 600 \end{array}$$

Write a fraction and simplify. $\dfrac{600 \div 200}{400 \div 200} = \dfrac{3}{2}$

Change $\frac{3}{2}$ to a percent. $2\overline{)3.0}$ $1.5 = 150\%$

Answer: The percent of increase is 150%.

EXAMPLE 4 A dance school had 1,000 students. The enrollment decreased to 400 students. Find the percent of decrease.

Find the amount of decrease.

$$\begin{array}{r} 1{,}000 \\ -400 \\ \hline 600 \end{array}$$

Write a fraction and simplify. $\dfrac{600 \div 200}{1{,}000 \div 200} = \dfrac{3}{5}$

Change $\frac{3}{5}$ to a percent. $5\overline{)3.0}$ $.60 = 60\%$

Answer: The percent of decrease is 60%.

CLASS EXERCISES

1. Find the percent of increase.

a. 50 to 70 **b.** 25 to 40 **c.** 45 to 72 **d.** 3.5 to 14 **e.** 20 to 21.1 **f.** 10 to $14\frac{1}{2}$

2. Find the percent of decrease.

a. 24 to 12 **b.** 64 to 16 **c.** 75 to 50 **d.** 20 to 15 **e.** 20 to 15.5 **f.** 30 to 3

3. In January, Martha sold 20 cars. The next month, she sold 35 cars. Find the percent of increase in the number of cars that Martha sold.

4. During a sale, the price of a shirt was changed from $20 to $16. Find the percent of decrease in the price of the shirt.

5. In one hour, the temperature went from 25° to 27°. Find the percent of increase in temperature.

6. Before going on a diet, Mr. Coleman's weight was 200 pounds. After dieting, he reached a weight of 170 pounds. What was the percent of decrease in his weight?

HOMEWORK EXERCISES

1. Find the percent of increase.
 a. 25 to 38 b. 50 to 52 c. 20 to 20.2
 d. 10 to 67 e. $4\frac{1}{2}$ to 9 f. 5 to $6\frac{1}{2}$

2. Find the percent of decrease.
 a. 100 to 90 b. 75 to 25 c. 50 to 49
 d. 200 to 148 e. 200 to 180 f. 200 to 0

3. The price of a dress went from $35 to $42. Find the percent of increase.

4. When a store owner prices an item by adding a percent increase to the amount that the item cost him, the percent is called a *markup*. Mr. Andrews paid $20 for a shirt and sold it for $28. What was the markup?

5. The number of people in a chorus went from 27 to 18. Find the percent of decrease.

6. In one year, Mark's height went from 60 inches to 63 inches. What was the percent of increase in Mark's height?

7. The number of student organization members increased from 400 to 500. Find the percent of increase.

8. Mr. James sold 40 cars in June and 25 cars in July. Find the percent of decrease in sales.

9. In one year, the number of cases of measles reported in Southside High School dropped from 24 to 8. What was the percent of decrease?

10. Before exercising, Martha's pulse rate was 75. After exercising, her pulse rate was 90. Find the percent of increase.

11. The price of a share of stock went from $24 to $20. What was the percent of decrease?

12. The price of a computer dropped from $300 to $240. Find the percent of decrease.

13. The graph shown gives the average number of people per U.S. household for six selected years.
 The percent decrease from 1940 to 1990 is closest to
 (a) 1.1% (b) 30% (c) 70% (d) 142%

Average Size of U.S. Households

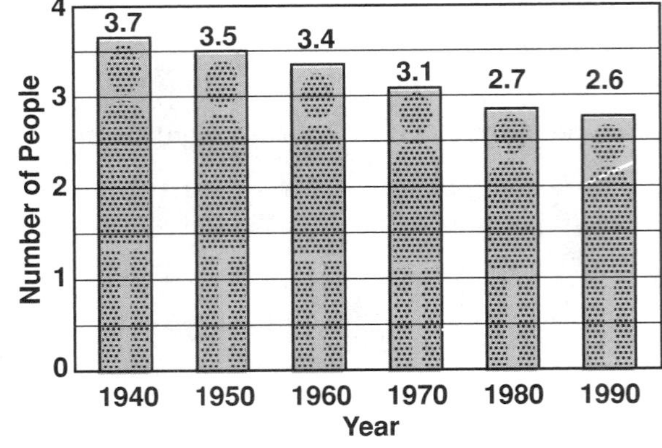

14. In the sequence below, each number is 10% greater than the previous number.
 1,000; 1,100; 1,210; 1,331; ...
 a. Find the fifth number in the sequence.
 b. Find the percent increase from the first number to the fifth.

SPIRAL REVIEW EXERCISES

1. A player for the Knicks missed 16 of the 24 shots he took in playoff games in New York. On what percent of his shots did he score?

2. At the end of the regular season, a member of the Boston Celtics team had made 348 of the 870 field goals he had attempted. What percent of the shots did he make?

3. Of the 1,600 pages in Jim's encyclopedia, 55.5% have illustrations. How many pages have illustrations?

4. Twenty-four percent of Mrs. Reed's employees work part-time. If 18 of the employees work part-time, how many employees does Mrs. Reed have?

5. Find $\frac{4}{5}$ of $\frac{15}{16}$.

6. Use your calculator to find the change from a $50 bill after making purchases of $17.98, $20.95, and $4.75.

7. Round 12,498 to the nearest thousand.

8. The number of tickets sold at the Bijou Theater each day of one week is as shown. What was the total number of tickets sold that week?

Sunday	505
Monday	68
Tuesday	47
Wednesday	59
Thursday	103
Friday	452
Saturday	651

9. Subtract 17.38 from 20.2.

10. Which number is *not* a multiple of 7?
(a) 7 (b) 17 (c) 42 (d) 70

11. The wholesale price of a three-inch bolt is $0.08. If the price marked on a box of these bolts by a wholesale supplier is $16.00, how many bolts does the box contain?

TEAMWORK

Look in the newspaper to find the opening and closing prices of five stocks. Calculate the percent increase or decrease for each stock. If your team owned 50 shares of each stock, how much would you have earned or lost on each stock from opening to closing? What would have been the percent increase or decrease on your team's total holdings? Prepare a team report that summarizes your findings.

UNIT 9-1 Using Percents to Find Discounts

THE MAIN IDEA

1. The amount by which an original price, or ***list price***, is lowered is called the ***amount of discount***. The new price is called the ***sale price***.
2. Percents are used to show the ***rate of discount***.
3. To find the amount of discount:
   a. Write the rate of discount as a decimal.
   b. Multiply the list price by this decimal.

   Amount of Discount = List Price × Rate of Discount
4. To find the sale price:
   a. Find the amount of discount.
   b. Subtract the amount of discount from the list price.

   Sale Price = List Price – Amount of Discount

EXAMPLE 1 A television set is on sale at 15% off the list price of $429. Find the amount of discount.

Write the percent as a decimal. 15% = 0.15

Multiply the list price by the decimal.

```
  $429  ←list price
 ×0.15  ←rate of discount
 21 45
 42 9
$64.35  ←amount of discount
```

KEYING IN

Use this key sequence on a calculator:

Key Sequence	***Display***
429 [×] 15 [%]	64.35

Answer: The amount of discount is $64.35.

EXAMPLE 2 Find the sale price of a camera that has a list price of $148.75 and is being sold at a 35% discount.

To find the amount of discount, multiply the list price by the rate of discount written as a decimal.

```
  $148.75  ← list price
   ×0.35   ← rate of discount
   74375
  44625
 $52.0625  ← amount of discount
```

Round the amount of discount to the nearest cent. $52.06(2)5 or $52.06

To find the sale price, subtract the amount of discount from the list price.

```
 $148.75  ← list price
  −52.06  ← amount of discount
  $96.69  ← sale price
```

KEYING IN

On a calculator, you can find the sale price directly with the following key sequence:

Key Sequence	***Display***
148.75 [−] 35 [%]	96.6875

Round the number on the display to the nearest cent. $96.68(7)5 or $96.69 *Ans.*

EXAMPLE 3 For Washington's Birthday Sale, everything at the Best Department Store was discounted 25%. Mr. Billings bought gloves listed at $15.90, socks listed at $8.75, and a shirt listed at $22.95. What was the total sale price for these items?

To find the total list price, add the list price of each of the items.

```
  $15.90
    8.75
  +22.95
  $47.60  ← total list price
```

To find the amount of discount, multiply the total list price by the rate of discount expressed as a decimal.

```
   $47.60  ← total list price
    ×0.25  ← rate of discount
   2 3800
   9 520
 $11.9000  ← amount of discount
```

To find the total sale price, subtract the amount of discount from the total list price.

```
  $47.60  ← total list price
  −11.90  ← amount of discount
  $35.70  ← total sale price
```

Answer: The total sale price was $35.70.

EXAMPLE 4 A man's suit was reduced from $220 to $187. What was the percent of discount?

THINKING ABOUT THE PROBLEM

Finding the percent of discount is like finding the percent of decrease.

Find the amount of discount.

$$\begin{array}{r} 220 \\ -187 \\ \hline 33 \end{array}$$

220 ← original price
−187 ← sale price
33 ← amount of discount

Write a fraction and simplify.

$$\frac{33 \div 11}{220 \div 11} = \frac{3}{20}$$

Write $\frac{3}{20}$ as a decimal, then as a percent.

$$20\overline{)3.00} = 0.15 = 15\%$$

Answer: The percent of discount was 15%.

EXAMPLE 5 Luxury Furniture Company is selling a $495 sofa at 45% off. Discount Dan's regular price for the same sofa is $425 and it is now on sale at $\frac{1}{3}$ off. Which store has the better buy?

Find the price at Luxury Furniture:

495 ← original price
×0.45 ← rate of discount
24 75
198 0
222.75 ← amount of discount

$495.00 ← original price
−222.75 ← amount of discount
$272.25 ← sale price

Find the price at Discount Dan:

$\frac{1}{3}$ off means divide by 3.

$$3\overline{)425.000} = 141.666 = 141.67$$

Round up to next cent.

$425.00 ← original price
−141.67 ← amount of discount
$283.33 ← sale price

Answer: Luxury Furniture at $272.25 has a better buy than Discount Dan at $283.33.

CLASS EXERCISES

1. A tie that has a list price of $18 is on sale at 10% off. Find the amount of discount?
2. A CD player that regularly sells for $240 is on sale at a 25% discount. Find the amount of discount.
3. A shirt that regularly sells for $18.99 is on sale at 35% off. What is the amount of discount?
4. Find the sale price of a baseball glove that has a list price of $35 and is being sold at a 15% discount.
5. Find the sale price of an $18,000 car that is being sold at a 12% discount.
6. Find the sale price of a bicycle that has a list price of $190.80 when it is discounted at each of the following rates.

   a. 10% **b.** 25% **c.** 30% **d.** 45%
7. A $45.75 dress is being sold at a 15% discount. What is the sale price?
8. Virginia bought a pen listed at $3.90, a notebook listed at $1.40, and a desk calendar listed at $5.95 at a store that discounts all items at 20%. What was the total amount that she paid?
9. During a 40% discount sale, Mr. Waters bought a toaster oven regularly priced at $35.90, a clock regularly priced at $29, and a radio regularly priced at $45.50.

   a. How much did Mr. Waters save?

   b. What was the total amount that he paid?
10. Louise bought 3 books originally priced at $3.49 each at a 15%-off book sale. What was the total amount that she spent?
11. The price of a one-year subscription to Luck Magazine was discounted from $28 to $25.50. Find the percent of discount.
12. Merchandise Mart and Super Savings Store both sell the Browning toaster oven at $42.95. One week, both stores have the item on sale, Merchandise Mart selling it for $\frac{1}{3}$ off, and Super Savings Store selling it at a 30% discount.

    a. Which store is offering the better deal?

    b. How much more can be saved by buying at the better sale price?
13. At a 20%-off sale, Isaac bought a sweater for $30.36. What was the original price of the sweater?

HOMEWORK EXERCISES

1. A jacket that has a list price of $80 is on sale at 20% off. Find the amount of discount.

2. A radio that regularly sells for $39 is on sale at a 15% discount. What is the amount of discount?

3. A pair of shoes that usually costs $23.99 is on sale at 25% off. How much money is saved by buying a pair of shoes on sale?

4. How much will the discount be on a suit listed at $185 and sold at a discount of 30%?

5. A game listed at $19.50 was sold at a discount of 15%. What was the sale price?

6. Find the sale price of a $22,500 station wagon that is being sold at a 10% discount.

7. A $499 video recorder is being sold at a 20% discount. What is the sale price?

8. Find the sale price of a television set that has a list price of $369 when it is discounted at the rate of:

   a. 10% **b.** 15% **c.** 25% **d.** 50%

9. A shirt listed at $25.99 is being sold at a 25% discount. What is the sale price?

10. Ms. Jewel bought a blouse that is regularly $15.75, a skirt regularly $24.50, and a handbag regularly $18.95 at a 30%-off sale.

    a. What was the total amount that she saved?

    b. What was the total amount that she spent?

11. Benjamin bought 4 CDs listed at $12.99 each at a 15%-off sale. How much did he spend?

12. Roberta bought a stereo listed at $319 and a CD player listed at $119.80 at a 20%-off sale. What was the total amount that she spent?

13. Albert bought 2 pens regularly priced at $1.99 each and 3 pens regularly priced at $2.50 each, at a store that discounts all items at 15%. What was the total amount that he spent?

14. Ron bought 2 shirts listed at $28.70 each, 4 ties listed at $10.50 each, and a pair of slacks listed at $31, all at a 20%-off sale. What was the total amount that he spent?

15. One store advertised a $20 toy reduced to $16.70. Another store advertised a $40 toy reduced to $35.20. Which store offered the greater percent of discount?

16. Store A sells a swim suit for $34.95, and store B sells the same model for $32.25. At their end-of-summer sales, store A offers $\frac{3}{5}$ off, and store B gives a 55% discount. What is the better sale price?

17. Mr. Valentine saw a \$179 power mower on sale in two stores. One store was selling the mower at 45% off, and the other was selling it at $\frac{2}{5}$ off. How much more can he save by buying at the better sale price?

18. A computer game with a list price of \$24 is first discounted by 20% and then by an additional 10%.

a. Find the final sale price.

b. How do these two discounts compare to a single discount of 30%?

19. The sale price of a chair is \$416.50 after a 15% discount has been given. What was the original price of the chair?

20. Darlene bought a down jacket at an end-of-season 55%-off sale. If she paid \$47.25, what was the original price?

SPIRAL REVIEW EXERCISES

1. Of the 600 trees on the Dyson property, 210 are apple trees. What percent are apple trees?

2. The temperature decreased from 90°*F* to 75°*F*. What was the percent of decrease?

3. Use your calculator to multiply:

$$\frac{1{,}800}{420} \times \frac{1{,}050}{600}$$

4. Mary earns \$7.50 an hour. How much will she earn in 8 hours?
(a) \$54 (b) \$60 (c) \$56 (d) \$52

5. Which decimal has the largest value?
(a) 0.356 (b) 0.298 (c) 0.9 (d) 0.87

6. If 3 gallons of gasoline cost \$4.20, the cost of 7 gallons is
(a) \$29.40 (b) \$9.80
(c) \$8.40 (d) \$7.70

7. The value of $5 \times 4 + 2 \times 9$ is
(a) 198 (b) 48 (c) 38 (d) 270

8. Three stores are featuring the same folding chairs, with prices as shown in the display signs. What is the lowest price for four chairs?

ALLEN's Best Price only \$14.25 each	BONDIO's Special 4 for \$59	2 for \$27 Buy at CORMAN's

9. In a period of five years, the number of cars and trucks built in the United States and Canada by foreign automobile manufacturers increased from 467,678 to 1.8 million. The percent of increase is closest to
(a) 100% (b) 200% (c) 300% (d) 400%

UNIT 9-2 Using Percents to Find Sales Tax

THE MAIN IDEA

1. *Taxes* are monies collected by city, state, and federal governments to help pay for schools, highways, buildings, and other public services.
2. A tax on something bought in a store is called a *sales tax*. The rate of sales tax is given as a percent.
3. To find the amount of sales tax:
   a. Write the percent as a decimal.
   b. Multiply the price by this decimal.

   Amount of Sales Tax = Price × Rate of Sales Tax
4. To find the total cost when a sales tax is collected:
   a. Find the sales tax.
   b. Add the sales tax to the price.

   Total Cost = Price + Amount of Sales Tax

EXAMPLE 1 Iris bought a jacket for \$70. If the sales tax is 8%, how much tax must she pay?

Write the percent as a decimal. $8\% = 0.08$

Multiply the price by this decimal.

\$70	← price
×0.08	← rate of sales tax
\$5.60	← amount of sales tax

KEYING IN

On a calculator:

Key Sequence	*Display*
70 [×] 8 [%]	5.6

Answer: The sales tax is \$5.60.

EXAMPLE 2 Saleem bought a pair of slacks for \$29.95. If the sales tax is $7\frac{1}{2}\%$, how much tax must he pay?

Write the percent as a decimal. $7\frac{1}{2}\% = 7.5\% = 0.075$

Multiply the price by this decimal.

\$29.95	← price
×0.075	← rate of sales tax
14975	
2 0965	
\$2.24625	← amount of sales tax

Round the sales tax to the nearest cent. \$2.24(6)25 or \$2.25

Answer: The sales tax is \$2.25.

EXAMPLE 3 Marvin bought a camera for \$390 and film for \$12. If the sales tax is 7%, how much tax must he pay?

To find the total price, add the individual prices.

$$\begin{array}{r} \$390 \\ +12 \\ \hline \$402 \end{array} \leftarrow \text{total price}$$

Write the percent as a decimal.

$7\% = 0.07$

Multiply the total price by this decimal.

$$\begin{array}{rl} \$402 & \leftarrow \text{total price} \\ \times 0.07 & \leftarrow \text{rate of sales tax} \\ \hline \$28.14 & \leftarrow \text{amount of sales tax} \end{array}$$ *Ans.*

EXAMPLE 4 A computer sells for \$499. If the rate of sales tax is $8\frac{1}{4}\%$, find the total cost.

Write the percent as a decimal. $8\frac{1}{4}\% = 8.25\% = 0.0825$

Multiply the price by this decimal.
Round up.

$$\begin{array}{rl} \$499 & \leftarrow \text{price} \\ \times 0.0825 & \leftarrow \text{rate of sales tax} \\ \hline 2495 & \\ 998 & \\ 39\ 92 & \\ \hline \$41.1675 & \leftarrow \text{amount of sales tax} \end{array}$$

$= \$41.17$

To find the total cost, add the sales tax to the price.

$$\begin{array}{rl} \$499.00 & \leftarrow \text{price} \\ +41.17 & \leftarrow \text{sales tax} \\ \hline \$540.17 & \leftarrow \text{total cost} \end{array}$$ *Ans.*

KEYING IN

Key Sequence	***Display***
499 [+] 8.25 [%]	540.1675 or \$540.17

EXAMPLE 5 While on vacation, the Florio family bought a fishing rod priced at $32. The cash register total, including the sales tax, was $34.56. What was the rate of sales tax in that state?

THINKING ABOUT THE PROBLEM

Finding the rate of sales tax is like finding the percent of increase.

Key Sequence	*Display*
34.56 [−] 32 [÷] 32 [%]	8.

Answer: The rate of sales tax was 8%.

EXAMPLE 6 The Salengo Discount Department Store charges its customers the county sales tax rate of 7%. Shown here is part of the table used by store clerks to determine how much tax must be charged.

Sally Sims bought a hair bow for $2.39. What was the total cost of the purchase, including sales tax?

Refer to the tax table to find the amount of the tax. The amount of the sale, $2.39, is between the table entries of 2.36 and 2.49. The tax shown is 0.17.

Add the sales tax to the price.

$2.39 ← price
+0.17 ← sales tax
$2.56 ← total cost *Ans.*

Sales Tax Table: 7% Tax

Amount of Sale *From*	*To*	Tax	Amount of Sale *From*	*To*	Tax
0.08	0.21	0.01	2.22	2.35	0.16
0.22	0.35	0.02	2.36	2.49	0.17
0.36	0.49	0.03	2.50	2.64	0.18
0.50	0.64	0.04	2.65	2.78	0.19
0.65	0.78	0.05	2.79	2.92	0.20
0.79	.92	0.06	2.93	3.07	0.21
0.93	1.07	0.07	3.08	3.21	0.22
1.08	1.21	0.08	3.22	3.35	0.23
1.22	1.35	0.09	3.36	3.49	0.24
1.36	1.49	0.10	3.50	3.64	0.25
1.50	1.64	0.11	3.65	3.78	0.26
1.65	1.78	0.12	3.79	3.92	0.27
1.79	1.92	0.13	3.93	4.07	0.28
1.93	2.07	0.14	4.08	4.21	0.29
2.08	2.21	0.15	4.22	4.35	0.30

CLASS EXERCISES

1. If the rate of sales tax is 5%, find the amount of tax for each price.
   a. \$15 **b.** \$38 **c.** \$105 **d.** \$240.50 **e.** \$19.95
2. Frank bought a suit for \$95. If the rate of sales tax is 7%, how much tax must he pay?
3. What is the tax on a cassette recorder costing \$120.50 if the rate of sales tax is 8%?
4. The rate of sales tax is 6%. How much tax will Roberta pay if she buys a book for \$12 and a pen for \$5?
5. Fred spent \$50.95 on cassette tapes and \$38 for stereo headphones. If the rate of sales tax is 8%, how much tax did Fred pay?
   (a) \$7.11 (b) \$7.12 (c) \$0.71 (d) \$0.72
6. Rosa bought 3 cassettes at \$2.99 each. If the rate of sales tax is 8%, how much tax did she pay?
7. Find the total cost of each purchase if the rate of sales tax is 6% and the price is:
   a. \$120 **b.** \$210 **c.** \$325 **d.** \$590.25 **e.** \$14.99
8. Jason bought a clock radio for \$38.50. What is the total amount that he paid if the rate of sales tax is 7%?
9. How much would you pay for a jacket that costs \$90 if the rate of sales tax is 8%?
   (a) \$5.40 (b) \$95.40 (c) \$7.20 (d) \$97.20
10. Mr. and Mrs. Altman had dinner at a local restaurant. The bill came to \$40, and the waiter added \$2 for sales tax. What was the rate of sales tax?
11. The City Council of Brooktown voted to increase the sales tax from 6% to 7%. How much would this add to the cost of a \$130 radio?
12. If the rate of sales tax is 7%, find the amount of tax on a purchase of
    a. 57¢ **b.** \$4 **c.** \$2.50 **d.** \$1.75
13. Find the total cost of each purchase if the rate of sales tax is 7%.
    a. \$1.25 **b.** \$3.98 **c.** \$0.49 **d.** \$2
14. Mrs. Ascher paid for a \$56 motel room with her credit card and found that with sales tax she had been charged \$59.64. What was the rate of sales tax?
15. Mr. Gregory paid for a sweater marked \$39 with a \$50 bill. After adding sales tax to the price of the sweater, the cashier gave him \$7.88 change. What rate of sales tax was used?

HOMEWORK EXERCISES

1. Find the amount of sales tax on each price **a.** \$18 **b.** \$25 **c.** \$80.50 **d.** \$19.99

   (1) if the rate of sales tax is 6% **(2)** if the rate of sales tax is $8\frac{1}{2}\%$

2. Phil bought shoes that sold for \$50. If the rate of sales tax is 7%, how much sales tax did he pay?
3. Mrs. Williams bought a coat costing \$120 and a pair of shoes costing \$50. How much tax did she pay if the rate of sales tax is 8%?
4. Elizabeth bought a handbag for \$24.99, a pair of gloves for \$9.95, and a scarf for \$7.49. If the rate of sales tax is 6%, how much tax did she pay?
5. Find the total cost of each purchase if the rate of sales tax is 7% and the price is:

   a. \$20 **b.** \$60 **c.** \$19 **d.** \$86 **e.** \$14.50 **f.** \$36.25 **g.** \$42.99 **h.** \$112.68

6. Monty bought 2 suits at \$179.95 each. If the rate of sales tax is 8%, how much did he spend?
7. Jill bought a baseball glove costing \$28.49 and a bat costing \$19.99. If the rate of sales tax is 6%, how much did she spend?
8. Mary added \$700 in options to a car that sells for \$8,000. If the rate of sales tax is 5%, how much must Mary pay for the car?
9. Luis bought 5 pairs of socks that cost \$2.79 for each pair and 2 belts that cost \$8.99 each. If the rate of sales tax is 7%, how much did Luis pay?
10. A librarian ordered 12 tapes. Three of the tapes cost \$3.99 each, 3 cost \$4.99 each, 4 cost \$5.49 each, and 2 cost \$7 each. If the rate of sales tax is 8%, what was the total cost?
11. The McCoy's are vacationing in a state with a 7% sales tax. Their own state tax is $7\frac{1}{2}\%$. How much money will they save if they buy a \$700 television set while on vacation?
12. If the total cost for a coat that sells for \$80 is \$84.80, then the rate of sales tax is
    (a) 5% (b) 6% (c) 7% (d) 8%

In 13 and 14, use the tax table on page 300.

13. Find the amount of tax on a purchase of

    a. 89¢ **b.** \$3.25 **c.** \$4.19 **d.** \$1

14. Find the total cost of each purchase.

    a. \$1.85 **b.** \$3 **c.** 50¢ **d.** \$4.29

15. Liz went shopping with \$80 in her handbag. After paying for a pair of sneakers that had a \$47 price tag, she had \$30.18 left. What was the rate of sales tax?

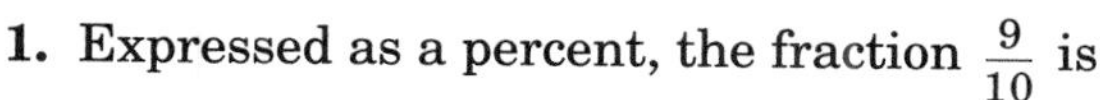

1. Expressed as a percent, the fraction $\frac{9}{10}$ is

(a) 9% (b) 29%
(c) 90% (d) 0.09%

2. 85% written as a fraction is

(a) $\frac{85}{1000}$ (b) $\frac{4}{5}$ (c) $\frac{15}{17}$ (d) $\frac{17}{20}$

3. $\frac{3}{5} + \frac{8}{5}$ equals

(a) $\frac{11}{10}$ (b) $2\frac{1}{5}$ (c) $\frac{24}{25}$ (d) $\frac{3}{8}$

4. $\frac{1}{3} + \frac{2}{9}$ equals

(a) $\frac{3}{12}$ (b) $\frac{1}{2}$ (c) $\frac{5}{9}$ (d) $\frac{15}{18}$

5. $\frac{3}{4}$ of 80 is

(a) 60 (b) 6 (c) 48 (d) $20\frac{3}{4}$

6. Find the sale price of a sofa bed that has a list price of $699 and is being sold at a 15% discount.

7. From the sum of 5 and −7, subtract −2.

8. Use your calculator to add:

$$7\frac{1}{2} + 3\frac{1}{4} + 5\frac{3}{5}$$

9. Over a ten-year period, airlines doubled the average full fare on domestic flights and increased the average discount fare by 30%. If the average discount fare at the beginning of the ten years was $184.90, what was that fare at the end of that period?

10. Find the next number in the sequence:

100, 20, 4, $\frac{4}{5}$, ?

11. The graph shows the total number of inches of rain each year in Clearport during a five-year period.

The percent of decrease in rainfall from the first year to the fifth year was

(a) 300% (b) 150%
(c) 75% (d) 25%

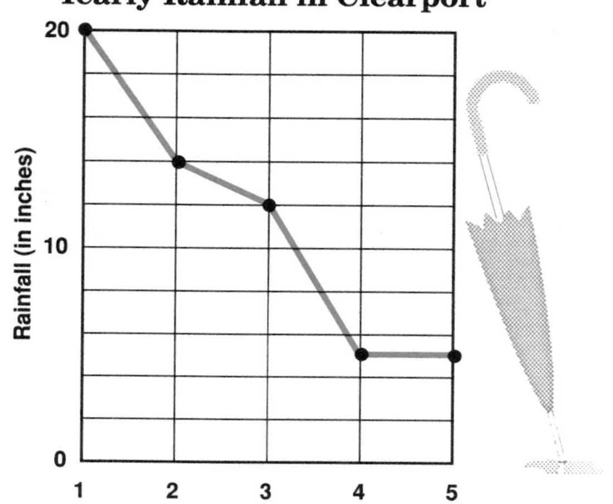

UNIT 9-3 Using Percents to Find Simple Interest

FINDING THE AMOUNT OF INTEREST

THE MAIN IDEA

1. ***Interest*** is a charge for money that is borrowed. For example, when you deposit money in a savings account, the bank pays you interest because it borrows your money to make investments. When you take a loan from a bank, you pay the bank interest for borrowing the bank's money.
2. Interest rates are given per year and are written as percents.
3. The amount of money that is invested or borrowed is called the ***principal***.
4. To find the amount of interest for one year:
   a. Write the rate of interest as a decimal.
   b. Multiply the principal by this decimal.

   $$\text{Interest} = \text{Principal} \times \text{Rate}$$

5. To find the total amount that a borrower must repay or to find the total amount in a savings account, add the interest to the principal.

   $$\text{Amount} = \text{Principal} + \text{Interest}$$

EXAMPLE 1 Mr. Smith deposited \$300 into his savings account. How much interest will he receive at the end of one year if the interest rate is 5%?

Write the interest rate as a decimal. 5% = 0.05

Multiply the principal, \$300, by this decimal (the rate).

$$\begin{array}{rl} \$300 & \leftarrow \text{principal} \\ \times 0.05 & \leftarrow \text{rate of interest} \\ \hline \$15.00 & \leftarrow \text{amount of interest} \end{array}$$

KEYING IN

On a calculator with a [%] key, this key sequence displays the amount of interest directly:

Key Sequence	***Display***
300 [×] 5 [%]	15.

Answer: The interest is \$15.

EXAMPLE 2 Jerry has $250 in his savings account. If the bank pays $6\frac{1}{2}\%$ interest, how much money will Jerry have in his account a year from now?

Write the rate as a decimal. $6\frac{1}{2}\% = 6.5\% = 0.065$

To find the amount of interest, multiply the principal, $250, by the rate, 0.065.

$$\begin{array}{r l} \$250 & \leftarrow \text{principal} \\ \times 0.065 & \leftarrow \text{rate of interest} \\ \hline 1\ 250 & \\ 15\ 00\ \ & \\ \hline \$16.250 & \\ = \$16.25 & \leftarrow \text{amount of interest} \end{array}$$

Add the amount of interest, $16.25, to the principal.

$$\begin{array}{r l} \$250.00 & \leftarrow \text{principal} \\ +16.25 & \leftarrow \text{amount of interest} \\ \hline \$266.25 & \leftarrow \text{amount at end of one year} \end{array}$$

KEYING IN

On a calculator with a [%] key, the following key sequence displays the sum of the principal and interest:

Key Sequence	***Display***	
250 [+] 6.5 [%]	266.25	← principal + interest

Answer: At the end of one year, Jerry will have $266.25 in his savings account.

CLASS EXERCISES

1. Find the interest for one year for each of the following:

a. $80 at 6% **b.** $200 at 6% **c.** $420 at 5% **d.** $1,500 at 5.5%

e. $5,000 at $6\frac{1}{2}\%$ **f.** $250,000 at $6\frac{1}{4}\%$ **g.** $100 at 6.5% **h.** $200 at 5.5%

2. What is the interest on $500 paid by a bank at the end of one year if the rate of interest is 5%?

3. What is the yearly interest on $2,000 invested at $5\frac{1}{2}\%$?

4. What is the interest on a one-year loan of $800 if the interest rate is 9%?

5. Carol borrowed $120 for one year at 9.25% interest. What is the amount of interest that Carol must pay?

6. Jason invested \$250 at 8%. How much interest will Jason earn at the end of one year?
7. Roberto borrowed \$300 for one year at 11%. What is the total amount of money that Roberto must repay?
8. Della deposits \$1,000 in a savings account. If the interest rate is 5.25%, how much money will be in Della's account at the end of one year?

FINDING SIMPLE INTEREST FOR TIME OTHER THAN ONE YEAR

THE MAIN IDEA

1. To find the amount of simple interest when the amount of time is other than a year:
   a. Change the rate of interest to a decimal.
   b. Multiply the principal by this decimal.
   c. Multiply the product by the time expressed in years.

   Simple Interest = Principal × Rate × Time

2. Interest calculated in this way is called ***simple interest***. With simple interest, the principal and the amount of interest earned each year stay the same.

EXAMPLE 3 Don invested \$500 at 7% for 3 years. How much interest will Don earn?

Multiply the principal by the rate of interest written as a decimal.

\$500	← principal
×0.07	← rate of interest
\$35.00	← amount of interest for one year

Multiply the amount of interest for 1 year, \$35, by the time in years.

\$35	
×3	← time in years
\$105	

KEYING IN

The following key sequence can be used on a calculator with a [%] key:

Key Sequence	***Display***	
500 [×] 7 [%]	35.	← interest for 1 year
[×] 3 [=]	105.	← interest for 3 years

Answer: Don will earn \$105 in 3 years.

EXAMPLE 4 Mr. Walsh invested \$8,000 at an 8% annual (yearly) interest. What will be the value of his investment 4 years later?

Multiply the principal by the rate of interest written as a decimal.

\$8,000	← principal
×0.08	← rate of interest
\$640.00	← amount of interest for 1 year

Multiply by the time in years.

\$640	
×4	← time in years
\$2,560	← amount of interest for 4 years

Add the amount of interest to the principal.

\$8,000	← principal
+2,560	← amount of interest
\$10,560	← value of investment 4 years later

KEYING IN

Begin by storing the principal in memory, since it will be needed in more than one calculation.

Key Sequence	***Display***	
8000 [M+]	M 8000.	← principal
[×] 8 [%]	M 640.	← interest for 1 year
[×] 4 [+]	M 2560.	← interest for 4 years
[MR] [=]	M 10560.	← amount after 4 years

Answer: In 4 years, the value of the investment will be \$10,560.

EXAMPLE 5 Adele invested \$3,000 at 6% for 3 months. How much interest did she earn?

Find the amount of interest for one full year.

\$3,000	← principal
×0.06	← rate of interest
\$180.00	← amount of interest for 1 full year

Express 3 months as part of a year.

$$\frac{3 \text{ months}}{12 \text{ months}} = \frac{1}{4} \text{ year}$$

Find the interest for $\frac{1}{4}$ of the year.

$$\frac{1}{\cancel{4}_{1}} \times \cancel{180}^{45} = \$45$$

Key Sequence	***Display***
3000 [×] 6 [%] [×] 3 [÷] 12 [=]	45.

Answer: Adele earned \$45 interest in 3 months.

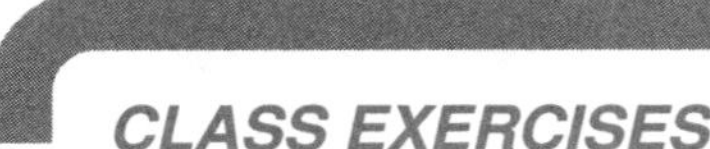

CLASS EXERCISES

1. Sam invested \$1,000 at 5%. How much interest will he earn after:
   a. one year **b.** 5 years **c.** 10 years **d.** 6 months
2. Ms. Santiago invested \$4,000 at 6% interest. How much is her investment worth 5 years later?
3. Write the key sequence used if Class Exercise 2 is solved using a calculator.
4. If \$1,000 is invested at 5%, the amount of interest earned after 9 months is
   (a) \$100 (b) \$900 (c) \$90 (d) \$37.50

HOMEWORK EXERCISES

1. Find the interest for each investment: **a.** in one year **b.** in 3 months
   (1) \$100 at 7% **(2)** \$350 at 5% **(3)** \$500 at 6% **(4)** \$4,000 at $6\frac{1}{2}\%$ **(5)** \$10,000 at $5\frac{1}{2}\%$
2. Ben borrowed \$500 for one year at 9%. How much interest must he pay?
3. Karen invests \$200 at a yearly interest rate of 6%. What will be the total value of Karen's investment at the end of one year?
4. Find the total value of each investment: **a.** after one year **b.** after 6 months
   (1) \$10,000 at 7% **(2)** \$600 at 4.5% **(3)** \$1,000 at 6.5% **(4)** \$500 at 5.25%
5. David invested \$1,000 at $7\frac{1}{2}\%$ interest for 5 years. How much interest will he earn?
6. Kay invested \$500 at 5%. How much is her investment worth at the end of 8 years?
7. William invested \$2,500 at 6%. What is the value of his investment 6 months later?

8. Write a key sequence that could be used on a calculator to solve Homework Exercise 7.

SPIRAL REVIEW EXERCISES

1. With a sales tax of 4%, what is the total cost of a book priced at $15?

2. The balance in Fred's checking account was $133.78 before he made a deposit of $75 and wrote a check for $39.50. What was his new balance?

3. By 1990, approximately 1% of the 13,000 professional baseball players had been inducted into the Hall of Fame. How many players were in the Hall of Fame by 1990?

4. If 3 pairs of sneakers cost $93, what is the cost of 5 pairs of sneakers?

5. How much greater is 8.39 than 7.86?

6. Find the product of $\frac{5}{8}$ and $\frac{24}{35}$.

7. A stock valued at $68 a share had a 12% increase in value followed by a 5% decrease.
   a. What was the final value of a share of the stock?
   b. How does the final value compare to the value of another stock that started at $68 and increased in value by 7%?

8. Round 2,563,419 to the nearest hundred thousand.

9. Evaluate: $-3 + (-8) + (+10)$

10. The graph shows alumni donations for a five-year period.

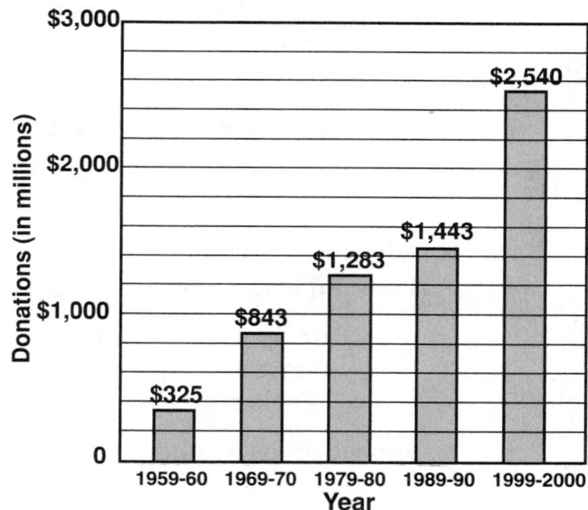

The percent increase from the year 1979–1980 to the year 1999–2000 is approximately

(a) 25% (b) 50% (c) 100% (d) 200%

UNIT 9-4 Using Percents to Find Compound Interest

THE MAIN IDEA

1. *Compound interest* is a way of computing interest that enables an investment to grow more rapidly than with simple interest.
2. Before finding the value of an investment with compound interest:
   a. Determine the total number of *periods* (number of times that interest will be compounded):

   Number of Periods = Number of Years × Number of Periods in One Year

   b. Find the interest rate per period:

   Interest Rate Per Period = Annual Interest Rate ÷ Number of Periods in One Year

3. To find the value of an investment with compound interest:
   a. Find the interest for the first period:

   Amount of Interest for a Period = Principal at Start of the Period × Interest Rate per Period

   b. Add the interest to the principal at the beginning of the period. The sum is the new principal for the next period.

   Principal at Start of Period + Interest for Period = Principal at End of Period = Principal at Start of Next Period

   c. Repeat Steps 3a and 3b for each period, one at a time, until the interest is calculated and added for every period (the number determined in Step 2a).

EXAMPLE 1 Find the total number of periods for each investment time.

	Number of Years	***Interest Compounded***	***Number of Periods In One Year***	***Total Number of Periods***
a.	3	Semiannually	2	$3 \times 2 = 6$
b.	$2\frac{1}{2}$	Quarterly	4	$2\frac{1}{2} \times 4 = 10$
c.	4.5	Monthly	12	$4.5 \times 12 = 54$
d.	2	Daily	365	$2 \times 365 = 730$

EXAMPLE 2 Find the interest rate per period, both as a decimal and as a percent, for an interest rate of 6% per year.

	Interest Compounded	Number of Periods Per Year	Interest Rate Per Period: As a Decimal	Interest Rate Per Period: As a Percent
a.	Annually	1	0.06	6%
b.	Semiannually	2	$0.06 \div 2 = 0.03$	3%
c.	Quarterly	4	$0.06 \div 4 = 0.015$	1.5%
d.	Monthly	12	$0.06 \div 12 = 0.005$	0.5%

EXAMPLE 3 Find the interest for the first period on an investment of $1,000 at 6% per year.

Use the interest rate per period from Example 2.

	Interest Compounded	Interest Rate Per Period	Interest for 1 Period
a.	Annually	6% or 0.06	$\$1{,}000 \times 0.06 = \$60$
b.	Semiannually	3% or 0.03	$\$1{,}000 \times 0.03 = \$30$
c.	Quarterly	1.5% or 0.015	$\$1{,}000 \times 0.015 = \$15$
d.	Monthly	0.5% or 0.005	$\$1{,}000 \times 0.005 = \$5.00$

EXAMPLE 4 Bill invested $500 at 10% per year, compounded annually. Find the value of his investment after 4 years.

Find the number of periods:

$$4 \text{ years} \times 1 \text{ period per year} = 4 \text{ periods}$$

The interest rate per period is 10%, or 0.1 when written as a decimal.

THINKING ABOUT THE PROBLEM

The strategy of making a table helps keep track of the growth of the principal from one period to the next. Complete one line of the table before going on to the next line.

Period	Principal	Interest	New Principal
1st (after 1 yr.)	$500	$0.1 \times 500 = 50$	$500 + 50 = 550$
2nd (after 2 yr.)	550	$0.1 \times 550 = 55$	$550 + 55 = 605$
3rd (after 3 yr.)	605	$0.1 \times 605 = 60.50$	$605 + 60.50 = 665.50$
4th (after 4 yr.)	665.50	$0.1 \times 665.50 = 66.55$	$665.50 + 66.55 = 732.05$

Answer: After 4 years, the value of Bill's investment will be $732.05.

EXAMPLE 5 Use a calculator to find the value of an $800 investment at the end of one year, if the rate of interest is 6% per year, compounded quarterly.

Interest compounded quarterly is added to the principal 4 times each year. There are 1×4, or 4 periods in all, with an interest rate of $6\% \div 4$, or 1.5%, per period.

On the calculator, add 1.5% four times.

Key Sequence	***Display***	
800	800.	← initial principal
[+] 1.5 [%]	812.	← value after first period
[+] 1.5 [%]	824.18	← value after second period
[+] 1.5 [%]	836.5427	← value after third period
[+] 1.5 [%]	849.09084	← value after fourth period

Answer: The value of the investment after one year will be $849.09.

EXAMPLE 6 Marsha invested $10,000 at 8% interest per year, compounded annually, and Rachel invested the same amount at 8% simple interest. After 3 years, how much greater than Rachel's investment will Marsha's investment be?

For Marsha's investment there are 3 time periods. The interest rate for each period is 8%.

Key Sequence	***Display***	
10000 [+] 8 [%]	10800.	← value after 1 year
[+] 8 [%]	11664.	← value after 2 years
[+] 8 [%]	12597.12	← value after 3 years

After 3 years, Marsha's value will be $12,597.12.

For Rachel's investment the amount of interest each year will be the same.

Interest = Principal × Rate × Time

Key Sequence	***Display***	
10000 [M+]	M 10000.	← principal
[×] 8 [%]	M 800.	← interest for 1 year
[×] 3 [M+]	M 2400.	← interest for 3 years
[MRC]	M 12400.	← value after 3 years

After 3 years, Rachel's value will be $12,400.

Find the difference:

$12,597.12	← Marsha's investment
−12,400.00	← Rachel's investment
$ 197.12	← difference

Answer: After 3 years, the value of Marsha's investment will be $197.12 greater than Rachel's.

EXAMPLE 7 Lee invested \$10,000 at 8% interest, compounded quarterly. What was the percent of increase in the value of his investment after one year?

There are four time periods, with an interest rate of 8% ÷ 4, or 2%, for each period. The growth of the investment is shown in the table.

Time Period	Principal at the Beginning of the Period	Value at the End of the Period
1	\$10,000	\$10,200.00
2	10,200	10,404.00
3	10,404	10,612.08
4	10,612.08	10,824.321 (round to 10,824.32)

To find the percent of increase, first find the amount of increase.

\$10,824.32	← value after 1 year
−10,000.00	← amount invested
\$ 824.32	← amount of increase

Then, find what percent the amount of increase is of the amount invested.

$$\frac{\text{Amount of increase}}{\text{Amount invested}} = \frac{824.32}{10{,}000.00}$$

$$= 0.082432 = 8.2432\%, \text{ or about } 8\tfrac{1}{4}\%$$

The percent of increase is about $8\frac{1}{4}$%. This rate, the percent of growth of an investment in one year, is called the ***Annual Percentage Rate (APR)***.

Answer: After one year, the percent of increase was about $8\frac{1}{4}$%.

CLASS EXERCISES

1. Find the total number of periods for a 4-year investment, if interest is compounded:
a. annually **b.** semiannually **c.** quarterly **d.** monthly

2. Find the interest rate per period, both as a decimal and as a percent, when the interest rate is 6% per year and interest is compounded:
a. annually **b.** semiannually **c.** quarterly

3. Find the interest for the first period on an investment of \$2,000 at 6% per year, if interest is compounded semiannually.

4. Mr. Holmes invested \$1,000 at $6\frac{1}{2}$% per year, compounded annually. Find the value of his investment after three years.

In 5 and 6, use a calculator.

5. A sum of $5,000 is invested at 3% per year, compounded semiannually. What is the value of the investment after two years?

6. Mrs. Winston invested $2,000 at 6.5% per year, compounded annually.
   a. At the end of five years, what was the value of her investment?
   b. What is the percent increase in the value of her investment over the five years?

7. A friend asks your advice about two investments available to him. He can invest $5,000 at a rate of 5%, compounded semiannually, or at a simple interest rate of 6%. If he wants to invest for one year, explain how you would advise him.

8. Mrs. McKenzie invested $20,000 at 8%, compounded quarterly. Find the annual percentage rate (the percent increase in the investment in one year).

HOMEWORK EXERCISES

1. Find the total number of periods for a $2\frac{1}{2}$ year investment, if interest is compounded:

   a. annually **b.** semiannually **c.** quarterly **d.** monthly

2. Find the interest rate per period, both as a decimal and as a percent, when the interest rate is 5% per year and interest is compounded:

   a. annually **b.** semiannually **c.** quarterly

3. Find the interest for the first period on an investment of $6,000 at 6% per year if interest is compounded quarterly.

In 4–10, use a calculator.

4. Find the final value of each investment.
   a. $1,100 at 8% per year, compounded annually for 4 years.
   b. $3,000 at 6% per year, compounded semiannually for 2 years.
   c. $2,500 at 5% per year, compounded quarterly for 2 years.

5. Mr. Williams invested $5,000 at 6.5% per year, compounded annually. Find the percent of increase in the value of his investment after 3 years.

6. A newspaper shows two advertisements on the same page for investments of $10,000. One offers a simple interest rate of 6%, and the other offers an interest rate of 5%, compounded quarterly.
   a. Which investment will have the greater value after one year?
   b. How much greater will it be?

7. Tamara invested $1,000 at 7% per year, compounded annually. How many years will it take for her investment to at least double in value?

8. Workman's Savings Bank is offering 6% per year, compounded semiannually, for money invested in its special fund. Edward invests $500 with the goal of having it grow to at least $750. How long must he keep his money in the investment to reach his goal?

9. How much more would you earn in one year if you invested $10,000 at 5% compounded semiannually rather than at 5% simple interest?

10. Find the annual percentage rate for an investment that earns 5% interest compounded quarterly.

SPIRAL REVIEW EXERCISES

1. Which group of decimals is in order from least to greatest?

   (a) 0.7, 0.09, 0.68, 0.625, 0.375
   (b) 0.375, 0.68, 0.625, 0.7, 0.09
   (c) 0.375, 0.625, 0.68, 0.7, 0.09
   (d) 0.09, 0.375, 0.625, 0.68, 0.7

2. In the past decade, the population of the state of Texas grew 20% to 17 million people. If the same trend continues, what will the population of Texas be a decade from now?

3. A large-screen television set that sells for $750 is on sale for "30% off." One way to find the amount of money saved when the set is bought at the sale price is to multiply $750 by

   (a) $\frac{1}{5}$ (b) $\frac{2}{5}$ (c) $\frac{3}{10}$ (d) $\frac{1}{2}$

4. The next number in the sequence 2, −4, 8, −16, 32, ... is

   (a) −16 (b) −32 (c) −64 (d) 64

5. Jack represented deposits in his checking account with positive numbers and withdrawals with negative numbers. He began the month with a balance of $685 and recorded the following transactions: +50, +75, −25, −60, −95, −115. What was the balance in his account at the end of the month?

6. Maria invested $500 for 3 years at 6% simple interest. How much was her investment worth after 3 years?

7. Last year, 40 students took Advanced Placement History. This year, 50 students are taking the course. What is the percent of increase in the number of students taking the course?

8. Which key sequence would have the effect of increasing $34.60 by 12%?

   (a) 34.6 [+] 88 [%]
   (b) 34.6 [×] 112 [%]
   (c) 34.6 [+] .12 [%]
   (d) 34.6 [+] .12 [=]

9. Divide: $\frac{-24}{25} \div 1\frac{3}{5}$

10. The best estimate for 5.38 × 6.87 from the given choices is

    (a) 120 (b) 25 (c) 30 (d) 35

11. The number 0.008 written as a fraction in simplest form is

    (a) $\frac{4}{5}$ (b) $\frac{2}{5}$ (c) $\frac{2}{25}$ (d) $\frac{1}{125}$

UNIT 9–5 Using Percents in a Budget

THE MAIN IDEA

1. A ***budget*** is a plan for spending money. Each item in a budget is allowed a certain part of the total income.
2. The parts of the income that are allowed for individual items are written as percents.
3. To find the amount of money that a budget allows for a particular item:
   a. Write the percent budgeted for that item as a decimal.
   b. Multiply the total income by this decimal.

   $$\text{Amount Budgeted} = \text{Total Income} \times \text{Percent Budgeted}$$
4. The sum of the percents budgeted for all the items is 100%.
5. The sum of the individual amounts of money budgeted for all the items equals the total income.

EXAMPLE 1 The Vivaldi family budget allows 35% for rent and utilities, 30% for food, 23% for clothing and entertainment, and 12% for savings. If their monthly income is $3,500, how much money does their budget allow for each item?

Find the amount of money allowed for each item by changing the percent budgeted to a decimal and multiplying $2,500 by this decimal.

	Item	*Percent Budgeted*	*Amount Budgeted*
	Rent and Utilities	35%	$1225
	Food	30%	$1050
	Clothing and Entertainment	23%	$805
	Savings	12%	$420
Check:	Total	100%	$3,500

EXAMPLE 2 The graph shown gives the Huang family weekly budget. Calculate the percent of the budget that was planned for clothing.

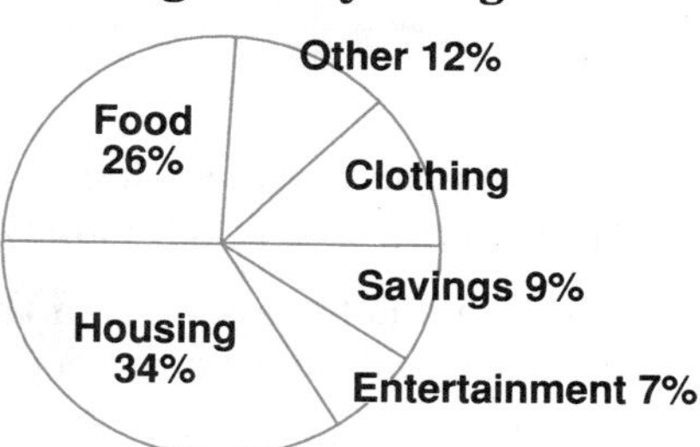

THINKING ABOUT THE PROBLEM

The sum of all the individual percents budgeted is 100%. To find the missing percent, subtract the sum of all the other percents from 100%.

Add the percents planned for food, housing, entertainment, savings, and other.

$$\begin{array}{r} 26\% \\ 34\% \\ 7\% \\ 9\% \\ +12\% \\ \hline 88\% \end{array}$$

Subtract 88% from 100%, the whole amount.

$$\begin{array}{r} 100\% \\ -88\% \\ \hline 12\% \end{array}$$

Answer: 12% of the budget was planned for clothing.

EXAMPLE 3 The Carter budget allows 35% for housing, 25% for food, 15% for clothing, 9% for transportation, 10% for entertainment, and 6% for savings. If the Carter monthly income is $3,950, how much should they plan to spend on both food and clothing?

Since 25% is budgeted for food and 15% is budgeted for clothing, 25% + 15% or 40% is budgeted for both food and clothing.

Find 40% of $3,950.

$$\begin{array}{r} \$3{,}950 \\ \times 0.40 \\ \hline \$1{,}580.00 \end{array}$$

Answer: The Carters should plan to spend $1,580 on food and clothing.

EXAMPLE 4 Mr. Jerome earns $4,000 a month. His monthly expenses for housing are $1,600. What percent of his income should he budget for housing?

Write as a fraction the part of Mr. Jerome's income that he spends for housing. $\frac{1{,}600}{4{,}000}$

Simplify the fraction. $\frac{1{,}600 \div 800}{4{,}000 \div 800} = \frac{2}{5}$

Change $\frac{2}{5}$ to a decimal. $5\overline{)2.00}$ with quotient 0.4

$\frac{2}{5} = 0.4$

Write 0.4 as a percent. $0.4 = 40\%$

Answer: Mr. Jerome should budget 40% of his income for housing.

CLASS EXERCISES

1. Ms. Johnson earns $3,500 a month. In her budget, she allows 30% for the rental of her apartment. How much money has she budgeted for rent?
2. Ethan plans to save 40% of the money that he earns baby-sitting. If Ethan earned $80, how much should he save to keep to his budget?
3. The Aho family budget allows 15% of the weekly income for clothing. How much money can be spent on clothing if the total weekly income is $780?
4. John's budget allows 20% for clothing, 40% for food, and 26% for entertainment. If savings is the only other item on the budget, what percent is allowed for savings?
5. Copy and complete the table shown for the Solomon family budget if the monthly income is $3,200.

Monthly Budget for the Solomon Family

Item	Percent Budgeted	Amount Budgeted
Food	20%	
Housing	25%	
Clothing	22%	
Savings	6%	
Entertainment	12%	
Other	15%	

6. The graph shown gives Jane's weekly budget.
   a. If Jane's weekly income was $670, how much money did she budget altogether for clothing and savings?
   b. What percent of the budget was planned for "other" expenses?

Weekly Budget

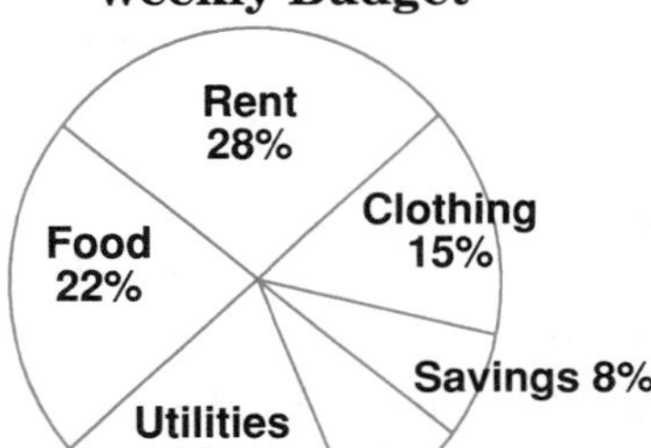

7. The Wu family budget allows 28% for housing, 22% for food, 18% for clothing, 11% for transportation, 12% for entertainment, and 9% for savings. If the Wus' weekly income is $950, how much money is budgeted for entertainment and savings?
8. Mr. Read earns $960 a week. If his weekly expenses for food are $240, what percent of his income should he budget for food?
9. Ms. Inoue earns $4,800 a month. What percent of her income should she budget for rent if she pays $1,920 a month for rent?

HOMEWORK EXERCISES

1. Caru works part time as a cashier for $80 a week. If she budgets 35% for savings, how much should she save each week?

2. Mr. George earns $3,250 a month. His budget allows 15% for entertainment. How much money has he budgeted for entertainment?

3. The Jones family budget allows 20% of the weekly income for food. How much money can be spent on food if the total weekly income is $920?

4. A guide for a family budget recommends that 30% of the monthly income be budgeted for housing. How much money should be budgeted for housing if the monthly income is $4,200?

5. Using the data in the graph below, find the amount budgeted for each item if the weekly income is $900.

Weekly Budget

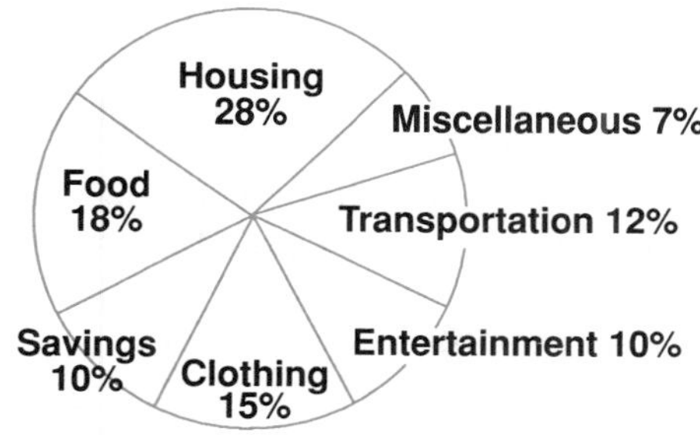

6. The Johnson family budget is shown in the chart.

   a. What percent of the budget was planned for food?

   b. If the Johnson monthly income was $4,960, how much money was budgeted for transportation and miscellaneous?

Item	Percent Budgeted
Housing	29%
Food	?
Clothing	18%
Savings	10%
Transportation	12%
Miscellaneous	8%

7. Roberta's weekly budget allows 30% for food, 25% for clothing, 12% for savings, and 20% for entertainment. If transportation is the only other item on her budget, what percent is allowed for transportation?

8. The Reliable Delivery Company budgets 30% for transportation, 25% for repairs, and 45% for the workers' salaries. If the total weekly expenses are $3,500, how much is budgeted for repairs and salaries?

9. The utility bills for the Levine family are $180 a month. If the Levine family's monthly income is $3,600, what percent of the income should be budgeted for utility expenses?

10. Mary earns $120 a week on her part-time job. What percent of her income should she budget for savings if she wants to save $30 a week?

11. Mr. Rome earns $1,000 a week. If his weekly expenses for transportation are $80, what percent of his weekly income should he budget for transportation?

12. From Martin's $950 weekly paycheck, he budgets 20% for food. Last week he spent $210 for food. How much more than the amount budgeted did he spent for food?

SPIRAL REVIEW EXERCISES

1. How much tax is paid on a $20 shirt if the sales tax is 6%?

2. If 4 pounds of tomatoes cost $4.80, what is the cost of 5 pounds of tomatoes?

3. Joe invested $500 at 7% interest. What is the value of his investment one year later?

4. Ms. Joyce had a balance of $785.52 in her checking account. What is her new balance after she wrote checks for $119.80 and $285.40?

5. Mr. Phillips spent $17.20 on groceries. How much change will he receive from a $20 bill?
(a) $37.20 (b) $3.80
(c) $2.80 (d) $1.80

6. Doris bought two picture frames regularly priced at $9.95 each on sale at 20% off. How much did she pay for the frames?

7. Use your calculator to find the sum of the reciprocals of 250 and 625.

8. The rates for a telephone call from a coin box are:

$1.20 for the first minute
$0.15 for each additional minute

The cost of a 10-minute call is
(a) $1.85 (b) $2.55
(c) $0.95 (d) $0.65

9. The value of $-7 + (-2) + (-9)$ is
(a) -81 (b) -18 (c) -14 (d) 0

10. At a professional golf tournament, out of 367 golfers who were competing, 24 were left-handed. The percent of left-handed golfers in the tournament was closest to
(a) 0.3% (b) 7% (c) 30% (d) 50%

UNIT 9-6 Using Percents in Installment Buying; Using Percents With Credit Cards

USING PERCENTS IN INSTALLMEMT BUYING

THE MAIN IDEA

1. ***Installment buying*** is sometimes used instead of paying the entire price at the time of purchase. In an installment plan, a ***down payment*** is usually made at the time of purchase, followed by equal monthly ***installments***. The down payment is often a percent of the purchase price.
2. To find the cost of an item bought on an installment plan:
   a. Multiply the amount of each installment by the number of installments.
   b. Add the amount of the down payment.

   $$\text{Cost of Item on Installment Plan} = \text{Amount of Each Installment} \times \text{Number of Installments} + \text{Down Payment}$$
3. The cost of an item bought on an installment plan is more than the original price. This increase in cost is called the ***carrying charge*** of the installment plan.
4. To find the carrying charge, subtract the original price from the installment plan cost.

   $$\text{Carrying Charge} = \text{Installment Plan Cost} - \text{Original Price}$$

EXAMPLE 1 Find the cost of a television set that is bought on an installment plan by making a \$50 down payment and paying 10 monthly installments of \$40 each.

Multiply the amount of each installment by the number of installments.

$$\begin{array}{r} \$40 \\ \times 10 \\ \hline \$400 \end{array}$$

\$40 ← amount of each installment
×10 ← number of installments

Add the amount of the down payment.

$$\begin{array}{r} \$400 \\ +50 \\ \hline \$450 \end{array}$$

+50 ← down payment

Answer: The cost of the television set is \$450.

EXAMPLE 2 A vacation that is advertised for $780 can be paid for with a 10% down payment and 15 monthly installments of $60 each.

a. Find the total cost of the vacation on the installment plan.

To find the amount of the down payment, multiply $780 by 10% written as a decimal.

$$\begin{array}{r} \$780 \\ \times 0.10 \\ \hline \$78.00 \end{array} \leftarrow \text{down payment}$$

Multiply the amount of each installment by the number of installments.

$$\begin{array}{rl} \$60 & \leftarrow \text{amount of each installment} \\ \times 15 & \leftarrow \text{number of installments} \\ \hline 300 & \\ 60 & \\ \hline \$900 & \end{array}$$

Add the down payment.

$$\begin{array}{rl} \$900 & \\ +78 & \leftarrow \text{down payment} \\ \hline \$978 & \end{array}$$

Answer: The total cost of the vacation is $978.

b. What is the approximate percent increase over the advertised cost of the vacation?

The cost increased from $780 to $978, an increase of $198. To find the percent of increase, find the percent that 198 is of 780.

$$\frac{198}{780} = \frac{198 \div 6}{780 \div 6} = \frac{33}{130} \qquad 130\overline{)33.0000}\ \text{quotient } 0.2538 = 25.38\%$$

Answer: The percent increase is approximately 25%.

EXAMPLE 3 A $110 coat is bought on an installment plan by making a $20 down payment and paying 8 installments of $15 each. Find the carrying charge.

Multiply the amount of each installment by the number of installments.

$$\begin{array}{rl} \$15 & \leftarrow \text{amount of each installment} \\ \times 8 & \leftarrow \text{number of installments} \\ \hline \$120 & \end{array}$$

Add the $20 down payment.

$$\begin{array}{rl} \$120 & \\ +20 & \leftarrow \text{down payment} \\ \hline \$140 & \leftarrow \text{installment plan cost} \end{array}$$

To find the carrying charge, subtract the original price from the installment plan cost.

$$\begin{array}{rl} \$140 & \leftarrow \text{installment plan cost} \\ -110 & \leftarrow \text{original price} \\ \hline \$30 & \leftarrow \text{carrying charge} \end{array}$$

Answer: The carrying charge is $30.

CLASS EXERCISES

1. The Collins family buys a refrigerator by making a $400 down payment and paying 12 monthly installments of $25 each. How much does the refrigerator cost?
2. Find the installment price of a stereo if a down payment of $200 and 20 payments of $40 were made.
3. Mr. James bought a used car by making a down payment of $500 and 12 monthly payments of $300. How much did Mr. James pay for the car?
4. A $400 ring is bought on installment by making a 15% down payment and paying 10 installments of $38 each. Find the amount paid for the ring.
5. A $650 washing machine is bought on an installment plan with a 20% down payment. What will be paid for the washing machine if 24 installments of $25 each will be paid?
6. A $450 television set was purchased by making a down payment of $\frac{1}{3}$ of the price and arranging for 12 installments of $34.95 each. What is the installment price of the television set?
7. You can buy a $300 color television set by making a $60 down payment and paying 12 installments of $25 each. What is the carrying charge?
8. Mr. Jackson bought a $400 video recorder by making a $60 down payment and paying 20 installments of $22 each. How much more did Mr. Jackson pay for the video recorder by using the installment plan than if he had paid cash?
9. You can buy a $1,500 boat by paying $300 down and 20 installments of $80 each. The carrying charge is
   (a) $400 (b) $500 (c) $1,600 (d) $1,900

NO DOWN PAYMENT
24 months to pay

Credit same as cash

USING PERCENTS WITH CREDIT CARDS

THE MAIN IDEA

1. Buying with a ***credit card*** is another way of paying for something over time.
2. No down payment is required with a credit card purchase, but a ***finance charge*** is added to the monthly credit-card bill. This finance charge is calculated as a percent of the *average daily balance*, which takes into account the unpaid balance and the new purchases.
3. To find the new balance each month:
   a. Find the unpaid balance:

      Old Balance – Payments = Unpaid Balance

   b. Add the unpaid balance and the cost of additional purchases.
   c. Calculate the finance charge.
   d. The new balance is the sum of the unpaid balance, the cost of new purchases, and the finance charge.

      Unpaid Balance + Additional Purchases + Finance Charge = New Balance

EXAMPLE 4 Mrs. Post had a balance of $400 in her credit-card account last month. She made a payment of $100. If the monthly finance charge is 1.2% of the unpaid balance and if she doesn't use her credit card again this month, what will be the new balance next month?

Find the unpaid balance. $\qquad$ $400 – $100 = $300

Find the finance charge, which is 1.2% of $300.

$$\begin{array}{r} 300 \\ \times 0.012 \\ \hline 600 \\ 300 \\ \hline 3.600 \end{array} = \$3.60$$

Add the unpaid balance and the finance charge to find the new balance.

$$\begin{array}{rl} \$300.00 & \leftarrow \text{unpaid balance} \\ 3.60 & \leftarrow \text{finance charge} \\ \hline \$303.60 & \leftarrow \text{new balance} \end{array}$$

Answer: The new balance next month will be $303.60.

EXAMPLE 5 Mr. Amodeo had an unpaid balance of $620 on his credit-card account at the beginning of the month. He used his credit card to make additional purchases of $63.50, $29.80, and $114. Each month, his credit-card company adds a finance charge that is 1.5% of the total unpaid balance and new purchases. What will be the new balance on Mr. Amodeo's statement next month?

First, find the sum of the unpaid balance and the additional purchases.

$$\begin{array}{rl} \$620.00 & \leftarrow \text{unpaid balance} \\ 63.50 & \\ 29.80 & \leftarrow \text{additional purchases} \\ +114.00 & \\ \hline \$827.30 & \end{array}$$

Then, find the finance charge, which is 1.5% of $827.30.

$$\begin{array}{rl} \$827.30 & \\ \times 0.015 & \\ \hline \$12.4095 & \leftarrow \text{finance charge} \end{array}$$

The new balance on the next statement will be the sum of $827.30 and $12.41.

$$\begin{array}{rl} \$827.30 & \\ +12.41 & \\ \hline \$839.71 & \leftarrow \text{new balance} \end{array}$$

KEYING IN

These calculations can be done on a calculator by using the following key sequence:

Key Sequence	***Display***
620 [+] 63.5 [+] 29.8 [+] 114 [+] 1.5 [%]	839.7095

Answer: The new balance on Mr. Amodeo's next credit-card statement will be $839.71.

CLASS EXERCISES

1. Mr. Rodgers has an unpaid balance of $172.50 on his credit-card account. The monthly finance charge is 1.4% of the unpaid balance. What will the new balance be on his next statement if he makes no additional purchases?

2. Melanie had a balance of $242.70 on her credit-card account. She made a payment of $150. If the monthly finance charge was 1.5% of the unpaid balance, what will be the new balance on her next statement, assuming that she makes no further payments or purchases?

3. Mrs. Gentry had an unpaid balance of $480.50 on her last credit-card statement. She used her credit card to purchase a dress for $42.80 and a pair of shoes for $24.90. If the monthly finance charge is 1.8% of the total unpaid balance and new purchases, what will be the new balance on her next statement?

4. Evan had a balance of $376.90 on his credit-card account. He made a payment of $250 and only one additional purchase of $75.80. If the monthly finance charge is 1.6% of the sum of the unpaid balance and the new purchases, what will Evan's new balance be on the next statement?

HOMEWORK EXERCISES

1. A television set is bought by paying $50 down and 10 monthly installments of $38. How much did the television set cost?

2. Find the installment price of an air conditioner if a down payment of $45 and 12 payments of $30 were made.

3. Mr. Reynolds bought a new car by making a down payment of $5,500 and 24 monthly payments of $200.95. What is the total cost of the car?

4. Copy and complete the table shown.

Item	Down Payment	Amount of Each Installment	Number of Installments	Installment Plan Cost
Camera	$20	$10	15	
Stereo	$55	$22	10	
Tape Recorder	$35	$18	12	
Moped	$28	$20	14	
Used Car	$1,200	$150	24	
Boat	$4,000	$280	48	

5. A $500 video recorder is bought on an installment plan with a 10% down payment. What is the total paid for the video recorder if 20 installments of $28 are made?

6. Albert bought a $180 bicycle by paying 20% down and 10 installments of $19.50 each. What was the installment plan cost of the bicycle?

7. Marcia bought a $1,200 personal computer by making a $100 down payment and 20 payments of $65 each. What was the carrying charge?

8. Mr. Cobb bought a $5,000 used car by paying $1,800 down and 36 installments of $120.25 each. How much more than the $5,000 price did he pay for the car by using the installment plan?

9. A $625 sofa is bought by paying $125 down and 12 installments of $50 each. The carrying charge is

   (a) $200 (b) $100 (c) $725 (d) $600

10. Copy and complete the chart shown.

Item	Original Price	Down Payment	Amount of Each Installment	Number of Installments	Installment Plan Cost	Carrying Charge
Guitar	$250	$25	$22	12		
Kitchen Set	$528	$100	$30	18		
Typewriter	$198	$50	$28	6		
New Car	$9,800	$2,900	$190	42		

11. Mrs. Margulies has an unpaid balance of $78.80 on her credit-card account. The monthly finance charge is 1.5% of the unpaid balance. If she makes no additional purchases, what will the new balance be on her next statement?

12. Tom had an unpaid balance of $1,250 on his credit-card account. He made a payment of $350. If the monthly finance charge is 1.3% of the unpaid balance, what will be the new balance on his next statement?

13. Ms. Fenner owed $950 on her credit-card account last month. She made a payment of $400 and then used her credit card for purchases of $257.50, $94.32, and $195.

a. If the monthly finance charge is 1.6% of the total unpaid balance and new purchases, what will be the new balance on her next statement?

b. A different credit-card company computes the finance charge of 1.6% only on the unpaid balance, and not on new purchases. How much less would the finance charge be on Ms. Fenner's account if her account was with this credit-card company?

14. Mr. King had a balance of $175.40 on his credit-card account. He used his card to buy a lamp for $85.50 and made a payment of $100. If the monthly finance charge is 1.8% of the sum of the unpaid balance and new purchases, what will be the new balance on his next statement?

15. One month, Ms. James had a balance of $350 on her credit-card account. On the next statement, after making no purchases or payments, a finance charge of $4.20 was added to the unpaid balance.

a. Write a key sequence that can be used on a calculator to find the rate of interest that was used to compute the finance charge.

b. What was the interest rate used to compute the finance charge?

16. Using her credit card, Mrs. Ardo bought a $300 stereo and a $100 cassette recorder at a discount of 15%. She paid for her purchases with three monthly payments of $40, $100, $100, and a final payment of the unpaid balance in the fourth month. Each month, a finance charge of 1.5% was added to the unpaid balance.

a. How much did Mrs. Ardo actually pay for her purchases?

b. What percent of the original price did she actually save?

17. Mr. Crane wants to buy a $600 refrigerator. Which of the two following payment plans results in a smaller total cost?

Plan A: Make a 10% down payment and 6 monthly installment payments of $98.64.

Plan B: Use a credit card that has a 1.5% monthly finance charge, and pay $90 each month until the total balance is paid.

SPIRAL REVIEW EXERCISES

1. If sales tax is charged at a rate of 7%, what is the tax on a jacket priced at \$65?

2. If Mrs. Jolson bought two items priced at \$7.50 and \$3.10 and sales tax is charged at the rate of 6%, what was the total cost of her purchases?

3. 18% of 400 is
(a) 720 (b) 72 (c) 7.2 (d) .72

4. $9\frac{3}{8}$ is equal to
(a) $\frac{75}{8}$ (b) $\frac{66}{8}$ (c) $\frac{35}{8}$ (d) $\frac{12}{8}$

5. The fraction $\frac{2}{5}$ is equal to
(a) 20% (b) 0.2 (c) 0.4 (d) 25%

6. 125.862 rounded to the nearest hundred is
(a) 100 (b) 125.86
(c) 200 (d) 125.87

7. Thirteen thousand, forty-eight written as a numeral is
(a) 130,048 (b) 1,348
(c) 13,480 (d) 13,048

8. Use your calculator to help decide which number is greater than $\frac{3}{49}$ and less than $\frac{1}{12}$.
(a) 0.06 (b) 0.0605
(c) 0.0615 (d) 0.084

9. The total monthly income for the Wolfson family is \$4,000. One month, they spent \$360 on clothing. Use the circle graph to determine by how much they exceeded their budget for clothing that month.

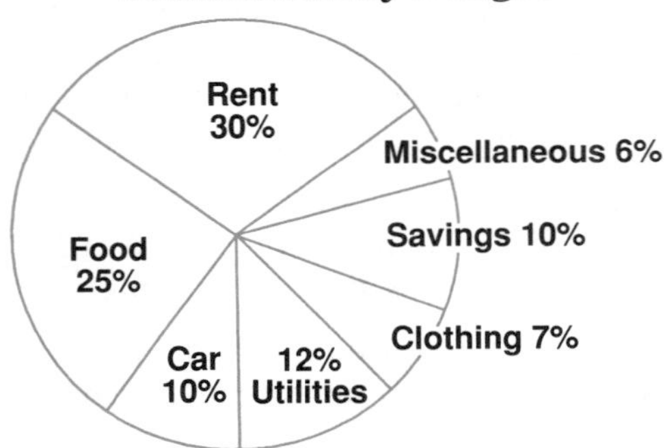

(a) \$40 (b) \$80 (c) \$120 (d) \$160

THEME 5

Statistics and Measurement

MODULE 10

MODULE 11

MODULE 12

UNIT 10-1 Frequency Tables; Histograms

FREQUENCY TABLES

THE MAIN IDEA

1. ***Statistics*** is the branch of mathematics that deals with organizing numerical facts, or ***data***, in order to study them.
2. One way we organize data is to record the number of times each value appears, the ***frequency***. This count, or ***tally***, produces a list of values that is called a ***frequency table***.
3. Data in a frequency table may be grouped in ***intervals***, sets of values between two numbers.

EXAMPLE 1 The frequency table shows the number of long-distance phone charges paid by the Jefferson Accounting Office during the month of March. The charges are grouped in $1 intervals.

Telephone Charges for March

Charge	Number of Calls (Frequency)
$1.00–$1.99	151
$2.00–$2.99	82
$3.00–$3.99	63
$4.00–$4.99	18
$5.00–$5.99	29
$6.00 and over	30

a. How many long-distance calls were made by the office during the month of March?

The table shows all the calls for March. Add to find the total number of calls.

151 + 82 + 63 + 18 + 29 + 30 = 373

Answer: There were 373 calls made in the month of March.

b. Which interval had the least number of calls? the greatest number of calls?

The least number of calls is 18. The greatest number of calls is 151.

Answer: The interval that had the least number of calls was $4.00–$4.99.
The interval that had the greatest number of calls was $1.00–$1.99.

EXAMPLE 2 Make a frequency table for these weights of 30 high school students. Group the data in 10-pound intervals.

Weights of 30 High School Students
(Weights rounded to the nearest pound)

128	118	140	116	97	132	121	107	99	96
106	103	110	99	112	108	126	130	122	119
118	111	117	143	113	123	114	98	120	141

First, note the highest and lowest weights: 143, and 96. Select intervals of 10 pounds that include these weights. List the intervals.

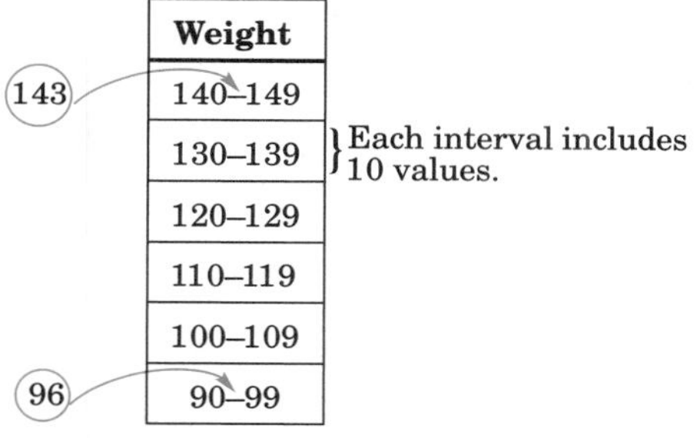

Second, tally (count, using stroke marks in groups of 5) the weights to show how many times weights are found in each interval.

Weight	Tally
140–149	𝍷𝍷𝍷
130–139	𝍷𝍷
120–129	𝍸 𝍷
110–119	𝍸 𝍸
100–109	𝍷𝍷𝍷𝍷
90–99	𝍸

Finally, count the tally marks and write the totals, or frequencies. Then write a title for the frequency table.

Weights of 30 High School Students

Weight	Tally	Frequency
140–149	𝍷𝍷𝍷	3
130–139	𝍷𝍷	2
120–129	𝍸 𝍷	6
110–119	𝍸 𝍸	10
100–109	𝍷𝍷𝍷𝍷	4
90–99	𝍸	5

To check, find the sum of all the frequencies and see if all 30 of the given weights were tallied.

$$\begin{array}{r} 3 \\ 2 \\ 6 \\ 10 \\ 4 \\ \underline{5} \\ 30 \checkmark \end{array}$$

Observe that the intervals were listed with the highest interval, 140–149, at the top of the table, and the lowest interval, 90–99, at the bottom. This arrangement is a matter of choice. Reversing the order, with 90–99 at the top, would be just as good.

EXAMPLE 3 The numbers listed are the heights, in inches, of 50 schoolchildren. Make a frequency table of these heights, grouping the data in 2-inch intervals.

60 48 52 55 50 61 51 51 53 49
61 59 49 48 48 49 55 57 60 53
49 58 61 50 54 56 55 59 61 57
56 48 54 60 61 54 49 60 48 51
59 50 55 59 51 53 57 54 49 59

To label the intervals, begin with the smallest height (48 in.) and count off (48–49, 50–51, etc.) until you have included the largest height (61 in.). Then, tally the heights and write the frequencies.

Check the total of the frequencies. Write a title.

Heights of 50 Schoolchildren

Height (inches)	Tally	Frequency
48–49	𝍸 𝍸 𝍷	11
50–51	𝍸 𝍷𝍷	7
52–53	𝍷𝍷𝍷𝍷	4
54–55	𝍸 𝍷𝍷𝍷	8
56–57	𝍸	5
58–59	𝍸 𝍷	6
60–61	𝍸 𝍷𝍷𝍷𝍷	9

Check: 50 ✓

CLASS EXERCISES

1. This is a frequency table of the average number of repairs made on different brands of automobiles during the first two years after purchase.
   a. Which brand had the worst record of repairs (the greatest number)?
   b. Which brand had the best record of repairs?
   c. Which brands had equal frequencies of repair?

Auto Repairs

Brand	Number of Repairs
A	2
B	7
C	3
D	5
E	8
F	3
G	1

2. a. Make a frequency table of the weights (in pounds), shown at the right, of airplane luggage. Use the intervals 14–16, 17–19, 20–22, 23–25, 26–28, and 29–31.
   b. Which interval has the greatest frequency?
   c. Which interval has the least frequency?
   d. What is the sum of all the frequencies?

22 16 22 16 21 18
22 27 29 31 30 28
18 17 17 30 29 30
27 22 19 15 18 15
28 22 15 14 27 22

3. a. Make a frequency table for the lengths of wire, given in inches. Use the intervals 4–5.9, 6–7.9, up through 14–15.9.
   b. Which interval has the greatest frequency?
   c. Which interval has the least frequency?
   d. What is the sum of all the frequencies?

4.5 6.0 14.4 15.8 7.3
8.0 12.0 14.0 8.0 5.3
8.0 5.0 11.0 15.0 11.1
7.9 6.5 4.9 14.5 11.5

HISTOGRAMS

THE MAIN IDEA

1. A ***histogram*** is like a bar graph of a frequency table. The horizontal scale shows the intervals, and the vertical scale shows the frequency. There is no space between bars.
2. Each bar of a histogram represents one interval of a frequency table. The height of each bar represents the frequency of data values in that interval.

EXAMPLE 4 The histogram below was made from a frequency table in which the students at New City High School were tallied according to their academic averages, rounded to the nearest tenth.

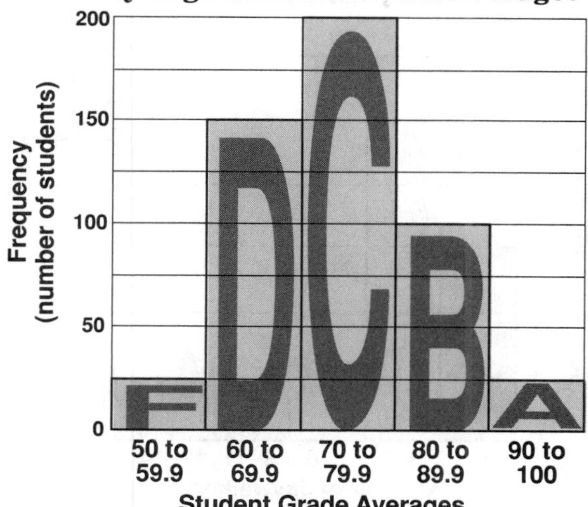

a. How many students are represented by the data in this histogram?

Add the five frequencies to find the total number of students:

$$25 + 150 + 200 + 100 + 25 = 500 \ \textit{Ans.}$$

b. How many more students had averages in the interval from 80 to 89.9 than in the interval from 50 to 59.9?

Find the numbers of students with averages in each interval, and then subtract to find the difference.

Students with averages from 80 to 89.9	→	100
Students with averages from 50 to 59.9	→	−25
Difference	→	75 *Ans.*

c. Students with averages of 80 or better are on the honor roll. What percent of the students at New City High School are on the honor roll?

Find the number of honor roll students.

Number of students with averages from 80 to 89.9	→	100
Number of students with averages from 90 to 100	→	+25
Total honor-roll students	→	125

Find what percent 125 is of the total.

$$\frac{125}{500} = 0.25 = 25\%$$

Answer: 25% of the students at New City are on the honor roll.

EXAMPLE 5 In a safety study conducted by a traffic department, the numbers of people who jaywalked from 10 A.M. to 11 A.M. at each of 12 different intersections near a school were recorded. The results are shown in the table below.

Results of Observations at 12 Intersections

Intersection	Number of Jaywalkers	Intersection	Number of Jaywalkers
Elm and 1st	3	Main and Oak	0
Elm and 2nd	1	Main and Ash	4
2nd and Ash	7	Main and 4th	5
3rd and Ash	4	Oak and 1st	8
Elm and Ash	11	Oak and 2nd	2
Oak and Ash	14	Oak and 3rd	3

a. Make a histogram of these observations, grouping the intersections with 0 to 2, 3 to 5, 6 to 8, 9 to 11, and 12 to 14 jaywalkers.

A frequency table is needed to draw a histogram. The frequency table shows that the histogram has 5 bars.

Number of Jaywalkers	Tally	Frequency (intersections)
0 to 2	|||	3
3 to 5	~~||||~~	5
6 to 8	||	2
9 to 11			1
12 to 14			1

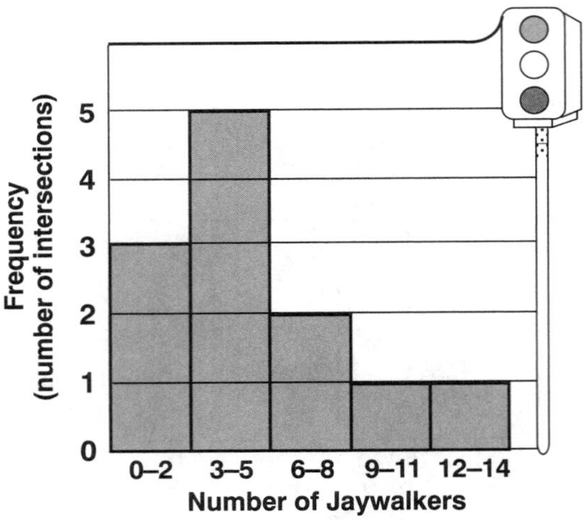

b. Is the traffic department justified in saying that there were at least 6 jaywalkers from 10 A.M. to 11 A.M. at the majority of the intersections studied? Explain.

Add the frequencies for intersections with 6 or more jaywalkers: $2 + 1 + 1 = 4$

Thus, of the 12 intersections studied, only 4 had 6 or more jaywalkers.

Answer: No, the statement is not justified, since fewer than half of the intersections studied had 6 or more jaywalkers.

CLASS EXERCISES

1. The histogram below shows the lengths of the songs most often requested by listeners of radio station WWXZ, tallied according to the playing times in seconds.

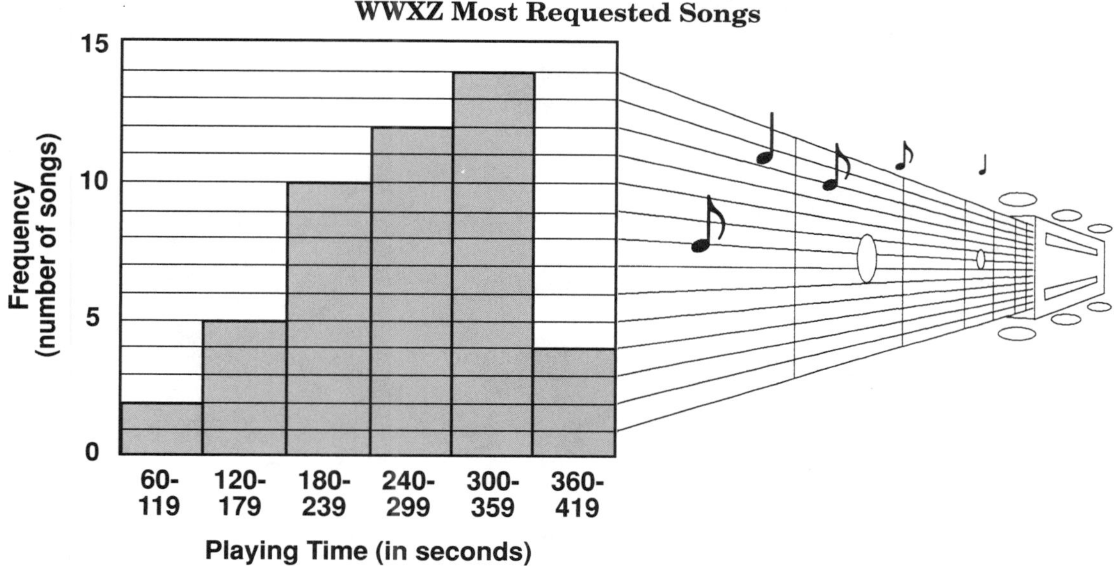

a. How many songs have playing times from 300–359 seconds?
b. How many more songs are there in the interval with the longest playing times than in the interval with the shortest playing times?
c. Which interval has half as many songs as the interval from 180–239 seconds?
d. How many songs are included in this histogram?
e. How many songs have a playing time of 240 seconds or more?
f. How many songs have a playing time that is less than 3 minutes?

2. The Best Friend Pet Clinic treated dogs with the following weights, in pounds, in one day:

22, 35.5, 16.3, 58, 42.5, 37, 18.8, 82.5,

25, 29.5, 51.4, 39.8, 8.5, 14, 19.6

a. Make a histogram of these weights, using intervals of 0 to 19.9 pounds, 20 to 39.9 pounds, 40 to 59.9 pounds, and so on.
b. How would the histogram change if 10-pound intervals were used?
c. Would 50-pound intervals be appropriate? Why, or why not?

1. This is a frequency table showing numbers of patients and their systolic blood pressures.
   a. In which group was the number of patients the greatest?
   b. In which group was the number of patients the least?
   c. How many patients had their blood pressure measured?

Systolic Blood Pressure	Number of Patients
110–119	10
120–129	27
130–139	23
140–149	12
150–159	3

2. This is a summary of the number of typing errors made one day by 6 secretaries.
   a. Who had the best record (the least number of errors)?
   b. Who had the worst record?
   c. How many typing errors did this group of secretaries make that day?

Secretary	Number of Errors
Miss A	5
Mr. B	8
Miss C	3
Mrs. D	11
Mr. E	10
Ms. F	1

3. The highest temperatures, in degrees Fahrenheit, recorded in a city on 20 consecutive days are as listed.

87	90	85	87	89	85	91	87	88	84
86	88	86	90	85	85	86	85	84	87

   a. Make a frequency table for these data.
   b. Which temperatures have the lowest frequencies? the highest frequency?
   c. What is the sum of the frequencies?

4. The ages of 30 teachers at Adams High School are given.

40	53	50	30	53	39	45	48	30	27
45	29	28	41	32	25	41	32	31	25
43	46	33	48	33	26	29	51	32	50

   a. Make a frequency table of these ages, grouping them in 4-year intervals, starting with 25–28.
   b. Which age group has the highest frequency?

5. Mary made a frequency table of the 100 books that she sold. How do you know that she counted wrong?

Type of Book	Frequency
Fiction	13
Travel	15
Humor	17
History	11
Science	14
Cookbooks	18
Home Repair	17

6. The captain of a charter fishing boat kept records of the number of pounds of fish caught by his clients one week. The results are shown in the histogram below.

Deep Sea Charter Fishing Expeditions

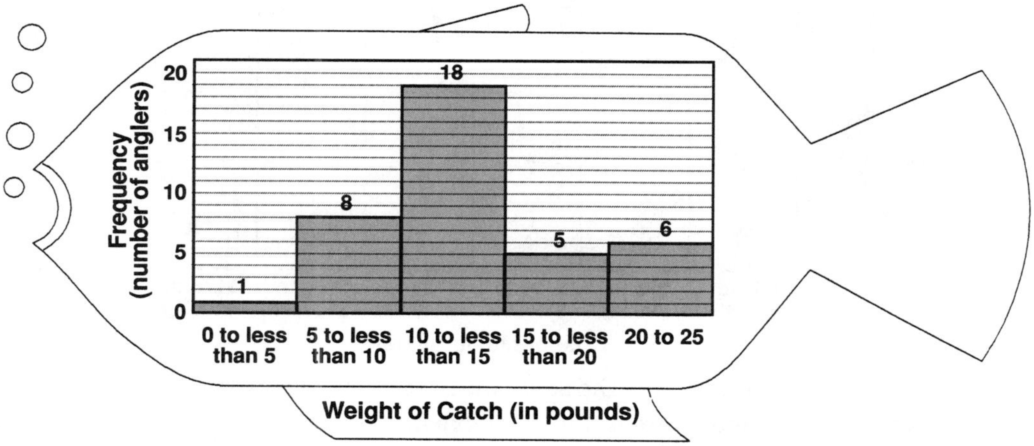

a. How many anglers caught from 20 to 25 pounds of fish?

b. In which single weight range did most of the catches fall?

c. Which two intervals combined had the same number of anglers as the interval 20 to 25?

d. How many anglers caught less than 10 pounds of fish? at least 10 pounds of fish?

e. How many anglers were included in this summary?

f. Explain why the intervals are given as 0 to *less than* 5, 5 to *less than* 10, etc. instead of 0–5, 5–10, etc.? In the last interval, why is it not necessary to state "20 to *less than* 25?"

g. Answer *true*, *false*, or *can't tell*. If you are unable to tell whether a statement is true or false, explain why.

(1) More than half of the anglers caught from 10 to less than 15 pounds of fish.

(2) Most of the anglers caught 10 pounds or more.

(3) There were no anglers without a catch that week.

7. The histogram shows data about the costs of some music systems.

Discount Music Systems ($200 or less)

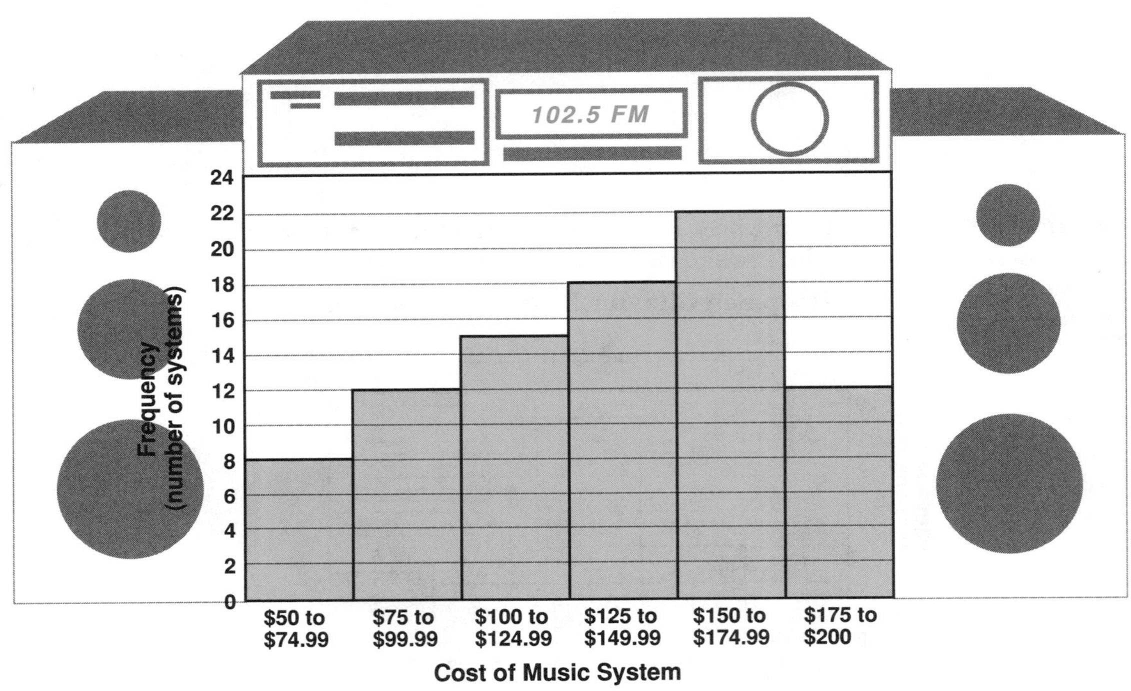

a. Which two price intervals have the same number of music systems?

b. How many more systems were in the most expensive category than in the least expensive?

c. Which price interval had the greatest number of systems?

d. How many systems were included in this summary?

e. How many systems were priced less than $100? between $100 and $200?

f. What percent of the systems in this histogram cost $150 or more?

8. The frequency table shows the birth weights (rounded to the nearest ounce) of babies born at North Central Hospital in one week. Make a histogram of this data, using the intervals shown in the frequency table.

Weight of Babies Born in One Week at North Central Hospital

Weight (ounces)	Number of Babies	Weight (ounces)	Number of Babies
96–100	2	116–120	8
101–105	4	121–125	3
106–110	3	126–130	1
111–115	6		

9. Visitors to a library in one afternoon borrowed the following numbers of books:

0, 2, 5, 1, 1, 0, 8, 3, 2, 4, 1, 3, 6, 1, 2,
2, 3, 2, 3, 1, 4, 9, 0, 3, 7, 0, 1, 6, 4, 3

a. Make a histogram of this data, using intervals of 0–1 books, 2–3 books, etc.

b. Write a sentence that describes the borrowing patterns that can be seen in the histogram.

SPIRAL REVIEW EXERCISES

1. Round 25,371,846 to the nearest hundred thousand.

2. Divide: $1.2\overline{)12.36}$

3. The number 0.25 written as a fraction is

(a) $\frac{1}{4}$ (b) $\frac{3}{10}$ (c) $\frac{1}{25}$ (d) $\frac{4}{5}$

4. When $\frac{1}{20}$ is divided by $\frac{1}{2}$, the result is

(a) $\frac{1}{10}$ (b) $\frac{1}{40}$ (c) $\frac{2}{10}$ (d) $\frac{2}{40}$

5. The least common denominator of the fractions $\frac{3}{4}$ and $\frac{7}{8}$ is

(a) 2 (b) 4 (c) 8 (d) 16

6. What is $\frac{2}{3}$ of 48?

7. Jim ordered a shirt for $19.75, a sweater for $32.80, and a pair of pants for $28.50 from a mail-order catalog. He added $9.08 to cover the sales tax and shipping costs. What was the total cost of his order?

8. Mary earned $6.70 per hour and worked $3\frac{1}{2}$ hours each day for 4 days. Use your calculator to find how much she earned.

9. In August 1990, before the Mideast conflict, the price of a barrel of crude oil was $23.48. By November 1990, the price of a barrel of oil had risen to $32.91. What was the increase in the price of 100 barrels?

10. From a piece of wood trim $8\frac{3}{4}$ inches long, a piece $2\frac{3}{16}$ inches long is removed. How long is the remaining piece?

TEAMWORK

Your teacher will give you a list of raw data collected from your class for one variable. Prepare histograms to summarize this data. For the first histogram, use 6 class intervals. For the second histogram, use 12 class intervals. Which histogram shows a better summary of the data? Prepare a team report describing your two histograms.

UNIT 10–2 Finding the Range; Stem-and-Leaf Displays

FINDING THE RANGE

THE MAIN IDEA

1. ***Ranked data*** is data that is arranged in order and numbered to show each item's position in the order from first to last. The order may be either increasing or decreasing.
2. The difference between the greatest value and the least value in the given data is called the ***range***. The range tells how far apart the two extremes are from each other.

   Range = Greatest Value – Least Value
3. A ***line plot*** is a display of a set of data on a number line. Each value is represented by an X above the corresponding point on the number line.

   Data: 2, 3, 3, 3, 5

   Line Plot

```
            X
            X
        X   X       X
  <-+---+---+---+---+---+->
    0   1   2   3   4   5
```

4. A line plot shows the range and ***distribution*** of the data. That is, it shows how the data is spread out over the range and how it is clustered within the range.

EXAMPLE 1 The five counties in the United States with the greatest populations in 1988 are listed below.

County	*Population*
Cook County, Illinois	5,284,343
Harris County, Texas	2,786,728
Kings County, New York	2,314,263
Los Angeles County, California	8,587,822
San Diego County, California	2,370,412

a. Make a table that ranks these five counties, by population, in decreasing order.

Order the populations from greatest to least. Then write the name of each county next to its population. Finally, number the counties from 1 to 5, beginning with the greatest.

U.S. Counties with Greatest Populations in 1988

Rank	County	Population
1	Los Angeles, CA	8,587,822
2	Cook, Il	5,284,343
3	Harris, TX	2,786,728
4	San Diego, CA	2,370,412
5	Kings, NY	2,314,263

b. What is the range of the populations of the five counties in the list above?

To find the range, subtract the least value from the greatest.

Range = 8,587,822 – 2,314,263
= 6,273,559 *Ans.*

EXAMPLE 2 The table below gives information about the regular varsity players on Eastend High School's women's softball team for the 1992 season. The data is ranked in decreasing order of hits.

Players Ranked by Number of Hits

Rank	*Player*	*Hits*	*Home Runs*	*Strikeouts*
1	Julie N.	35	6	9
2	Latoya W.	31	8	7
3	Sherry M.	29	5	3
4	Roberta S.	28	4	4
5	Melanie A.	21	3	6
6	Kin L.	15	0	2
7	Teresa D.	14	4	19
8	Tasha J.	14	6	14
9	Yoko M.	13	2	10

a. Rank the five players with the least number of strikeouts, in increasing order.

Order the numbers of strikeouts from least to greatest. Write the name of each player next to the number of strikeouts. Number the names.

Answer:

Rank	*Player*	*Strikeouts*
1	Kin L.	2
2	Sherry M.	3
3	Roberta S.	4
4	Melanie A.	6
5	Latoya W.	7

b. What is the range of hits for team members?

Range = Greatest Value – Least Value

Range = 35 – 13, or 22 *Ans.*

c. Make a line plot of the number of home runs made by all the regular varsity players.

Answer:

Number of Home Runs

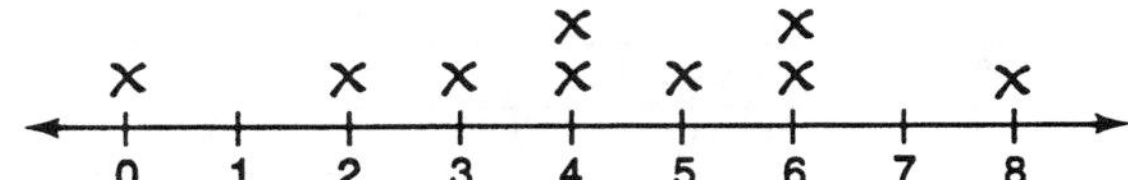

The line plot shows that the data is clustered between the values 2 to 6.

EXAMPLE 3 A quality control engineer measured the thickness of six samples of machine parts. The measurements, in inches, were 0.215, 0.234, 0.208, 0.239, 0.211, and 0.218. What was the range of these measurements?

Range = Greatest Value – Least Value
= 0.239 – 0.208, or 0.031 *Ans.*

CLASS EXERCISES

1. The weights of nine members of a Sta-Trim exercise class were:

135, 162, 128, 117, 146, 173, 128, 150, 161

   a. Rank these weights in decreasing order.
   b. Find the range of these weights.
   c. If everyone loses 4 pounds in two weeks, what effect will this have on the range?

2. In the New Jersey area in 1990, the newspapers with the largest circulations were:

Newspaper	*Circulation*
Newsday	714,128
The Asbury Park Press	159,629
The North Jersey Herald and News	88,113
The Record	159,550
The Star-Ledger	476,257
The Times Herald-Record	85,607

   a. Make a table in which the newspapers are ranked in decreasing order of circulation.
   b. Find the range of the data.
   c. If an advertising agency wants to know where its ads will be most effective, would it look at the top or at the bottom of your table?

3. The table below gives information about Mr. Ortega's mathematics class. It is ranked by latenesses in increasing order.

Rank	Student	Latenesses	Test Average	Absences
1	Evan M.	0	91	1
2	Dwayne C.	1	98	0
3	Jill D.	2	94	3
4	Becky D.	3	87	5
5	Hok L.	4	79	4
6	Annmarie B.	7	89	8
7	Ahmed M.	10	73	2
8	Thomas W.	14	68	5

Using the data in this table:

   a. Make a table that shows the top five students ranked by test average in decreasing order.
   b. Make a table that shows the five students with the greatest numbers of absences ranked in increasing order.
   c. Make a line plot of the number of absences for these students.
   d. How can you find the range from the line plot?

4. A construction company purchased seven house lots of the following sizes: $2\frac{1}{2}$ acres, $1\frac{3}{4}$ acres, $1\frac{3}{8}$ acres, $2\frac{1}{8}$ acres, $1\frac{1}{8}$ acres, $1\frac{3}{4}$ acres, and $2\frac{1}{4}$ acres. What is the range of these lot sizes?

STEM-AND-LEAF DISPLAYS

THE MAIN IDEA

1. A ***stem-and-leaf display*** is another way to organize and present data.
2. In a stem-and-leaf display, numbers that start with the same digit or digits, called the ***stem***, are listed on the same line. Each data value is represented by a ***leaf***, which is a single digit that distinguishes it from other values with the same stem.

Data
18, 24, 25

Stem-and-Leaf Display

Stems		Leaves
1	|	8
2	|	4 5

Key: 1 | 8 represents 18

3. To make a stem-and-leaf display:
   a. List the stems.
   b. Write a leaf for each data value.
   c. Rewrite the display, writing the leaves in increasing order.
   d. Write a key for the display.

EXAMPLE 4 The student averages for Mr. Jasper's homeroom are shown in this stem-and-leaf display.

Student Averages

Stem	Leaves
6	1 4 8
7	2 3 3 5 6 6 6 7 7 8 9 9
8	0 2 2 5 8 8 8
9	0 1 4

Key: 6 | 1 represents an average of 61.

a. How many students are there?

Count the leaves, the numbers to the right of the vertical line.

Answer: There are 25 students.

b. What is the range of the averages?

Range = Greatest Value – Least Value
Range = 94 – 61, or 33 *Ans.*

c. What percent of the students had averages in the eighties?

There are 7 leaves with a stem of 8, representing 7 scores in the eighties. Find what percent 7 is of the total:

$$\frac{7}{25} = 0.28 = 28\% \textit{ Ans.}$$

EXAMPLE 5 Fifty people were asked to complete a puzzle for a psychology experiment. Use the data at the right to make a stem-and-leaf display of the number of seconds required by each of these people.

Number of Seconds Needed to Complete Puzzle

135	156	176	144	158	152	161	159	168	170
136	141	139	146	163	166	171	140	138	150
163	157	178	177	143	137	155	151	143	156
169	172	170	134	153	159	164	149	137	150
143	136	140	150	151	166	159	148	157	134

First, draw a vertical line and list the stems vertically on one side of the line. Each stem will represent the set of values that begin with those digits.

13
14
15
16
17

Observe that, in this case, the stems have two digits, because a leaf may only be a single digit.

Next, write the ones digit of each data value on the horizontal line next to the correct stem.

For example, the first value, 135, is indicated by writing the digit 5 on the line with a stem of 13.

Stem	Leaves
13	569874764
14	4160339308
15	6829075163900197
16	18363946
17	6018720

Finally, rewrite the display, rearranging the leaves in order from least to greatest. Write a key for the display.

Number of Seconds Needed to Complete Puzzle

Stem	Leaves
13	445667789
14	0013334689
15	0001123566778999
16	13346689
17	0012678

Key: 13 | 4 represents 134 seconds.

CLASS EXERCISES

1. The approximate weights of roasting turkeys sold at Food Mart one day are shown in the stem-and-leaf display.

   Weight of Turkeys Sold at Food Mart

Stem	Leaves
0	99
1	012224689
2	0236

   Key: 1 | 2 represents 12 pounds.

   a. How many turkeys were sold that day?

   b. What is the range of the weights?

   c. What percent of the turkeys weighed 12 pounds?

2. The distance from the target for 30 parachute jumps to the actual landing point was measured in inches.

   a. Make a stem-and-leaf display of the data:

145	122	108	138	133	126
140	135	101	119	132	127
130	138	143	139	100	115
121	119	125	136	111	145
128	135	123	143	128	130

   b. Which interval of distances has the greatest frequency?

   c. What is the least distance recorded in the last interval?

1. Seven stores charged different prices for a particular brand and model of videocassette recorder. The prices were $459, $519, $499, $435, $475, $530, and $485.
   a. Rank the prices in decreasing order.
   b. Find the range of the prices.
   c. Find the range of the middle three ranked prices.
2. On one Spring day in 1999, the pollen counts in six selected cities were: 1,259 in Oklahoma City, 32 in Washington, D.C., 168 in Seattle, 1,049 in St. Louis, 76 in New Orleans, and 2,045 in Atlanta.
   a. Make a table in which the cities are ranked in increasing order of pollen counts.
   b. Find the range of the pollen counts for the six cities listed.
3. The table gives baseball data for the Eastern Division of the American League in 1999.

American League–Eastern Division (1999)

Team	Hits	At Bats	Home Runs
Boston	1,551	5,579	176
New York	1,568	5,568	193
Toronto	1,581	5,642	212
Baltimore	1,572	5,637	203
Tampa Bay	1,531	5,586	145

   Using the data:
   a. Make a table showing the top three teams ranked by the number of home runs.
   b. Make a table showing the three teams with the least number of times at bats, ranked in increasing order.
4. Rank the following in decreasing order: $\frac{3}{8}, \frac{1}{2}, \frac{7}{16}, \frac{13}{24}, \frac{3}{4}$
5. The prices of a quart of milk at six different stores are $1.19, $1.43, $1.26, $1.15, $1.57, and $1.35. What is the range of these prices?

6. The number of visitors for each day of an art exhibit is shown in the stem-and-leaf display below.

a. What is the range of the numbers of visitors?

b. On how many days did more than 280 people view the exhibit?

c. On what percent of the days were there fewer than 280 visitors?

Visitors to Art Exhibit

Stem	Leaf
25	4
26	0 8
27	1 2 2 7 8 8
28	0 5 6 6 6 9
29	3 4 5
30	1
31	0

Key: 25 | 4 represents 254 visitors.

7. The number of phone calls received by the switchboard of an investment company was recorded for the same time period on 40 consecutive business days. The results are given below:

Number of Phone Calls Received in One Day

52	14	33	27	41	12	54	18	36	23
40	26	24	11	39	25	17	50	43	29
31	45	35	41	13	22	18	14	51	31
10	20	17	29	13	34	22	16	18	23

a. Make a stem-and-leaf display of this data.

b. On a typical day during that time period, which of the following intervals best describes the number of phone calls received?

(a) 20–29 calls (b) 30–39 calls (c) 40–49 calls (d) 50–59 calls

c. Within the interval with the greatest frequency, which number of phone calls had the greatest frequency?

d. Describe the general shape of the display, and explain what the shape reveals about the numbers of phone calls.

8. The stocks of twelve companies in a certain industry showed the following price changes between the opening and closing of the stock market one day.

a. Write a signed number to represent each change.

b. Make a line plot to show the price changes.

c. What is the range of the price changes?

d. What percent of the stocks increased in value that day?

e. Can you tell from this data which stock had the greatest value at the end of the day?

Price Changes of Selected Stocks ($)

GTCo	increased 2	PaxCo	gained 2
IntRub	lost 3	MarCorp	increased 1
IndSp	gained 1	ActiCo	stayed the same
DynBf	lost 7	Supra	lost 1
PConsol	gained 5	ApexCo	decreased 2
MaxGen	decreased $1\frac{1}{2}$	XYZCo	gained $1\frac{1}{2}$

SPIRAL REVIEW EXERCISES

1. Find the remainder when 350 is divided by 12.

2. Multiply the sum of 4.6 and 3.9 by the difference between 6.2 and 5.7.

3. The population of Greenfalls is 12,762 and the population of Boxboro is 42,089. Find the total population of the two towns, rounded to the nearest ten thousand.

4. At the veterinarian's office, Mr. Simpson stepped on the scale while holding his dog, Ginger. The scale read 230 pounds. If Mr. Simpson weighs 197 pounds, what is Ginger's weight?

5. Jerry the Pinball Wizard scored 12,500 points more than the previously listed champion, who scored 35,650 points. What was Jerry's score?

6. Donna spent $2.50 on bus fare for 5 schooldays last week. This week the fare was raised, and she spent $3.00 for 5 days. How much did the fare go up per day?

7. What is the next number in the sequence 1, 2, 3, 5, 8, 13, 21, ___?

8. Which expression can be used to represent the total cost of seven books that cost $5.50 each and eight more books that cost $5.50 each?
   (a) $5.5 \times (7 + 8)$ (b) $5.5 + 7 \times 8$
   (c) $8 + 7 \times 5.5$ (d) $5.5 \times 7 \times 8$

9. Roseanne took a mathematics exam on which there were 45 questions of equal value, totaling 100%. She answered 37 questions correctly. Use your calculator to compute her score.

10. $\frac{12}{40}$ has the same value as
    (a) $\frac{1}{4}$ (b) $\frac{24}{20}$ (c) $\frac{4}{10}$ (d) $\frac{3}{10}$

11. $3\frac{7}{8}$ written as an improper fraction is
    (a) $\frac{10}{8}$ (b) $\frac{24}{8}$ (c) $\frac{21}{8}$ (d) $\frac{31}{8}$

TEAMWORK

Your teacher will give you two lists of raw data for one variable, one collected from the girls in your class and one collected from the boys. Make two stem-and-leaf displays, back-to-back, for the two sets of data. Compare the results for the boys and girls. Prepare a team report that describes the two displays.

UNIT 10–3 Finding the Mean; Finding the Mode

FINDING THE MEAN

THE MAIN IDEA

1. The ***mean*** of a set of data is the ***average***.
2. To find the mean:
   a. Find the sum of the items.
   b. Divide the sum by the number of items.
3. The mean of a set of data can be shown on a line plot. For example,

Data

7, 3, 5, 8, 3, 8, 9, 4, 3, 5

The sum of the 10 data items is 55.

mean = 55 ÷ 10 = 5.5

Line Plot

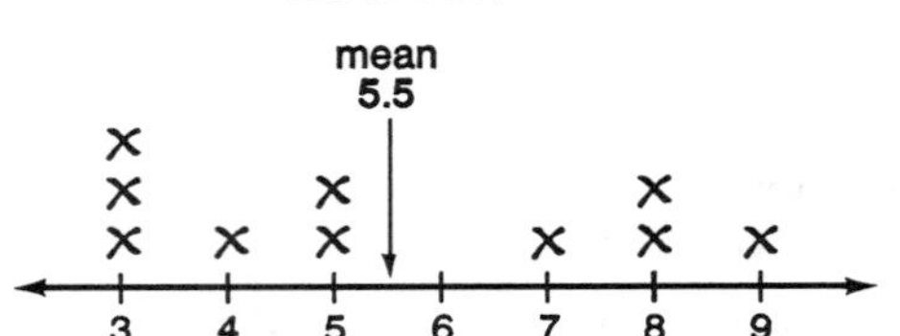

EXAMPLE 1 John's grades for 8 exams were 74, 76, 76, 78, 82, 82, 82, and 86. Make a line plot of John's grades, showing the mean grade.

To find the mean of the 8 grades:

74 + 76 + 76 + 78 + 82 + 82 + 82 + 86 = 636

636 ÷ 8 = 79.5 mean

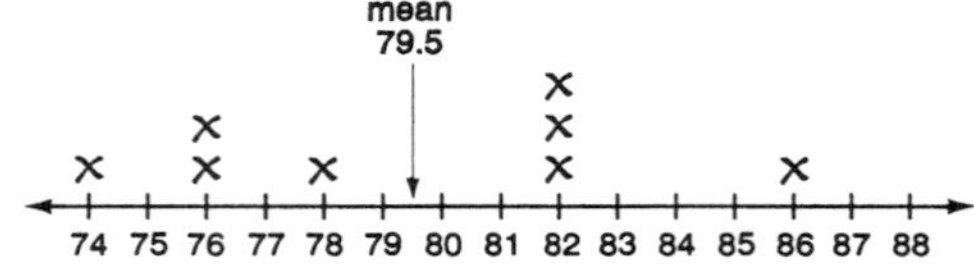

EXAMPLE 2 The table shows auto production in the U.S.A. Find the mean number of autos produced per year, rounded to the nearest hundred.

Year	Number of Autos
1982	422,800
1983	565,100
1984	635,200
1985	666,800
1986	626,300
1987	590,400

Key Sequence:

422800 [+] 565100 [+] 635200 [+] 666800 [+] 626300 [+] 590400 [÷] 6 [=]

Display: 584433.33

Answer: Rounded to the nearest hundred, the mean number of autos produced per year for this 6-year period was 584,400.

EXAMPLE 3 Stephanie has grades of 78, 92, and 84 in 3 math tests. What grade must she get on a fourth test in order to have an average grade of 85?

THINKING ABOUT THE PROBLEM

The mean is given and one of the grades must be found. This is a good place to guess and check. Guess a value for the missing grade and check by computing the mean. If the resulting mean is higher than 85, try a lower guess. If the resulting mean is lower than 85, try a higher guess.

80 ← guess	85 ← higher guess	86 ← higher guess
78	78	78
92	92	92
84	84	84
334	339	340
$334.0 \div 4 = 83.5$ ← mean (too low)	$339.00 \div 4 = 84.75$ ← mean (still too low, but closer)	$340 \div 4 = 85$ ← mean (correct)

Answer: Stephanie needs a grade of 86 on the fourth test.

CLASS EXERCISES

1. Find the mean of each group of numbers.

a. 27, 27, 27, 27, 27 **b.** 11, 17, 13, 16, 18 **c.** 52, 61, 59, 63, 57

d. 36, 48, 52, 37, 50, 46, 38, 51 **e.** 79, 89, 87, 78, 88, 89 **f.** 1.1, 1.3, 1.2, 0.4

g. 26.1, 23.4, 28.6, 20.4, 18.0, 21.1 **h.** 63.02, 48.12, 52.35, 60.03 **i.** 3, $2\frac{3}{4}$, $2\frac{1}{4}$

2. Which set of numbers has the greatest mean?

(a) 45, 46, 46, 47 (b) 45, 46, 46, 47, 45, 46 (c) 47, 46, 45, 44 (d) 48, 42, 46, 47

3. The average height of four students was 67 inches. The heights of three students were 64 inches, 68 inches, and 70 inches. What was the height of the fourth student?

4. Mrs. Harris, the English teacher, thought that the average number of students in her classes was greater than the average number of students in Mr. Smith's classes. Was she right?

Number of Students in Mrs. Harris' Classes	Number of Students in Mr. Smith's Classes
35	34
36	35
42	35
28	37
30	34

5. The table below gives the areas of six New England states in square miles. Find the mean of these areas.

State	Area (sq. mi.)
Connecticut	4,872
Maine	30,995
Massachusetts	7,824
New Hampshire	8,993
Rhode Island	1,058
Vermont	9,273

FINDING THE MODE

THE MAIN IDEA

1. The *mode* of a set of data is the value that has the *greatest frequency.*
2. To find the mode of a set of data:
   a. Tally the values and write the frequencies.
   b. Choose the value with the greatest frequency.
3. On a line plot, the mode is the value that has the greatest number of X's above it.
4. For some sets of data, there may be more than one mode, or no mode.

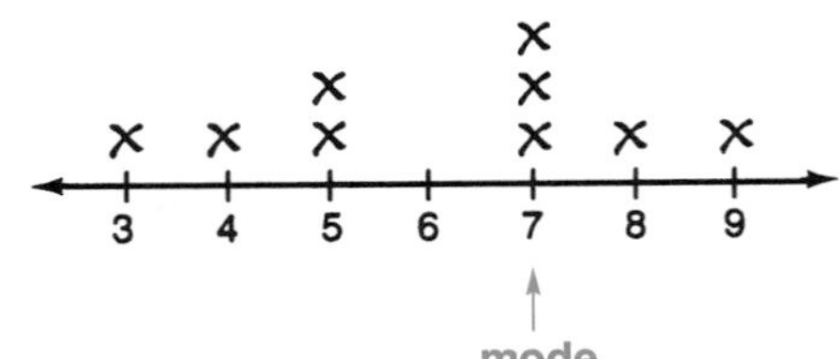

EXAMPLE 4 Find the mode for each data set.

	Data Set	***Mode(s)***
a.	72, 23, 35, 23, 34, 23	23
b.	72, 23, 72, 23, 34	72 and 23
c.	72, 23, 35, 34, 71	no mode

EXAMPLE 5 For this list of 25 prices of grocery orders, find the mode by making a frequency table.

$ 3.50	12.75	4.25	9.00	10.00
8.00	9.00	8.00	8.75	10.25
11.00	7.50	11.50	8.50	7.00
9.50	9.00	10.00	8.00	7.50
8.50	9.50	9.00	9.00	11.00

Price	Tally	Frequency	Price	Tally	Frequency
$ 3.50	|	1	9.00	~~||||~~	5
4.25	|	1	9.50	||	2
7.00	|	1	10.00	||	2
7.50	||	2	10.25	|	1
8.00	|||	3	11.00	||	2
8.50	||	2	11.50	|	1
8.75	|	1	12.75	|	1

From the frequency table, read the value that has the highest frequency.

Answer: The mode is $9.00.

EXAMPLE 6 For this set of average temperatures (in degrees Fahrenheit) of 30 cities, find the mode by making a stem-and-leaf display.

54	48	31	46	40	72
32	54	40	48	31	46
70	31	46	34	46	72
32	40	68	31	46	32
31	46	31	45	46	31

Average Temperatures (°F) in 30 Cities

Stem	Leaf
3	1 1 1 1 1 1 1 2 2 2 4
4	0 0 0 5 6 6 6 6 6 6 6 8 8
5	4 4
6	8
7	0 2 2

Key: 3 | 1 means 31° F

From the display, you can see that there are two different temperatures, 31 and 46, that are written the greatest number of times. Therefore, there are two modes.

Answer: The modes are 31° and 46°.

CLASS EXERCISES

1. Find the mode of each group of numbers.
   a. 12, 27, 12, 14, 21, 12
   b. 13, 16, 31, 14, 15
   c. 57, 75, 62, 61, 75, 56, 61, 75, 57
   d. 102, 105, 205, 103, 105, 102, 105
   e. 62, 59, 62, 59, 59, 62
   f. $2\frac{3}{4}, 2\frac{1}{2}, 2\frac{3}{4}, 2, 2\frac{1}{8}, 2\frac{3}{4}, 2\frac{1}{2}$
   g. 7.1, 7.4, 7.5, 7.4, 7.6, 7.4, 7.4, 7.9
   h. 78.5, 114.6, 98.3, 89.3, 78.6, 89.3, 94.3, 89.3, 93.8

2. For each frequency table, find the mode.

a.

Height (in.)	Frequency
62	7
63	11
64	12
65	10
66	5

b.

Price	Frequency
$10.00	7
11.00	6
12.00	6
13.00	5
14.00	7
15.00	6

c.

Distance (mi.)	Frequency
10	7
15	12
20	21
25	17
30	29
35	25
40	19
45	16

3. Make a frequency table and find the mode for this set of 20 shoe sizes.

$10\frac{1}{2}$ $8\frac{1}{2}$ $9\frac{1}{2}$ 9 8 8 11 8 $9\frac{1}{2}$ 10
9 $9\frac{1}{2}$ $9\frac{1}{2}$ $10\frac{1}{2}$ $8\frac{1}{2}$ $9\frac{1}{2}$ $8\frac{1}{2}$ $8\frac{1}{2}$ 11 $9\frac{1}{2}$

4. Make a frequency table and find the mode for this set of distances, in miles.

12.4 10.1 5.6 3.1 5.6 6.3 2.4 12.3
7.1 6.9 9.7 6.5 2.4 3.1 4.2 4.2
6.9 7.5 5.6 3.1 2.4 7.1 3.1 6.5

5. For the data on the line plot shown, find:
   a. the range b. the mean c. the mode

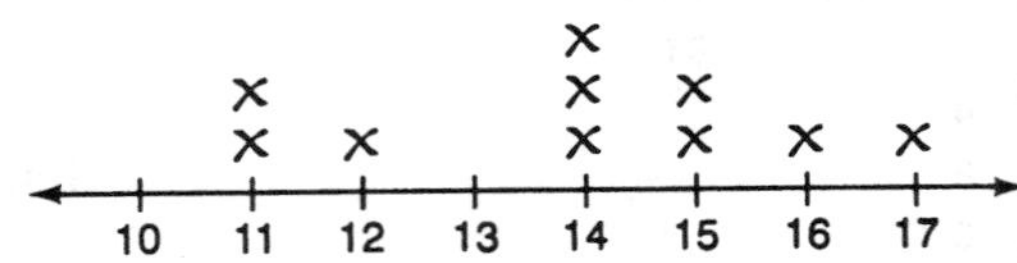

6. Does every set of data have:
   a. a mean? b. a mode? c. a range?

HOMEWORK EXERCISES

1. Find the mean of each group of numbers.

   a. 20, 30, 40 **b.** 28, 32, 47, 63 **c.** 15, 16, 14, 17, 18, 16 **d.** 38, 42, 56, 26

   e. 30, 50, 60, 80, 100 **f.** 28, 30, 26, 38, 22 **g.** 46, 39, 54, 18, 71, 92 **h.** 5, 5.5, 6.2, 5.3

   i. 9.6, 8.3, 2.4, 11.7 **j.** 2, 5, $5\frac{1}{2}$, 6, $6\frac{1}{2}$ **k.** 3, $3\frac{1}{2}$, $3\frac{1}{4}$, $3\frac{3}{4}$, 3, 4 **l.** –2, 5, 1, –1

2. Mark's grades on 4 exams were 78, 83, 85, and 90. What was the mean of his grades?

3. Mary's temperatures (°*F*) for the past 5 mornings were 99°, 100°, 100°, 101°, 100°.

   a. What was her mean temperature? **b.** What was the mode of the temperatures?

   b. Make a line plot that shows both the mean and the mode.

   c. If, on the sixth morning, her temperature was 98.6°, explain how this would affect the mean and the mode.

4. Which set of numbers has the greatest mean? Explain how you can solve this problem without actually calculating the mean of each set.

   (a) 23, 27, 28, 30 (b) 32, 22, 26, 22 (c) 22, 23, 24, 24 (d) 26, 25, 24, 23

5. These are the numbers of miles run by 3 joggers over a period of 5 days. Which jogger had the highest mean mileage? Which jogger had the lowest mean mileage?

Jane	Carlo	Milly
3	1	5
6	9	4
2	2	5
3	2	6
5	3	5

6. These are the tolls that Ms. Reed and Mr. Sweeney each paid over a period of 5 days. Which person had the higher mean toll?

Ms. Reed	Mr. Sweeney
\$2.00	\$2.00
\$2.50	\$3.00
\$1.50	\$2.50
\$3.00	\$1.50
\$2.50	\$3.00

7. Find the mean of this group of numbers. 63, 68, 64, 68, 68, 65, 62

8. The mean of three numbers is 120. If two of the numbers are 90 and 150, what is the third number?

9. The average yardage gained by four running backs in a football game was 42 yards. If three of the running backs gained 21 yards, 38 yards, and 50 yards, how many yards did the fourth running back gain?

10. The average noontime temperature for four days in December was 8°*F*. The noontime temperatures for the first three days were 9°*F*, 10°*F*, and 20°*F*. Find the noontime temperature for the fourth day.

11. The table shows the number of performances for 5 hit shows on New York City's Broadway. Find the mean number of performances.

Show Title	Number of Performances
A Chorus Line	5,832
Forty-Second Street	3,486
Grease	3,388
Fiddler on the Roof	3,242
My Fair Lady	2,717

12. For the set of numbers 5, 15, 25, 5, 35, 50, 5, 15, 5, 25, the mode is
(a) 4 (b) 5 (c) 15 (d) 50

13. In each case, find the mode of the given set of numbers.
a. 11, 32, 27, 33, 32, 12, 32 **b.** 26, 49, 55, 38, 26, 19, 67
c. 59, 60, 61, 61, 61, 70, 72 **d.** 101, 117, 103, 132, 115, 119
e. 125, 156, 198, 165, 156, 120, 156 **f.** $3\frac{1}{2}$, 4, $3\frac{1}{2}$, $4\frac{1}{2}$, $3\frac{1}{2}$, 5, $3\frac{1}{2}$, $5\frac{1}{2}$, $4\frac{1}{2}$
g. 10.4, 10.5, 10.3, 10.2, 10.3, 10.4, 10.3, 10.3 **h.** 98.7, 97.8, 99.3, 93.7, 97.8, 98.3, 97.8, 98.7

14. For the frequency table, find the mode.

Age	Number of Students
13	35
14	68
15	129
16	134
17	112
18	52

15. For the stem-and-leaf display, find the mode.

Exam Scores

Stem	Leaf
6	2 5 8 8
7	2 2 2 5 8
8	3 5 5 5 5 7
9	1 2 4 4
10	0 0

Key: 7 | 2 represents 72%

16. Make a frequency table and find the mode for this group of lengths (in cm) of electrical wire.

11.4	10.1	9.6	9.5	11.5	9.4	11.0	9.6	6.9
9.4	9.6	11.0	25.3	11.4	25.2	11.4	9.6	9.5
9.5	9.6	10.1	9.7	10.1	9.6	6.9	10.2	10.1

17. **a.** Make a frequency table for this group of shirt sizes.

$16\frac{1}{2}$	14	$16\frac{1}{2}$	15	16	16	$14\frac{1}{2}$
$14\frac{1}{2}$	$16\frac{1}{2}$	17	$15\frac{1}{2}$	$16\frac{1}{2}$	16	$16\frac{1}{2}$
14	$16\frac{1}{2}$	$15\frac{1}{2}$	15	16	$16\frac{1}{2}$	$15\frac{1}{2}$

b. Find the mode for the shirt sizes.
c. Explain how a store manager can use the mode when ordering more shirts.
d. Why is the mean of this data not very helpful to the store manager?

18. The men and women in a college ski club are to be arranged in order of height for a group photograph. Their heights are listed.

6'0"	5'5"	5'11"	6'1"	5'11"	5'4"	5'7"	6'4"
5'1"	4'11"	5'11"	5'8"	6'1"	5'9"	5'2"	5'4"
5'6"	5'11"	5'5"	6'3"	5'9"	5'5"	5'4"	5'10"
5'8"	5'2"	5'3"	5'8"	5'6"	5'11"	5'4"	5'4"

a. How many modes are shown in the data?
(a) 0 (b) 1 (c) 2 (d) 3

b. How can you explain the number of modes?

19. Find the mean of the exam scores shown in the frequency table below.

Math Exam Scores

Score	Tally	Frequency
70	|||	3
75	~~||||~~	5
80	~~||||~~	5
85	||||	4
90	||||	4
95	||	2
100	||	2

SPIRAL REVIEW EXERCISES

1. The table shows Mary's absences for 5 months. In which month did she have the greatest frequency of absence?

Month	Number of Absences
Sept.	0
Oct.	2
Nov.	5
Dec.	1
Jan.	4

2. Find the sum of 4.86, 11.9, and 28.09.

3. Find the product of $\frac{30}{39}$ and $\frac{78}{105}$.

4. Copy and complete the chart to express each number in two other ways.

Fraction	Decimal	Percent
$\frac{3}{4}$		
	0.7	
		45%

5. 18 is 90% of what number?

6. Before buying a CD player, Jill compared prices at 5 stores. The range of the prices was $59.75. If the highest price was $229.00, what was the lowest price?

7. James received a gift certificate for $100. He bought 4 sweaters on sale at $18.99 each. How much money did he still have?

8. Which number should come next in this pattern? 10, 9, 7, 4, 0, ___
(a) 4 (b) −2 (c) −4 (d) −5

9. The value of $12 + 6 \times 5 + 4$ is
(a) 160 (b) 94 (c) 46 (d) 32

10. It is estimated that every dollar spent on measles vaccinations saves $11.90 in medical costs. A plan to immunize more children would cost $90.9 million. How much would the plan save?

UNIT 10–4 Finding the Median; Comparing the Mean, Median, and Mode

FINDING THE MEDIAN

THE MAIN IDEA

1. The ***median*** of a group of numbers is the middle value when the numbers are arranged in order.
2. There are as many values greater than the median as there are less than the median.
3. To find the median, count the number of items.
   a. If the number of items is odd, there is one middle value, which is the median.
   b. If the number of items is even, there are two middle values. The mean of these two middle values is the median. In this case, the median will be a value that is not one of the original values.

EXAMPLE 1 Find the median of this group of numbers:

24.4, 26.1, 27.5, 29.2, 30.7, 34.9, 36.8

Since these numbers are already in order, count them. There are 7, which is an odd number. The middle number is the fourth number.

24.4, 26.1, 27.5, 29.2, 30.7, 34.9, 36.8

There are 3 numbers less than the median. (24.4, 26.1, 27.5)

The middle number is the median. (29.2)

There are 3 numbers greater than the median. (30.7, 34.9, 36.8)

Answer: The median is 29.2.

EXAMPLE 2 Find the median of this group of numbers: 21, 23, 24, 26, 28, 34, 36, 39

Since the numbers are already in order, count them. There are 8, which is an even number. The median is halfway between the two middle numbers.

21, 23, 24, 26, 28, 34, 36, 39

two middle numbers (26, 28)

There are 4 numbers less than the median. (21, 23, 24, 26)

median (between 26 and 28)

There are 4 numbers greater than the median. (28, 34, 36, 39)

Find the mean of the two middle numbers.

$$\begin{array}{r} 26 \\ +28 \\ \hline 54 \end{array} \qquad 54 \div 2 = 27$$

Answer: The median is 27 (not one of the original values).

EXAMPLE 3 Find the median of this group of numbers: 26, 34, 14, 13, 27, 33, 32, 28, 29

First, put the numbers in order.

13, 14, 26, 27, 28, 29, 32, 33, 34

Then, locate the middle number by counting. Since there are 9 numbers, the middle number is the fifth number.

13, 14, 26, 27, 28, 29, 32, 33, 34

There are 4 numbers less than the median.

There are 4 numbers greater than the median.

The middle number is the median.

Answer: The median is 28.

EXAMPLE 4 Find the median of this set of numbers: 3, 10, 5, 12, 4, 2

Order the numbers and locate the middle.

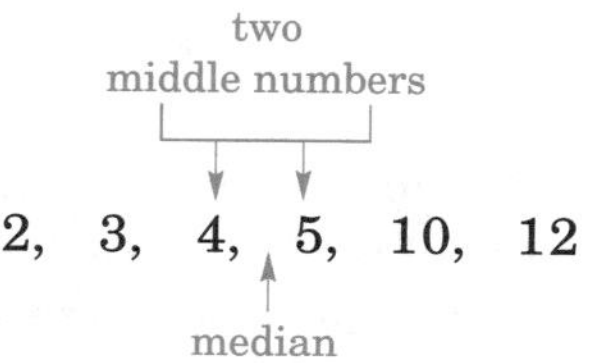

Find the mean of the two middle numbers.

$$\begin{array}{r} 4 \\ +5 \\ \hline 9 \end{array} \qquad 2\overline{)9.0} = 4.5$$

Answer: The median is 4.5.

CLASS EXERCISES

1. Find the median of each group of numbers.

a. 47, 53, 38, 21, 50 **b.** 102, 98, 112 **c.** 12, 3, 9, 15, 11, 8, 4 **d.** 27, 18, 14, 6

e. 22, 28, 20, 26 **f.** 75, 80, 67, 92 **g.** 97, 96, 95, 99, 92, 94, 93

h. 54, 55, 55, 54, 54, 55, 55, 54, 54, 54, 55 **i.** 18.6, 17.8, 18.7, 19 **j.** 32.4, 32.6, 31.8, 32.2

2. These are the prices of 5 automobiles: $8,300 $9,240 $8,210 $10,335 $8,877
What is the median price?

3. Find the median of the following group of prices of houses in Brickville:
$52,000 $48,500 $63,000 $71,350

4. The number of cars sold in each of the first 5 months of the year by the Acme Car Lot are shown in the table. What was the median number of sales?

Month	Number of Cars Sold
Jan.	20
Feb.	24
March	18
April	40
May	52

COMPARING THE MEAN, MEDIAN, AND MODE

THE MAIN IDEA

1. The mean, median, and mode are called ***measures of central tendency***. Each one, in its own way, gives information about the data.
   a. The mean is like the "balancing point" for the data.
   b. The median is in the middle position of the ordered values.
   c. The mode tells which value appears most often, when there is such a value.
2. The mean or the median may be, but need not be, equal to one of the given data values. The mode must be one of the data values.
3. A value that is unusually low or high, called an ***extreme value***, or ***outlier***, can greatly decrease or increase the mean, but will not affect the median or mode more than any other value.

EXAMPLE 5 A survey asked the top honor students how many hours they had spent using the library last week. The results were: 2, 3, 4, 6, and 10 hours, with a mean of 5 hours.

Show that the mean is a "balancing point" by doing the following:

(1) Find the difference between each data value and the mean value 5.

(2) Find the total of the differences for

a. values below the mean

Values Below the Mean	Difference Mean – Value
2	$5 - 2 = 3$
3	$5 - 3 = 2$
4	$5 - 4 = 1$
TOTAL	6

b. values above the mean

Values Above the Mean	Difference Value – Mean
6	$6 - 5 = 1$
10	$10 - 5 = 5$
TOTAL	6

(3) Compare the totals. Since the total difference above the mean equals the total difference below the mean, the mean is a balancing point for the data.

EXAMPLE 6 The rents for a group of apartments are \$450, \$500, \$500, \$570, \$650, \$700 and \$1,950. Which better represents this data, the mean or median?

$$\text{Mean} = \frac{450 + 500 + 500 + 570 + 650 + 700 + 1,950}{7}$$

$$= 760$$

Median = 570 (the middle number)

THINKING ABOUT THE PROBLEM

Except for one very high rent, all rents are less than the mean, which is so high only because one of the values, \$1,950, is an extreme value. In this case, the mean of \$760 is misleading because it gives the impression that the rents are generally much higher than they really are.

There are three rents less than the median of \$570, and three greater than \$570. The median, in this case, gives a better general picture of the actual data.

EXAMPLE 7 Mr. Wallace, the manager of a shoe store, kept a record of the sizes of all the pairs of shoes sold in his store one morning:

$$8, 8, 8\tfrac{1}{2}, 9, 9, 9, 9, 9\tfrac{1}{2}, 10,$$

$$10, 10\tfrac{1}{2}, 10\tfrac{1}{2}, 11\tfrac{1}{2}, 11\tfrac{1}{2}$$

Which measure of central tendency will be most useful in helping him decide what quantities of the different sizes he should order in the future?

Mean = 9.5714285

Median = $9\frac{1}{4}$ Mode = 9

THINKING ABOUT THE PROBLEM

Since neither the mean nor the median is an actual shoe size, the mode is most useful.

EXAMPLE 8 A teacher is trying a new grading system. Five test scores for each of three students are given. Under the new grading system, Jose received a lower grade than did Jeff and Yves. Why might Jose have received the lower grade?

Jose	Jeff	Yves
100	0	70
98	62	71
90	90	69
62	98	71
0	100	69

Find the mean, median, range, and general trend of the scores for each student.

	Jose	Jeff	Yves
Mean	70	70	70
Median	90	90	70
Range	100	100	2
Trend	Scores are decreasing.	Scores are increasing.	Scores are stable.

Observe that the mean score of all 3 students is the same. Jose's and Jeff's scores also have the same median and range, but they still received different grades. Jose's scores show a decreasing trend, while Jeff's scores are increasing, and Yves' scores are stable.

Answer: The teacher must be evaluating the trend of the scores, as well as the mean, median, and range, to determine the grade.

CLASS EXERCISES

1. The annual salaries of seven workers at the Health Bran Bread Company were $30,000, $32,500, $35,200, $42,000, and $48,500.

   a. Find the mean salary. b. Find the median salary.

   c. Which is a better description of these salaries? Explain your answer.

2. In one day, the A&B Hardware Store sold packages of light bulbs with the following numbers of watts: 25, 40, 50, 50, 60, 60, 60, 75, 75, 100, 100, 100, 100, and 200 watts.

   a. Find the mean. b. Find the median. c. Find the mode.

   d. Which is the most useful to the store owner for the purpose of analyzing light bulb sales and ordering more? Explain your reasoning.

3. In a contest, the 5 winners received cash prizes of $50, $50, $50, $500, and $50,000. After calculating the mean, median, and mode for these values, tell which value most fairly reflects the "average" winning.

4. Two of Dr. Blake's patients kept records of their systolic blood pressure for one week.

	Mr. Cody	Mrs. Sanchez
Mon.	140	116
Tue.	135	131
Wed.	128	118
Thurs.	139	138
Fri.	136	160
Sat.	141	145
Sun.	137	158

   a. Calculate the mean systolic blood pressure for each patient.

   b. Why might Dr. Blake be more concerned about one patient than the other after looking at these readings?

5. The number of people using the services of a walk-in medical clinic each day for one week were as follows: 16, 19, 12, 22, 18, 24, and 8 patients

   a. Find the mean number of walk-in patients per day.

   b. Make a line plot of the data, showing all the data values and the mean.

   c. Show that the mean is a "balancing point" by doing the following:

      (1) Find the difference between each data value and the mean.

      (2) Find the total of the differences for:
         (a) values below the mean
         (b) values above the mean

      (3) Compare the totals.

HOMEWORK EXERCISES

1. Find the median of each group of numbers.
   a. 37, 22, 54, 23, 27 b. 11, 9, 8, 12, 10 c. 126, 117, 152 d. 5, 7, 7, 12, 6, 6, 5
   e. 42, 48, 52, 50 f. 33, 34, 34, 33, 33, 34, 34, 33, 33 g. 66, 67, 68, 69, 68, 67
   h. 13.1, 12.6, 16.1, 14.5 i. 36.7, 37.5, 37.6 j. $8\frac{1}{8}, 6\frac{3}{8}, 9\frac{1}{2}, 3$ k. $3\frac{1}{3}, 3\frac{5}{6}, 3\frac{1}{6}, 3\frac{2}{3}, 3\frac{1}{2}, 3$
2. Find the median of this group of book prices. $15.25 $11.30 $10.00 $14.75 $13.25
3. In the group of annual salaries, find the median.
   $134,000 $27,000 $21,750 $30,000 $31,000 $32,500
4. The table shows the numbers of new subscriptions sold by Bobby on his paper route last week. What was the median number of new subscriptions?

Day	Monday	Tuesday	Wednesday	Thursday	Friday
Number of New Subscriptions	2	8	3	4	2

5. The graph shows the number of reported cases of Lyme disease for the years 1984 through 1990.

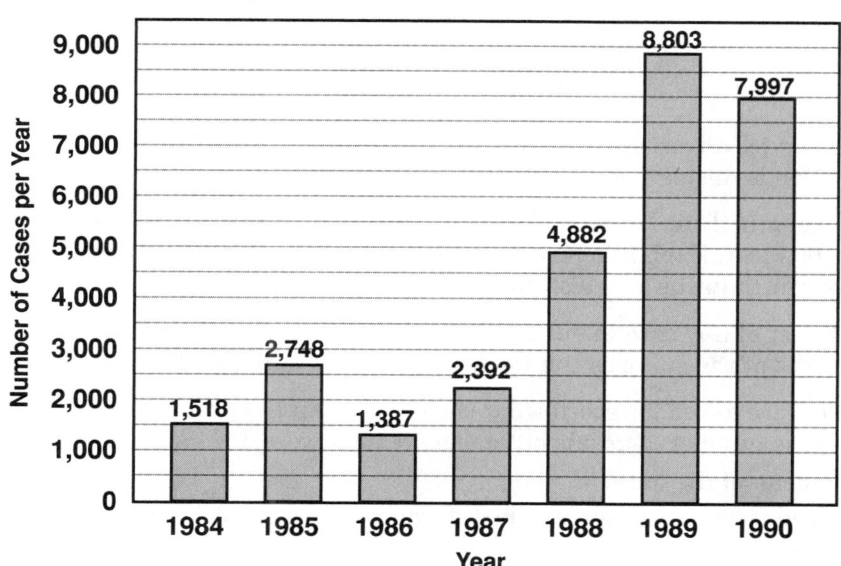

   a. Find the median number of cases reported. b. Find the mean number of cases reported.
6. During a sale at La Shoppe, dresses of the following sizes were sold: 6, 6, 8, 8, 10, 10, 10, 10, 10, 12, 12, 12, 12, 14, 14, 16. Which measure of central tendency is most useful in understanding the needs of customers? Explain your answer.

7. There is one position open on the Weber High women's bowling team. Mei and Jenny each bowled six games for Coach Williams and got the scores shown.

   a. Find the mean score for each bowler.

   b. Explain why Coach Williams picked Jenny.

Mei	Jenny
155	145
115	135
115	158
185	132
190	140
110	160

8. The selling prices of the five homes in a development were \$180,000, \$182,000, \$185,000, \$186,000, and \$235,000.

   a. Which is a better description of these selling prices, the median or the mean?

   b. If the selling price of the most expensive home increased to \$255,000 because of remodeling, what effect would this have on the median and the mean?

9. A tutoring service prepares students for a college entrance examination. Scores of clients who took the exam last year, both before and after tutoring, are shown in the table.

Student	Score Before Tutoring	Score After Tutoring	Increase
A	480	510	30
B	540	560	20
C	490	490	0
D	560	740	180
E	530	540	10
F	610	760	150
G	600	630	30

   The tutoring service advertised that, on the average, scores of clients increased by 60 points. Explain why this is misleading.

10. The average depth of the Atlantic Ocean is 12,880 feet, and the average depth of the Indian Ocean is 13,002 feet. Explain how it is possible that the greatest known depth of the Atlantic Ocean is greater than the greatest known depth of the Indian Ocean.

11. The median cost of day care in suburbs of large cities is \$88 per week. A speaker claimed that in three-fourths of the suburbs the weekly cost was less than \$88 a week. Explain his error.

12. Eight people were asked to record what they had eaten the previous day. Their records were analyzed for the number of grams of fat. The results were 40, 48, 56, 62, 75, 83, 91, 95 grams.

    a. Find the mean of the data.

    b. Use a signed number to express the difference between each data value and the mean, with a negative number representing a value below the mean.

    c. Add the differences found in part **b** and explain the result.

SPIRAL REVIEW EXERCISES

1. Find the mode for this group of numbers.

   48, 44, 41, 47, 48

2. Kim's restaurant bill came to $12. She wanted to leave the waiter a 15% tip. How much money should she have left for the tip?

3. $5\frac{1}{2}\%$ written as a decimal is

   (a) 5.5 (b) 0.55 (c) 0.055 (d) 0.0505

4. The sum of two numbers is −12. If one of the numbers is $-8\frac{3}{8}$, what is the other number?

5. The odometer on the Jackson family car read 23986.5 before they left on vacation. When the family returned, the reading was 24863.2. How many miles did they travel while on vacation?

6. What number should come next in this pattern?

   1, −3, 9, −27, 81, ____

7. Ted borrowed $97.86 from his father to buy a set of weights. To repay his debt, he worked at his father's store for 7 hours at $6.25 an hour. How much did he still owe?

8. Each of five friends wants to pay an equal share of a restaurant bill of $58.30. How much should each pay?

9. Mark received grades of 82, 89, and 75 on the first three exams. What grade must he get on the fourth exam to have an 85 average?

10. If 3,000 people attended a concert and one-fifth bought programs that cost $2.50 each, how much money was spent on programs?

11. Use your calculator to find the value of a $370 investment that earns 7% interest compounded annually for 3 years.

12. Each of 1,285 runners paid a $10 entry fee to run in a five-kilometer race that was held to raise money for a charity. If 54 sponsors contributed a total of $12,206 and the total expenses of organizing the race were $11,653, how much money was raised for the charity?

13. The table below gives the average full fare and average discount fare on a 2,160-mile round-trip domestic flight for each year from 1994 through 2000.

Year	Average Full Fare	Average Discount Fare
1994	$490	$230
1995	510	230
1996	540	220
1997	580	210
1998	620	230
1999	650	230
2000	710	250

The greatest difference between the average cost of a full fare and the average cost of a discount fare in any given year was:

(a) $710 (b) $250 (c) $460 (d) $960

UNIT 10–5 Percentiles; Box-and-Whisker Plots

PERCENTILES

THE MAIN IDEA

1. If data is ranked in increasing order,
   a. the median divides the data into 2 sets of values with equal frequency;
   b. ***quartiles*** divide the data into 4 sets of values with equal frequency;
   c. ***percentiles*** divide the data into 100 sets of values with equal frequency.

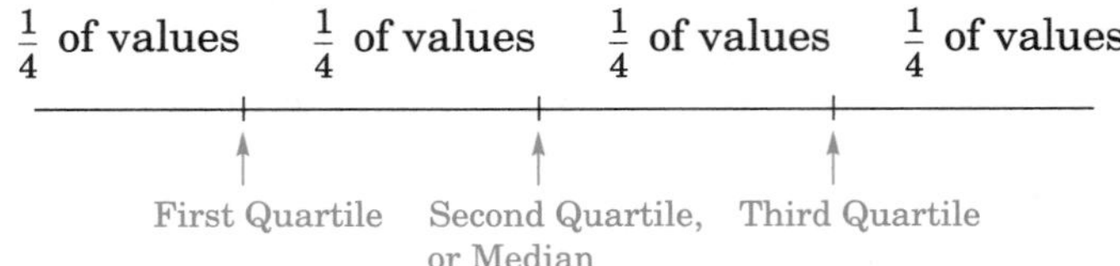

2. The 25th percentile is another name for the first quartile; it is greater than or equal to 25% of the values. The 50th percentile is another name for the second quartile, or median; it is greater than or equal to 50% of the values. The 75th percentile, or third quartile, is greater than or equal to 75% of the values.
3. Percentiles are used to describe where a particular value is found in a ranked set of data. If, in a set of data, the value 58 is the 80th percentile, then 58 is greater than or equal to 80% of the values, and less than 20% of the values.

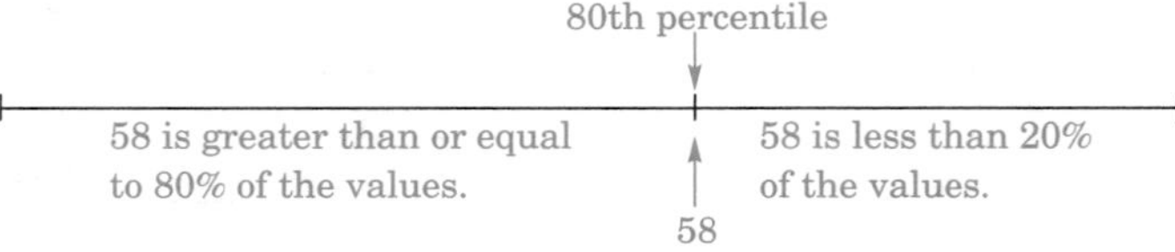

EXAMPLE 1 A newspaper surveyed its readers about the amount of time they spent reading the paper on a given day. The results are shown.

Maximum Number of Minutes	20	30	45
Percentile	25th	50th	75th

If 116 readers reported reading the paper for 20 minutes or less, how many readers were included in the survey in all?

The 25th percentile is 20 minutes. That is, 25% of the readers reported reading the paper for no more than 20 minutes that day. Since 25%, or $\frac{1}{4}$, of the readers surveyed was 116, the total number of readers surveyed was 4×116, or 464. *Ans.*

EXAMPLE 2 On a nationwide reading exam, Julio's score was the 95th percentile.

a. What portion of all the students taking the exam had scores less than or equal to Julio's? greater than Julio's?

The fact that Julio's score was the 95th percentile means that 95% of all the scores were less than or equal to Julio's. The remaining 5% had scores greater than Julio's.

b. If 250,000 students took the exam, how many had scores that weren't any better than Julio's?

The *number* of students with scores that weren't any better than Julio's was 95% of 250,000: $0.95 \times 250{,}000 = 237{,}500$ *Ans.*

EXAMPLE 3 Last year, when Eugene took his college entrance exam, he scored 1140, which was the 89th percentile for his group. This year his younger brother Michael took his exam and received a score of 1160. Which brother placed better in his group?

Although Michael's score of 1160 is greater than Eugene's score, 1140, we do not know the percentile for Michael's score. It is possible that, even though the numerical value of his score (his "raw score") is greater than Eugene's, it may have a lower rank in all of the scores for this year.

Answer: Without knowing the percentile for Michael's score, we do not have enough information to tell who did better.

CLASS EXERCISES

1. In a group of 1,000 marathon runners, Tien's final time was the 54th percentile.

a. How many runners were faster than Tien?

b. How many runners were no faster than Tien?

2. A group of drivers took a written driving exam. Mrs. Stein's score was the 10th percentile. What part of the group taking the exam knew no more about driving than Mrs. Stein?

3. In a set of exam scores, the following percentiles were determined:

Score	40	63	66	72	75	81	86	90	95	98
Percentile	10th	20th	25th	40th	50th	60th	70th	75th	90th	95th

a. Which score was the 1st quartile?

b. Which score was the 2nd quartile?

c. Which score was the median?

d. Which score was the 3rd quartile?

e. The prefix "quart-" means "four", as in the words "quarter" (one-fourth) and "quartet" (a group of four). Explain why there are only three quartiles.

BOX-AND-WHISKER PLOTS

THE MAIN IDEA

1. A useful way to show the distribution of a set of data is to make a kind of graph called a ***box-and-whisker plot***. In a box-and-whisker plot, the range, the first quartile, the median, the third quartile, and the lowest and highest values can be seen at a glance.
2. To make a box-and-whisker plot:
   a. Rank the data and find the median and the first and third quartiles.
   b. Draw a number line that extends over the range of values.
   c. Draw a box by drawing vertical line segments at the first and third quartiles and connecting them with two horizontal line segments.
   d. Draw a vertical line inside the rectangle to represent the median.
   e. Place a point at the extreme low and extreme high values, and draw line segments (whiskers) from the points to the box.

   For example, for the data 2, 5, 5.5, 6, 7, 8, 10, 13, a box-and-whisker plot is:

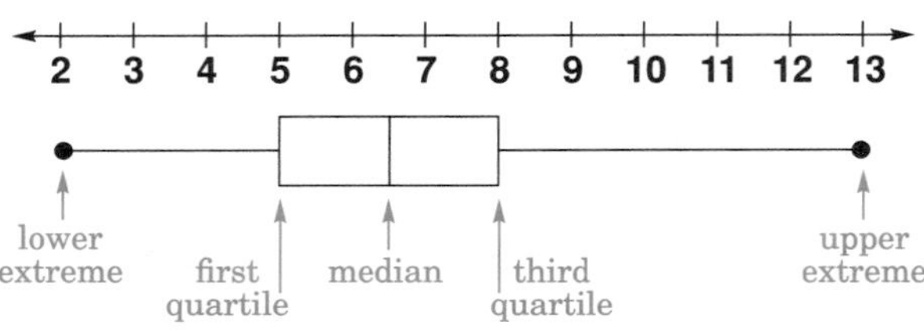

EXAMPLE 4 Make a box-and-whisker plot for grades on a science exam:

78, 81, 89, 75, 77, 91, 100, 53, 80, 87, 83, 79, 95, 91, 90, 88

Rank the grades in increasing order, and locate the necessary values:

53, 75, 77, 78, 79, 80, 81, 83, 87, 88, 89, 90, 91, 91, 95, 100

First Quartile 78.5 — Median 85 — Third Quartile 90.5

Draw a scale from 50 to 100. Then draw a box to represent values from 78.5 to 90.5, and draw a line inside the box at 85 to represent the median. Finally, draw whiskers from the box to the two extreme values, 53 and 100.

Grades Scored on a Science Exam

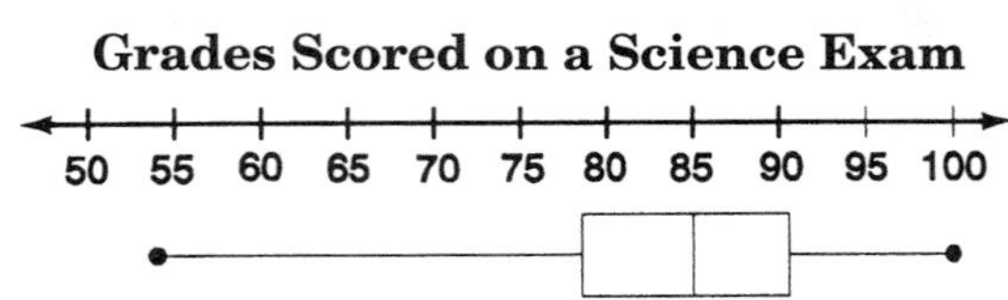

EXAMPLE 5 A survey asked people how many times during one month they watched the news on television. Read the box-and-whisker plot to answer the questions.

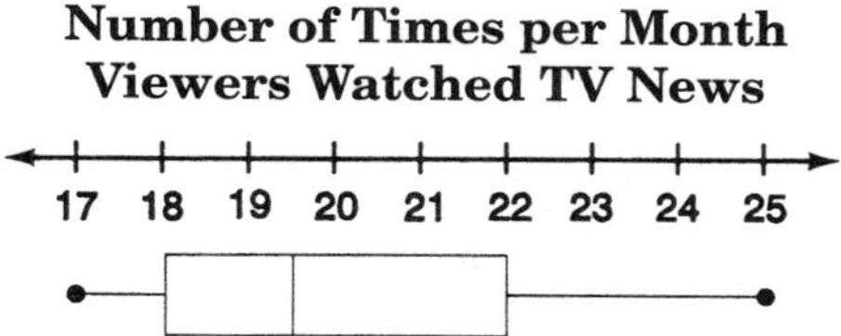

a. What is the range of the data?

The whiskers extend from 17 to 25. The range is the difference between the greatest value and the least value: 25 – 17 = 8

Answer: The range is 8.

b. What is the median of this data?

The line inside the box is positioned at the median, which is 19.5.

Answer: The median is 19.5

c. What is the first quartile?

The first quartile is where the box begins, at 18.

Answer: The first quartile is 18.

d. What is the third quartile?

The third quartile is where the box ends, at 22.

Answer: The third quartile is 22.

e. What does the box-and-whisker plot tell you about people's viewing habits?

Answer: 50% of all viewers watched TV news 18–22 times per month.

CLASS EXERCISES

1. From the given box-and-whisker plots, determine the range, the first quartile, the median, and the third quartile of each set of data.

a.

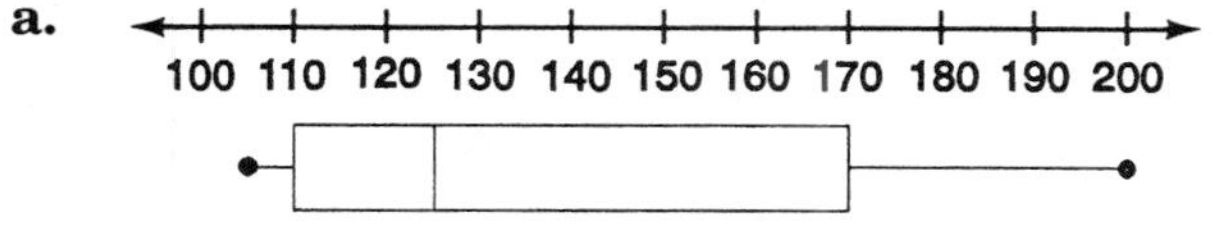

b.

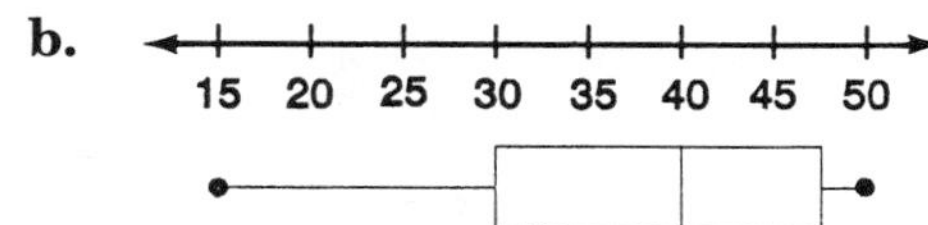

2. Make a box-and-whisker plot of the following data:

27 35 42 32 29 25 48 31 39 49 42 36 39 50 42 40 28 30

HOMEWORK EXERCISES

1. Athena's total score on a physical fitness test was the 87th percentile. What part of the student population was more physically fit than Athena?

2. When the working lives of 540 washing machines were recorded, it was found that 7 years was the 80th percentile. How many washing machines in this group lasted 7 years or less?

3. In a group of 250 ice skaters, those scoring above the 88th percentile in compulsory exercises will go on to the next level of competition. How many skaters will go on to compete further?

4. A track coach determined the percentiles for the broad-jump distances of his team members:

Distance (inches)	58	60	63	65	70	72	75	80	83
Percentile	10th	20th	25th	40th	50th	60th	70th	75th	90th

   a. What was the first quartile?
   b. Give three different names for the distance 70 inches.
   c. What was the third quartile?
   d. Which is a reasonable estimate of the 65th percentile?
   (a) 40 inches (b) 65 inches (c) 71 inches (d) 73 inches

   e. Charlene jumped farther than 40% of the other team members. Describe how far she jumped.

5. Read each box-and-whisker plot below, and give the range, median, first quartile, and third quartile of each set of data.

a.

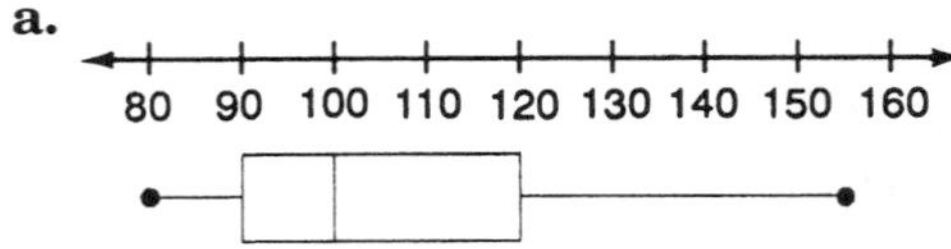

b.

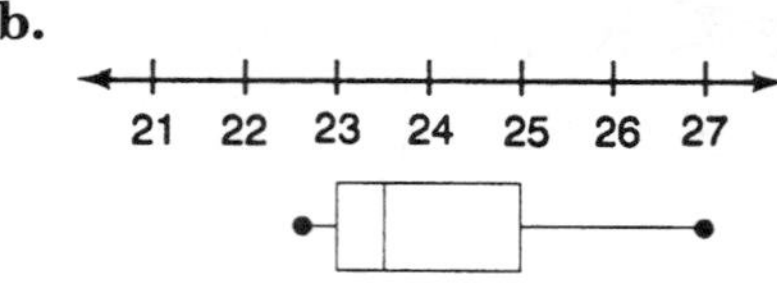

c.

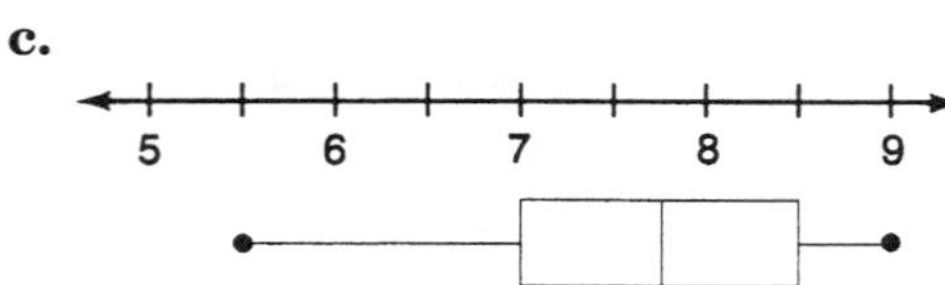

6. Make a box-and-whisker plot of the given lengths, in inches, of fish netted by a conservationist.
8 21 17 18.5 20.5 10 12.5 11.25 8.5 22 29.5 13.25 11 10.25 23.5 19 18 21.5 22 23

SPIRAL REVIEW EXERCISES

1. For the group of numbers given:

   25, 27, 25, 26, 24, 25, 27

   a. Find the mode. **b.** Find the mean.

2. The rainfall, in inches, for 7 days in New City was: 2, 0, 1, 2, 0, 1, 1

   Find the average rainfall for the 7 days.

3. Add: $\frac{2}{5} + \frac{3}{20}$

4. Multiply: 0.18×0.02

5. If the sales tax is charged at the rate of 8%, what is the final cost of a $20 dress?

6. The mean of the set of numbers 12, 12, 15, 20, and 36 is

   (a) 12 (b) 15 (c) 19 (d) 20

7. Which key sequence will display the selling price of a $35 sweater that is being sold at a 15% discount?

   (a) 35 [×] 85 [%]
   (b) 35 [×] .15 [=]
   (c) 35 [−] 15 [=]
   (d) 35 [−] .15 [=]

8. Write a key sequence that can be used to find the number 16% of which is 37.22.

9. From a $12\frac{1}{2}$-pound roast, Mr. Bruno trimmed $\frac{3}{4}$ pound of fat. How much meat was left?

10. What is the mode of the group of numbers shown?

    $1\frac{1}{2}, 1\frac{2}{3}, 2\frac{1}{4}, 2\frac{1}{2}, 2\frac{1}{2}, 2\frac{1}{2}, 3\frac{1}{3}$

11. The smallest of the values shown is

    (a) 9% (b) $\frac{1}{10}$ (c) 0.08 (d) $7\frac{3}{4}\%$

12. 318.234 rounded to the nearest tenth is

    (a) 310 (b) 320
    (c) 318.2 (d) 318.3

13. Replace each ? by $<$, $=$, or $>$ to make a true statement.

    a. $\frac{5}{8}$? $\frac{5}{9}$ **b.** $\frac{4}{5}$? $\frac{3}{5}$

    c. $2\frac{1}{2}$? $\frac{2}{5}$ **d.** $3\frac{3}{4}$? $\frac{30}{8}$

14. The cost of a coin-box telephone call from New York City to Miami is $2.00 for the first minute and $0.20 for each additional minute. What is the cost of a 15-minute call?

15. Mr. Loman had a balance of $1,247.58 in his checking account. He wrote checks for $198.50, $250.00, and $37.95. What was his new balance?

UNIT 11-1 Customary Measures of Length

THE MAIN IDEA

1. The most common *customary (English) units of length* are the ***inch*** (in.), the ***foot*** (ft.), the ***yard*** (yd.), and the ***mile*** (mi.).

   An inch is about the length across a bottle cap.

   A foot is about the length of an average man's foot.

   A yard is about the distance from a doorknob to the floor.

   A mile is about as long as 18 football fields.

2. These units have the following relationships to each other:

   12 inches = 1 foot (12" = 1')   3 feet = 1 yard   5,280 feet = 1 mile

3. In general: To change from a smaller unit to a larger unit, divide.
   To change from a larger unit to a smaller unit, multiply.

EXAMPLE 1 Make each of the required unit changes.

	Required Unit Change	*Conversion Information*	*Arithmetic*
a.	Change 66 inches to feet. *Answer:* 66 inches = $5\frac{1}{2}$ feet	To change from the smaller unit (inches) to the larger unit (feet), divide by 12.	$66 \div 12 = 5\frac{6}{12} = 5\frac{1}{2}$
b.	Change 7 feet to inches. *Answer:* 7 feet = 84 inches	To change from the larger unit (feet) to the smaller unit (inches), multiply by 12.	$7 \times 12 = 84$
c.	Change 14 feet to yards. *Answer:* 14 feet = $4\frac{2}{3}$ yards	To change from the smaller unit (feet) to the larger unit (yards), divide by 3.	$14 \div 3 = 4\frac{2}{3}$
d.	Change 8 yards to feet. *Answer:* 8 yards = 24 feet	To change from the larger unit (yards) to the smaller unit (feet), multiply by 3.	$8 \times 3 = 24$

EXAMPLE 2 Complete the table:

Distance in Feet	10,560	31,680	12,672	3,696	55,440
Distance in Miles					

KEYING IN

To change from feet to miles, divide by 5,280. Since there are five calculations of this type to do, use the constant feature of the calculator.

Key Sequence	***Display***
10560 ÷ 5280 =	2.
31680 =	6.
12672 =	2.4
3696 =	0.7
55440 =	10.5

Answer:

Distance in Feet	10,560	31,680	12,672	3,696	55,440
Distance in Miles	2	6	2.4	0.7	10.5

CLASS EXERCISES

1. Change to feet;

a. 72 inches **b.** 144 inches

c. 30 inches **d.** 54 inches

2. Change to inches:

a. 12 feet **b.** 100 feet

c. $8\frac{1}{4}$ feet **d.** $18\frac{3}{4}$ feet

3. Helen needed 48 inches of trim for her skirt. How many feet of trim did she need?

4. A golf ball landed $5\frac{3}{4}$ feet from the flag. How many inches away from the flag was it?

5. Change to yards:

a. 36 feet **b.** 300 feet

c. 144 feet **d.** 13 feet

6. Change to feet:

a. 4 yards **b.** 100 yards

c. $10\frac{2}{3}$ yards **d.** $\frac{1}{6}$ yard

7. Rafi ran 17 yards for a touchdown. How many feet did he run?

8. Mrs. Rivera measured her windows for curtains and found that she needed 45 feet of material. How many yards of material did she need?

9. Change to feet:

a. 1 mile **b.** 3 miles **c.** 10 miles **d.** $2\frac{1}{2}$ miles **e.** 4.5 miles

10. Marcia and Bill walked 4 miles to get help for their disabled car. How many feet did they walk?

11. Ms. Anderson needed $3\frac{1}{2}$ yards of material for curtains that she was making. The fabric shop has a piece 120 inches long on sale. Explain why Ms. Anderson should or should not purchase this piece for her curtains.

HOMEWORK EXERCISES

1. Change to feet:
   a. 156 inches **b.** 1,440 inches
   c. 111 inches **d.** 9 inches

2. Change to inches:
   a. 120 feet **b.** 45 feet
   c. $9\frac{1}{2}$ feet **d.** $19\frac{3}{4}$ feet

3. A shelf is 48 inches above the floor. How many feet above the floor is it?

4. Mavis is 63 inches tall. What is her height in feet?

5. Roberta broad-jumped $7\frac{3}{4}$ feet. How many inches did she jump?

6. Change to yards:
   a. 300 feet **b.** 480 feet **c.** 10 feet
   d. 101 feet **e.** 10.5 feet

7. Change to feet:
   a. 9 yards **b.** 42 yards **c.** 100 yards
   d. $8\frac{2}{3}$ yards **e.** 9.5 yards

8. A telephone installer found that she needed 75 feet of wire. How many yards did she need?

9. A racetrack is 440 yards long. What is its length in feet?

10. The distance from third base to home plate is 90 feet. How many yards is this?

11. Copy and complete the table, changing the measures in inches to yards.

Length in Inches	180	126	45	504	309.6
Length in Yards					

12. Jesse biked for 12 miles in a bike-a-thon. How many feet did he travel?

13. Change to the indicated unit.
    a. 6 feet = [?] inches **b.** 48 inches = [?] feet **c.** 17 yards = [?] feet
    d. 35 feet = [?] yards **e.** 10 miles = [?] feet **f.** $9\frac{1}{3}$ yards = [?] feet
    g. 20 inches = [?] feet **h.** 48 feet = [?] yards **i.** 19 feet = [?] inches

14. Joan and Phyllis kept logs of the distances they jogged for five days. Their records are shown at the right. Which runner jogged the greater total distance for the five days?

15. A telephone repair team has 10 miles of cable. How many 100-yard lengths of cable can they cut?

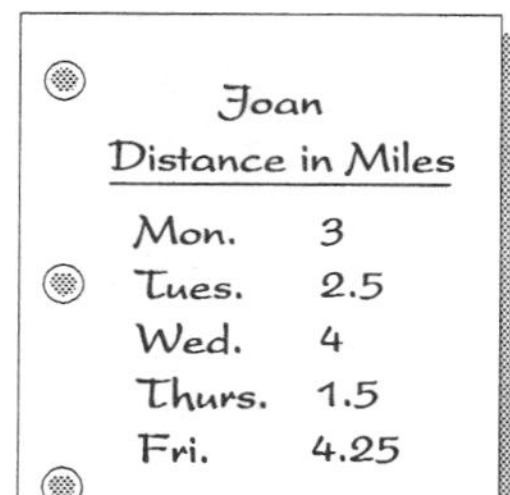

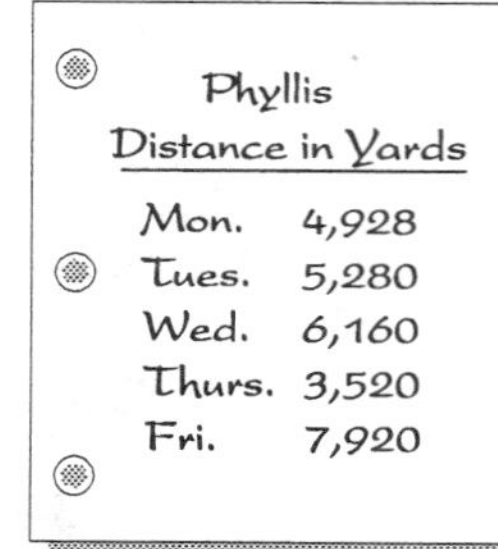

SPIRAL REVIEW EXERCISES

1. If Debra's grades on her mathematics tests are 75, 85, 65, 90, and 80, find the average of her grades.

2. Find the median for the following test scores: 55, 60, 75, 75, 80, 80, 95

3. What is the mode of the given set of numbers?

   23, 29, 29, 35, 46, 47, 47, 47, 50

4. How much would a $5\frac{1}{2}\%$ city sales tax add to the cost of a $399.99 stereo? Round the answer to the nearest cent.

5. A plumber charges $36 for the first hour and $12.50 for each half-hour after the first hour. If the plumber worked for $2\frac{1}{2}$ hours to fix a sink, how much was his bill?

6. 10% of $256 is

   (a) $256 (b) $25.60 (c) $2.56 (d) $221.40

7. 32.39 rounded to the nearest tenth is

   (a) 30 (b) 32 (c) 32.3 (d) 32.4

8. The original price of a handbag was reduced by 20% to be sold for $38. Use your calculator to find the original price.

9. The next number in the pattern 80, 78, 74, 68, 60, ... is

   (a) 52 (b) 50 (c) 48 (d) 46

10. The bar graph below shows the number of athletes who competed for the chance to represent the United States in the Olympic games in four sports and the number who were actually selected. About what percent of the athletes who tried out for these sports were actually selected?

    (a) 40% (b) 20% (c) 7% (d) 4%

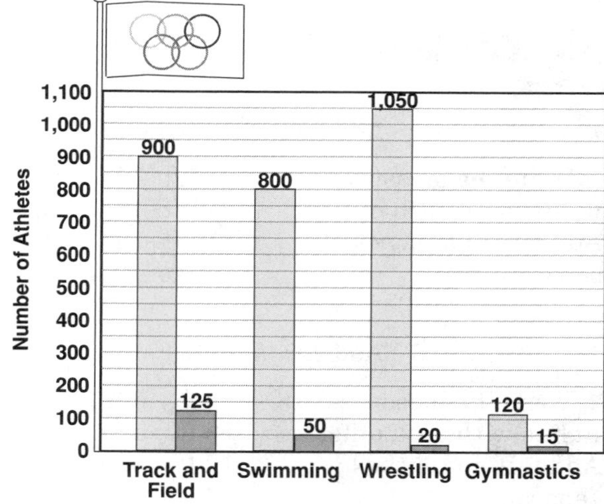

11. Multiply −5 by −3.

12. Evaluate: $11 + 4 \times 2 \times 3$

UNIT 11-2 Adding and Subtracting Measures

THE MAIN IDEA

1. To add or subtract measures that are given in two or more related units, such as feet and inches:
   a. Write the measures one below the other so that feet are in one column and inches are in another column.
   b. Add or subtract the feet and inches separately, renaming units as needed.
2. To add or subtract measures that are given in different units (for example, one in inches and another in feet):
   a. Write one of the given measures with the same unit as the other.
   b. Add or subtract the two measures that are now expressed in terms of the same unit and label the result with this unit.

EXAMPLE 1 Add 7 ft. 3 in. to 5 ft. 4 in.

Write the measures in line with each other. Add feet and inches separately.

```
  7 ft. 3 in.
 +5 ft. 4 in.
 12 ft. 7 in.  Ans.
```

EXAMPLE 2 Add 4 ft. 9 in. to 8 ft. 5 in.

Write the measures in line with each other. Add feet and inches separately.

```
  4 ft.  9 in.
 +8 ft.  5 in.
 12 ft. 14 in.
```

Since the number of inches is greater than 12, divide by 12.

```
     1
 12)14
    12
     2
```

The quotient 1 gives one more foot, and the remainder 2 stays as 2 inches.

12 ft. 14 in.
= 12 ft. + 1 ft. 2 in.
= 13 ft. 2 in. *Ans.*

EXAMPLE 3 Add 10 yd. 2 ft. 8 in. to 4 yd. 2 ft. 9 in.

Add yards, feet, and inches separately.

```
 10 yd. 2 ft.  8 in.
 +4 yd. 2 ft.  9 in.
 14 yd. 4 ft. 17 in.
```

Since the number of inches is greater than 12, divide 17 by 12.

```
     1
 12)17
    12
     5
```

The quotient 1 gives one more foot, and the remainder 5 stays as 5 inches.

14 yd. 4 ft. 17 in.
= 14 yd. 4 ft. + 1 ft. 5 in.
= 14 yd. 5 ft. 5 in.

Since the number of feet is greater than 3, divide 5 by 3.

```
   1
 3)5
   3
   2
```

The quotient 1 gives one more yard, and the remainder 2 stays as 2 feet.

14 yd. 5 ft. 5 in.
= 14 yd. + 1 yd. 2 ft. 5 in.
= 15 yd. 2 ft. 5 in. *Ans.*

EXAMPLE 4 Subtract 4 ft. 7 in. from 10 ft. 11 in.

Write the measures in line with each other. Subtract feet and inches separately.

$$\begin{array}{r} 10 \text{ ft. } 11 \text{ in.} \\ -4 \text{ ft. } 7 \text{ in.} \\ \hline 6 \text{ ft. } 4 \text{ in.} \end{array} \quad \textit{Ans.}$$

EXAMPLE 5 Subtract 22 in. from $2\frac{1}{2}$ yd.

Change from yards to inches using 1 yard = 36 inches.

$$2\frac{1}{2} \times 36 = \frac{5}{\cancel{2}_1} \times \frac{\cancel{36}^{18}}{1} = 90$$

Subtract the two measures in inches.

$$\begin{array}{r} 90 \text{ in.} \\ -22 \text{ in.} \\ \hline 68 \text{ in.} \end{array} \quad \textit{Ans.}$$

EXAMPLE 6 Subtract 8 ft. 6 in. from 14 ft. 3 in.

Write the measures in line with each other. We cannot subtract 6 in. from 3 in.

$$\begin{array}{r} 14 \text{ ft. } 3 \text{ in.} \\ -8 \text{ ft. } 6 \text{ in.} \\ \hline \times \end{array}$$

In the minuend, rename 1 foot as 12 inches.

$$\begin{aligned} &14 \text{ ft. } 3 \text{ in.} \\ &= 13 \text{ ft. } 12 \text{ in.} + 3 \text{ in.} \\ &= 13 \text{ ft. } 15 \text{ in.} \end{aligned}$$

Now we can subtract.

$$\begin{array}{r} 13 \text{ ft. } 15 \text{ in.} \\ -8 \text{ ft. } 6 \text{ in.} \\ \hline 5 \text{ ft. } 9 \text{ in.} \end{array} \quad \textit{Ans.}$$

CLASS EXERCISES

1. Add:
   a. 6 ft. 8 in. and 5 ft. 3 in.
   b. 11 ft. 6 in. and 1 ft. 8 in.
   c. 10 ft. and 3 ft. 9 in.
   d. 17 ft. 8 in. and 11 in.
   e. 4 yd. 1 ft. 10 in. and 5 yd. 1 ft. 1 in.
   f. 7 yd. 1 ft. 9 in. and 4 yd. 2 ft. 6 in.
   g. 4 ft. 11 in. and 6 yd. 2 ft. 5 in.
   h. 2 yd. 2 in. and 10 ft. 10 in.
   i. 6 mi. 4,176 ft. and 10 mi. 943 ft.
   j. 4 mi. 5,100 ft. 4 in. and 3 mi. 1,500 ft. 10 in.

2. Sarah wishes to fence in her vegetable garden, which is shaped in the form of a triangle. If the measures of the sides of the triangle are 6 ft. 3 in., 5 ft. 4 in., and 7 ft. 5 in., how many feet of fencing must Sarah buy?

3. If Sarah were to enlarge her garden so that the measurements were 11 ft. 11 in., 10 ft. 10 in., and 9 ft. 9 in., how many feet of fencing would she have to buy?

   (a) 30 ft. (b) 31 ft. (c) 32 ft. (d) $32\frac{1}{2}$ ft.

4. If Sarah were to rearrange her garden into the shape of a quadrilateral (4-sided figure) with sides measuring 11 ft. 11 in., 10 ft. 10 in., 9 ft. 9 in., and 8 ft. 8 in., how many feet of fencing would she have to buy? (a) 39 ft. (b) 41 ft. (c) 42 ft. (d) 43 ft.

5. Subtract:
a. 11 ft. 7 in. from 19 ft. 9 in.
b. 7 ft. 11 in. from 9 ft. 9 in.
c. 9 ft. from 16 ft. 6 in.
d. 5 ft. 4 in. from 8 ft.
e. 10 yd. 1 ft. 4 in. from 14 yd. 2 ft. 7 in.
f. 8 yd. 1 ft. 3 in. from 15 yd. 2 ft. 1 in.
g. 2 yd. 2 ft. 7 in. from 8 yd. 1 ft. 4 in.
h. 12 mi. 4,444 ft. from 15 mi. 4,700 ft.
i. 3 mi. 3,300 ft. from 7 mi. 2,000 ft.
j. $10\frac{1}{2}$ mi. from 20 mi. 2,800 ft.

6. A tailor has $8\frac{1}{3}$ yd. of material from which to make:
a. two dresses. If the first dress requires 3 yd. 2 ft. 7 in., how much material is left for the second?
b. a three-piece suit. If the jacket requires 2 yd. 1 ft. 9 in. and the vest requires 1 yd. 4 in., how much material is left for the skirt?

7. Subtract the first measure from the second.
a. 42 inches; 5 feet
b. 0.5 mile; 4,920 feet
c. 27 inches; $\frac{3}{4}$ yard

HOMEWORK EXERCISES

1. Add:
a. 8 ft. 1 in. and 3 ft. 9 in.
b. 10 ft. 4 in. and 7 ft. 11 in.
c. 16 ft. and 4 ft. 8 in.
d. 7 ft. 3 in. and 9 in.
e. 3 yd. 1 ft. 7 in. and 5 yd. 1 ft. 4 in.
f. 9 yd. 2 ft. 4 in. and 3 yd. 1 ft. 8 in.
g. 4 mi. 2,000 ft. and 20 mi. 1,000 ft.
h. 19 mi. 3,300 ft. and 9 mi. 2,200 ft.

2. Ted wishes to frame a tapestry to hang on a wall. If two sides of the tapestry measure 4 ft. 2 in. each and the other two sides measure 5 ft. 8 in. each, how much framing does he need?

3. Mrs. Elkins wishes to trim a blouse with ribbon. If each sleeve requires 2 ft. 4 in. of ribbon and the front of the blouse requires 4 ft. 4 in., how many yards of ribbon must Mrs. Elkins buy?
(a) 3 yd. (b) 4 yd. (c) 6 yd. (d) 9 yd.

4. In an athletic contest, Jane runs 2,500 feet, and bikes for 1.5 miles. She qualifies for the finals if she has covered a distance of 10,000 feet. Does Jane qualify?

5. Subtract:
a. 7 ft. 4 in. from 10 ft. 8 in.
b. 11 ft. 9 in. from 15 ft. 1 in.
c. 5 ft. from 12 ft. 8 in.
d. 6 ft. 7 in. from 10 ft.
e. 6 yd. 1 ft. 8 in. from 10 yd. 2 ft. 11 in.
f. 8 yd. 2 ft. 6 in. from 12 yd. 1 ft. 4 in.
g. 16 mi. 100 ft. from 20 mi. 1,000 ft.
h. 4 mi. 1,000 ft. from 10 mi. 500 ft.

6. A tailor has $4\frac{2}{3}$ yd. of material from which to make:

   a. a skirt and a blouse. If the blouse requires 1 yd. 2 ft. 7 in., how much material is left for the skirt?

   b. a three-piece outfit. If the blouse requires 1 yd. 1 ft. 11 in. and the skirt requires 2 yd. 2 ft. 6 in., how much material is left for the sash?

7. Subtract the first measure from the second: **a.** $3\frac{1}{2}$ yards; 135 inches **b.** 8,940 feet; 2 miles

8. Which is closest to 200 feet? (a) 2,350 inches (b) 66 yards (c) 190 feet (d) 0.04 mile

9. Mrs. Veley has a piece of ribbon 10 ft. 3 in. long. She wants to cut it into three pieces of equal length. How long should each piece be?

SPIRAL REVIEW EXERCISES

1. Use the graph to estimate the number of miles traveled by Mr. S.

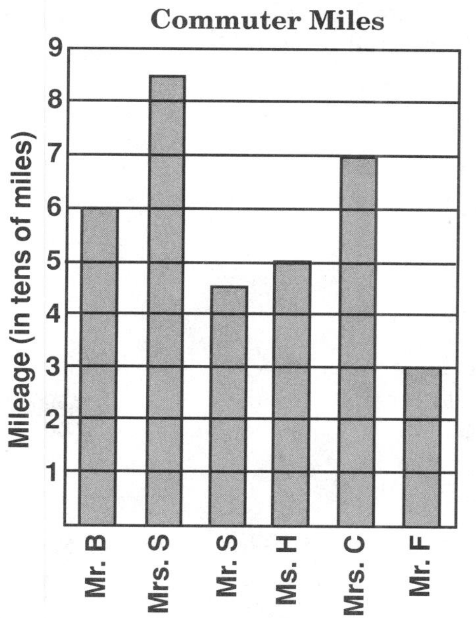

2. The cost of a steak dinner at five different restaurants is as follows: $11, $12, $9, $18, and $16. What is the mean cost?

3. Of 5 ft. 6 in., 5 ft. 8 in., 5 ft. 9 in., 5 ft. 11 in., and 6 ft., what is the median height?

4. Sam needed 84 feet of webbing to reweb a chair. How many yards did he need?

5. $6\frac{2}{3}$ yards equals (a) 21 feet (b) 20 feet (c) 19 feet (d) 18 feet

6. Change $1\frac{1}{4}$ miles to feet.

7. A library lends videos at the rate of $1.50 for the first day and $0.25 for each day thereafter. If Camille borrowed a video for 7 days, how much must she pay?

8. Write $\frac{7}{8}$ as a decimal.

9. Which is the larger value, $\frac{1}{4}$ or $\frac{2}{9}$?

10. Write a key sequence to display the cost of a $12 meal plus a 15% tip.

11. Which is the smaller value, 0.1 or 0.01?

12. Add +7 and −11.

13. 42.7 rounded to the nearest ten is
    (a) 40 (b) 42 (c) 43 (d) 50

14. 20% of $10,000 is
    (a) $2 (b) $200 (c) $20 (d) $2,000

UNIT 11-3 Customary Measures of Weight

THE MAIN IDEA

1. The most common *customary units of weight* are the ***ounce*** (oz.), the ***pound*** (lb.), and the ***ton*** (T.).
   An ounce is about the weight of a spoon.
   A pound is about the weight of a loaf of bread.
   A ton is about the weight of a small car.
2. These units have the following relationships to each other:
   16 ounces = 1 pound   2,000 pounds = 1 ton

EXAMPLE 1 Make each of the required unit changes.

	Required Unit Change	***Conversion Information***	***Arithmetic***
a.	Change 48 ounces to pounds. *Answer:* 48 ounces = 3 pounds	To change from the smaller unit (ounces) to the larger unit (pounds), divide by 16.	$48 \div 16 = 3$
b.	Change $2\frac{3}{4}$ pounds to ounces. *Answer:* $2\frac{3}{4}$ pounds = 44 ounces	To change from the larger unit (pounds) to the smaller unit (ounces), multiply by 16.	$2\frac{3}{4} \times 16$ $= \frac{11}{\cancel{4}_1} \times \frac{\cancel{16}^4}{1} = 44$
c.	Change 3,500 pounds to tons. *Answer:* 3,500 pounds = $1\frac{3}{4}$ tons	To change from the smaller unit (pounds) to the larger unit (tons), divide by 2,000.	$3{,}500 \div 2{,}000$ $= 1\frac{1{,}500}{2{,}000} = 1\frac{3}{4}$
d.	Change $2\frac{1}{2}$ tons to pounds. *Answer:* $2\frac{1}{2}$ tons = 5,000 pounds	To change from the larger unit (tons) to the smaller unit (pounds), multiply by 2,000.	$2\frac{1}{2} \times 2{,}000$ $= \frac{5}{\cancel{2}_1} \times \frac{\cancel{2{,}000}^{1{,}000}}{1}$ $= 5{,}000$

EXAMPLE 2 Combine the measures.

a.
10 lb. 5 oz.
+7 lb. 8 oz.
17 lb. 13 oz. *Ans.*

b.
16 T. 500 lb.
−4 T. 200 lb.
12 T. 300 lb. *Ans.*

No renaming of units was necessary here.

EXAMPLE 3 Combine the measures.

a.
14 lb. 11 oz.
+3 lb. 7 oz.
17 lb. 18 oz.
= 17 lb. + 1 lb. 2 oz.
= 18 lb. 2 oz. *Ans.*

Rename 18 oz.
16) 18 = 1 lb.
16
2 oz.

b.

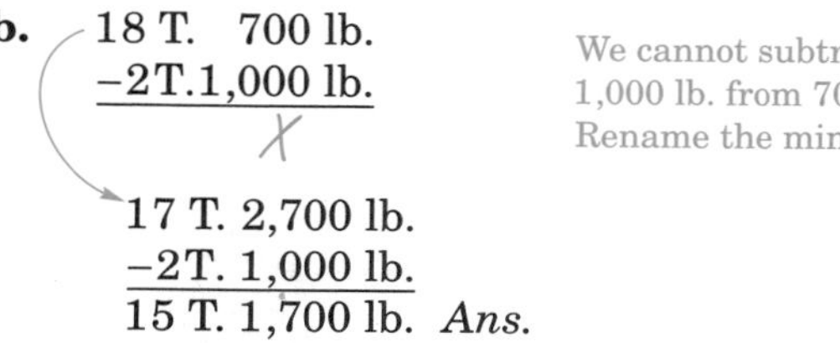

CLASS EXERCISES

1. Change to pounds: **a.** 160 oz. **b.** 48 oz. **c.** 68 oz. **d.** 8 oz.
2. Change to ounces: **a.** 5 lb. **b.** 20 lb. **c.** $\frac{1}{4}$ lb. **d.** $2\frac{1}{2}$ lb.
3. Change to tons: **a.** 4,000 lb. **b.** 10,000 lb. **c.** 5,000 lb. **d.** 1,000 lb.
4. Change to pounds: **a.** 3 tons **b.** 10 tons **c.** $8\frac{3}{4}$ tons **d.** $4\frac{1}{2}$ tons
5. Joseph collected 196 ounces of scrap aluminum. How many pounds of aluminum did he collect?
6. Mrs. Green used $2\frac{3}{4}$ pounds of butter in her baking. How many ounces of butter did she use?
7. If a loaded shipping crate weighs 7,500 pounds, what is its weight in tons?
8. What is the weight, in pounds, of a $4\frac{1}{4}$-ton truck?
9. Combine the measures as indicated.

a.	**b.**	**c.**	**d.**
8 lb. 1 oz.	18 T. 1,000 lb.	17 lb. 14 oz.	100 T. 900 lb.
+3 lb. 9 oz.	−12 T. 100 lb.	+3 lb. 8 oz.	−27 T. 1,200 lb.

10. 14 oz. + 22 oz. + 10 oz. is equal to
(a) 1 lb. 12 oz. (b) 2 lb. 6 oz. (c) 2 lb. 14 oz. (d) 3 lb. 2 oz.

11. Copy and complete the table.

Weight in Pounds	6,000	7,000	1,600	20,000	200
Weight in Tons					

HOMEWORK EXERCISES

1. Change to pounds: **a.** 480 oz. **b.** 176 oz. **c.** 40 oz. **d.** 20 oz.

2. Change to ounces: **a.** 4 lb. **b.** 15 lb. **c.** $3\frac{1}{2}$ lb. **d.** $\frac{3}{4}$ lb.

3. Change to tons: **a.** 8,000 lb. **b.** 120,000 lb. **c.** 9,000 lb. **d.** 10,200 lb.

4. Change to pounds: **a.** 2 tons **b.** 25 tons **c.** $\frac{1}{4}$ ton **d.** $9\frac{3}{10}$ tons

5. How many pounds of turkey are there in 240 ounces?

6. Helen packed 228 ounces of canned goods. How many pounds did she pack?

7. A recipe calls for $1\frac{1}{4}$ pounds of chicken. How many ounces of chicken is this?

8. An airplane had a 6,000-pound cargo. How many tons was this?

9. What is the weight, in tons, of a truck that weighs 5,500 pounds?

10. There is a 3-ton limit on a certain stretch of road. What is the limit in pounds?

11. A truck dumped $\frac{3}{4}$ of a ton of gravel. How many pounds of gravel was this?

12. Combine the measures as indicated.

a.
```
  11 lb.  4 oz.
+10 lb. 10 oz.
```
b.
```
  19 T. 900 lb.
 −7 T. 100 lb.
```
c.
```
   1 lb.  1 oz.
+15 lb. 15 oz.
```
d.
```
  20 T. 100 lb.
−10 T. 500 lb.
```

13. At birth, the Johnson twins weighed 5 lb. 14 oz. and 6 lb. 3 oz.

a. What was the combined birth weight?

b. During the first month, the smaller twin gained 1 lb. 10 oz. and the larger twin gained 11 oz. What was the new combined weight?

c. At a checkup, the doctor put both twins on the scale at the same time and the recorded weight was 20 lb. 3 oz. If one twin weighed 9 lb. 8 oz., what was the weight of the other twin?

14. 15 oz. + 21 oz. + 28 oz. is equal to (a) 2 lb. (b) 3 lb. (c) $3\frac{1}{2}$ lb. (d) 4 lb.

15. A cook added 11 ounces of vegetables and 5 ounces of chicken to 5 pounds of rice. What was the total weight of the mixture?

16. Mr. Barnes' truck can haul a maximum gross weight (weight of truck + weight of load) of 12,500 pounds. The truck, when empty, weighs $2\frac{1}{2}$ tons. What is the heaviest load that Mr. Barnes can safely carry in the truck?

17. If a 1-foot section of a beam weighs 42 pounds, what is the weight, in tons, of an 80-foot section of the same beam?

1. This bar graph shows the number of times five students were absent from school in one year.

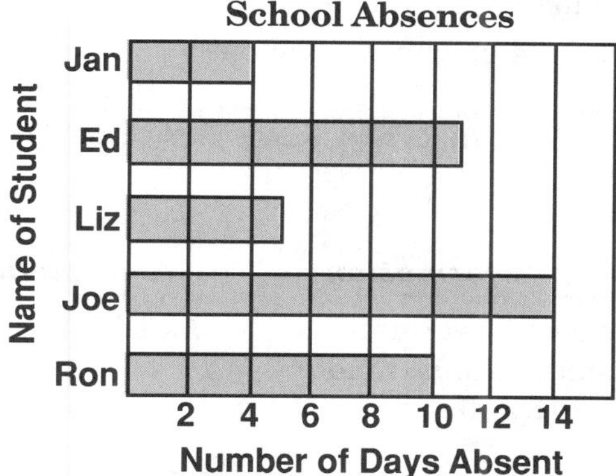

How many more days than Liz was Joe absent?

(a) 20 (b) 9 (c) 11 (d) 8

2. Find the mode of these temperatures.

56°, 59°, 59°, 60°, 63°, 64°, 70°

3. Find the mean of these weekly salaries.

\$150, \$185, \$225, \$280

4. Change the given measures of length to the indicated units.

a. 16 feet = ? inches

b. 10 miles = ? feet

c. 15 yards = ? feet

5. The quarterback of the Clinton High School football team completed 60% of his passes. How many passes did he complete if he threw a total of 225 passes?

6. If Cheryl buys 3 bottles of nail polish at \$1.75 each, how much change should she receive from a \$10 bill?

7. For a long-distance call from a pay telephone, Wanda has \$4.75 in quarters. How many quarters does she have?

8. What percent of 40 is 8?

9. A \$60 jacket is on sale for \$30. The percent of discount is

(a) 25% (b) $33\frac{1}{3}\%$ (c) 50% (d) 75%

10. Find the total amount of sales tax on purchases of \$15.75, \$28.95, and \$39.99 if the rate of sales tax is $5\frac{1}{2}\%$.

11. Mr. Sachs withdrew \$125 from his savings account, which had a balance of \$2,515. What is the new balance?

12. Add: $2.7 + 15.06 + 7.93 + 0.8$

13. Multiply: $5\frac{1}{3} \times 3\frac{3}{8}$

14. What is the greatest common factor of 36 and 45?

UNIT 11-4 Customary Measures of Liquid Volume

THE MAIN IDEA

1. ***Volume*** is a measure of how much room something takes up.
2. The common *customary units of liquid volume* are the ***fluid ounce*** (fl. oz.), the ***cup*** (c.), the ***pint*** (pt.), the ***quart*** (qt.), and the ***gallon*** (gal.).

   The fluid ounce, which is a measure of volume, is not the same as the ounce that is a measure of weight.

   A fluid ounce is the size of the contents of a bottle of iodine.
   A cup is about as much as a coffee mug holds.
   A pint is about as much as a tall spray-paint can holds.
   A quart is the size of the contents of a tall milk container.
   A gallon is about as much as a small sink holds.
3. These units have the following relationships to each other:

   16 fluid ounces = 1 pint 8 fluid ounces = 1 cup
   2 cups = 1 pint 2 pints = 1 quart 4 quarts = 1 gallon

EXAMPLE 1 Make each of the required unit changes.

	Required Unit Changes	*Conversion Information*	*Arithmetic*
a.	Change 3 pints to fluid ounces. *Answer:* 3 pints = 48 fluid ounces	To change from the larger unit (pints) to the smaller unit (fluid ounces), multiply by 16.	$3 \times 16 = 48$
b.	Change 80 fluid ounces to pints. *Answer:* 80 fluid ounces = 5 pints	To change from the smaller unit (fluid ounces) to the larger unit (pints), divide by 16.	$80 \div 16 = 5$
c.	Change 5 pints to quarts. *Answer:* 5 pints = $2\frac{1}{2}$ quarts	To change from the smaller unit (pints) to the larger unit (quarts), divide by 2.	$5 \div 2 = 2\frac{1}{2}$
d.	Change $5\frac{1}{2}$ quarts to pints. *Answer:* $5\frac{1}{2}$ quarts = 11 pints	To change from the larger unit (quarts) to the smaller unit (pints), multiply by 2.	$5\frac{1}{2} \times 2$ $= \frac{11}{\not{2}_1} \times \frac{\not{2}^1}{1} = 11$

EXAMPLE 2 Combine the measures as indicated.

a.

$$\begin{array}{r} 10 \text{ gal. } 3 \text{ qt. } 1 \text{ pt.} \\ +9 \text{ gal. } 2 \text{ qt. } 1 \text{ pt.} \\ \hline 19 \text{ gal. } 5 \text{ qt. } 2 \text{ pt.} \end{array}$$

Change 2 pints to 1 quart.

= 19 gal. 6 qt.

Rename 6 quarts.

$$\begin{array}{r} 1 \text{ gal.} \\ 4\overline{)6} \\ \underline{4} \\ 2 \text{ qt.} \end{array}$$

= 19 gal. + 1 gal. 2 qt.

= 20 gal. 2 qt. *Ans.*

b.

$$\begin{array}{r} 3 \text{ pt. } 8 \text{ fl. oz.} \\ -1 \text{ pt. } 12 \text{ fl. oz.} \\ \hline \end{array}$$

We cannot subtract 12 fl. oz. from 8 fl. oz.

$$\begin{array}{r} 2 \text{ pt. } 24 \text{ fl. oz.} \\ -1 \text{ pt. } 12 \text{ fl. oz.} \\ \hline 1 \text{ pt. } 12 \text{ fl. oz.} \end{array}$$ *Ans.*

Rename the minuend.

CLASS EXERCISES

1. Change to fluid ounces: **a.** 11 pints **b.** 6 pints **c.** $3\frac{1}{2}$ pints
2. Change to pints: **a.** 32 fl. oz. **b.** 56 fl. oz. **c.** 172 fl. oz.
3. Change to pints: **a.** 5 quarts **b.** 2 quarts **c.** $3\frac{1}{2}$ quarts
4. Change to quarts: **a.** 2 pints **b.** 1 pint **c.** 14 pints
5. Copy and complete the table. Change each given quantity to four other equivalent quantities.

Gallons	Quarts	Pints	Cups	Fluid Ounces
8				
	20			
		40		
			104	
				320

6. Mom's recipe called for 3 cups of milk. How many fluid ounces of milk were needed?
7. At the service station, Mr. Jackson discovered that his car needed 2 quarts of oil. How many pints of oil did his car need?
8. The Russo family wanted a half-gallon of ice cream. They decided to get separate pints instead of one large container, so that they could order a different flavor for each pint. How many different flavors could they order?
9. A bottle of milk contained 2 quarts, and a bottle of juice contained 72 fluid ounces. Which bottle contained more?
10. Mr. Rose's recipe for pizza dough called for $3\frac{1}{2}$ ounces of olive oil. He wanted to make 10 times the usual amount of dough. How many cups of oil should he have used?

11. Combine the measures as indicated.

a. 15 gal. 2 qt. 1 pt.
+ 5 gal. 1 qt. 1 pt.

b. 7 pt. 9 fl. oz.
−2 pt. 14 fl. oz.

HOMEWORK EXERCISES

1. Change to fluid ounces: **a.** 3 pints **b.** 8 pints **c.** $4\frac{1}{2}$ pints

2. Change to pints: **a.** 64 fl. oz. **b.** 40 fl. oz. **c.** 108 fl. oz.

3. Change to pints: **a.** 3 quarts **b.** 10 quarts **c.** $1\frac{1}{2}$ quarts

4. Change to quarts: **a.** 4 pints **b.** 12 pints **c.** 27 pints

5. Copy and complete the table. Change each given quantity to four other equivalent quantities.

Gallons	Quarts	Pints	Cups	Fluid Ounces
$1\frac{1}{2}$				
	16			
		60		
			160	
				288

6. A batch of bread dough requires 4 cups of water. How many fluid ounces is this?

7. At lunch, a group of children drank 3 quarts of milk. How many pints of milk did they drink?

8. For the school fair, 4 gallons of ice cream were bought. If the ice cream was packed into 1-pint containers, how many containers were needed?

9. Which contains more, a 2-gallon picnic jug or a container that holds 260 fluid ounces?

10. If a jar contains 12 fluid ounces, how many cups do 16 of these jars hold?

11. Combine the measures as indicated.

a. 12 gal. 3 qt.
+5 gal. 2 qt.

b. 8 pt. 3 fl. oz.
−1 pt. 9 fl. oz.

12. A gourmet grocer had a total of 14 gal. 2 qt. 1 pt. of its special salad dressing. The grocer sold the following amounts: 3 gal. 3 qt., 1 qt. 1 pt., and 1 pt. How much salad dressing remained in the store?

13. Get Well Pharmacy pays $7.99 for a gallon of cough syrup. It repackages the syrup in 8-ounce bottles that sell for $1.39 each. How much profit does the pharmacy make?

SPIRAL REVIEW EXERCISES

1. Change each of the given weights to the indicated unit.

   a. 80 oz. = [?] lb. **b.** 10,000 lb. = [?] T.

   c. 9 lb. = [?] oz. **d.** 10 oz. = [?] lb.

   e. $3\frac{1}{4}$ T. = [?] lb. **f.** $\frac{1}{2}$ lb. = [?] oz.

2. A bowling alley is 60 feet long. What is its length in yards?

3. The distance from the floor to the ceiling is 8 feet. How many inches is this?

4. A truck weighs 11,000 lb. How many tons does it weigh?

5. The highest temperature recorded for each of the first 5 days of July was 81°, 83°, 84°, 85°, and 88°. What was the median high temperature for these days?

6. The cost of a pound of chopped meat at 3 different butcher shops is $2.20, $2.60, and $3.00. What is the average cost per pound?

7. Find the mode of this set of numbers.

   9, 9, 11, 14, 15, 15, 15, 17, 19

8. Ms. Fine, a real estate agent, earns a 7% commission on her home sales. How much does she earn on selling a home that costs $85,000?

9. Multiply $1\frac{2}{3}$ by 18.

10. A $225 television set is on sale at 25% off. If the rate of sales tax is 8%, the key sequence that can be used to calculate the total cost is

    (a) 225 [×] 25 [%]
    (b) 225 [×] 25 [%] [+] 225 [=]
    (c) 225 [+] 25 [%] [+] 8 [%]
    (d) 225 [−] 25 [%] [+] 8 [%]

11. Find the sum of 12.8, 17.05, 26, and 2.11.

12. Calculate: $2 \times 8 + 5 \times 3$

13. Subtract −7 from −9.

14. From the line graph below, estimate the number of sit-ups that Charlie did on Wednesday.

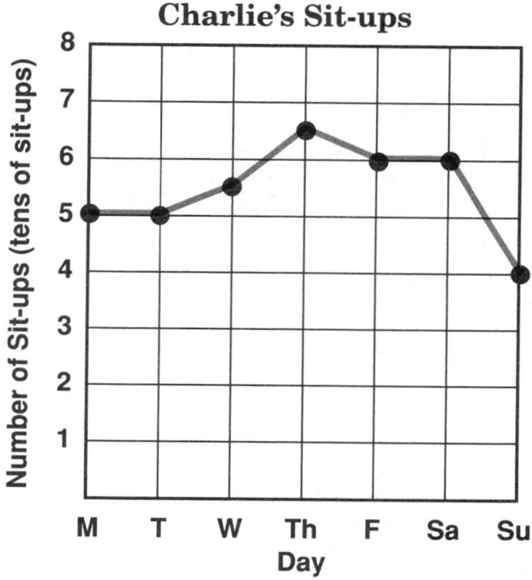

UNIT 11-5 Measuring Time; Time Zones

MEASURING TIME

THE MAIN IDEA

1. The *units of time* have the following relationships to each other:

   60 seconds (sec.) = 1 minute (min.)
   60 minutes = 1 hour (hr.)
   24 hours = 1 day (da.)

2. The time of day is divided into 2 parts:

   A.M. before noon
   P.M. after noon

3. To compute the *elapsed* time (time gone by) between two given times:
   a. If both times are A.M. or both times are P.M. in the same day, subtract the earlier time from the later time.
   b. If the first time is A.M. and the second is P.M. within the same day, subtract the A.M. time from 12:00 noon and add the result to the P.M. time.
   c. If the first time is P.M. and the second is A.M. within a 24-hour period, subtract the P.M. time from 12:00 midnight and add the result to the A.M. time.

EXAMPLE 1 How much time has elapsed between 3:15 P.M. and 7:40 P.M.?

Since both times are P.M., subtract directly.

$$\begin{array}{r} 7{:}40 \\ -3{:}15 \\ \hline 4{:}25 \end{array}$$

Answer: 4 hr. 25 min.

EXAMPLE 2 How much time has elapsed from the first clock to the second?

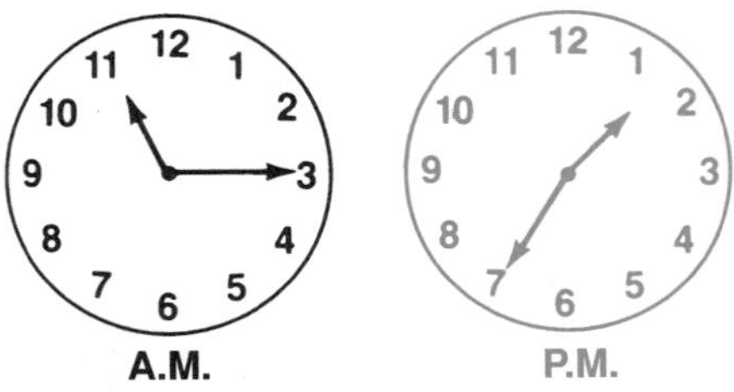

First, read the clocks to tell the times they show. (11:15 A.M. to 1:35 P.M.)

Subtract 11:15 from 12:00 noon.

$$\begin{array}{r} 12{:}00 \\ -11{:}15 \\ \hline \end{array}$$

To do the subtraction, rename 12:00 noon as 11:60.

$$\begin{array}{r} 11{:}60 \\ -11{:}15 \\ \hline 00{:}45 \end{array}$$

Add this result to the P.M. time.

$$\begin{array}{r} 1{:}35 \\ +00{:}45 \\ \hline 1{:}80 \end{array}$$

Rename 80 minutes as 1 hour 20 minutes.

1:80 = 1 hour + 1 hour 20 minutes
= 2 hours 20 minutes *Ans.*

EXAMPLE 3 Rose must get to work by 8:10. If her trip from home takes 35 minutes, by what time must she leave home in order to get to work on time?

Subtract 35 minutes from 8:10.

```
  8:10
 −:35
```

Change 1 hour to 60 minutes.

```
  7:70
 −:35
  7:35
```

Answer: She must leave by 7:35.

EXAMPLE 4 Combine the measures as indicated.

a.

```
  4 da. 17 hr. 25 min.
 +8 da. 19 hr. 45 min.
 12 da. 36 hr. 70 min.
```

Since the number of minutes > 60, rename.

= 12 da. 36 hr. + 1 hr. 10 min.
= 12 da. 37 hr. 10 min.

Since the number of hours > 24, rename.

= 12 da. + 1 da. 13 hr. 10 min.
= 13 da. 13 hr. 10 min. *Ans.*

b.

```
 15 hr. 20 min. 10 sec.
 −2 hr. 30 min. 15 sec.
```

We cannot subtract 15 sec. from 10 sec. Rename.

```
 15 hr. 19 min. 70 sec.
 −2 hr. 30 min. 15 sec.
```

We cannot subtract 30 min. from 19 min. Rename.

```
 14 hr. 79 min. 70 sec.
 −2 hr. 30 min. 15 sec.
 12 hr. 49 min. 55 sec.   Ans.
```

EXAMPLE 5 A commuter train going from Bay Village to Central City took 28 minutes in actual running time and also made five 4-minute local stops on the way. If the train left Bay Village at 7:03 A.M., at what time did it arrive in Central City?

THINKING ABOUT THE PROBLEM

Find the number of minutes needed for the entire trip; then add that total to the starting time.

```
  28 minutes    running time
 +20 minutes    five 4-minute local stops
  48 minutes
```

```
 7:03           starting time
 +48 minutes
 7:51           arrival time
```

Answer: The train arrived in Central City at 7:51 A.M.

CLASS EXERCISES

1. How much time elapsed between the first time and the second time?
   a. 5:00 A.M. to 10:25 A.M. **b.** 2:15 P.M. to 4:25 P.M. **c.** 3:45 A.M. to 5:15 A.M.
   d. 8:52 P.M. to 11:12 P.M. **e.** 3:37 A.M. to 5:00 A.M. **f.** 9:30 A.M. to 12:00 noon
   g. 10:30 A.M. to 1:15 P.M. **h.** 6:45 A.M. to 2:40 P.M. **i.** 10:15 A.M. to 3:45 P.M.
   j. 9:52 P.M. to 2:55 A.M. **k.** 10:45 A.M. to 12:00 midnight

2. How much time has elapsed from the first clock to the second?

a.

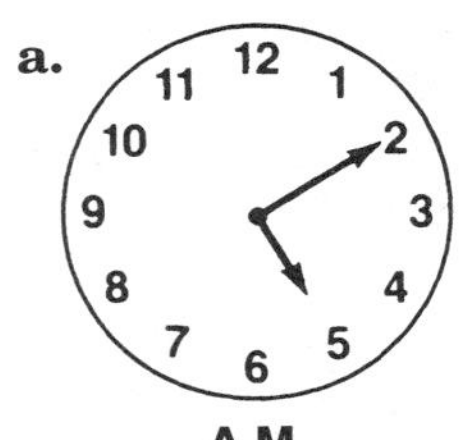

A.M.

A.M.

b.

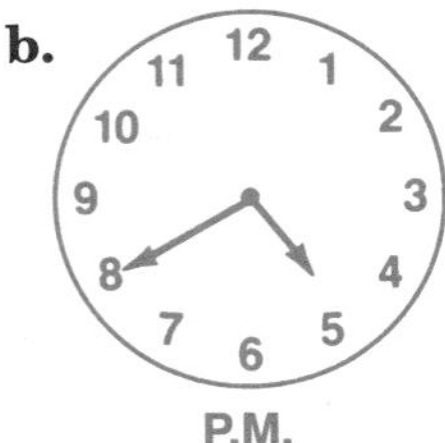

P.M.

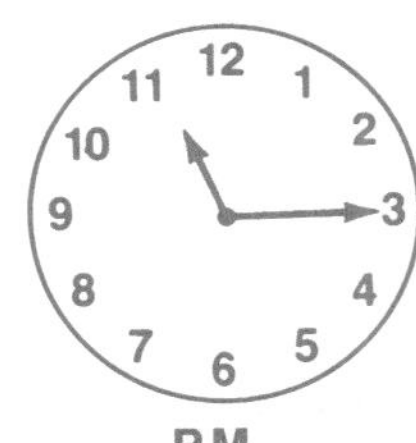

P.M.

c.

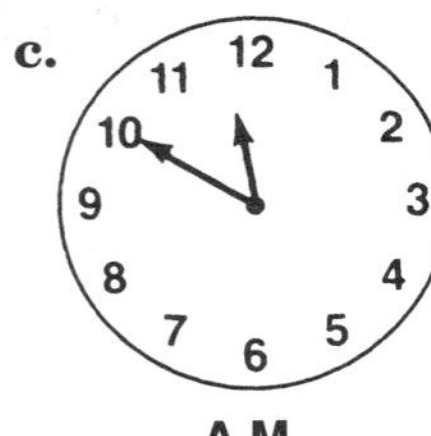

A.M.

P.M.

d.

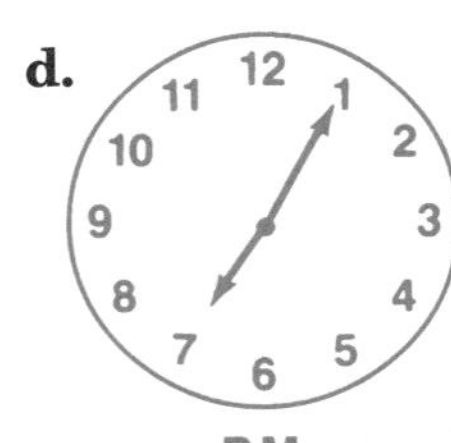

P.M.

A.M.

3. How much time elapsed between sunrise at 6:01 A.M. and sunset at 7:29 P.M.?

4. Mike slept from 11:17 P.M. to 6:15 A.M. How long did he sleep?

5. Combine the measures as indicated.

a.
```
  7 da. 19 hr. 30 min.
+1 da. 15 hr. 35 min.
```

b.
```
 17 hr. 10 min. 12 sec.
 −5 hr. 15 min. 26 sec.
```

6. Mr. Garcia worked 8 hours and 10 minutes on Monday and 7 hours and 35 minutes on Tuesday. What was the total amount of time that he worked?

7. A two-part standardized test was given to the junior class. The students had 1 hour and 45 minutes for Part *A* and 40 minutes for Part *B*. If the test was begun at 8:30 A.M., and there was a 10-minute break between the two parts, at what time did the test end?

8. Ken started doing his homework at 7:15 P.M. He spent 20 minutes on English, 15 minutes on Italian, 18 minutes on mathematics, and 35 minutes on science. If he worked without interruptions, at what time did he finish?
   (a) 7:59 P.M. (b) 8:27 P.M. (c) 8:43 P.M. (d) 9:23 P.M.

TIME ZONES

THE MAIN IDEA

1. The time of day is related to the position of the sun at a given location.
2. *Time zones* divide the world into regions that have the same time of day at a given moment.
3. Most of the United States is divided into four time zones, named in order from West to East: Pacific Time, Mountain Time, Central Time, Eastern Time
4. Starting at the West Coast, the consecutive time zones are each one hour later.

The map below shows four time zones in the United States. The clocks above each zone show the time in each zone at the same moment. For example, when it is 1:00 A.M. in the Pacific Time Zone, it is 4:00 A.M. in the Eastern Time Zone.

EXAMPLE 6 When it is 1:00 P.M. in Minnesota, what time of day is it in:

a. Wyoming **b.** Oregon **c.** Vermont

a. One time zone line is crossed in traveling West from Minnesota to Wyoming. Subtract 1 hour from 1:00 P.M.

Answer: It is 12:00 noon in Wyoming.

b. Oregon is two time zones west of Minnesota. Subtract 2 hours from 1:00 P.M.

Answer: It is 11:00 A.M. in Oregon.

c. Vermont is one time zone to the east of Minnesota. Add one hour to 1:00 P.M.

Answer: It is 2:00 P.M. in Vermont.

EXAMPLE 7 A plane left Los Angeles, California at 8:00 A.M. and landed in New York City $6\frac{1}{2}$ hours later. What was the time of day in New York City when it landed?

When the plane landed in New York, the time in Los Angeles was 8:00 A.M. + $6\frac{1}{2}$ hours, or 2:30 P.M.

Traveling East from Los Angeles to New York requires crossing 3 time zone lines. The time in New York City was 2:30 P.M. + 3 hours, or 5:30 P.M.

Answer: The plane landed at 5:30 P.M. in New York City.

CLASS EXERCISES

Exercises 1–3 refer to the time zone map on page 389.

1. Find the time in each of the given states when it is 10:00 A.M. in Utah.
   a. California **b.** South Dakota **c.** Florida
2. If it is 11:00 P.M. in the Eastern Time Zone, what time is it in the Mountain Time Zone?
3. A plane leaves Chicago, Illinois at 7:00 P.M. and travels 3 hours and 45 minutes to Seattle, Washington. What is the time of day in Seattle when it lands?

HOMEWORK EXERCISES

1. How much time elapsed between the first time and the second time?
   a. 5:10 P.M. to 7:35 P.M. **b.** 8:50 A.M. to 11:15 A.M. **c.** 7:47 P.M. to 10:22 P.M.
   d. 8:42 A.M. to 12:05 P.M. **e.** 2:42 P.M. to 4:50 P.M. **f.** 11:12 A.M. to 12:17 P.M.
   g. 10:39 A.M. to 2:15 P.M. **h.** 10:40 P.M. to 5:35 A.M. **i.** 12:00 noon to 2:35 A.M.

2. How much time elapsed from the first clock to the second?

a.

A.M.

A.M.

b.

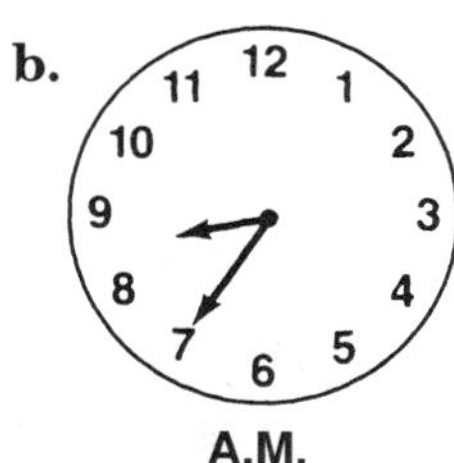

A.M.

P.M.

3. Mrs. Miller wants to set the timer on her stove to ring 25 minutes before 3:15 P.M. For what time should she set the timer?

4. The Wang family left their home at 5:50 P.M. and drove for 47 minutes to pay a visit. At what time did they arrive?

5. Harry left for work at 7:20 A.M. and Joan left an hour and 45 minutes later. At what time did Joan leave?

6. The #6 bus arrived at the bus terminal at 5:39 P.M. If the trip took 2 hours and 45 minutes, at what time did the bus start the trip?

7. Combine the measures as indicated.

a.
```
 15 da. 18 hr. 47 min.
+11 da. 10 hr. 20 min.
```

b.
```
 16 hr.  4 min. 1 sec.
 −5 hr. 10 min. 5 sec.
```

8. Peter practiced the trumpet for 1 hr. and 40 min. in the afternoon and 2 hr. and 20 min. in the evening.

   a. What was the total amount of time that Peter practiced?

   b. How much longer did Peter practice in the evening?

9. Robert worked for 3 hr. 20 min. on Monday, 2 hr. 45 min. on Tuesday, 4 hr. 50 min. on Thursday, and 1 hr. 35 min. on Friday. What are Robert's total earnings if his salary is $6.50 an hour?

10. Barbie needs 30 min. for morning jogging, 35 min. for washing and dressing, and 25 min. for breakfast. If she must leave for work at 7:40 A.M., for what time should she set her alarm?

11. The Moros left home at 6:05 A.M. and drove to a campsite, arriving at 3:50 P.M. Along the way, they made three 5-minute stops and one 20-minute stop. What was their actual travel time?

    (a) 9 hr. (b) 9 hr. 10 min. (c) 9 hr. 45 min. (d) 13 hr. 40 min.

Exercises 12–14 refer to the time zone map on page 389.

12. Find the time in each of the given states when it is 11:30 P.M. in Oklahoma.

    a. Maine **b.** California **c.** Pennsylvania **d.** Nevada

13. If it is 3:10 A.M. in the Central Time Zone, what time is it in the Pacific Time Zone?

14. A train leaves Boston, Massachusetts at 8:30 A.M. and travels 21 hours to Chicago, Illinois. What is the time of day in Chicago when the train arrives?

15. Explain why the lines that separate the time zones are irregular instead of straight lines.

16. At 4:00 P.M. in San Francisco, California, Mark called a computer company in Boston, Massachusetts to place an order. Explain why he was unable to reach a salesperson.

17. Generally, if two cities are in two neighboring time zones, the city to the west will have an earlier time of day than the city to the east, at the same moment. Find a place on the time zone map where this rule is reversed.

18. **a.** Write a signed number to represent the time change when a person travels from one time zone to another:

(1) From the Eastern Time Zone to the Central Time Zone

(2) From the Mountain Time Zone to the Eastern Time Zone

(3) From the Central Time Zone to the Pacific Time Zone

b. Write a rule for using signed numbers to find the time in one time zone, given the time in a different time zone.

1. Change each measure of volume to the indicated unit.

a. 160 fl. oz. = [?] quarts

b. 11 gallons = [?] quarts

2. How many ounces in 2 pounds?

3. What is the length in inches of a 9-foot rope?

4. A loaded cargo crate weighed 9,500 pounds. How many tons did it weigh?

5. A group of 850 students took a Spanish Placement Examination. Inez's score was the 92nd percentile. How many students scored below Inez on the exam?

6. Felipe purchased items for \$3.75, \$1.98, and \$1.57. How much change did he receive from a \$10 bill?

7. The product of 0.02 and 50 is

(a) 0.1 (b) 1.0 (c) 10 (d) 50.02

8. The quotient of 25 divided by 0.05 is

(a) 5 (b) 50 (c) 500 (d) 5,000

9. A poll asked 1,000 people how often they ate salad. Use the graph to tell how many people ate salad at least once a week.

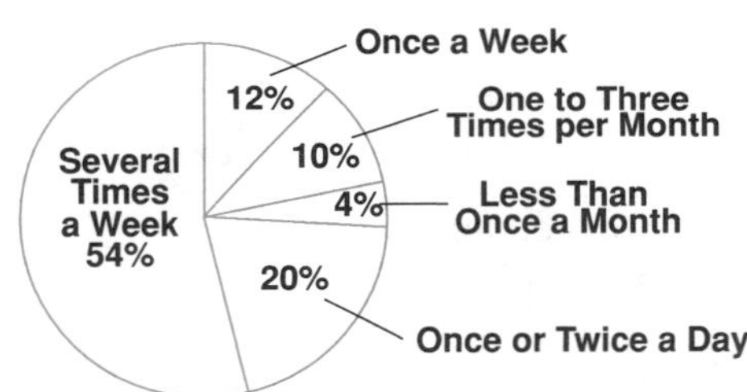

10. Mr. Blank had \$2,475 in his savings account and made a withdrawal of \$298. What is the new balance?

11. The mean of five numbers is 27.5. Four of the numbers are 30, 18, 23, and 32. Write a key sequence to find the fifth number.

12. The selling price of a \$30 sweater that is sold at a 15% discount is

(a) \$34.50 (b) \$26.50

(c) \$15.00 (d) \$25.50

13. 0.35 equals (a) $\frac{35}{1,000}$ (b) $\frac{7}{20}$ (c) $\frac{1}{4}$ (d) $\frac{2}{5}$

MODULE 12

UNIT 12-1 Metric Measures of Length

THE MAIN IDEA

1. The most common *metric units of length* are the ***millimeter*** (mm), the ***centimeter*** (cm), the ***meter*** (m), and the ***kilometer*** (km).

   A millimeter is about the thickness of a dime.

   A centimeter is about the width of a telephone push button.

   A meter is about the length of a baseball bat.

   A kilometer could be the distance between two towns.

2. In metric measurement, the *prefix* in the name of a unit tells you how that unit compares to a basic unit. Some of the most common prefixes are:

Prefix	*Meaning*	*Name of Unit*	*Relationship to a Meter*
milli	$\frac{1}{1,000}$	*milli*meter	$\frac{1}{1,000}$ of a meter
centi	$\frac{1}{100}$	*centi*meter	$\frac{1}{100}$ of a meter
kilo	1,000	*kilo*meter	1,000 meters

3. The customary and metric units of length compare in the following way:

   A meter is slightly longer than a yard. 1 m = 1.1 yd.

   A centimeter is slightly less than half an inch. 1 cm = 0.4 in.

   A kilometer is slightly more than half a mile. 1 km = 0.6 mi.

EXAMPLE 1 Which would be the most convenient metric unit of length to use in measuring each item? What would be the corresponding customary unit of length?

	Item to be Measured	*The Most Convenient Metric Unit of Length*	*The Corresponding Customary Unit*
a.	a length of carpeting	the meter	the yard
b.	the length of an auto trip	the kilometer	the mile
c.	the length of a page	the centimeter	the inch
d.	the thickness of a coin	the millimeter	the inch

EXAMPLE 2 Which is the most likely length of a home television screen?
(a) 40 mm (b) 40 cm (c) 40 m (d) 40 km

Choice (b). Since a centimeter is roughly the thickness of a finger, 40 of these units would be a reasonable size for a television screen. 40 mm would be only about the height of a stack of 40 dimes—too small for a television screen. 40 m would be longer than most rooms in a house, and 40 km could be the distance between two cities.

EXAMPLE 3 Make each of the required unit changes.

	Required Unit Change	***Conversion Information***	***Arithmetic***
a.	Change 47 millimeters to centimeters.	To change from the smaller unit (millimeters) to the larger unit (centimeters), divide by 10. The easy way to divide by 10 is to move the decimal point one place to the left.	$47 \div 10 = 4.7$
	Answer: 47 millimeters = 4.7 centimeters		
b.	Change 11.3 centimeters to millimeters.	To change from the larger unit (centimeters) to the smaller unit (millimeters), multiply by 10 (move the decimal point one place to the right).	$11.3 \times 10 = 113$
	Answer: 11.3 centimeters = 113 millimeters		
c.	Change 480 meters to kilometers.	To change from the smaller unit (meters) to the larger unit (kilometers), divide by 1,000 (move the decimal point three places to the left).	$480 \div 1{,}000 = 0.48$
	Answer: 480 meters = 0.48 kilometer		

CLASS EXERCISES

1. Which would be the most convenient metric unit of length to use in measuring each item? What would be the corresponding customary unit of length?

a. the length of window curtains **b.** the distance from the earth to the moon
c. the length of a swimming pool **d.** the thickness of a piece of sheet metal
e. the height of a tree **f.** the length of a train track **g.** the width of a tooth
h. the width of a piece of movie film **i.** the width of a record album cover

2. The most likely height of a basketball player is: (a) 2 mm (b) 2 cm (c) 2 m (d) 2 km

3. The most likely length of a playing card is: (a) 85 mm (b) 85 cm (c) 85 m (d) 85 km

4. Change to centimeters: **a.** 68 mm **b.** 10 mm **c.** 300 mm

5. Change to millimeters: **a.** 8 cm **b.** 48.5 cm **c.** 13 cm

6. Change each measurement to an equivalent measurement, using the given units.
a. 32 cm to m **b.** 7.2 m to cm **c.** 2,420 m to km **d.** 9.6 km to m

7. Which measure is not equivalent to the others?
(a) 5,400 mm (b) 54 cm (c) 5.4 m (d) 0.0054 km

8. Tom's locker is 2 m tall and Kit's locker is 210 cm tall. Which locker is taller?

HOMEWORK EXERCISES

1. Which would be the most convenient metric unit of length to use in measuring each item? What would be the corresponding customary unit of length?
a. the thickness of a magazine **b.** the width of a doorway **c.** the height of a house
d. the length of a garden hose **e.** the length of a dollar bill **f.** a person's neck size
g. the space between two teeth **h.** the width of a hair ribbon
i. the distance from Los Angeles to San Francisco **j.** the length of the equator

2. Choose the most likely measurement.

a. the length of a person's arm:	(a) 1 mm	(b) 1 cm	(c) 1 m	(d) 1 km
b. the diameter of a nickel:	(a) 20 mm	(b) 20 cm	(c) 20 m	(d) 20 km
c. the thickness of a nickel:	(a) 0.2 mm	(b) 0.2 cm	(c) 0.2 m	(d) 0.2 km
d. the height of the Empire State Building:	(a) 0.466 mm	(b) 0.446 cm	(c) 0.446 m	(d) 0.446 km

3. Change to centimeters:
a. 54 mm **b.** 1,482 mm **c.** 97 mm

4. Change to millimeters:
a. 37 cm **b.** 2.05 cm **c.** 0.03 cm

5. Change each measurement to an equivalent measurement using the given units.
a. 620 cm to m **b.** 8.04 m to cm **c.** 6,540 m to km **d.** 0.48 km to m

6. Which measure is not equivalent to the others?
(a) 0.6 m (b) 60 cm (c) 0.006 mm (d) 0.0006 km

7. What is the difference in the lengths of two metal bars that are 2.1 m and 2.35 cm long?

8. On a bar graph, each millimeter in the height of a bar represents $50. What is the amount of money represented by a bar that is 3.6 cm high?

9. Combine the measures as indicated.

a.
```
  15 km 600 m 30 cm
 +50 km 800 m 90 cm
```

b.
```
  40 m 10 cm 1 mm
 −30 m 20 cm 5 mm
```

SPIRAL REVIEW EXERCISES

1. Add:
```
 5 hr. 18 min.
 2 hr. 47 min.
```

2. Subtract:
```
 8 hr. 21 min.
 3 hr. 38 min.
```

3. Philip left for Chicago at 7:35 A.M. and arrived on the same day at 2:05 P.M. (same time zone). How long was the trip?

4. The data for the heights of 72 children showed that 48 inches was the 45th percentile. How many children in the group were shorter than 48 inches?

5. The graph shows the average number of days of vacation given to employees in several countries after a year of service. The mean of these numbers is
(a) 10 (b) 20 (c) 22.2 (d) 24

Vacation Days after 1 Year of Employment

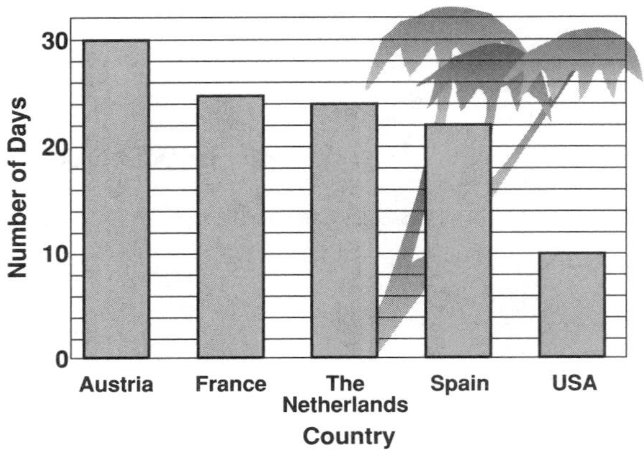

6. Last year, Mark's height was 50 inches. If his height increased by 10%, what is his new height?

7. 50 is 20% of what number?

8. $\frac{9}{12}$ written as a decimal is
(a) 0.25 (b) 0.50 (c) $0.\overline{6}$ (d) 0.75

9. Helen purchased items totaling $7.83. The amount of change that she should get from a $20 bill is (a) $2.17 (b) 12.17 (c) $3.17 (d) $13.17

10. The product of 0.9 and 0.07 is
(a) 6.3 (b) 0.63 (c) 0.063 (d) 0.97

11. Which has the greatest value?
(a) $\frac{1}{4}$ (b) $\frac{2}{5}$ (c) .35 (d) 30%

12. Use your calculator to determine which statement is true.
(a) $\frac{15}{23} < \frac{13}{21}$ (b) $\frac{15}{23} = \frac{13}{21}$
(c) $\frac{15}{23} > \frac{13}{21}$ (d) $\frac{13}{21} > \frac{15}{23}$

13. Use a signed number to show:
a. a $50 deposit **b.** 10° below zero

UNIT 12–2 Using a Ruler: The Inch Scale and The Metric Scale

THE INCH SCALE

THE MAIN IDEA

1. On a ruler, an inch scale is marked in the following way:

   The inches are numbered next to the longest marks.

   The half inches are shown by the second longest marks, which divide each inch into two equal parts.

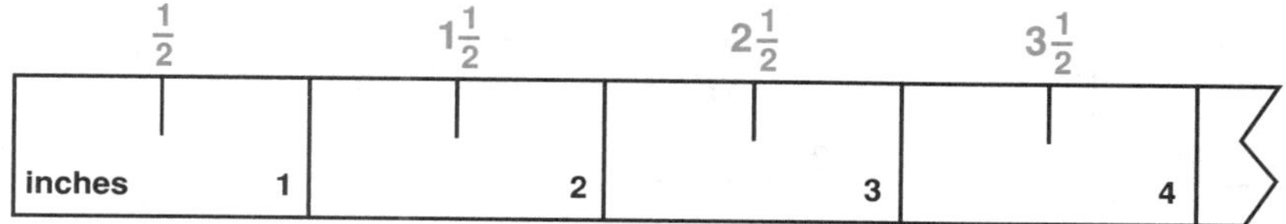

   Each inch is broken into a larger number of equal parts by marks that decrease in length.

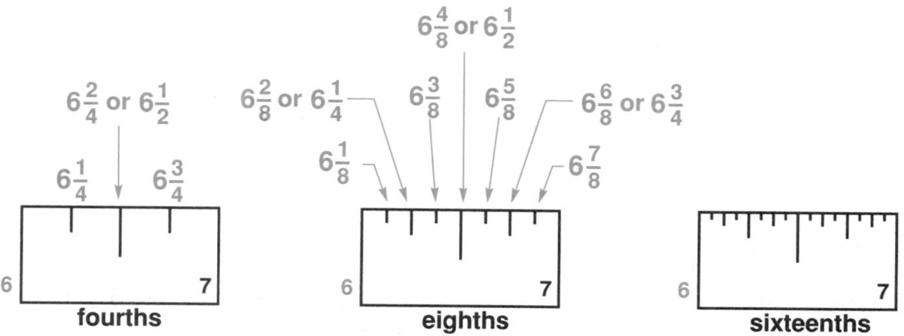

2. To measure a line segment with the inch scale on a ruler, line up one end of the line segment with the beginning of the ruler. Then read:
   a. the number of the inch just to the left of the other end of the line segment, and
   b. the fraction mark closest to that end of the line segment.

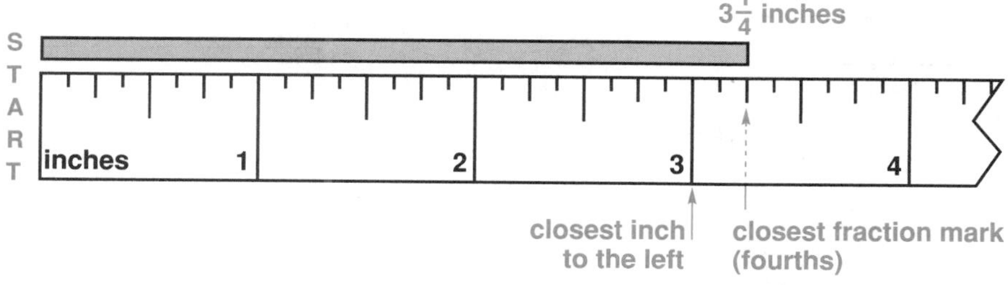

EXAMPLE 1 Tell how the inch is divided.

a.

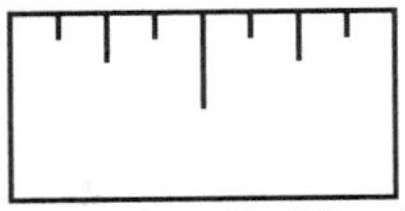

Since the marks divide the inch into 8 equal parts, they show eighths.

b.

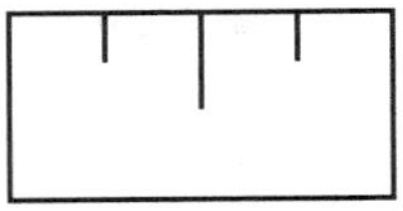

Since the marks divide the inch into 4 equal parts, they show fourths.

EXAMPLE 2 Draw a line segment $4\frac{3}{4}$ inches long.

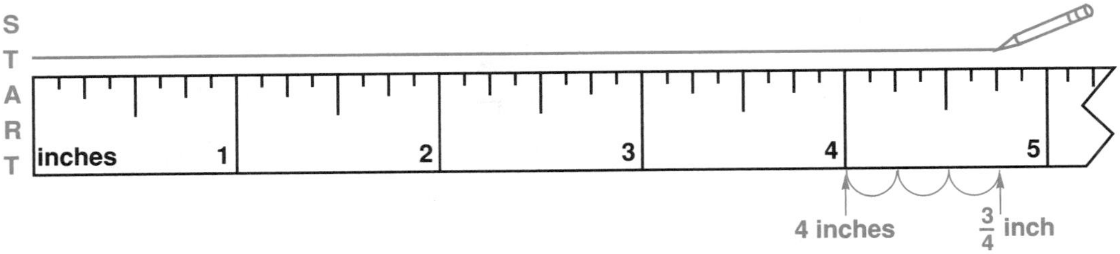

START

inches 1 2 3 4 5

4 inches $\frac{3}{4}$ inch

CLASS EXERCISES

1. Tell how each inch is divided. **a.** **b.** 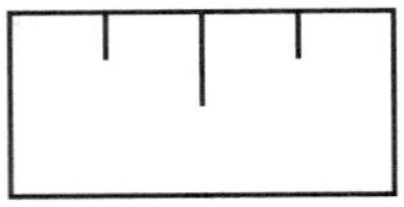 **c.**

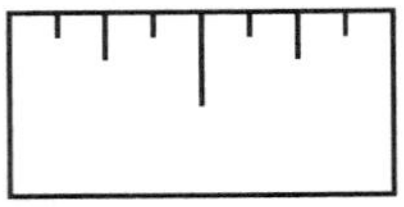

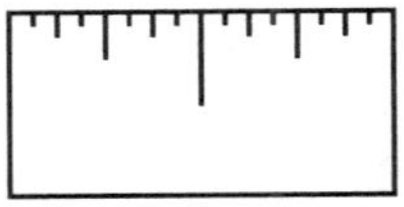

2. Draw a line segment having each length.

a. 5 inches **b.** $5\frac{1}{4}$ inches **c.** $4\frac{1}{2}$ inches **d.** $3\frac{3}{4}$ inches **e.** $5\frac{1}{8}$ inches

f. $3\frac{5}{8}$ inches **g.** $5\frac{6}{8}$ inches **h.** $5\frac{3}{4}$ inches **i.** $4\frac{7}{8}$ inches

3. Measure the length, in inches.

a.

b.

c.

d.

e.

f. **g.**

h. **i.**

4. The figure shown is to be made of decorative wire that costs $2.50 per inch. Measure to find how much the necessary wire will cost.

THE METRIC SCALE

THE MAIN IDEA

1. On a ruler, a metric scale is marked in the following way:

   The centimeters are numbered next to the longest marks.

   Each centimeter is divided into 10 equal parts, 10 millimeters. The second longest marks show 5-millimeter intervals (half centimeters).

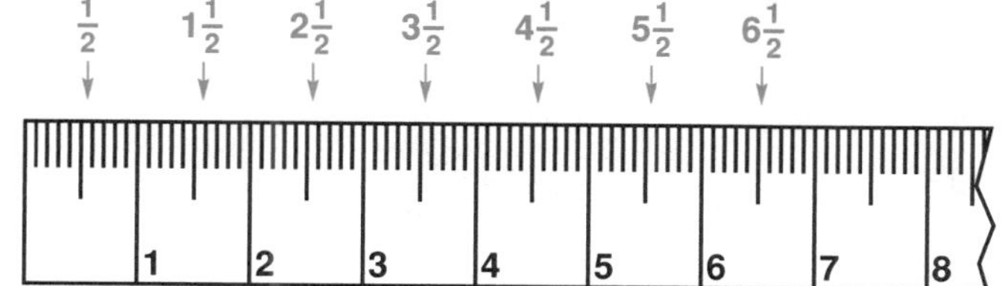

2. To measure a line segment with the metric scale on a ruler, line up one end of the line segment with the beginning of the ruler. Then read:
   a. the number of the centimeter just to the left of the other end of the line segment, and
   b. the millimeter mark closest to that end of the line segment.

   Since 1 mm = $\frac{1}{10}$ cm, the millimeters may be written as a decimal.

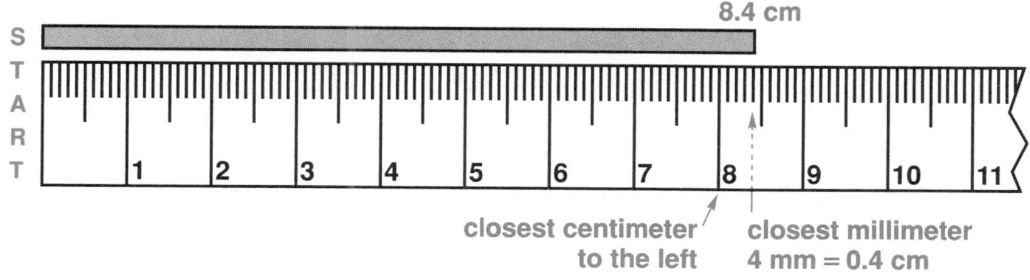

EXAMPLE 3 Write the number of millimeters shown as a decimal fraction of a centimeter.

8 mm = 0.8 cm

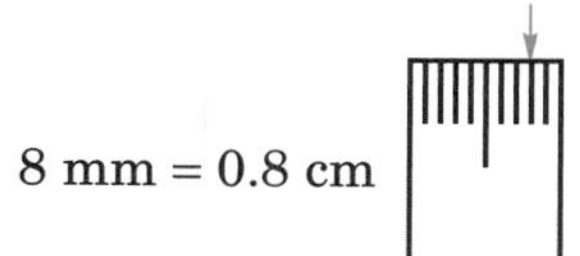

EXAMPLE 4 Draw a line segment 8.6 cm long.

Since 0.6 cm = 6 mm, count 6 mm past 8 cm.

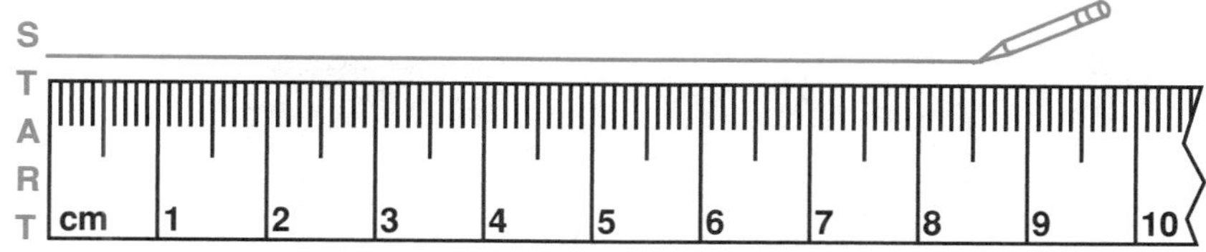

CLASS EXERCISES

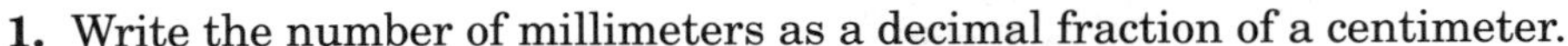

1. Write the number of millimeters as a decimal fraction of a centimeter.

a. 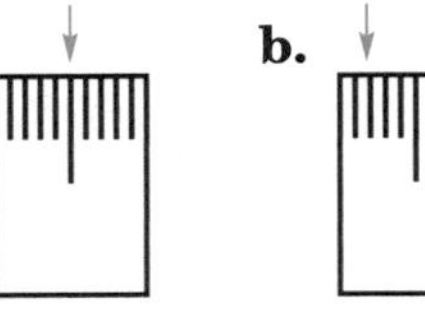 b. 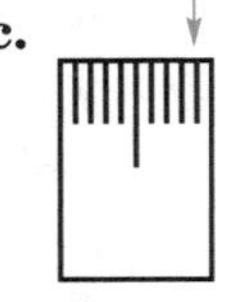 c. d. 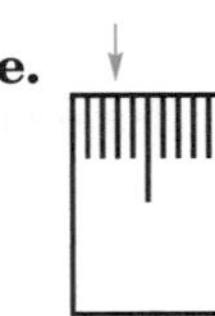 e.

2. Draw a line segment having each length.

   a. 4.5 cm b. 3.2 cm c. 6.8 cm d. 10.1 cm e. 11 mm
   f. 0.7 cm g. 9 mm h. 9.9 cm i. 6.0 cm

3. Measure the length, in centimeters, of each line segment.

   a. ______ b. ______
   c. ______ d. ______ e. ______
   f. ______ g. ______
   h. ______ i. ______

HOMEWORK EXERCISES

1. Tell how each inch is divided.

   a. b. c. d.

2. Draw a line segment having each length.

   a. 4 inches b. $4\frac{1}{4}$ inches c. $6\frac{1}{2}$ inches d. $4\frac{3}{8}$ inches e. $2\frac{5}{8}$ inches

3. Measure the length of each line segment in inches.

   a. ______
   b. ______
   c. ______
   d. ______
   e. ______
   f. ______ g. ______
   h. ______
   i. ______ j. ______

4. Write the numbers of millimeters as a decimal fraction of a centimeter.

a. **b.** **c.** **d.** **e.**

5. Draw a line segment having each length.

a. 3.8 cm	**b.** 5.2 cm	**c.** 1.4 cm	**d.** 4.1 cm	**e.** 1.9 cm
f. 8 mm	**g.** 0.2 cm	**h.** 8.8 cm	**i.** 3.0 cm	**j.** 15 mm

6. Measure the length of each line segment in centimeters.

a.

b. **c.**

d.

e. **f.**

g.

h. **i.**

j.

7. On the map below, each centimeter stands for 400 miles. What is the length, in miles, of the airplane flight from New York to Chicago to Los Angeles, as shown on the map?

8. On this road map, each centimeter stands for 15 miles. If Ed's car can go 21 miles on a gallon of gasoline, which is the closest estimate of the number of gallons of gasoline he needs to drive from Lowville to Hightown?

(a) $4\frac{1}{2}$ gallons (b) 5 gallons

(c) $3\frac{3}{4}$ gallons (d) $2\frac{1}{2}$ gallons

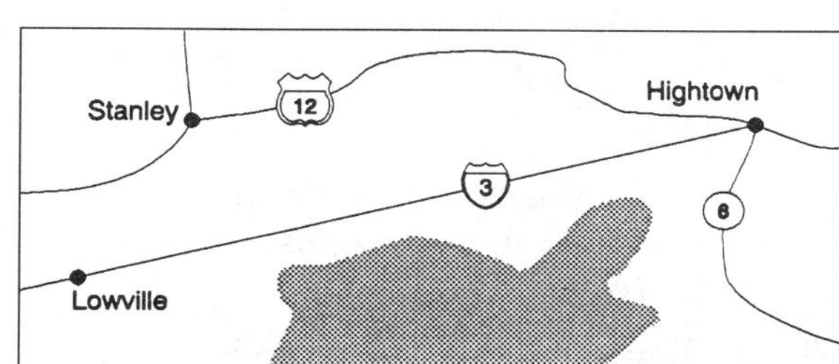

9. In this drawing of a field, each centimeter represents 3 meters. If fencing costs \$21 per meter, what would be the total cost of fencing the four sides of the field?

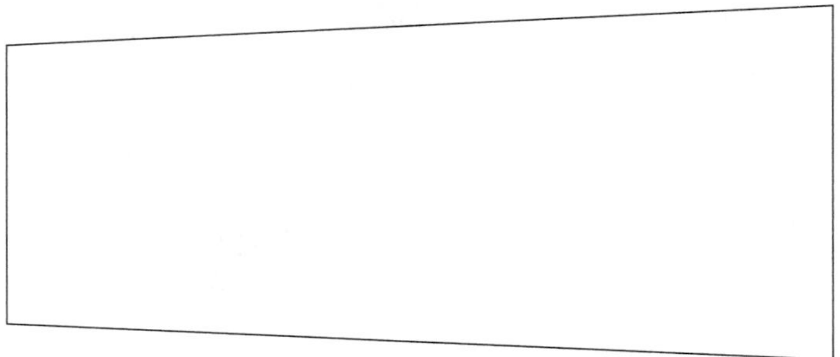

SPIRAL REVIEW EXERCISES

1. Change each given measure of length to the indicated unit.

   a. 11 km = ? m **b.** 0.6 cm = ? mm

   c. 1.8 m = ? cm **d.** 250 cm = ? m

2. The width of a loose-leaf album cover is closest to

   (a) 30 mm (b) 30 cm
   (c) 3 m (d) 30 m

3. The length of a football field is closest to

   (a) 92 mm (b) 92 cm
   (c) 92 m (d) 92 km

4. Combine the measures as indicated.

   a. 4 yd. 2 ft. 10 in.
   +6 yd. 1 ft. 8 in.

   b. 7 gal. 1 qt.
   −4 gal. 2 qt. 1 pt.

5. If it takes Mary 25 minutes to get to school, at what time should she leave in order to get there at 8:10 A.M.?

6. Mr. Johnson earns \$12.80 an hour. One week, he worked 40 hours and 2 hours overtime. If he is paid time and one-half for overtime, how much did he earn?

7. Ms. Byron invested \$2,000 at 6% annual interest. How much interest did she earn the first year?

8. Use your calculator to help you find the LCM (least common multiple) of 8, 12, and 18.

9. Round 1,475,892 to the nearest ten thousand.

10. Add: $\frac{3}{8} + \frac{3}{4}$

11. Subtract 19.48 from 28.07.

12. $\frac{23}{4}$ written as a mixed number is

    (a) $4\frac{3}{4}$ (b) $5\frac{1}{4}$ (c) $5\frac{3}{4}$ (d) $7\frac{1}{2}$

13. Divide 0.27 by 1,000.

14. 16 is 20% of what number?

15. The greatest common factor of 24 and 48 is

    (a) 6 (b) 12 (c) 24 (d) 48

UNIT 12–3 Metric Measures of Weight

THE MAIN IDEA

1. The most common *metric units of weight* are the ***milligram*** (mg), the ***gram*** (g), the ***kilogram*** (kg), and the ***metric ton*** (t).

   A milligram is about the weight of a few grains of salt.

   A gram is about the weight of a small paperclip.

   A kilogram could be the weight of a dozen tomatoes.

   A metric ton might be the weight of a small car.

2. The prefixes in the names of the most common metric units of weight have the following meanings:

Prefix	*Meaning*	*Name of Unit*	*Relationship to a Gram*
milli	$\frac{1}{1{,}000}$	*milli*gram	$\frac{1}{1{,}000}$ of a gram
kilo	1,000	*kilo*gram	1,000 grams

3. A metric ton is 1,000 times a kilogram. (A metric ton is not identified by a prefix.)

4. The customary and metric units of weight compare in the following way:

A kilogram is slightly more than two pounds.	A metric ton is slightly more than a customary ton.	A gram is a very small part of an ounce.
1 kg = 2.2 lb.	1 t = 1.1 T.	1 g = $\frac{1}{28}$ oz.

EXAMPLE 1 Which would be the most convenient metric unit of weight to use in measuring each item? What would be the corresponding customary unit of weight?

	Item to be Measured	*The Most Convenient Metric Unit of Weight*	*The Corresponding Customary Unit*
a.	the weight of a bicycle	the kilogram	the pound
b.	the weight of a wristwatch	the gram	the ounce
c.	the amount of fat in a breadstick	the milligram	the ounce
d.	the weight of a truckload of sand	the metric ton	the ton

EXAMPLE 2 Make each of the required unit changes.

	Required Unit Change	*Conversion Information*	*Arithmetic*
a.	Change 1,450 milligrams to grams.	To change from the smaller unit (milligrams) to the larger unit (grams), divide by 1,000 (move the decimal point 3 places left).	$1{,}450 \div 1{,}000 = 1.45$
	Answer: 1,450 milligrams = 1.45 grams		
b.	Change 3.2 grams to milligrams.	To change from the larger unit (grams) to the smaller unit (milligrams), multiply by 1,000 (move the decimal point three places to the right).	$3.2 \times 1{,}000 = 3{,}200$
	Answer: 3.2 grams = 3,200 milligrams		
c.	Change 3,200 grams to kilograms.	To change from the smaller unit (grams) to the larger unit (kilograms), divide by 1,000.	$3{,}200 \div 1{,}000 = 3.2$
	Answer: 3,200 grams = 3.2 kilograms		
d.	Change 1.75 metric tons to kilograms.	To change from the larger unit (metric tons) to the smaller unit (kilograms), multiply by 1,000.	$1.75 \times 1{,}000 = 1{,}750$
	Answer: 1.75 metric tons = 1,750 kilograms		

CLASS EXERCISES

1. To weigh each item, which metric unit is most convenient? Which customary unit?

a. the weight of a person **b.** the weight of a truck **c.** the weight of a suitcase
d. the weight of 2 slices of bread **e.** the weight of a truckload of bricks
f. the amount of Vitamin C in an orange **g.** the weight of a heavy gold ring

2. How many grams are in each?
a. 3,200 mg **b.** 7,050 mg **c.** 850 mg

3. How many milligrams are in each?
a. 5 g **b.** 7.4 g **c.** 13.64 g

4. Change to kilograms: **a.** 4,800 g **b.** 5,675 g **c.** 80 g

5. Change to grams: **a.** 8.9 kg **b.** 502.7 kg **c.** 0.002 kg

6. Change to metric tons: a. 10,500 kg b. 650 kg c. 91.8 kg
7. Change to kilograms: a. 3 t b. 4.6 t c. 0.75 t
8. 5.6 g is equivalent to (a) 5,600 mg (b) .056 kg (c) 560 mg (d) 5.6 kg
9. Mrs. Smith's diamond earrings weigh 3.6 grams. How many milligrams do they weigh?
10. A truck dumped 3.6 metric tons of concrete. How many kilograms of concrete was this?
11. A loaf of bread contains 28,000 mg of protein. What is this weight in grams?
12. Which measure is unequal to the others? (a) 3,500 mg (b) 3.5 kg (c) 3,500 g (d) 0.0035 t

HOMEWORK EXERCISES

1. To weigh each item, which metric unit is most convenient? Which customary unit?
   a. the weight of a television set
   b. the weight of a pen
   c. the weight of a postage stamp
   d. the weight of a full cement mixer
   e. the weight of a deck of playing cards
   f. the amount of iron in a vitamin pill
   g. a baby's weight
   h. the weight of a small insect
2. How many grams are in each?
   a. 4,800 mg b. 998 mg c. 34 mg
3. How many milligrams are in each?
   a. 4.1 g b. 29.02 g c. 0.007 g
4. Change to kilograms: a. 6,720 g b. 205 g c. 24 g
5. Change to grams: a. 12 kg b. 3.01 kg c. 0.5 kg
6. Change to metric tons: a. 12,000 kg b. 725 kg c. 683.4 kg
7. Change to kilograms: a. 7 t b. 8.3 t c. 0.403 t
8. 22.2 g equals (a) 2.22 mg (b) 222 mg (c) 22.2 mg (d) 0.0222 kg
9. A loaf of bread contained 5 grams of fat. How many kilograms of fat were in the loaf?
10. Mr. James used 2.4 metric tons of sand to mix concrete. How many kilograms did he use?
11. Which of the following measures is not equivalent to the others?
    (a) 2,750 mg (b) 2.75 g (c) 0.0275 kg (d) 0.00275 kg
12. The Ace Refrigerated Truck Company charges the following rates to ship dairy products:
    $1.95 for the first 50 kg $0.35 for each additional 50 kg or portion thereof
    Find the total shipping cost for 450 kg of cottage cheese, 280 kg of butter, and 610 kg of milk.

13. Mr. Samuels must limit his sodium intake to 1,000 mg a day. His breakfast contained 350 mg of sodium. For lunch he had 6 ounces of shrimp, 1 cup of broccoli, and 2 tomatoes. Use the table to find the number of milligrams of sodium he can consume for dinner without going over the 1,000-milligram limit.

Food	Sodium Content
3 oz. shrimp	137 mg
1 cup broccoli	35 mg
1 raw carrot	34 mg
1 tomato	14 mg
1 tbsp. ketchup	156 mg
4 oz. cottage cheese (unsalted)	14 mg

SPIRAL REVIEW EXERCISES

1. Change each given measure of length to the indicated unit.

   a. 30 cm = [?] mm **b.** 5 m = [?] cm
   c. 720 mm = [?] cm **d.** 8 km = [?] m

2. Find the measure of each line segment, using the indicated unit of measure.

   a. ____________
   centimeters

   b. ______________________________________
   inches

3. In 1945, the United Nations had 51 member countries. By 1999, the number of member countries increased to 188. Admission of new members must be approved by at least $\frac{2}{3}$ of the member nations. How many more votes were needed to admit a new member in 1999 than in 1945?

4. Add: 5 hr. 29 min.
   2 hr. 37 min.

5. What time is it 2 hours and 15 minutes after 11:15 A.M.?

6. If there is a 5% sales tax, what will be the total cost of a $38 radio?

7. Evaluate: $12 + 18 \times 5 \div 6$

8. Philip got 18 hits in 40 times at bat. What percent of his times at bat did he get a hit?

9. To the product of −7 and −3, add −6.

10. Use your calculator to help you decide whether 101 is prime or composite.

11. The line graph shows the average price of self-serve regular unleaded gasoline over a five-month period. What is the range of the data?

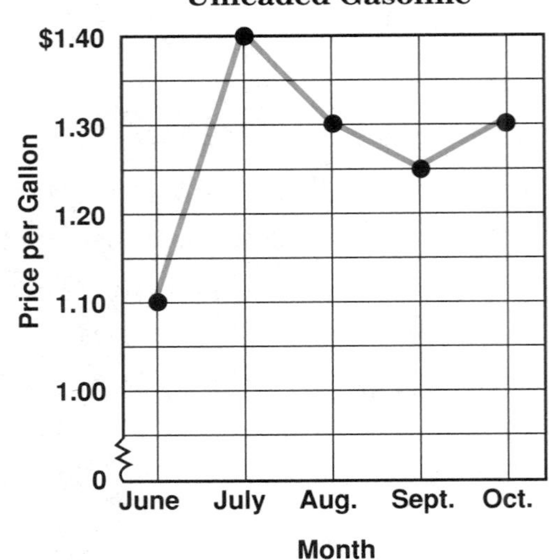

UNIT 12-4 Metric Measures of Volume

THE MAIN IDEA

1. The most common *metric units of volume* are the ***milliliter*** (mL), the ***liter*** (L), and the ***kiloliter*** (kL).

   A milliliter is about the size of a small sugar cube. It is the same as a cubic centimeter, which is the volume contained in a small box all of whose sides are 1 cm long.

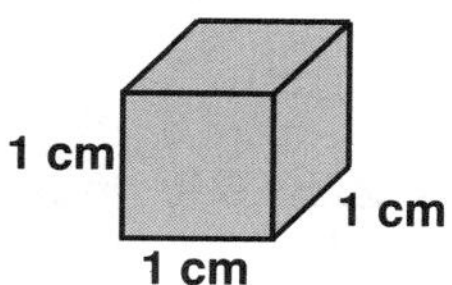

   A liter is just a little larger than a one-quart milk container.

   A kiloliter could be the size of a large carton. It is the same as a cubic meter, which is the volume contained in a box all of whose sides are 1 m long.

2. The prefixes in the names of the most common metric units of volume have the following meanings:

Prefix	*Meaning*	*Name of Unit*	*Relationship to a Liter*
milli	$\frac{1}{1,000}$	*milli*liter	$\frac{1}{1,000}$ of a liter
kilo	1,000	*kilo*liter	1,000 liters

3. The customary and metric units of volume compare in the following way:
   A liter is slightly more than a quart.

$$1 \text{ L} = 1.1 \text{ qt.}$$

EXAMPLE 1 Name the metric unit of volume that would be most convenient to use in measuring each item.

Item to be Measured	*The Most Convenient Metric Unit of Volume*
a. the capacity of a bathtub	the kiloliter
b. the amount of juice in a large picnic cooler	the liter
c. the amount of medicine in a bottle	the milliliter

EXAMPLE 2 Make the required unit change.

Required Unit Change	***Conversion Information***	***Arithmetic***
Change 7,520 milliliters to liters.	To change from the smaller unit (milliliters) to the larger unit (liters), divide by 1,000 (move the decimal point 3 places left).	$7{,}520 \div 1{,}000 = 7.52$

Answer: 7,520 milliliters = 7.52 liters

CLASS EXERCISES

1. Name the metric unit of volume that would be most convenient to use in measuring each item.
  a. the amount of cider in a large jug **b.** the capacity of a refrigerator
  c. the capacity of a car trunk **d.** the volume of a room

2. Change to liters: **a.** 3,450 mL **b.** 785 mL **c.** 124.3 mL

3. Change to milliliters: **a.** 9 L **b.** 32.5 L **c.** 0.0385 L

4. Find the number of kiloliters in: **a.** 8,000 L **b.** 660 L **c.** 8.25 L

5. Find the number of liters in: **a.** 17 kL **b.** 8.5 kL **c.** 0.95 kL

6. The number of cubic centimeters in 5 milliliters is (a) 0.05 (b) 0.5 (c) 5 (d) 50

7. Dr. Gomez gave Anita a 2-cubic-centimeter injection of vaccine. How many milliliters was it?

8. Joe and Barbara each drank a half liter of juice with lunch. How many milliliters altogether?

9. How many cubic meters of sand are there in 750 liters?

10. Mr. Gallo bought 4 liters of gasoline. How many milliliters was this?

11. Which measure is unequal to the others? (a) 85 L (b) 8,500 mL (c) 85,000 mL (d) 0.085 kL

HOMEWORK EXERCISES

1. Name the metric unit of volume that would be most convenient to use in measuring each item.
  a. the contents of a tea bag **b.** the contents of a spoon
  c. the contents of a swimming pool **d.** the contents of an automobile gas tank

2. Change to liters: **a.** 2,005 mL **b.** 583 mL **c.** 326.25 mL
3. Change to milliliters: **a.** 20 L **b.** 49.4 L **c.** 0.0036 L
4. Find the number of kiloliters in: **a.** 7,750 L **b.** 11 L **c.** 95.4 L
5. Find the number of liters in: **a.** 48 kL **b.** 1.04 kL **c.** 0.008 kL
6. How many cubic meters are there in 500 liters? (a) 0.5 (b) 5 (c) 50 (d) 500
7. Rosalie's bottle of perfume contains 3.5 cubic centimeters. How many milliliters is that?
8. At the track team's open house reception, the guests drank $8\frac{1}{2}$ liters of bottled water. How many milliliters was this?
9. A department store put a 250-mL bottle of J'Aime perfume on the counter for customers to sample. At the end of the day, there were 75 mL left in the bottle. If the perfume is worth $0.22 per milliliter, what was the value of the perfume that was used for samples?
10. Five liters of juice concentrate are mixed with 8.5 liters of water to make a juice drink that sells for $1.10 a liter. What is the total value of the resulting juice drink?

SPIRAL REVIEW EXERCISES

1. Change each of the given measures of weight to the indicated unit.
   a. 540 g = [?] kg **b.** 72 g = [?] mg
   c. 4,000 g = [?] t **d.** 25 mg = [?] g
2. A cake weighs 1.5 kilograms. What is its weight in grams?
3. If it costs $1.20 to ship one kilogram of freight, what is the cost of shipping a package weighing 3.5 kilograms?
4. If a bus ride takes 3 hr. 35 min. and the bus arrives at its destination at 8:15 P.M., at what time did it start out?
5. Subtract $3\frac{5}{8}$ from $6\frac{1}{2}$.
6. Find 250% of 64.
7. Round 241.7 to the nearest ten.
8. Carol bowled 3 games. If each game cost $1.35, how much change did she receive from a $5 bill?
9. Jim buys a watch by making a $25 down payment and paying 15 installments of $12 each. What is the total cost of the watch?
10. Subtract –10 from –40.
11. Which key sequence could be used to change 15 inches to centimeters?
    (a) 15 [÷] 2.54 [=] (b) 15 [×] 2.54 [=]
    (c) 15 [+] 2.54 [=] (d) 15 [−] 2.54 [=]
12. By 1997, 49 of the 50 states had passed seat-belt laws. The percent of the states that had adopted seat-belt laws by 1997 was about
    (a) 50% (b) 67% (c) 80% (d) 98%
13. Rounded to the nearest thousandth, the number 2,746.9752 is
    (a) 3,000 (b) 2,746.976
    (c) 2,746.975 (d) 2,000

UNIT 12-5 Temperature

THE MAIN IDEA

1. Two different scales for measuring temperature are the ***Celsius*** *(C)* scale and the ***Fahrenheit*** *(F)* scale.
2. On both scales, the unit of measure is called the ***degree*** (°), but a Celsius degree is almost twice as large as a Fahrenheit degree.

   Because Celsius degrees are larger units than Fahrenheit degrees, it takes fewer of them to measure the same amount of heat. A Celsius temperature above zero will always be a smaller number than the equivalent Fahrenheit temperature.
3. Here is a comparison of some familiar temperatures on both scales.

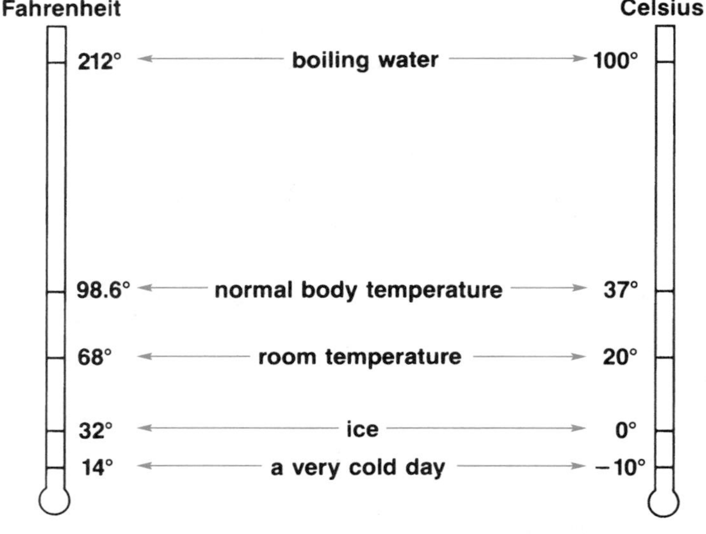

EXAMPLE 1 On a very hot day, Jerry measured the temperature with a Celsius thermometer and a Fahrenheit thermometer at the same time. He jotted down the two temperatures, 35° and 95°, but forgot which was which. Which temperature was measured with a Celsius thermometer and which with a Fahrenheit thermometer?

For the same amount of heat above zero, the Celsius temperature is less than the Fahrenheit temperature.

Answer: The temperature is 35°*C* or 95°*F*.

EXAMPLE 2 State whether each temperature is probably Fahrenheit or Celsius.

a. A fall on an ice patch: 10°

This is probably 10°*F*, since 10°*C* is well above the melting point of ice. *Ans.* Fahrenheit

b. Swimming in the ocean on a summer day: 25°

This is probably 25°*C*, since 25°*F* is colder than the temperature of ice. *Ans.* Celsius

c. A person's temperature during a fever: 102°

This is 102°*F*, since 102°*C* would be hotter than boiling water. *Ans.* Fahrenheit

EXAMPLE 3 Choose the more reasonable temperature.

a. A snowy day: 30°*C* or 30°*F*

30°*F* is around the temperature of ice.
30°*C* is more than normal room temperature.

Answer: 30°*F*

b. A cup of hot soup: 90°*C* or 90°*F*

90°*C* is a bit lower than the temperature of boiling water.
90°*F* is cooler than normal body temperature.

Answer: 90°*C*

EXAMPLE 4 Determine whether the temperature change represents an increase or a decrease, and state the amount of change.

To determine the change in temperature, subtract the first value from the second.

	Temperature Change	*Subtraction*	*Result*
a.	40°*F* to 31°*F*	31 – 40 = –9	decrease of 9°
b.	54.8°*C* to 56.7°*C*	56.7 – 54.8 = 1.9	increase of 1.9°
c.	–10°*F* to –3°*F*	–3 – (–10) –3 + 10 = +7	increase of 7°
d.	–4°*C* to 15°*C*	15 – (–4) = 15 + 4 = 19	increase of 19°
e.	20°*F* to –5°*F*	–5 – 20 = –25	decrease of 25°

CLASS EXERCISES

1. Two scientists measured the temperature of the same body of water at the same time, one with a Celsius thermometer and one with a Fahrenheit thermometer. If their readings were 25° and 77°, which was the Fahrenheit temperature?

2. Mrs. White gave Mrs. Ross her recipe for apple pie. The recipe said, "Heat the oven to 150°," but it didn't tell whether this was 150° Celsius or Fahrenheit. Which was it, probably?

3. Ada's pen pal in France wrote to her and said that the temperature in her town was 30°*C*. Was it a hot day or a cold day?

4. In Montreal one day, the temperature had dropped 10 Celsius degrees and in New York it had dropped 10 Fahrenheit degrees. In which city did the temperature drop more?

5. State whether each temperature is probably Fahrenheit or Celsius.

   a. a warm spring day: 25° **b.** an iced drink: 35° **c.** a person's body temperature: 37°
   d. room temperature in a school: 70° **e.** a cup of warm coffee: 150°

6. Choose the more reasonable temperature.
   a. a cup of hot chocolate: 180°*C* or 180°*F*
   b. swimming pool water: 25°*C* or 25°*F*
   c. a person's forehead: 99°*C* or 99°*F*
   d. a chilly autumn day: 5°*C* or 5°*F*

7. Find each change in temperature.
   a. from 15°*C* to 27°*C*
   b. from 85°*F* to 73°*F*
   c. from 48.6°*C* to 37.8°*C*
   d. from 101.5°*F* to 102.6°*F*
   e. from 72.5°*F* to 70.4°*F*
   f. from 10.1°*C* to 20°*C*

8. Find each change in temperature.
   a. from +18°*C* to +24°*C*
   b. from –32°*F* to +8°*F*
   c. from +11°*C* to –8°*C*
   d. from –21°*F* to –18°*F*
   e. from –5°*C* to –12°*C*
   f. from –14°*F* to 0°*F*

9. Find the change in temperature from the first thermometer reading to the second.

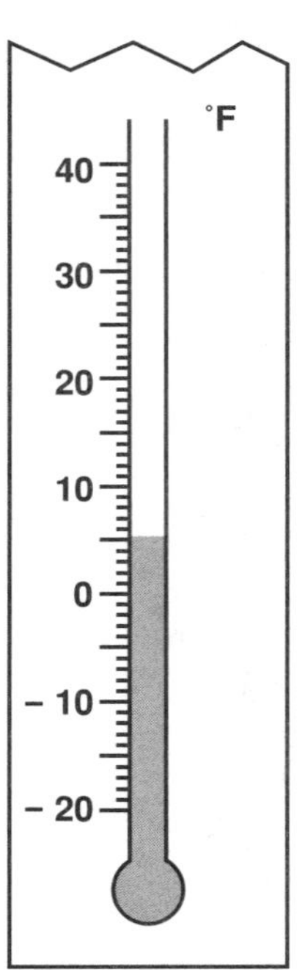

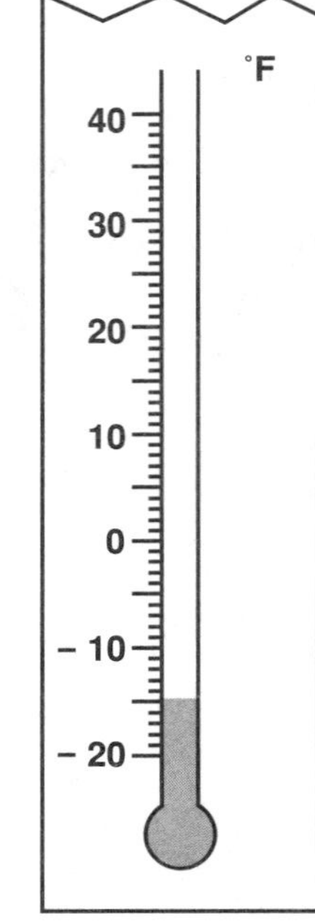

1. A display sign outside a bank gave the temperature in both Celsius and Fahrenheit. If the sign showed 30° and 86°, which was the Celsius temperature?

2. The instruction manual for Mr. Henry's new oil burner told him to set his thermometer to 68°. Is it more likely that this meant 68° Celsius or 68° Fahrenheit?

3. The temperature of Joe's coffee was 40°*C*. Was the coffee warm or cold?

4. Which is a greater increase of heat, a rise of 5 Celsius degrees or a rise of 5 Fahrenheit degrees?

5. State whether each temperature is probably Fahrenheit or Celsius.

   a. the inside of a freezer: 10° **b.** the inside of a refrigerator: 40°
   c. a glass of iced tea: 38° **d.** a hot summer day: 35°

6. In each case, choose the more reasonable temperature.

   a. a pizza oven: 150°*C* or 150°*F* **b.** bath water: 30°*C* or 30°*F*
   c. aquarium water: 75°*C* or 75°*F* **d.** a hot meal: 65°*C* or 65°*F*

7. The temperature of Mrs. Sweeney's car engine changed from 75°*F* to 150°*F*. Find the change in temperature.

8. In each case, find the change in temperature.

   a. from 32°*F* to 15°*F* **b.** from 100°*C* to 117°*C* **c.** from 0°*C* to 15°*C*
   d. from 20.5°*C* to 11.4°*C* **e.** from 101.5°*F* to 102.6°*F* **f.** from 98°*F* to 89°*F*

9. In each case, find the change in temperature.

   a. from +15°*C* to +7°*C* **b.** from –12°*F* to +8°*F* **c.** from –34°*F* to –27°*F*
   d. from +14°*C* to –26°*C* **e.** from –28°*C* to –37°*C* **f.** from 0°*F* to –16°*F*

10. Find the change in temperature from the first thermometer reading to the second.

a.

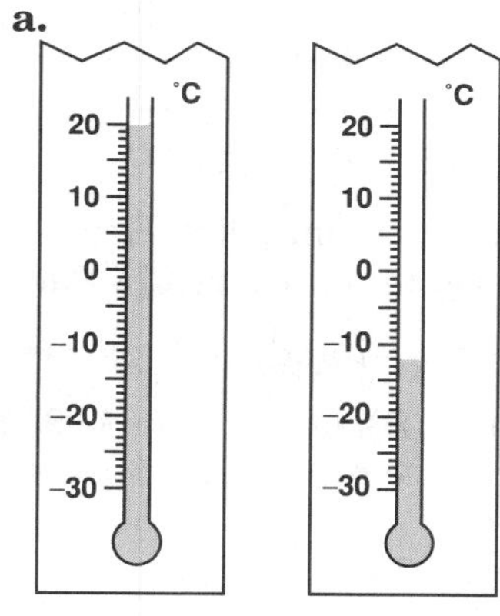

b.

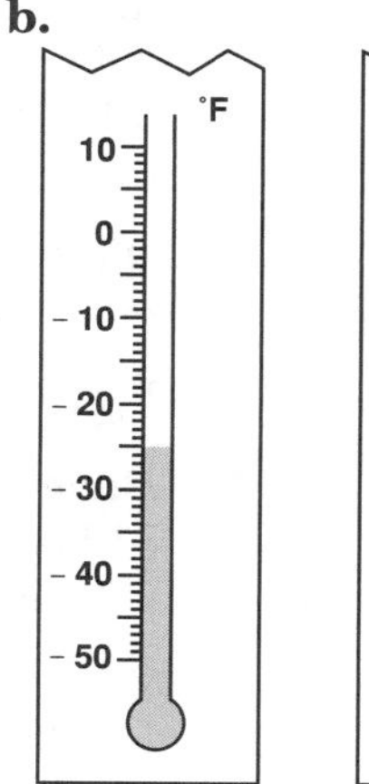

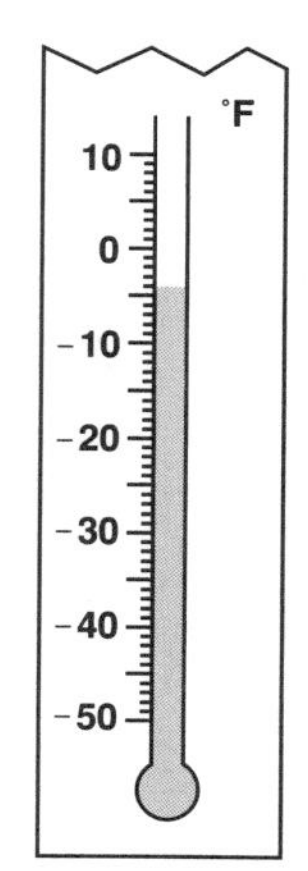

11. The graph shows the average monthly temperature (°F) for Fairbanks, Alaska over a 10-month period.
    - **a.** What is the range of the temperatures shown on the graph?
    - **b.** What is the mean of these temperatures?
    - **c.** Which of these temperatures is closest to the freezing point of water?

Average Monthly Temperature – Fairbanks, Alaska

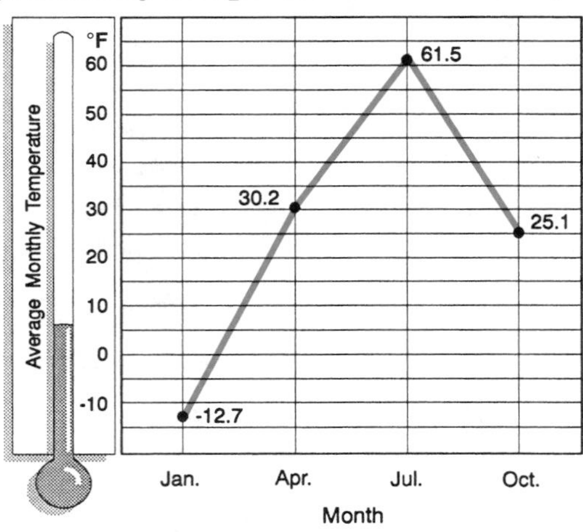

12. The temperature of a chemical solution drops 2 Celsius degrees every 30 minutes. The temperature at 11:25 A.M. was 55°C. What will the temperature of the solution be at 2:55 P.M.?

13. While on location for a special assignment, a reporter wrote that the temperature in degrees Celsius was equal to the temperature in degrees Fahrenheit. Describe the type of clothing that the reporter probably would have worn while on this assignment. Explain your reasoning.

SPIRAL REVIEW EXERCISES

1. Change each of the given measures of volume to the indicated unit.
   - **a.** 75 mL = [?] L **b.** 7.2 kL = [?] L
   - **c.** 14 L = [?] mL **d.** 2,800 L = [?] kL

2. Find 1% of 500.

3. Add: $5\frac{3}{4} + 2\frac{3}{16}$

4. Subtract 11.7 from 19.2.

5. Divide: $\frac{7}{8} \div \frac{1}{2}$

6. The cost of a home computer went from $800 to $600. What was the percent decrease in price?

7. Write 0.93 as a percent.

8. $\frac{3}{8}$ written as a decimal is
   (a) 0.125 (b) 0.25 (c) 0.375 (d) 0.5

9. Marla earns $5\frac{1}{2}$% annual interest on $500 savings. How much interest did she earn at the end of one year?

10. In 1970, the average number of gallons of gasoline used annually by a car driven in the United States was 1,034. By 1990, this figure had dropped to 870. Use your calculator to find the percent decrease to the nearest tenth of a percent.

11. On a map, 4 cm represent 1 km. How many cm are used to represent 275 km?

12. Find the product of –7 and –9.

13. Rhonda bought 7 CD's priced from $9.95 to $14.95. A reasonable price range for the 7 CD's is between
    (a) $180 and $200 (b) $98 and $104
    (c) $62 and $98 (d) $9.95 and $13.95

THEME 6

Ratio, Proportion, and Probability

MODULE 13

MODULE 14

UNIT 13–1 The Meaning of Ratio; Using Ratios to Express Rates

THE MEANING OF RATIO

THE MAIN IDEA

1. A *ratio* is a comparison between two numbers.
2. A ratio can be written in three ways:
   a. by using the word "to" 2 to 3
   b. by using a colon (:) instead of the word "to" 2 : 3
   c. by writing a fraction in which the fraction bar replaces the word "to" $\frac{2}{3}$
3. A ratio is read using the word "to."

2 to 3, 2 : 3, and $\frac{2}{3}$ are all read "2 to 3."

EXAMPLE 1 In an English class, there are 16 girls and 14 boys. In three different ways, write the ratio of: **a.** girls to boys **b.** boys to girls **c.** girls to students in the class

Write the number of the item mentioned first as the first number of the ratio.

Ways to Write a Ratio	*Ratio of Girls to Boys*	*Ratio of Boys to Girls*	*Ratio of Girls to Class*
Using the word "to"	16 to 14	14 to 16	16 to 30
Using a colon	16 : 14	14 : 16	16 : 30
Using a fraction	$\frac{16}{14}$	$\frac{14}{16}$	$\frac{16}{30}$

CLASS EXERCISES

1. Copy the chart shown and write each ratio in two other ways.

Using "to"	Using ":"	As a Fraction
5 to 8		
	3 : 7	
		$\frac{4}{11}$

2. At Carla's party, there were 8 girls and 6 boys. What was the ratio of boys to girls?
3. Fieldston High School's football team won 7 games and lost 5 games. What was the ratio of the number of games won to the number of games lost?
4. In Bill's aquarium, there are 7 angelfish and 6 guppies. Find:
   a. the ratio of angelfish to the total number of fish
   b. the ratio of angelfish to guppies
   c. the ratio of guppies to angelfish
5. In Professor Wilson's library, there are 30 history books and 40 mathematics books. Find:
   a. the ratio of the number of history books to the number of mathematics books
   b. the ratio of the number of mathematics books to the total number of history books and mathematics books
6. On a tray of fruit, there are 5 oranges, 3 apples, and 4 pears. Find:
   a. the ratio of the number of oranges to the number of apples
   b. the ratio of the number of oranges to the number of pears
   c. the ratio of the number of oranges to the total number of fruits
7. Linda bought 9 yellow balloons, 12 red balloons, and 20 white balloons to decorate her home for her party. Write a ratio for:
   a. the number of yellow balloons to the number of red balloons
   b. the number of red balloons to the number of white balloons
   c. the number of yellow balloons to the total number of balloons

8. A mobile health clinic offered free blood-pressure screening. At Location A, 121 out of 986 participants were advised to see a doctor. At Location B, 98 were advised to see a doctor out of 672 participants. Write the ratio of participants advised to see a doctor to total participants at each location.
   a. Location A **b.** Location B **c.** For which location was the ratio greater?

USING RATIOS TO EXPRESS RATES

THE MAIN IDEA

1. When a ratio is used to compare two different kinds of quantities, it is called a *rate*. Such ratios may use words like *for*, *in*, or *per* instead of *to*. For example:

   A car traveling at a *rate* of 55 miles *per* hour compares the number of miles traveled, 55, to the number of hours traveled, 1.

2. In simplest form, a rate expresses the quantity of one item per *single unit* of the other. For example:

$$\frac{300 \text{ miles}}{6 \text{ hours}} = \frac{50 \text{ miles}}{1 \text{ hour}} \text{ or 50 miles per hour}$$

EXAMPLE 2 Express each ratio as a rate in simplest form.

	Ratio	***Rate***	***Rate in Simplest Form***
a.	135 words in 3 minutes	$\frac{135 \text{ words}}{3 \text{ minutes}}$	$\frac{45}{1}$ or 45 words per minute
b.	\$2.80 for 14 pencils	$\frac{\$2.80}{14 \text{ pencils}}$ or $\frac{280¢}{14 \text{ pencils}}$	$\frac{20}{1}$ or 20 cents per pencil

CLASS EXERCISES

1. Express each ratio as a rate in simplest form.
   a. 234 kilometers in 6 hours **b.** 400 words in 10 minutes **c.** \$2.70 for 9 rolls
   d. \$23.75 for 5 hours **e.** 275 students for 11 teachers
2. If Suki can do 30 push-ups in 6 minutes, what is Suki's exercise rate?
3. If Ernesto can read 350 words in 7 minutes, what is Ernesto's reading rate?
4. If Evan earns \$38.40 for 6 hours of work, what is Evan's rate of pay?
5. Mr. Hall drove 208 miles using 16 gallons of gasoline. At what rate did he use gasoline?
6. In 120 minutes of television viewing, Marcie counted 40 commercials. At what rate did the commercials appear?

HOMEWORK EXERCISES

1. In Mr. Paulson's class, there are 17 boys and 15 girls. What is the ratio of the number of boys to the number of girls?

2. Newton's soccer team won 12 games and lost 7 games. What was the ratio of the number of games lost to the number of games won?

3. On a class test, 19 students passed and 7 students failed. Find the ratio of:
   a. the number of students who passed to the number of students who failed
   b. the number of students who failed to the total number of students who took the test

4. In a pet store, there are 15 dogs and 10 cats. Find the ratio of the number of:
   a. cats to the number of dogs **b.** dogs to the total number of dogs and cats

5. In a basket of fruit, there were 7 apples, 5 pears, and 9 oranges. Write a ratio for:
   a. the number of apples to the number of oranges
   b. the number of oranges to the number of pears
   c. the number of pears to the total number of pieces of fruit

6. Mary sold 17 student tickets, 24 adult tickets, and 11 senior-citizen tickets for the school play. Write a ratio for:
   a. the number of student tickets to the number of adult tickets
   b. the number of student tickets to the number of senior-citizen tickets
   c. the number of senior-citizen tickets to the total number of tickets

7. Express each ratio as a rate in simplest form.
   a. 249 miles in 5 hours **b.** 2 defective tires in every 100 tires produced
   c. 5 teachers for 60 students **d.** 1,000 miles using 40 gallons of gasoline
   e. 50 strikeouts in 35 innings **f.** 85 baskets in 100 attempts

8. Milton can type 240 words in 8 minutes. What is his typing rate?

9. If Florence earns $48.60 for 12 hours of work, what is her rate of pay?

10. An accountant takes 48 hours to complete 8 tax forms. At what rate did the accountant complete the tax forms?

11. In one year, 17 million men and 31 million women were on diets in the United States. Write a ratio for:
    a. the number of men on diets to the number of women on diets that year
    b. the number of men on diets to the total number of dieters that year

12. At Ana's party, there were 8 girls and 6 boys. At Maria's party, there were 12 girls and 9 boys. How did the ratio of the number of boys to the number of girls at Ana's party compare to the ratio of the number of boys to the number of girls at Maria's party?

13. Lincoln High School evaluated the physical fitness of its students. The results are shown in the table. Write a ratio of the number of students with each rating compared to the total number of students evaluated.

Poor/Not Fit	132
Physically Fit	252
Superior Fitness	96

a. Poor/Not Fit b. Physically Fit c. Superior Fitness

d. Find the sum of the ratios for all three ratings and explain the result.

SPIRAL REVIEW EXERCISES

1. Write each fraction in simplest form.

a. $\frac{25}{35}$ b. $\frac{14}{32}$ c. $\frac{62}{62}$

2. Mr. Homer's bank pays him 6% interest per year on his special savings account. If he has invested \$320 in this account, how much interest will he earn in one year?

3. The Diaz family budget allows 75% for rent and food. If their monthly income is \$2,800, how much does the budget allow for rent and food?

4. At its white sale, Mancy's department store is offering a 20% discount on designer sheets. If the original selling price of a queen size sheet is \$16.99, how much will a customer pay after the discount?

5. Janice plans to save 25% of the money she earns baby-sitting. This week Janice worked 12.5 hours and was paid \$4.50 per hour. How much money should she save?

6. Mr. Math bought a computer by making a down payment of \$100 and paying 6 installments of \$90 each. The total cost of the computer was

(a) \$600 (b) \$540 (c) \$640 (d) \$960

7. Which is equal to $\frac{1}{2}$?

(a) 0.5 (b) 5% (c) 0.05 (d) 0.5%

8. To the product of 8 and −3, add −9.

9. The number of grams in 3.8 kilograms is

(a) 0.38 (b) 0.038 (c) 380 (d) 3,800

10. An investment of \$500 is to earn interest at the rate of 6%, compounded semi-annually. Use your calculator to find the total value at the end of 2 years.

11. The circle graph shows the origins of the nearly 400,000 foreign students who are studying in colleges in the United States. The number of foreign students who come from either Africa, the Middle East, or Europe is

(a) 40,000 (b) 48,000
(c) 72,000 (d) 112,000

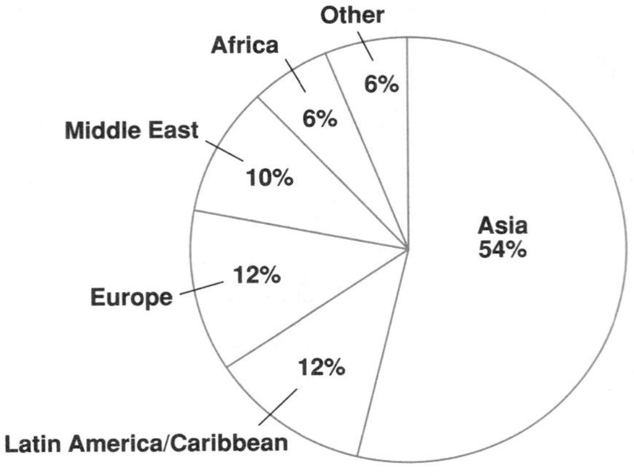

UNIT 13-2 The Meaning of Proportion; Forming a Proportion

THE MEANING OF PROPORTION

THE MAIN IDEA

1. A *proportion* is a statement that two ratios are equal.
2. A proportion tells that, in two different ratios, the numbers compare to each other in the same way.

   The proportion $\frac{3}{4} = \frac{9}{12}$ is read "3 *is to* 4 as 9 *is to* 12."

   This means 3 compares to 4 in the same way that 9 compares to 12.

3. There are four numbers in a proportion. The first and last numbers are called the *extremes*. The second and third numbers are called the *means*.

$$\frac{\overset{\text{extreme}}{\text{1st number}}}{\underset{\text{mean}}{\text{2nd number}}} = \frac{\overset{\text{mean}}{\text{3rd number}}}{\underset{\text{extreme}}{\text{4th number}}}$$

EXAMPLE 1 Write each proportion in words.

	In Numbers	*In Words*
a.	$\frac{3}{7} = \frac{9}{21}$	3 is to 7 as 9 is to 21.
b.	5 : 4 = 20 : 16	5 is to 4 as 20 is to 16.

EXAMPLE 2 Write as a proportion.

	Statement	*Proportion*
a.	8 is to 16 as 12 is to 24.	$\frac{8}{16} = \frac{12}{24}$
b.	9 is to 3 as 15 is to 5.	$\frac{9}{3} = \frac{15}{5}$

EXAMPLE 3 In each proportion, name the means and the extremes.

	Proportion	*Means*	*Extremes*
a.	$\frac{9}{18} = \frac{1}{2}$	18 and 1	9 and 2
b.	20 : 4 = 5 : 1	4 and 5	20 and 1

CLASS EXERCISES

1. Write each proportion in words.

a. $\frac{7}{28} = \frac{2}{8}$ **b.** $1 : 3 = 6 : 18$

2. Name the means and the extremes.

a. $\frac{4}{20} = \frac{2}{10}$ **b.** $9 : 27 = 2 : 6$

3. Write each statement as a proportion. Name the means and the extremes.

a. 4 is to 24 as 2 is to 12 **b.** 50 is to 6 as 25 is to 3 **c.** 1 is to 2 as 3 is to 6

FORMING A PROPORTION

THE MAIN IDEA

Two ratios form a proportion if:

1. the two fractions are equivalent.

$\frac{3}{12} = \frac{8}{32}$ is a proportion

$\frac{3 \div 3}{12 \div 3}$ $\frac{8 \div 8}{32 \div 8}$ fractions are equivalent

$\frac{1}{4} = \frac{1}{4}$ ✔

or

2. the ***cross products*** are equal.

$\frac{3}{12} = \frac{8}{32}$ is a proportion

$\frac{3}{12} \times \frac{8}{32}$ cross products are equal

$96 = 96$ ✔

EXAMPLE 4 Tell whether each statement is a proportion.

	Statement	*Are the Fractions Equivalent?*	*Are the Cross Products Equal?*	*Answer*
a.	$\frac{2}{5} = \frac{12}{30}$	$\frac{2}{5}$; $\frac{12 \div 6}{30 \div 6} = \frac{2}{5}$ yes	$\frac{2}{5} \times \frac{12}{30}$ $5 \times 12 \stackrel{?}{=} 2 \times 30$ $60 = 60$ yes	proportion
b.	$\frac{3}{7} = \frac{2}{5}$	$\frac{3}{7}$; $\frac{2}{5}$ no	$\frac{3}{7} \times \frac{2}{5}$ $7 \times 2 \stackrel{?}{=} 3 \times 5$ $14 = 15$ no	not a proportion

EXAMPLE 5 Use the numbers 1, 3, 5, and 15 to write a proportion.

Try an arrangement.

$$\frac{1}{3} \stackrel{?}{=} \frac{5}{15}$$

Test the cross products.

$$\frac{1}{3} \times \frac{5}{15}$$

$$3 \times 5 \stackrel{?}{=} 1 \times 15$$

$$15 = 15 \checkmark$$

$\frac{1}{3} = \frac{5}{15}$ is a proportion.

There are other arrangements of the four numbers that also form proportions.

$$\frac{1}{5} = \frac{3}{15}, \frac{3}{1} = \frac{15}{5}, \frac{15}{3} = \frac{5}{1}$$

EXAMPLE 6 Mona and Amy were comparing their monthly food budgets. Mona spent $15 for snacks and $125 for basic groceries. Amy spent $12 for snacks and $100 for basic groceries. Do the ratios for the cost of snacks to groceries form a proportion?

$$\frac{\text{Mona's snacks}}{\text{Mona's groceries}} \stackrel{?}{=} \frac{\text{Amy's snacks}}{\text{Amy's groceries}}$$ Write the "skeleton" of a proportion that follows the word pattern.

$$\frac{15}{125} \stackrel{?}{=} \frac{12}{100}$$ Substitute the given numbers.

$$125 \times 12 \stackrel{?}{=} 15 \times 100$$ Test the cross products.

$$1{,}500 = 1{,}500 \checkmark$$

Answer: The comparison of Mona's and Amy's food budgets results in a proportion.

EXAMPLE 7 In a survey of 2,865 television viewers in Chicago, 382 of them reported that they watched a program on the environment. In a survey of 4,320 viewers in Cleveland, 576 reported that they watched the same program. Is the ratio of viewers of this program to total viewers the same in both cities?

KEYING IN

For each city, write the ratio for the total number of viewers to the number who watched the program on the environment. See whether the ratios form a proportion.

Chicago *Cleveland*

total number of viewers → / viewers of the program →

$$\frac{2{,}865}{382} = \frac{4{,}320}{576}$$

Method 1 Show the cross products are equal.

Key Sequence	***Display***	
2865 [×] 576 [=]	1650240.	← cross product 1
382 [×] 4320 [=]	1650240.	← cross product 2

Method 2 Show the ratios are equal.

Key Sequence	***Display***	
2865 [÷] 382 [=]	7.5	← ratio 1
4320 [÷] 576 [=]	7.5	← ratio 2

Answer: The ratio is the same in both cities.

CLASS EXERCISES

1. Tell whether each statement is a proportion.

   a. $\frac{7}{5} = \frac{28}{20}$ b. 100 : 5 = 50 : 2 c. $\frac{40}{30} = \frac{8}{6}$ d. 11 : 66 = 4 : 24 e. $\frac{2}{8} = \frac{8}{34}$

   f. $\frac{4}{9} = \frac{8}{16}$ g. $\frac{27}{36} = \frac{24}{32}$ h. 15 : 10 = 10 : 6 i. 90 : 60 = 2 : 3

2. Use each set of numbers to write four proportions. a. 2, 7, 10, and 35 b. 45, 39, 30, and 26
3. Jose and Cheryl made photos of their families. Jose's photo was 8" wide and 10" long. Cheryl's photo was 11" wide and 14" long. Were the dimensions of the photos in proportion?
4. Mr. Rand rode 200 miles using 16 gallons of gas. Mr. Burton rode 150 miles using 14 gallons of gas. Are these rates in proportion?
5. Chung is 60 inches tall and weighs 120 pounds. Ralph is 72 inches tall and weighs 144 pounds. Are these heights and weights in proportion?
6. The height of a building is 300' and its width is 200'. In Martha's model of the building, the height is 6" and the width is 5". Is Martha's model in proportion to the building?
7. In a bakery, Mr. Wu increased a recipe that called for 37 lb. of flour and 115 c. of milk to 296 lb. of flour and 966 c. of milk. Determine if Mr. Wu increased the two ingredients in proportion.

HOMEWORK EXERCISES

1. Write each proportion in words.

   a. $\frac{3}{8} = \frac{9}{24}$ b. 2 : 5 = 14 : 35

   c. $\frac{12}{48} = \frac{3}{12}$ d. $\frac{16}{48} = \frac{2}{6}$

2. Name the means and the extremes.

   a. 19 : 38 = 1 : 2 b. $\frac{55}{33} = \frac{10}{6}$

   c. 12 : 20 = 15 : 25 d. $\frac{20}{16} = \frac{10}{8}$

3. Write each statement as a proportion. Name the means and the extremes.

   a. 2 is to 3 as 8 is to 12 b. 22 is to 16 as 33 is to 24 c. 90 is to 36 as 135 is to 54

4. Tell whether each statement is a proportion.

   a. $\frac{3}{9} = \frac{6}{18}$ b. 28 : 7 = 64 : 16 c. 95 : 100 = 17 : 20 d. $\frac{60}{80} = \frac{14}{16}$ e. $\frac{20.0}{300} = \frac{24}{36}$

5. Use each set of numbers to write four proportions.

   a. 3, 5, 15, and 25 b. 81, 63, 36, and 28 c. 18, 24, 27, and 36

6. Carol walked 8 miles in 3 hours. John walked 15 miles in 6 hours. Are these rates in proportion?

7. Seiji bought 6 apples for $1.80. Jane bought 5 apples for $1.50. Are these rates in proportion?

8. In a basketball game, Maxine made 12 shots out of 25 attempts, and Rosa made 6 shots out of 13 attempts. Are these rates in proportion?

9. Elizabeth is 64" tall and weighs 120 lb. Joan is 60" tall and weighs 100 lb. Are these heights and weights in proportion?

10. On a map, 3 inches represent 90 miles. On a second map, 2 inches represent 60 miles. Are the two maps in proportion?

11. The population of Alter is 13,850, including 1,177 senior citizens. The population of Copley is 30,477, including 2,592 senior citizens. Determine if the numbers of senior citizens in the two towns are in proportion.

SPIRAL REVIEW EXERCISES

1. In Mr. Benson's class, 5 students have blond hair, 8 students have brown hair, and 13 students have black hair. Write a ratio for:
   a. the number of brown-haired students to the number of black-haired students
   b. the number of black-haired students to the number of blond-haired students
   c. the number of blond-haired students to the total number of students

2. Pat drove 320 miles on 25 gallons of gas. What was the rate of gasoline use?

3. Carl bought 2 cassette tapes at $4.99 each. How much change did Carl receive from a $20 bill?

4. Mr. Stein bought a dishwasher having a list price of $380 on an installment plan. If he made a 20% down payment and paid 15 installments of $25 each, what were the carrying charges?

5. With a sales tax of 8%, what is the total cost of a suit priced at $84?

6. 36 is what percent of 48?

7. Bananas are 75.7% water. If a bunch of bananas weighs about 4 pounds, find the number of pounds of water.

8. Which key sequence could be used to change 17 hours to minutes?

   (a) 60 [÷] 17 [=] (b) 17 [÷] 60 [=]
   (c) 17 [×] 60 [=] (d) 60 [×] 17 [×] 24 [=]

9. The sum of 5.8 and 0.23 is

   (a) 81 (b) 8.1 (c) 6.03 (d) 5.93

10. Evaluate: $12 + \frac{1}{2} \times 10 \times 40$

11. The percent of the figure that is shaded is

   (a) 25% (b) 50% (c) $66\frac{2}{3}$% (d) 75%

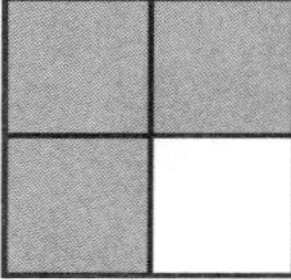

UNIT 13-3 Solving Proportions; Unit Pricing

SOLVING PROPORTIONS

THE MAIN IDEA

1. If one of the numbers in a proportion is unknown, represent it by using a letter. For example: $\frac{5}{20} = \frac{n}{48}$
2. To find an unknown number in a proportion:
   a. Set the cross products equal to each other. (In a proportion, the product of the means is equal to the product of the extremes.)
   $20 \times n = 5 \times 48$
   $20 \times n = 240$
   b. Divide the multiplier of n into the other cross product.
   $n = \frac{240}{20}$
   The result is the unknown number.
   $n = 12$
   c. Check this value by substituting it into the original proportion.
   $\frac{5}{20} = \frac{n}{48}$
   The two fractions should be equivalent.
   $\frac{5}{20} \stackrel{?}{=} \frac{12}{48}$ $\quad \frac{5}{20} = \frac{5 \cdot 1}{5 \cdot 4}, \frac{12}{48} = \frac{12 \cdot 1}{12 \cdot 4}$
   $\frac{1}{4} = \frac{1}{4}$ ✔

EXAMPLE 1 Find the value of n: $\frac{5}{7} = \frac{n}{42}$

$7 \times n = 5 \times 42$ Set the cross products equal.
$7 \times n = 210$

$n = \frac{210}{7}$ Divide 7 into 210.
$n = 30$

Check in the original proportion.

$\frac{5}{7} = \frac{n}{42}$

$\frac{5}{7} \stackrel{?}{=} \frac{30}{42}$ $\quad \frac{30}{42} = \frac{6 \cdot 5}{6 \cdot 7}$

$\frac{5}{7} = \frac{5}{7}$ ✔

Answer: $n = 30$

EXAMPLE 2 Use a calculator to solve the proportion: $\frac{15{,}845}{n} = \frac{12{,}676}{659}$

KEYING IN

To solve a proportion on a calculator, divide the numerical cross product by the multiplier of n.

Key Sequence	***Display***
15845 [×] 659 [÷] 12676 [=]	823.75

To check the value of n, the values of the cross products must be equal.

Key Sequence	***Display***	
15845 [×] 659 [=]	10441855.	← cross product 1
12676 [×] 823.75 [=]	10441855.	← cross product 2

Answer: $n = 823.75$

EXAMPLE 3 From a class of 30 girls and 18 boys, a committee will be formed. The number of boys and girls on the committee should be proportional to the number of boys and girls in the class. If the committee will contain 3 boys, how many girls should be on the committee?

Use the words of the problem to think out a pattern.

$$\frac{\text{number of girls in class}}{\text{number of boys in class}} = \frac{\text{number of girls on committee}}{\text{number of boys on committee}}$$

Three of the four numbers are given. Let n represent the missing number, the number of girls on the committee.

$$\frac{30}{18} = \frac{n}{3}$$

Set the cross products equal.

$$18 \times n = 30 \times 3$$
$$18 \times n = 90$$

Divide by the multiplier of n.

$$n = \frac{90}{18}$$
$$n = 5$$

Check the value of n in the original proportion.

$$\frac{30}{18} = \frac{n}{3}$$
$$\frac{30}{18} \stackrel{?}{=} \frac{5}{3}$$
$$\frac{6 \cdot 5}{6 \cdot 3} \stackrel{?}{=} \frac{5}{3}$$
$$\frac{5}{3} = \frac{5}{3} \checkmark$$

Answer: There should be 5 girls on the committee.

EXAMPLE 4 If Julie can type 200 words in 4 minutes, how many words can she type in 10 minutes? At what rate does Julie type?

Use the words of the problem to write a proportion, letting n represent the number of words in 10 minutes.

Solve the proportion.

$$\frac{200 \text{ words}}{4 \text{ minutes}} = \frac{n \text{ words}}{10 \text{ minutes}}$$

$$4 \times n = 2{,}000$$

$$n = \frac{2{,}000}{4}$$

$$n = 500$$

Check the value of n in the original proportion.

$$\frac{200}{4} = \frac{n}{10}$$

$$\frac{200}{4} \stackrel{?}{=} \frac{500}{10}$$

$$\frac{50}{1} = \frac{50}{1} \checkmark$$

This is the rate at which Julie types.

Answer: Julie can type 500 words in 10 minutes.
She types at the rate of 50 words per minute.

CLASS EXERCISES

1. Find each value of n: **a.** $\frac{6}{9} = \frac{n}{72}$ **b.** $8 : 3 = n : 15$ **c.** $\frac{n}{55} = \frac{3}{5}$ **d.** $\frac{16}{n} = \frac{28}{35}$

e. $n : 12 = 5 : 4$ **f.** $33 : n = 11 : 5$ **g.** $0.7 : 1.4 = 7 : n$ **h.** $\frac{8}{3} = \frac{56}{n}$

2. A recipe calls for 4 cups of flour and 2 teaspoons of cinnamon. Mrs. James wants to increase the recipe so that it will use 28 cups of flour. How much cinnamon should she use?

3. In a certain country, the banks exchange 5 buckos for every 4 U.S. dollars. How many buckos will they exchange for 52 U.S. dollars?

4. The ratio of boys to girls in a school is 9 : 10. If there are 1,368 boys in the school, how many girls are there?

5. Mr. Harris knew that his car could go 14 miles for every gallon of gasoline. How far could he travel using 12 gallons?

6. If 100 children eat 20 pounds of cheese for lunch in the school cafeteria, how much cheese should the dietician order for 250 children?

7. Solve each proportion: **a.** $\frac{108}{67} = \frac{n}{6{,}298}$ **b.** $\frac{n}{38{,}844} = \frac{943}{83}$ **c.** $\frac{154}{215} = \frac{31{,}764}{n}$

8. A speed of 60 miles per hour is equivalent to 88 feet per second. How many feet per second is equivalent to 35 miles per hour?

UNIT PRICING

THE MAIN IDEA

1. *Unit pricing* helps to compare the prices of two different sizes of a product.
2. The unit price of a product is its price per ounce, per quart, or any unit of measure. To compare two unit prices, they must both be in the same unit.
3. To compute unit price, solve the proportion: $\frac{\text{Unit Price}}{\text{1 Unit}} = \frac{\text{Price of Package}}{\text{Number of Units in Package}}$

   Since the denominator of the first ratio is 1: $\text{Unit Price} = \frac{\text{Price of Package}}{\text{Number of Units in Package}}$
4. The unit price of a product is usually rounded to the nearest tenth of a cent.

EXAMPLE 5 A 22-ounce bottle of liquid detergent sells for $1.65. Find the unit price.

$$\text{Unit Price} = \frac{\text{Price of Package}}{\text{Number of Units in Package}}$$

$$= \frac{1.65}{22} = 0.075$$

Answer: The unit price is $0.075, or 7.5¢, per ounce.

EXAMPLE 6 A 64-oz. bottle of juice costs $1.79, and a 40-oz. bottle costs $1.07. Determine the better buy.

Key Sequence	***Display***	***Unit Price***
1.79 [÷] 64 [=]	0.0279687	2.8¢ per oz.
1.07 [÷] 40 [=]	0.02675	2.7¢ per oz.

Answer: The 40-oz. bottle is the better buy.

CLASS EXERCISES

1. Find the unit price for: **a.** 5 lbs. of roast beef for $17.50 **b.** 8 lbs. of potatoes for $4.72 **c.** 18 oz. of olive oil for $3.20 **d.** a 64-oz. bottle of bleach for $0.79
2. Determine the better buy.
   a. a 5-pound bag of dog food for $3.59 or a 20-pound bag for $12.99
   b. a 24-ounce jar of applesauce for $1.25 or a 16-ounce jar for $0.89
   c. a 14-ounce jar of spaghetti sauce for $1.19 or a 20-ounce jar for $1.60

HOMEWORK EXERCISES

1. Find each value of n:

   a. $\frac{n}{12} = \frac{21}{36}$ b. $\frac{20}{n} = \frac{4}{5}$ c. $\frac{45}{50} = \frac{n}{20}$ d. $\frac{8}{12} = \frac{14}{n}$

   e. $\frac{5}{8} = \frac{n}{72}$ f. $12 : 15 = 30 : n$ g. $5 : 4 = n : 28$ h. $\frac{n}{1.5} = \frac{36}{27}$

   i. $\frac{30}{n} = \frac{3}{0.5}$ j. $18 : 12 = 24 : n$ k. $2.4 : n = 1.2 : 3.6$ l. $\frac{200}{120} = \frac{n}{150}$

2. If a recipe for 4 loaves of bread uses 3 tablespoons of honey, how much honey will be needed for 6 loaves?
3. A psychologist formed 2 groups of subjects in which the numbers of men and women were proportional. In one group, she placed 8 men and 5 women. In the second group, she placed 40 men. How many women were in the second group?
4. A nurse prepares a certain mixture by using 3 cc of water with 7 cc of medicine. How much water should he use with 98 cc of medicine?
5. If Paul can read 6 pages in 24 minutes, how many pages can he read in 60 minutes?
6. If it rains at the rate of 0.14 inch per hour, how much will it rain in 24 hours?
7. Solve each proportion.

   a. $\frac{n}{4{,}522} = \frac{416}{119}$ b. $\frac{8{,}160}{n} = \frac{96}{324}$ c. $\frac{57}{39} = \frac{n}{13{,}299}$ d. $\frac{637}{62} = \frac{54{,}782}{n}$

8. A speed of 60 miles per hour is equivalent to 88 feet per second. What speed in miles per hour is equivalent to 77 feet per second?
9. The ratio of boys to girls at a school basketball game is 7 : 3. If there are 87 girls attending the game, how many boys are there?
10. The ratio of adults to children at a circus is 7 : 8. If there are 408 children at the circus, what is the total number of people there?
11. Explain why you cannot enlarge a 4-inch by 5-inch photograph to an 11-inch by 14-inch size.
12. Find the unit price for:

    a. 64 oz. of orange juice for \$2.59 b. 5 lb. of bananas for \$2.25

    c. 6 lb. of apples for \$5.94 d. 12 oz. of cheese for \$2.49

13. Determine the better buy.

a. a 5-pound bag of flour for \$1.29 or a 2-pound bag for \$0.59
b. a 3-liter tin of olive oil for \$8.99 or a 1-liter bottle for \$3.20
c. a 6-ounce container of yogurt for \$0.79 or an 8-ounce container for \$0.85
d. five 8-ounce cans of tomato sauce at \$0.20 each or a 32-ounce can at \$0.89
e. 7 ounces of shampoo for \$2.59 or $8\frac{1}{2}$ ounces for \$3.10
f. a 32-ounce jar of salsa for \$1.29, an 18-ounce jar for \$2.49, or a 10-ounce jar for \$1.59
g. a 7-ounce tube of toothpaste at \$2.69 plus a 50¢-off coupon or an 8.2-ounce tube for \$2.39
h. an 11-ounce can of condensed soup at \$1.89 (to which you must add 1 can of water) or a 20-ounce can of ready-to-serve soup at \$1.59

SPIRAL REVIEW EXERCISES

1. If there are 24 American-made cars and 14 imported cars, what is the ratio of imported cars to the total number of cars?

2. Use the numbers 28, 5, 4, and 35 to write four proportions.

3. I bought 4 compact discs at \$12.50 each and received a fifth one free. What was the average cost of one compact disc?

4. $5\frac{1}{2}\%$ written as a decimal is

(a) 0.55 (b) 0.05 (c) 0.055 (d) 0.0055

5. $\frac{3}{20}$ written as a percent is

(a) 35% (b) 25% (c) 15% (d) 5%

6. If the sales tax rate is 6%, what is the total cost of 3 items priced at \$4.80, \$11.20, and \$6.40?

7. Determine how many years it will take for an investment of \$250 that earns 6.25% interest compounded annually to double in value.

8. What is the total cost of a car if the down payment is \$6,000 and there are 24 installments of \$300 each?

9. What is the price of each pen if a dozen pens cost \$8.28?

10. The first word printed on a T-shirt costs \$3 and each additional word costs \$0.75. What is the cost of a 5-word message?

11. Which is a composite number?

(a) 2 (b) 7 (c) 11 (d) 15

12. Find the sum of $11\frac{2}{3}$ and $\frac{1}{6}$.

13. How many 18-inch-long pieces can be cut from 6,300 inches of cable?

14. 28 is 50% of

(a) 14 (b) 56 (c) 2.8 (d) 50

15. The reciprocal of –1 is

(a) –1 (b) 1 (c) 0 (d) $-\frac{1}{2}$

UNIT 13-4 Using Proportions in Scale Drawings

THE MAIN IDEA

1. A ***scale drawing*** is a drawing in which all the dimensions of the actual objects are reduced or enlarged proportionally. For example, a map is a scale drawing.
2. The ratio between any actual length and the corresponding length in the drawing always has the same value.
3. To find an unknown length in a scale drawing, write and solve a proportion.

EXAMPLE 1 On a map, the 30-km distance between Avon and Boyle is represented by a line segment 2 cm long, and the distance between Boyle and Carson by a segment of 3 cm. What is the actual distance between Boyle and Carson?

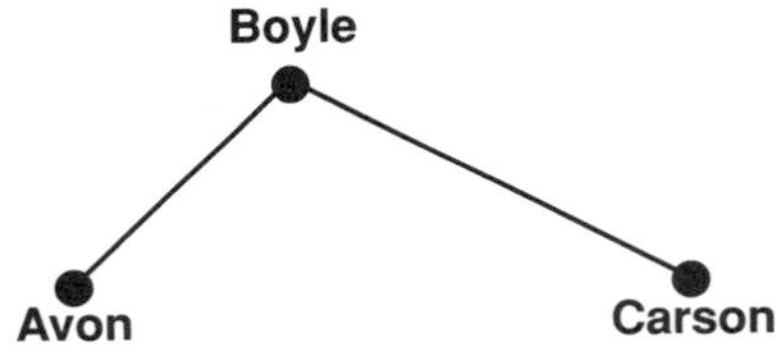

Use the words of the problem to write a proportion, letting n represent the unknown length, the second actual distance.

$$\frac{\text{1st actual distance}}{\text{1st map length}} = \frac{\text{2nd actual distance}}{\text{2nd map length}}$$

$$\frac{30 \text{ km}}{2 \text{ cm}} = \frac{n \text{ km}}{3 \text{ cm}}$$

$$2 \times n = 90$$

$$n = \frac{90}{2}$$

$$n = 45$$

Answer: The actual distance between Boyle and Carson is 45 kilometers.

EXAMPLE 2 What is the approximate distance between Acton and Bream?

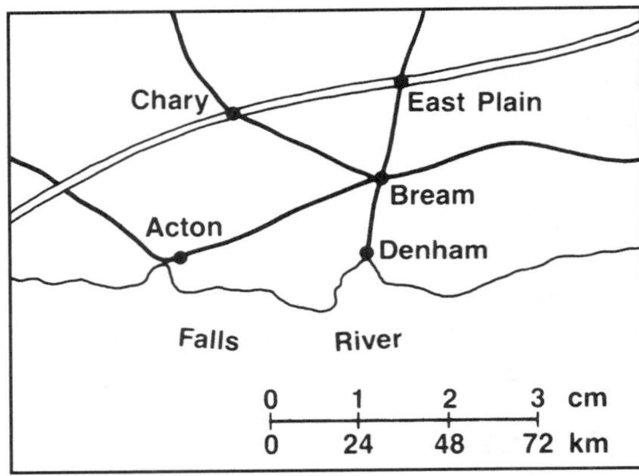

Use the map scale to estimate the map length between Acton and Bream, about 2.5 cm. The scale shows that 1 cm on the map represents 24 km. Write a proportion.

$$\frac{\text{scale length}}{\text{actual distance}} = \frac{\text{map length}}{\text{unknown distance}}$$

$$\frac{1 \text{ cm}}{24 \text{ km}} = \frac{2.5 \text{ cm}}{n \text{ km}}$$

$$1 \times n = 24 \times 2.5$$

$$n = 60$$

Answer: The actual distance between Acton and Bream is about 60 km.

EXAMPLE 3 On a diagram, a doorway is represented by a rectangle of width 2 inches and height 4.5 inches. If the width of the actual doorway is 4 feet, what is its height?

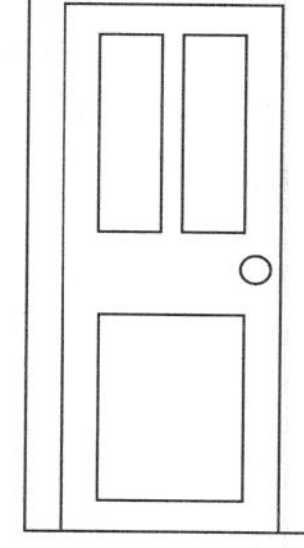

Write a proportion, letting n represent the actual height.

$$\frac{\text{diagram width}}{\text{diagram height}} = \frac{\text{actual width}}{\text{actual height}}$$

$$\frac{2 \text{ in.}}{4.5 \text{ in.}} = \frac{4 \text{ ft.}}{n \text{ ft.}}$$

$$2 \times n = 18$$

$$n = \frac{18}{2}$$

$$n = 9$$

Answer: The height of the actual doorway is 9 ft.

EXAMPLE 4 A stamp of length 20 mm and width 30 mm is enlarged to be reproduced in a book. If the length of the enlargement is $1\frac{1}{3}$ in., what is its width?

Write a proportion, letting n represent the width of the enlargement.

$$\frac{\text{stamp length}}{\text{stamp width}} = \frac{\text{enlarged length}}{\text{enlarged width}}$$

$$\frac{20 \text{ mm}}{30 \text{ mm}} = \frac{1\frac{1}{3} \text{ in.}}{n \text{ in.}}$$

$$20 \times n = 30 \times 1\frac{1}{3}$$

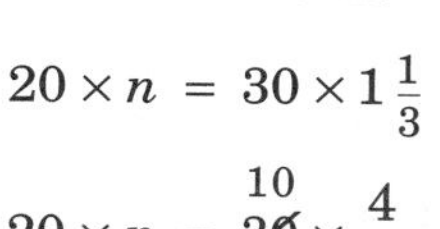

$$20 \times n = \overset{10}{\cancel{30}} \times \frac{4}{\underset{1}{\cancel{3}}}$$

$$20 \times n = 40$$

$$n = \frac{40}{20}$$

$$n = 2$$

Answer: The width of the enlargement is 2 in.

EXAMPLE 5 On a certain scale drawing, all the dimensions are shown $\frac{1}{16}$ actual size. What is the actual length of a desk that is 3 inches long on the drawing?

Write a proportion, letting n represent the missing length.

$$\frac{\text{drawing length}}{\text{actual length}} = \frac{1}{16}$$

$$\frac{3 \text{ in.}}{n \text{ in.}} = \frac{1}{16}$$

$$1 \times n = 48$$

$$n = 48$$

Answer: The actual length is 48 inches.

CLASS EXERCISES

1. A length of 15 miles is represented on a scale drawing by a line segment 3 centimeters long. What length is represented by a line segment 3.5 centimeters long?
2. The width of an island that is actually 100 mi. wide is represented on a map by a line segment 2 in. long. If the length is represented by a line segment 3 in. long, how long is the island?
3. How would you represent a distance of 75 kilometers on a map that represents 25 kilometers by a line segment 4 centimeters long?
4. A scale drawing shows all dimensions $\frac{1}{20}$ actual size. How long is a steel bar that is represented by a line segment 2.5 inches long?
5. A truck is 32 feet long and 8 feet high. In a photograph, the length of the truck is 5 inches. How high is the truck in the photograph?
6. Stephanie made an accurate model of an airplane. In her model, the length of the airplane is 11 inches and the wingspread is 9 inches. If the wingspread of the actual airplane is 27 feet, what is its length?
7. For his science project, Mark made a giant model of a bacterium. The actual length of the bacterium is 0.05 inch and its width is 0.02 inch. Mark wants his model to be 24 centimeters long. How wide should he make it?
8. A computer chip is $\frac{1}{4}$ inch wide and $\frac{1}{2}$ inch long. The diagram for the chip is 36 inches long. How wide is the diagram?

HOMEWORK EXERCISES

1. On a map, a distance of 200 miles is represented by a line segment 5 inches long. What distance is represented by a line segment 4 inches long on this map?
2. A length of 35 cm is shown as a line segment 2 cm long on a scale drawing. If a length is shown as a line segment 7 cm long on this drawing, what is its actual size?
3. On an architect's blueprint, a room that is 12 feet wide and 20 feet long is represented by a rectangle that is 5 cm long. How wide is the rectangle?
4. On a scale drawing, every length is shown $\frac{1}{12}$ actual size. What is the length of a line segment that represents a length of 30 feet?
5. How would you represent a 42-foot length on a scale drawing that represents a 20-foot length by a line segment that is 2 centimeters long?

6. A sheet of paper that is $8\frac{1}{2}$ inches wide and 11 inches long is to be photographed and reduced so that the width of the photograph will be $4\frac{1}{4}$ inches. How long will the photograph be?

7. A man is 6 feet tall and his daughter is 4 feet tall. In a photograph, the man is $2\frac{1}{2}$ inches tall. What is the height of his daughter in the photograph?

8. On a sign, a bottle that is actually 15 inches tall is shown 10 feet tall. If the bottle is actually 6 inches wide, how wide is it on the sign?

SPIRAL REVIEW EXERCISES

1. In each proportion, find the value of n.

   a. $\frac{n}{32} = \frac{33}{24}$ **b.** $6 : n = 9 : 3$

   c. $\frac{27}{9} = \frac{9}{n}$ **d.** $1.5 : 15 = n : 40$

2. Ms. Phillips makes a salary of $265 a week and a 2% commission on her gasoline sales. Last week, she sold $8,580 worth of gasoline. Find her total earnings.

3. The baby-sitter at the health club charges $2.00 for the first hour and $0.50 for each additional hour or part of an hour. Joy left her daughter with the baby-sitter for $2\frac{1}{2}$ hours. How much did this cost her?

4. If the population of Trenton, New Jersey, was about 105,000 one year and about 94,500 ten years later, by what percent did the population decrease in those 10 years?

5. Joel got 15 strikes in 60 frames of bowling. What percent of his frames bowled were strikes?

6. 16 is 25% of

   (a) 4 (b) 8 (c) 32 (d) 64

7. Mr. Lopez buys a diamond ring by making a $300 down payment and 24 installment payments of $50 each. The total cost of the ring is

   (a) $1,200 (b) $1,500
   (c) $900 (d) $1,600

8. If the rate of sales tax is 4%, the amount of tax charged on a $15 purchase is

   (a) $6.00 (b) $0.42
   (c) $15.60 (d) $0.60

9. 24% written as a fraction is

   (a) $\frac{12}{100}$ (b) $\frac{13}{50}$ (c) $\frac{6}{25}$ (d) $\frac{24}{50}$

10. Find the product of $\frac{28}{39}$ and $\frac{26}{35}$.

11. Ms. Ortega drove 795 miles and paid a total of $47.50 for gas. Use your calculator to find, correct to the nearest cent, her cost of gas per mile.

12. Find the greatest common factor of 60 and 45.

13. When 58 is divided by 7, the remainder is

    (a) 2 (b) 3 (c) 5 (d) 6

UNIT 13-5 Pictographs

THE MAIN IDEA

1. In a *pictograph*, quantities are represented by symbols.
2. There are two types of pictographs:
   a. In some pictographs, each symbol stands for a given amount. The *number* of symbols is proportional to the quantity represented.

1,000 Stamps

3,500 Stamps

   b. In other pictographs, different amounts are shown with proportionately sized symbols. The *size* of each symbol is proportional to the quantity represented.

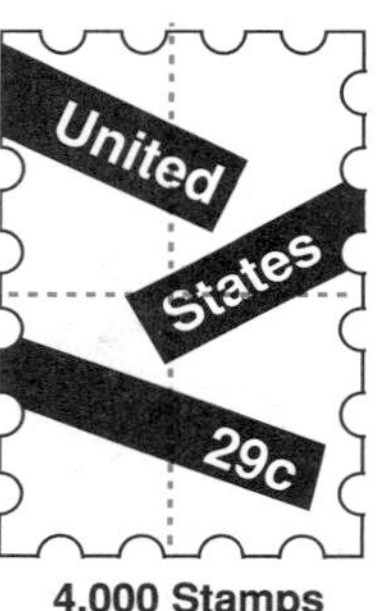

4,000 Stamps

EXAMPLE 1 This pictograph shows the number of books circulated by the local library over a period of six months. How many more books were circulated in October than in September?

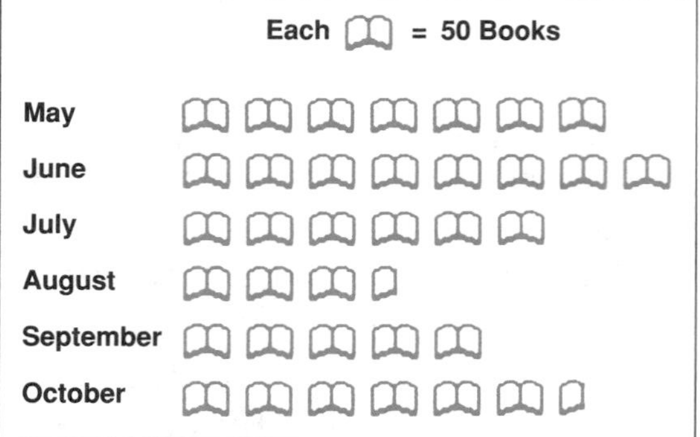

Count the number of symbols for September and for October. Each symbol represents 50 books.

$6\frac{1}{2} \times 50$ or $6.5 \times 50 = 325$ ← books in October

$5 \times 50 = 250$ ← books in September

75 ← difference

Answer: 75 more books were circulated in October than in September.

EXAMPLE 2 In a pictograph that shows the number of cars sold by a dealer each month, 2 symbols represent 32 cars. How many cars are represented by 5 symbols?

Write a proportion, letting n represent the unknown value.

$$\frac{\text{Jan. actual cars}}{\text{Jan. car symbols}} = \frac{\text{Feb. actual cars}}{\text{Feb. car symbols}}$$

$$\begin{aligned} \frac{32}{2} &= \frac{n}{5} \\ 2 \times n &= 5 \times 32 \\ 2 \times n &= 160 \\ n &= 80 \end{aligned}$$

Answer: 5 symbols represent 80 cars.

EXAMPLE 3 If 5 bottle symbols represent 200 babies, which of the following choices is the best estimate of the number of babies represented by 4 bottle symbols?
(a) 40 (b) 80 (c) 100 (d) 160

For an estimate, there is no need to work out a proportion.

Think: Since 4 symbols are more than half the number of symbols in the given set of 5 bottles, the number of babies they represent should be more than half of 200.

Answer: (d)

EXAMPLE 4 This pictograph represents the number of homes in Sunnyside that local planners hope to heat by solar power each year from 1995 to 2000.

Number of Homes to Be Heated by Solar Power

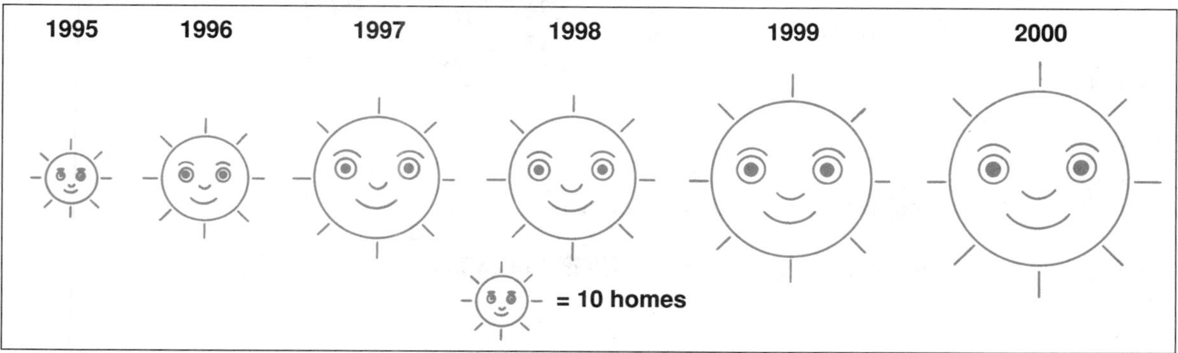

To answer the questions that follow, consider the *sizes* of the symbols.

a. Will the number of homes heated by solar power increase or decrease from 1995 to 2000?

Since the size of the sun symbol grows larger, this shows an increase.

Answer: Increase

b. In which year will there be no change from the previous year in the number of homes with solar heat?

Look for 2 consecutive years in which the symbols are the same size.

Answer: 1998

c. If the number of solar homes in 1998 is 8 times the number in 1995, and the number in 2000 is 10 times the number in 1995, how many more solar homes are there in 2000 than in 1998?

10	←	in 1995
10×10 or 100	←	in 2000
8×10 or 80	←	in 1998
20	←	difference

Answer: There are 20 more solar homes in 2000 than in 1998.

EXAMPLE 5 In a pictograph that shows the growth of camera sales, a small camera symbol represents 10,000 cameras.

a. How many cameras are represented by a symbol that is 3 times as high and 3 times as wide?

THINKING ABOUT THE PROBLEM

Method 1 Draw a diagram. Observe that 3×3, or 9, original symbols can fit into one large symbol that is 3 times as high and 3 times as wide.

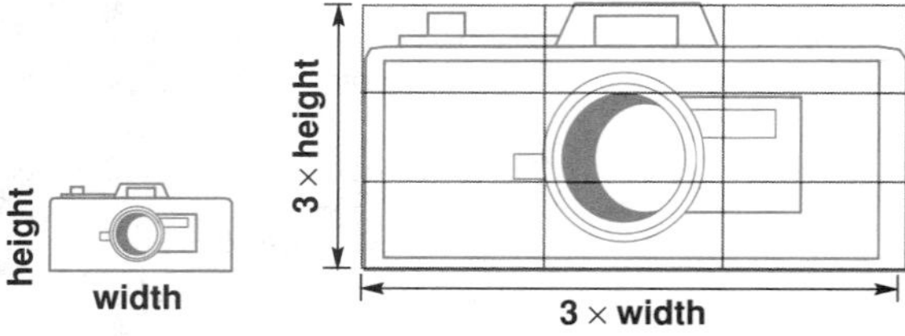

Method 2 Write a proportion.

$$\frac{10{,}000}{1} = \frac{90{,}000}{9}$$

← number of cameras
← number of times symbol is enlarged

Answer: The new symbol represents 90,000 cameras.

Observe that Method 2 is used in parts **b** and **c** that follow.

b. How many cameras are represented by a symbol that is 1.5 times as high and 1.5 times as wide as the small camera symbol?

The symbol is 1.5×1.5, or 2.25 times, as large as the original.

Solve the proportion:

$$\frac{10{,}000}{1} = \frac{n}{2.25} \begin{array}{l} \leftarrow \text{number of cameras} \\ \leftarrow \text{number of times symbol is enlarged} \end{array}$$

$$1 \times n = 10{,}000 \times 2.25$$
$$n = 22{,}500$$

Answer: The symbol represents 22,500 cameras.

c. The symbol that represents 10,000 cameras is about 0.5 cm high and 1.5 cm wide. How many cameras are represented by a symbol that is 1 cm high and 3 cm wide?

Find the ratio of the new dimensions to the original dimensions:

$$\begin{array}{l} \text{new dimensions} \rightarrow \\ \text{original dimensions} \rightarrow \end{array} \quad \overset{\text{Height}}{\frac{1 \text{ cm}}{0.5 \text{ cm}}} = \overset{\text{Width}}{\frac{3 \text{ cm}}{1.5 \text{ cm}}} = \frac{2}{1}$$

Thus, this new symbol is 2×2, or 4 times, as large as the original.

$$\frac{10{,}000}{1} = \frac{n}{4} \begin{array}{l} \leftarrow \text{number of cameras} \\ \leftarrow \text{number of times symbol is enlarged} \end{array}$$

$$1 \times n = 10{,}000 \times 4$$
$$n = 40{,}000$$

Answer: This new symbol represents 40,000 cameras.

EXAMPLE 6 Make a pictograph to show the information given in the table.

Number of Trees Planted in New Township	
March	400
April	500
May	700
June	250
July	150
August	50
September	450

Decide upon a symbol related to the subject of the graph, and decide how many units it represents.

Determine how many whole symbols and parts of a symbol are needed to picture the given information.

Number of Trees Planted in New Township

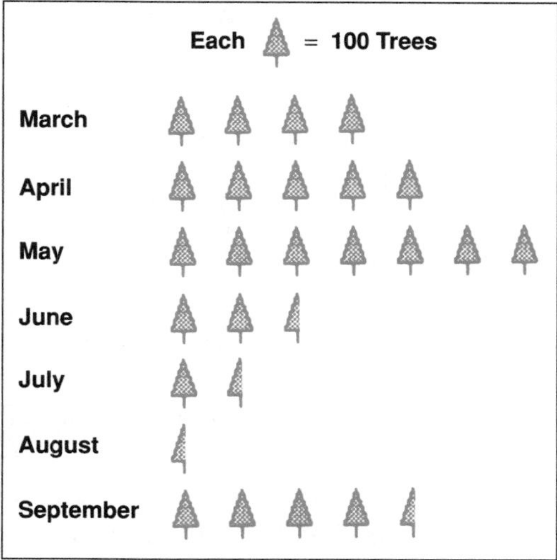

CLASS EXERCISES

1. This pictograph shows the number of frozen yogurt cones sold by the Dipit Yogurt Parlor over a period of six days.
   a. How many more cones were sold on Friday than on Thursday?
   b. On which day were twice as many cones sold as on Monday?
   c. On which day were 30 fewer cones sold than on Thursday?
   d. What was the total number of cones sold during the six days?

Number of Cones Sold

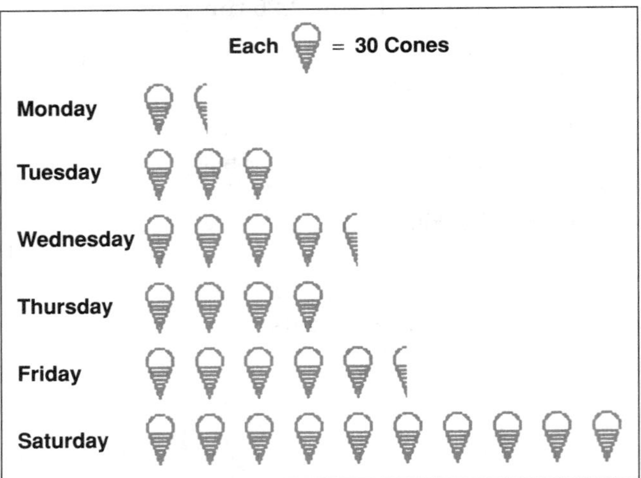

2. In a pictograph that shows the number of campers visiting a state park each month, 4 tent symbols represent 600 campers. How many tent symbols are needed to represent 900 campers?

3. If $2\frac{1}{2}$ apple symbols represent 10,000 bushels of apples, which choice is the best estimate of the number of bushels represented by 10 apple symbols?
   (a) 20,000 (b) 40,000 (c) 50,000 (d) 100,000

4. This pictograph represents the amount of rainfall over a 4-month period.

Amount of Rainfall

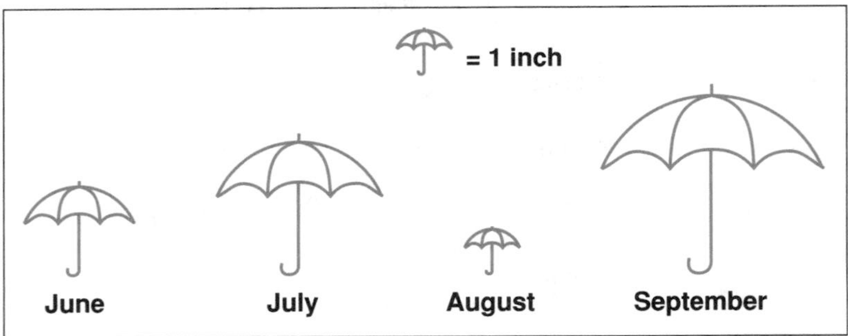

   a. Which month had the smallest amount of rainfall? the greatest?
   b. Which month(s) had more rainfall than June?
   c. Which month(s) had less rainfall than July?
   d. Which month(s) had less rainfall than August?

5. If this symbol represents $5,000, how much money is represented by a symbol that is:

a. 2 times as high and 2 times as wide?
b. 5 times as high and 5 times as wide?
c. 1.8 times as high and 1.8 times as wide?
d. $\frac{1}{2}$ as high and $\frac{1}{2}$ as wide?

6. Make a pictograph to illustrate the information given in each table.

a.

Number of Cats Rescued by the Animal Shelter	
January	300
February	500
March	200
April	250
May	100
June	50

b.

Number of Hours of Exercise per Week	
Mary	14
Gina	28
Melanie	8
Alexia	20
Marion	12
Hope	18

HOMEWORK EXERCISES

1. This pictograph shows the number of traffic accidents in Laurelton for a period of five months.

Number of Traffic Accidents

Each [car] = 10 Accidents

March
April
May
June
July

a. How many accidents were there in April?
(a) 2 (b) 3 (c) 20 (d) 25
b. How many more accidents were there in June than in July?
c. For which two months combined was the total number of accidents about equal to the number for June?

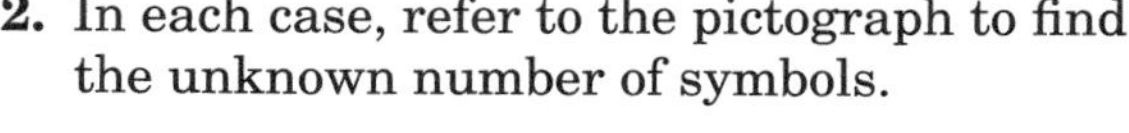

2. In each case, refer to the pictograph to find the unknown number of symbols.

48 days of warm weather

a. [?] symbols for 32 days

600 pints of milk

b. [?] symbols for 960 pints

2,600 theater goers

c. [?] symbols for 1,625 theater goers

3. In a pictograph, symbols of $4\frac{1}{2}$ fish represent 1,800 tons of fish. Of the given choices, the best estimate of the number of tons of fish represented by 2 symbols is

(a) 825 (b) 1,050 (c) 1,200 (d) 1,650

4. This pictograph represents the number of pies sold by Pie Palace during a 5-month period.

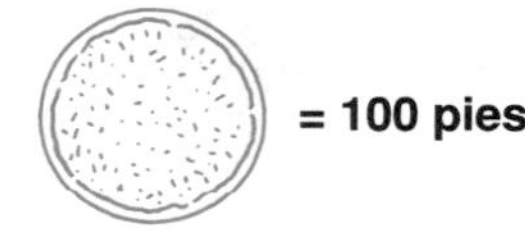

Number of Pies Sold

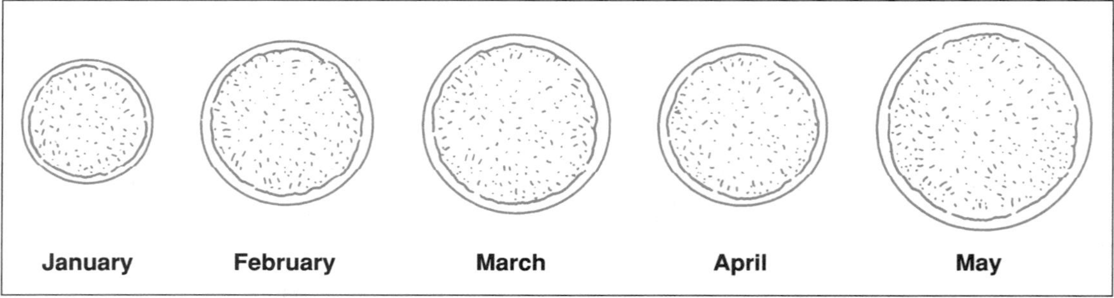

a. In which month were the most pies sold? the fewest?

b. In which 2 months were sales about the same?

c. Which symbol better represents the average number of pies sold per month during this period, February or May?

5. The symbol shown represents movie-ticket sales of 200,000.

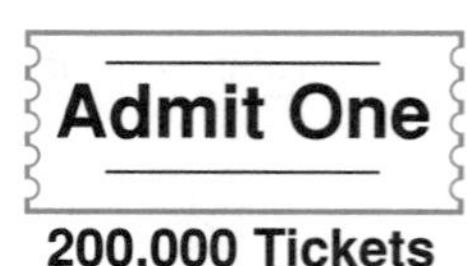

a. How many ticket sales are represented by a symbol that is 4 times as wide and 4 times as high?

b. How many ticket sales are represented by a symbol $2\frac{1}{2}$ times as wide and $2\frac{1}{2}$ times as high?

c. How many times longer should the height and width of the symbol be to represent 1,800,000 ticket sales?

d. If the symbol for 200,000 ticket sales has a height of 1.2 cm and width of 2 cm, what should be the dimensions of the symbol to represent 800,000 tickets?

6. Make a pictograph to show the given information.

Bushels of Apples Shipped by Citrus Coop	
June	3,000
July	4,000
August	1,500
September	7,500
October	8,000
November	500

7. Mark wants to make a pictograph to show the given information about the number of families in his neighborhood with 2 or more jobs in the family.

Year	Number of Families with 2 or More Jobs
1995	23
1996	38
1997	47
1998	52
1999	42
2000	50

He decides to use the symbol to stand for one family. What advice would you give Mark before he begins to make the graph?

SPIRAL REVIEW EXERCISES

1. In its first four games, a high school basketball team scored 288 points. If the team continues scoring at the same rate, how many total points will it have scored in all after 11 games?

2. Mrs. Chase weighs 120 pounds, and her daughter weighs 80 pounds. The ratio of Mrs. Chase's weight to her daughter's weight is (a) 3 : 2 (b) 5 : 2 (c) 2 : 5 (d) 2 : 3

3. On a map, 1 inch represents 5 miles. How long a line segment would be used to represent 35 miles?

4. Jonathan bought a pair of shoes priced at $39.75. Which is the best estimate of a 5% sales tax on the shoes?

   (a) $1.50 (b) $2.00 (c) $2.20 (d) $3.00

5. For every 500 calories burned in weekly exercise, the risk of diabetes for men drops 6%. If a man burns 2,500 calories each week by exercising, by what percent will he reduce his chances of getting diabetes?

6. If 8 out of 160 seniors had overdue books, what percent of the seniors was this?

7. The Russos drove 1,740 miles on a trip. They stopped six times for gasoline, making purchases of 11.6 gal., 12.8 gal., 10.7 gal., 11.1 gal., 9.9 gal., and 11.2 gal. Find, correct to the nearest tenth, the number of miles per gallon of gasoline that they averaged.

8. Which is not equivalent to the others?

   (a) 50 m (b) 0.05 km
   (c) 500 cm (d) 5,000 cm

9. Jason has to drive from Canton to Newton. If he has already traveled $\frac{7}{8}$ of the distance, how many more miles does he have left?

10. Felix plans to cut a piece of paper that is 5 feet 4 inches long into 6 equal pieces. Which of the following expressions could he use to find the length of each piece in inches?

    (a) $(5 \times 36 + 4) \div 12$ (b) $(5 \times 6 + 4) \div 12$
    (c) $(5 \times 12 + 4) \div 6$ (d) $(5 + 4 \times 12) \div 6$

UNIT 14-1 The Meaning of Probability

THE MAIN IDEA

1. ***Probability*** describes how certain it is that a particular event, or ***outcome***, will occur.
2. Probability is the *ratio* of the number of successful outcomes to the total number of possible outcomes.
3. The set of all possible outcomes is called the ***sample space***.
4. To find a probability:
   a. Count the number of successful outcomes, or ways a given event can happen.
   b. Count the number of possible outcomes in the sample space.
   c. Form a ratio that compares the number of successful outcomes to the total number of possible outcomes.

$$\text{probability} = \frac{\text{number of successful outcomes}}{\text{number of possible outcomes}}$$

EXAMPLE 1 What is the probability that a card picked at random from a standard deck of playing cards will be a king?

THINKING ABOUT THE PROBLEM

"At random" means to pick without looking. In a standard deck, there are 52 cards, divided into 4 suits (hearts, diamonds, clubs, and spades). There are 13 cards in each suit (2, 3, 4, 5, 6, 7, 8, 9, 10, jack, queen, king, and ace). Two suits are red: hearts and diamonds; two suits are black: clubs and spades.

Write the number of kings in a deck. 4 ← number of successful outcomes

Write the total number of cards. 52 ← number of possible outcomes

Write the ratio.

$$\text{probability} = \frac{\text{number of successful outcomes}}{\text{number of possible outcomes}}$$

$$= \frac{4}{52}$$

Answer: The probability of picking a king at random from a standard deck of cards is $\frac{4}{52}$, or $\frac{1}{13}$.

EXAMPLE 2 The arrow on the spinner is equally likely to stop in any of the 4 spaces.

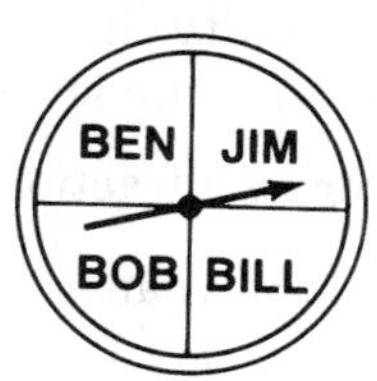

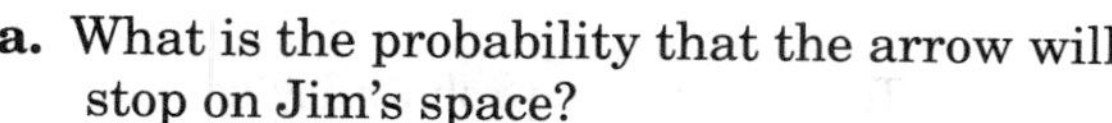

a. What is the probability that the arrow will stop on Jim's space?

Write the number of spaces labeled "Jim."

1 ← number of successful outcomes

Count the total number of spaces.

4 ← number of possible outcomes

Write the probability as a ratio.

$$\text{probability} = \frac{\text{number of successful outcomes}}{\text{number of possible outcomes}}$$

$$= \frac{1}{4}$$

Answer: The probability that the arrow will stop on Jim's space is $\frac{1}{4}$.

b. What is the probability that the arrow will stop on a space with a name that begins with the letter "B"?

$$\text{probability} = \frac{\text{number of successful outcomes}}{\text{number of possible outcomes}}$$

$$= \frac{3}{4} \begin{array}{l} \leftarrow \text{number beginning with “B”} \\ \leftarrow \text{total number of spaces} \end{array}$$

Answer: The probability that the arrow will stop on a space with a name that begins with the letter "B" is $\frac{3}{4}$.

EXAMPLE 3 A box contains 3 red marbles and 2 yellow marbles. What is the probability that a marble chosen at random will be red?

$$\text{probability} = \frac{\text{number of successful outcomes}}{\text{number of possible outcomes}}$$

$$= \frac{3}{5} \begin{array}{l} \leftarrow \text{number of red marbles} \\ \leftarrow \text{total number of marbles} \end{array}$$

Answer: The probability that a red marble will be chosen is $\frac{3}{5}$.

EXAMPLE 4 The ratio of the number of boys to the total number of students in a class is 4 : 5. If a student is chosen at random from that class, what is the probability that the student will be a boy?

The probability of choosing a boy is the same as the ratio of the number of boys to the total number of students.

Answer: The probability of choosing a boy is $\frac{4}{5}$.

EXAMPLE 5 What is the probability that a month of the year chosen at random will have a name beginning with the letter "J"?

Write the answer as a fraction, as a decimal, and as a percent.

List the names of the months of the year (the sample space):

January, February, March, April, May, June, July, August, September, October, November, December

$$\text{probability} = \frac{\text{number of successful outcomes}}{\text{number of possible outcomes}}$$

$$= \frac{3}{12} \begin{array}{l} \leftarrow \text{January, June, July} \\ \leftarrow \text{total number of months} \end{array}$$

Answer: The probability of randomly picking a month beginning with "J" is $\frac{1}{4}$, or 0.25, or 25%.

KEYING IN

To express the probability as a decimal:

Key Sequence	***Display***
3 [÷] 12 [=]	0.25

To express the probability as a percent:

Key Sequence	***Display***
3 [÷] 12 [%]	25.

EXAMPLE 6 In a dealer's lot of 50 cars, 30 are used cars. If a car is chosen at random from that lot, what is the probability that it will be a new car?

Find the number of successful outcomes, which is the number of new cars.

$$\begin{array}{rl} 50 & \leftarrow \text{total number of cars} \\ -30 & \leftarrow \text{number of used cars} \\ \hline 20 & \leftarrow \text{number of new cars} \end{array}$$

Write the probability of choosing a new car.

$$\frac{20}{50} \begin{array}{l} \leftarrow \text{number of new cars} \\ \leftarrow \text{total number of cars} \end{array}$$

Simplify the ratio.

$$\frac{20 \div 10}{50 \div 10} = \frac{2}{5}$$

Answer: The probability of choosing a new car is $\frac{2}{5}$.

CLASS EXERCISES

1. A box contains 25 coins, of which 7 are nickels. If a coin is chosen at random from that box, what is the probability of choosing a nickel?

2. Find the probability that the arrow on the spinner shown will stop on the space marked with:

   a. "A" **b.** "B" **c.** "C"

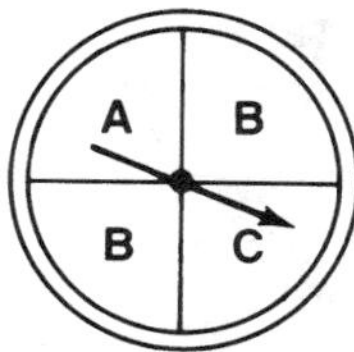

3. In a room containing 32 people, 11 persons are wearing blue shirts. What is the probability that the first person leaving the room will be wearing a blue shirt?

4. In our alphabet, there are 26 letters, of which 5 are vowels, and the rest are consonants. Each letter is written on a separate card and placed in a box. If a card is chosen at random from the box, what is the probability that a vowel is chosen?

5. There are 4 aces in a standard deck of 52 cards. If the cards are shuffled, what is the probability that an ace will be the top card?

6. A box contains only 3 red marbles, 5 yellow marbles, and 4 green marbles. If a marble is chosen at random from the box, what is the probability of choosing:

   a. a red marble **b.** a green marble **c.** a yellow marble

7. What is the probability that a card chosen at random from a standard deck of cards will be:

   a. a black card **b.** a heart

   c. a picture card (count only jacks, queens, and kings as picture cards)

   d. a number card (count an ace as the number 1)

8. What is the probability that a pet chosen at random from among all the pets in a certain animal shelter will be a dog if the ratio of the number of dogs to the total number of pets is 3 : 5? Write the answer as a fraction, as a decimal, and as a percent.

9. In a fish tank, there are 16 fish. Twelve of these are guppies. What is the probability that a fish chosen at random from the tank will not be a guppy?

10. Find the probability of choosing:

    a. a red card from a standard deck

    b. a 3 from the digits 0, 1, 2, 3, 4, 5, 6, 7, 8, 9

    c. an "m" from the letters of the word "mummy"

    d. a girl from a class of 12 girls and 24 boys

    e. an even number from the integers 1 through 25

1. There are 13 hearts in a standard deck of 52 cards. If a card is chosen at random from the deck, find the probability of choosing a heart.

2. A box contains 14 marbles, of which 7 are red. If a marble is chosen at random from the box, what is the probability of choosing a red marble?

3. A drawer contains 10 pairs of socks, of which 6 pairs are brown. If a pair of socks is drawn at random from the drawer, what is the probability of drawing a pair of brown socks?

4. Find the probability that the arrow on the spinner will stop:
   a. on an even number
   b. on a multiple of 3

5. In a random draw from {names of months of the year}, find the probability of selecting a month with:
   a. exactly 4 letters
   b. more than 6 letters

6. The ratio of the number of boys to the total number of guests at Judd's party is 3 : 8. If a person is chosen at random from these guests, what is the probability that the person will be a boy?

7. The ratio of parakeets to the total number of birds in a pet store is 4 : 9. If a bird is chosen at random, what is the probability that it will be a parakeet?

8. If a marble is chosen at random from a bag containing 8 red marbles and 10 green marbles, what is the probability of choosing a green marble?

9. Mrs. Jones has 5 nickels and 10 quarters in her purse. If she picks a coin at random, the probability that it will be a nickel is

   (a) $\frac{5}{10}$ (b) $\frac{1}{3}$ (c) $\frac{10}{15}$ (d) $\frac{10}{5}$

10. A record collection contains 8 jazz records and 24 classical records. The probability that a record chosen at random from this collection is a classical record is

    (a) 75% (b) 50% (c) $33\frac{1}{3}$% (d) 25%

11. If a letter is chosen at random from the letters of the word "SLEEP," the probability of choosing an "E" is (a) 0.2 (b) 0.4 (c) 0.5 (d) 0.6

12. As a fraction in simplest form, as a decimal, and as a percent, write the probability of choosing:
    a. 5 from the numbers 1, 5, 10, 15, and 20
    b. a "T" from the letters of the word "TURTLE"
    c. a boy from a group of 14 girls and 6 boys
    d. a black card from a standard deck of cards

13. There is a 15% chance that it will rain. Express this probability as a fraction in simplest form.

14. Change each of the given probabilities, which are expressed as percents, into fractions in simplest form. **a.** 30% **b.** 80% **c.** 50% **d.** 41% **e.** 32%

15. The probability of Carl's getting a base hit is 0.250. Express this probability as a fraction in simplest form.

16. Change each of the given probabilities, which are expressed as decimals, into fractions in simplest form. **a.** 0.9 **b.** 0.75 **c.** 0.37 **d.** 0.005 **e.** 0.85

17. If one card is drawn at random from a standard deck, what is the probability that it is an ace or a red card?

SPIRAL REVIEW EXERCISES

1. 2.4 m is the same as

   (a) 24 cm (b) 240 cm
   (c) 2,400 cm (d) 0.24 cm

2. Mr. Fein began work at 8:25 A.M. and finished at 4:55 P.M. How many hours did he work?

3. From a group of 40 students, 12 of them went to summer school. What percent of them went to summer school?

4. Round 135.967 to the nearest:

   a. hundred **b.** hundredth

5. Find the product of $2\frac{3}{5}$ and 20.

6. Find the sum of 8.9, 0.06, and 12.39.

7. Write a key sequence that can be used to find the median of the numbers 14.75, 14.83, 14.92, 15.36, 15.36, and 15.64.

8. The largest value of the choices below is

   (a) $\frac{1}{7}$ (b) 7 (c) 7.7 (d) 77%

9. The table gives the average number of complaints about lost or damaged baggage per 1,000 passengers for six airlines. What is the median of this data?

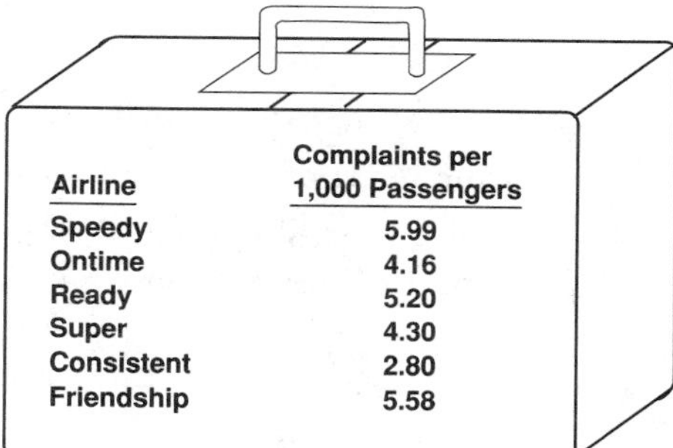

Airline	Complaints per 1,000 Passengers
Speedy	5.99
Ontime	4.16
Ready	5.20
Super	4.30
Consistent	2.80
Friendship	5.58

10. A handbag priced at $38.80 is being sold at a half-price sale. If the sales tax is 5%, then the total cost is

    (a) $1.93 (b) $19.40
    (c) $20.37 (d) $40.74

11. Joseph was absent from school 18 days and present 162 days. The ratio of the number of days present to the total number of days is

    (a) 1 : 9 (b) 9 : 1 (c) 9 : 10 (d) 10 : 9

UNIT 14-2 Probabilities of 0 and 1

THE MAIN IDEA

1. If there are *no successful ways* for an event to happen, then the probability is 0.

$$\text{probability} = \frac{\text{number of successful outcomes}}{\text{number of possible outcomes}}$$

$$= \frac{0}{\text{total}} = 0$$

   This means that the event is *impossible*.

2. If *every way is a successful way* for an event to happen, then the probability is 1.

$$\text{probability} = \frac{\text{number of successful outcomes}}{\text{number of possible outcomes}}$$

$$= \frac{\text{total}}{\text{total}} = 1$$

   This means that the event is *certain*.

3. The least possible probability is 0. The greatest probability is 1. All other probabilities are numbers between 0 and 1.

EXAMPLE 1 What is the probability that the arrow on this spinner will stop on a space marked "John"?

$$\text{probability} = \frac{\text{number of successful outcomes}}{\text{number of possible outcomes}}$$

There are no spaces marked "John," and there are 4 spaces in all.

$$\text{probability} = \frac{0}{4} = 0$$

Answer: The probability that the arrow will stop on a space marked "John" is 0. (The event is impossible.)

EXAMPLE 2 What is the probability that the arrow on this spinner will stop on an odd number?

$$\text{probability} = \frac{\text{number of successful outcomes}}{\text{number of possible outcomes}}$$

All 4 of the numbers are odd.

$$\text{probability} = \frac{4}{4} = 1$$

Answer: The probability that the arrow will stop on an odd number is 1. (The event is certain.)

EXAMPLE 3 What is the probability that if a day of the week is chosen at random its name will have 4 letters?

List the names of the days of the week:

Monday, Tuesday, Wednesday, Thursday, Friday, Saturday, Sunday

There are no days with a 4-letter name. probability $= \frac{0}{7} = 0$

EXAMPLE 4 What is the probability that a month of the year chosen at random will have fewer than 32 days?

The number of days in a month can only be 28, 29, 30, or 31.

All the months have fewer than 32 days. probability $= \frac{12}{12} = 1$

CLASS EXERCISES

1. What is the probability that the arrow on this spinner will stop on:

a. an even number?

b. an odd number?

2. From a box that contains only 11 blue marbles, what is the probability in a random draw of picking one that is:

a. green?

b. blue?

3. What is the probability that a day of the week chosen at random will begin with "X"?

4. If a bag contains only 5 nickels and 7 dimes, find the probability of picking a quarter.

5. Find the probability that the arrow on this spinner will stop on a name that:

a. begins with "A" **b.** begins with "O" **c.** contains "N"

d. has exactly 3 letters **e.** has at least 3 letters

f. has no more than 3 letters **g.** has more than 5 letters

HOMEWORK EXERCISES

1. What is the probability that the arrow on this spinner will stop on:

a. a vowel?

b. a consonant?

2. From a box that contains only 15 red marbles, what is the probability in a random draw of picking one that is:

a. red?

b. blue?

3. What is the probability that a day of the week chosen at random will have a name:
   a. ending with "y"? **b.** beginning with a vowel? a consonant?

4. There are 80 coins in a box, including pennies, nickels, dimes, and quarters. Each of the coins is dated 1990. Ramon concluded that the probability of selecting a silver-colored coin was 1. Explain the error in his reasoning.

5. A box contains 9 marbles, numbered from 1 to 9. Find the probability of picking at random a marble having: **a.** the number 9 **b.** an even number **c.** an odd number **d.** a number less than 10 **e.** a number greater than 20 **f.** a prime number

6. A box contains only 15 red marbles and 21 yellow. One marble is chosen at random, and it is yellow. This yellow marble is not put back into the box, and a second marble is chosen. Find the probability that the second marble is: **a.** yellow **b.** red

SPIRAL REVIEW EXERCISES

1. Last year, Mr. Halston saved 15% of the $40,000 that he earned. At the end of the year, he invested this money at 5% simple interest. How much interest will this money earn in one year?

2. One box of cereal has 22 oz. and costs $3.45; another has 16 oz. and costs $2.65. Which is the better buy?

3. Multiply: $\frac{5}{12} \times \frac{3}{10}$

4. $\frac{19}{7}$ equals (a) $2\frac{5}{7}$ (b) $2\frac{3}{7}$ (c) $1\frac{5}{7}$ (d) $\frac{7}{19}$

5. 842 mL has the same value as
   (a) 0.842 L (b) 8.42 L
   (c) 84.2 L (d) 842,000 L

6. Find the sum of −20 and +42.

7. Evaluate: 38 − 18 ÷ 2 + 7

8. Use the graph below to determine the average number of minutes that Alice spends on the phone each day:
   (a) 3 (b) 30 (c) 25 (d) $2\frac{1}{2}$

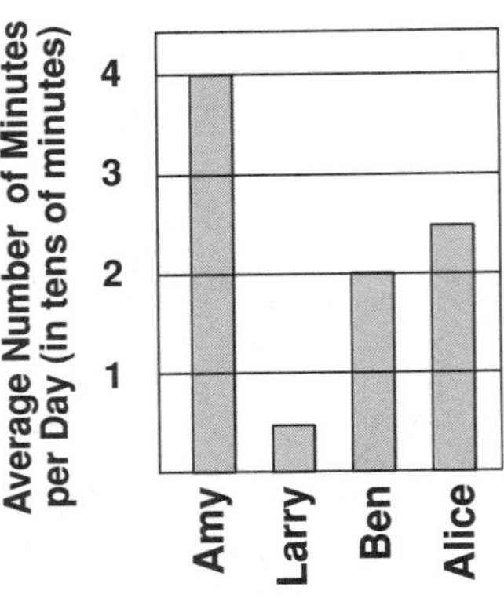

UNIT 14-3 Tree Diagrams; The Counting Principle

TREE DIAGRAMS

THE MAIN IDEA

1. To draw a ***tree diagram*** that illustrates a sample space:
   a. From a single point, draw a branch for each of the possible outcomes for the first activity and list the outcomes next to the branches.
   b. For each outcome of the first activity, draw as many branches as there are outcomes for the second activity, listing these outcomes.
   c. The sample space is the set of all possible outcomes.

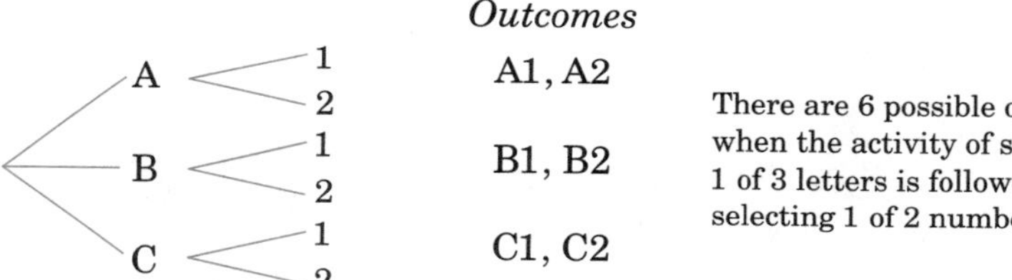

There are 6 possible outcomes when the activity of selecting 1 of 3 letters is followed by selecting 1 of 2 numbers.

   d. If there are more than two activities, continue in the same way.

EXAMPLE 1 On two flights each day from City X to City Y, an airline offers choices of class, time, and meal.

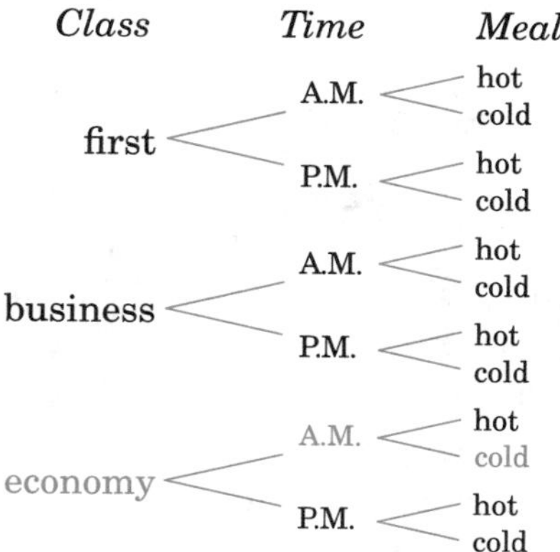

a. Describe the flight shown in color.

Answer: economy class – A.M. – cold meal

b. How many different choices are there?

Count the number of branches in the tree diagram, starting with first class – A.M. – hot meal, and ending with economy – P.M. – cold.

Answer: There are 12 combinations.

EXAMPLE 2 A portfolio for a Social Studies project must include a picture and background information on one item from each set of subjects listed.

Set 1	***Set 2***	***Set 3***
Scenic Views	*National Leader*	*News Event*
Yellowstone Park	Winston Churchill	Hindenburg Explosion
Niagara Falls	Harry Truman	Royal Wedding

Make a tree diagram to show the sample space. List the sample space.

Scenic Views — *National Leaders* — *News Event*

- Yellowstone Park
  - Churchill
    - Hindenburg Explosion
    - Royal Wedding
  - Truman
    - Hindenburg Explosion
    - Royal Wedding
- Niagara Falls
  - Churchill
    - Hindenburg Explosion
    - Royal Wedding
  - Truman
    - Hindenburg Explosion
    - Royal Wedding

Yellowstone-Churchill-Hindenburg	Yellowstone-Churchill-Wedding
Yellowstone-Truman-Hindenburg	Yellowstone-Truman-Wedding
Niagara-Churchill-Hindenburg	Niagara-Churchill-Wedding
Niagara-Truman-Hindenburg	Niagara-Truman-Wedding

CLASS EXERCISES

1. Josette's options for Saturday night include going to either of two concerts with any one of three friends, as shown in the tree diagram.

- The Jazz
  - Meg
  - Kim
  - Jo
- Rhythm & Co.
  - Meg
  - Kim
  - Jo

a. How many options are in the sample space?

b. How many items in the sample space include:

(1) The Jazz **(2)** Kim **(3)** The Jazz and Kim

2. A bag contains a blue marble, a red marble, and a green marble. One marble is selected, its color is recorded, and it is returned to the bag. This process is repeated one more time. Draw a tree diagram and list the sample space for the marble pairs.

3. A group of friends always meets on the 1st, 2nd, or 3rd day of a month with 30 days. Make a tree diagram to show the sample space of these choices.

4. Lena has school clothing as listed:

Skirts	*Blouses*	*Shoes*
black	white	black loafers
red	beige	brown loafers
plaid	black	tan loafers
white		black heels
		brown heels

Suppose you were making a tree diagram to find all the possible school outfits that Lena can put together. Answer each question without actually drawing the tree diagram. Then explain how you arrived at your answer.

a. How many branches would there be at the end of the diagram?

b. How many paths would include the red skirt?

c. How many paths would include the red skirt and the tan loafers?

THE COUNTING PRINCIPLE

THE MAIN IDEA

1. The ***counting principle*** is a way of determining the *number* of items in a sample space for a series of activities without actually making a tree diagram or listing and counting the items.
2. The counting principle is:

Number of items in the sample space	=	Number of outcomes for 1st activity	×	Number of outcomes for 2nd activity	×	Number of outcomes for 3rd activity	× etc.

EXAMPLE 3 A committee of 4 students is to be chosen to represent a school. The committee must contain 1 freshman, 1 sophomore, 1 junior, and 1 senior. The choices must be made from a group of 5 freshmen, 4 sophomores, 6 juniors, and 10 seniors. How many different committees could be formed?

Total number of committees	=	Number of freshmen	×	Number of sophomores	×	Number of juniors	×	Number of seniors
	=	5	×	4	×	6	×	10
	=	1,200 different committees						

CLASS EXERCISES

1. Mrs. Curcio, an English teacher, wants to assign a novel, a play, an essay, and a short story to be read by her 10th-grade class. There are 5 novels, 12 plays, 7 essays, and 8 short stories on the reading list for 10th grade. How many different assignments can Mrs. Curcio put together?
2. In the town of Lyons, the town council wants to choose a month and a day of the week that will be publicized as "Pride of Lyons Day." How many different selections of a month and a day of the week are possible?
3. A soup factory labels cases of canned soup with codes that consist of a letter of the alphabet followed by one digit from 0 to 9. How many cases can be labeled without repeating a code?
4. On a quiz with 5 multiple-choice questions, each question had choices (a), (b), (c), and (d) as possible answers. How many different ways are there of answering the five questions?

HOMEWORK EXERCISES

1. Barbara has rye, wheat, and soda crackers. She has made egg salad, tuna salad, onion dip, and guacamole. Make a tree diagram to show how many different kinds of party snacks she can prepare by putting one mix on each cracker. List the sample space.
2. In a word puzzle, Sean must choose 1 of the 5 vowels followed by the letter "T" or "L".
   a. Make a tree diagram, and list the sample space for these selections.
   b. How many of these selections start with the letter "U"? end with the letter "L"?
3. Philip, Martha, and Julia write their names on slips of paper and drop them into a hat. A name is chosen at random, the slip is replaced in the hat, and a second name is chosen.
   a. Make a tree diagram, and list the sample space for these selections.
   b. How many selections of 2 names result in a choice of 2 different persons?
4. Each morning and afternoon on Monday through Friday, 2 secretaries, Al and Bev, take turns going to the post office.
   a. Make a tree diagram, and list the sample space for the selection of a secretary for the morning and afternoon of each day.
   b. In how many of these selections does: **(1)** Bev go on Wednesday? **(2)** Al go in the afternoon? **(3)** either Al or Bev go in the morning?
5. The new Turbo car can be ordered in the colors listed. How many different color combinations are possible?

Upholstery	*Roof*	*Body*
pumpkin	butterscotch	grape
plum	tomato	almond
raspberry		toast

6. Cara has a 12-volume set of encyclopedias. Each volume is divided into 5 chapters, each chapter contains 4 sections, and each section contains 6 paragraphs. How many paragraphs are there?

7. An ice-cream store advertises 31 different flavors. If a two-scoop ice-cream cone can have either two scoops of the same flavor or of 2 different flavors, how many different kinds of two-scoop cones can be made?

8. Identification numbers for a fleet of trucks will have a digit from 1 to 3 followed by a letter of the alphabet followed by the digit 2. How many trucks can be labeled, each with a different ID?

9. A new one-hour television show on Channel X will be scheduled once a week, either immediately before, during, or immediately after the daily 6:00 P.M. news hour on Channel Y. How many different time slots are being considered for the new show on Channel X?

10. How many different results can there be of a spin of the spinner and a toss of the coin?

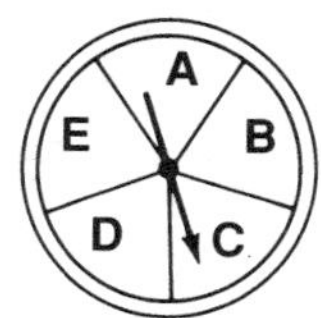

SPIRAL REVIEW EXERCISES

1. What is the probability that a month chosen at random will have a name that ends with the letter "r"?

2. In choosing a day of the week at random, what is the probability that its name will have fewer than 10 letters?

3. In a bag, there are 13 red marbles and 17 green marbles. The probability that a marble picked at random will be red is

(a) $\frac{13}{30}$ (b) $\frac{17}{30}$ (c) $\frac{13}{17}$ (d) $\frac{17}{13}$

4. Mr. Syms buys a television set by making a \$50 down payment and paying 20 installments of \$18 each. Find the cost of the television set.

5. To the quotient of −20 and 4, add −6.

6. Which key sequence will change 1,452 mm to meters?

(a) 1452 [÷] 1000 [=] (b) 1452 [×] 1000 [=]
(c) 1452 [÷] 100 [=] (d) 1452 [×] 100 [=]

7. Ms. Jennings needs maps for her delivery company. The chart below gives the bulk discount that is available for the maps. The least amount of money that she must spend to get at least 80 maps is

(a) \$60 (b) \$75.75 (c) \$80 (d) \$100

Bulk Discount Rates

1–4 maps	\$2.50 each
5–10	1.50 each
11–50	1.25 each
51–100	1.00 each
101–500	0.75 each
501 or more	0.60 each

8. In 1980, there were approximately 121.6 million cars on the road in the United States. By the year 2000, it is estimated that the number of cars on the road increased to 190 million. The percent increase is approximately

(a) 25% (b) 36% (c) 56% (d) 70%

UNIT 14-4 Probability with Number Cubes; Compound Events

PROBABILITY WITH NUMBER CUBES

THE MAIN IDEA

1. Number cubes (dice) each have six sides with numbers or dots that represent the numbers 1 through 6.
2. A toss of one *die* can have any of 6 possible results:

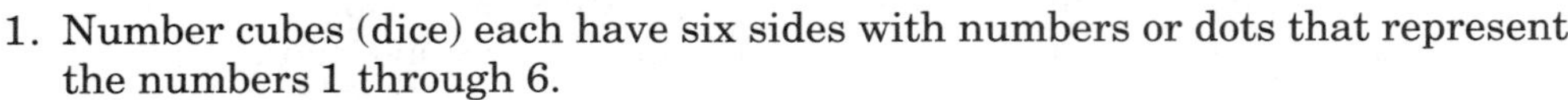

3. A toss of two *dice* can have any of 36 possible results. The sample space is:

1, 1 → 2	2, 1 → 3	3, 1 → 4	4, 1 → 5	5, 1 → 6	6, 1 → 7
1, 2 → 3	2, 2 → 4	3, 2 → 5	4, 2 → 6	5, 2 → 7	6, 2 → 8
1, 3 → 4	2, 3 → 5	3, 3 → 6	4, 3 → 7	5, 3 → 8	6, 3 → 9
1, 4 → 5	2, 4 → 6	3, 4 → 7	4, 4 → 8	5, 4 → 9	6, 4 → 10
1, 5 → 6	2, 5 → 7	3, 5 → 8	4, 5 → 9	5, 5 → 10	6, 5 → 11
1, 6 → 7	2, 6 → 8	3, 6 → 9	4, 6 → 10	5, 6 → 11	6, 6 → 12

EXAMPLE 1 If one die is rolled, what is the probability that it will show:

a. the value "5"? Of the 6 ways the die can come up, only 1 of these ways is "5." probability $= \frac{1}{6}$

b. an even value? Of the 6 ways the die can come up, 3 ways are even: "2," "4," "6" probability $= \frac{3}{6}$, or $\frac{1}{2}$

EXAMPLE 2 When rolling a pair of dice, what is the probability of getting:

a. a sum of 7? Of the 36 ways in which a pair of dice can come up, 6 ways show a sum of 7. probability $= \frac{6}{36}$, or $\frac{1}{6}$

b. a sum of 13? Since the greatest sum is 12, there is no way to get 13. probability $= \frac{0}{36}$, or 0

EXAMPLE 3 Which sums from 2 through 12 have the lowest probability of coming up in a roll of a pair of dice?

THINKING ABOUT THE PROBLEM

A key word in the problem is "lowest." This suggests comparing the probabilities of all the possible sums, from 2 through 12, to find the lowest.

Of the 36 ways in which a pair of dice can fall, only 1 way is a sum of 2: $1 + 1$ probability of getting a sum of $2 = \frac{1}{36}$

Only 1 way is a sum of 12: $6 + 6$ probability of getting a sum of $12 = \frac{1}{36}$

All other sums can be formed in more than 1 way (for example, $5 = 2 + 3$ or $4 + 1$).

Thus, all other sums have probabilities that are greater than $\frac{1}{36}$.

Answer: The sums with the lowest probabilities are 2 and 12.

CLASS EXERCISES

1. In rolling one die, find the probability that it will show:

a. "1" **b.** "2" **c.** "3" **d.** "4" **e.** "5" **f.** "6" **g.** "7"
h. an even number **i.** an odd number **j.** a number less than 4

2. In rolling a pair of dice, find the probability of getting a sum of:

a. 2 **b.** 3 **c.** 4 **d.** 5 **e.** 6 **f.** 7 **g.** 8 **h.** 9 **i.** 10 **j.** 11 **k.** 12 **l.** 1

3. When a pair of dice is rolled, what is the probability that:

a. the sum will be an odd number? **b.** the sum will be greater than 9?

4. When a pair of dice is rolled, what sum has the same probability as a sum of 5?

COMPOUND EVENTS

THE MAIN IDEA

1. If the outcome of one event in no way influences the outcome of a second event, the two events are ***independent events***. Events that do influence each other are called ***dependent events***.
2. A ***compound event*** consists of two or more events.
3. To find the probability of a compound event made up of independent events, multiply the probabilities of the independent events.
4. To find the probability of a compound event made up of two dependent events:
   a. Find the probability of the first event.
   b. Assuming a successful outcome for the first event, find the probability of the second event *after the first event has occurred*.
   c. Multiply the two probabilities.

EXAMPLE 4 Tell whether the events described are independent or dependent.

	Events	***Dependent or Independent?***
a.	Rolling a "6" on a die; getting "heads" on a coin toss	Independent
b.	Drawing an ace from a deck of cards; then drawing another ace without replacing the first ace.	Dependent; the number of successful outcomes for the second event is related to the result of the first.
c.	Drawing a king from a deck of cards and replacing it; drawing a king again.	Independent

EXAMPLE 5 A white die and a green die are rolled. Find the probability of obtaining:

a. "6" on the white and "6" on the green

The events are independent. probability of "6" on white = $\frac{1}{6}$

probability of "6" on green = $\frac{1}{6}$

Multiply the probabilities: $\frac{1}{6} \times \frac{1}{6} = \frac{1}{36}$

b. an odd number on the white die and an odd number on the green

The events are independent.

probability of odd on white = $\frac{3}{6}$, or $\frac{1}{2}$

probability of odd on green = $\frac{3}{6}$, or $\frac{1}{2}$

Multiply the probabilities: $\frac{1}{2} \times \frac{1}{2} = \frac{1}{4}$

c. an odd number on the white and the same odd number on the green

The events are dependent.

probability of odd on white = $\frac{3}{6}$, or $\frac{1}{2}$

probability of green showing the one outcome that was on the white = $\frac{1}{6}$

Multiply the probabilities: $\frac{1}{2} \times \frac{1}{6} = \frac{1}{12}$

CLASS EXERCISES

1. Tell whether the events described are independent or dependent.

a. A coin that is tossed shows "tails," and a card drawn from a standard deck is a queen.

b. A white die and a green die are rolled. The white die shows an odd number and the green die shows an even number.

c. From a class of 9 boys and 7 girls, a student is selected at random. Then a second student is selected at random from the remaining students. A girl is selected for both.

2. A die is rolled two times in a row. What is the probability that it will show a "5" both times?

3. The Samson triplets Marcia, Mary, and Howard are all in the same class. The class has 12 girl and 15 boys. One boy and one girl are selected at random. What is the probability that Howard and one of his sisters will be selected?

4. A bag contains 3 red and 2 blue marbles. A marble is selected at random, its color is recorded, and a second marble is selected.

a. If the first marble is put back in the bag before the second marble is selected, what is the probability that the first marble is red and the second is blue?

b. If the first marble is red and is not put back in the bag, what is the probability that the second marble is blue?

c. If the first marble is not replaced, what is the probability that the first marble is red and the second marble is blue?

d. If the first marble is not replaced, what is the probability that both marbles are blue?

HOMEWORK EXERCISES

1. In rolling a die, find the probability that it will show:
   a. "2" **b.** "1" **c.** "9" **d.** a number less than 5 **e.** a number less than or equal to 6

2. The probability of rolling a die and having it show the value "3" is

   (a) $\frac{1}{5}$ (b) $\frac{1}{6}$ (c) $\frac{5}{6}$ (d) $\frac{1}{2}$

3. The probability of rolling a number that is a multiple of 3 with a die is

   (a) $\frac{1}{6}$ (b) $\frac{1}{3}$ (c) $\frac{1}{2}$ (d) $\frac{2}{3}$

4. When a pair of dice is rolled, find the probability that the sum is: **a.** 7 **b.** 2

5. In rolling a pair of dice, the probability that the sum will be 8 is

   (a) $\frac{1}{8}$ (b) $\frac{5}{36}$ (c) $\frac{8}{36}$ (d) $\frac{1}{11}$

6. In rolling a pair of dice, the probability of getting a sum less than 15 is

   (a) 0 (b) $\frac{15}{36}$ (c) $\frac{1}{15}$ (d) 1

7. A coin is tossed at the same time that a die is rolled. What is the probability of getting "heads" on the coin and "4" or higher on the die?

8. Tell whether the situation described involves independent or dependent events.

   a. The spinner shown stops at "4", and a queen is drawn from a standard deck of cards.
   b. The number shown on the spinner and the number shown on a die that is rolled have a sum less than 4.
   c. A bag contains 4 red apples and 3 green apples. Two friends each remove an apple from the bag and find that both apples are the same color.

9. A die is rolled three times in a row. What is the probability that it will show "3" on the first roll, "4" on the second roll, and "5" on the third roll?

10. Which is less likely to happen on three rolls of a die, getting three "1's" in a row or getting "4", "5", and "6" in that order? Explain your reasoning.

11. What is the probability that these two spinners will stop at:
    a. the number "3" and the letter "t"?
    b. an even number and a vowel?
    c. a prime number and a consonant?
    d. an odd number and the letter "y"?
    e. a number less than 10 and a letter in the word "ate"?

12. A fair coin tossed four times results in four "heads." What is the probability that a fifth toss will also show "heads"? Explain.

13. Two cards are drawn at random from a standard deck of cards. If the first card is a queen and is replaced before the second card is drawn, is the probability of drawing 2 queens greater than, less than, or equal to the probability of drawing 2 queens if the first card is not replaced? Explain.

SPIRAL REVIEW EXERCISES

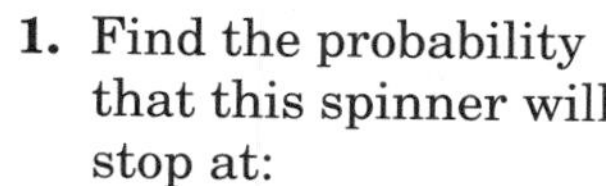

1. Find the probability that this spinner will stop at:

   a. "3" **b.** "7"
   c. a number less than 5
   d. an even number
   e. a number less than 7

2. If a card is drawn at random from a standard deck of playing cards, what is the probability that the card chosen is:
   a. a club **b.** a red card
   c. a picture card **d.** a red queen
   e. a black picture card

3. From the sum of $\frac{1}{2}$ and $\frac{2}{5}$, subtract $\frac{7}{10}$.

4. Write $\frac{37}{24}$ as a mixed number.

5. Express $\frac{5}{8}$ as a decimal.

6. Find the product of (–4) and (+14).

7. If a temperature rises from –3° to +7°, how many degrees has it risen?

9. A $240 bike is on sale at 10% off. The sale price is
   (a) $208 (b) $200
   (c) $196 (d) $216

10. Predict the display on an entry-order calculator after the key sequence:

    147 [−] 147 [÷] 3 [=]

    (a) 98 (b) Error (c) 49 (d) 0

11. Mr. Jameson purchased four compact discs at $11.50 each. If the sales tax rate is 5%, how much change did he receive from a $50 bill?

12. Johnson's Market sells 5 pounds of apples for $1.95, while the Farmers' Market sells 8 apples for $1.20. If the average apple weighs $\frac{3}{8}$ pound, which market is offering the better price per pound?

13. John spent $4.38 on a tape. How much change did he receive from a $10 bill?
    (a) $15.62 (b) $5.72
    (c) $5.62 (d) $6.62

8. A sandwich shop offers a luncheon special in which customers can choose 1 meat, 1 topping, and a type of bread from the menu below.

   How many different kinds of sandwiches can be made from this menu?

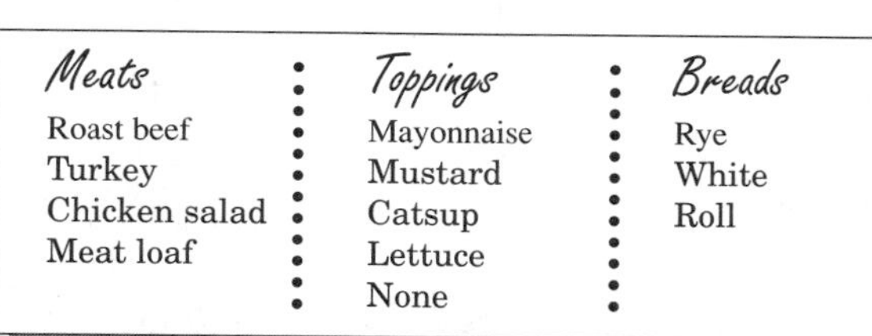

Meats	Toppings	Breads
Roast beef	Mayonnaise	Rye
Turkey	Mustard	White
Chicken salad	Catsup	Roll
Meat loaf	Lettuce	
	None	

UNIT 14-5 Permutations; Combinations

PERMUTATIONS

THE MAIN IDEA

1. An arrangement of a set of items is called a ***permutation***. There are 6 permutations of the 3 letters A, B, and C:

   ABC BAC CAB ACB BCA CBA

2. The symbol 5! represents the product $5 \times 4 \times 3 \times 2 \times 1$. Read the symbol 5! as "five ***factorial***."
3. To find the number of permutations of a set of n items, use *one* of these methods:
   a. List all the arrangements.
   b. Make a tree diagram of the arrangements.
   c. Find the value of $n!$.
4. When a group of items is to be selected from a larger set, the larger set is called the ***selection set***.

EXAMPLE 1 Evaluate each factorial expression.

a. $6! = 6 \times 5 \times 4 \times 3 \times 2 \times 1 = 720$

b. $2! \times 3! = (2 \times 1) \times (3 \times 2 \times 1) = 2 \times 6 = 12$

c. $$\frac{5!}{3!} = \frac{5 \times 4 \times 3 \times 2 \times 1}{3 \times 2 \times 1} = \frac{5 \times 4 \times \overset{1}{\cancel{3}} \times \overset{1}{\cancel{2}} \times 1}{\underset{1}{\cancel{3}} \times \underset{1}{\cancel{2}} \times 1} = \frac{5 \times 4}{1} = 20$$

d. $\frac{9!}{6!}$ Find the value of the denominator and store it in memory. Then, find the value of the numerator and divide it by the denominator recalled from memory.

Key Sequence	***Display***	
6 [×] 5 [×] 4 [×] 3 [×] 2 [M+]	M 720.	← denominator
9 [×] 8 [×] 7 [×] 6 [×] 5 [×] 4 [×] 3 [×] 2 [=]	M 362880.	← numerator
[÷] [MR] [=]	M 504.	

EXAMPLE 2 The manager of The Sports Store wants to show equipment for football, basketball, hockey, and soccer in 4 separate window displays.

a. Make a list to find the number of different arrangements of the 4 displays.

Use the first letter of each sport as an abbreviation.

Work in an organized way to be sure you list all of the possibilities.

FBHS	BFHS	HFBS	SFBH
FBSH	BFSH	HFSB	SFHB
FHBS	BHFS	HBFS	SBFH
FHSB	BHSF	HBSF	SBHF
FSBH	BSFH	HSFB	SHFB
FSHB	BSHF	HSBF	SHBF

24 arrangements

b. Make a tree diagram to show the number of arrangements.

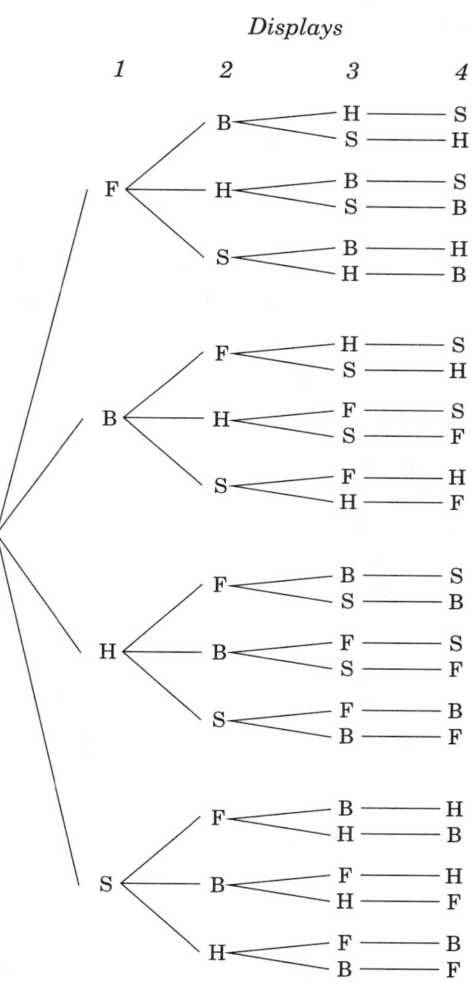

c. Evaluate 4! to show the number of different arrangements of the 4 displays.

$$4 \times 3 \times 2 \times 1 = 24$$

EXAMPLE 3 The director of an art gallery wants to select 2 paintings from a set of 5 for display, one on the north wall, and one on the east wall.

a. In how many ways can a "north" and an "east" painting be selected?

The paintings are to be selected from a larger set. There are 5 items in the selection set, and so, 5 ways of selecting the first painting. After the first painting is chosen, there are 5 – 1, or 4, ways to select the second painting.

$$\text{number of permutations of 2 paintings out of 5} = \underbrace{5 \times 4}_{2 \text{ factors}} = 20$$

The number of factors is equal to the number of paintings selected.

Answer: There are 20 ways of selecting a "north" and an "east" painting out of the 5 paintings.

b. If the 5 paintings are represented by A, B, C, D, and E, what is the probability that an arrangement chosen at random will be first A, then B?

Of the 20 possible arrangements, only 1 will have A first, then B.

$$\text{probability} = \frac{1}{20}$$

CLASS EXERCISES

1. Find the value of **a.** $9!$ **b.** $3! \times 2!$ **c.** $\frac{8!}{3!}$ **d.** $\frac{7!}{4!}$ **e.** $\frac{10!}{9!}$ **f.** $\frac{87!}{86!}$

2. Maxine needs to call Art, Beth, Carla, and Dan. Make a list to show the possibilities for the order in which she could make the calls.

3. At a mall, a group wants to go to a video store, a snack bar, and a sporting goods store. Make a tree diagram to show their choices for the order in which they could make all of these stops.

4. How many different arrangements can be made using all of the letters in the word "SPOT"? How many of these arrangements are English words?

5. Five students are making an oral presentation to their history class. In how many ways can they seat themselves in a row at the front of the room?

6. A musical director must put 8 songs in sequence for a winter concert. How many different sequences can be made?

7. A baseball coach wants to choose 2 out of 5 players for the positions of manager and captain. In how many ways can the coach assign 2 players to these positions?

8. In a word game, Yolanda has the letters A, E, S, and T on individual plastic tiles. In how many ways can she select 2 letters, one at a time, from this set?

9. What is the probability of selecting the number 385 at random from all possible arrangements of the digits 3, 5, and 8?

10. How many 2-digit house numbers can be formed from the digits 1 through 9 if both digits are different?

11. The names of 4 students—Charles, Monica, Stephanie, and Andre—are to be drawn in sequence from slips of paper in a hat to determine the order in which they will make their oral reports to the class. What is the probability that

a. the sequence Monica, Charles, Stephanie, Andre will be chosen?

b. the sequence that is chosen will begin with Andre?

COMBINATIONS

THE MAIN IDEA

1. A selection of items, without arranging them in any particular order, is called a ***combination***. The combinations of 2 letters chosen from among A, B, and C are: AB, BC, and AC. There is only 1 combination of all 3 letters.
2. Combinations do not include different permutations (arrangements) of the same selections. AB and BA are the same combination.
3. For a given set of 2 or more items, there are always more permutations than combinations because for each *combination* of n items there are $n!$ *permutations* (ways to arrange them in order).
4. If n items are to be chosen from a larger selection set, the number of combinations of those n items can be found as follows:
   a. Find the number of ways to choose and arrange those n items (permutations).
   b. Divide by $n!$, which is the number of permutations (arrangements), of those n items.

EXAMPLE 4 Two students are to be selected from a group of 5 students.

a. List the permutations of 2 students chosen from 5 students. Let the letters A, B, C, D, and E represent the students.

There are 5 ways to choose the first student, then 4 ways to choose the second:

$5 \times 4 = 20$ ways to choose

AB	BA	CA	DA	EA
AC	BC	CB	DB	EB
AD	BD	CD	DC	EC
AE	BE	CE	DE	ED

Answer: There are 20 permutations.

b. Find the number of different committees of 2 students that can be formed out of the 5 students.

From the 20 permutations, eliminate the duplications. As a committee choice, AB is the same as BA.

AB	~~BA~~	~~CA~~	~~DA~~	~~EA~~
AC	BC	~~CB~~	~~DB~~	~~EB~~
AD	BD	CD	~~DC~~	~~EC~~
AE	BE	CE	DE	~~ED~~

Answer: There are 10 combinations.

Note the relation between the number of combinations and the number of permutations of selecting 2 of 5 items.

number of permutations of 2 out of 5 ⟶ 5×4
number of permutations of 2 out of 2 ⟶ 2×1

$$\frac{5 \times 4}{2 \times 1} = \frac{20}{2} = 10$$

⟵ number of combinations of 2 out of 5

EXAMPLE 5 Eli, Jamie, Sue, Tony, Maria, and Hua are divided equally into two teams. What is the probability that Eli, Jamie, and Maria will form one team?

First, find the number of combinations of 3 children out of 6.

$$\text{number of combinations of 3 out of 6} = \frac{\text{number of permutations of 3 out of 6}}{\text{number of permutations of 3 out of 3}}$$

$$= \frac{6 \times 5 \times 4}{3 \times 2 \times 1} = 20 \text{ combinations}$$

Of the 20 possible teams of 3 children, only 1 will include Eli, Jamie, and Maria.

Answer: The probability that Eli, Jamie, and Maria form a team is $\frac{1}{20}$.

CLASS EXERCISES

1. In a spelling game, Angela has to pick a random set of 3 tiles from a set of 4 tiles. In how many different ways can she make her selection?
2. With dinner, a restaurant offers a choice of 2 vegetables: carrots, string beans, and peas.
   a. Is the order in which a diner chooses the vegetables important to the result?
   b. How many combinations of 2 vegetables does the restaurant offer?
3. The letters of the word "PAINT" are written on individual slips of paper and 3 are chosen at random. What is the probability that the letters "P," "A," and "T," will be chosen?
4. Four months are to be chosen at random in order to award gifts to those workers with birthdays in those months. In how many ways can the 4 months be chosen?
5. A social studies committee must choose 2 of the United States to study as a special project. In how many ways can the committee make its choice?
6. Marty needed his Spanish book and science book for school one morning. In his rush out of the house, he grabbed the top two books from a random pile of 6 books on his desk. What is the probability that he grabbed the books he needed?

HOMEWORK EXERCISES

1. Find the value of: **a.** $7!$ **b.** $4! \times 3!$ **c.** $\frac{8!}{7!}$ **d.** $\frac{40!}{39!}$ **e.** $\frac{7!}{4!}$ **f.** $\frac{11!}{7!}$
2. For a demonstration tape, a rock band will choose 3 songs from a group of 7 songs and record them in order. How many different arrangements can they make?
3. Three different letters of the alphabet are to be chosen and arranged as a 3-letter identification code. How many different codes can be made?
4. Mr. Loman has to visit the towns of Quenton, Patcher, Dunkle, and Waters on his sales route. In how many different ways can he arrange the sequence of his visits to these towns?
5. Five cards, an ace, a 10, a king, a queen, and a jack are shuffled. What is the probability that they will turn up in the order ace, queen, 10, jack, king?
6. The Vixens have 4 pitchers on the softball team, Myra, Nicole, Abby, and Teresa. The coach wants to use all 4 pitchers in the game.
   a. In how many different ways can the coach put the 4 pitchers in sequence?
   b. What is the probability that the sequence used will be Teresa, Myra, Nicole, Abby?
   c. What is the probability that the sequence used will begin with Abby?

7. How many combinations of 3 puppies can be chosen from a litter of 7 to be given away?

8. In how many ways can Will select 2 coins from his collection of 10 Canadian coins?

9. Each student in a literature class must write about 3 poems chosen from the following list:

Annabel Lee	*Home-Thoughts, From Abroad*
Elegy Written in a Country Churchyard	*Crossing the Bar*
I Hear America Singing	*Sonnets from the Portuguese*

How many different selections of 3 poems can be made?

10. The digits 1, 2, 3, 5, and 7 are written on individual slips of paper. Two are selected at random. What is the probability that the sum of the 2 digits is

a. greater than 10? **b.** less than 7? **c.** odd? **d.** prime?

11. Four United States senators are to be selected at random for a special committee. What is the probability that the 2 senators from Iowa and the 2 from Colorado will be chosen?

12. Toni's CD player has a random feature that selects songs from a disk in the machine and plays them in random order. If she now has a disk with 12 songs in the machine, what is the probability that the next 4 songs will be songs 6, 2, 8, and 1, in any order?

SPIRAL REVIEW EXERCISES

1. When a pair of dice is tossed, what is the probability that the sum will be 14?

2. Kim's restaurant bill is $12. She wants to leave the waiter a 15% tip. How much money should she leave for the tip?

3. What is the probability that a die will show a number less than 3 when it is tossed?

4. A stock valued at $50 a share had a 12% increase in price followed by a 5% decrease. What was the value of the stock after these two changes in price?

5. Find the mean of these weekly salaries:

$425, $480, $650, $685

6. There are 9 roads leading from Potsdam to York and 4 roads leading from York to Clark. How many possible routes can be taken from Potsdam to York and then to Clark?

7. Which numeral represents one million, fifty-two thousand?

(a) 1,520,000 (b) 105,200
(c) 1,052,000 (d) 1,005,200

8. Which key is unnecessary to find half the product of 75 and 38?

75 [×] 38 [=] [÷] 2 [=]

9. An inspector found 7 defective parts in a batch of 750 parts. At this rate, how many defective parts should the inspector expect to find in a batch of 20,000 parts?

10. Mr. Crane bought a $600 refrigerator by making a 10% down payment and 12 payments of $48 each. What percent of the list price of the refrigerator did he pay in carrying charges?

UNIT 14-6 Using Probability to Predict Outcomes

THE MAIN IDEA

If you know the probability of an event and you know the number of trials that will take place, you can predict the *number of times* that you can expect the event to happen:

1. Find the probability that the event will happen.
2. Multiply this probability by the number of trials.

expected number of successes = probability of success × number of trials

EXAMPLE 1 About how many times can you expect to get the value "4" if you roll a die 180 times in a row?

Find the probability of getting the value "4" on 1 roll of a die.	probability $= \frac{1}{6}$
Multiply the probability by the number of trials.	$\frac{1}{\cancel{6}_1} \times \cancel{180}^{30} = 30$

Answer: You can expect to get "4" about 30 times in 180 rolls of a die.

EXAMPLE 2 In 144 rolls of a pair of dice, about how many times can you expect a sum of 8?

Find the probability of getting a sum of 8 with a pair of dice.	probability $= \frac{5}{36}$
Multiply the probability by the number of trials.	$\frac{5}{\cancel{36}_1} \times \cancel{144}^{4} = 20$

Answer: You can expect a sum of 8 about 20 times in 144 rolls of a pair of dice.

CLASS EXERCISES

1. If a die is rolled 120 times in a row, how many times would you expect to roll the value "1"?

2. If you rolled a pair of dice 180 times in a row, tell how many times you would expect to roll:
a. a sum of 10 **b.** a sum of 7 **c.** a sum of 11 **d.** a sum of 3 **e.** a sum of 16

3. With a *fair* coin, either side is equally likely to occur in a toss. A *weighted* coin makes one side more likely to occur. If the probability of getting "heads" is $\frac{2}{3}$ when a certain weighted coin is tossed, how many "heads" would you expect to get in 60 tosses of this coin?

4. In basketball practice, the probability that Warren will make a basket is $\frac{2}{3}$. How many baskets can Warren be expected to make in 330 attempts?

5. In 150 spins of this spinner, tell about how many times you can expect it to stop on:

   a. the letter "a" **b.** the letter "c" **c.** the letter "z"

6. A magician shuffles a deck of cards, chooses a card at random, and then puts the chosen card back into the deck. If he does this 32 times in a row, how many hearts would you expect him to choose?

7. How many times would you expect to get each of the following in 104 individual random selections of a card from a standard deck, if the chosen card is returned to the deck each time?

   a. a king **b.** a picture card **c.** a red card **d.** a number card (count an ace as number 1)

HOMEWORK EXERCISES

1. In 300 rolls of a die, how many times would you expect the die to show the value "4"?

2. In 500 tosses of a fair coin, how many times would you expect the coin to land heads up?

3. If you rolled a pair of dice 108 times in a row, state how many times you would expect it to roll:

   a. a sum of 5 **b.** a sum of 2 **c.** a sum of 8 **d.** a sum of 14

4. In 3,000 rolls of a die, how many times would you expect the die to show a number less than 7?

5. The probability that it will rain on a given day is 0.2. In the next 100 days, how many rainy days can you expect?

6. The probability that Reggie will hit a home run is $\frac{1}{15}$. How many home runs can Reggie be expected to hit in 450 times at bat?

7. In 100 spins of this spinner, tell about how many times you can expect the arrow to stop on:

   a. the number "1" **b.** the number "5"
   c. the number "3" **d.** the number "7"

8. A marble is picked at random from a box containing 3 red marbles, 2 blue marbles, and 5 yellow marbles. The marble is then placed back into the box. If this is done 150 times, how many times can you expect a red marble to be picked?

9. In 180 rolls of a die, the number of times you would expect it to show a number less than 10 is
   (a) 30 (b) 90 (c) 120 (d) 180

10. Copy and complete the chart shown.

Probability of Success	Number of Trials	Expected Number of Successes
0.3	50	
20%	90	
	200	100
$\frac{3}{4}$		75

SPIRAL REVIEW EXERCISES

1. A jar contains only 4 red marbles, 5 white marbles, and 6 blue marbles. If a marble is chosen at random from this jar, find the probability that the marble chosen is:

   a. red **b.** green **c.** not red
   d. not green **e.** red or blue

2. Given: the set of positive integers that are less than 20 and that are multiples of 3. If a number is chosen at random from this set, find the probability that the number chosen is:

   a. a two-digit number **b.** an odd number
   c. a prime number **d.** a multiple of 4
   e. a multiple of 7 **f.** less than 30

3. If a die is rolled, find the probability that it will show:

   a. the value "5"
   b. a value that is not "5"
   c. a value that is less than 5
   d. a value that is not less than 5

4. 80% is equivalent to

   (a) $\frac{7}{10}$ (b) $\frac{4}{5}$ (c) $\frac{8}{100}$ (d) $\frac{20}{100}$

5. The sum of $\frac{1}{2}$ and $\frac{2}{3}$ is

   (a) $\frac{3}{5}$ (b) $\frac{1}{3}$ (c) $\frac{3}{4}$ (d) $\frac{7}{6}$

6. The quotient of +3 and –6 is

   (a) –18 (b) +18 (c) $+\frac{1}{2}$ (d) $-\frac{1}{2}$

7. Find, to the nearest cent, the amount of a 7% sales tax on an item that sells for $10.99.

8. The Altman family has a monthly income of $3,750. They budget $2,850 for all items except food. Write a key sequence that can be used to find the percent of their monthly income that the Altman's budget for food.

9. If 3 apples cost $0.99, then the cost of 5 apples is

   (a) $0.33 (b) $1.32 (c) $1.65 (d) $2.97

10. In the United States, about 6 billion pounds of dog food and 2 billion pounds of cat food are purchased each year. The circle graphs show store sales of the different types of dog foods and cat foods. The total number of pounds of dry dog food and dry cat food sold in one year is approximately

    (a) 0.86 billion lbs. (b) 4.0 billion lbs.
    (c) 1.02 billion lbs. (d) 5.06 billion lbs.

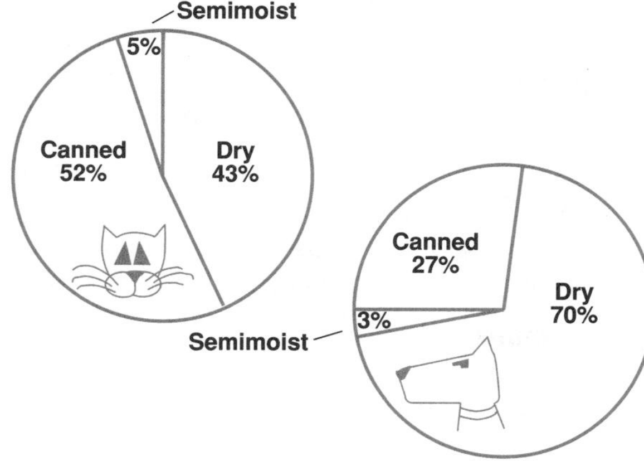

11. In a typical hour of children's programming on network television, approximately 12 minutes is devoted to commercials. What percent of the time is devoted to commercials?

UNIT 14-7 Finding the Probability That an Event Will Not Happen

THE MAIN IDEA

1. The sum of the probability that an event will happen and the probability that it will not happen is 1.

   probability of happening + probability of not happening = 1

2. To find the probability that an event will not happen, subtract the probability that the event will happen from 1.

   probability of not happening = 1 − probability of happening

EXAMPLE 1 The probability of getting "heads" with a certain weighted coin is $\frac{3}{8}$. What is the probability of not getting "heads"?

probability of not getting "heads"
= 1 − probability of getting "heads"

$= 1 - \frac{3}{8}$

$= \frac{8}{8} - \frac{3}{8} = \frac{5}{8}$

EXAMPLE 2 What is the probability of not getting a heart when choosing a card at random from a standard deck?

Find the probability of getting a heart.

number of hearts → 13
total number of cards → 52

$\frac{13}{52}$, or $\frac{1}{4}$

probability of not getting a heart
= 1 − probability of getting a heart

$= 1 - \frac{1}{4} = \frac{4}{4} - \frac{1}{4} = \frac{3}{4}$

EXAMPLE 3 If the probability of choosing a girl from a certain class is 0.45, what is the probability of not choosing a girl?

probability of not choosing a girl
= 1 − probability of choosing a girl
= 1 − 0.45 = 1.00 − 0.45 = 0.55

EXAMPLE 4 The probability that a train will be on time is 72%. What is the probability that the train will not be on time?

probability of not being on time
= 1 − probability of being on time
= 1 − 72% = 100% − 72% = 28%

CLASS EXERCISES

1. Each number represents the probability that an event will happen. Find the probability that the event will not happen: **a.** $\frac{5}{6}$ **b.** 0.36 **c.** 85% **d.** 1 **e.** 0.01 **f.** 0
2. The probability of getting "heads" with a certain weighted coin is $\frac{3}{10}$. What is the probability of not getting "heads" with this coin?
3. The probability of selecting a vowel from the letters of our alphabet is $\frac{5}{26}$. What is the probability of selecting a consonant?
4. The probability that it will not rain tomorrow is 60%. What is the probability that it will rain?
5. If a card is picked at random from a standard deck, what is the probability that the card will not be a picture card?

HOMEWORK EXERCISES

1. Each number represents the probability that an event will happen. Find the probability that the event will not happen: **a.** 37% **b.** $\frac{3}{8}$ **c.** 0 **d.** 0.45 **e.** $\frac{9}{16}$ **f.** 100%
2. The probability of getting "heads" with a certain weighted coin is $\frac{4}{9}$. What is the probability of not getting "heads" with this coin?
3. The probability of John's winning the race is 0.29. What is the probability of John's not winning?
4. The probability that a card chosen at random from a standard deck of playing cards will not be a club is (a) $\frac{1}{4}$ (b) $\frac{1}{2}$ (c) $\frac{3}{4}$ (d) 1
5. The probability that it will rain today is 40%. What is the probability that it will not rain?
6. The probability of picking a red ball at random from a bag is $\frac{7}{35}$. What is the probability of not picking a red ball at random from the bag?
7. The probability of a candidate's winning an election is 0.52. What is the probability of the candidate's not winning that election?
8. James said that the probability of drawing a heart at random from a standard deck is $\frac{1}{2}$. He reasoned that there are only two possibilities: either the card will be a heart or it will not be a heart. Explain why James' reasoning is wrong.

SPIRAL REVIEW EXERCISES

1. Jim has a $\frac{3}{10}$ probability of getting a hit. What is the number of hits Jim can be expected to get in 200 times at bat?

2. The names of the days of the week are written on separate slips of paper and placed in a hat. One slip of paper is chosen at random from the hat. Find the probability of:
   - **a.** picking a day whose name begins with the letter "T"
   - **b.** picking a day whose name ends with the letter "y"
   - **c.** picking a day whose name has 7 letters
   - **d.** picking a day whose name starts with the letter "Z"

3. Write each probability as a fraction in simplest form.

   a. 83% **b.** 0.95 **c.** 0.125 **d.** 19%
   e. 58% **f.** 25% **g.** 0.007 **h.** 0.98

4. What percent of 40 is 8?

5. Find the probability that the arrow on the spinner will stop on:

   - **a.** the number "6"
   - **b.** an odd number
   - **c.** an even number
   - **d.** a prime number
   - **e.** a number less than 10
   - **f.** a two-digit number

6. Mr. Kartov bought a shirt for \$24 and a tie for \$11. If the sales tax is 5%, how much change should he receive from two \$20 bills?

7. The greatest common factor of 12 and 18 is
   (a) 2 (b) 3 (c) 6 (d) 9

8. Subtract –15 from 35.

9. Round 0.9827 to the nearest thousandth.

10. The largest value of the choices below is
    (a) $\frac{5}{8}$ (b) 0.67 (c) 68% (d) $\frac{9}{16}$

11. The product of 0.05 and 0.9 is
    (a) 4.5 (b) 0.45 (c) 0.045 (d) 0.0045

12. Find the cost of each inch of a lace border if 8 yards of the border cost \$109.44.

13. Divide: $3\frac{3}{4} \div \frac{5}{16}$

14. Evaluate: $3(17 + 5) - 2 \times 6$

15. Ernesto earns \$8.40 an hour, and he makes time-and-a-half for any time over 35 hours a week. If he worked 41 hours during a week, how much did he earn?

16. $\frac{199}{200} \times \frac{23}{50}$ is about
    (a) 500 (b) 50 (c) 5 (d) 0.5

THEME 7

Introduction to Algebra

MODULE 15

MODULE 16

MODULE 15

UNIT 15-1 The Meaning of Variable

THE MAIN IDEA

1. In mathematics, a ***sentence*** is a *complete statement*.

   $13 + 7 = 4 \times 5$ is a sentence.

2. A mathematical sentence uses one of these symbols:

$=$ is equal to	$<$ is less than	$\leq$ is less than or equal to
$\neq$ is not equal to	$>$ is greater than	$\geq$ is greater than or equal to

   Think of these symbols as the *verbs* of the sentences.

3. A sentence that includes an unknown quantity is called an ***open sentence***.

   $3 \times \boxed{?} = 45$ is an open sentence.

4. The unknown quantity in an open sentence is represented by a ***variable***. Variables are shown as follows:
   a. as a blank space with a question mark $\boxed{?} + 10 = 21$
   b. as a symbol, such as an empty square or triangle $\triangle + 10 = 21$
   c. as a letter $x + 10 = 21$

5. In algebra, variables are shown by using letters.

6. The variable in an open sentence usually stands for the words "a number" or "an unknown number."

EXAMPLE 1 Which are sentences?

	"Verb"	*Sentence?*
a. $7 \times 3 = 16 + 5$	$=$	yes
b. $6 + 3 - 5 < 2 \times 3$	$<$	yes
c. $6 + 5 \times 2$	none	no
d. $45 \div 9 \leq 10$	$\leq$	yes

EXAMPLE 2 Which are open sentences?

	Variable	*Open Sentence?*
a. $5 + \boxed{?} = 12$	$\boxed{?}$	yes
b. $9 \div 3 = 2 + 1$	none	no
c. $20 \times \square > 100$	$\square$	yes
d. $15 \times a = 60$	a	yes
e. $12 + 7 - 19 < 25 - 12 - 3$	none	no

CLASS EXERCISES

1. Which are sentences?

   a. $100 + 20 - 40$ b. $5 \times 3 > 4 \times 3$ c. $144 \div 12 = 12$ d. $100 - 2 \geq 50 + 40$
   e. 5×60 f. $5 \times 60 = 3 \times 100$ g. $2 \times 2 \times 2 = 8$ h. $40 \div 8 \neq 4$ i. $40 \div 8 + 4$

2. Which are open sentences?

   a. $6 \times m = 24$ b. $\triangle - 12 = 36$ c. $144 \div \boxed{?} = 6$ d. $120 \div 6 = 20$
   e. $5 + \bigcirc < 20$ f. $x - 10 \geq 40$ g. $4 \times 3 \leq 6 \times 8$ h. $25 + 15 - 5 > \boxed{?}$
   i. $3 \times 3 \times 3 < 30$ j. $4 \times 3 + 2 \neq 4 + 3 \times 2$

3. If x stands for "a number," which expression means "a number minus 2"?
   (a) $x - 2$ (b) $2 - x$ (c) $8 - 2$ (d) $2 - 8$

4. Which expression means "4 times an unknown number"?
   (a) 4×12 (b) $x + 4$ (c) $7 + 7 + 7 + 7$ (d) $4 \times n$

5. Which sentence means "a number plus 12 is less than 20"?
   (a) $8 + 12 < 20$ (b) $x + 12 > 20$ (c) $a + 12 < 20$ (d) $n + 12 \leq 20$

6. Which sentence means "2 times a number is equal to the number plus 8"?
   (a) $2 \times a = 4 + 8$ (b) $2 \times 8 = 8 + 8$ (c) $2 \times 5 = 2 + 8$ (d) $2 \times b = b + 8$

7. Which sentence means "an unknown number minus 5 is greater than or equal to 5"?
   (a) $m - 5 \leq 5$ (b) $y - 5 \geq 5$ (c) $10 - 5 \geq 5$ (d) $10 - 5 \leq 5$

HOMEWORK EXERCISES

1. Which are sentences?

   a. $15 \times \frac{2}{3} = 14 - 4$ b. $96 \div 12$ c. $34 \times 3 > 14 \times 5$ d. $5 + 9 \neq 8 \times 2$
   e. 32×7 f. $50 \div 2 + 17 \times 5$ g. $30 \div 3 = 15 - 5$ h. $32 - 7 \leq 40$

2. Which are open sentences?

   a. $9 + 7 = 16$ b. $x - 2 = 6$ c. $12 \times 3 < 40$ d. $\square + 4 = 15$
   e. $19 + 17 = \triangle$ f. $60 \div 5 \neq 42$ g. $25 \div 5 + 8 \times 3 < 100$ h. $\boxed{?} = 50$
   i. $3 \times y = 36$ j. $1{,}001 > 50 \times 20$ k. $\square \div 6 = 8$ l. $5 \times a > 40$

3. Which expression means "a number increased by 5"? (a) $8 + 5$ (b) $y - 5$ (c) $y + 5$ (d) $8 - 5$

4. Which means "9 subtracted from a number"? (a) $9 - n$ (b) $n - 9$ (c) $9 \times n$ (d) $n \div 9$

5. Which sentence means "5 times a number is equal to 15"?
   (a) $5 \times n = 15$ (b) $5 + n = 15$ (c) $15 \times n = 5$ (d) $n - 5 = 15$

6. Which sentence means "a number decreased by 7 is greater than 10"?
(a) $x + 7 \geq 10$ (b) $x + 7 < 10$ (c) $x - 7 > 10$ (d) $x + 7 > 10$

7. Which sentence means "20 is greater than an unknown number plus 6"?
(a) $20 + n > 6$ (b) $20 \times n < 6$ (c) $20 > n - 6$ (d) $20 > n + 6$

8. Which sentence means "a number divided by 3 is equal to 5 less than the number"?
(a) $3 \times n = 5 - n$ (b) $3 \times n = n - 5$ (c) $n \div 3 = n - 5$ (d) $3 \div n = 5 - n$

9. Explain why you cannot tell whether the sentence $n + 17 = 15$ is true, but you know that the sentence $n - n = 0$ is always true.

SPIRAL REVIEW EXERCISES

1. This graph has data about the graduates at Apex High School for a 5-year period.

Graduates of Apex High School

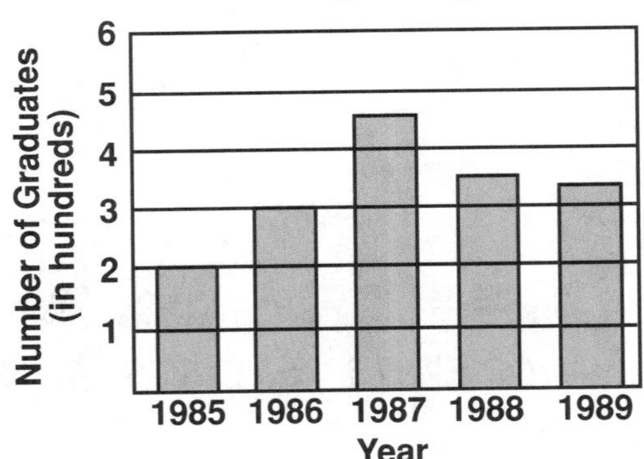

The number of graduates in 1989 was about
(a) 3 (b) 4 (c) 350 (d) 325

2. Monica was given the job of counting the spectators as they arrived at a basketball game. On her calculator she entered the key sequence 0 [+] 1 and then pressed the [=] key every time a spectator passed through the turnstile. Explain how this process would count the number of spectators.

3. A box contains 3 red marbles and 7 green marbles. The probability of picking a red marble at random is
(a) 3 (b) 7 (c) $\frac{3}{7}$ (d) $\frac{3}{10}$

4. Find the mean of 8.9, 4.6, 12.1, and 9.8.

5. Find the number of centimeters in 25 meters.

6. Add: $-10 + (-4)$

7. What is the sum of $\frac{2}{3}$ and $\frac{3}{5}$?

(a) $\frac{2}{5}$ (b) $1\frac{4}{15}$ (c) $\frac{5}{8}$ (d) $\frac{6}{15}$

8. 0.14 has the same value as

(a) 14% (b) 1.4% (c) $\frac{14}{10}$ (d) $\frac{14}{1,000}$

9. If the price of one share of Starline Industries stock is \$50 and its earnings are \$2.50 per share, then the ratio of the stock's price to its earnings is:
(a) 20 : 1 (b) 2 : 1 (c) 1 : 2 (d) 1 : 20

UNIT 15-2 Writing Algebraic Expressions

THE MAIN IDEA

1. You ***translate*** word phrases into algebraic expressions by writing symbols for the words.
2. You have been using the symbols +, –, ×, and ÷ for the operations of arithmetic, and you know certain key words that are associated with these operations.
3. Since it could be confusing to write $2 \times x$, there are other ways to write "2 times the number x": $2 \cdot x$ $2(x)$ $(2)(x)$ $2x$
4. Parentheses ( ) are used to group numbers. A fraction bar is also used.

EXAMPLE 1 Translate each word phrase into an algebraic expression.

Word Phrase	*Answer*	*Word Phrase*	*Answer*
a. the sum of x and 2 (the sum of → +)	$x + 2$	**e.** a number increased by 7	$x + 7$
b. the difference of n and 4 (the difference of → –)	$n - 4$	**f.** 4 more than a number	$y + 4$
c. the product of y and 3 (the product of → ×)	$3y$, $3 \cdot y$, $3(y)$, or $(3)(y)$	**g.** Al's age decreased by 2 years	$a - 2$
d. the quotient of b and 7 (the quotient of → ÷)	$\frac{b}{7}$ or $b \div 7$	**h.** 5 pounds less than Bo's weight	$b - 5$
		i. the difference of 4 and n	$4 - n$

Note: The unknown quantity can be represented by any variable.

EXAMPLE 2 Use parentheses or a fraction bar to group numbers.

Answer

a. twice the difference of y and 3
$2 \cdot (y - 3)$ — Answer: $2(y - 3)$

b. the sum of 7 and m, divided by 3
$(7 + m) \div 3$ — Answer: $(7 + m) \div 3$ or $\frac{7 + m}{3}$

CLASS EXERCISES

1. Use the variable n to translate each word phrase into an algebraic expression.
   a. a number increased by 11 **b.** the product of 6 and a number
   c. a number divided by 10 **d.** 7 less than a number
   e. 7 decreased by a number **f.** the sum of a number and 15
   g. the difference between a number and 8
   h. $\frac{3}{4}$ of a number
   i. 20% of a number **j.** 24 divided by a number

2. Write each expression algebraically.
   a. 2 times the sum of a number and 5
   b. 6 more than 4 times a number
   c. the product of 5 and a number, decreased by 11
   d. the difference between a number and 5, divided by 3
   e. the sum of a number and 6, multiplied by 7 more than the number

3. Which expression means "10 more than a number"?
   (a) $n + 10$ (b) $10n$ (c) $(10)(n)$ (d) $10 \cdot n$

4. Which expression means "8 decreased by a certain number"?
   (a) $x \div 8$ (b) $\frac{8}{x}$ (c) $n - 8$ (d) $8 - a$

5. Which expression means "7 more than twice a number"?
   (a) $7 \cdot (2y)$ (b) $2y + 7$ (c) $n + (2)(7)$ (d) $7y + 2$

6. Which expression means "the sum of 5 times a number and 9"?
   (a) $x(9 + 5)$ (b) $5b + 9$ (c) $(9)(5a)$ (d) $y + (a)(5)$

7. Which expression means "the sum of a number and 3, multiplied by the difference between the number and 2"?
   (a) $(a - 3)(a + 2)$ (b) $(a + 3)(a - 2)$ (c) $(a - 2) + (a + 3)$ (d) $3a - 2$

8. Which expression means "4 less than the product of 3 and a number"?
   (a) $4 - (3 + n)$ (b) $4 - 3n$ (c) $(3 + n) - 4$ (d) $3n - 4$

9. Write an algebraic expression for each word phrase.
   a. Mary's age decreased by 3 **b.** twice Mr. Jones' salary
   c. the sum of the price of a book and \$3 **d.** 4 inches more than Mike's height
   e. $\frac{1}{5}$ of the number of students in the class

HOMEWORK EXERCISES

1. Use the variable n to translate each word phrase into an algebraic expression.
   - a. a number divided by 7
   - b. a number diminished by 30
   - c. a number increased by 75
   - d. the product of -15 and a number
   - e. 5 times a number
   - f. 85 subtracted from a number
   - g. 35 divided by a number
   - h. the sum of a number and 37.5
   - i. $\frac{5}{8}$ of a number
   - j. the difference between 20 and a number

2. Write each expression algebraically.
   - a. 8 increased by 4 times a number
   - b. 10 subtracted from 9 times a number
   - c. 3 times the sum of a number and -4
   - d. 7 less than 5 times a number
   - e. 15 more than a number, divided by 2
   - f. -50 divided by the sum of a number and 2
   - g. the difference of 15 and a number, multiplied by 3
   - h. the difference of a number and 7, multiplied by 5 more than the number
   - i. 15 more than twice a number, multiplied by 3 less than the number

3. Which expression means "a number decreased by 7"?

   (a) $n + 7$ (b) $n - 7$ (c) $7n$ (d) $\frac{n}{7}$

4. Which expression means "5 less than twice a number"?

   (a) $5 - 2n$ (b) $5 + 2n$ (c) $2n - 5$ (d) $2(n - 5)$

5. Which expression means "the product of 3 and 7 less than a number"?

   (a) $n - 21$ (b) $3(7 - n)$ (c) $3(7 + n)$ (d) $3(n - 7)$

6. Which expression means "10 more than a number, divided by 5"?

   (a) $\frac{n+10}{5}$ (b) $\frac{n-10}{5}$ (c) $n + 10 \div 5$ (d) $n - 10 \div 5$

7. Which expression means "a number increased by 7, multiplied by 2 more than the number"?

   (a) $(a + 7)(2a)$ (b) $(a - 7)(2a)$ (c) $(a + 7)(a + 2)$ (d) $(a - 7)(a + 2)$

8. Write an algebraic expression for each word phrase.
   - a. June's wages increased by $30
   - b. Joseph's weight decreased by 5 pounds
   - c. 3 times the distance that Carlos walked
   - d. $\frac{1}{2}$ the price of a movie ticket
   - e. The sum of the cost of a jacket and a $20 shirt

9. Maria earned d dollars on Monday and $28 on Tuesday. Write algebraically the total amount of money Maria earned.

10. James has x dollars and spends \$50 for a radio. Write algebraically the amount of money that James has left.

11. Mr. Vivaldi divides w dollars evenly among his 5 children. Write algebraically the amount of money given to each child.

12. Explain why you cannot tell which is the greater quantity, $x + 5$ or $n + 20$.

SPIRAL REVIEW EXERCISES

1. Which is a sentence?

(a) $3(5 + 4)$ (b) $5(3 + 4)$
(c) $5 + 4 = 9$ (d) $5 + 4 \times 9$

2. Which expression means "3 less than a number"?

(a) $3 - x$ (b) $x - 3$ (c) $3 \div x$ (d) $x \div 3$

3. James bought 2 quarts of milk. How many fluid ounces did he buy?

4. What is the mode of $1\frac{1}{8}$, $2\frac{1}{4}$, $1\frac{1}{2}$, $1\frac{1}{8}$, $2\frac{5}{16}$, $1\frac{1}{4}$, $1\frac{1}{8}$, and $2\frac{1}{4}$?

5. The key sequence 12.5 [÷] 36 [=] will change

(a) 12.5 inches to feet
(b) 12.5 feet to inches
(c) 12.5 inches to yards
(d) 12.5 yards to inches

6. The remainder when 225 is divided by 17 is

(a) 0 (b) 1 (c) 13 (d) 4

7. $\frac{24}{15}$ equals (a) $1\frac{7}{15}$ (b) $1\frac{2}{3}$ (c) $1\frac{3}{5}$ (d) $1\frac{11}{15}$

8. $4 - (-5)$ equals (a) -1 (b) 9 (c) 1 (d) -9

9. Andy spent an average of 140 minutes a day on homework. The circle graph shows the average number of minutes spent on each subject.

Time Spent on Homework

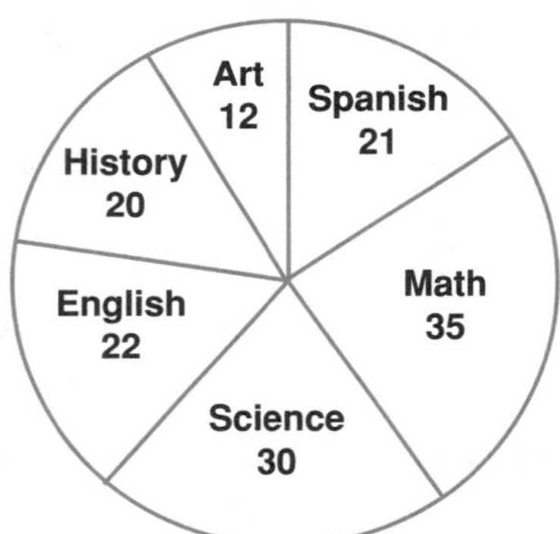

What percent of Andy's homework time was spent on Math?

(a) 15% (b) 20% (c) 25% (d) 35%

10. If the sales tax is 6%, how much tax is paid on a \$35 purchase?

11. If 7 boxes of cereal cost \$9.10, what is the cost of 3 boxes of cereal?

UNIT 15-3 Evaluating Algebraic Expressions

THE MAIN IDEA

1. To ***evaluate*** (find the value of) an algebraic expression, replace the variables by their numerical values. Then perform the operations that are shown.
2. If there is more than one operation, the operations follow the order of operations that you learned.
   a. Do the operations inside parentheses first.
   b. Do the multiplications and divisions next. Work from left to right.
   c. Do the additions and subtractions last. Work from left to right.

EXAMPLE 1 Find the value of $3x$ when $x = 12$.

THINKING ABOUT THE PROBLEM

$3x$ means "3 times x." If x is 12, $3x$ means "3 times 12," written as:

3×12, $3(12)$, or $3 \cdot 12$

Answer: When $x = 12$, $3x = 36$.

EXAMPLE 2 Evaluate $2 + 3a$ when $a = -5$.

Replace the variable a by the numeral -5.	$2 + 3a$ $= 2 + 3(-5)$
Multiply first.	$= 2 + -15$
Do the addition.	$= -13$ *Ans.*

EXAMPLE 3 Find the value of $5(a + b)$ when $a = 2$ and $b = 7$.

Replace the variables by the numerals.	$5(a + b)$ $= 5(2 + 7)$
Work inside the parentheses first.	$= 5(9)$
Do the multiplication.	$= 45$ *Ans.*

EXAMPLE 4 Evaluate $2x - 5y + 3z$ when $x = 10$, $y = 3$, and $z = -1$.

Replace the variables.	$2x - 5y + 3z$ $= 2(10) - 5(3) + 3(-1)$
Multiply first.	$= 20 - 15 + (-3)$
Add and subtract from left to right.	$= 5 + (-3)$ $= 2$ *Ans.*

CLASS EXERCISES

1. Evaluate each expression, using the given values of the variables.
   a. $3 + w$ when $w = 11$ **b.** $a - 12$ when $a = 27$ **c.** $6x$ when $x = 25$
   d. $\frac{b}{3}$ when $b = 144$ **e.** $23 - y + 2$ when $y = -19$ **f.** $\frac{6z}{18}$ when $z = -12$
2. Evaluate each expression, using the values $x = 13$ and $y = 7$.
   a. $x + y$ **b.** $x - y$ **c.** xy **d.** $\frac{x}{y}$ **e.** $2x + y$ **f.** $x + 2y$
   g. $2x + 2y$ **h.** $4x - 3y$ **i.** $x + y - x - y$ **j.** $7x - 13y$
3. Use the values $a = 5$ and $b = 3$ to evaluate each expression.
   a. $3(a + b)$ **b.** $(a + b) - 7$ **c.** $4(2a + b)$ **d.** $6(a - b) + 5$
   e. $(a + b)(a - b)$ **f.** $(2a + b)(2a - b)$ **g.** $3(b + a)$ **h.** $4(b + 2a)$
4. Find the value of $2\ell + 2w$, using the given values of the variables.
   a. $\ell = 2, w = 3$ **b.** $\ell = 3, w = 2$ **c.** $\ell = 2\frac{1}{2}, w = 3\frac{1}{2}$ **d.** $\ell = 4, w = 1$
5. If $m = 12$ and $p = -1$, the value of $3m - 2p$ is (a) 33 (b) 34 (c) 36 (d) 38

HOMEWORK EXERCISES

1. Evaluate each expression, using the given values of the variables.
   a. $x + 19$ when $x = 20$ **b.** $y - 26$ when $y = 54$ **c.** $31 + z$ when $z = 29$
   d. $20x$ when $x = -5$ **e.** $5y$ when $y = 6$ **f.** $\frac{w}{7}$ when $w = 28$
   g. $35 - a$ when $a = -20$ **h.** $11c$ when $c = 9$ **i.** $-14 + x$ when $x = 10$
2. Find the value of $85 - w$ when w is: **a.** 50 **b.** 0 **c.** -85 **d.** 12.9 **e.** $80\frac{5}{8}$ **f.** -5.5
3. Find the value of $30x$ when x is: **a.** 3 **b.** -9 **c.** 0 **d.** $\frac{1}{2}$ **e.** 5.8 **f.** $\frac{2}{3}$ **g.** $\frac{-5}{6}$
4. Find the value of $50 + 3y$ when y is: **a.** 2 **b.** -7 **c.** 16 **d.** 0 **e.** $5\frac{1}{3}$ **f.** 10.7 **g.** $\frac{-1}{3}$
5. If $x = 20$ and $y = 30$, evaluate: **a.** $4x - 2y$ **b.** $\frac{60}{x} + \frac{30}{y}$ **c.** $9(y - x)$ **d.** $9(x - y)$
6. If $a = 12$ and $b = 4$, evaluate: **a.** $3a - b$ **b.** $3(a - b)$ **c.** $3b - a$ **d.** $(a - b)(a + b)$
   e. $5b + 2a$ **f.** $\frac{1}{3}a + b$ **g.** $10(a + b)$ **h.** $10(a - 3b)$

7. Find the value of $4y - 3x$, using the given values of the variables.

 a. $y = 2, x = 1$ **b.** $y = 5, x = 6$ **c.** $y = 0, x = 0$ **d.** $y = \frac{1}{2}, x = \frac{1}{3}$

8. If $d = 10$ and $t = 5$, then the value of $2d + 4t$ is (a) 120 (b) 40 (c) 28 (d) 255

9. If $k = 20$ and $m = 15$, then the value of $3k - 4m$ is (a) 0 (b) 15 (c) 5 (d) 35

10. If $x = 2$ and $y = -1$, the value of $3x - y$ is (a) 5 (b) 6 (c) 7 (d) 8

11. If $a = 5$, $b = 2$, and $c = 3$, explain why $\frac{a+b}{c-3}$ cannot be evaluated.

SPIRAL REVIEW EXERCISES

1. Which expression means "5 decreased by twice a number"?
 (a) $5 + 2n$ (b) $2n - 5$
 (c) $5 - 2n$ (d) $2 - 5n$

2. Which sentence means "twice a number is greater than or equal to 12"?
 (a) $2x > 12$ (b) $2x \geq 12$
 (c) $2x \leq 12$ (d) $2x < 12$

3. The number of liters in 4,500 mL is
 (a) 450 (b) 45 (c) 4.5 (d) 0.45

4. $-7 + (-4)$ is equal to
 (a) -11 (b) $+11$ (c) -3 (d) $+3$

5. Which key sequence will produce the same result as the key sequence: 57 [×] .23 [=]
 (a) 57 [÷] 23 [=]
 (b) 57 [÷] 23 [×] 100 [=]
 (c) 57 [×] .23 [%]
 (d) 57 [×] 23 [÷] 100 [=]

6. Find the median of 35, 41, 52, 58, 63, 65, and 68.

7. Jerry bought a \$30 shirt on sale for \$15. If the rate of sales tax is 8%, how much change did he receive from a \$20 bill?

8. The balance in Ms. Lorenzo's checking account was \$950.75. If she made a deposit of \$432.85 and wrote a check for \$173.40, what was her new balance?

9. Multiply: 2.9 and 0.7

10. This graph shows data about Mandy's temperature.

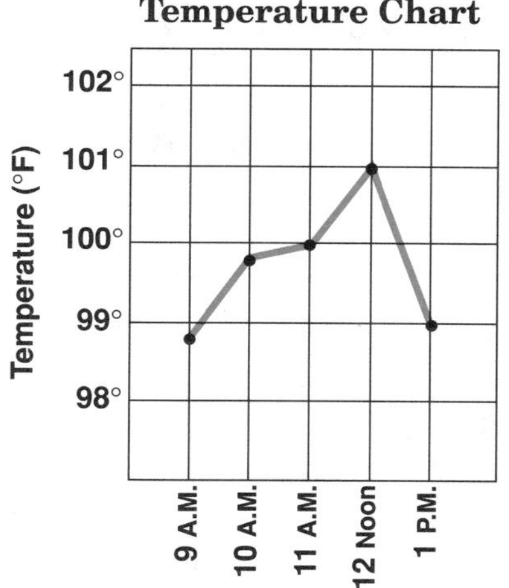

How much did Mandy's temperature drop from 12 Noon to 1 P.M.?

11. On a mathematics test, Sandy answered 20 of 25 questions correctly. What percent of the questions did Sandy get correct?

UNIT 15-4 Exponents; Powers of Ten; Scientific Notation

EXPONENTS

THE MAIN IDEA

1. An *exponent* shows how many times a *base* is used as a factor.

   6^3 is read

   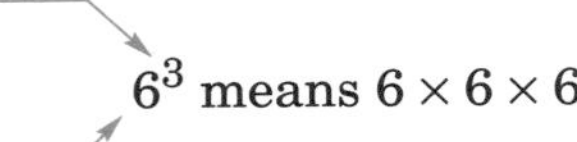

   "6 raised to the third power," "6 to the third power," or "6 cubed."

2. To include exponents in the order of operations:
   a. Work inside parentheses.
   b. Work with exponents.
   c. Multiply and divide.
   d. Add and subtract.

EXAMPLE 1

a. The number 5 is to be used 4 times as a factor. Write this product in two ways.

Answer: $5 \times 5 \times 5 \times 5$ or 5^4

b. The number 4 is to be used 5 times as a factor. Write this product in two ways.

Answer: $4 \times 4 \times 4 \times 4 \times 4$ or 4^5

EXAMPLE 2

a. Write the expression $a \cdot a \cdot a \cdot b \cdot b$, using exponents.

Answer: $a^3 \cdot b^2$ or a^3b^2

b. Write the expression x^4y^2z without exponents.

Answer: $x \cdot x \cdot x \cdot x \cdot y \cdot y \cdot z$

EXAMPLE 3 Write each expression in words.

	Expression	*Answer*
a.	5^2	five raised to the second power or five squared
b.	a^3	a to the third power or a cubed

EXAMPLE 4 Evaluate 7^6 on a calculator.

Key Sequence	*Display*
7 [×] [=] [=] [=] [=] [=] (5 multiplications)	117649.

Answer: $7^6 = 117,649$

EXAMPLE 5 Evaluate each expression.

	Expression	*Calculation*
a.	$(-5)^2 \cdot 2^3$	$-5 \times -5 \times 2 \times 2 \times 2$ $= 25 \times 8$ $= 200$ *Ans.*
b.	$5 \cdot 3^2$	$5 \times 3 \times 3$ $= 15 \times 3$ $= 45$ *Ans.*
c.	$(5 \cdot 3)^2$	$(15)^2$ $= 15 \times 15$ $= 225$ *Ans.*
d.	$5 + 3^2$	$5 + 3 \times 3$ $= 5 + 9 = 14$ *Ans.*

EXAMPLE 6 Evaluate each expression when $x = 3$ and $y = 4$.

	Expression	*Calculation*
a.	$x(y)^2$	$3(4)^2$ $= 3(4 \times 4) = 3(16) = 48$ *Ans.*
b.	$(xy)^2$	$(3 \times 4)^2$ $= (12)^2 = 12 \times 12 = 144$ *Ans.*
c.	x^2y^2	$3^2 \times 4^2$ $= 3 \times 3 \times 4 \times 4$ $= 9 \times 16 = 144$ *Ans.*
d.	x^2y	$3^2 \times 4$ $= 3 \times 3 \times 4 = 9 \times 4 = 36$ *Ans.*

CLASS EXERCISES

1. Write each product in two ways, one of which uses exponents.
   a. 6 used as a factor 3 times **b.** x multiplied by x **c.** w times w times w
2. Write each expression, using exponents.
   a. $10 \cdot y \cdot y \cdot y \cdot y$ **b.** $(p)(p)(q)(q)(q)$ **c.** $a \cdot a \cdot a \cdot b \cdot b \cdot b$ **d.** $3(r)(r)(r)(r)(s)(s)$
3. Write each expression without exponents: **a.** a^3b^2 **b.** a^2b^3 **c.** x^3y^3z **d.** xy^3z^3
4. Write in words: **a.** 6^3 **b.** $(-4)^8$ **c.** $5 \cdot x^4$ **d.** $m^2 \cdot n^3$ **e.** $a^4 \cdot b^2 \cdot c^6$
5. Evaluate: **a.** 3^2 **b.** 2^3 **c.** 4^3 **d.** $(-3)^4$ **e.** 10^2 **f.** $(-10)^3$ **g.** 2^6 **h.** $\left(\frac{1}{2}\right)^2$ **i.** $(0.1)^3$
6. Evaluate: **a.** 12^6 **b.** 2^{12} **c.** 8^8 **d.** $(-5)^7$
7. Evaluate each expression.
   a. 2×3^2 **b.** $2^2 \times 3^2$ **c.** $3 \times (-2)^3$ **d.** $3^3 \times 2^3$ **e.** $3^2 \times 2^3$ **f.** $(3 \times 2)^3$
   g. $2^2 + 3^2$ **h.** $(-3)^3 + 4^2$ **i.** $10^2 - 7^2$ **j.** $10^2 + 7^2$ **k.** $10^2 + (-7)^2$ **l.** $(-10)^2 + 7^2$
8. Evaluate each expression when $x = 2$.
   a. $2 \cdot x^2$ **b.** $x^2 \cdot x^3$ **c.** x^5 **d.** $5x^2$ **e.** $5^2 \cdot x^2$ **f.** $(5x)^2$ **g.** $x^2 + 3$ **h.** $3x^2 + x$
9. The value of $2 \times 10^3 + 8 \times 10^2 + 5 \times 10^1 + 1$ is (a) 28.51 (b) 285.1 (c) 2,851 (d) 28,510
10. Evaluate: **a.** $3 \times 10^3 + 4 \times 10^2 + 7$ **b.** $9 \times 10^4 + 7 \times 10^2 + 5 \times 10^1 + 8$

POWERS OF TEN

THE MAIN IDEA

1. When an exponent is reduced by 1, the expression has 1 less factor.

$2^4 = 2 \times 2 \times 2 \times 2$
$2^3 = 2 \times 2 \times 2$

2. When the exponent of a power of 10 is reduced by 1, the value is divided by 10.

Power of Ten	*Meaning*	*Division*	*Value*
10^4	$10 \times 10 \times 10 \times 10$		10,000
10^3	$10 \times 10 \times 10$	$10{,}000 \div 10$	1,000
10^2	10×10	$1{,}000 \div 10$	100
10^1	10	$100 \div 10$	10

3. The pattern of dividing by 10 can be continued to include an exponent of 0 and negative exponents.

Power of Ten		*Value*
10^1		10
10^0	$10 \div 10$	1
10^{-1}	$1 \div 10$	0.1
10^{-2}	$0.1 \div 10$	0.01

EXAMPLE 7 Find the value of 10^{-7}.

Continue the pattern in item 3 of The Main Idea.

Power of Ten		*Value*
10^{-3}	$0.01 \div 10$	0.001
10^{-4}	$0.001 \div 10$	0.0001
10^{-5}	$0.0001 \div 10$	0.00001
10^{-6}	$0.00001 \div 10$	0.000001
10^{-7}	$0.000001 \div 10$	0.0000001

Observe that the number of decimal places matches the exponent.

Answer: $10^{-7} = 0.0000001$

SCIENTIFIC NOTATION

THE MAIN IDEA

1. A number in the form 2.5×10^3 is in ***scientific notation***. The first factor is a number between 1 and 10, and the second factor is a power of 10.
2. To change a number *from* scientific notation to standard notation, move the decimal point of the first factor according to the exponent of 10.

   $2.5 \times 10^3 = 2.500.$ Move 3 places to the right since the exponent is positive.
   $2.5 \times 10^3 = 2{,}500$

   $2.5 \times 10^{-3} = .002.5$ Move 3 places to the left since the exponent is negative.
   $2.5 \times 10^{-3} = 0.0025$

3. To change a number *to* scientific notation:
   a. Place a decimal point so that the first factor is a number between 1 and 10.
   b. Count how many places the decimal point must be moved to change the new first factor into the original number
      The number of places is the exponent of 10, positive if the decimal point must be moved to the right, negative if moved left.
   c. Show the two factors as a product.

EXAMPLE 8 Write each number in scientific notation.

a. 56,000

Place a decimal point so that the first factor is a number between 1 and 10.

5.6000

5.6 is between 1 and 10.

Count how many places the decimal point must be moved to the right to change 5.6 to 56,000.

5. 6 0 0 0 .

Move 4 places to the right.

Use 4 as the exponent of 10.

10^4 ← exponent

Show the product.

5.6×10^4 *Ans.*

b. 0.0000296

Place a decimal point so that the first factor is a number between 1 and 10.

0.0000296

2.96 is between 1 and 10.

Count the number of places the decimal point must be moved to the left to change 2.96 to 0.0000296.

0.0 0 0 0 2 .9 6

Move 5 places to the left.

Use –5 as the exponent of 10.

Show the product.

2.96×10^{-5} *Ans.*

CLASS EXERCISES

1. Find the value of each power of 10: **a.** 10^4 **b.** 10^0 **c.** 10^{-3} **d.** 10^{-5} **e.** 10^7
2. Write each number in standard notation.
   a. 1.8×10^2 **b.** 2.45×10^4 **c.** 8.3×10^6 **d.** 7.06×10^{-4} **e.** 1.3×10^{-1} **f.** 9.8×10^{-7}
3. Write in scientific notation: **a.** 5,700 **b.** 620,000 **c.** 1,250 **d.** 0.00045 **e.** 0.83 **f.** 0.000003
4. The standard notation for 2.5×10^3 is (a) 2,500 (b) 25^3 (c) 2.5000 (d) 250
5. The scientific notation for 175,000 is (a) 175×10^3 (b) 17.5×10^4 (c) 1.75×10^5 (d) 1.75×10^3
6. The scientific notation for 0.00457 is (a) 4.57×10^{-4} (b) 4.57×10^{-3} (c) 4.57×10^3 (d) 4.57×10^4
7. The time needed by some modern computers to do simple calculations is measured in *nanoseconds* or *picoseconds*.
   a. A picosecond is 10^{-12} second. Write 10^{-12} in standard notation.
   b. A nanosecond is 10^{-9} second. Write 10^{-9} in standard notation.
8. Under ideal conditions, some modern microscopes can be used to view objects as small as 0.00002 cm long. Write this number in scientific notation.

HOMEWORK EXERCISES

1. Write each product in two ways, one of which uses exponents.
   a. 10 used as a factor 4 times **b.** a times a times a **c.** m multiplied by itself
2. Write each expression, using exponents.
   a. $x \cdot x \cdot x \cdot x$ **b.** $a \cdot a \cdot b \cdot b$ **c.** $5 \cdot y \cdot y \cdot y$ **d.** $4(a)(b)(b)(b)$ **e.** $12m \cdot m \cdot m \cdot r \cdot r$
3. Write an equivalent expression without exponents: **a.** x^4y^2 **b.** x^2y^4 **c.** a^2b^3c **d.** ab^2c^3
4. Write in words: **a.** 8^2 **b.** 5^3 **c.** 9^5 **d.** 5^9 **e.** w^6 **f.** y^3 **g.** $3x^3$ **h.** x^2y^4 **i.** $x^3y^4z^5$

5. Evaluate: **a.** 6^2 **b.** 5^3 **c.** $(-3)^5$ **d.** 10^4 **e.** 10^5 **f.** 2^5 **g.** $(-11)^2$ **h.** $\left(\frac{1}{3}\right)^3$ **i.** $(0.1)^4$

6. Evaluate: **a.** 6^5 **b.** 11^6 **c.** 9^7 **d.** 2^{20}

7. Evaluate: **a.** 5×3^2 **b.** $4^2 \times 5^2$ **c.** $(4 \times 5)^2$ **d.** $6 \times (-2)^3$ **e.** $6^2 + 8^2$ **f.** $(-5)^2 + 12^2$

8. Evaluate each expression when $a = 3$.

a. a^3 **b.** $2a^2$ **c.** $a^2 \cdot a^3$ **d.** a^5 **e.** $(4 + a)^2$ **f.** $a^2 - a$ **g.** $2a^2 + 5$

9. The value of $9 \times 10^3 + 6$ is (a) 906 (b) 960 (c) 9,006 (d) 9,060

10. Evaluate: **a.** $4 \times 10^3 + 7 \times 10^2 + 1 \times 10^1 + 5$ **b.** $8 \times 10^2 + 7$ **c.** $7 \times 10^3 + 1 \times 10^2 + 4$

11. Find the value of each power of 10. **a.** 10^6 **b.** 10^{-6} **c.** 10^{-1} **d.** 10^1 **e.** 10^0

12. Write each number in standard notation.

a. 7.3×10^2 **b.** 3.1×10^3 **c.** 5.42×10^5 **d.** 3.7×10^{-3} **e.** 8.92×10^{-1} **f.** 9.6×10^{-6}

13. Write in scientific notation: **a.** 98,000 **b.** 870,000 **c.** 4,720 **d.** 0.034 **e.** 0.000026 **f.** 0.000892

14. The standard notation for 1.2×10^4 is (a) 1.20000 (b) 1,200 (c) 12^4 (d) 12,000

15. The scientific notation for 28,400 is (a) 0.284×10^5 (b) 2.84×10^4 (c) 28.4×10^3 (d) 284×10^2

16. A *byte* is the amount of computer memory needed to store one letter. There are 8 *bits* in one byte and 2^{10} bytes in one kilobyte.

a. Evaluate 2^{10}. **b.** Write 2^{10} in scientific notation.

c. Express in scientific notation the number of bits in one kilobyte.

17. The galaxies closest to our own galaxy are the Large Magellanic Cloud and the Small Magellanic Cloud.

a. The Large Magellanic Cloud is 1.5×10^5 light years from us. Write the underlined expression in standard notation.

b. The Small Magellanic Cloud is 173,000 light years from us. Write the underlined number in scientific notation.

c. Which is closer to us, the Large or the Small Magellanic Cloud?

TEAMWORK

Look in newspapers and magazines and find five articles that mention very large or very small numbers. Rewrite these numbers in scientific notation. Prepare a sheet to which you have attached excerpts from these articles showing the numbers you have chosen and how you have rewritten them.

SPIRAL REVIEW EXERCISES

1. $-8 - (-5)$ is equal to
(a) -13 (b) $+13$ (c) -3 (d) $+3$

2. Write an algebraic expression that means "a person's age decreased by 12 years."

3. If the probability that it will rain is 0.4, what is the probability that it will not rain?

4. In a group of 20 athletes, 8 were professionals and 12 were amateurs. The ratio of professional athletes to the total number of athletes was
(a) $\frac{12}{20}$ (b) $\frac{2}{5}$ (c) $\frac{2}{3}$ (d) $\frac{4}{5}$

5. The price of a shirt was reduced from \$30 to \$24. The percent of decrease is
(a) 40% (b) $66\frac{2}{3}\%$ (c) 8% (d) 20%

6. Find $\frac{4}{5}$ of $\frac{2}{7}$.

7. Divide: $1.04 \div 0.02$

8. An inspector will reject a metal sheet manufactured by a machine if its thickness differs from 0.32 inches by more than 5%. Find the range of acceptable thicknesses.

9. What is the probability that the first of these spinners will stop on "blue" and the other on an even number?

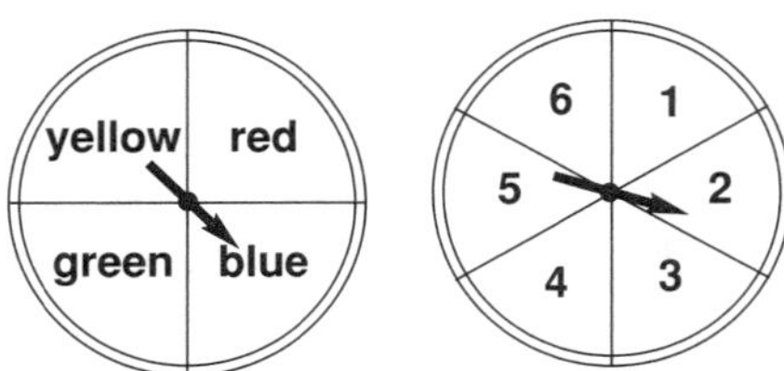

10. This graph shows the kinds of vehicles passing through a toll booth in one day.

Kinds of Vehicles

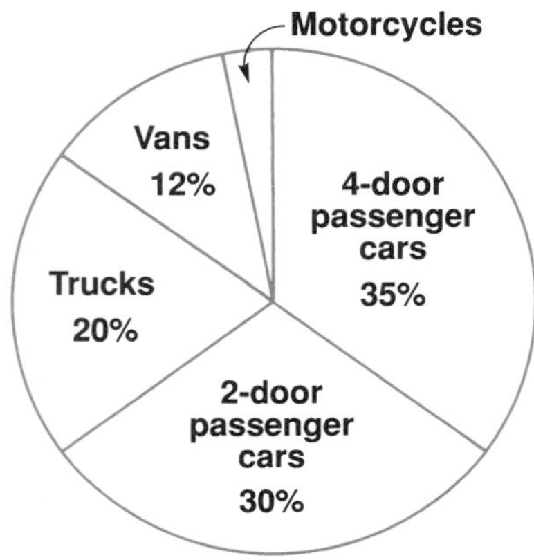

100% = 1,300 Vehicles

How many of the vehicles passing through the toll booth were motorcycles?

11. A flight from London to Tokyo stops in Anchorage, Alaska for 1.5 hours. The flight from London to Anchorage takes 8 hours 55 minutes, and from Anchorage to Tokyo, 7 hours 10 minutes. How much time will be saved by taking a direct flight from London to Tokyo that takes 11 hours 30 minutes?

12. Evaluate: 10!

UNIT 15-5 Squares and Square Roots

THE MAIN IDEA

1. a. When a number is raised to the second power, the number is *squared*. — 5^2 is 5 squared.
   b. The result is the ***square*** of the original number. — 25 is the square of 5.
   c. The original number is the ***square root*** of the result. — 5 is the square root of 25.
2. The symbol $\sqrt{}$ means "square root." — $5 = \sqrt{25}$
3. Whole numbers with square roots that are also whole numbers are ***perfect squares***. — 25 is a perfect square.
4. You should become familiar with the perfect squares through 12^2.

$1^2 = 1$	$1 = \sqrt{1}$	$5^2 = 25$	$5 = \sqrt{25}$	$9^2 = 81$	$9 = \sqrt{81}$
$2^2 = 4$	$2 = \sqrt{4}$	$6^2 = 36$	$6 = \sqrt{36}$	$10^2 = 100$	$10 = \sqrt{100}$
$3^2 = 9$	$3 = \sqrt{9}$	$7^2 = 49$	$7 = \sqrt{49}$	$11^2 = 121$	$11 = \sqrt{121}$
$4^2 = 16$	$4 = \sqrt{16}$	$8^2 = 64$	$8 = \sqrt{64}$	$12^2 = 144$	$12 = \sqrt{144}$

5. The square root of a number that is not a perfect square is an ***irrational number***. Its value can only be *approximated*. — $\sqrt{28}$ is about 5.3.

EXAMPLE 1 Write equivalent sentences for: $8^2 = 64$

Answer: 64 is the square of 8;
8 is the square root of 64; $8 = \sqrt{64}$

EXAMPLE 2 Find the square of 11.

The square of 11 means 11^2

Answer: $11^2 = 11 \times 11 = 121$

EXAMPLE 3 What number is $\sqrt{81}$?

Answer: Since $9^2 = 9 \times 9 = 81$, then $\sqrt{81} = 9$

EXAMPLE 4 Which number is a perfect square? (a) 90 (b) 144 (c) 50 (d) 2

Answer: (b)

EXAMPLE 5 Find: $\sqrt{128{,}881}$

Key Sequence	***Display***
128881 [√]	359.

Answer: $\sqrt{128{,}881} = 359$

EXAMPLE 6 Approximate $\sqrt{70}$ to the nearest tenth.

THINKING ABOUT THE PROBLEM

Since the closest perfect squares to 70 are 64 and 81, $\sqrt{70}$ must be between $\sqrt{64}$, which is 8, and $\sqrt{81}$, which is 9. To approximate $\sqrt{70}$, use the strategy of guessing and checking.

$$64 < 70 < 81$$
$$\sqrt{64} < \sqrt{70} < \sqrt{81}$$
$$8 < \sqrt{70} < 9$$

Guess for $\sqrt{70}$	*Check by squaring.*		*Compare the square to 70.*
8.2	$(8.2)^2 = 8.2 \times 8.2 = 67.24$	←	Too small. Try a greater number.
8.3	$(8.3)^2 = 8.3 \times 8.3 = 68.89$	←	Still too small. Try a greater number.
8.4	$(8.4)^2 = 8.4 \times 8.4 = 70.56$	←	Too big.

Because $(8.3)^2$ is less than 70 and $(8.4)^2$ is greater than 70, $\sqrt{70}$ is between 8.3 and 8.4. To decide which is closer, compare their squares to 70.

$$\begin{array}{r} 70.00 \\ -68.89 \leftarrow (8.3)^2 \\ \hline 1.11 \end{array} \qquad \begin{array}{r} 70.56 \leftarrow (8.4)^2 \\ -70.00 \\ \hline 0.56 \end{array}$$

Since $0.56 < 1.11$, $(8.4)^2$ is closer to 70 than is $(8.3)^2$.

$\sqrt{70} \approx 8.4$ *Ans.*

On a calculator:

Key Sequence	***Display***
70 [√]	8.3666003

Round the result on the display to the nearest tenth: 8.4

CLASS EXERCISES

1. For each sentence, write equivalent sentences, using the word "square," the words "square root," and the symbol $\sqrt{\ }$: **a.** $10^2 = 100$ **b.** $121 = 11^2$ **c.** $1^2 = 1$ **d.** $169 = 13^2$ **e.** $20^2 = 400$
2. Evaluate: **a.** 6^2 **b.** 4^2 **c.** 14^2 **d.** 32^2 **e.** 1.2^2 **f.** $\left(\frac{1}{3}\right)^2$ **g.** $(0.03)^2$
3. Find: **a.** $\sqrt{81}$ **b.** $\sqrt{100}$ **c.** $\sqrt{144}$ **d.** $\sqrt{1{,}600}$ **e.** $\sqrt{484}$ **f.** $\sqrt{625}$ **g.** $\sqrt{\frac{1}{4}}$
4. Which numbers are perfect squares? (a) 1,000 (b) 40 (c) 49 (d) 121 (e) 5 (f) 0
5. Approximate to the nearest tenth: **a.** $\sqrt{17}$ **b.** $\sqrt{46}$ **c.** $\sqrt{130}$ **d.** $\sqrt{53}$ **e.** $\sqrt{700}$
6. Find: **a.** $\sqrt{3{,}249}$ **b.** $\sqrt{14{,}641}$ **c.** $\sqrt{790{,}321}$ **d.** $\sqrt{509{,}796}$
7. Approximate to the nearest hundredth: **a.** $\sqrt{95}$ **b.** $\sqrt{128}$ **c.** $\sqrt{878}$ **d.** $\sqrt{1{,}426}$

HOMEWORK EXERCISES

1. For each sentence, write equivalent sentences, using the word "square," the words "square root," and the symbol $\sqrt{\ }$: **a.** $9^2 = 81$ **b.** $0^2 = 0$ **c.** $50^2 = 2{,}500$ **d.** $10{,}000 = 100^2$
2. Evaluate: **a.** 1^2 **b.** 20^2 **c.** 12^2 **d.** 13^2 **e.** 22^2 **f.** $\left(\frac{3}{5}\right)^2$ **g.** 100^2
3. Find: **a.** $\sqrt{9}$ **b.** $\sqrt{121}$ **c.** $\sqrt{900}$ **d.** $\sqrt{225}$ **e.** $\sqrt{\frac{9}{100}}$ **f.** $\sqrt{1{,}000{,}000}$
4. Which numbers are perfect squares? (a) 64 (b) .64 (c) 640 (d) 6,400 (e) .064
5. Approximate to the nearest tenth: **a.** $\sqrt{21}$ **b.** $\sqrt{3}$ **c.** $\sqrt{10}$ **d.** $\sqrt{107}$
6. Find: **a.** $\sqrt{3{,}969}$ **b.** $\sqrt{19{,}881}$ **c.** $\sqrt{273{,}529}$ **d.** $\sqrt{4{,}225}$
7. Approximate to the nearest thousandth: **a.** $\sqrt{52}$ **b.** $\sqrt{158}$ **c.** $\sqrt{19}$ **d.** $\sqrt{278}$
8. When is the square of a number less than the number?
9. Explain why $\sqrt{-16}$ is not -4.

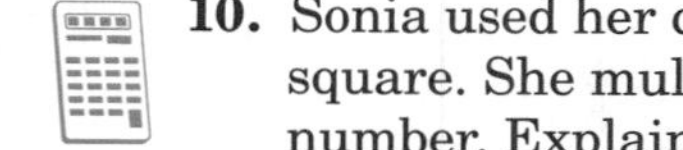

10. Sonia used her calculator to approximate the square root of a number that is not a perfect square. She multiplied the number in the display by itself, but did not get back to her original number. Explain what happened.
11. **a.** Find 4^2. **b.** Find $(-4)^2$.
    c. Nancy said that 16 has two square roots, +4 and −4. Explain how this can be true.
    d. How many numbers have squares equal to 25? **e.** How many square roots does 25 have?
    f. Do all numbers have two square roots? Explain, using examples.

SPIRAL REVIEW EXERCISES

1. Write 9.52×10^6 in standard notation.
2. The value of $7 \times 10^2 + 3 \times 10 + 5$ is
   (a) 7,035 (b) 735 (c) 375 (d) 7,350
3. Find the value of $50 - x^2$ when x is 7.
4. The opposite of 5 is
   (a) $\frac{1}{5}$ (b) $-\frac{1}{5}$ (c) -5 (d) 0

5. Find a number which, when used to replace ? in the following key sequence, will result in the display of a greater number: ? [√]
6. If no digit can be used more than once, the number of 4-digit numbers that can be formed using only the digits 1, 3, 5, 7, and 9 is: (a) 30 (b) 12 (c) 120 (d) 125
7. Find the value of $16 + 4 \times 8 \div 2$.
8. The mode for the set of numbers 5, 5, 10, 15, 35 is (a) 5 (b) 10 (c) 12 (d) 14

UNIT 15-6 Combining Like Terms; Adding and Subtracting Polynomials

THE MAIN IDEA

1. A ***term*** may contain 3 parts:

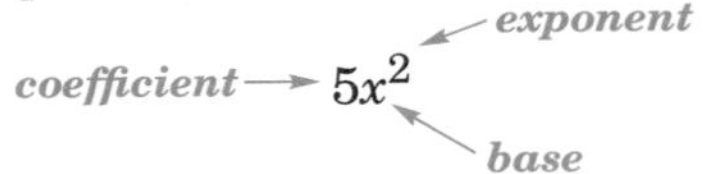

2. With no variable base, the term is a ***constant***, such as 5.
3. A coefficient of 1 or –1 is understood, so that x^2 means $1x^2$ and $-x^2$ means $-1x^2$.
4. An exponent of 1 is understood, so that $5x$ means $5x^1$.
5. In a single term, the only operations are multiplication and division.

   $-4a^2b^3$ and $\frac{5d^3}{2c}$ are each terms. $x + y$ is not one term.

6. ***Like terms*** have exactly the same variables with the same exponents.

   $7x^2$ and $5x^2$ are like terms.
   $5x^2$ and $5y^2$ are not like terms.

7. Like terms can be *combined* by addition or subtraction.

   $5x^2 + 7x^2 = 12x^2$
   $5x^2 - 7x^2 = -2x^2$
   $5x^2 + 5y^2$ cannot be combined.

EXAMPLE 1 Is the expression a term? Describe it if it is a term.

If the expression is not a term, explain why.

	Term?	*Description or Reason*
a. 24	yes	constant
b. w	yes	variable
c. $x + 2$	no	the *sum* of two terms
d. $5a$	yes	product of a constant and a variable
e. $6w - 1$	no	the *difference* of two terms
f. $-4x^3y$	yes	product of a constant, two variable bases and their exponents

EXAMPLE 2 Combine like terms as indicated.

a. Add $3a$ and $7a$.

$3a + 7a = (3 + 7)a$
$= 10a$

b. Find the sum of $-3y^3$ and $-7y^3$.

$-3y^3 + (-7y^3) = [-3 + (-7)]y^3$
$= -10y^3$

c. Find the sum of $3x^2$, $-5x^4$, $6x^2$, and $4x^4$.

Arrange the like terms together.

$3x^2 + 6x^2 + (-5x^4) + 4x^4$
$= (3 + 6)x^2 + [(-5) + 4)]x^4$
$= 9x^2 + (-1x^4)$
$= 9x^2 - x^4$

d. Subtract $-6a^5$ from $9a^5$.

To subtract, add the opposite.

$9a^5 - (-6a^5) = 9a^5 + (+6a^5)$
$= (9 + 6)a^5$
$= 15a^5$

EXAMPLE 3 Express the sum of the lengths of the sides of this figure.

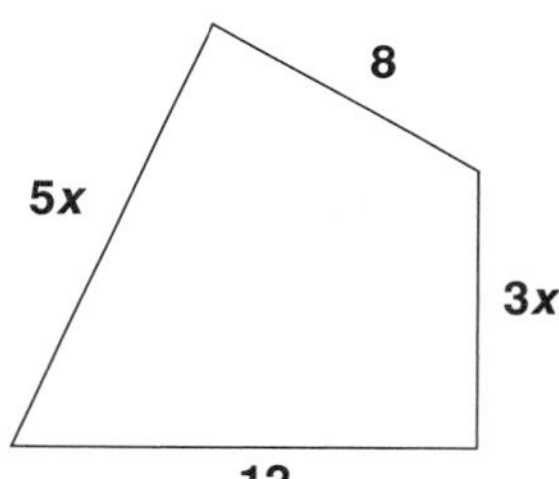

Express the sum and combine the terms.

$5x + 12 + 3x + 8 = (5 + 3)x + (12 + 8)$
$= 8x + 20$

CLASS EXERCISES

1. Which are terms? (a) 5 (b) p^3 (c) $x + 2$ (d) $5ab$ (e) $-3x^2y^2$ (f) $3 - y$ (g) $-\frac{11x}{20y}$ (h) $-5a^4bc^3$

2. Identify the like terms: 4 $5xy$ $8x^2$ 100 $-9x$ $-xy$ $14y^2$ $-3x^2$ $2x$ $2xy^2$ $-11y$ $4x^2$ xy

3. Combine like terms as indicated.

a. Add $6xy$ and $-4xy$. **b.** Find the sum of $-2x^3$ and $-7x^3$. **c.** Add: $4a^2b$, $-5a^2b$, a^2b, $3a^2b$

d. From $8x^2y^3$ subtract x^2y^3. **e.** Subtract $-10m^2n$ from $20m^2n$. **f.** Add: $12de^2$, -14, $-4de^2$, 13, de^2

4. The weights (in ounces) of five packages are represented as $6x$, $3x$, $5x$, 12, and 9. Express the total weight of the five packages.

ADDING AND SUBTRACTING POLYNOMIALS

THE MAIN IDEA

1. A polynomial is a sum or difference of terms. Special names are given to polynomials with

1 term: ***monomial***	2 terms: ***binomial***	3 terms: ***trinomial***
$5a^2$	$5a^2 + b$	$5a^2 + b + 3c$

2. Polynomials can be combined by adding or subtracting like terms.

EXAMPLE 4 Classify each polynomial according to the number of terms.

	Polynomial	***Number of Terms***	***Name***
a.	$-15w^2xy^3$	1 term	monomial
b.	$36x^2 + 12x - 9$	3 terms	trinomial
c.	$5xy + 3ab - 6bc - 12 + x$	5 terms	polynomial
d.	$5a - b$	2 terms	binomial

EXAMPLE 5 Add $4y^2 - 3y + 12$ and $6y^2 + 2y - 8$.

Express the sum of the polynomials and add the like terms.

$$\begin{array}{l} 4y^2 \ + (-3y) + 12 \\ \underline{6y^2 \ + \ 2y \ + (-8)} \\ 10y^2 + \ (-y) \ + 4 \end{array}$$

$$= 10y^2 - y + 4$$

EXAMPLE 6 From $5a^2 - 3ab + 15$ subtract $-2a^2 - 3ab + 14$.

To subtract, add the opposite.

$$5a^2 - 3ab + 15 - (-2a^2 - 3ab + 14)$$

$$= 5a^2 - 3ab + 15 + (2a^2 + 3ab - 14)$$

Combine like terms.

$$= 5a^2 + 2a^2 - 3ab + 3ab + 15 - 14$$

$$= (5 + 2)a^2 + (-3 + 3)ab + (15 - 14)$$

$$= 7a^2 + 0ab + 1$$

$$= 7a^2 + 1$$

CLASS EXERCISES

1. Classify each polynomial according to the number of terms.
   a. $3ab + 5bc - 7cd$ **b.** $5x^2y - 11xy^2$ **c.** $w + x + y + z$ **d.** $-2de^2f^4$ **e.** $2z$ **f.** 2
2. Add the polynomials.
   a. $x^2 + 5$ and $5x^2 + 20$ **b.** $3a - 2b$ and $-a + 5b$
   c. $(15y + 10z) + (y - 2z) + (12z)$ **d.** $(8a^2 + 6a + 2) + (-a^2 - a + 1)$
3. Express the total length of the line segment as a binomial.

3a + b | a + b | 5a − b | 7a − 3b

4. The map shows a triangular lot where three streets intersect. Express the total distance around the lot as a binomial.

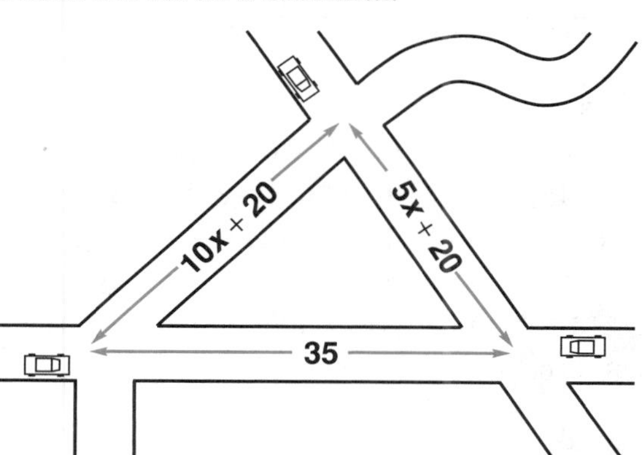

5. Subtract the polynomials.
   a. From $8x + 6$, subtract $5x + 5$.
   b. Subtract $8a - 2b$ from $12a - 3b$.
   c. $(2x^2 + 3x - 10) - (x^2 + x + 10)$
   d. $12xyz^2 - 12xyz^2$ **e.** $10a^2b^3 - 9a^2b^3$
6. Simplify each expression by combining like terms and arranging the terms so that the exponents are in decreasing order.
   a. $5x + 7 + 3x - 2$
   b. $9x^3 - 2x^2 + 7x - 9 + 8x^3 - 9x^2 - 16x$
   c. $15y^3 - 3y^2 + 6 - 8y^3 + 9y^2 + 6y + 9$
   d. $15x^2y - 13xy^2 + 9x^2y - 8xy^2$
7. Express the distance from A to B as a binomial.

A B C
7x − 2
10x − 5

8. From the sum of $9p^2 - 7pq - q^2$ and $4p^2 + pq + 2q^2$, subtract $6p^2 + q^2$.
9. **a.** Add: $(3x + 1) + (3x + 1) + (3x + 1) + (3x + 1) + (3x + 1)$
   b. Use the distributive property to multiply: $5(3x + 1)$
   c. Explain how the expressions in part **a** and part **b** are related.
10. Gift fruit baskets include x pieces of fruit and 3 jars of preserves.
    a. Write an expression for the number of items in 6 fruit baskets.
    b. How many items are there in 6 fruit baskets if the value for x is 8?

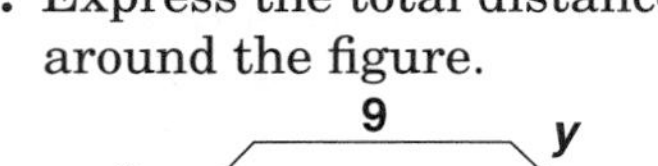

HOMEWORK EXERCISES

1. Identify the like terms: $5ab$ -12.5 $7xy^2$ 1 $3xy$ $5b$ xy^2 $5a$ ab $\frac{5}{12}$ $12x^2y$

2. Combine like terms as indicated.

a. $-5xy + (-3xy)$ **b.** Add $5a^2b$ and $7a^2b$.

c. Find the sum of $-6wxy$ and $3wxy$.

d. Find the sum of $10pq^2$, -15, $5pq^2$, and 32.

e. Add $3.5xy$ and $-3.5xy$. **f.** Subtract $45a^3$ from $30a^3$.

g. $30ax^2 - 29ax^2$ **h.** From $8mn$, subtract $-2mn$.

3. Express the total distance around the figure.

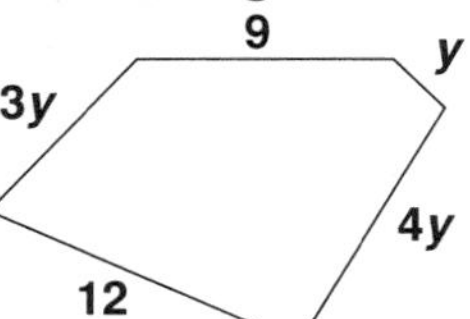

4. Express the total distance from A to E in the least number of terms possible.

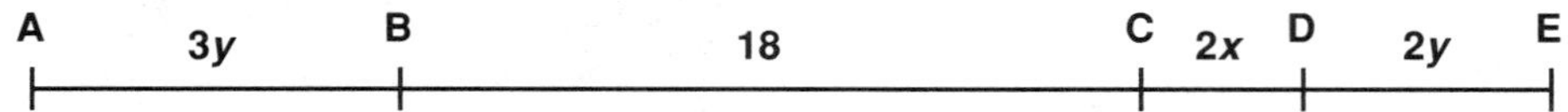

5. Classify each polynomial according to the number of terms.

a. $6a^2 + 12$ **b.** $x + y + 2$

c. $-12x^2y^3$ **d.** $-12x^2 + y^3$

e. $3a - 2b + c - 4d$

6. Add the polynomials.

a. $5x + 12$ and $x + 20$ **b.** $18y - 6$ and $12y - 8$

c. $(-6a + 60) + (5a - 45)$ **d.** $(19x - 14y) + (-10x - 14y)$

e. $(5a + 7b - 3c) + (-2a + 3b - c)$ **f.** $(9x^2 + 11) + (3x^2 - 11)$

g. $4a^2b^3 + (-5a^2b^3)$ **h.** $(5x - 3y) + (2x + 2y) + (-4x - y)$

i. To the sum of $(51a + b)$ and $(-48a + 50b)$, add $(a - 50b)$.

j. Find the sum of $(30x^2 + y^2 + z^3)$, $(8x^2 - 3y^2 + z^3)$, and $(-4y^2 - z^3)$.

7. Express the distance around the lawn as a binomial.

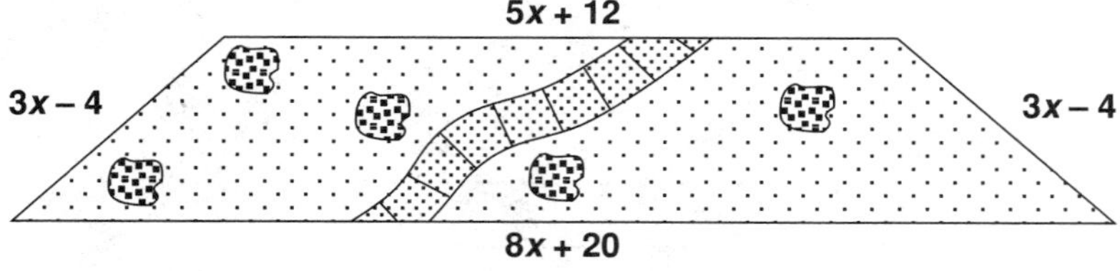

8. Express the total distance from A to F, using as few terms as possible.

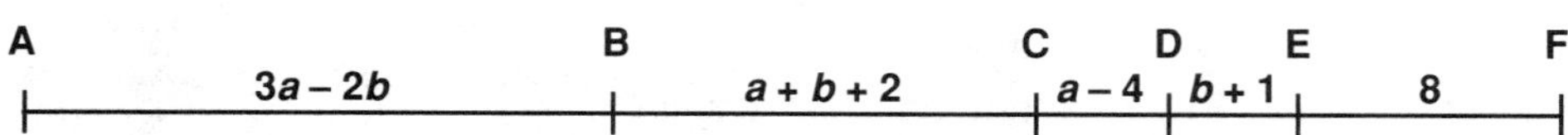

9. The shipping weight of a printer ribbon is x ounces, and the shipping weight of a loose-leaf binder is y ounces. The chart gives data about the contents and weights of 5 packages that must be shipped.

Package Number	Number of Ribbons	Number of Binders	Weight of Ribbons	Weight of Binders	Total Weight
1	2	3	$2x$	$3y$	$2x + 3y$
2	4	7	$4x$	$7y$	$4x + 7y$
3	1	6	x	$6y$	?
4	5	1	?	?	?
5	?	?	$3x$	?	$3x + 8y$

a. Copy and complete the chart.

b. Express the total weight of all five packages as a binomial.

10. Subtract the polynomials.

a. $(7m + 3p) - (5m + 2p)$ **b.** $(3x - 4y) - (2x + 5y)$

c. $(-8a + 9b) - (7a - b)$ **d.** $(2p + 3q + 5r) - (p + q + r)$

e. From $10x^2 + 20y^2$, subtract $-3x^2 + y^2$.

f. Subtract $-25a^2 - 16b^2$ from $16a^2 - 25b^2$.

g. From the sum of $125m^3$, $78n^2$, and $-39p$, subtract $125m^3 + 78n^2 - 39p$.

11. Simplify each expression by combining like terms and arranging the terms so that the exponents are in decreasing order.

a. $5x^4 + 9x^2 - 9x - 17 + 8x^3 - 9x^2 + 7x$

b. $12y^3 + 3y^2 - 9y + 17y^3 - 9y^2 - 7y - 8$

c. $10xy^3 + 3x^2y^2 - 9xy^3 - 4x^2y^2 - 4xy^3$

12. The number of acres of land in a park is expressed as $3x^2 + 7x - 2$. The number of acres used for tourist services is represented as $x^2 - 10$. Represent the number of acres used for other purposes.

13. Express the distance from B to C, using the least number of terms possible.

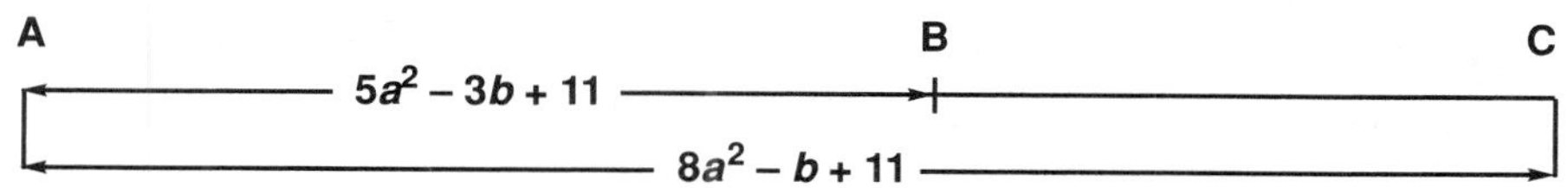

14. The amount of copper (in grams) in a square sheet is represented by $5p^2 + 7q^2$. The amount of copper removed to make a circular hole, as shown in the figure, is represented by $3p^2 - q^2$. Represent the amount of copper that remains after the circular piece is removed.

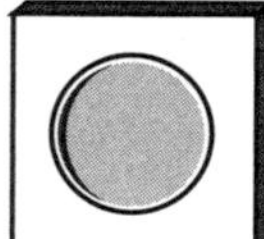

15. The volume of a laboratory solution is represented by $(5a + 14b)$ cubic cm. This is combined with another solution of volume $(7a - 2b)$ cubic cm. From this mixture, a volume of $(6a - 2b)$ cubic cm is removed. Represent the volume of solution that remains.

16. Cara bought 5 pairs of socks and returned 3 pairs. The cost (in dollars) of one pair is represented by x.
   a. Write an expression for each of the following:
      (1) the cost of the five pairs of socks that Cara bought
      (2) the cost of the 3 pairs of socks that Cara returned
      (3) the cost of the socks that Cara did not return
   b. Use this example to explain why $5x - 3x$ equals $2x$.

17. a. Add: $(x + 7) + (x + 7) + (x + 7) + (x + 7)$
   b. Use the distributive property to multiply: $4(x + 7)$
   c. Using $x = 6$, verify that the expressions given in parts **a** and **b** are equivalent. Is this true for all values of x?

18. A magazine has x pages in black and white and 16 pages in color. Write an expression for the total number of pages in 10 copies of the magazine.

SPIRAL REVIEW EXERCISES

1. Which number is not a perfect square?
   (a) 10 (b) 100 (c) 25 (d) 49

2. Find the value of each square root.
   a. $\sqrt{1}$ **b.** $\sqrt{9}$ **c.** $\sqrt{81}$ **d.** $\sqrt{169}$

3. Evaluate: $5 \times 10^3 + 6 \times 10 + 9$

4. What is the value of $4x^3$ when $x = 2$?

5. What is the opposite of -3?
   (a) $+3$ (b) $\frac{1}{3}$ (c) $-\frac{1}{3}$ (d) 0

6. What is the probability that this spinner will stop on a multiple of 3?

7. Ribbon costs \$0.19 per foot. If David bought 4 yards of ribbon, how much did he spend?

8. Subtract $11\frac{5}{8}$ from $18\frac{1}{4}$.

9. Write the number 25,415 as a product of prime numbers.

10. Eight basketball teams are in a tournament. If there are no ties, in how many different ways can first, second, and third place be determined?

11. On a plane flight, the ratio of seats that are filled to the total number of seats is 4 to 9. If there are 63 seats, how many passengers are on the plane?

UNIT 16-1 Solving Equations by Adding; Using an Equation to Solve a Problem

SOLVING EQUATIONS BY ADDING

THE MAIN IDEA

1. An ***equation*** is a mathematical sentence stating that *two quantities are equal*.

   $22 - 15 = 7$

2. When an equation contains a variable, *a value of the variable that makes the equation true* is called a ***solution*** of the equation.

   If $x - 15 = 7$, then $x = 22$ is the solution.

3. To ***solve*** an equation like this, add 15 (the opposite of –15), to *both* sides, thus getting the variable x to stand alone while keeping the equation "in balance."

   $$x - 15 = 7$$
   $$x \underbrace{- 15 + 15}_{0} = 7 + 15$$
   $$x = 22$$

4. To check a solution, substitute the number for the variable in the original equation and see if a true sentence results.

   $$x - 15 = 7$$
   $$22 - 15 \stackrel{?}{=} 7$$
   $$7 = 7 \checkmark$$

EXAMPLE 1 Tell whether 3 is a solution of the equation $7x + 12 = 33$.

In the original equation, substitute 3 for x. $\quad 7(3) + 12 \stackrel{?}{=} 33$

Evaluate, following the order of operations. $\quad 21 + 12 \stackrel{?}{=} 33$

Since a true sentence results, 3 is a solution. $\quad 33 = 33 \checkmark$

EXAMPLE 2 Solve for q: $17 + q = -20$

Add the opposite of 17 to both sides.

$$17 + q = -20$$
$$\underbrace{(-17) + 17}_{0} + q = -20 + (-17)$$
$$q = -37$$

Check: In the original equation, substitute –37 for q and evaluate.

$$17 + q = -20$$
$$17 + (-37) \stackrel{?}{=} -20$$
$$-20 = -20 \checkmark$$

Answer: $q = -37$

CLASS EXERCISES

1. Tell whether the given value of the variable is a solution of the given equation.
   a. $x + 7 = 28;\ x = 21$ **b.** $x - 13 = 13;\ x = 26$ **c.** $5 + a = 21;\ a = 26$
   d. $3n = 120;\ n = 40$ **e.** $2x + 11 = 27;\ x = 15$ **f.** $3x - 10 = 17;\ x = 9$

2. Solve and check.
   a. $x + 14 = 23$ **b.** $y - 20 = 60$ **c.** $-35 + x = 43$ **d.** $a + 1 = -1$
   e. $z - 18 = -3$ **f.** $y + 19 = 19$ **g.** $101 = m - 87$ **h.** $73 + b = 98$
   i. $x - 25 = -25$ **j.** $y + 19 = -19.2$ **k.** $13 = 31 + z$ **l.** $n - 42 = -39$

USING AN EQUATION TO SOLVE A PROBLEM

THE MAIN IDEA

1. Writing and solving an equation is a useful problem-solving strategy.
2. To use an equation to solve a word problem:
   a. Choose a variable to represent the unknown quantity.
   b. Translate the words of the problem into an equation.
   c. Solve the equation.
3. To check a solution, substitute the solution for the words that describe it in the original problem. If a true statement results, the solution checks.

EXAMPLE 3 Write an equation for each sentence. Use n to represent the unknown number.

	Sentence	*Equation*
a.	A number increased by 17 equals 32. (n / $+17$ / $= 32$)	$n + 17 = 32$
b.	3 times a number equals 42. ($3 \cdot$ / n / $= 42$)	$3n = 42$
c.	13 less than a number equals 73. ($n - 13$ / $= 73$)	$n - 13 = 73$

EXAMPLE 4 The Sears Tower is 1,454 feet tall. It is 104 feet taller than the World Trade Center. How tall is World Trade?

Represent the unknown quantity.

Let W = height of the World Trade Center.

Translate the words of the problem into an equation.

Tower is 104′ taller than World Trade.

$$\underbrace{1{,}454}_{\text{Tower}} \quad \underbrace{=}_{\text{is}} \quad \underbrace{104 + W}_{\text{104′ taller than World Trade}}$$

To solve, get W alone by adding the opposite of 104.

$$1{,}454 + (-104) = 104 + (-104) + W$$

$$1{,}350 = W$$

Check: Substitute in the original problem. Is 1,454 really 104 more than 1,350? Yes.

Answer: The World Trade Center is 1,350' tall.

EXAMPLE 5 Mr. Yuen's age 9 years ago was 21. How old is he now?

Represent the unknown quantity.

Translate the words of the problem into an equation.

Solve the equation. To get the variable alone, add an opposite.

Let y = Mr. Yuen's age now.
Then $y - 9$ = his age 9 years ago.

Age 9 years ago was 21.

$$y - 9 = 21$$

Rewrite subtraction. $y + (-9) = 21$

Add opposite. $y + (-9) + (9) = 21 + (9)$

$$y = 30$$

Check: Substitute in the original problem. If he is 30 years old now, was he 21 years old 9 years ago? Yes.

Answer: Mr. Yuen is now 30 years old.

CLASS EXERCISES

1. Write an equation for each sentence. Use n to represent the unknown number.

a. A number increased by 31 equals 48.
b. Twelve more than a number is 81.
c. 56 decreased by a number equals 29.
d. 31 less than a number is 147.
e. 6 times a number is 90.
f. The sum of 16 and a number is 112.

2. Write an equation to solve each word problem. Solve and check.

a. If a number is increased by 15, the result is 42. Find the number.
b. If the price of a coat is reduced by \$15, the new price is \$75. Find the original price.
c. The musical *Hello, Dolly!* was performed on Broadway 1,496 more times than *Funny Girl*. There were 2,844 performances of *Hello, Dolly!*. How many were there of *Funny Girl*?
d. In 12 years, Sam will be 52. How old is he now?

HOMEWORK EXERCISES

1. Tell whether the given value of the variable is a solution of the given equation.
   a. $x + 17 = 24;\ x = 7$ **b.** $a - 9 = 27;\ a = 36$ **c.** $8 + y = 21;\ y = 17$
   d. $3m = 33;\ m = 11$ **e.** $z + 15 = 15;\ z = 0$ **f.** $n - 14 = 1;\ n = 13$
   g. $35 + x = 25;\ x = -10$ **h.** $x - 1 = 1;\ x = 1$ **i.** $x + 3 = 5;\ x = -2$

2. Solve each equation and check.
   a. $x + 13 = 25$ **b.** $x - 19 = 11$ **c.** $y - 25 = -35$ **d.** $1 + x = 1.5$
   e. $y - 13 = 13$ **f.** $14 + y = -14$ **g.** $x + 67 = 93$ **h.** $100 = n - 47$
   i. $y - 39 = -38$ **j.** $100 + a = 5$ **k.** $x - 15 = -32$ **l.** $y + 26 = -51$
   m. $y - 26 = 51\frac{3}{4}$ **n.** $44\frac{1}{2} + x = 22$ **o.** $-2 = b - 150$ **p.** $x - 25 = 74$

3. Write an equation for each sentence. Use n to represent the unknown number.
   a. A number increased by 43 equals 92. **b.** Nine more than a number is –34.
   c. A number decreased by 47 equals 194. **d.** 48 less than a number is 67.
   e. 4 times a number is 1,488. **f.** The sum of a number and 26.4 is 105.

4. Write an equation to solve each word problem. Solve and check.
   a. Ms. McGowan put 50 pounds of cargo on her truck. The resulting weight was 6,000 pounds. Find the weight of the truck without the cargo.
   b. If a number is deceased by 12, the result is –30. Find the number.
   c. Charles Lindbergh's flight from New York to Paris in 1927 was 1,584 miles longer than Amelia Earhart's flight from Newfoundland to Ireland five years later. If Lindbergh's flight was 3,610 miles long, how long was Earhart's flight?
   d. If the present temperature increases by 5°, the temperature will be –8°*F*. Find the present temperature.
   e. Rachel has $12 less than her brother Joshua to spend on a gift for their mother. If Joshua has $32, how much does Rachel have?
   f. Six years ago, Samantha was 3 years younger than her cousin Alex, who was then 14. How old is Samantha now?

SPIRAL REVIEW EXERCISES

1. Find the sum of $2x^2y$ and $(-30x^2y)$.
2. Subtract: $(9x^2 - 3x + 9) - (6x^2 + 7x - 12)$
3. What is the value of $2x^3 + 3x^2 - 2$ when $x = 3$?
4. What is the probability that the first spinner will stop on an even number and the second spinner will stop on a vowel?

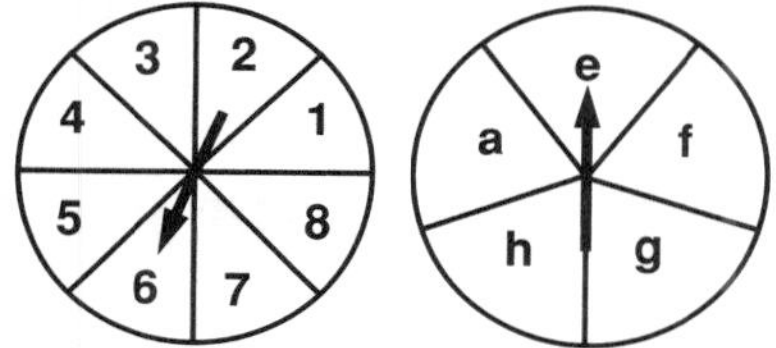

5. The spacecraft Mercury, piloted by Alan Shepard in 1961, had a thrust of 78,000 pounds. A modern space shuttle has a thrust of 7.8 million pounds. How many times as great as the thrust of the Mercury is that of the modern shuttle?

   (a) 10^2 (b) 10^3 (c) 10^4 (d) 10^6

6. A 75-foot maple tree has between 500,000 and 700,000 leaves, In the autumn, when the leaves fall, this is enough to fill ten 32-gallon plastic bags. The approximate number of leaves in each bag is between

   (a) 5,000 and 7,000 (b) 6,667 and 9,333
   (c) 15,625 and 21,875 (d) 50,000 and 70,000

7. Between which two consecutive integers does $\sqrt{131}$ lie?

8. A diner's special lunch price is $3.89. Use the constant feature of your calculator to make a chart that shows the total prices for purchases of from 1 to 6 lunch specials.

9. The graph below gives the average number of miles traveled annually by drivers in the United States for each of the given years.

Average Annual Miles Traveled Per Driver

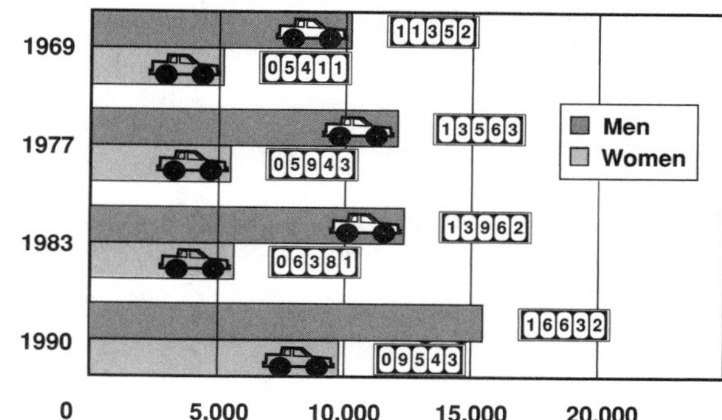

Based on the data, the greatest percent of increase is between

(a) 1977 and 1983 for male drivers
(b) 1969 and 1977 for female drivers
(c) 1969 and 1977 for male drivers
(d) 1983 and 1990 for female drivers

10. One grocery store is selling 3 cans of soup for $0.99. Another store is advertising 8 cans of the same soup on sale for $2.71. Which is the better buy?

UNIT 16-2 Solving Equations by Multiplying

THE MAIN IDEA

To solve an equation that involves multiplication, like

$$9x = 72 \qquad \text{or} \qquad \frac{1}{7}x = -5$$

multiply each side of the equation by the reciprocal of the coefficient.

EXAMPLE 1 Solve for x: $9x = 72$

$9x = 72$

Multiply each side of the equation by the reciprocal of 9.

$\frac{1}{9}(9x) = \frac{1}{9}(72)$

$x = 8$

(Remember: $9 \times \frac{1}{9} = 1$)

Check: In the original equation, substitute 8 for x and evaluate.

$9x = 72$

$9(8) \stackrel{?}{=} 72$

$72 = 72$ ✔

Answer: $x = 8$

EXAMPLE 2 Solve for x: $\frac{x}{7} = -5$

Rewrite the division as multiplication.

$\frac{1}{7} \cdot x = -5$

Multiply each side of the equation by the reciprocal of $\frac{1}{7}$.

$7(\frac{1}{7} \cdot x) = 7(-5)$

$x = -35$

Check: In the original equation, substitute -35 for x and evaluate.

$\frac{x}{7} = -5$

$\frac{-35}{7} \stackrel{?}{=} -5$

$-5 = -5$ ✔

Answer: $x = -35$

EXAMPLE 3 If the value of a coin is multiplied by 6, the result is \$3. Find the value of the coin.

THINKING ABOUT THE PROBLEM

Use the problem-solving strategy of writing and solving an equation.

Let y = the value of the coin.

Write an equation. $6y = 3$

Multiply both sides of the equation by the reciprocal of 6.

$\frac{1}{6}(6y) = \frac{1}{6}(3)$

$y = \frac{3}{6}$

$y = \frac{1}{2}$

Check: Replace the words "the value of a coin" with the number $\frac{1}{2}$ to see if a true statement results.

If $\frac{1}{2}$ is multiplied by 6, the result is 3.

$6 \times \frac{1}{2} \stackrel{?}{=} 3$

$3 = 3$ ✔

Answer: The value of the coin is $\frac{1}{2}$ of a dollar, or 50 cents.

CLASS EXERCISES

1. Solve each equation and check.

a. $8x = 48$ **b.** $7a = 52.5$ **c.** $-11x = 132$ **d.** $84 = 12b$ **e.** $9x = -36$
f. $10y = -70$ **g.** $-45a = 15$ **h.** $14x = 7$ **i.** $42x = 0$ **j.** $18x = 18$

2. Solve each equation and check.

a. $\frac{x}{8} = 4$ **b.** $96 = \frac{y}{3}$ **c.** $\frac{a}{5} = 1$ **d.** $\frac{x}{9} = -4$ **e.** $\frac{b}{-7} = -6$ **f.** $\frac{y}{2} = -1$

3. Solve each equation and check.

a. $x + 4 = 12$ **b.** $-4x = 12$ **c.** $\frac{1}{4}x = 12$ **d.** $12x = 4$ **e.** $12 - x = 4$

f. $12 + x = 4$ **g.** $\frac{x}{12} = 4$ **h.** $\frac{1}{12}x = 4$ **i.** $x - 12 = 4$ **j.** $4 - x = 12$

4. Write an equation to solve each word problem. Solve and check.

a. If a number is multiplied by 7, the result is –98. Find the number.
b. If Muriel can do twice as many sit-ups as Jack, who does 75, how many sit-ups can she do?
c. If the weight of a package is multiplied by $\frac{1}{3}$, the result is 15 pounds. Find the weight.
d. One-fifth of the length of a rope is 8 feet. Find the length of the rope.

HOMEWORK EXERCISES

1. Solve each equation and check.

a. $5x = 85$ **b.** $3x = -48$ **c.** $-7x = 98$ **d.** $11x = -99$ **e.** $4x = -24$
f. $6x = 54$ **g.** $-72 = 6x$ **h.** $10x = 5$ **i.** $15x = -5$ **j.** $3x = -15$

2. Solve each equation and check.

a. $\frac{x}{4} = 80$ **b.** $\frac{y}{6.5} = 48$ **c.** $\frac{a}{5} = 105.5$ **d.** $\frac{x}{7} = 14\frac{1}{3}$ **e.** $33 = \frac{n}{2}$

f. $\frac{x}{-9} = 1$ **g.** $\frac{x}{8} = -1$ **h.** $-96 = \frac{y}{8}$ **i.** $\frac{m}{12} = -144$ **j.** $\frac{-144}{x} = 12$

3. Solve each equation and check.

a. $5x = 10$ **b.** $5.7 + x = 10$ **c.** $x - 5 = 10\frac{2}{3}$ **d.** $\frac{1}{5}x = 10$ **e.** $10x = -5$

4. Write and solve an equation that says: "3 times a number is 4.5"

5. Write an equation to solve each word problem. Solve and check.

a. Emily donates $\frac{1}{2}$ of her baby-sitting money to charity. Last week, she donated \$17. How much money did she earn baby-sitting?

b. If a number is divided by 15, the result is –18. Find the number.

c. A theater sold \$872 worth of tickets. If each ticket cost \$4, how many tickets were sold?

SPIRAL REVIEW EXERCISES

1. Solve each equation and check.

a. $x + 34 = 52$ **b.** $y - 11 = 74$

c. $a - 13 = -12$ **d.** $y + 25 = -34$

2. Which statement is correct?

(a) 16 is the square root of 4.

(b) 25 is the square of 5.

(c) $\frac{1}{4}$ is the square root of $\frac{1}{2}$.

(d) 100 is the square root of 1,000.

3. Which number is a perfect square?

(a) 5 (b) 55 (c) 9 (d) 99

4. 36 is what percent of 50?

5. The value of $5 \times 10^3 + 2 \times 10 + 9$ is

(a) 5,290 (b) 5,209

(c) 5,029 (d) 52,009

6. The mean of 12, 12, 20, 30, and 56 is

(a) 12 (b) 20 (c) 24 (d) 26

7. The product of –18 and 3 added to the quotient of –25 and –5 is:

(a) –1 (b) –11 (c) –49 (d) –59

8. The key sequence 7 [×] [=] [=] [=] [=] [=] can be used to calculate

(a) 7^2 (b) 7^5 (c) 7^6 (d) 6^7

9. As shown in the bar graph, the number of cars sold in March is closest to

(a) 2 (b) 150 (c) 175 (d) 200

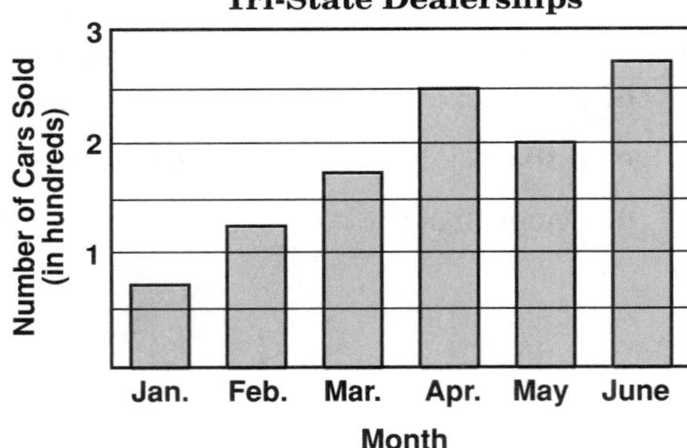

10. The number of meters in 3 kilometers is

(a) 3 (b) 30 (c) 300 (d) 3,000

11. Jack, Carlos, Ramón, and Hector each wrote his name on a slip of paper and placed the slip in a box. The slips were then removed from the box one at a time. What is the probability that the slips were picked in alphabetical order?

UNIT 16-3 Solving Equations With Two Operations

THE MAIN IDEA

To solve an equation with two operations, like $5x + 8 = 38$:

1. First, work on the additions by adding an opposite to both sides of the equation.
2. Then, multiply both sides by the reciprocal of the coefficient.

EXAMPLE 1 Solve for x: $5x + 8 = 38$

Add the opposite of 8 to both sides.

$$5x + 8 = 38$$
$$5x + 8 + (-8) = 38 + (-8)$$
$$5x = 30$$

Multiply both sides by the reciprocal of 5.

$$\frac{1}{5}(5x) = \frac{1}{5}(30)$$
$$x = 6$$

Check: Substitute 6 for x in the original equation, and evaluate.

$$5x + 8 = 38$$
$$5(6) + 8 \stackrel{?}{=} 38$$
$$30 + 8 \stackrel{?}{=} 38$$
$$38 = 38 \checkmark$$

Answer: $x = 6$

EXAMPLE 2 Solve for x: $2x - 3 = -13$

Rewrite the subtraction.

$$2x - 3 = -13$$
$$2x + (-3) = -13$$

Add 3 to both sides.

$$2x + (-3) + (3) = -13 + (3)$$
$$2x = -10$$

Multiply both sides by $\frac{1}{2}$.

$$\frac{1}{2}(2x) = \frac{1}{2}(-10)$$
$$x = -5$$

Check: Substitute -5 for x in the original equation, and evaluate.

$$2x - 3 = -13$$
$$2(-5) - 3 \stackrel{?}{=} -13$$
$$-10 - 3 \stackrel{?}{=} -13$$
$$-13 = -13 \checkmark$$

Answer: $x = -5$

EXAMPLE 3 If one-fourth of a number is increased by 6, the result is 5. Find the number.

Let n = the number.

$$\frac{1}{4}n + 6 = 5$$
$$\frac{1}{4}n + 6 + (-6) = 5 + (-6)$$
$$\frac{1}{4}n = -1$$
$$4\left(\frac{1}{4}n\right) = 4(-1)$$
$$n = -4$$

Check: If $\frac{1}{4}$ of -4 is increased by 6, is the result 5?

$$\frac{1}{4}(-4) + 6 \stackrel{?}{=} 5$$
$$-1 + 6 \stackrel{?}{=} 5$$
$$5 = 5 \checkmark$$

Answer: The number is -4.

EXAMPLE 4 Solve and check: $4(2x + 1) = 60$

$$4(2x + 1) = 60$$
$$8x + 4 = 60 \quad \text{Distributive property.}$$
$$8x + 4 + (-4) = 60 + (-4) \quad \text{Add opposite.}$$
$$8x = 56$$
$$\frac{1}{8} \cdot 8x = \frac{1}{8} \cdot 56 \quad \text{Multiply by reciprocal.}$$
$$x = 7$$

Check: Substitute 7 for x.

$$4(2x + 1) = 60$$
$$4(2 \cdot 7 + 1) \stackrel{?}{=} 60$$
$$4(14 + 1) \stackrel{?}{=} 60$$
$$4(15) \stackrel{?}{=} 60$$
$$60 = 60 \checkmark$$

Answer: $x = 7$

CLASS EXERCISES

1. Solve each equation and check.

a. $3x + 4 = 10$ **b.** $6 + 2x = -12$ **c.** $5y + 7 = 27.5$ **d.** $\frac{a}{4} + 2 = 34$

e. $9 + \frac{1}{3}z = 3$ **f.** $5x + 15 = 0$ **g.** $5x + 6 = 32.5$ **h.** $3 + 8x = 51$

i. $\frac{y}{7} + 7 = -21$ **j.** $57 + 3q = 39$ **k.** $\frac{1}{2}x + 21 = 3$ **l.** $11m + 9 = -79$

2. Solve each equation and check.

a. $2x - 6 = 10$ **b.** $3x - 5 = -11$ **c.** $8y - 12 = 40$ **d.** $5a - 1 = 14$

e. $7m - 9 = -2$ **f.** $10x - 14 = 36$ **g.** $3m - 3 = 3$ **h.** $5y - 2 = -2$

3. Write an equation to solve each word problem. Solve and check.

a. If three times a number is decreased by 8, the result is 22. Find the number.

b. For a class trip, 3 buses were filled and 6 other students went by car. If there were 126 students on the trip, how many were in each bus?

c. At a half-price sale, a $2 sales tax was added to the sale price of a radio. If the resulting cost was $36, what was the original price of the radio?

d. Together, Sandra and Lou are planning to spend $54 on an anniversary gift for their parents. Lou has $16. Sandra needs to contribute twice as much money as she originally planned. How much did Sandra originally plan to give?

4. Solve each equation and check.

a. $2(x + 1) = 18$ **b.** $5(3x + 2) = 55$ **c.** $4(x - 3) = 60$ **d.** $3(2x - 1) = 39$

HOMEWORK EXERCISES

1. Solve each equation and check.

a. $2x + 7 = 15$ **b.** $6y + 8 = 38$ **c.** $1 + 9z = 82$ **d.** $8a + 5 = 53$
e. $3b + 14 = 29$ **f.** $4q + 8 = 12$ **g.** $5x + 5 = 45$ **h.** $2m + 22 = 30$
i. $7 + 12a = 31$ **j.** $8y + 17 = 97$ **k.** $26 + 6x = 14$ **l.** $5y + 53 = 108$
m. $4z + 20 = 0$ **n.** $15 + 12x = -9$ **o.** $3x + 2 = 2$ **p.** $6y + 8 = 14$

2. Solve each equation and check.

a. $3x - 4 = 11$ **b.** $2y - 8 = 18$ **c.** $5a - 15 = 5$ **d.** $6m - 6 = 0$
e. $9x - 3 = -21$ **f.** $10y - 10 = 100$ **g.** $11a - 44 = -22$ **h.** $-3 + 14t = -3$

3. Solve each equation and check.

a. $\frac{x}{4} + 2 = 5$ **b.** $\frac{1}{5}y - 3 = 1$ **c.** $\frac{a}{3} + 12 = 9$ **d.** $\frac{1}{7}x + 5 = 5$ **e.** $\frac{n}{4} - 4 = 1$

f. $\frac{1}{3}x - 1 = -1$ **g.** $\frac{y}{10} - 5 = -45$ **h.** $\frac{1}{12}x + 1 = -3$ **i.** $\frac{1}{3}x + 6 = 12$ **j.** $\frac{n}{2} - 3 = -2$

4. Solve each equation and check.

a. $3x + 1 = 10$ **b.** $\frac{x}{3} + 5 = 14$ **c.** $8m - 3 = 21$ **d.** $\frac{y}{3} - 12 = -2$

e. $11 + 4r = 15$ **f.** $5x + 12 = -13$ **g.** $\frac{a}{2} + 17 = 12$ **h.** $\frac{1}{4}x - 1 = 0$

i. $2 + 2a = 16$ **j.** $\frac{1}{3}x + 9 = 9$ **k.** $\frac{n}{5} - 1 = -2$ **l.** $\frac{y}{10} - 8 = -16$

5. Write an equation to solve each word problem. Solve and check.

a. When a number is multiplied by 6 and the product is increased by 10, the result is 130. Find the number.

b. If 15 less than one-third of a number is –25, find the number.

c. Carl said, "Think of a number between 10 and 20. Multiply it by 5. Subtract 17." If Mary's result was 58, what was her original number?

d. Mike's Bikes rents a bicycle for $15 plus $2 an hour. JoAnn paid $23. For how many hours did she rent the bike?

e. If 3 times a number decreased by 5 is the same as 2 times the number increased by 8, find the number.

6. Solve each equation and check.

a. $3(x + 2) = 27$ **b.** $2(2x + 1) = 94$ **c.** $5(x - 1) = 55$ **d.** $4(5x - 2) = 72$

SPIRAL REVIEW EXERCISES

1. Tell whether the given value of the variable is a solution of the given equation.

   a. $x + 13 = -13;\ x = -1$

   b. $a - 17 = 27;\ a = 44$

   c. $3m = -18;\ m = -6$

   d. $\frac{y}{8} = -4;\ y = -2$

2. Evaluate $5x + 2y$ when $x = 4$ and $y = 6$.

3. An electric company charges:

   $0.18 per day for a service charge
   $0.14 per kilowatt-hour for the first 250 kwhr.
   $0.15 per kilowatt-hour after the first 250 kwhr.

   How much is the cost of 1,000 kilowatt-hours used over a 30-day period?

4. Jason deposited checks in the amounts of $34.80, $96.50, and $111.95 into his checking account, which had a balance of $492.75 before the deposits. What was his new balance?

5. If a pair of shoes that sells for $45 is on sale for 30% off, what is the sale price?

6. Find $\frac{3}{4}$ of the difference of 360 and 80.

7. Solve the proportion:

$$\frac{3}{7} = \frac{21}{x}$$

8. Subtract 81.9 from 90.82.

9. Evaluate: 5!

10. Find the number that, when used to replace [?] in the following key sequence, would result in a calculator display equal to 2.3×10^5.

    920 [×] [?] [=]

11. In 1918, when the Boston Red Sox won the World Series, each player received $1,102 as the winners' share. In 1990, when the Cincinnati Reds won the World Series, each player received $112,533. The winners' share for the 1990 Reds is about how many times as much as the winners' share for the 1918 Red Sox?

    (a) 10^0 (b) 10^1 (c) 10^2 (d) 10^3

12. Combine like terms:
    $3x^2y - 9xy^2 + 4xy^2 - 7x^2y + 12x^2y$

13. 20 kilometers is equal to

    (a) 200 m (b) 2,000 m
    (c) 20,000 m (d) 200,000 m

14. Find the probability that the first of these spinners will stop on "red" and the other on a vowel.

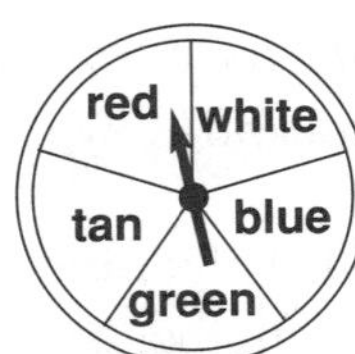

UNIT 16–4 Formulas

WORKING WITH FORMULAS

THE MAIN IDEA

1. A ***formula*** is a mathematical sentence that shows how variables are related to each other. The variable that stands alone is the ***subject***.

   $\underbrace{P}_{\text{subject}} = 2\ell + 2w$

2. You can evaluate the subject of a formula for particular values of the other variables.

   When $\ell = 5$ and $w = 6$,
   $P = 2(5) + 2(6)$
   $P = 10 + 12 = 22$

3. You can write a formula if you are given either a word sentence or a table of values that shows how the variables are related.

EXAMPLE 1 Write a formula that says: "The distance is equal to the product of the rate and the time."

Choose variables to stand for the quantities. — Let D represent Distance, R represent Rate, T represent Time.

"=" for "is equal to" — $D =$

Write "the product of the rate and the time." — $R \times T$ or $R(T)$, or RT

Answer: $D = RT$

EXAMPLE 2 Mrs. Burns buys some meat at $3 a pound for a total cost of x dollars. Write a formula to find the number of pounds (n) of meat she bought.

Answer: $3n = x$ or $n = \dfrac{x}{3}$

EXAMPLE 3 Write a formula that states the relationship between x and y.

x	3	4	5	6
y	21	28	35	42

THINKING ABOUT THE PROBLEM

The table shows the values of y for different values of x. Look for a pattern. Is the same number always added to x? Is y always multiplied by the same number?

Note that in this table, x is always multiplied by 7 to produce the value of y. Thus, y is always equal to 7 times x, as in $28 = 7(4)$.

Also, y is always divided by 7 to produce the value of x. Thus, x is always equal to y divided by 7, as in $5 = \frac{35}{7}$.

Answer: $y = 7x$ or $x = \dfrac{y}{7}$

CLASS EXERCISES

1. Name the subject of each formula: **a.** $A = \pi r^2$ **b.** $I = PRT$ **c.** $\frac{1}{2}bh = A$

2. Find the value of the subject of each formula, for the given values of the variables.

   a. $D = rt$ $r = 45, t = 3$ **b.** $I = prt$ $p = 2{,}000, r = .05, t = 2$

   c. $2\ell + 2w = p$ $\ell = 7, w = 11$ **d.** $A = s^2$ $s = 12$

   e. $A = \frac{1}{2}bh$ $b = 8, h = 5$ **f.** $P = a + b + c$ $a = 2.3, b = 5, c = 4.3$

   g. $S = 6e^2$ $e = 2$ **h.** $C = P + .05P$ $P = 50$

3. The formula $S = 200 + 10t$ gives Mr. Smith's salary S, in dollars, for a week in which he works t hours overtime. Find S when t equals: **a.** 0 **b.** 5 **c.** 10 **d.** 18

4. Write a formula for each sentence.

   a. The time of a train trip is equal to 400 miles divided by the train's rate.
   b. The number of quarts is equal to 4 times the number of gallons.
   c. A father's age is 22 years more than his son's age.
   d. The total price of a bag of apples is equal to 0.49 times the number of pounds of apples.
   e. The cost of a cab ride is equal to the number of miles driven multiplied by the rate per mile.

5. Write a formula that states the relationship between the variables in each table of values.

a.

x	7	8	9	10	11
y	11	12	13	14	15

b.

m	20	22	24	26	28
n	27	29	31	33	35

c.

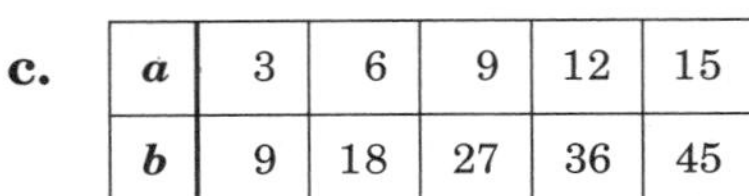

a	3	6	9	12	15
b	9	18	27	36	45

d.

q	30	60	90	120	150
r	5	10	15	20	25

e.

k	1	2	3	4	5
l	1	4	9	16	25

f.

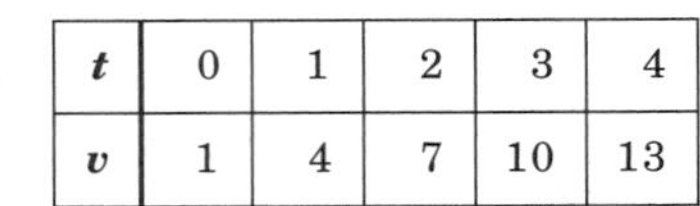

t	0	1	2	3	4
v	1	4	7	10	13

6. Write a formula for the relationship between:

   a. the total cost of a box of pastries and the number of pastries if each pastry costs $0.80
   b. a train's speed and its traveling time on a 300-mile trip
   c. liters and milliliters if there are 1,000 milliliters in every liter
   d. weeks and the number of days in them

TRANSFORMING FORMULAS

THE MAIN IDEA

1. A formula can be rewritten so that the subject is any one of the variables of the formula. Changing the subject of a formula is called ***transforming*** the formula.
2. To transform a formula, solve the equation for the new subject as though all other variables were constants.

EXAMPLE 4 Transform the formula $d = rt$ so that the subject is t.

To *isolate* t (get it to stand alone), "undo" the multiplication by r: multiply both sides of the formula by the reciprocal of r.

$$d = rt$$
$$\frac{1}{r} \cdot d = \frac{1}{r} \cdot (rt)$$
$$\frac{d}{r} = t$$
$$\text{or } t = \frac{d}{r} \quad \textit{Ans.}$$

EXAMPLE 5 Given the formula $x = y - 5$, solve for the variable y.

Rewrite the subtraction.

$$x = y - 5$$
$$x = y + (-5)$$

Then, add 5 to each side in order to isolate y.

$$x + 5 = y + (-5) + 5$$
$$x + 5 = y$$
$$\text{or } y = x + 5 \quad \textit{Ans.}$$

CLASS EXERCISES

1. Transform $A = \ell w$ so that w is the subject.
2. Make d the subject of $R = \frac{d}{t}$.
3. Solve for the variable s: $P = 4s$
4. Solve for the variable t: $S = 16t + 34$

HOMEWORK EXERCISES

1. Name the subject of each formula: **a.** $A = \ell w$ **b.** $P - D = C$ **c.** $C = 2\pi r$ **d.** $R = \frac{d}{t}$
2. The formula $C = 70 + 30(m - 3)$ gives the cost, C, of a long-distance telephone call of m minutes, when $m \geq 3$. Find C when m equals: **a.** 3 **b.** 5 **c.** 10

3. Find the value of the subject of each formula for the given values of the variables.

a. $V = \ell wh$ $\quad \ell = 10, w = 8, h = 6$

b. $A = \frac{1}{2}bh$ $\quad b = 8, h = 6$

c. $P = 2(\ell + w)$ $\quad \ell = 15, w = 5$

d. $V = e^3$ $\quad e = 5$

e. $C = 2\pi r$ $\quad \pi = \frac{22}{7}, r = 14$

f. $C = \frac{5}{9}(F - 32)$ $\quad F = 32$

g. $A = \frac{1}{2}h(b + c)$ $\quad h = 4, b = 6, c = 10$

h. $F = \frac{9}{5}C + 32$ $\quad C = 100$

4. Write a formula for each sentence.
   a. The number of centimeters is equal to 100 times the number of meters.
   b. The number of days is equal to 7 times the number of weeks.
   c. The salary is equal to 10.25 times the number of hours worked.
   d. The number of feet is equal to the number of inches divided by 12.

5. Write a formula that states the relationship between the variables in each table of values.

a.

x	4	5	6	7	8
y	9	10	11	12	13

b.

m	22	20	16	12	8
n	18	16	12	8	4

c.

a	1	2	3	4	5
b	60	30	20	15	12

d.

r	18	27	36	45	54
s	6	9	12	15	18

e.

c	7	9	11	13	15
d	7	9	11	13	15

f.

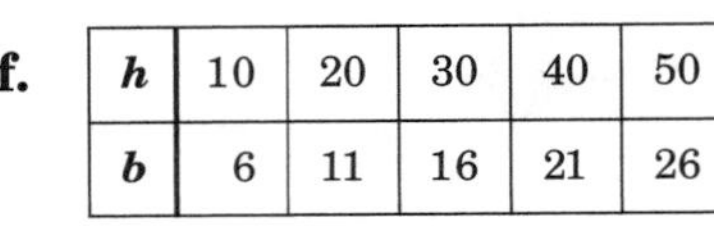

h	10	20	30	40	50
b	6	11	16	21	26

6. Write a formula for the relationship between:
   a. the total cost of a set of texts and the number of texts if each costs $27.50
   b. an airplane's speed and its traveling time on a 2,000-mile trip
   c. grams and kilograms if there are 1,000 grams in every kilogram
   d. years and the number of months in them

7. Transform $S = p - d$, so that p is the subject.

8. Make t the subject of $I = prt$.

9. Solve the variable b: $P = a + b + c$

10. Solve for the variable h: $W = 6h + e$

11. Explain why a scientist might want to transform the formula $F = \frac{9}{5}C + 32$, which converts a temperature in Celsius to a temperature in Fahrenheit, so that its subject is C.

SPIRAL REVIEW EXERCISES

1. Solve each equation and check.

a. $8x = 96$ **b.** $-6x = 66$

c. $5x = -35$ **d.** $21a = 7$

2. Solve each equation and check.

a. $y + 53 = 91$ **b.** $a - 51 = 50$

c. $x - 17 = 34$ **d.** $18 + x = 22$

3. Solve each equation and check.

a. $\frac{a}{9} = 7$ **b.** $\frac{x}{5} = -6$

c. $\frac{-m}{8} = 64$ **d.** $\frac{y}{4} = -144$

4. Tell whether the given value of the variable is a solution of the given equation.

a. $x + 47 = 53;\ x = -6$

b. $3m = -12;\ m = -4$

c. $x - 15 = 15;\ x = 0$

d. $3x - 7 = 17;\ x = 8$

5. The sum of –9 and –3 is

(a) 27 (b) 3 (c) –6 (d) –12

6. The expression a^3 means

(a) $a \cdot a \cdot a$ (b) $3 \cdot a$

(c) $a + 3$ (d) $\frac{3}{a}$

7. If 4 pounds of potatoes cost $1.38, what is the cost of 9 pounds?

8. If 9 out of 25 students are absent, what percent of the students are present?

9. Which key sequence would display the square of 49?

(a) 49 [×] 2 [=] (b) 49 [√]

(c) 49 [×] 49 [×] 49 [=] (d) 49 [×] [=]

10. The table below give the average number of children for families in the U.S.A. for each of three years.

Year	Average Number of Children in a U. S. Family
1970	2.3
1980	1.9
1990	1.8

Which of the following statements is true?

(a) The average number of children per family increased by about 50% from 1970 to 1990.

(b) The average number of children per family increased by about 20% from 1970 to 1990.

(c) The average number of children per family decreased by about 20% from 1970 to 1990.

(d) The average number of children per family decreased by about 50% from 1970 to 1990.

UNIT 16-5 Graphing Open Sentences on a Number Line

THE MAIN IDEA

1. Recall that a number is represented by a point on a number line.
   a. All numbers greater than the number are to its right.
   b. All numbers less than the number are to its left.

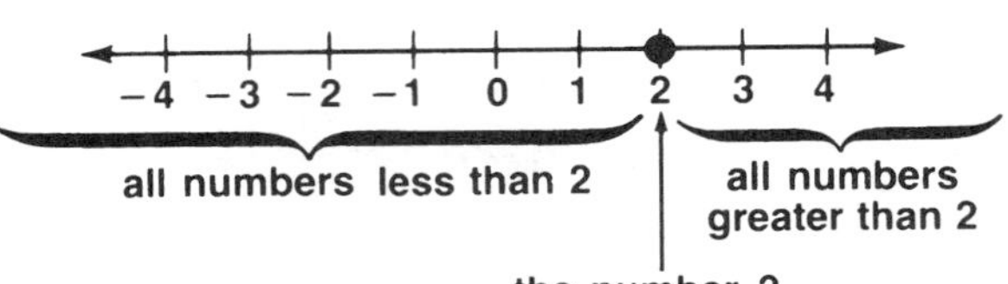

2. Here are some typical graphs.

	The graph shows	*Open Sentence*	*Graph*
a.	a given number.	$x = 2$	
b.	all numbers greater than a given number.	$x > 2$	The *open circle* at 2 shows that the value 2 *is not included* in the graph.
c.	all numbers greater than or equal to a given number.	$x \geq 2$	The *closed circle* at 2 shows that the value 2 *is included* in the graph.
d.	all numbers less than a given number.	$x < 2$	
e.	all numbers less than or equal to a given number.	$x \leq 2$	
f.	all numbers between two given numbers. The sentence tells whether or not to include the end values.	$-1 \leq x \leq 2$	
		$-1 < x < 2$	
		$-1 \leq x < 2$	
		$-1 < x \leq 2$	

CLASS EXERCISES

1. Graph each of the given open sentences on a number line.

a. $x < -5$ **b.** $x \geq -2$ **c.** $x > 7$ **d.** $x \leq -3$ **e.** $-4 \leq x < 1$

f. $-10 < x \leq -5$ **g.** $x > -1$ **h.** $x \leq -5$ **i.** $-1 \leq x < 1$ **j.** $x < -3$

k. $x > -5$ **l.** $-2 \leq x \leq 0$ **m.** $x \geq -2$ **n.** $-1 < x < 2$ **o.** $-7 < x \leq -5$

2. What is the inequality that is represented by the graph? 
(a) $x > -3$ (b) $x < -3$ (c) $x \geq -3$ (d) $x \leq -3$

3. Which open sentence is represented by the graph? 
(a) $x \leq 4$ (b) $x > -3$ (c) $-3 \leq x < 4$ (d) $-3 < x \leq 4$

4. The graph that represents the open sentence $-6 < x \leq -3$ is

(a) 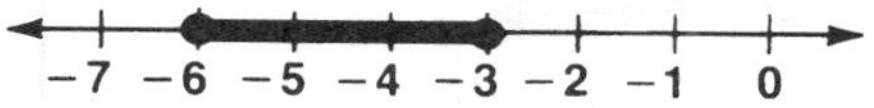 (b) 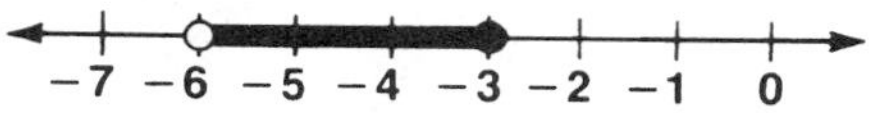 (c) 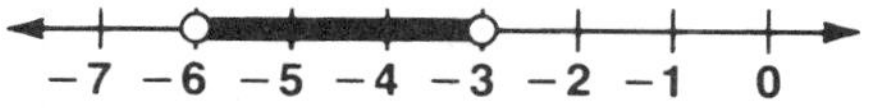 (d)

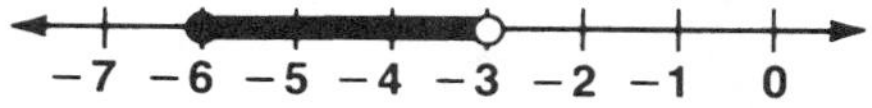

HOMEWORK EXERCISES

1. Graph each of the given open sentences on a number line.

a. $x > -3$ **b.** $x < 2$ **c.** $x \geq -4$ **d.** $x < -5$ **e.** $x > -7$

f. $x \leq 0$ **g.** $x \geq -6$ **h.** $x \geq 3$ **i.** $x \leq -7$ **j.** $x > -6$

k. $0 < x < 5$ **l.** $-3 \leq x < 1$ **m.** $-2 \leq x \leq 3$ **n.** $-5 < x \leq -1$ **o.** $-6 \leq x \leq 0$

2. What is the inequality that is represented by the graph? 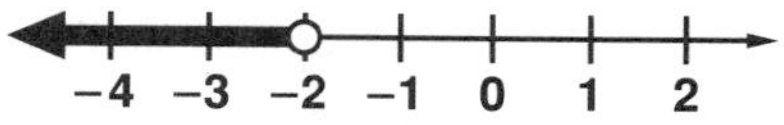
(a) $x > -2$ (b) $x < -2$ (c) $x < 2$ (d) $x \leq -2$

3. What is the inequality that is represented by the graph?

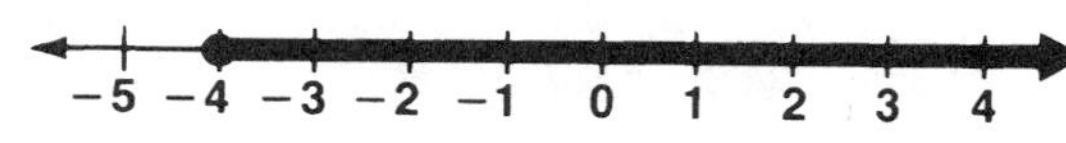

(a) $x \geq -4$ (b) $x \leq -4$ (c) $x > -4$ (d) $x < 4$

4.  The graph represents (a) $x > -7$ (b) $x < 1$
(c) $-7 < x < 1$ (d) $-7 \leq x < 1$

5.  The graph represents (a) $x \geq -5$ (b) $x \leq 0$
(c) $-5 < x < 0$ (d) $-5 \leq x \leq 0$

6. The graph that represents the open sentence $x < 3$ is

(a) 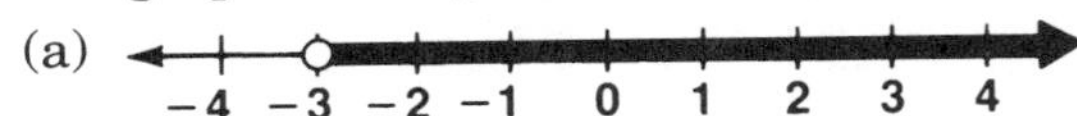

(b)

(c) 

(d) 

7. The graph that represents the open sentence $-2 < x \leq 3$ is

(a)

(b) 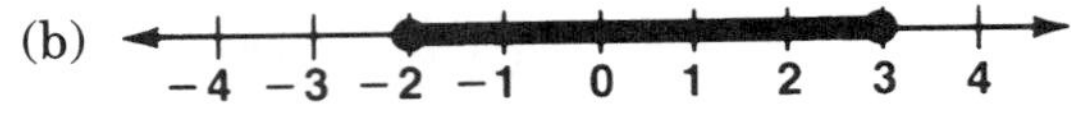

(c) 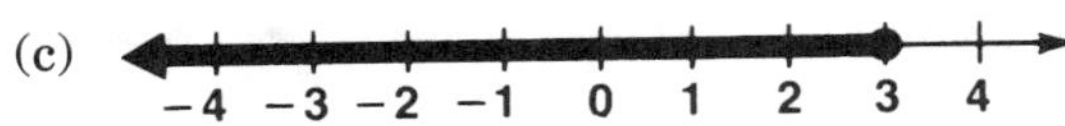

(d) 

8. On a number line, show all the values of x that make the statement true.

a. $x \leq -2$ or $x > 3$ **b.** $x \leq -2$ and $x > 3$

SPIRAL REVIEW EXERCISES

1. Evaluate $5x + 2y$ when $x = 4$ and $y = 6$.

2. If $m = 3$ and $p = -2$, find the value of $m^2 + p^3$.

3. Replace [?] with < or > to make a true comparison.

a. 0 [?] -6 **b.** -26 [?] -10

c. -112 [?] 3 **d.** -40 [?] -100

e. -8.4 [?] -9 **f.** $\frac{-1}{4}$ [?] $\frac{-1}{3}$

4. Name the opposite of each signed number.

a. $\frac{-3}{4}$ **b.** $+9$ **c.** -11.5

d. 0 **e.** $-2\frac{1}{2}$ **f.** 3.4

5. Mr. Graham invests $500 at 5.5% simple interest. How much interest does he earn after one year?

6. Solve the equation $5x - 12 = 93$ and check.

7. Which number is a solution of the equation $2x + 5 = -1$?

(a) -3 (b) 3 (c) $1\frac{1}{2}$ (d) $-1\frac{1}{2}$

8. Add: $\frac{2}{3} + \frac{3}{5}$

9. Multiply: $\frac{2}{3} \times \frac{3}{5}$

10. What percent of 40 is 8?

11. Solve the proportion: $\frac{127.5}{35.7} = \frac{n}{16.8}$

12. If in a class there are 20 boys and 15 girls, the ratio of the number of boys to the total number of students in the class is

(a) $\frac{3}{4}$ (b) $\frac{4}{3}$ (c) $\frac{3}{7}$ (d) $\frac{4}{7}$

13. From a kindergarten class of 20 children, the teacher wants to choose 2 children. The first chosen will be paper monitor and the second will be block monitor. In how many different ways can this be done?

UNIT 16-6 Solving Inequalities

THE MAIN IDEA

1. An ***inequality*** is a sentence that states that *two quantities are unequal*. A *solution* is any value for the variable that makes the sentence true.
2. Inequalities often have many solutions. To solve an inequality means to describe the ***solution set***, which is the set of all the solutions of the inequality.
3. The following properties are used to solve inequalities:
   a. ***Addition Property of Inequality***

   If two numbers are unequal and the same number is added to each of them, the results are unequal in the *same order* as the original numbers.

   $$5 > -2$$
   $$5 + 3 > -2 + 3$$
   $$8 > 1$$

   The symbol ">" remains unchanged.

   b. ***Multiplication Property of Inequality***

   If two numbers are unequal and each is multiplied by the same *positive* number, the results are unequal in the *same order* as the original numbers.

   $$-3 < 1$$
   $$-3(+6) < 1(+6)$$
   $$-18 < 6$$

   The symbol "<" remains unchanged.

   If two numbers are unequal and each is multiplied by the same *negative* number, the results are unequal in the *opposite order* to the original numbers.

   $$-3 < 5$$
   $$-3(-1) > 5(-1)$$
   $$3 > -5$$

   The symbol "<" must be reversed to ">" for the resulting inequality to be true.

EXAMPLE 1 Is -4 a solution of $-2x \geq 14$?

Substitute -4 for x in the inequality.

$$-2x \geq 14$$
$$-2 \cdot (-4) \stackrel{?}{\geq} 14$$

Simplify and compare the two sides of the sentence.

$$+8 \geq 14$$
false

Since the sentence is false $(8 < 14)$, -4 is not a solution of $-2x \geq 14$.

EXAMPLE 2 Is 11 in the solution set of the inequality $a - 24 > -20$?

Substitute 11 for a in the inequality.

$$a - 24 > -20$$
$$11 - 24 \stackrel{?}{>} -20$$

Simplify and compare the two sides of the sentence.

$$-13 > -20$$
true

Since the sentence is true, 11 is in the solution set of $a - 24 > -20$.

EXAMPLE 3 Solve $x - 17 \leq 21$ and graph the solution set.

Begin as you would when solving an equation; add +17 to each side of the number sentence. Adding a positive number to each side of an inequality does not reverse its order. Perform the operations shown.

$$x - 17 \leq 21$$
$$x - 17 + (+17) \leq 21 + (+17)$$
$$x \leq 38$$

Check: Substitute a value less than 38 in the original sentence.

$$x - 17 \leq 21$$
$$37 - 17 \overset{?}{\leq} 21$$
$$20 \leq 21$$
true

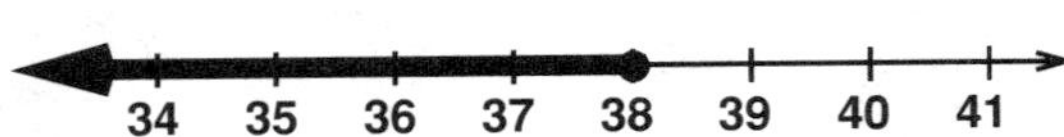

EXAMPLE 4 Solve $5x > -65$ and graph the solution set.

Multiply each side of the number sentence by $\frac{1}{5}$. Multiplying each side of an inequality by a positive number does not reverse its order.

$$5x > -65$$
$$\frac{1}{5} \cdot (5x) > \frac{1}{5} \cdot (-65)$$
$$x > -13$$

Check: Substitute a value greater than −13 in the original sentence.

$$5x > -65$$
$$5(-12) \overset{?}{>} -65$$
$$-60 > -65$$
true

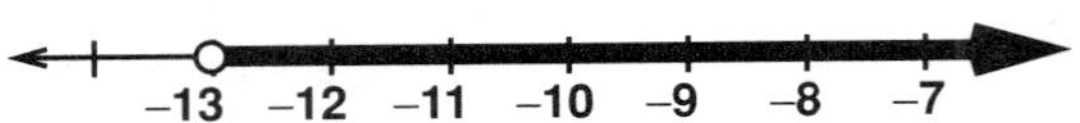

EXAMPLE 5 Solve $-2y \geq -20$ and graph the solution set.

Multiply each side by $\frac{-1}{2}$. Reverse the order of the inequality.

$$-2y \geq -20$$
$$\frac{-1}{2} \cdot (-2y) \leq \frac{-1}{2} \cdot (-20)$$
$$y \leq 10$$

Check: Substitute a value less than 10 in the original sentence.

$$-2y \geq -20$$
$$-2(9) \overset{?}{\geq} -20$$
$$-18 \geq -20$$
true

EXAMPLE 6 Solve $3x - 10 < 14$.

As with equations, add +10 to each side. Then, multiply each side by $\frac{1}{3}$.

$$3x - 10 < 14$$
$$3x - 10 + (+10) < 14 + (+10)$$
$$3x < 24$$
$$\frac{1}{3} \cdot 3x < \frac{1}{3} \cdot 24$$
$$x < 8$$

Check: Substitute a value less than 8 in the original sentence.

$$3x - 10 < 14$$
$$3(7) - 10 \overset{?}{<} 14$$
$$21 - 10 \overset{?}{<} 14$$
$$11 < 14$$
true

Answer: $x < 8$

EXAMPLE 7 The greatest load that Mr. Grant's truck can carry is 12,000 pounds. He wants to carry 4 crates of equal weight. Write and solve an inequality to express the number of pounds each crate may weigh.

THINKING ABOUT THE PROBLEM

As with equations, a problem-solving strategy is to use an inequality to solve a problem.

Let W = weight of each crate.	Choose a variable for the unknown.
weight of 4 crates $\leq 12{,}000$	Load cannot be > 12,000.
$4W \leq 12{,}000$	Write the words in symbols.
$\frac{1}{4} \cdot (4W) \leq \frac{1}{4} \cdot (12{,}000)$	Multiply by the reciprocal of 4.
$W \leq 3{,}000$	The order remains the same.

Check in the words of the problem.

Answer: The weight of each crate may be 3,000 pounds or less.

CLASS EXERCISES

1. For each inequality, tell whether the given value of the variable is a solution.

a. $x + 5 > 15;\ x = 1$ **b.** $a - 7 \leq 12;\ a = 12$ **c.** $2x \geq -40;\ x = -10$
d. $3y < 45;\ y = 16$ **e.** $4x + 8 > 22;\ x = 3$ **f.** $-2x + 2 \leq -12;\ x = -6$

2. Solve each inequality and graph the solution set on a number line.

a. $a + 9 < 23$ **b.** $b - 12 > -20$ **c.** $x - 3 \geq 42$ **d.** $y + 14 \leq -88$
e. $5x \geq 35$ **f.** $10y < -130$ **g.** $4x + 14 < 34$ **h.** $6a - 20 \geq -50$
i. $-5x > 16$ **j.** $10 - x \leq 15$ **k.** $3x + 2 > 5$ **l.** $3x + 2 < 5$

3. Each of the 6 third-grade classes may send some children to a special assembly program, but the total number of children sent from the third grade must not be more than 24. Write and solve an inequality to express the number of children each class may send if each third-grade class sends the same number of children.

4. Sam wants to buy some ties that are priced at $10 each. He has decided to buy a pair of shoes for $40, but the total cost of the ties and the shoes must be less than $100, or he won't have enough money for carfare to get home. Write and solve an inequality to express the number of ties he may buy.

5. You have seen the addition and multiplication properties of inequality. Explain why there is no need for a subtraction or division property.

6. There is a property for multiplying each side of an inequality by a positive number and a property for multiplying by a negative number. Why is there no property for multiplying each side of an inequality by 0? Use examples to explain your answer.

HOMEWORK EXERCISES

1. For each inequality, tell whether the given value of the variable x is a solution.

   a. $x - 3 < 12;\ 15$ **b.** $x + 17 \geq 20;\ 5$ **c.** $8x \leq -40;\ -6$ **d.** $5x > -100;\ -25$

   e. $x - 24 \geq 44;\ 72$ **f.** $x + \frac{1}{3} > 5;\ 4$ **g.** $3x + 2 < 20;\ 4$ **h.** $2x - 2 > -10;\ 4$

   i. $-3x \geq 60;\ -24$ **j.** $-4x - 5 \leq -23;\ -7$ **k.** $5 - 2x > -3;\ 4$ **l.** $x - 4.6 \leq -5.6;\ -2.4$

2. Solve each inequality and graph each solution set on a number line.

   a. $a + 7 > 35$ **b.** $m - 6 < 42$ **c.** $x - 15 \geq -20$ **d.** $y + 7.8 \leq 9.4$

   e. $3x < 63$ **f.** $9y \leq -81$ **g.** $-12a < 48$ **h.** $-20y \geq -100$

   i. $5x + 3 < 48$ **j.** $-2m \geq 42$ **k.** $-5b \leq -70$ **l.** $-3y + 12 > -72$

   m. $0.6 - 0.4y < -2.2$ **n.** $1.1a - 4 > 4.8$ **o.** $4x + 1 \geq 1$

3. Mrs. Roberts needs at least 34 buttons to finish some dresses she is making. Buttons are sold in packages of 8. Write and solve an inequality to express the number of packages she must buy in order to finish the dresses.

4. In a beaker with a 50-milliliter capacity, a chemist has 47.3 milliliters of glycerine. Write and solve an inequality to express the number of milliliters of glycerine that can still be added to the beaker without going over its capacity.

5. At the end of the third quarter of a basketball game, the score was Gophers 42, Chipmunks 64. In the fourth quarter, the Chipmunks did not score any points and the Gophers won the game. If each basket is worth 2 points, write and solve an inequality to express the number of baskets the Gophers could have scored in the fourth quarter.

6. If N is a number less than 12, explain with inequalities and graphs why the opposite of the number must be greater than –12.

SPIRAL REVIEW EXERCISES

1. If $P = 2\ell + 2w$, find the value of P when $\ell = 15.5$ and $w = 12$.

2. Which open sentence is represented by the graph?

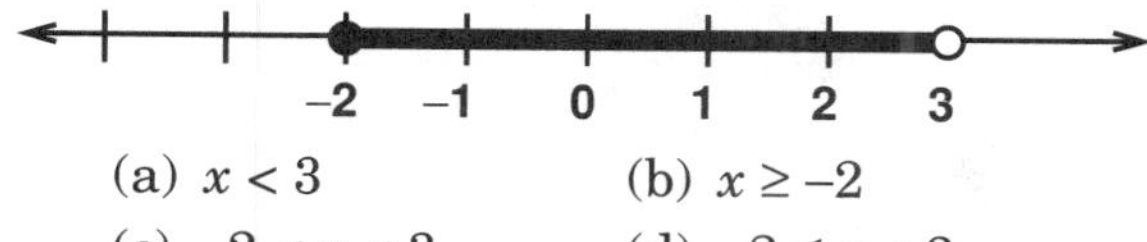

(a) $x < 3$ (b) $x \geq -2$
(c) $-2 < x < 3$ (d) $-2 \leq x < 3$

3. What is the value of $3x^2 - 5x$ when $x = -2$?

4. The cost of 5 posters is $12.80. Find the cost of 12 posters.

5. Each of the letters of the word ARKANSAS is written on a slip of paper and placed in a hat. If one slip of paper is chosen at random, what is the probability that an A will be selected?

6. A pair of jeans is on sale for 25% off the original price of $30. If the rate of sales tax is 5%, what is the total cost of the jeans?

7. Joel begins an 18-mile workout to train for a marathon. He walks one-half of the distance before resting; and then jogs one-third of the remaining distance before stopping again. How many miles will complete his workout?

8. Approximate $5 \times \sqrt{5}$, to the nearest hundredth.

9. Jay is decorating his living room. He is considering gray or tan for the carpeting, white or gray for the walls, and white, mauve, or gray for the drapes. If he chooses one color for carpeting, one for paint, and one for drapes, the number of decorating possibilities is:

(a) 4 (b) 8 (c) 12 (d) 16

10. A poll of 500 adults was taken to find out how they acquired their Halloween costumes. This graph shows the results.

Costume Customs

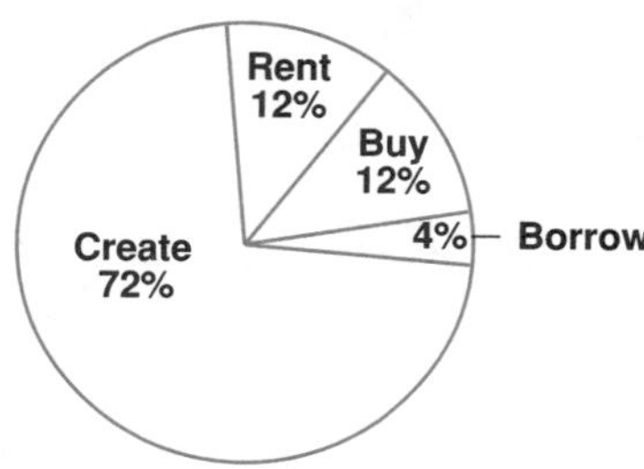

The difference in the number of adults who created their own costumes and those who borrowed theirs is

(a) 340 (b) 300 (c) 280 (d) 40

11. $\sqrt{78}$ is between which pair of consecutive integers?

(a) 7 and 8 (b) 8 and 9
(c) 9 and 10 (d) 39 and 40

12. The value of $6 \times 10^4 + 3 \times 10^2 + 4 \times 10^1$ is

(a) 634 (b) 6,340
(c) 63,340 (d) 60,340

13. The number 0.00091 written in scientific notation is

(a) 9.1×10^{-4} (b) 9.1×10^{-3}
(c) 9.1×10^3 (d) 9.1×10^4

14. In 1985, about 2.6 million personal computers were sold. By 1990, the sales had grown to 8.5 million. The percent of increase in sales is about

(a) 225% (b) 50% (c) 100% (d) 330%

THEME 8

Introduction to Geometry

MODULE 17

MODULE 18

MODULE 19

MODULE 17

UNIT 17-1 Basic Geometric Figures

THE MAIN IDEA

1. The 3 most basic geometric figures are:

point

Point Q holds a position in space.

line

Line AB, or $\overleftrightarrow{AB}$, extends without end in two opposite directions.

plane

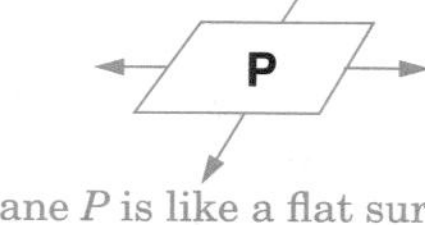

Plane P is like a flat surface that extends without end in all directions.

2. A *part* of a line is called a:

ray

Ray AB, or $\overrightarrow{AB}$, consists of one endpoint and all the points on the line on one side of that endpoint.

line segment

Line segment RS, or $\overline{RS}$, contains two endpoints and all the points on the line between these endpoints.

3. Of the figures listed above, only a line segment has a measurable length.

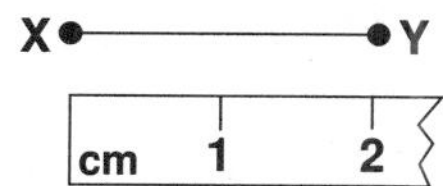

The length of $\overline{XY}$, written XY (without the bar), is 2 cm.

4. ***Congruent segments*** have equal length.

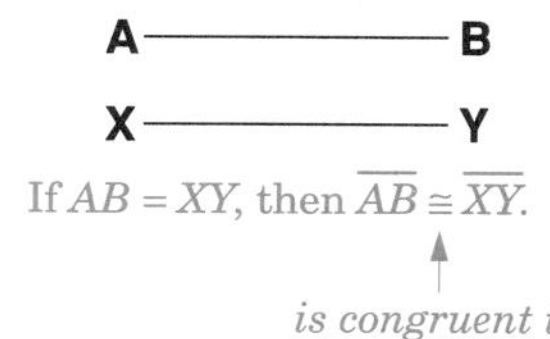

If $AB = XY$, then $\overline{AB} \cong \overline{XY}$.

is congruent to

5. The ***midpoint*** of a line segment is the point that divides it into two segments with equal lengths.

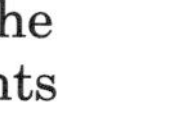

If M is the midpoint of $\overline{RS}$, then $RM = MS$ and $\overline{RM} \cong \overline{MS}$.

EXAMPLE 1 Name each colored figure in words, and write two different symbols for each figure.

	Figure		*Names/ Symbols*
a.	P Q R S T	To name a line, name any two points on the line. Write the word "line" or draw a double arrow.	Line *PS*; $\overleftrightarrow{QR}$; $\overleftrightarrow{SP}$
b.	P Q R S T	To name a ray, name the endpoint first, then any other point on the ray. Write "ray" or draw an arrow.	Ray *RP*; $\overrightarrow{RP}$; $\overrightarrow{RQ}$
c.	P Q R S T	This ray is the opposite of ray *RP* above.	Ray *RS*; $\overrightarrow{RS}$; $\overrightarrow{RT}$
d.	P Q R S T	To name a line segment, name the two endpoints in either order. Write "segment" or draw a line segment.	Segment *PR*; $\overline{PR}$; $\overline{RP}$

EXAMPLE 2 Referring to the diagram below, use symbols to name the following:

a. two lines that pass through *E*

Answer: $\overleftrightarrow{EA}$, $\overleftrightarrow{ED}$

b. two rays with endpoint *F*

Answer: $\overrightarrow{FD}$, $\overrightarrow{FE}$

c. a point on $\overleftrightarrow{CD}$ that is not on $\overrightarrow{CD}$

Answer: point *B*

d. three different segments on $\overleftrightarrow{AE}$

Answer: $\overline{EC}$, $\overline{CA}$, $\overline{EA}$

EXAMPLE 3 If $IM = 14$ in., find *IJ*.

Let $IJ = x$ in. Write an equation and solve.

$$\begin{aligned} IJ + JK + KL + LM &= IM \\ x + 2 + 5 + 3 &= 14 \\ x + 10 &= 14 \\ x + 10 + (-10) &= 14 + (-10) \\ x &= 4 \end{aligned}$$

Answer: $IJ = 4"$

EXAMPLE 4 If *M* is the midpoint of $\overline{CD}$, find *CD*.

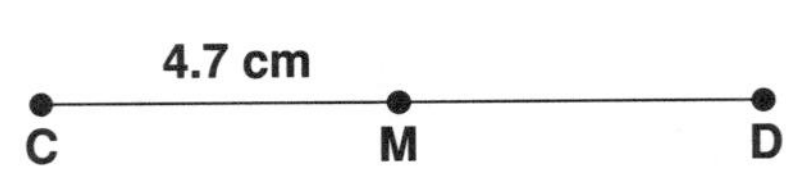

Since the midpoint creates two line segments of equal length, *MD* also equals 4.7 cm.

$$CD = CM + MD = 4.7 \text{ cm} + 4.7 \text{ cm} = CD = 9.4 \text{ cm} \quad \textit{Ans.}$$

CLASS EXERCISES

1. Name each colored figure in words, and write two different symbols for each figure.

a. A B C **b.** D E F **c.** G H K

2. Referring to the diagram at the right, use symbols to name each of the following:

a. five points

b. two line segments that have B as one endpoint

c. three rays that have D as the endpoint

d. two lines that pass through point A

e. a point on $\overleftrightarrow{DB}$ that is not on $\overrightarrow{DB}$

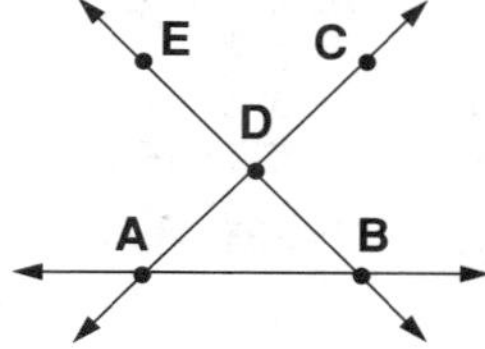

3. Use the diagram below to find:

a. BC **b.** AB **c.** LK

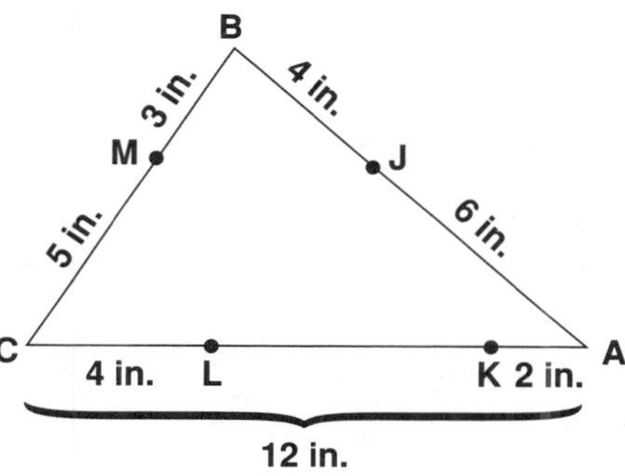

4. M is the midpoint of $\overline{QR}$. If $QR = 18$ ft., find MR.

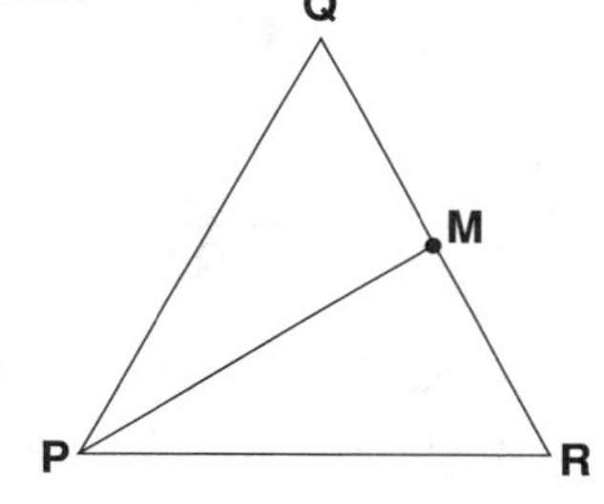

5. In the diagram, $AB = 20$ cm and $CB = 14$ cm. If M is the midpoint of $\overline{AC}$, find AM.

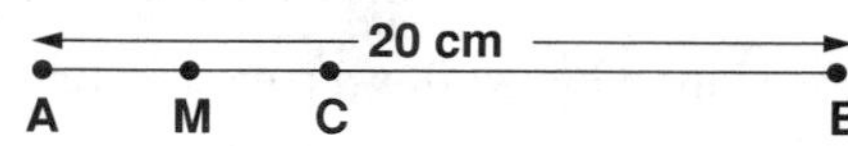

HOMEWORK EXERCISES

1. Name each colored figure in words, and write two different symbols for each figure.

a. P Q R **b.** S T W **c.** V J M

2. Referring to the diagram at the right, name each of the following:

a. five points **b.** the line in three different ways

c. two line segments that have D as one endpoint

d. two rays that have B as the endpoint

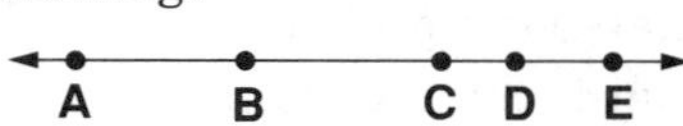

3. Use symbols to name each of the following in the diagram at the right:

a. six points **b.** two rays that have Z as the endpoint

c. two lines that pass through T

d. three line segments that have S as one endpoint

e. the point where $\overleftrightarrow{YS}$ and $\overleftrightarrow{ZV}$ meet

f. a point that is on $\overleftrightarrow{ST}$, but not on $\overline{ST}$

4. In the diagram at the right, $AF = 18$ m.

a. Find AC. **b.** Find BE. **c.** Find EF.

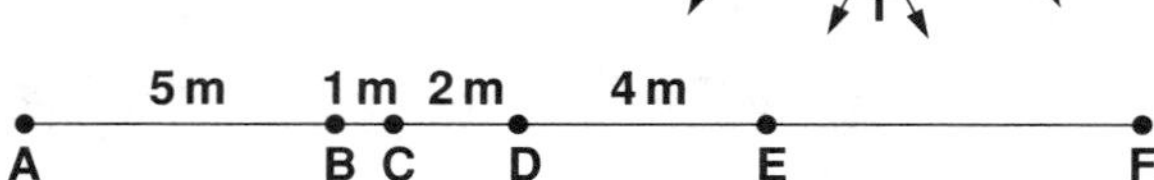

5. In the diagram below, $GM = MP = PQ = QK$ and $GK = 20$ in.

a. Find PQ. **b.** Find MK.

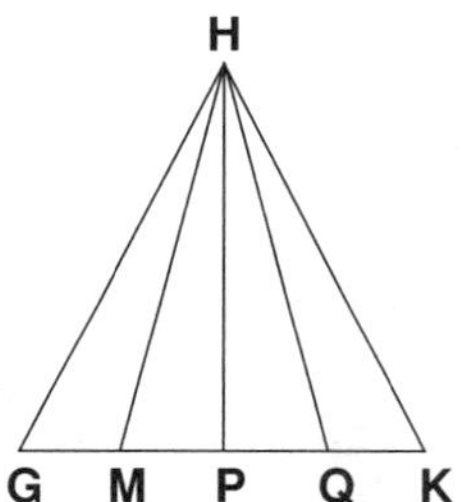

6. In the diagram below, P is the midpoint of $\overline{AC}$. If $AC = 8.7$ m, find PC.

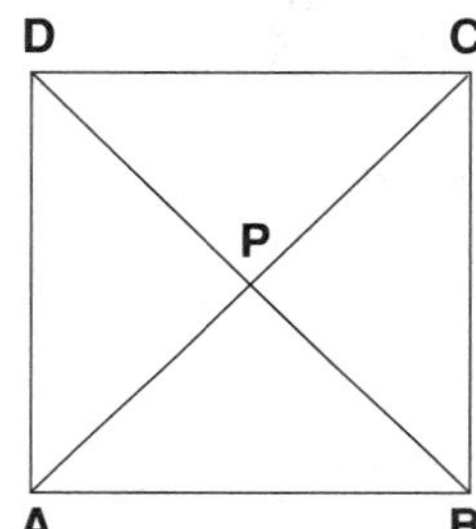

7. In the diagram, F is the midpoint of $\overline{EG}$, $GH = 15$ cm, and $EF = 8$ cm. Find EH.

8. Given five points, A, B, C, D, and E. Using two points for each figure:

a. How many different line segments can be drawn? Name them.

b. How many different rays can be drawn? Name them.

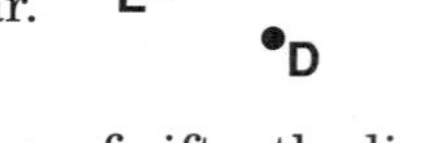

9. Five friends agree to give each other birthday gifts in the coming year.

a. How many gifts will be given?

b. Which part of Homework Exercise 8 better illustrates this exchange of gifts, the line segments in part **a** or the rays in part **b**? Explain.

10. Five people at a meeting each shake hands with every other person.

a. How many handshakes will there be?

b. Which part of Homework Exercise 8 better illustrates the handshakes at the meeting, the line segments in part **a** or the rays in part **b**? Explain.

11. A committee is made up of Ana, Bo, Carl, Dale, and Elena. Two people will be selected at random to represent the group.

a. What is the probability that the two chosen will be Ana and Bo?

b. What is the probability that either Ana or Bo will be chosen?

c. Explain how you can use the results of Homework Exercise 8 to find these probabilities.

SPIRAL REVIEW EXERCISES

1. Solve: $-4x > 3$

2. The graph of the solution set of $3x + 9 \leq -12$ is

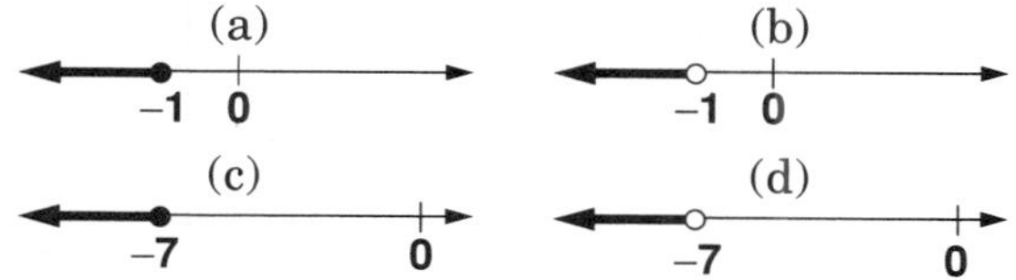

3. A length of 3 meters is equivalent to
(a) 30 cm (b) 30 mm
(c) 300 cm (d) 3,000 cm

4. If n is used to represent a number, then "5 less than n" is
(a) $5n$ (b) $n - 5$ (c) $5 - n$ (d) $\frac{n}{5}$

5. The value of $2x + 5$ when $x = 3$ is
(a) 28 (b) 16 (c) 8 (d) 11

6. If $I = prt$, find the value of I when $p = 1{,}000$, $r = 0.05$, and $t = 2$.

7. Evaluate: $\frac{1}{10} + \frac{2}{5} \times \frac{1}{2}$

8. The product of 0.8 and 0.7 is
(a) 1.5 (b) 0.1 (c) 0.56 (d) 5.6

9. The Student Council spent $200 for flowers that they sold to raise money. If they sold 150 roses at $1.25 each and 200 carnations at $0.75 each, how much profit did they make?

10. If one student is selected at random from a school club in which there are 12 sophomores, 14 juniors, and 26 seniors, the probability that the student will be a sophomore is
(a) $\frac{1}{12}$ (b) $\frac{3}{13}$ (c) $\frac{7}{26}$ (d) $\frac{1}{2}$

11. An inspector at a sweater factory found five imperfect sweaters in a sample of 95 sweaters. How many imperfect sweaters should the inspector expect to find in a lot of 570 sweaters?

12. Splat Dishwashing Liquid is sold in a 22-ounce bottle for $1.69 and in a 32-ounce bottle for $2.62. Which is the better buy?

13. The graph shows the estimated cost of a night out for a family of four.

Next month, Mr. and Mrs. Santana plan to take their two children with them to see two movies and one live theater production.

Using the data in the graph, the best estimate of the cost for these three evenings is:
(a) $96.50 (b) 142.00
(c) $164.00 (d) $200.50

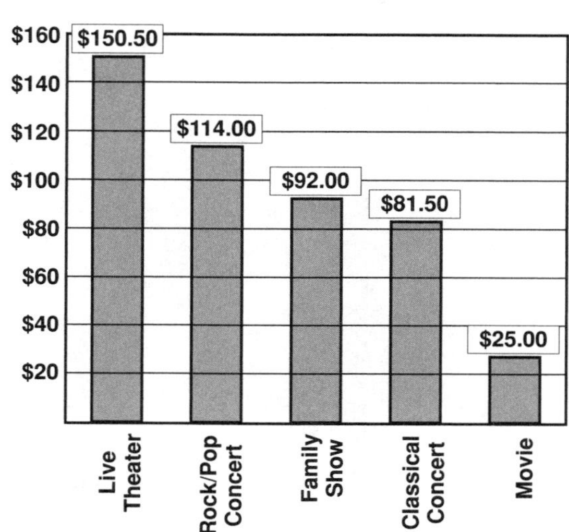

UNIT 17-2 Angles

THE MEANING OF AN ANGLE

THE MAIN IDEA

1. An ***angle*** (symbol ∠) is the figure formed by two ***rays*** that have the same endpoint called the ***vertex*** (plural ***vertices***).

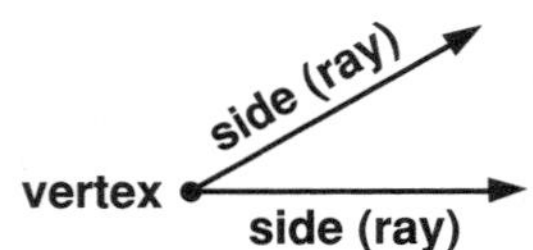

2. Ways of naming an angle are:

with a capital letter at its vertex

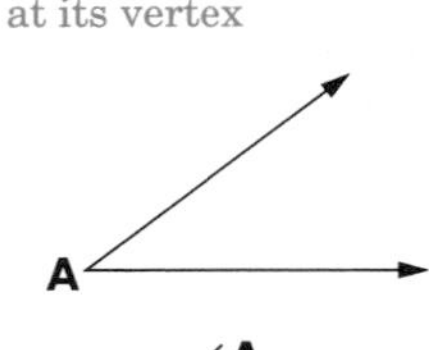

with a number inside the angle

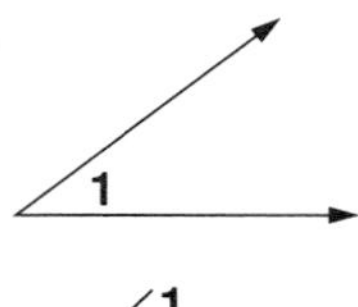

with three capital letters, of which the middle letter is the vertex

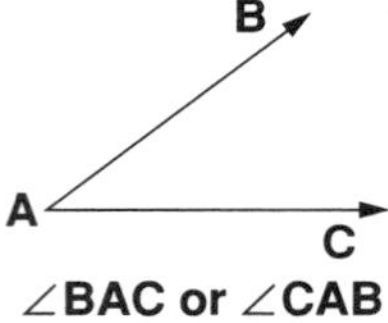

EXAMPLE 1 Name each angle by using a single letter and by using three letters.

a.

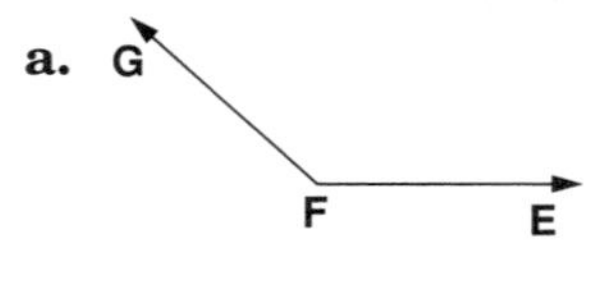

Answer: ∠*F*, ∠*GFE*, or ∠*EFG*

The vertex must be the middle letter.

b.

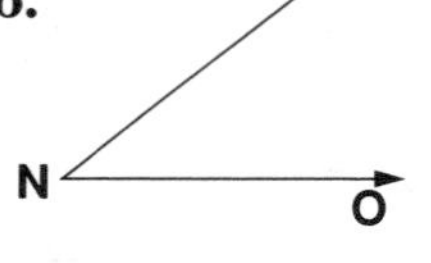

Answer: ∠*N*, ∠*MNO*, or ∠*ONM*

EXAMPLE 2 Use three letters to name ∠1, ∠2, and ∠3.

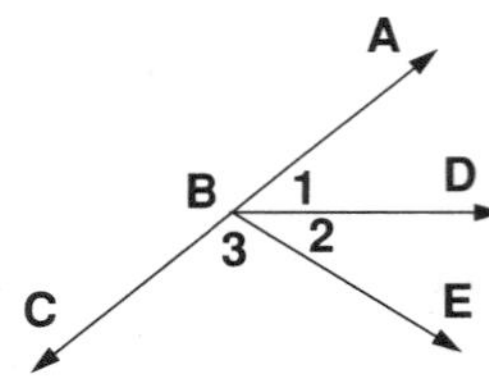

Answer: ∠1 is ∠*ABD* or ∠*DBA*.
∠2 is ∠*DBE* or ∠*EBD*.
∠3 is ∠*EBC* or ∠*CBE*.

MEASURING ANGLES

THE MAIN IDEA

1. The ***degree*** is the unit of measure for angles. There are 360 degrees, written 360°, in a ***complete rotation***.

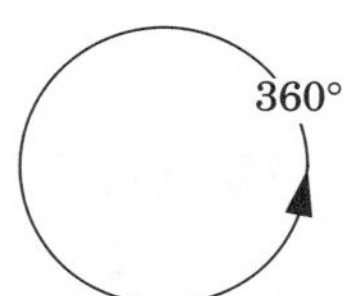

2. A ***protractor*** is an instrument used to measure angles. It has two scales, each starting at 0° and ending at 180°, to make it convenient to measure an angle regardless of its position.

   To find the measure of $\angle ABC$:

   Place the center of the protractor on the vertex of the angle.

   Line up the base line of the protractor with one side of the angle.

   Ray BC intersects the *inner* scale at 0°.

   Note where the other side cuts across the inner scale.

   Ray BA intersects the inner scale at 50°.

   The measure of $\angle ABC$ is 50°.

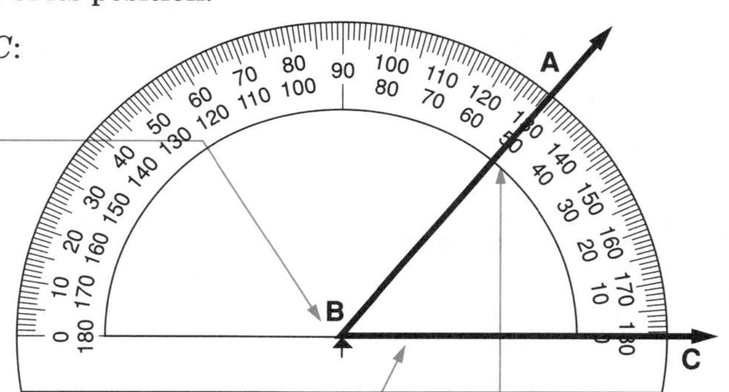

3. The measure of an angle is determined only by the number of degrees in the angle, not by the lengths of the sides.

EXAMPLE 3 Find the measure of $\angle LMN$.

Read the outer scale, because ray ML crosses the outer scale at 0°.

Answer: The measure of $\angle LMN$ is 110°.

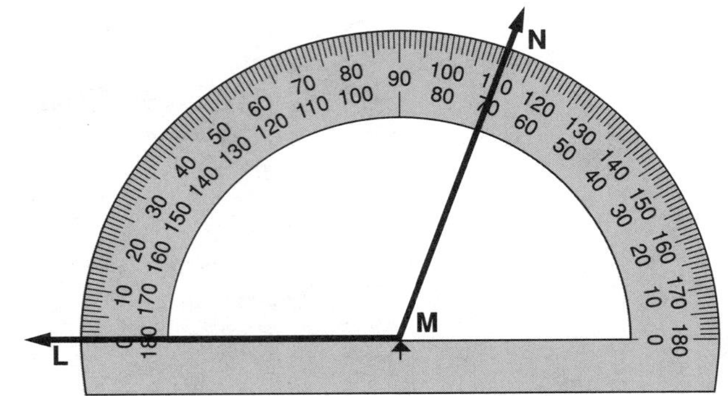

CLASS EXERCISES

1. Name each angle by using a single letter and by using three letters.

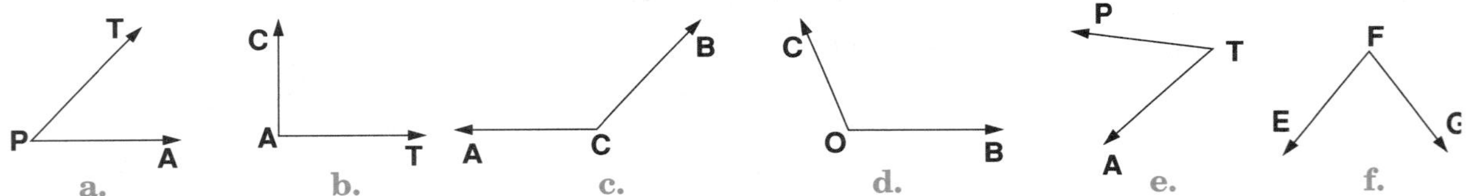

2. Use three letters to name the numbered angles.

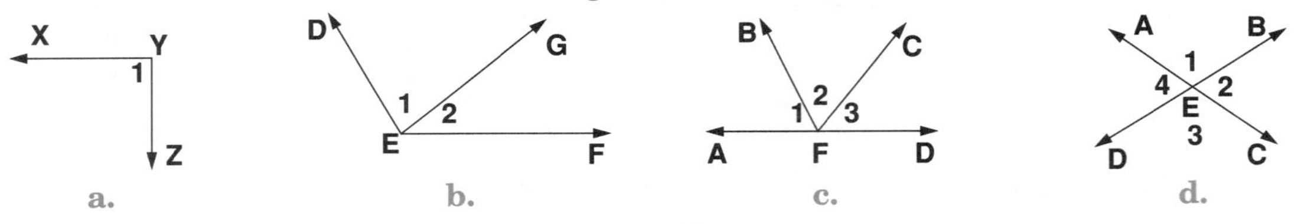

3.

a. How many angles are formed:
  (1) by 3 rays with the same endpoint?
  (2) by 4 rays with the same endpoint?

b. Explain why it is confusing to talk about $\angle P$.

4. Trace each angle on your paper, and use a protractor to find the measure of each angle. If necessary, extend the sides of the angle.

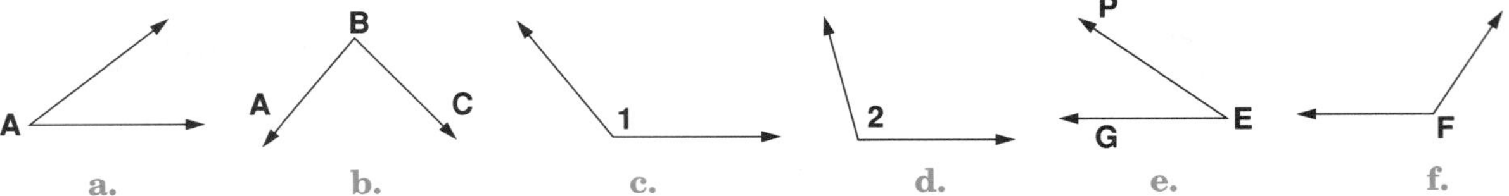

CLASSIFYING ANGLES

THE MAIN IDEA

1. Angles are classified according to their measure.

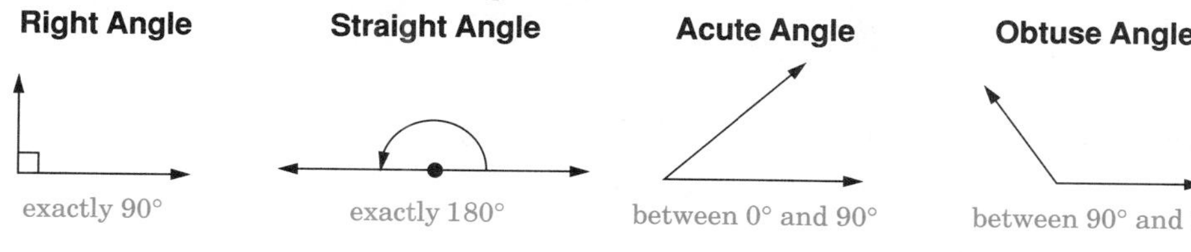

2. ***Congruent angles*** have equal measures. (The symbol $m\angle A$ means "the measure of $\angle A$.")

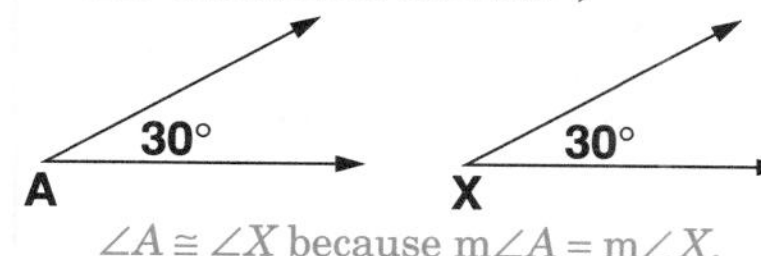

$\angle A \cong \angle X$ because $m\angle A = m\angle X$.

3. The ***bisector of an angle*** is a ray that divides an angle into two congruent angles.

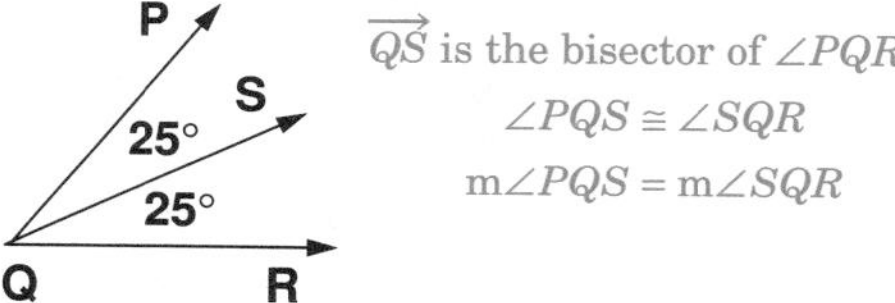

$\overrightarrow{QS}$ is the bisector of $\angle PQR$.

$\angle PQS \cong \angle SQR$

$m\angle PQS = m\angle SQR$

EXAMPLE 4 From its measure, classify the angle.

	Measure of the Angle	*Type of Angle*
a.	50°	acute angle
b.	98°	obtuse angle
c.	180°	straight angle
d.	90°	right angle

EXAMPLE 5 Evaluate:

	Given Angle	*Answer*
a.	$\frac{3}{10}$ of a right angle	$\frac{3}{10} \times 90° = 27°$
b.	$\frac{2}{9}$ of a straight angle	$\frac{2}{9} \times 180° = 40°$
c.	$\frac{1}{2}$ of an angle with measure of 100°	$\frac{1}{2} \times 100° = 50°$

CLASS EXERCISES

1. From its shape, classify each angle.

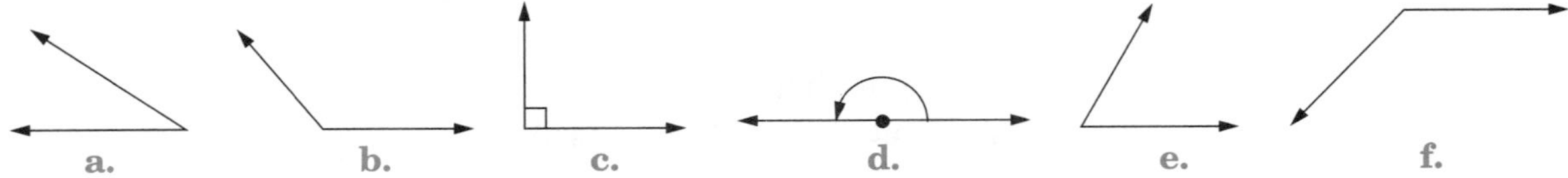

2. From its measure, classify each angle.

a. 25° **b.** 89° **c.** 146° **d.** 180° **e.** 79° **f.** 46° **g.** 90° **h.** 101°

3. Find the number of degrees in each given angle.

a. $\frac{1}{3}$ of a right angle **b.** $\frac{5}{6}$ of a straight angle **c.** $\frac{7}{12}$ of a complete rotation

4. Name two congruent angles.

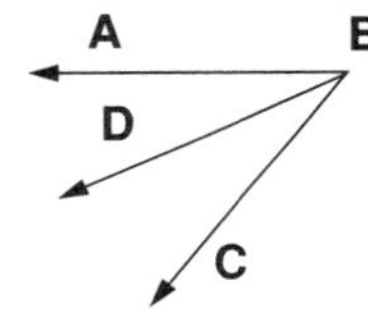

$\overrightarrow{BD}$ bisects $\angle ABC$.

5. Classify the two congruent angles formed by drawing the angle bisector of:

a. an acute angle **b.** an obtuse angle
c. a right angle **d.** a straight angle

PAIRS OF ANGLES

THE MAIN IDEA

1. ***Complementary angles*** are a pair of angles whose measures have a sum of 90°. Each angle is called the ***complement*** of the other.
2. ***Supplementary angles*** are a pair of angles whose measures have a sum of 180°. Each angle is called the ***supplement*** of the other.
3. When two lines intersect, two pairs of ***vertical angles*** are formed. Vertical angles are congruent.

EXAMPLE 6 Classify the angle pairs. Use measures to justify answers.

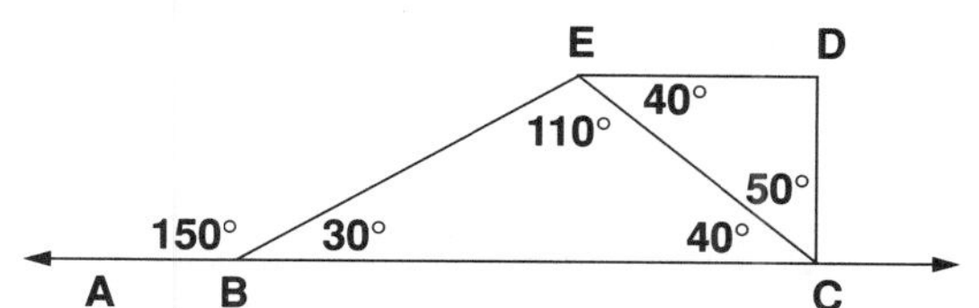

a. $\angle BCE$ and $\angle ECD$

$40° + 50° = 90°$

Answer: complementary angles

b. $\angle ABE$ and $\angle EBC$

$150° + 30° = 180°$

Answer: supplementary angles

c. $\angle EBC$ and $\angle BED$

$30° + (110° + 40°) = 180°$

Answer: supplementary angles

EXAMPLE 7 In the diagram, line *DE* and line *FG* intersect at *H*.

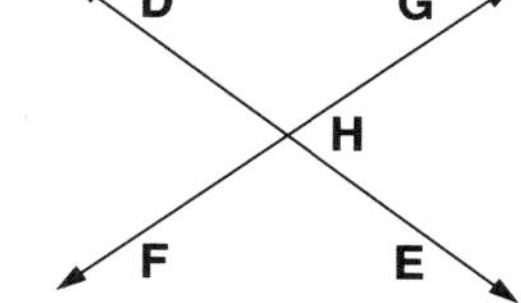

If m$\angle GHE$ is 72°, find:

a. m$\angle DHF$

Since $\angle GHE$ and $\angle DHF$ are vertical angles, they are equal in measure.

Answer: m$\angle DHF$ = 72°

b. m$\angle DHG$

Since $\angle GHE$ and $\angle DHG$ form a straight angle, they are supplementary.

Write and solve an equation.

Let x = m$\angle DHG$. $\quad x + 72 = 180$

Add −72 to each side. $\quad x + 72 + (-72) = 180 + (-72)$

$x = 108$

Answer: m$\angle DHG = 108°$

EXAMPLE 8 Two complementary angles have measures that are represented by $3x$ and $7x$. Find the measures of the angles.

Since the angles are complementary, the sum of their measures is 90°.

Write and solve an equation.

$3x + 7x = 90$ complementary angles

$10x = 90$ Add like terms.

$\frac{1}{10}(10x) = \frac{1}{10}(90)$ Multiply by reciprocal.

$x = 9$

To find the measure of each angle, substitute 9 for x in the expression for each angle.

$3x = 3(9) = 27 \qquad 7x = 7(9) = 63$

Check: Is the sum of the measures 90°?

$27° + 63° \stackrel{?}{=} 90°$

$90° = 90°$ ✔

Answer: The angles have measures of 27° and 63°.

CLASS EXERCISES

1. In the diagram shown, $m\angle BEC = 150°$.
   a. Find $m\angle AED$.
   b. Find $m\angle BED$.

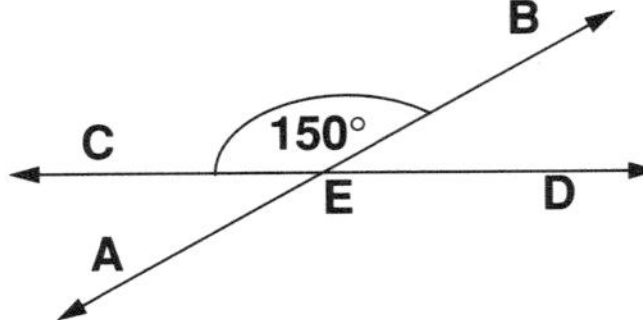

2. In the diagram, $m\angle ACD = 165°$. Find $m\angle DCB$.

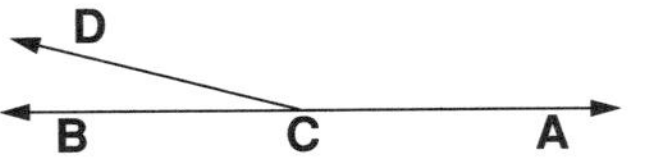

3. Find the complement of an angle whose measure is 17°.
4. Find the supplement of an angle whose measure is 97°.
5. In the diagram, $m\angle CAD = 36°$ and $m\angle CDB = 52°$.
   a. Find $m\angle CDA$.
   b. Find $m\angle ADE$.

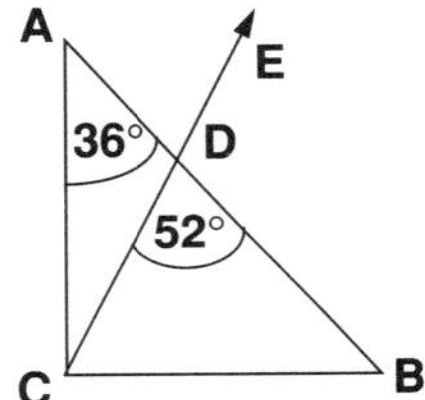

6. In the diagram, $\angle A$ is the supplement of $\angle ABC$.
   a. Find the value of x.
   b. Find $m\angle ABC$.
   c. Find $m\angle CBE$.

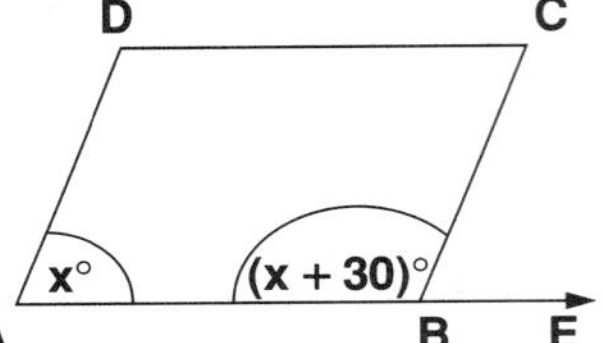

HOMEWORK EXERCISES

1. Name each angle by using a single letter and by three letters.

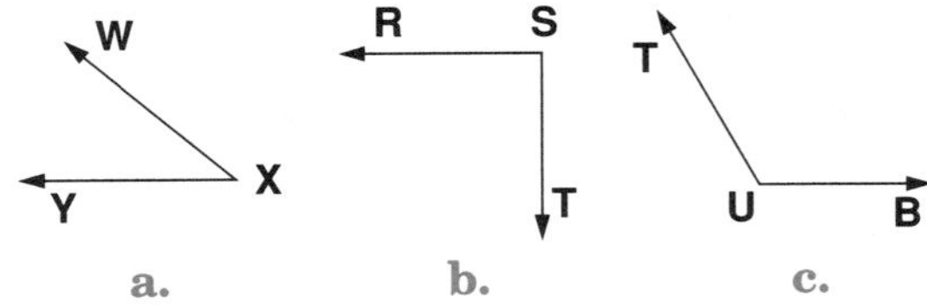

2. Use three letters to name the numbered angles.

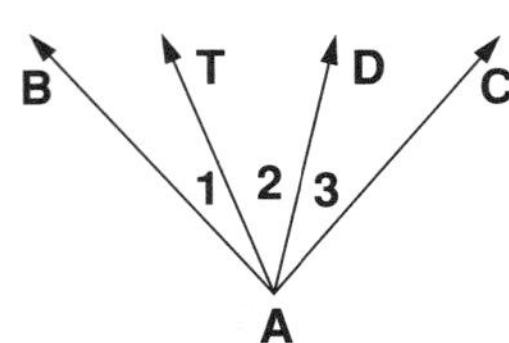

3. From the diagram, name six angles that have vertex P.

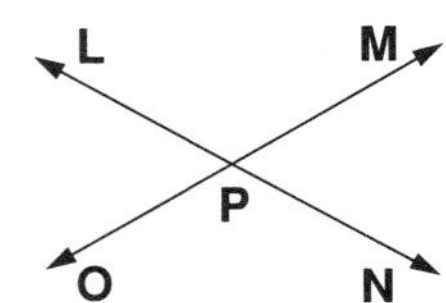

4. Trace each angle on your paper, and use a protractor to find the measure of each angle. If necessary, extend the sides of the angle.

a. b. 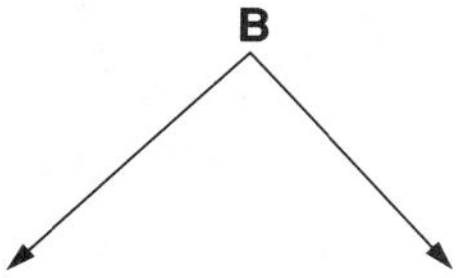 c.

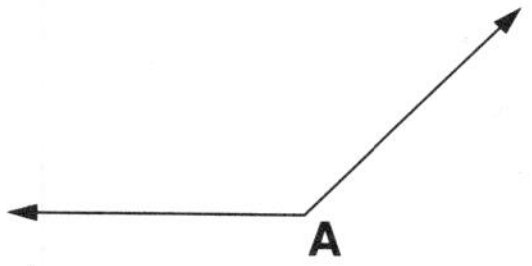

d. 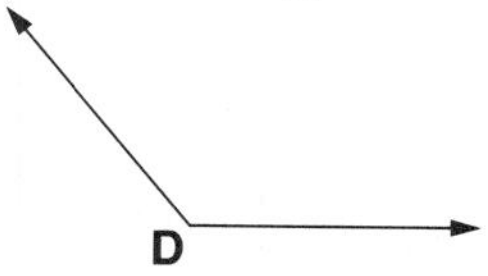 e. 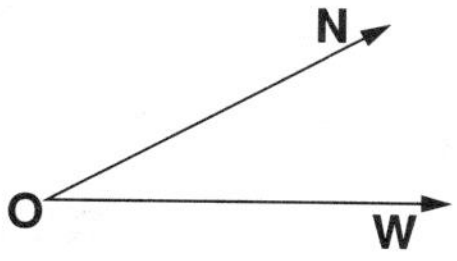 f.

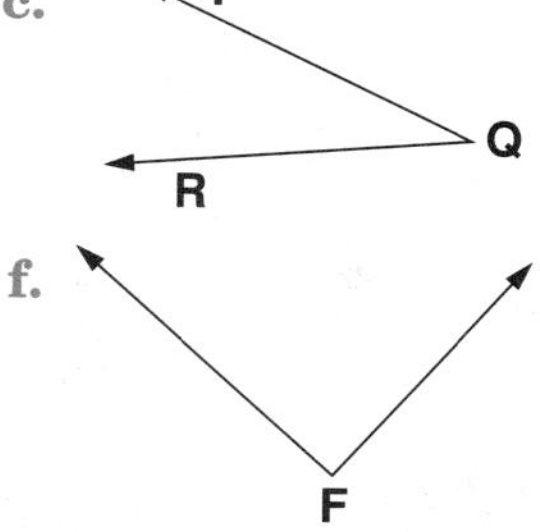

5. From its shape, classify each angle.

a. b. 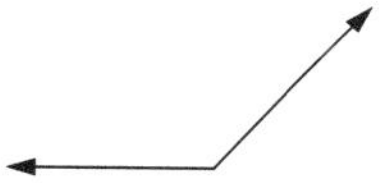 c. d. e.

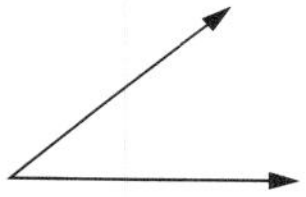

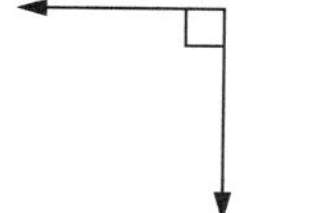

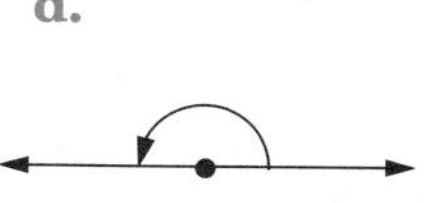

6. From its measure, classify each angle.

a. 90° **b.** 37° **c.** 105° **d.** 175° **e.** 180° **f.** 49° **g.** 68° **h.** 178°

7. The number of degrees contained in $\frac{2}{3}$ of a right angle is

(a) 30° (b) 60° (c) 45° (d) 120°

8. The number of degrees contained in $\frac{1}{4}$ of a straight angle is

(a) $22\frac{1}{2}°$ (b) 45° (c) 30° (d) 40°

9. The ratio of the number of degrees contained in a right angle to the number of degrees contained in a straight angle is

(a) 1 : 3 (b) 2 : 3 (c) 2 : 1 (d) 1 : 2

10. $\overrightarrow{BD}$ is the bisector of $\angle ABC$, and $\overrightarrow{YW}$ bisects $\angle XYZ$. If $\angle ABC \cong \angle XYZ$, how many other pairs of congruent angles can be named? Explain.

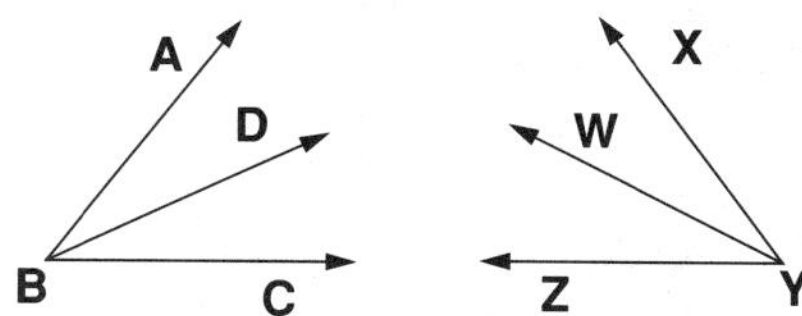

11. In the diagram, $\overrightarrow{QR}$ bisects $\angle AQD$ and $\angle BQC$. If $m\angle BQR = 25°$ and $m\angle CQD = 15°$, find the measure of $\angle AQD$.

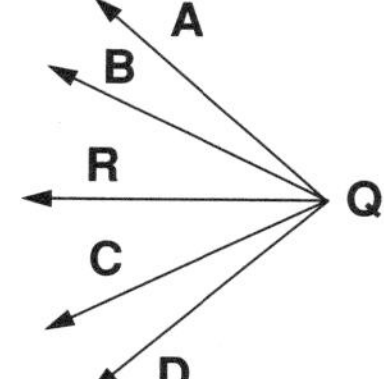

12. **a.** Using the 26 letters of the alphabet, print the capital letters that are usually drawn using only straight line segments, and that form: **(1)** just acute angles
(2) just right angles
(3) at least one obtuse angle

b. If a capital letter is chosen at random, what is the probability that it consists only of line segments that form right angles?

13. In the diagram, $\overline{HJ}$ and $\overline{KI}$ intersect at E.

a. Name two pairs of vertical angles.

b. Name a supplement of $\angle HEI$.

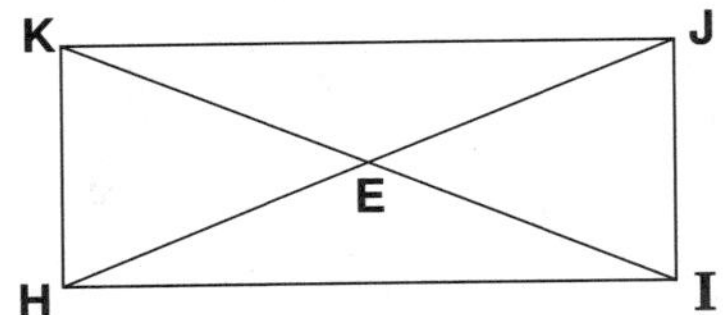

14. In the diagram, $m\angle PQT = 32°$.

a. Find $m\angle TQS$.

b. Find $m\angle RQS$.

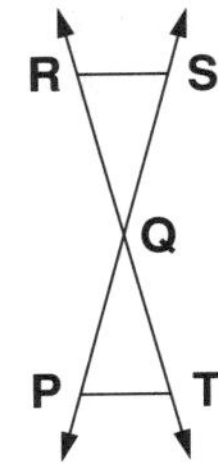

15. In the diagram, the measures of $\angle GEF$ and $\angle DEG$ are represented by x and $x + 50$. Find $m\angle DEG$.

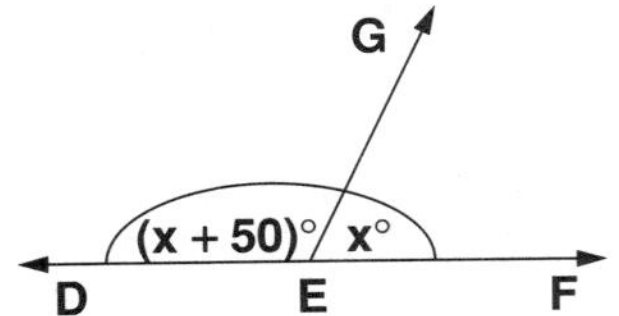

16. If $\overline{GD}$ and $\overline{DE}$ intersect at D to form an angle whose measure is 90° and $m\angle FDE = 34°$, find $m\angle GDF$.

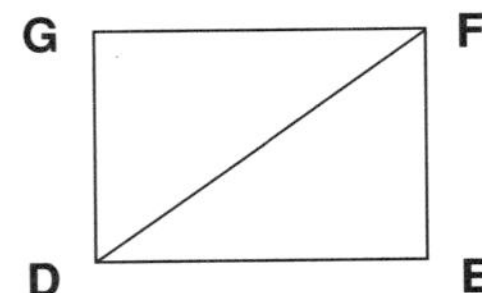

17. Find the complement of an angle whose measure is: **a.** 50° **b.** 27° **c.** 39° **d.** 89°

18. Find the supplement of an angle whose measure is: **a.** 111° **b.** 89° **c.** 42° **d.** 9°

19. Copy and complete the table. Describe the relationship between the complement of an angle and the supplement of the same angle.

Angle A	53°		28°		89°	5°
Its complement				45°		
Its supplement		117°				

20. Find the measure of an angle if it is equal in measure to its:
a. complement **b.** supplement

21. If the measures of two supplementary angles are represented by $2x$ and $7x$, find the measure of each of the angles.

22. In the diagram of two intersecting lines, what is the least amount of information you need to find the measures of $\angle FHD$, $\angle DHG$, $\angle GHE$, and $\angle FHE$?

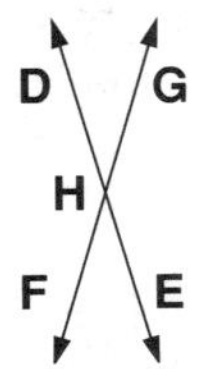

23. In the diagram, CED is a straight line, $m\angle AEB = 90°$, and $m\angle AED = 56°$.

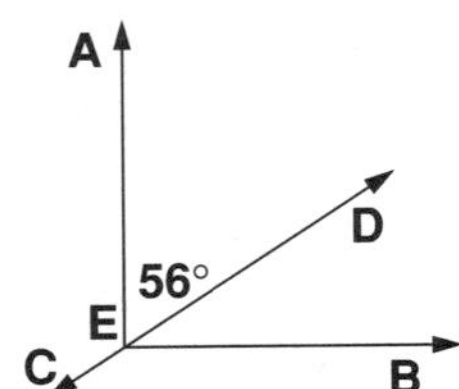

a. Find $m\angle DEB$.

b. Find $m\angle CEB$.

24. Draw angles X, Y, and Z so that $\angle X$ is the complement of $\angle Y$, and $\angle Y$ is the complement of $\angle Z$. What relationship is there between $\angle X$ and $\angle Z$? Explain.

SPIRAL REVIEW EXERCISES

1. Find the length of line segment AB.

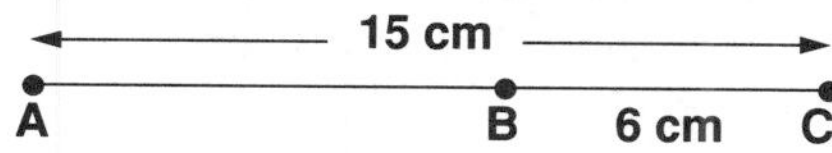

2. If $15 - 3n \leq 12$, then the solution set is

(a) $n = 1$ (b) $n \leq 1$
(c) $n \geq 1$ (d) $n \leq -1$

3. On a highway, the maximum legal speed limit is 55 miles per hour and the minimum legal speed limit is 40 miles per hour. If x represents the speed of a vehicle, which sentence best describes the legal speeds?

(a) $40 \leq x \leq 55$ (b) $40 < x < 55$
(c) $40 \leq x < 55$ (d) $40 < x \leq 55$

4. The cost of placing a classified ad in a newspaper is given by the formula $C = \$2.50 + \$0.50w$, where w is the number of words in the ad. Find the cost of an ad containing 18 words.

5. A dime and a quarter are tossed at the same time. What is the probability that the dime will show tails and the quarter will show heads?

(a) 1 (b) $\frac{1}{2}$ (c) $\frac{1}{4}$ (d) $\frac{1}{8}$

6. Which expression should come next in the pattern? $4x, 8x^2, 16x^3, 32x^4, \ldots$

(a) $48x^5$ (b) $64x^5$ (c) $40x^6$ (d) $48x^6$

7. 16 is what percent of 40?

8. Eight buses were needed to take 336 students on a field trip. If there were four adults on each bus, what was the ratio of adults to children?

9. Carla bought some CDs that cost \$9 each and a \$16 CD case. If the total cost was \$79, how many CDs did Carla buy?

10. Solve the proportion:

$$\frac{458}{n} = \frac{8{,}702}{24{,}244}$$

11. The average distance from the sun to the planet Venus is 67,200,000 miles. This distance written in scientific notation is

(a) 6.72×10^{-5} (b) 6.72×10^{5}
(c) 6.72×10^{-8} (d) 6.72×10^{7}

12. Ron bought a car that had a base price of \$11,926. He had a sun roof installed for an additional \$412. If the rate of sales tax was 6%, what was the total price of the car?

UNIT 17-3 Polygons and Triangles

CLASSIFYING POLYGONS AND TRIANGLES

THE MAIN IDEA

1. A *polygon* is a closed plane figure, with sides that are line segments. Polygons are classified by the number of sides, such as:

Number of Sides	*Name*
3	Triangle
4	Quadrilateral
5	Pentagon
6	Hexagon
8	Octagon
10	Decagon

2. To name a polygon, start with any vertex and name each vertex in order, going around the polygon in either direction.

pentagon *ABCDE* or pentagon *CBAED*

sides: $\overline{AB}$, $\overline{BC}$, $\overline{CD}$, $\overline{DE}$, $\overline{AE}$

angles: $\angle A$, $\angle B$, $\angle C$, $\angle D$, $\angle E$

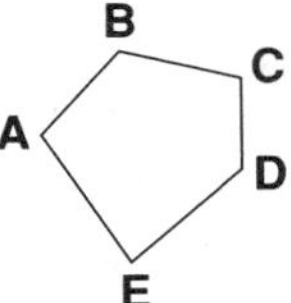

3. Triangles can be classified by the types of angles they contain.

Acute Triangle

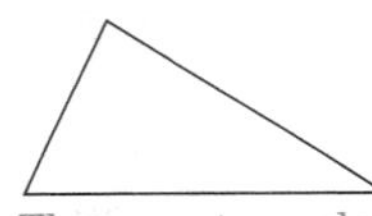

Three acute angles

Right Triangle

One right angle and two acute angles

Obtuse Triangle

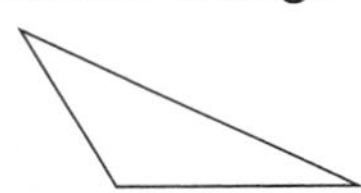

One obtuse angle and two acute angles

4. Triangles can also be classified by the number of congruent sides.

Scalene Triangle

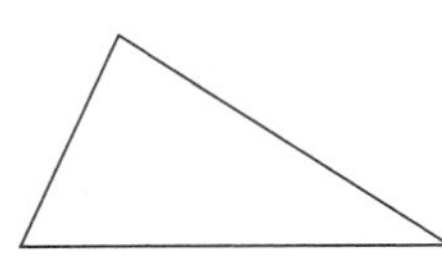

No congruent sides

Isosceles Triangle

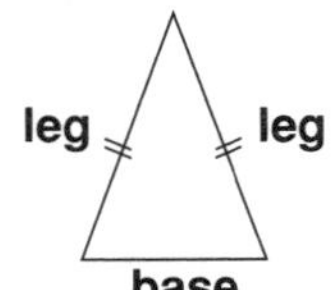

Two congruent sides

Equilateral Triangle

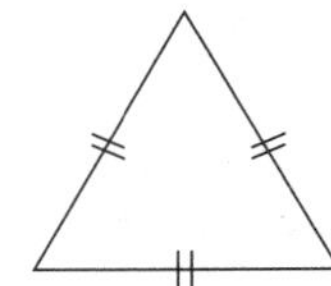

Three congruent sides

CLASS EXERCISES

1. Classify each polygon and name it in two ways.

a. 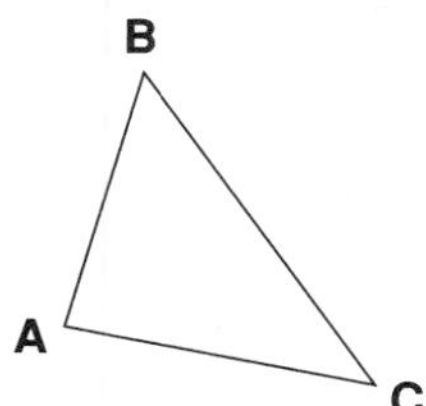

b. 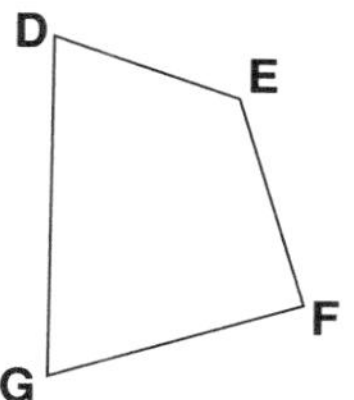

c. 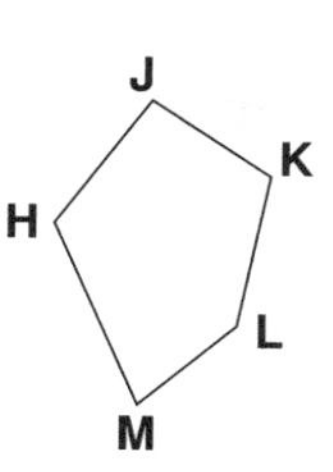

d. 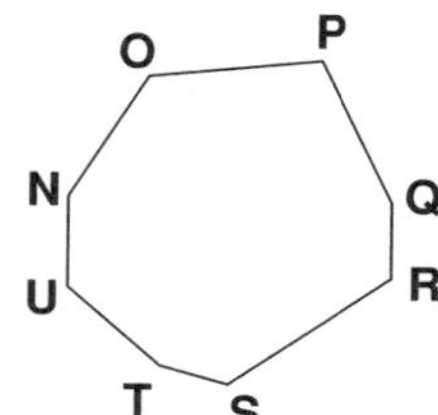

2. Name each triangle, its sides, and its angles.

a. 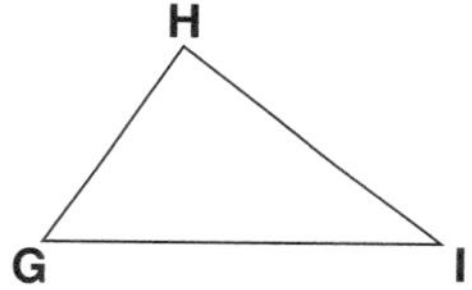

b. 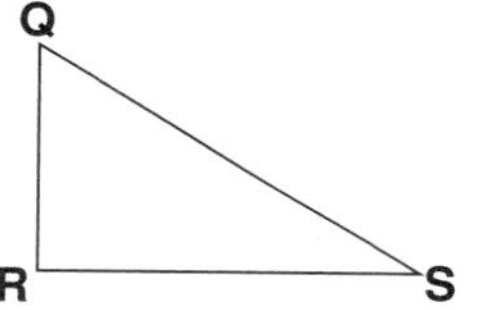

c. 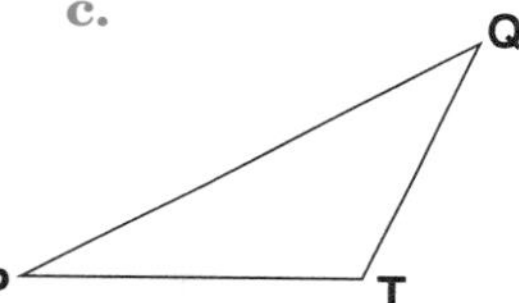

3. Classify each triangle according to its angles.

a. 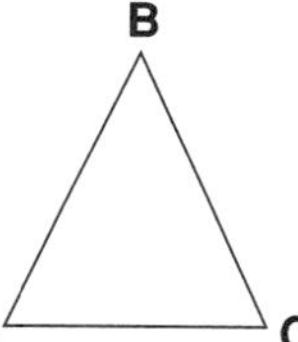

b. 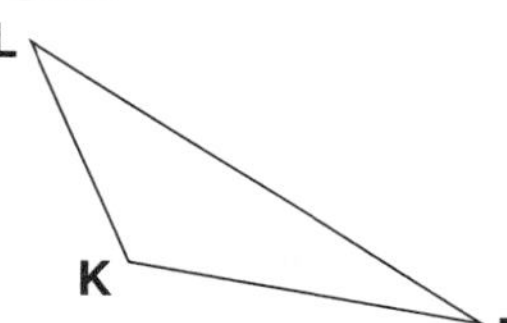

c. 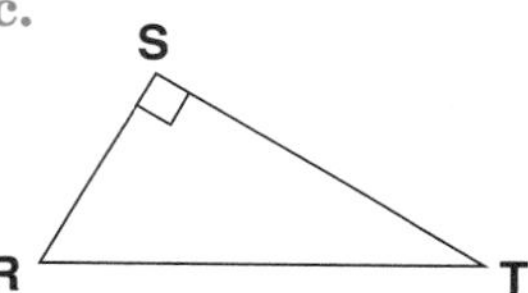

4. Classify each triangle according to its sides.

a. 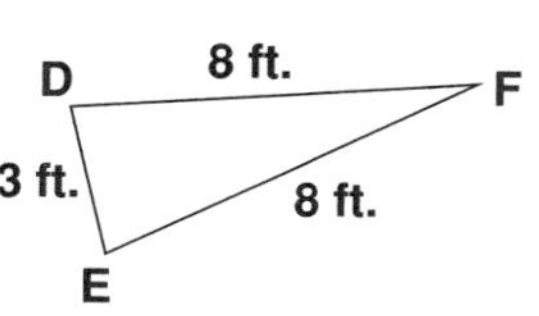

b. 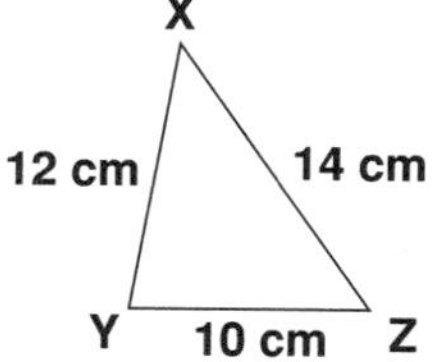

c. 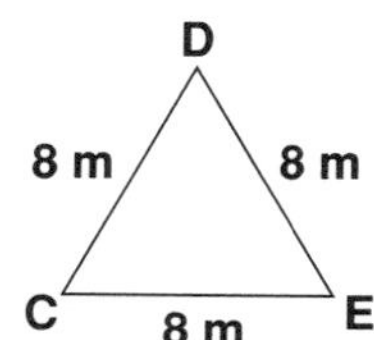

5. Classify each triangle according to its angles if the measures of the angles are:

a. 40°, 50°, 90° **b.** 125°, 25°, 30° **c.** 70°, 80°, 30°

6. Classify each triangle according to its sides if the measures of the sides are:

a. 7 in., 7 in., 4 in. **b.** 5 cm, 12 cm, 13 cm **c.** 9 cm, 9 cm, 9 cm

ANGLES OF A TRIANGLE

THE MAIN IDEA

1. The sum of the measures of the angles of any triangle is 180°.
2. An isosceles triangle has two congruent angles.
3. An equilateral triangle has three congruent angles.

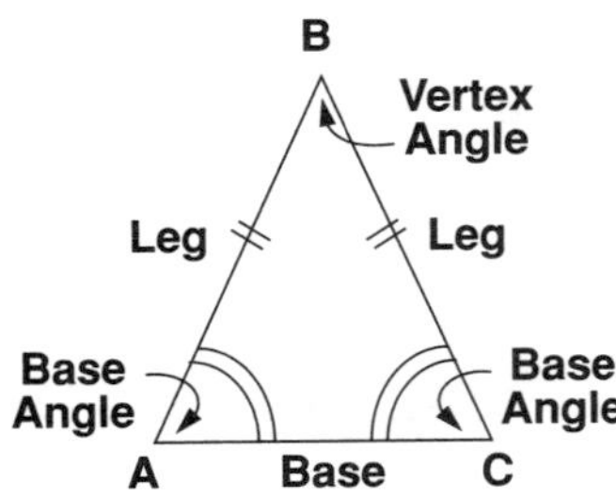

In isosceles triangle ABC, $\overline{AB} \cong \overline{BC}$ and $\angle A \cong \angle C$.

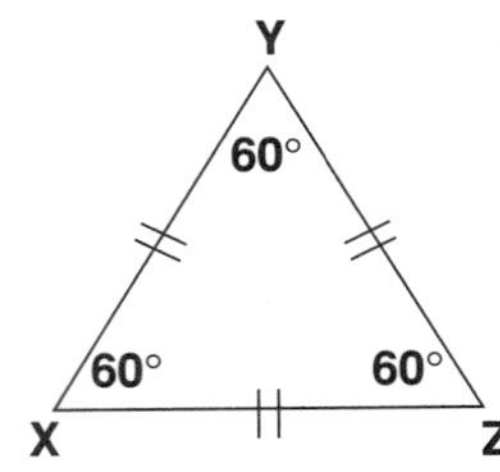

In equilateral triangle XYZ, $\overline{XY} \cong \overline{YZ} \cong \overline{XZ}$ and $\angle X \cong \angle Y \cong \angle Z$.

EXAMPLE 1 In isosceles triangle DEF, $DE = DF$, and $m\angle E = 29°$. Find $m\angle D$.

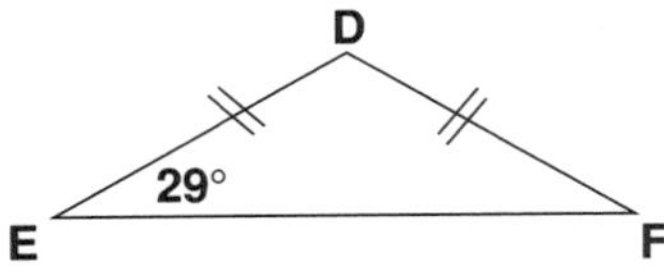

Since the base angles of an isosceles triangle are equal in measure, $m\angle F = 29°$.

Let x represent the measure of $\angle D$. The sum of the angles of the triangle is 180°.

$$\begin{aligned} x + 29 + 29 &= 180 \\ x + 58 &= 180 \\ x + 58 + (-58) &= 180 + (-58) \\ x &= 122 \end{aligned}$$

Answer: $m\angle D = 122°$

EXAMPLE 2 In an isosceles triangle, the vertex angle measures 30°. Find the measure of each base angle.

Let x = the measure of a base angle.

The base angles are equal in measure, and the sum of the 3 angles is 180°.

$$\begin{aligned} x + x + 30 &= 180 \\ 2x + 30 &= 180 \\ 2x + 30 + (-30) &= 180 + (-30) \\ 2x &= 150 \\ \tfrac{1}{2}(2x) &= \tfrac{1}{2}(150) \\ x &= 75 \end{aligned}$$

Check: Do the measures of the three angles total 180°?

$$\begin{aligned} 75° + 75° + 30° &\stackrel{?}{=} 180° \\ 180° &= 180° \checkmark \end{aligned}$$

Answer: The measure of each base angle is 75°.

CLASS EXERCISES

1. Find the measure of $\angle 1$ in each triangle.

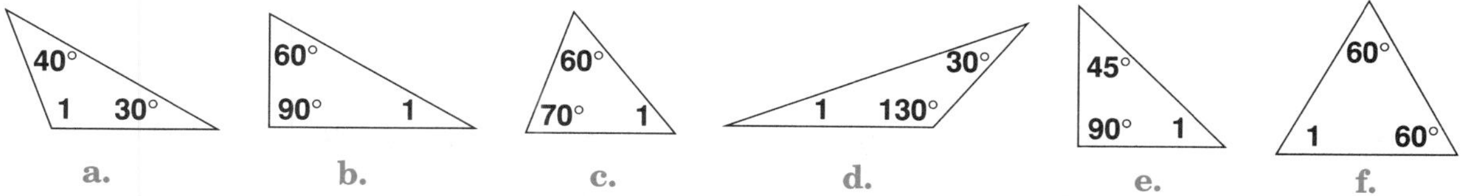

2. Find the measure of the third angle of a triangle if the measures of the other two angles are:
   a. 75° and 20° **b.** 15° and 105° **c.** 18° and 72° **d.** 47° and 90°
   e. 56° and 52° **f.** 165° and 5° **g.** 45° and 45° **h.** 100° and 30°
3. Tell whether or not you can draw a triangle with three angles that measure:
   a. 48°, 100°, 32° **b.** 65°, 65°, 50° **c.** 110°, 30°, 95° **d.** 60°, 50°, 40°
   e. 60°, 60°, 60° **f.** 90°, 90°, 1° **g.** 180°, 0°, 0° **h.** 90°, 90°, 0°
4. The maximum number of obtuse angles that a triangle can have is (a) 0 (b) 1 (c) 2 (d) 3
5. In isosceles triangle ABC, $\overline{BC} \cong \overline{CA}$ and $m\angle B = 50°$.
   a. Find the measure of $\angle A$. **b.** Find the measure of $\angle C$.
6. If the measure of an angle of an equilateral triangle is represented by $2x + 10$, find the value of x.
7. In an isosceles triangle, the measure of a base angle is 80°. Find the measure of the vertex angle.
8. Is it possible for a triangle to have more than one angle that is not acute? Explain.

HOMEWORK EXERCISES

1. Classify each polygon by the number of its sides.

a.

b.

c.

SPEED
LIMIT
50

d.

e.

2. Classify each triangle according to its angles.

a.

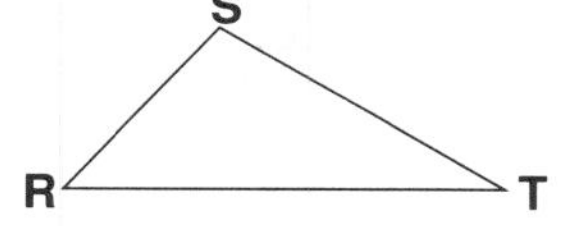

b.

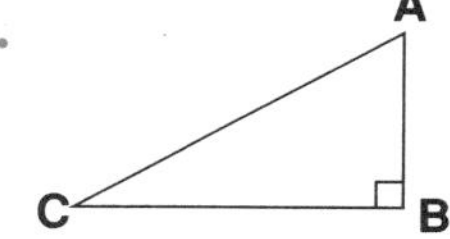

c.

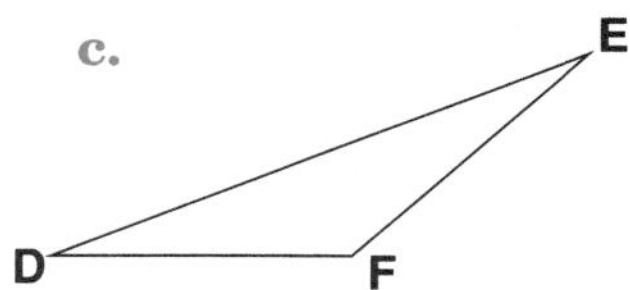

3. Classify each triangle according to its sides.

a. 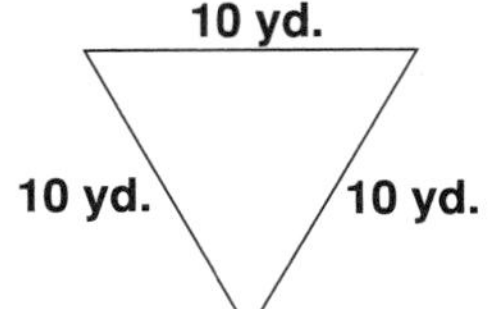

b. 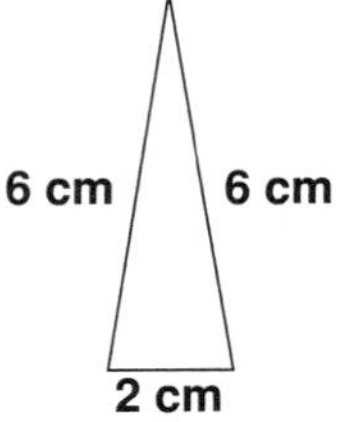

c. 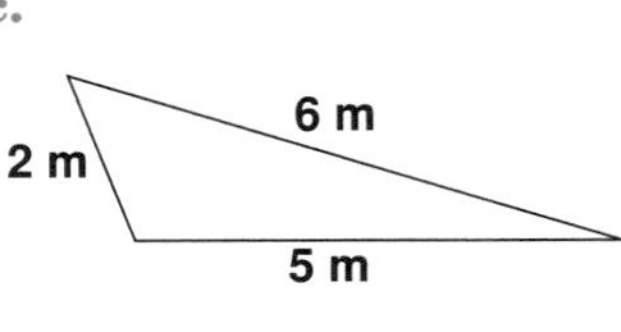

4. Classify each triangle according to its angles if the measures of the angles are:
   a. 20°, 75°, 85° **b.** 30°, 50°, 100° **c.** 30°, 60°, 90°

5. Classify each triangle according to its sides if the measures of the sides are:
   a. 3 in., 4 in., 5 in. **b.** 3 m, 3 m, 5 m **c.** 15 ft., 15 ft., 15 ft.

6. In isosceles triangle EFG, $EF = FG$ and $m\angle E = 55°$. **a.** Find $m\angle G$. **b.** Find $m\angle F$.

7. If the measure of one angle of an equilateral triangle is represented by $3x + 30$, find the value of x.

8. Find the measure of $\angle A$ in each triangle.

a. 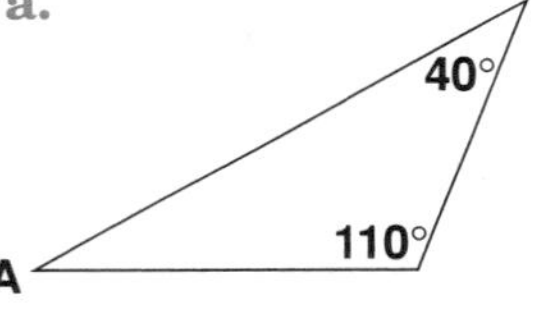

b. 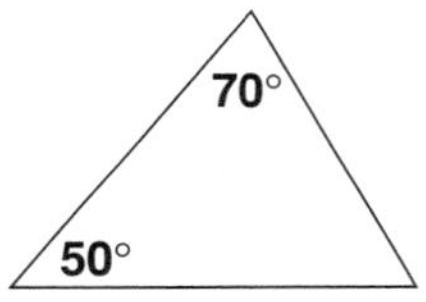

c. 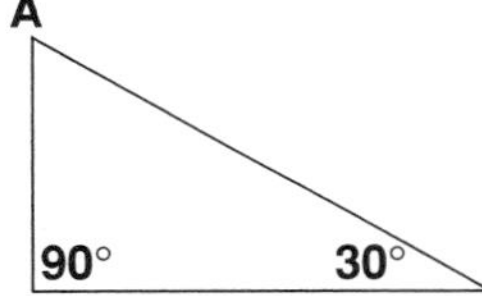

9. Find the measure of the third angle of a triangle if the measures of the other two angles are:
   a. 80° and 60° **b.** 58° and 109° **c.** 45° and 90° **d.** 30° and 60°

10. Tell whether or not you can draw a triangle with three angles that measure:
    a. 28°, 42°, 100° **b.** 50°, 50°, 60° **c.** 70°, 60°, 50° **d.** 110°, 35°, 35°

11. The maximum number of acute angles that a triangle can have is (a) 0 (b) 1 (c) 2 (d) 3

12. If the measure of a base angle of an isosceles triangle is 25°, find the measure of the vertex angle.

13. If the measure of the vertex angle of an isosceles triangle is 24°, find the measure of a base angle.

14. If a right triangle is also isosceles, what is the measure of each of its base angles?

15. If the measure of one acute angle of a right triangle is twice the measure of the other acute angle, find the measure of each angle.

16. Lee said that an equilateral triangle is also an isosceles triangle. Explain how you would defend his point of view.

17. Copy and complete the diagram by writing the following labels in the appropriate spaces:

Isosceles Triangles
Triangles
Equilateral Triangles
Right Triangles

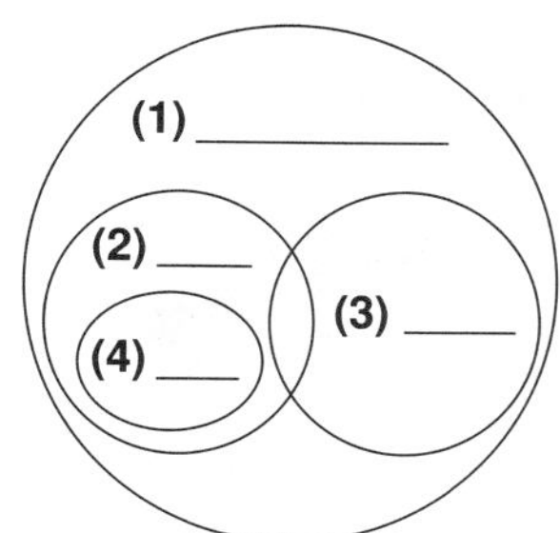

SPIRAL REVIEW EXERCISES

1. Find the supplement of an angle that has a measure of 58°.

2. If $\angle ABE$ and $\angle EBD$ are complementary, and $m\angle EBD = 40°$, find $m\angle ABE$.

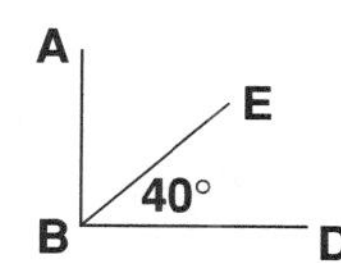

3. What percent of Mr. Jonah's paycheck is spent on entertainment?

Mr. Jonah's Expenses

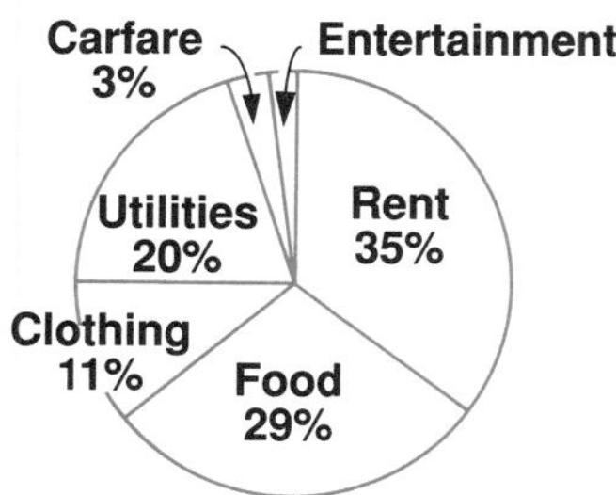

4. Solve the equation: $5x - 11 = -11$

5. Using the formula $A = \frac{1}{2}bh$, find A when $b = 4.2$ and $h = 20$.

6. Evaluate xy^2 when $x = 2$ and $y = 5$.

7. Divide 8,596 by 28.

8. Add: $\frac{1}{2} + \frac{1}{3}$

9. Which of the following is the graph of the inequality $-4 < x \leq 0$?

(a)

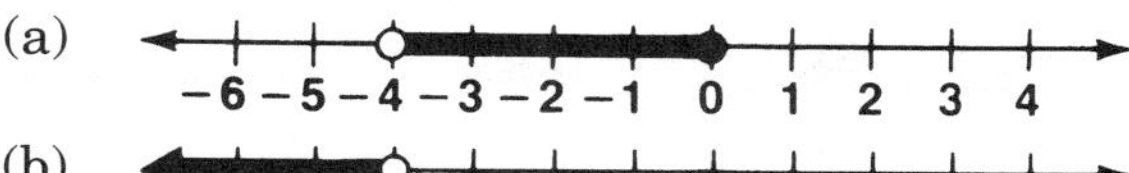

(b)

−6 −5 −4 −3 −2 −1 0 1 2 3 4

(c)

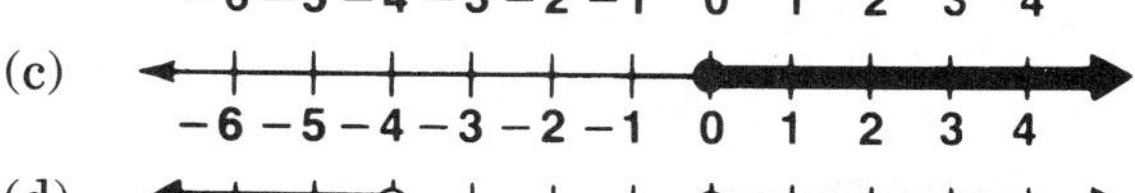

(d)

−6 −5 −4 −3 −2 −1 0 1 2 3 4

10. If the temperature increased from $-2°F$ to $+7°F$, by how many degrees did it increase?

11. A stereo system with a list price of \$900 was purchased for \$675. What was the percent of discount?

12. Find the probability of selecting the number 49 at random from a list of perfect squares that starts at 1 and ends at 256. Express your result as a percent.

13. In a stack of 20 playing cards, 6 are red. What is the probability that a card chosen at random from the stack will be black?

(a) 0.3 (b) 0.6 (c) 0.7 (d) 1.0

UNIT 17-4 Parallel Lines and Perpendicular Lines; Special Quadrilaterals

PARALLEL LINES AND PERPENDICULAR LINES

THE MAIN IDEA

1. ***Parallel lines*** are different lines in the same plane that go in exactly the same direction and, thus, never meet.
   ∥ means "is parallel to."

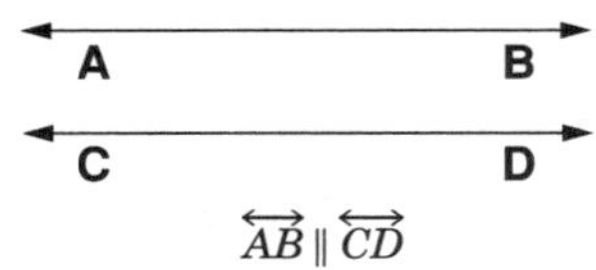

$\overleftrightarrow{AB} \parallel \overleftrightarrow{CD}$

2. ***Perpendicular lines*** are lines that meet at right angles.
   ⊥ means "is perpendicular to."

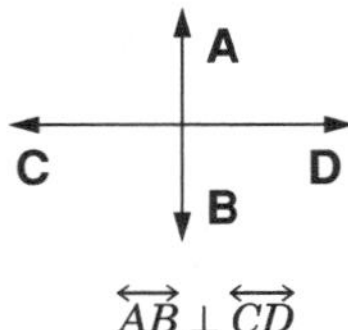

$\overleftrightarrow{AB} \perp \overleftrightarrow{CD}$

EXAMPLE 1 In each diagram, use words and symbols to tell which pairs of lines (or line segments) are drawn parallel and which pairs are drawn perpendicular.

Diagram	*Answer in Words*	*Symbols*
a.	Line *AB* and line *CD* are drawn parallel.	$\overleftrightarrow{AB} \parallel \overleftrightarrow{CD}$
	Line *JK* is drawn perpendicular to line *AB*.	$\overleftrightarrow{JK} \perp \overleftrightarrow{AB}$
	Line *JK* is drawn perpendicular to line *CD*.	$\overleftrightarrow{JK} \perp \overleftrightarrow{CD}$
b.	Line segment *AB* is drawn parallel to line segment *CD*.	$\overline{AB} \parallel \overline{CD}$
	Line segment *AE* is drawn perpendicular to line segment *CE*.	$\overline{AE} \perp \overline{CE}$

CLASS EXERCISES

1. In each diagram, use words and symbols to tell which pairs of lines (or line segments) are drawn parallel, and which pairs are drawn perpendicular.

a. 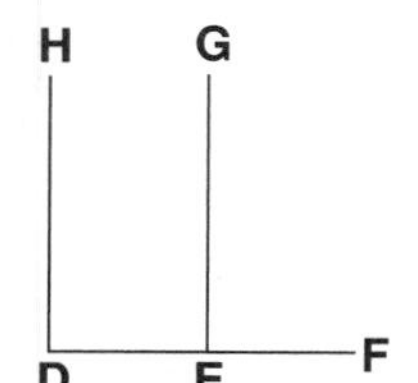

b. 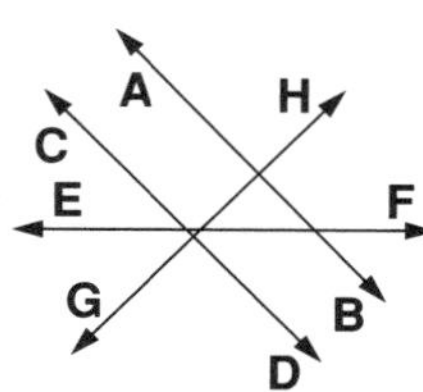

c. 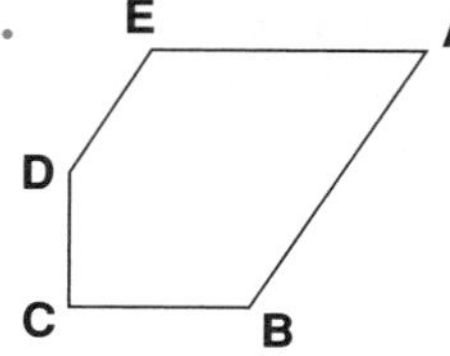

d. 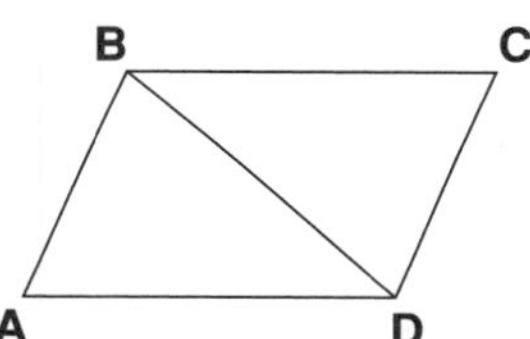

e. 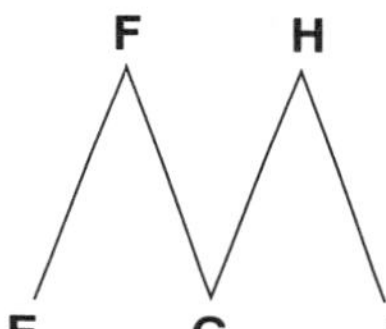

f. 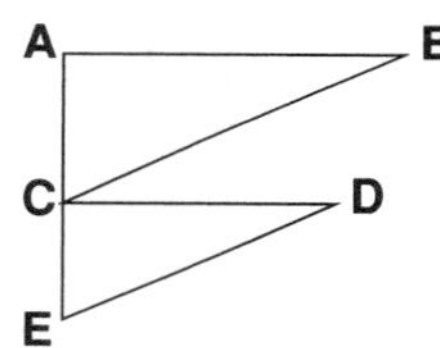

2. Using the parallel lines on a sheet of lined notebook paper, draw two parallel lines and a third line that intersects (crosses) them. Label the lines as shown in the diagram.

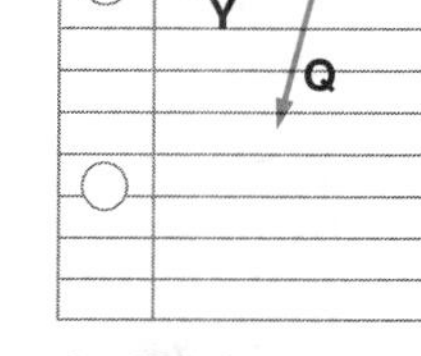

a. Measure all of the angles formed.
b. Name all the angles that have the same measure as $\angle PRX$.
c. Name all the angles that have the same measure as $\angle PRW$.
d. Name all the angles that are supplements of $\angle YSQ$.
e. Repeat parts **a** – **d**, using a new diagram in which $\overleftrightarrow{PQ}$ slants in the opposite direction.
f. Do the same angles have equal measures?
g. Are the same pairs of angles supplementary?

3. The walls of a room meet each other, the ceiling, and the floor to form the line segments shown.

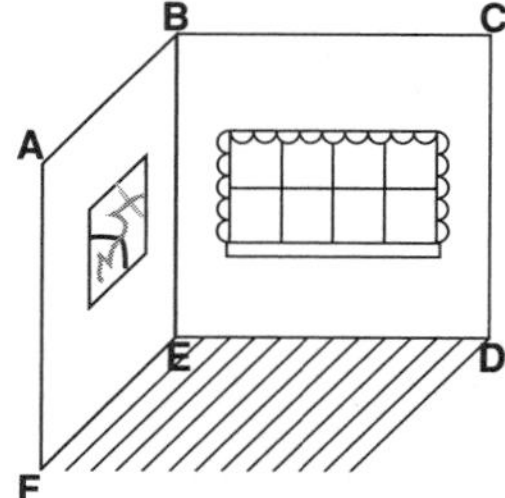

Name the following:
a. two pairs of parallel line segments
b. two pairs of perpendicular line segments
c. two line segments that are neither parallel nor perpendicular

SPECIAL QUADRILATERALS

THE MAIN IDEA

1. A ***trapezoid*** is a quadrilateral that has only one pair of opposite sides parallel.

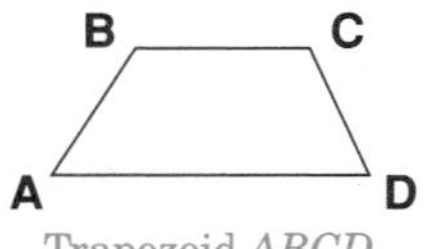

Trapezoid $ABCD$

$\overline{BC} \parallel \overline{AD}$

2. A ***parallelogram*** is a quadrilateral that has both pairs of opposite sides parallel. In a parallelogram, the opposite sides are congruent.

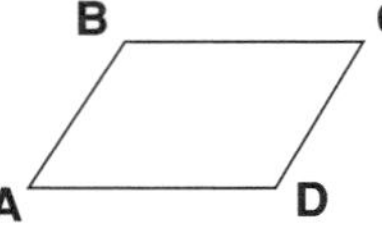

Parallelogram $ABCD$

$\overline{AB} \parallel \overline{DC}$, $\overline{BC} \parallel \overline{AD}$
$\overline{AB} \cong \overline{DC}$, $\overline{BC} \cong \overline{AD}$

3. A ***rhombus*** is a parallelogram with four congruent sides.

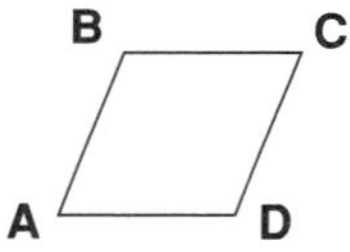

Rhombus $ABCD$

Has all properties of a parallelogram, and
$\overline{AB} \cong \overline{BC} \cong \overline{CD} \cong \overline{DA}$

4. A ***rectangle*** is a parallelogram that has four right angles.

Rectangle $ABCD$

Has all properties of a parallelogram, and
$\overline{AB} \perp \overline{AD}$, $\overline{AB} \perp \overline{BC}$
$\overline{DC} \perp \overline{BC}$, $\overline{DC} \perp \overline{AD}$

5. A ***square*** is a rectangle with four congruent sides.

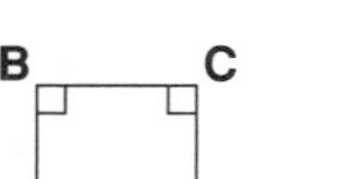

Square $ABCD$

Has all properties of a rectangle, and
$\overline{AB} \cong \overline{BC} \cong \overline{CD} \cong \overline{DA}$

EXAMPLE 2 In parallelogram $ABCD$, find the value of x.

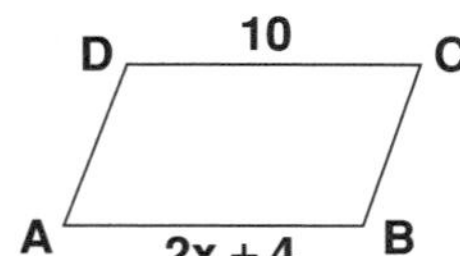

$$
\begin{aligned}
AB &= DC \\
2x + 4 &= 10 \\
2x &= 6 \\
x &= 3 \quad \textit{Ans.}
\end{aligned}
$$

CLASS EXERCISES

1. In each quadrilateral, tell which sides are drawn parallel and which are drawn perpendicular.

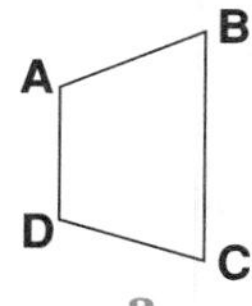

a.

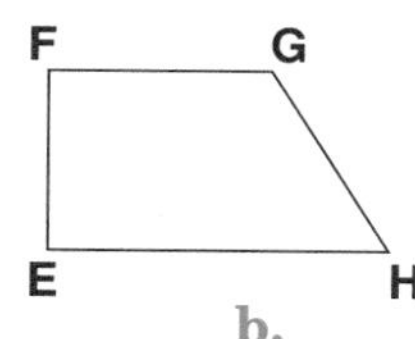

b.

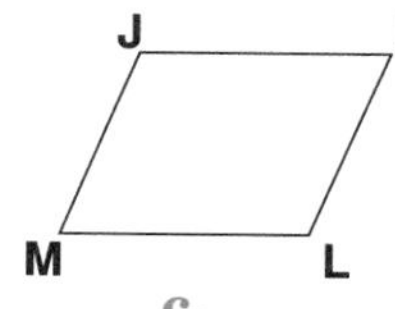

c.

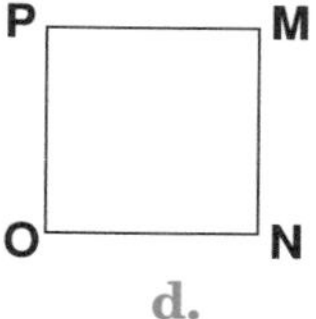

d.

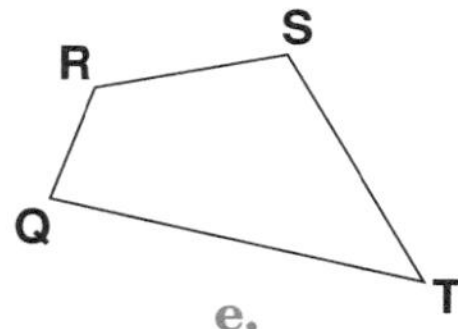

e.

2. In each quadrilateral, name the parallel sides, name the perpendicular sides, and find the missing lengths.

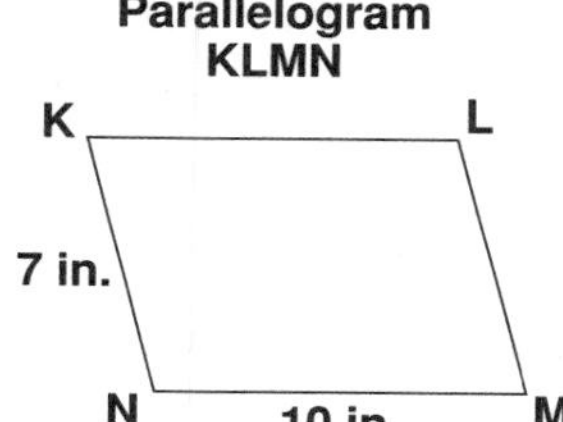

a.

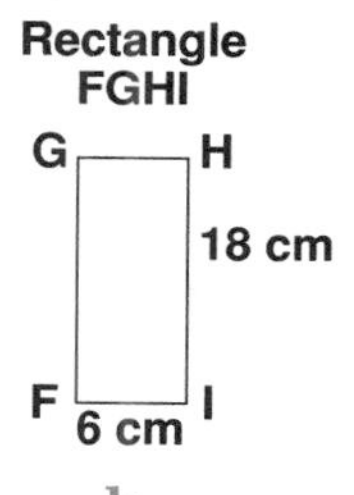

b.

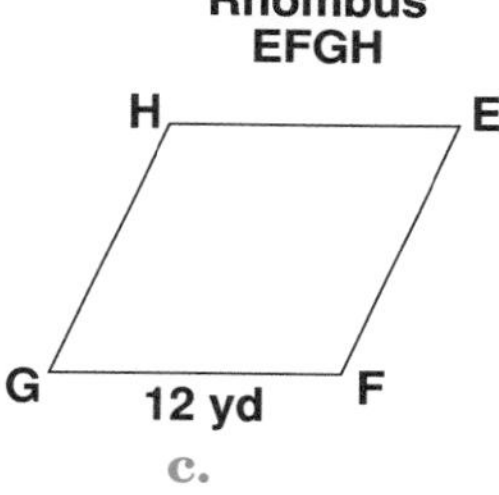

c.

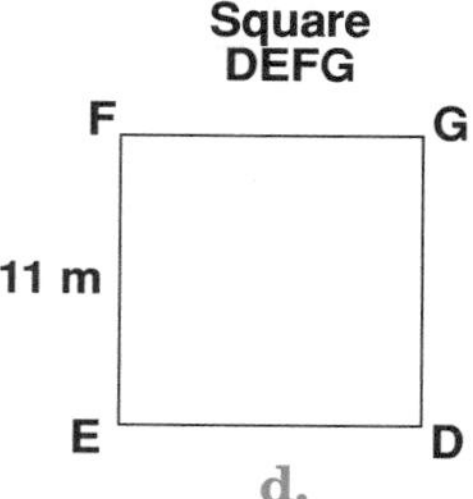

d.

3. In each quadrilateral, find the value of x.

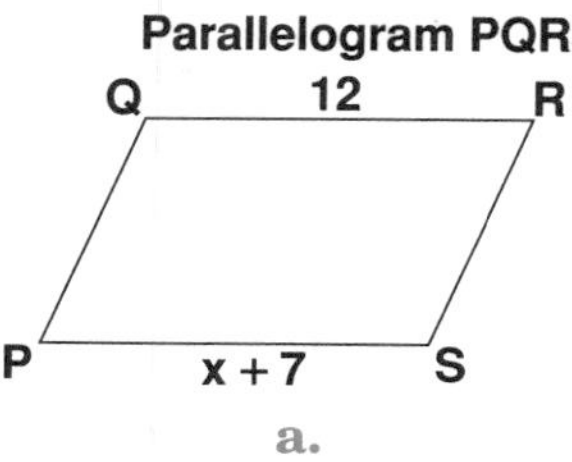

a.

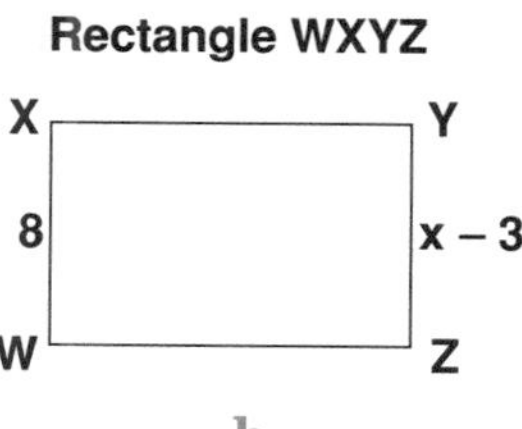

b.

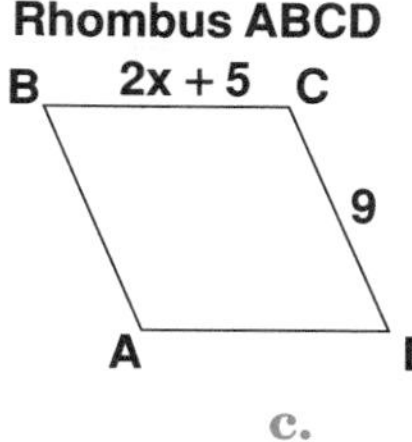

c.

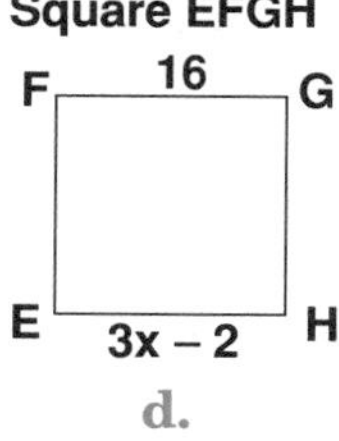

d.

4. Make a large copy of the diagram to the right. Referring to The Main Idea on page 554:

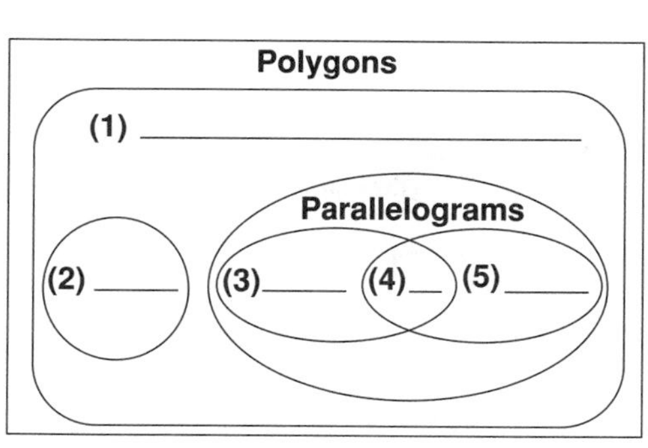

   a. Write the following names on the appropriate blank spaces:

      Trapezoids Rectangles Squares
      Rhombuses Quadrilaterals

   b. In each section of your diagram, draw a small quadrilateral of the type named in that section.

   c. Tell whether the statement is *always true*, *sometimes true*, or *never true*. Explain.

      (1) A square is a rhombus.
      (2) A rectangle is a square.
      (3) A parallelogram is a trapezoid.
      (4) A rhombus is a parallelogram.

1. Which pairs of lines (or segments) are drawn parallel and which are perpendicular?

a. L N Q P M O

b. S M R T

c. A D B E C

2. Which sides are drawn parallel and which are drawn perpendicular?

A B D C

a.

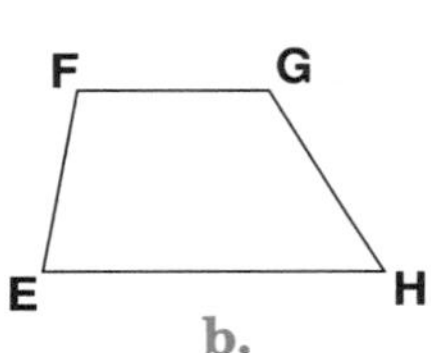

b.

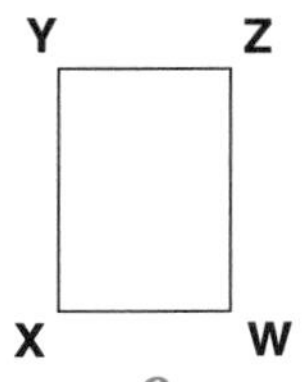

c.

d.

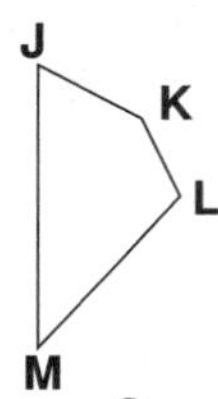

e.

3. Name the parallel sides and the perpendicular sides, and find the missing lengths.

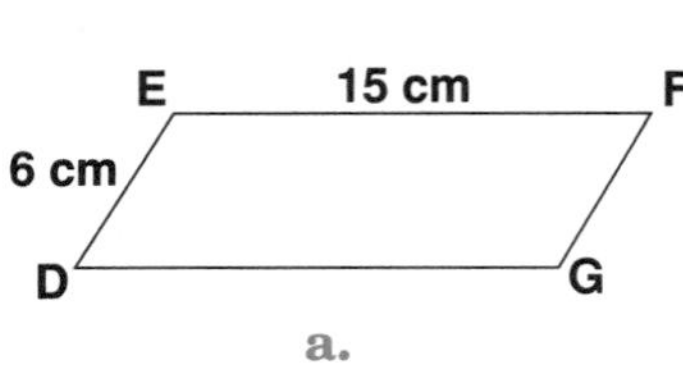

a.

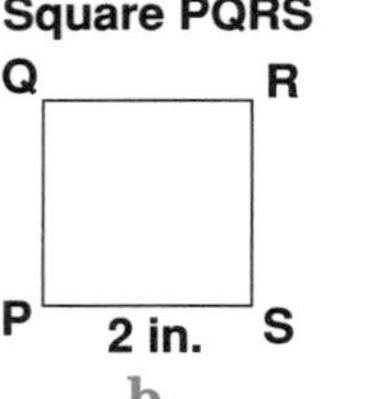

b.

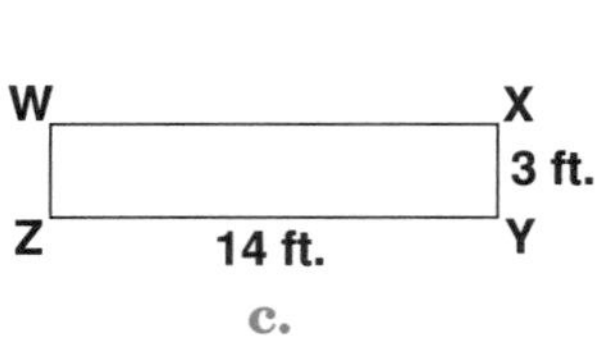

c.

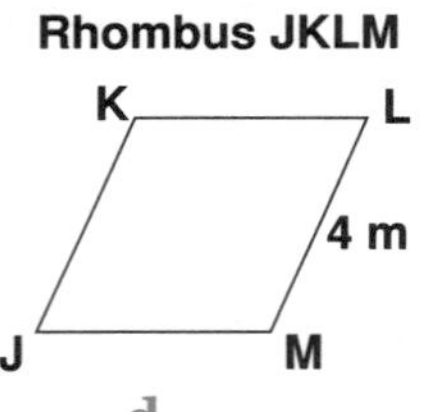

d.

4. In each quadrilateral, find the value of x.

Rectangle RSTU

S 21 T

R 4x + 1 U

a.

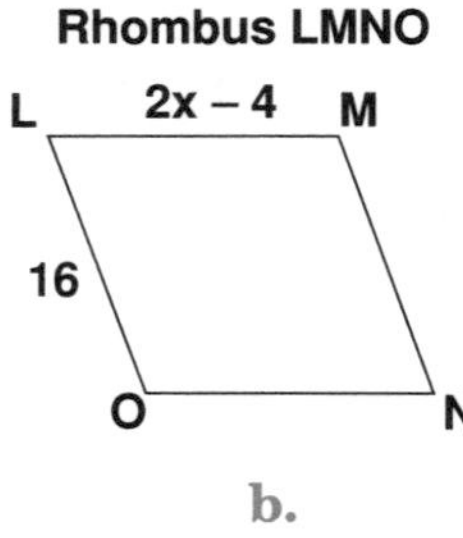

b.

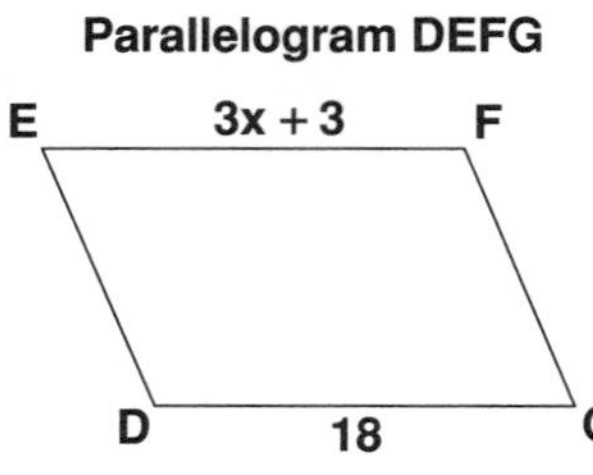

c.

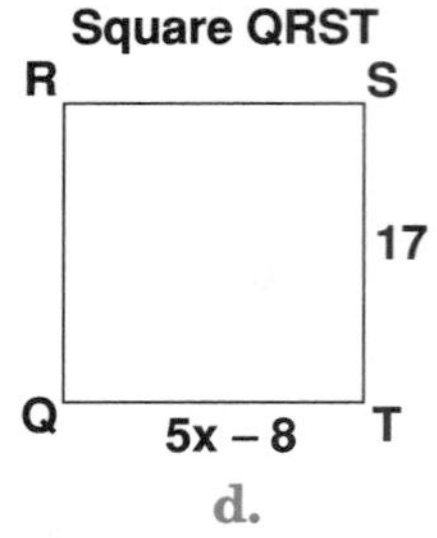

d.

5. a. Make a large copy of the diagram in Class Exercise 4 on page 555. Write the names of the special quadrilaterals in each section.
   b. Write the properties that make each type of polygon different from the polygons in the next larger section.
   c. Tell whether the statement is *always true*, *sometimes true*, or *never true*. Explain.
      (1) A quadrilateral has two parallel sides.
      (2) A square has two pairs of parallel sides.
      (3) A rectangle has two pairs of congruent sides.
      (4) A rhombus has no parallel sides.
      (5) A square has four congruent angles.
      (6) A rectangle is a polygon.

6. Suppose that you could see the set of stairs shown from all possible angles. How many edges (segments) would be parallel to $\overline{AB}$?

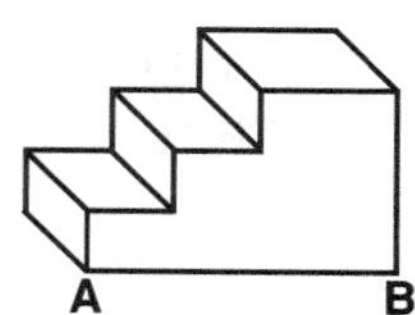

SPIRAL REVIEW EXERCISES

1. The number of degrees contained in $\frac{4}{5}$ of a right angle is
   (a) 18° (b) 36° (c) 72° (d) 144°

2. The kind of angle that has the smallest measure is
   (a) right (b) acute
   (c) obtuse (d) straight

3. Graph the solution set of $5x - 4 > 6$ on a number line.

4. A quart is equivalent to
   (a) $\frac{1}{4}$ gallon (b) 4 pints
   (c) 16 ounces (d) 64 ounces

5. 2,500 mL is equivalent to
   (a) 25 L (b) 2.5 L
   (c) 2.5 kL (d) 0.25 kL

6. If a represents a number, write an expression to represent "three more than the number."

7. How much time elapses between 11:52 P.M. and 6:12 A.M.?

8. If $A = \frac{1}{2}bh$, find the value of A when $b = 10$ and $h = 7$.

9. Find the difference between the mean and the median of the numbers 51.17, 52.03, 52.78, 54.63, and 56.33.

10. Solve: $3w - 4 = 7$

11. Multiply: 0.9×0.02

UNIT 17-5 The Pythagorean Relationship; Special Right Triangles

THE PYTHAGOREAN RELATIONSHIP

THE MAIN IDEA

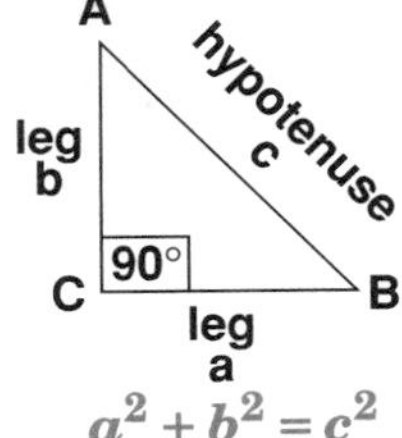

1. The ***hypotenuse*** of a right triangle is the side opposite the right angle. The hypotenuse is longer than either of the two ***legs*** (sides).
2. The Greek mathematician Pythagoras showed that, for the measures of the sides of any right triangle:

$$(\textit{leg})^2 + (\textit{leg})^2 = (\textit{hypotenuse})^2 \qquad a^2 + b^2 = c^2$$

3. A triangle is a right triangle if the measures of its three sides satisfy the Pythagorean relationship.

EXAMPLE 1 Find the length of the hypotenuse of a right triangle whose other two sides measure 6 mm and 8 mm.

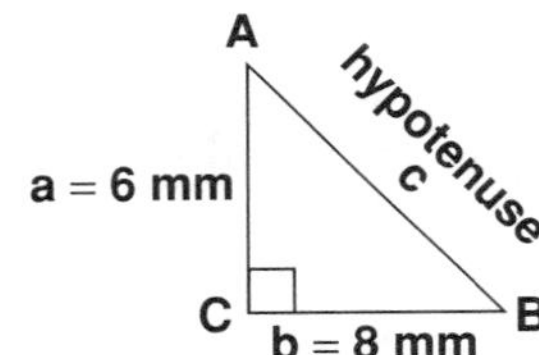

$a^2 + b^2 = c^2$	Pythagorean relation
$6^2 + 8^2 = c^2$	$a = 6$, $b = 8$
$36 + 64 = c^2$	Evaluate the squares.
$100 = c^2$	Combine.
$10 = c$	Take the square root.

Answer: The hypotenuse measures 10 mm.

EXAMPLE 2 Find the value of b in the right triangle shown.

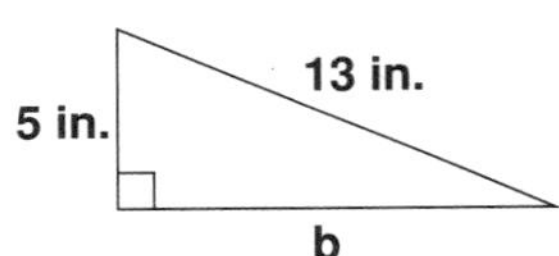

$a^2 + b^2 = c^2$	Pythagorean relation
$5^2 + b^2 = 13^2$	$a = 5$, $c = 13$
$25 + b^2 = 169$	Evaluate the squares.
$b^2 = 144$	Subtract 25.
$b = 12$	Take the square root.

Answer: $b = 12$ in.

EXAMPLE 3 Tell whether a triangle that has sides that measure 6 inches, 5 inches, and 9 inches is a right triangle.

See if the measures satisfy the Pythagorean relationship. $a^2 + b^2 = c^2$

Since 9 is the length of the longest side, substitute 9 for c. $6^2 + 5^2 \stackrel{?}{=} 9^2$

Evaluate on each side of the equals sign. $36 + 25 \stackrel{?}{=} 81$
$61 \neq 81$

Answer: Since the triangle does not satisfy the Pythagorean relationship, the triangle is not a right triangle.

EXAMPLE 4 The entrance to a building is 3 feet above ground level. A ramp 20 feet long is to be constructed so that people in wheelchairs may enter the building. How far from the base of the building (to the nearest tenth of a foot) must the base of the ramp be?

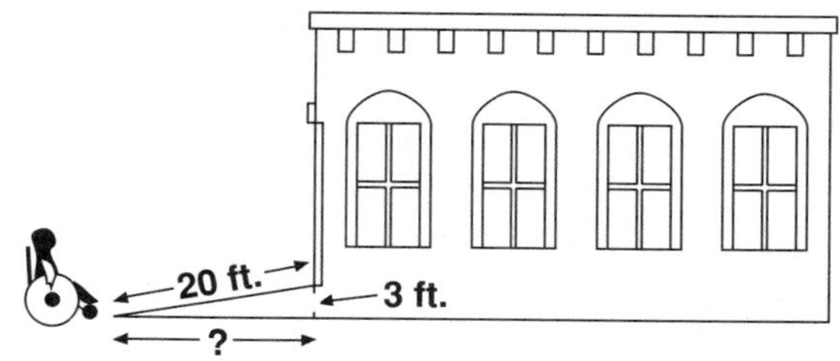

$a^2 + b^2 = c^2$
$3^2 + b^2 = 20^2$
$9 + b^2 = 400$
$b^2 = 391$

Key Sequence	***Display***
391 [√]	19.773719

Answer: The base of the ramp must be 19.8' from the base of the building.

CLASS EXERCISES

1. Find the length of the hypotenuse of a right triangle whose two legs measure:
a. 3 in. and 4 in. **b.** 5 cm and 12 cm **c.** 8 yd. and 15 yd. **d.** 12 m and 16 m

2. Find each missing length.

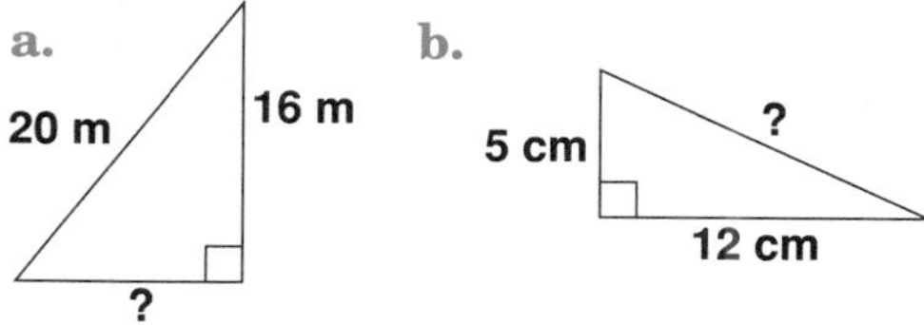

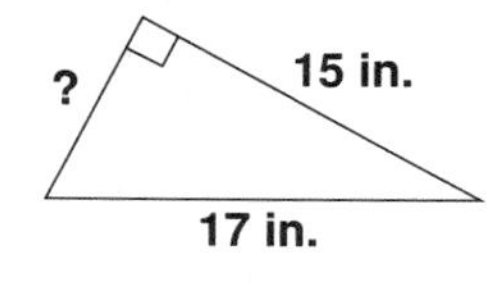

3. Tell whether each set of measurements can be the lengths of the sides of a right triangle.
a. 5 in., 12 in., 13 in. **b.** 3 ft., 4 ft., 6 ft. **c.** 9 cm, 12 cm, 15 cm **d.** 8 m, 15 m, 20 m

4. A ladder 25 feet long leans against a wall. If the foot of the ladder is 15 feet from the wall, how far up the wall does the ladder reach?

5. For right triangle ABC, with c the hypotenuse, find to the nearest tenth:
a. c if $a = 19$ and $b = 15$
b. a if $b = 10$ and $c = 16$

SPECIAL RIGHT TRIANGLES

THE MAIN IDEA

1. A right triangle that is also isosceles is often called a 45°–45°–90° triangle.

   Every 45°–45°–90° triangle has the following properties:

   a. The legs are congruent.

   b. The length of the hypotenuse is equal to the length of a leg multiplied by $\sqrt{2}$.

   hypotenuse $= \text{leg} \times \sqrt{2}$

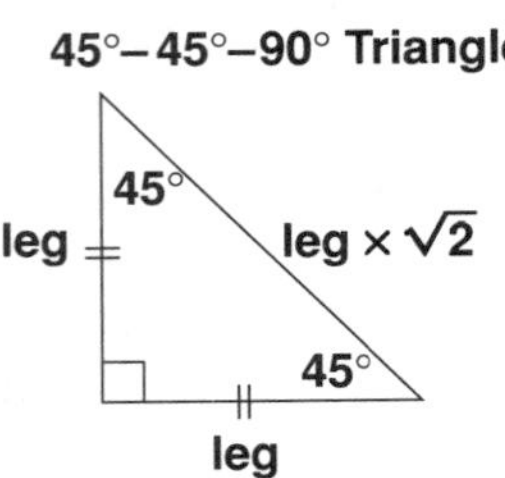

2. Another special right triangle is the 30°–60°–90° triangle.

   Every 30°–60°–90° triangle has the following properties.

   a. short leg $= \frac{1}{2}$ hypotenuse

   b. long leg $= \text{short leg} \times \sqrt{3}$

   $= (\frac{1}{2}\text{hypotenuse}) \times \sqrt{3}$

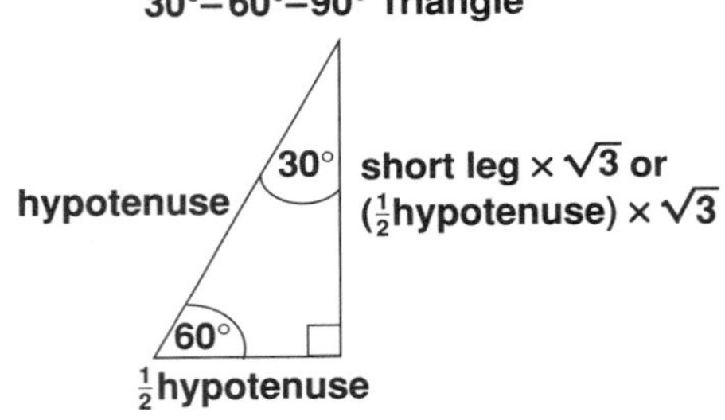

EXAMPLE 5 In isosceles right triangle ABC, angle C is a right angle and $AC = 9$ cm.

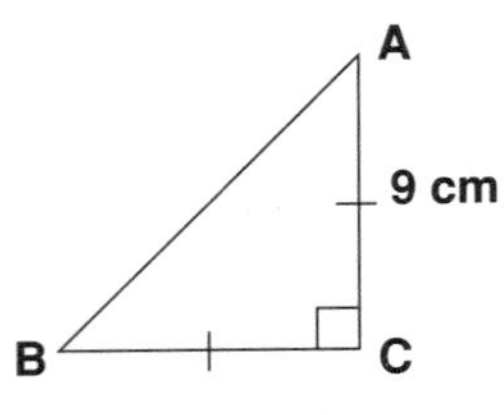

a. Find m$\angle A$.

An isosceles right triangle is a 45°–45°–90° triangle.

Answer: m$\angle A = 45°$

b. Find CB.

The legs of an isosceles triangle are equal in length.

Answer: $CB = 9$ cm

c. Find the length of $\overline{AB}$ to the nearest hundredth of a cm.

In a 45°–45°–90° triangle:
hypotenuse $= \text{leg} \times \sqrt{2}$
$= 9 \times \sqrt{2}$, or $9\sqrt{2}$

Key Sequence	***Display***
2 [√] [×] 9 [=]	12.727921

Answer: The length of $\overline{AB}$ is 12.73 cm.

EXAMPLE 6 In the diagram, $AB = 12$ units.

a. Find BC.

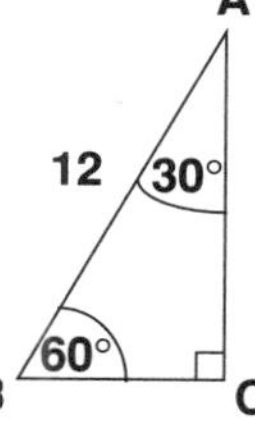

In a 30°–60°–90° triangle:

short leg = $\frac{1}{2}$hypotenuse

$BC = \frac{1}{2}(12)$

$BC = 6$ *Ans.*

b. Find AC to the nearest thousandth.

In a 30°–60°–90° triangle:

long leg = $\frac{1}{2}$hypotenuse $\times \sqrt{3}$

$AC = \frac{1}{2}(12)\sqrt{3}$

$AC = 6\sqrt{3}$

Key Sequence	***Display***
3 [√] [×] 6 [=]	10.392304

Answer: The length of $\overline{AC}$ is 10.392.

CLASS EXERCISES

1. In isosceles right triangle DEF, $\angle E$ is the right angle and the length of $\overline{DE}$ is 7 ft.

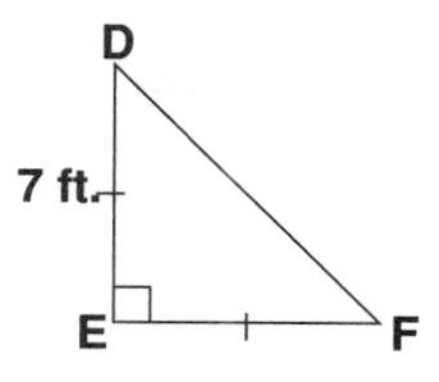

a. Find EF. **b.** Find DF.

c. Use a calculator to approximate DF correct to the nearest tenth of a foot.

2. In right triangle RST, m$\angle R = 30°$ and m$\angle T = 60°$. If $RT = 26$ in., find:

a. ST **b.** RS

c. Use a calculator to approximate the length of $\overline{RS}$ correct to the nearest whole inch.

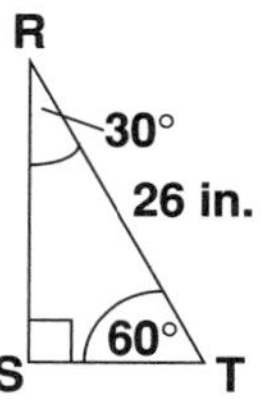

3. Find the value of x in each of the diagrams.

a.

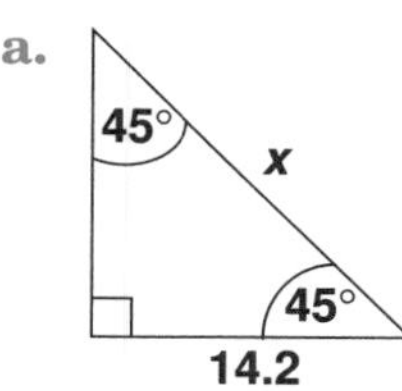

b.

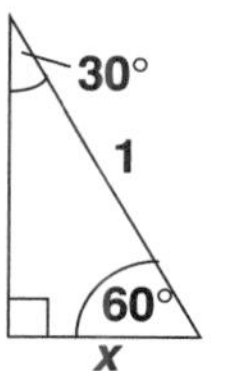

c.

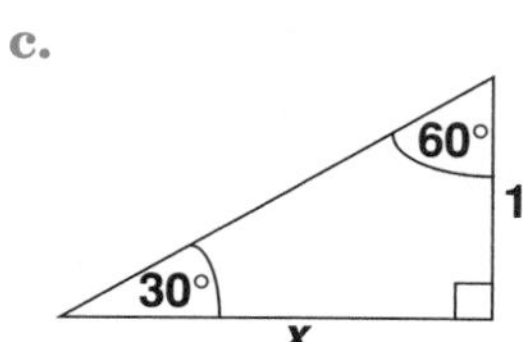

HOMEWORK EXERCISES

1. Find the length of the hypotenuse of a right triangle whose two legs measure:
   a. 4 cm and 3 cm **b.** 9 in. and 12 in. **c.** 16 m and 30 m **d.** 10 yd. and 24 yd.

2. Find each missing length.

   a. 17 m, ?, 8 m

   b. 6 m, ?, 8 m

   c. 12 cm, 20 cm, ?

   d. ?, 12 ft., 13 ft.

3. Tell whether each set of measurements can be the lengths of the sides of a right triangle.
   a. 2 cm, 3 cm, 4 cm **b.** 6 m, 8 m, 10 m **c.** 30 ft., 40 ft., 50 ft. **d.** 6 yd., 9 yd., 12 yd.

4. Find the length of the cable.

5. Find the distance between City A and City B.

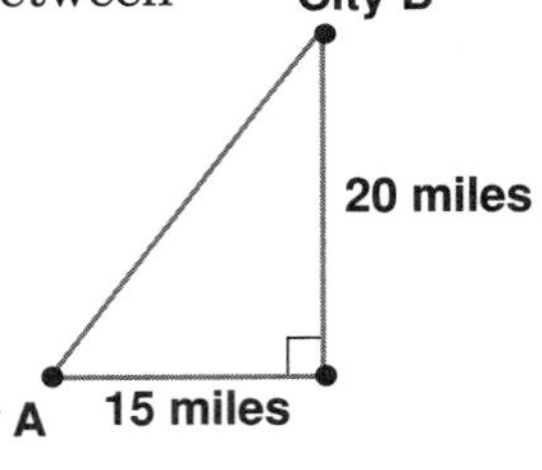

6. Use a calculator to find each missing length correct to the nearest tenth.

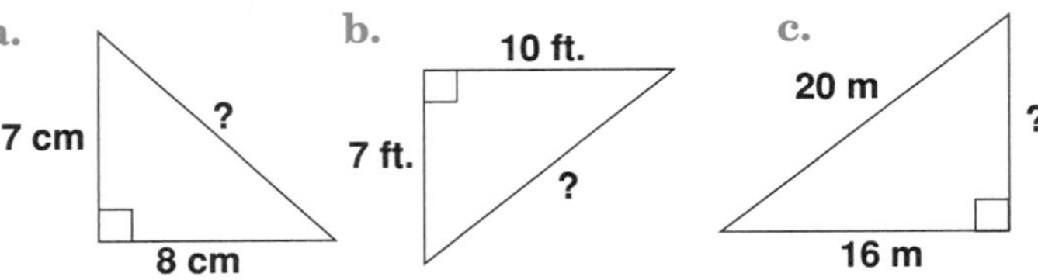

   d. 5 in., ?, 15 in.

7. In isosceles right triangle LMN, $\angle M$ is the right angle and $\overline{LM}$ is 100 meters long.

   a. Find $m\angle L$. **b.** Find $m\angle N$. **c.** Find LN.

   d. Use a calculator to approximate LN correct to the nearest hundredth of a meter.

N, M, L, 100 m

8. In triangle ABC, $\angle C$ is a right angle, $m\angle B = 30°$, and $AB = 36$ in.

   a. Find $m\angle A$. **b.** Find AC. **c.** Find BC.

   d. Use a calculator to approximate BC correct to the nearest tenth of an inch.

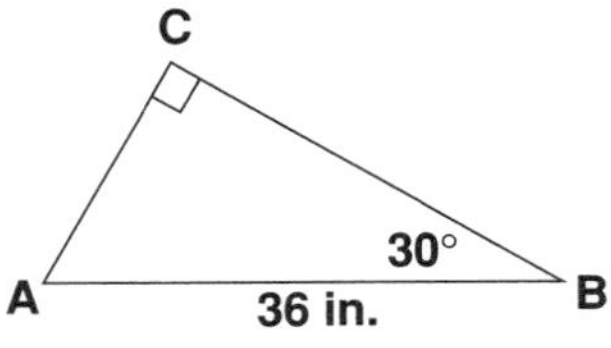

9. Find the value of x in each diagram.

a.

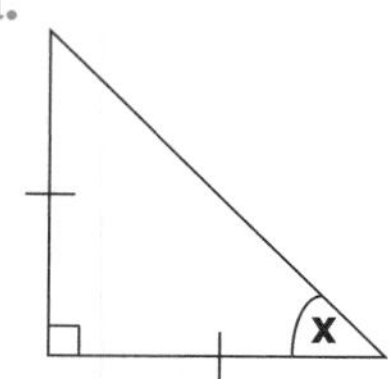

b.

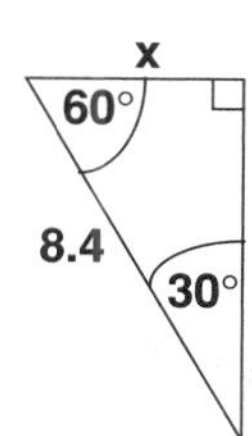

c.

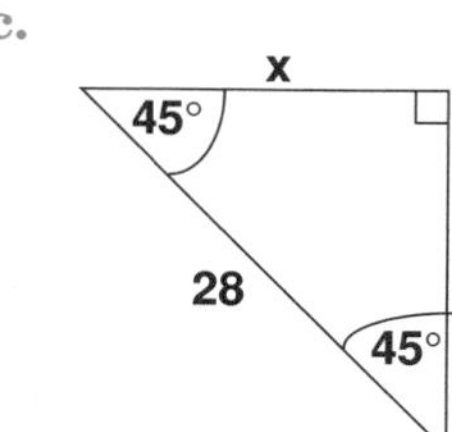

d.

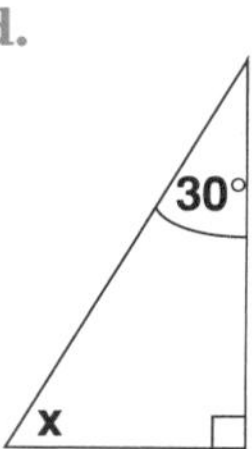

SPIRAL REVIEW EXERCISES

1. If the measures of two angles of a triangle are 50° and 80°, find the measure of the third angle.

2. If the measure of $\angle A$ is 89°, then $\angle A$ is
   (a) acute (b) right
   (c) obtuse (d) straight

3. Graph $-2 \leq x < 5$ on a number line.

4. The probability that the spinner will stop on a space numbered with a prime is

   (a) $\frac{1}{4}$ (b) $\frac{1}{2}$
   (c) $\frac{3}{4}$ (d) 1

5. If $A = bh$, the value of A when $b = 10$ and $h = 8$ is
   (a) 18 (b) 108 (c) 80 (d) $\frac{5}{4}$

6. Solve: $5x + 12 = 27$

7. The value of $3x - 7$ when $x = 9$ is
   (a) 6 (b) 32 (c) 36 (d) 20

8. 35 meters is equivalent to
   (a) 35 cm (b) 3.5 cm
   (c) 350 cm (d) 3,500 cm

9. The distance from Lawrence to Holtsville on a map is 3.6 inches, and the distance from Holtsville to Newtown is 1.8 inches. If the scale is one inch = 85 miles, what is the actual distance from Lawrence to Newtown by way of Holtsville?

10. After multiplying the number of rows by the number of seats per row in a theater and then subtracting this product from the number of tickets sold for one performance, Mel read the display [– 76.] on his calculator. This means:
    (a) There will be 76 empty seats.
    (b) There will be 76 ticket holders without seats.
    (c) There were 76 tickets sold.
    (d) The theater has 76 seats.

11. Mr. Roland's restaurant tab came to \$89.58. If he wants to add a 20% tip, then his total bill should be about
    (a) \$90.00 (b) \$99.00 (c) \$108.00 (d) \$118.00

12. This year, 260 students participated in the charity fund-raising drive. If this number is 125% of the number of students who participated last year, how many participated last year?

13. Evaluate: $\frac{7}{12} + \frac{3}{4} \times \frac{5}{6}$

UNIT 18-1 Similar Figures: Congruent Figures

SIMILAR FIGURES

THE MAIN IDEA

1. ***Similar figures*** are figures that have the same shape. The symbol ~ means "is similar to."
2. Similar figures have the following properties:
   a. A pair of sides in one figure compares in the same way as the corresponding pair of sides in the other figure. The ratios are equal.
   b. Corresponding angles are congruent (equal in measure).
3. To find an unknown length in a pair of similar figures, write and solve a proportion.

EXAMPLE 1 Quadrilateral *PQRS* is similar to quadrilateral *ABCD*.

a. Find the value of z.

Write a proportion comparing lengths of corresponding sides.

$$\frac{PQ}{PS} \longrightarrow \frac{6}{10} = \frac{9}{z} \longleftarrow \frac{AB}{AD}$$

$$6z = 90 \longleftarrow \text{Set the cross products equal.}$$

$$\frac{1}{6} \cdot 6z = \frac{1}{6} \cdot 90$$

$$z = 15 \quad \textit{Ans.}$$

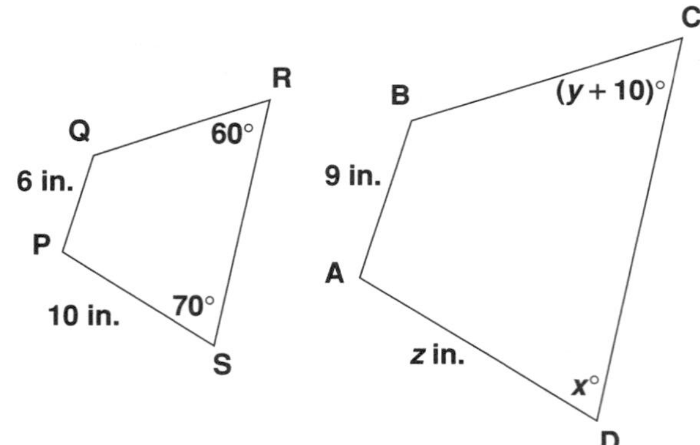

b. Find the values of x and y.

Because the quadrilaterals are similar, corresponding angles are equal in measure.

$$m\angle D = m\angle S = 70°$$
$$x = 70 \quad \textit{Ans.}$$

$$m\angle C = m\angle R$$
$$y + 10 = 60$$
$$y + 10 + (-10) = 60 + (-10)$$
$$y = 50 \quad \textit{Ans.}$$

EXAMPLE 2 Triangle *ABC* is similar to triangle *XYZ*.

Which of the following sets of lengths could be the lengths of $\overline{XY}$, $\overline{YZ}$, and $\overline{ZX}$?

(a) 10 in., 18 in., 20 in.
(b) 10 in., 18 in., 21 in.
(c) 10 in., 18 in., 22 in.
(d) 10 in., 18 in., 23 in.

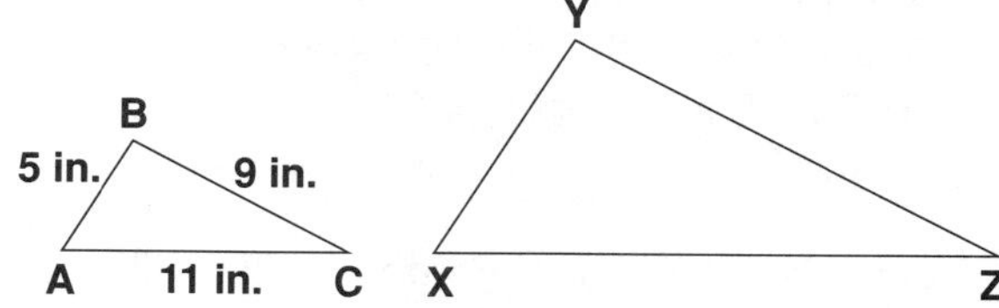

THINKING ABOUT THE PROBLEM

From the choices, you see that the measures of the two shorter sides of triangle *XYZ* are repeated: 10 and 18 appear in all the choices. Thus, the correct answer depends on the measure of the longest side.

From the original triangle, form a ratio that involves the longest side: $\frac{5}{11}$

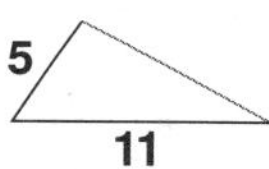

For each choice, consider the corresponding ratio to see which is equal to $\frac{5}{11}$: $\frac{10}{20}$ $\frac{10}{21}$ $\boxed{\frac{10}{22}}$ $\frac{10}{23}$

Answer: (c)

CLASS EXERCISES

1. Find the value of n in each pair of similar figures.

a.

6 mm
n mm
42 mm
14 mm

b.

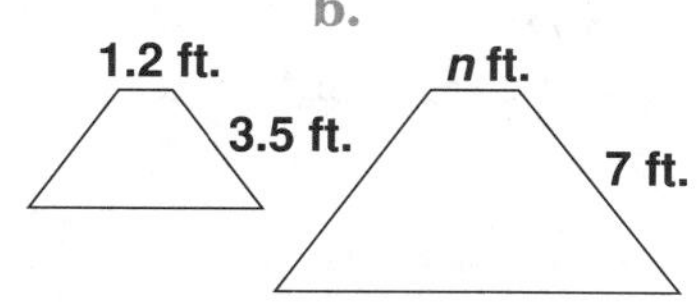

c.

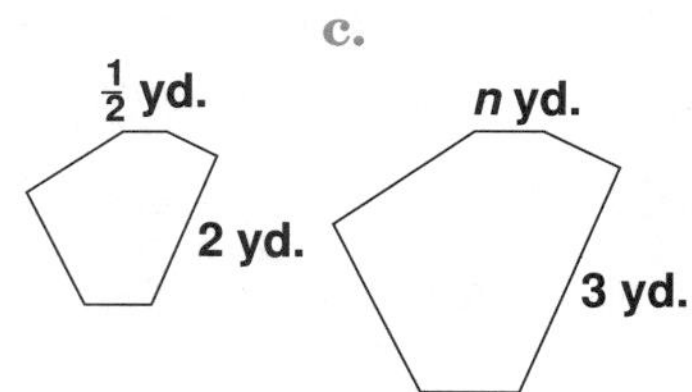

2. Which of the following sets of lengths could be the lengths of the sides of a triangle that is similar to triangle *ABC*?

(a) 2 cm, 3, cm, 4 cm (b) 2 cm, 3 cm, 3 cm
(c) 2 cm, 3 cm, 2 cm (d) 1 cm, 3 cm, 3 cm

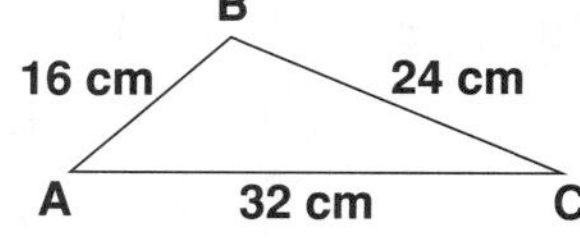

3. Rectangle *JKLM* is similar to rectangle *WXYZ*. Which of the following could be the length and width of rectangle *WXYZ*?

(a) 7 m, 2 m (b) 0.7 m, 2 m
(c) 0.07 m, 2 m (d) 0.007 m, 2 m

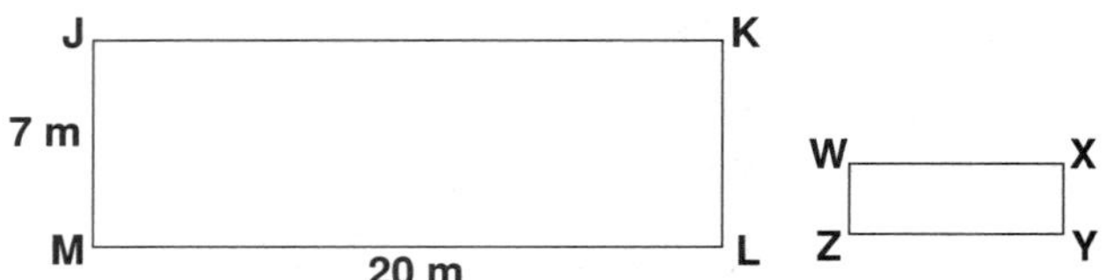

4. Pentagon *ABCDE* ~ Pentagon *PQRST*.

Find the value of: **a.** x **b.** y **c.** z

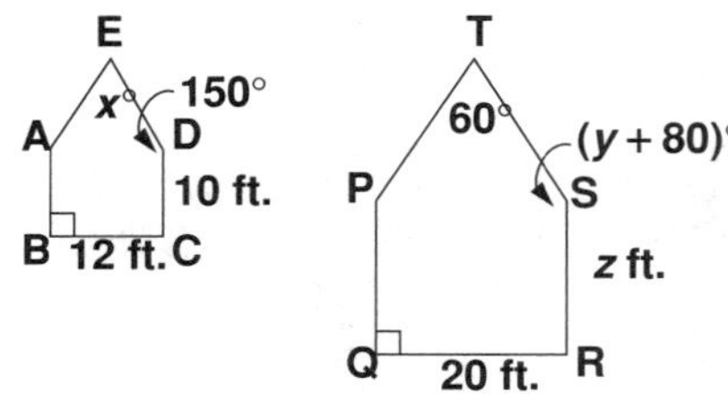

5. Triangle *ABC* ~ Triangle *DEF*.

Find the value of: **a.** x **b.** y **c.** z **d.** w

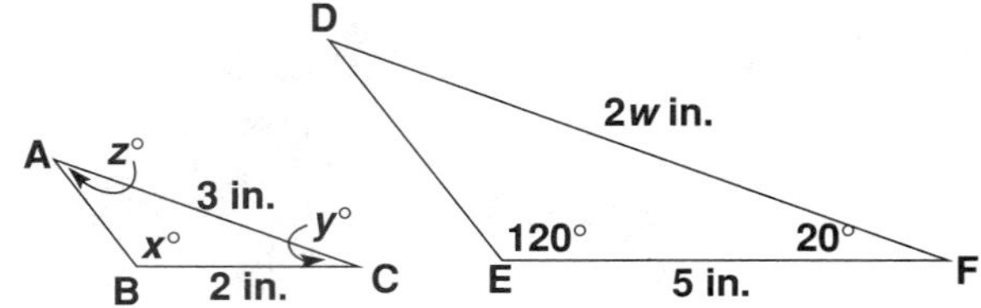

CONGRUENT FIGURES

THE MAIN IDEA

1. ***Congruent figures*** are figures that have the same size as well as the same shape. The symbol ≅ means "is congruent to."
2. Congruent figures have the following properties:
   a. Corresponding sides are congruent (equal in length).
   b. Corresponding angles are congruent (equal in measure).

EXAMPLE 3 Quadrilateral *ABCD* ≅ Quadrilateral *WXYZ*.

Find: **a.** AD **b.** WX **c.** $m\angle D$ **d.** $m\angle X$

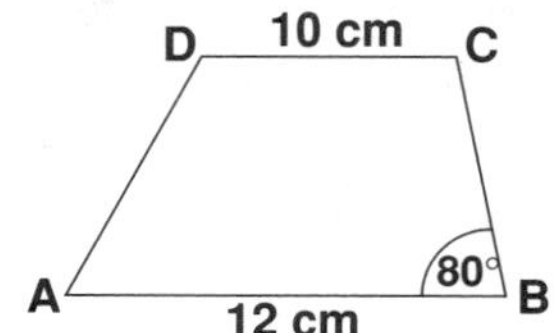

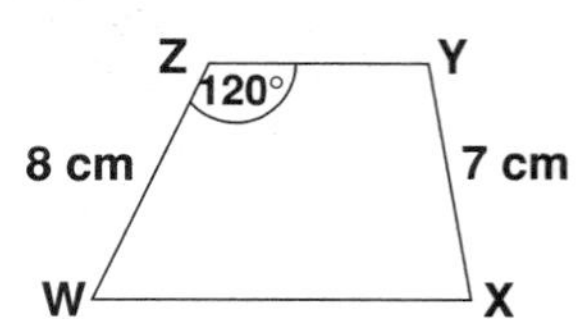

a. $\overline{AD} \cong \overline{WZ},\ AD = 8$ cm

b. $\overline{WX} \cong \overline{AB},\ WX = 12$ cm

c. $\angle D \cong \angle Z,\ m\angle D = 120°$

d. $\angle X \cong \angle B,\ m\angle X = 80°$

EXAMPLE 4 Triangle $SQR \cong$ Triangle HJK.

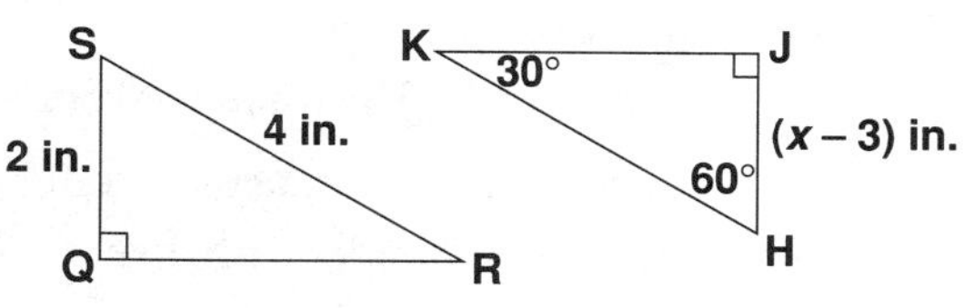

a. Find the measure of $\angle R$.

$\angle R$ is opposite $\overline{SQ}$, the shortest side of $\triangle SQR$. The angle in $\triangle HJK$ that corresponds to $\angle R$ is also opposite the shortest side, $\overline{HJ}$.

Therefore, $\angle K$ corresponds to $\angle R$, and their measures are equal. *Answer:* $m\angle R = 30°$

b. Find the value of x.

$\overline{SQ}$ and $\overline{HJ}$ are corresponding sides because they are both opposite the 30° angles.

$$SQ = HJ$$
$$x - 3 = 2$$
$$x - 3 + 3 = 2 + 3$$
$$x = 5 \quad \textit{Ans.}$$

CLASS EXERCISES

1. Tell why the figures in each pair are not congruent.

a.

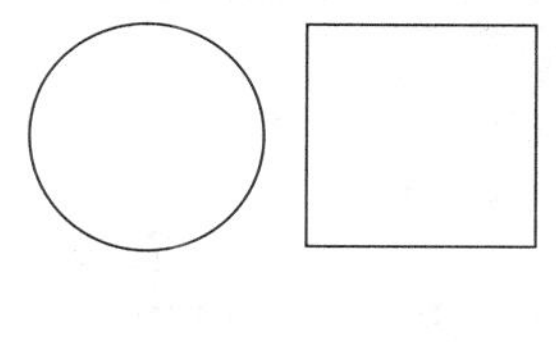

b.

2. If $\triangle ABC \cong \triangle XYZ$, list all the pairs of congruent sides and congruent angles.

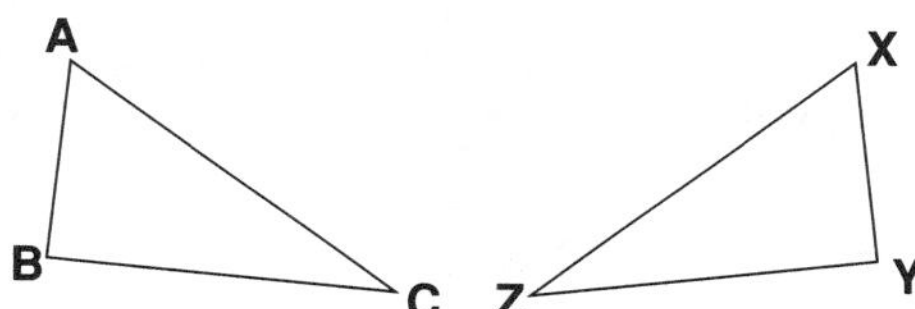

3. **a.** If $\triangle DEF \cong \triangle RST$, find the measure of each indicated angle or line segment.

   (1) $\overline{DE}$ **(2)** $\angle F$ **(3)** $\angle R$

   b. Is $\triangle DEF \sim \triangle RST$? Explain.

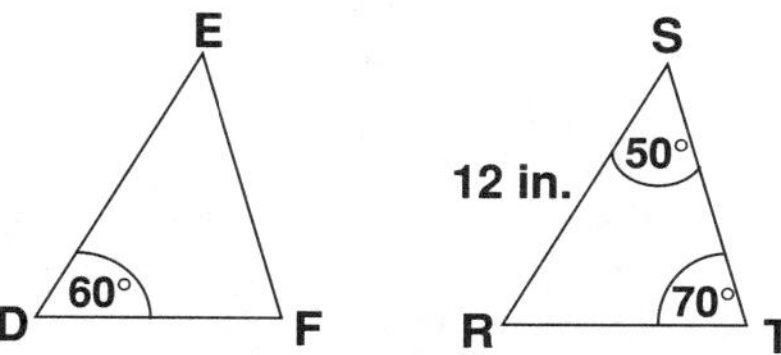

4. Parallelogram $ABCD \cong$ Parallelogram $EFGH$.
   Find the value of: **a.** x **b.** y **c.** z

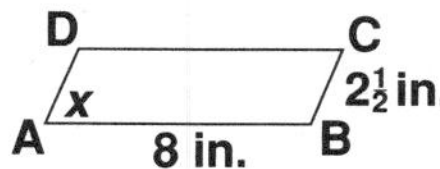

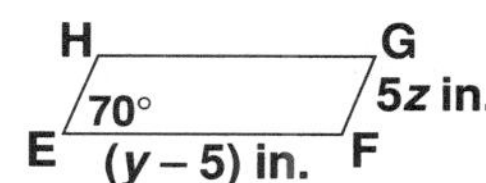

5. Triangle $JKM \cong$ Triangle NPQ.
   Find the value of: **a.** x **b.** y **c.** z

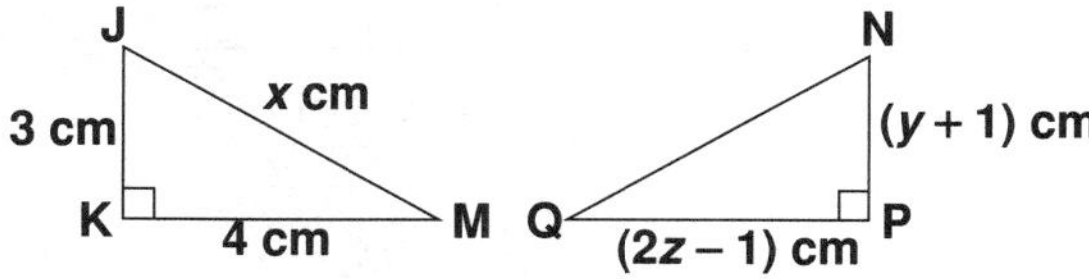

6. Two squares are
(a) never similar (b) always congruent (c) always similar (d) never congruent

7. An obtuse triangle and an acute triangle are
(a) sometimes similar (b) always similar (c) never similar (d) sometimes congruent

HOMEWORK EXERCISES

1. Find the value of n in each pair of similar figures.

a.

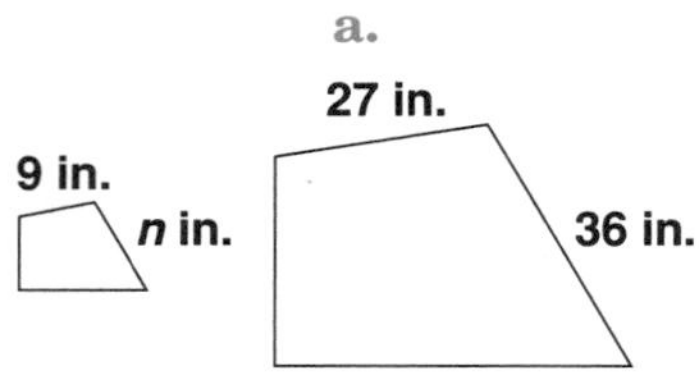

b.

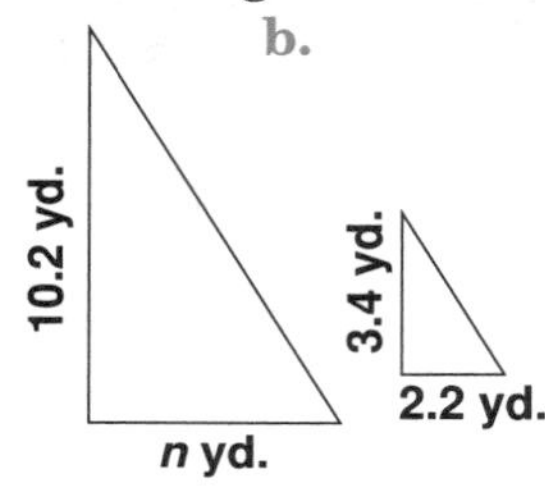

c.

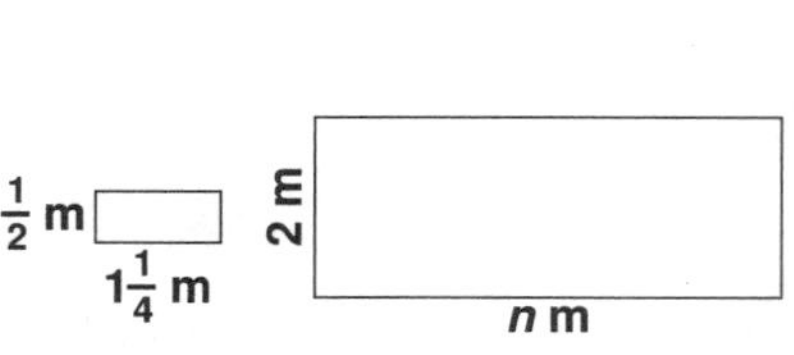

2. Which of the following sets of lengths could be the lengths of the sides of a triangle that is similar to $\triangle ABC$?
(a) 6 in., 10 in., 12.5 in. (b) 7 in., 10 in., 12.5 in.
(c) 6.5 in., 10 in., 12.5 in. (d) 7.5 in., 10 in., 12.5 in.

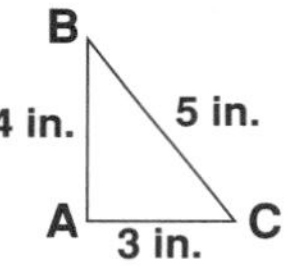

3. For each pair of figures, decide if they appear to be congruent. Explain your reasoning.

a.

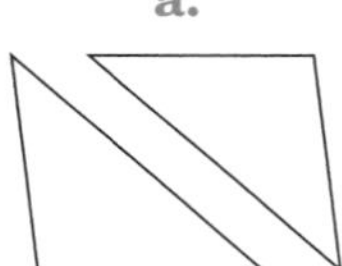

b.

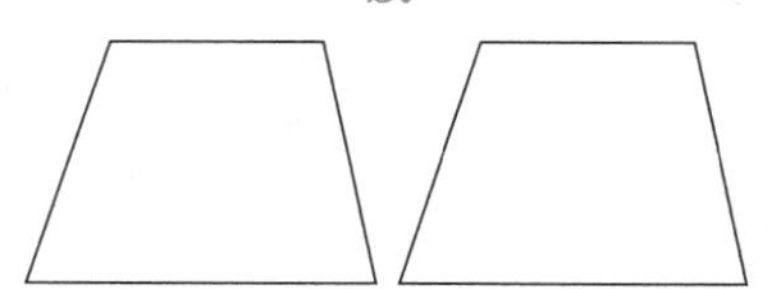

c.

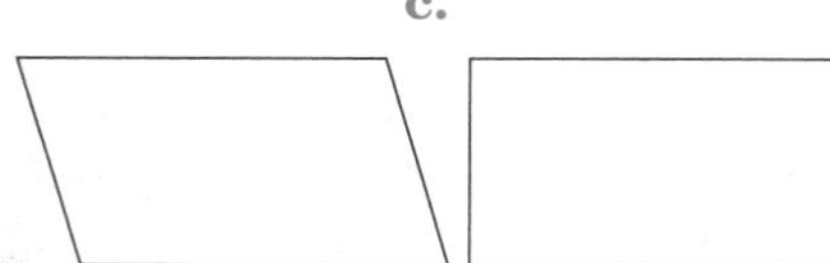

4. If $\triangle DEF \cong \triangle GHI$, list all pairs of congruent sides and congruent angles.

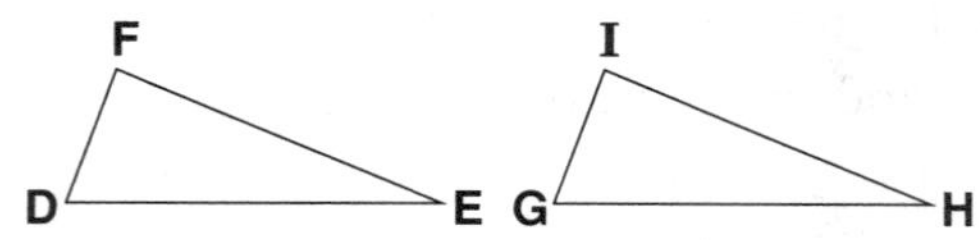

5. If $\triangle RST \cong \triangle XYZ$, find:
a. TS **b.** $m\angle X$ **c.** XZ

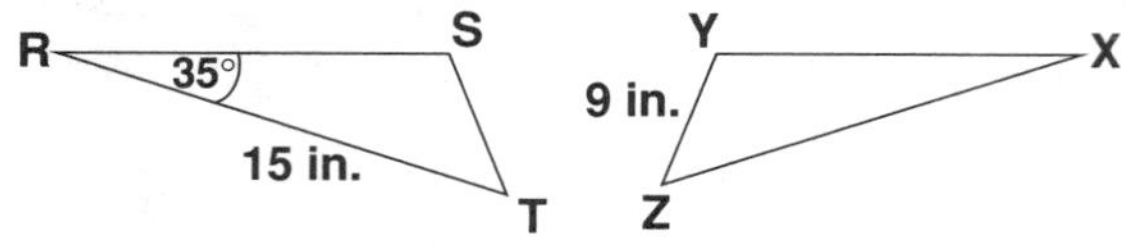

6. Given:

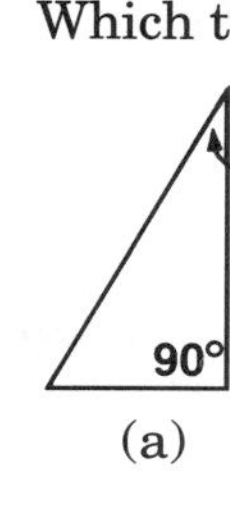

Which triangle is similar to triangle ABC?

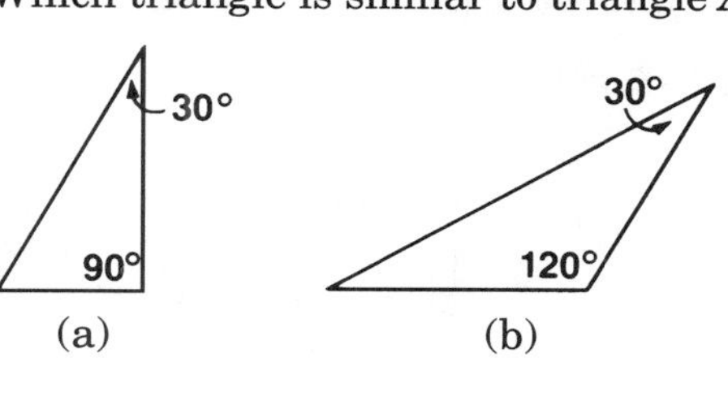

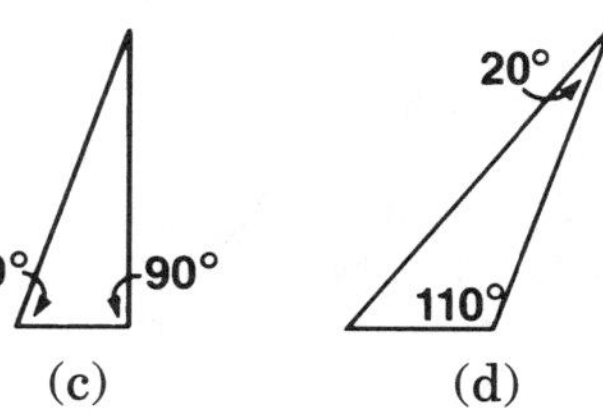

7. If $\triangle PQR \cong \triangle WXY$, find $m\angle P$.

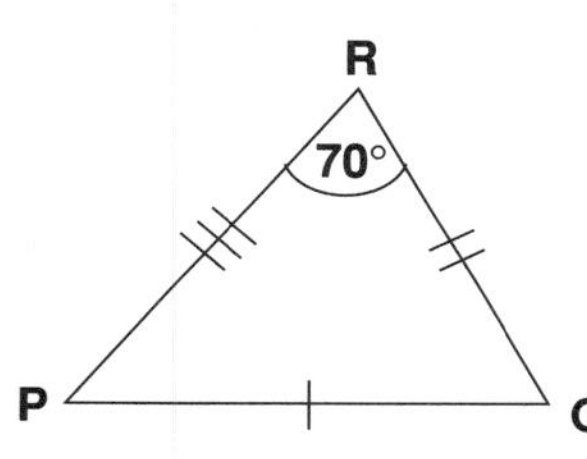

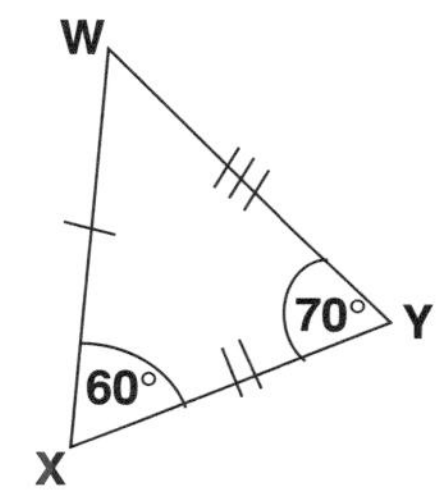

8. If $\triangle ABC \cong \triangle MNP$, find the value of:

a. x **b.** y **c.** z

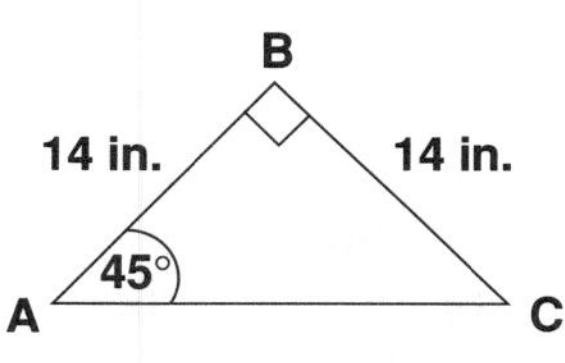

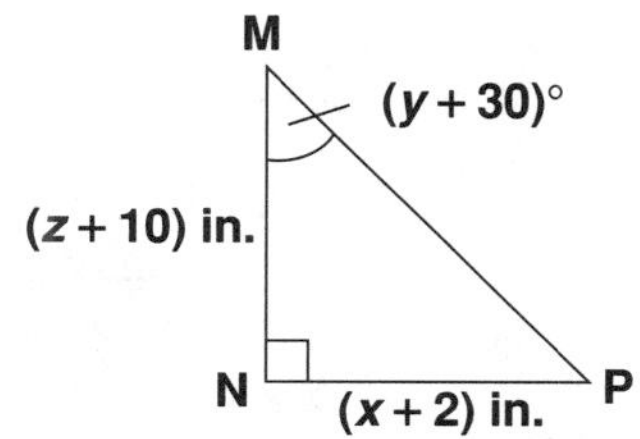

9. If the two triangles are similar, which of the following proportions can *not* be used to solve for n?

(a) $\frac{8}{12.5} = \frac{5}{n}$ (b) $\frac{8}{5} = \frac{n}{12.5}$

(c) $\frac{8}{5} = \frac{12.5}{n}$ (d) $\frac{n}{12.5} = \frac{5}{8}$

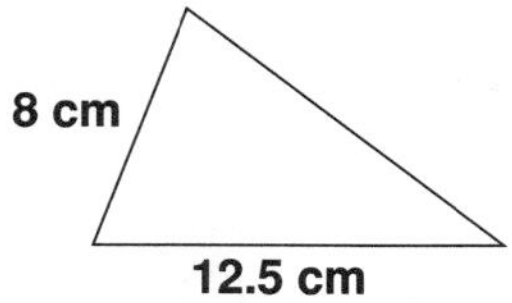

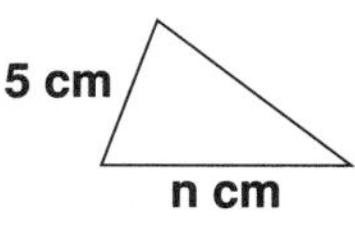

10. Tell if each statement is *sometimes true*, *always true*, or *never true*. Explain your answer.

a. Similar figures are congruent.

b. Congruent figures are similar.

c. Congruent figures have the same number of sides.

d. Pentagons are similar.

e. Right triangles are similar.

SPIRAL REVIEW EXERCISES

1. The shortest distance between the cities of Winston and Clark is
   (a) 125 miles
   (b) 130 miles
   (c) 150 miles
   (d) 200 miles

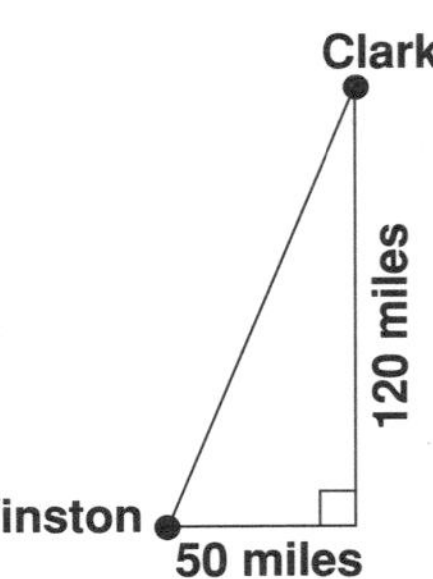

2. Which key sequence on an entry-order calculator will change 54 minutes 48 seconds to hours?
   (a) 54.48 [÷] 3600 [=]
   (b) 54 [×] 60 [+] 48 [÷] 3600 [=]
   (c) 54 [+] 48 [×] 60 [÷] 3600 [=]
   (d) 54 [÷] 3600 [+] 48 [=]

3. Change $7\frac{1}{2}\%$ to a decimal.

4. The measure of ∠ABC is
   (a) 65° (b) 115° (c) 75° (d) 125°

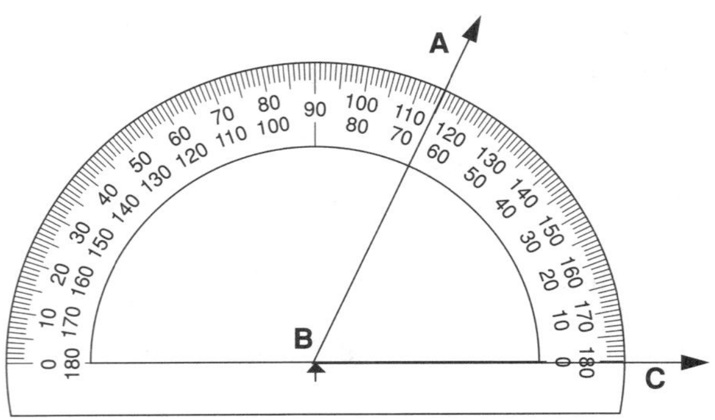

5. At 8 A.M., the temperature was $-5°F$, and at 12 noon the temperature was $20°F$. What was the increase in temperature?

6. Mr. Jackson left a 15% tip for the waiter. How much money did he leave if his bill was $30?

7. Graph $x < -3$ on a number line.

8. The measure of angle C is
   (a) 80° (b) 90°
   (c) 100° (d) 110°

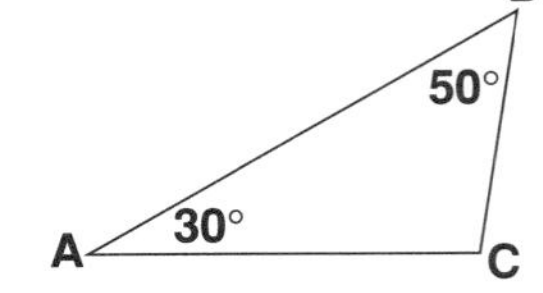

9. What number is 75% of 80?

10. Find 30% of 240.

11. Find the sum of $\frac{3}{4}$ and $\frac{7}{12}$.

12. In a poll, 1,000 adults were asked "Which is more important to your health, the food you eat, the exercise you get, or both equally?" The graph shows the results. How many more people felt that the food they eat was the most important factor than the number who felt that exercise was most important?

Factors Important to Health

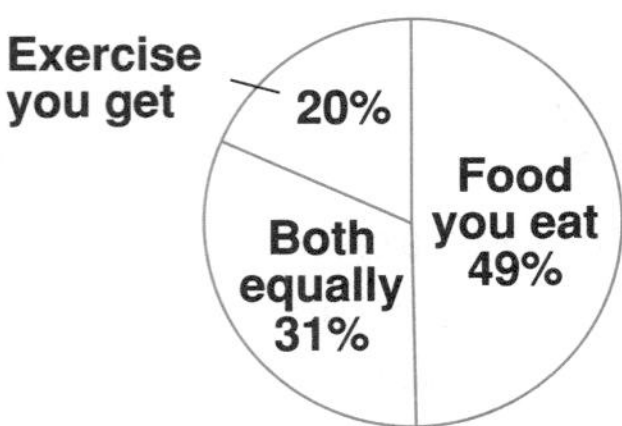

UNIT 18-2 Using Coordinates to Graph Points; Finding the Distance Between Two Points

USING COORDINATES TO GRAPH POINTS

THE MAIN IDEA

1. You can tell the location of points by using a horizontal number line and a vertical number line that cross at right angles to each other. For example, point P corresponds to –2 on the horizontal number line and +3 on the vertical number line.

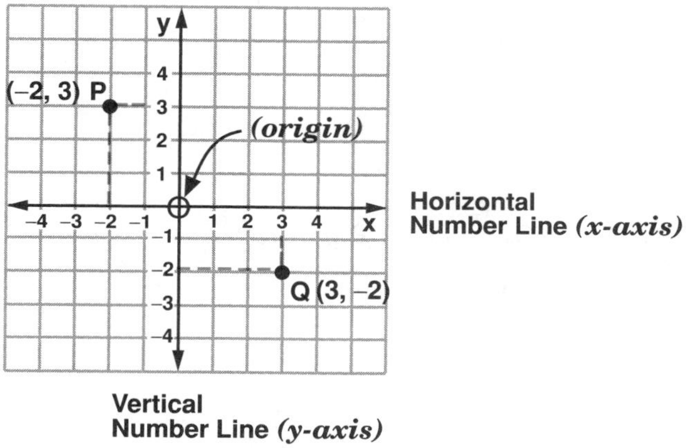

2. The numbers –2 and +3 are written as an ***ordered pair*** inside parentheses and separated by a comma. The numbers in the ordered pair (–2, 3) are called the ***coordinates*** of point P. The first number of an ordered pair gives the horizontal location and is called the ***x-coordinate***. The second number gives the vertical location and is called the ***y-coordinate***.

   Note that (3, –2) does not name the same location as (–2, 3).

3. The horizontal number line is called the ***x-axis***. The vertical number line is called the ***y-axis***. The point where the number lines cross is called the ***origin***. The coordinates of the origin are (0, 0).

4. To graph an ordered pair:
   a. Start at the origin, (0, 0).
   b. Read the x-coordinate and move to the left or to the right the number of units indicated.
   c. Read the y-coordinate and move up or down the number of units indicated.
   d. Put a dot at the location reached and label the point with a capital letter.

EXAMPLE 1 Use an ordered pair to name the location of point A.

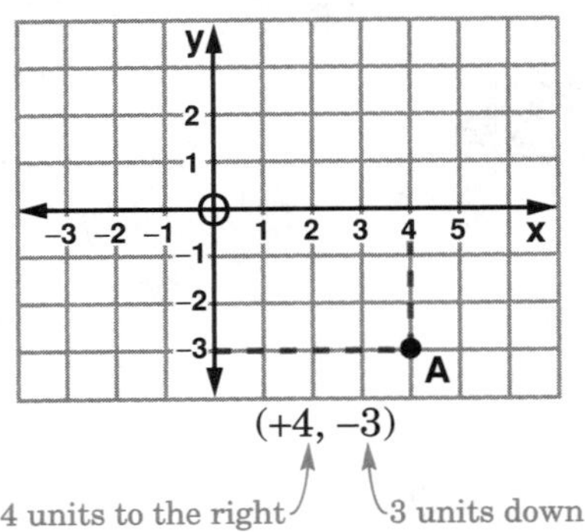

EXAMPLE 2 Graph and label point A (3, –2).

Start at (0, 0). Go 3 units to the right, 2 units down.

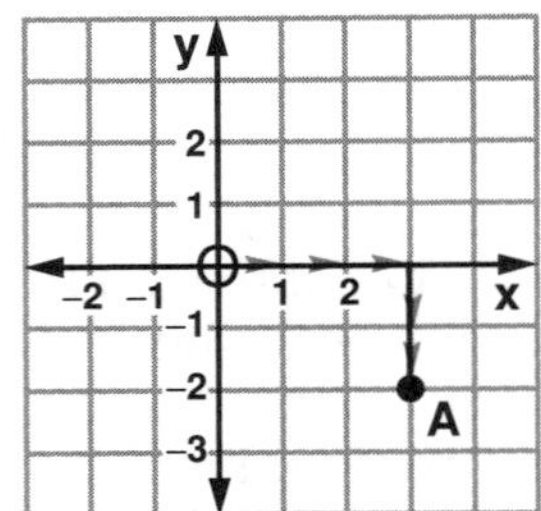

EXAMPLE 3 Use an ordered pair to tell the coordinates of each of the given points.

	Point	*Coordinates*
a.	A	(4, 1)
b.	B	(2, 0)
c.	C	(–3, 4)
d.	D	(–3, –2)
e.	E	(0, –3)
f.	F	(1, –4)

EXAMPLE 4 Name the point that has the given coordinates.

	Coordinates	*Point*
a.	(5, –3)	B
b.	(–2, 2)	E
c.	(4, 5)	D
d.	(0, 1)	C
e.	(–4, –2)	A

CLASS EXERCISES

1. Use an ordered pair to tell the coordinates of each of the given points.

   a. *A* **b.** *B* **c.** *C* **d.** *D* **e.** *E* **f.** *F*

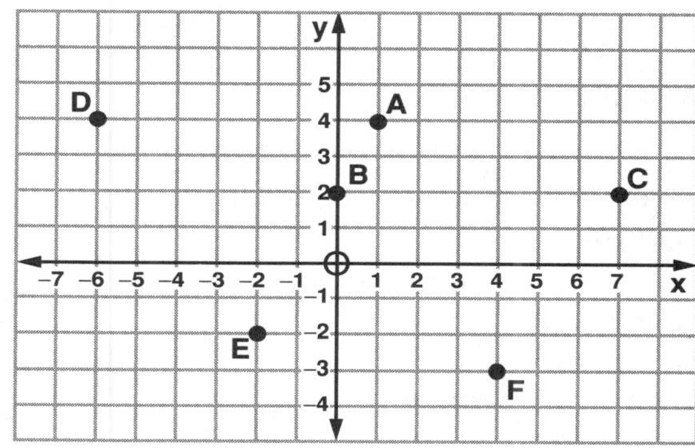

2. Tell the coordinates of each vertex of triangle *ABC*.

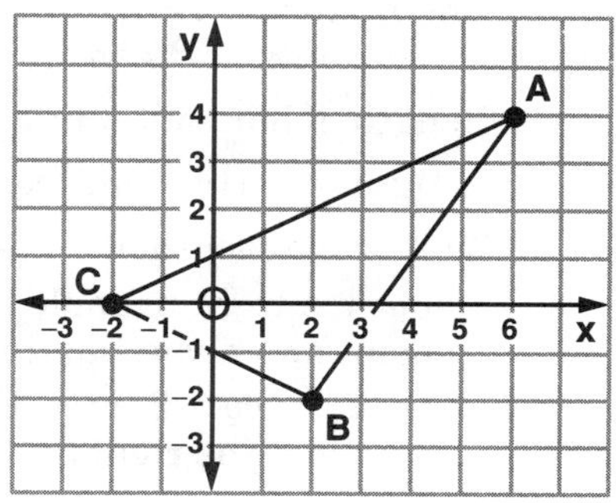

3. Name the point that has the given coordinates.

   a. (–1, 2) **b.** (–3, 0) **c.** (8, 2) **d.** (–5, –2)
   e. (2, 8) **f.** (5, –4) **g.** (0, –3) **h.** (–2 , –5)

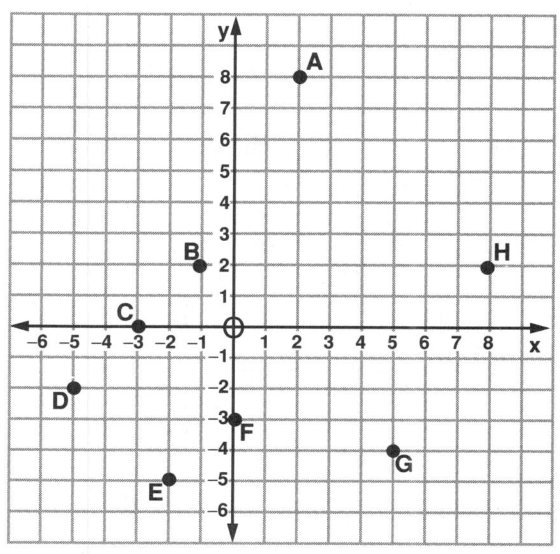

4. Graph and label the points having the given coordinates.

   a. *A*(2, 6) **b.** *B*(–3, 0) **c.** *C*(–4, 2)
   d. *D*(–6, 6) **e.** *E*(0, 2) **f.** *F*(2, 9)
   g. *G*(–7, –1) **h.** *H*(–1, –7)

5. **a.** Graph the points in each set, and join each set of points with line segments.

Set 1	*Set 2*	*Set 3*	*Set 4*
P(−2, 3)	*P*(−3, −4)	*X*(1, −2)	*X*(−2, −2)
Q(1, 3)	*Q*(0, −4)	*Y*(1, 0)	*Y*(−2, 1)
R(4, 3)	*R*(2, −4)	*Z*(1, 4)	*Z*(−2, 5)

   b. Describe the line segments in Set 1 and Set 2.

   c. Describe the line segments in Set 3 and Set 4.

   d. Describe $\overline{AB}$ if the endpoints are *A*(500, 100) and *B*(500, –600).

   e. Describe $\overline{XY}$ if the endpoints are *X*(–300, 400) and *Y*(750, 400).

FINDING THE DISTANCE BETWEEN TWO POINTS

THE MAIN IDEA

1. The ***distance between two points*** is the length of the line segment that has those points as its endpoints.
2. Length of horizontal line segment = |difference of x-coordinates|
3. Length of vertical line segment = |difference of y-coordinates|
4. To find the length of a line segment that is neither horizontal nor vertical:
   a. Form a right triangle by drawing a horizontal line segment through one point and a vertical line segment through the other point.
   b. Find the coordinates of the vertex of the right angle.
   c. Find the length of each leg (the horizontal and vertical sides of the triangle).
   d. Use the Pythagorean relationship to find the length of the segment (the hypotenuse of the right triangle).

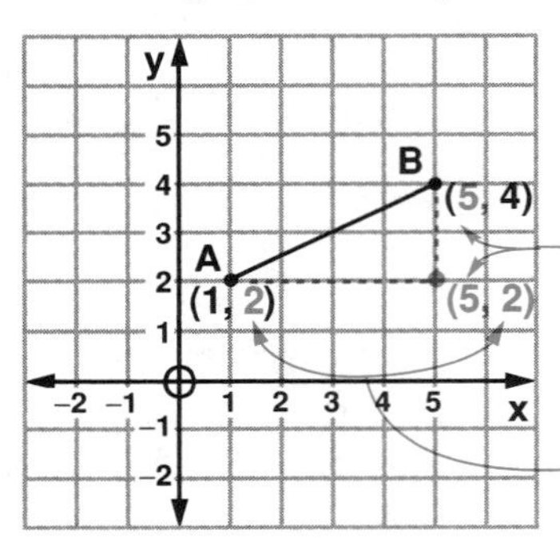

Points on the same vertical line have the same x-coordinate.

Points on the same horizontal line have the same y-coordinate.

EXAMPLE 5 Find the distance between (–1, 2) and (5, 2).

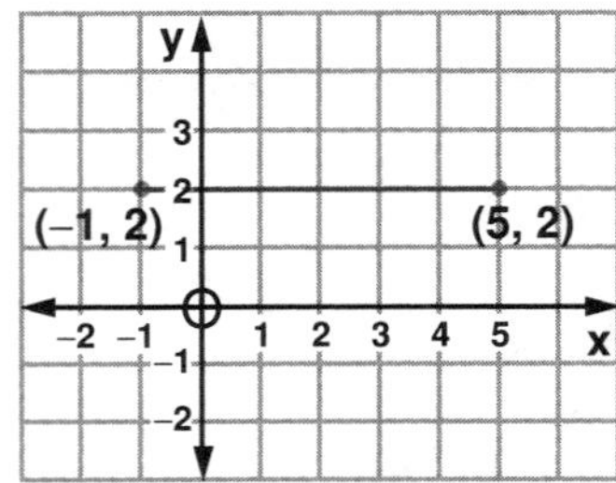

Length = |difference of x-coordinates|

$= |5 - (-1)| = |5 + (+1)| = |6| = 6$

EXAMPLE 6 Find the distance between (4, –2) and (4, 3).

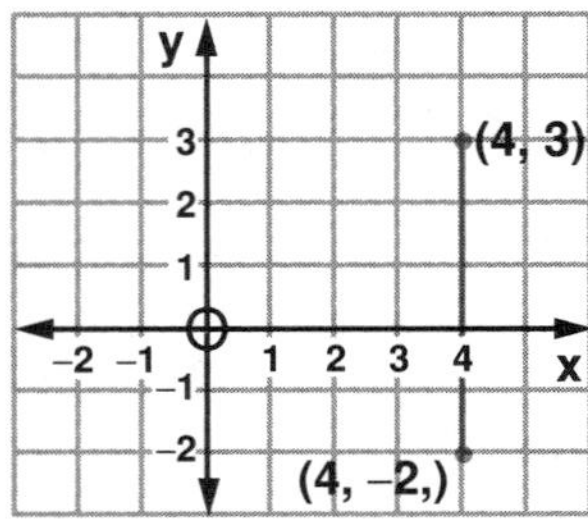

Length = |difference of y-coordinates|

$= |3 - (-2)| = |3 + (+2)| = |5| = 5$

EXAMPLE 7 Find the distance between $A(2, 3)$ and $B(5, 7)$.

Form right triangle ABC. The coordinates of C are $(5, 3)$.

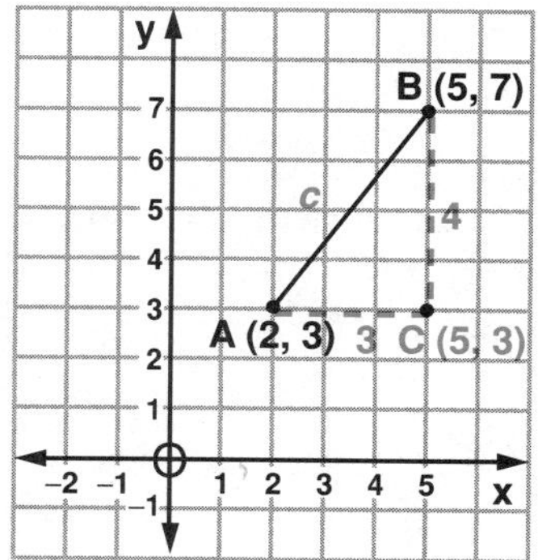

Find the length of each leg of the right triangle:

AC = length of horizontal leg = |difference of x-coordinates|
$= |5 - 2| = |3| = 3$

BC = length of vertical leg = |difference of y-coordinates|
$= |7 - 3| = |4| = 4$

Apply the Pythagorean relation: $c^2 = 3^2 + 4^2$

$c^2 = 9 + 16$

$c^2 = 25$

$c = 5$ *Answer:* The length of $\overline{AB}$ is 5.

CLASS EXERCISES

1. Find the distance between each of the given pairs of points.
a. (3, 6) and (3, 11) **b.** (–5, 4) and (5, 4) **c.** (–6, –1) and (–6, 5) **d.** (–3, –2) and (5, –2)

2. The coordinates of the vertices of rectangle $ABCD$ are $A(2, 5)$, $B(9, 5)$, $C(9, 0)$ and $D(2, 0)$. Find the length of each side of the rectangle.

3. Find the distance of each point from the origin: **a.** (4, 3) **b.** (–6, 8) **c.** (–3, –4) **d.** (5, –12)

4. Find the distance between each of the given pairs of points.
a. (2, 4) and (7, 16) **b.** (–3, 1) and (5, 7) **c.** (–1, –4) and (7, –10) **d.** (30, 15) and (0, –25)

5. A line segment has endpoints at (3, 7) and (5, 9). Approximate its length to the nearest tenth.

6. Quadrilateral $ABCD$ has vertices at $A(0, 0)$, $B(10, 0)$, $C(13, 4)$, and $D(3, 4)$.
a. Graph points A, B, C, and D and draw line segments to form quadrilateral $ABCD$.
b. Find the length of each side of quadrilateral $ABCD$.
c. What kind of quadrilateral is $ABCD$?

HOMEWORK EXERCISES

1. Use an ordered pair to tell the coordinates of each of the given points.

   a. A
   b. B
   c. C
   d. D
   e. E
   f. F
   g. G
   h. H

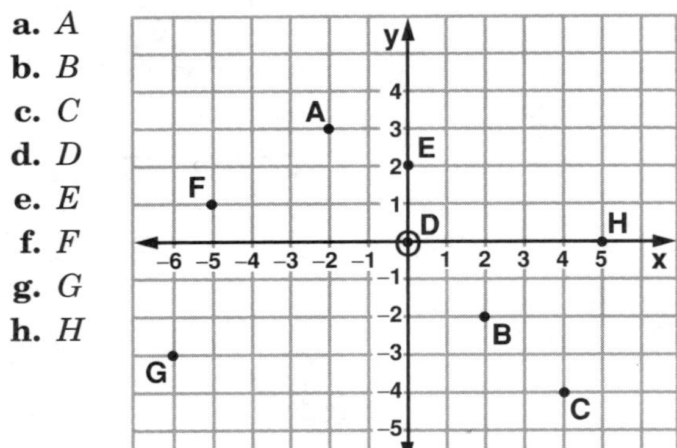

2. Tell the coordinates of each vertex of triangle DEF.

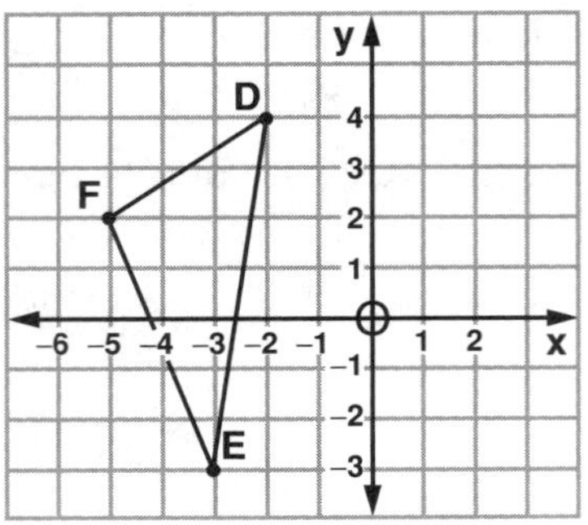

3. Name the point that has the given coordinates.

   a. (6, 1) b. (3, 2) c. (3, –2) d. (–4, 2)
   e. (–6, –3) f. (0, –3) g. (0, 3) h. (–8, 0)

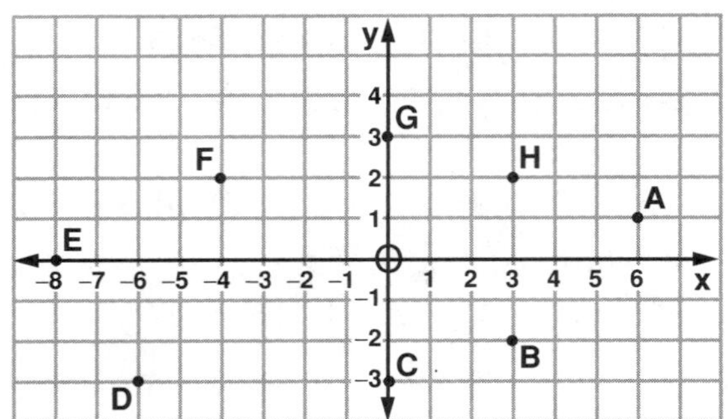

4. Graph and label:

   a. $A(7, 1)$ b. $B(0, 0)$ c. $C(-3, 5)$
   d. $D(5, -3)$ e. $E(-2, -6)$ f. $F(2, 6)$
   g. $G(-7, 2)$ h. $H(0, 6)$ i. $I(6, 0)$
   j. $J(-8, -4)$ k. $K(-5, 0)$ l. $L(-8, 6)$

5. If $(x, 7)$ and $(-3, 7)$ name the same point, what must x equal? Explain.

6. Find the distance between:

   a. (3, 2) and (8, 2) b. (4, –7) and (4, –9)
   c. (1, 4) and (7, 12) d. (–9, 1) and (6, 1)
   e. (2, 1) and (14, 6) f. (3, 7) and (7, 10)

7. The coordinates of the vertices of right triangle ABC are $A(9, 1)$, $B(6, 5)$ and $C(9, 5)$. Find the length of each side of the triangle.

8. A line segment has endpoints (–1, 2) and (5, –3). Approximate its length to the nearest tenth.

9. Approximate the distance of each point from the origin, to the nearest tenth.

   a. (3, 3) b. (5, –3) c. (–1, 4) d. (–1, 10)

10. Describe a procedure for finding the distance from any point to the origin.

11. The distance from Point P to the origin is 10. The coordinates of P are $(6, y)$. Find the value of y.

12. An ant started at point (0, 0) on a sheet of graph paper on which each space is 1 cm wide. It walked to point (0, 5), then to (3, 5), then to (3, 0), and then back to (0, 0). What was the total distance that the ant walked?

13. Graph points A, B, and C that are given. Then, find the coordinates for point D to form the type of quadrilateral named.

a. $A(2, 3)$, $B(1, 1)$, $C(6, 1)$; parallelogram $ABCD$ **b.** $A(-1, 1)$, $B(-1, -4)$, $C(1, -4)$; rectangle $ABCD$

c. $A(-3, 0)$, $B(0, 3)$, $C(3, 0)$; square $ABCD$ **d.** $A(-2, 0)$, $B(0, 3)$, $C(2, 0)$; rhombus $ABCD$

e. $A(-2, -1)$, $B(-1, 2)$, $C(1, 2)$; isosceles trapezoid $ABCD$
(In an isosceles trapezoid, nonparallel sides are congruent.)

SPIRAL REVIEW EXERCISES

1. If $\triangle RST \sim \triangle XYZ$, find XZ.

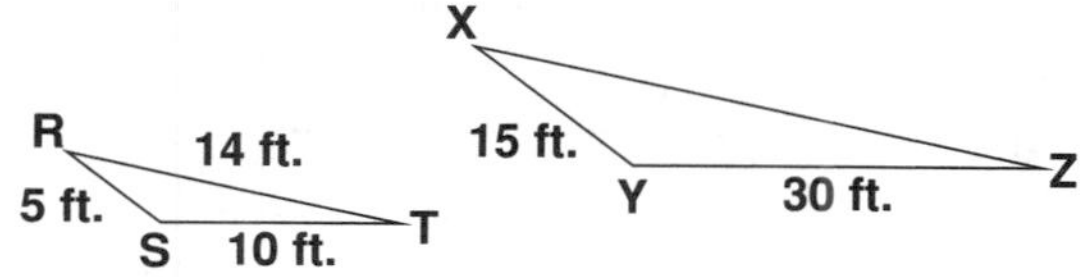

2. Tell whether each statement is *true* or *false*.

a. $-9 < 0$ **b.** $0 > -100$

c. $-5 > -9$ **d.** $-29 < 2$

e. $-2.6 < -2.8$ **f.** $-\frac{7}{8} > -\frac{5}{8}$

3. Which open sentence is represented by the graph shown?

(a) $x < -2$ (b) $x > -2$
(c) $x \leq -2$ (d) $x \geq -2$

4. Subtract –15 from 35.

5. Mr. Jackson left a 15% tip for the waiter in a restaurant. What was the total cost of the meal if his bill was $40?

6. The value of $7 \times 10^3 + 3 \times 10^2 + 5$ is

(a) 7,350 (b) 7,035
(c) 7,305 (d) 73,005

7. Divide: $\frac{8}{15} \div \frac{2}{5}$ 8. Subtract: $\frac{3}{4} - \frac{7}{20}$

9. A box contains 10 coins: 3 quarters, 2 dimes, and 5 nickels. If one coin is chosen at random from the box, find the probability of picking:

a. a quarter **b.** a dime **c.** a nickel
d. a half-dollar **e.** a dime or a nickel

10. If the sales tax is 5%, what is the total amount paid for a shirt that is priced at $14?

11. The greatest common factor of 12 and 18 is

(a) 2 (b) 3 (c) 6 (d) 9

12. A truck can carry a maximum load of 2,400 pounds safely. It has already been loaded with 27 cartons that weigh 52.8 pounds each. Find the maximum number of cartons weighing 47.25 pounds each that can be added safely to the load.

UNIT 18-3 Graphing a Linear Equation

THE MAIN IDEA

1. An equation whose graph is a straight line is called a ***linear equation***. Some examples of linear equations are: $y = 3x + 5$ $y = -2x$ $y + 2 = 48$
2. An ordered pair is a solution of a linear equation if a true statement results when the coordinates of the ordered pair are substituted for x and y in the equation. If a true statement results, then the point with those coordinates is on the graph of that line.
3. To find an ordered pair that is a solution of a linear equation:
   a. Select a value for x.
   b. Substitute that value into the equation.
   c. Solve to find the corresponding value of y.
4. The graph of a linear equation is a line that represents the infinite number of ordered pairs that are solutions of the equation.
5. To graph a linear equation:
   a. Find at least three ordered pairs that are solutions of the equation. Use a table to list the ordered pairs.
   b. Graph each ordered pair.
   c. Draw a line through these points.
   d. Label the graph with the equation.

x	y	(x, y)

EXAMPLE 1 Is (3, 11) a solution of the linear equation $y = 2x + 5$?

Substitute 3 for x and 11 for y. $11 \stackrel{?}{=} 2(3) + 5$

Does a true statement result? $11 \stackrel{?}{=} 6 + 5$

$11 = 11$ ✔

Answer: The point (3, 11) is on the line $y = 2x + 5$.

EXAMPLE 2 Find the ordered pair with x-coordinate of 2 that is on the line whose equation is $y = 3x - 6$.

Substitute 2 for x. $y = 3x - 6$

$y = 3(2) - 6$

Find the corresponding value of y. $y = 6 - 6$

$y = 0$

Answer: (2, 0)

EXAMPLE 3 Make a table of three solutions of $y = x + 4$ and graph the equation.

(1) Choose three values for x.

Try –2, 0, and 3.

x	y	(x, y)
–2		
0		
3		

(2) Substitute each value for x to find the corresponding value for y. Complete the table.

If $x = -2$:	*If* $x = 0$:	*If* $x = 3$:
$y = x + 4$	$y = x + 4$	$y = x + 4$
$y = -2 + 4$	$y = 0 + 4$	$y = 3 + 4$
$y = 2$	$y = 4$	$y = 7$

(3) Graph the 3 points, draw a line through them, and label the line with the equation.

Answer:

x	y	(x, y)
–2	2	(–2, 2)
0	4	(0, 4)
3	7	(3, 7)

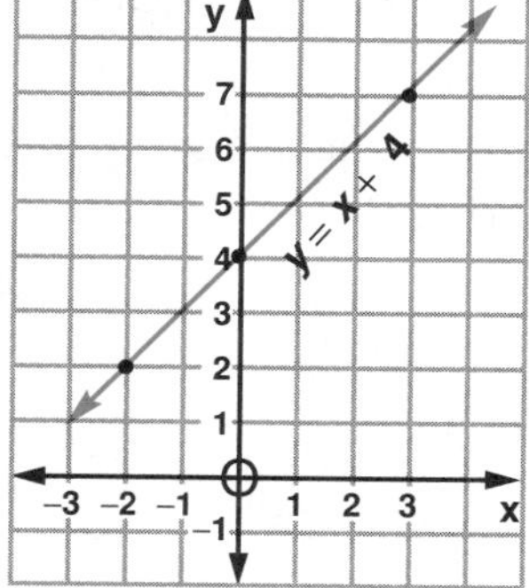

Observe that every ordered pair that makes the linear equation true is on the line. Furthermore, all points on the line, and only those points, have coordinates that make the equation true. Check some points on the graph above.

Try (–1, 3): $y = x + 4$
$3 = -1 + 4$
(on the line) $3 = 3$ true

Try (2, 7): $y = x + 4$
$7 = 2 + 4$
(not on line) $7 = 6$ false

EXAMPLE 4 Make a table of three solutions of $y = 4$ and graph the equation.

All values of x are paired with the same value of y. The graph is the horizontal line that is 4 units above the x-axis.

Answer:

x	y	(x, y)
–3	4	(–3, 4)
0	4	(0, 4)
4	4	(4, 4)

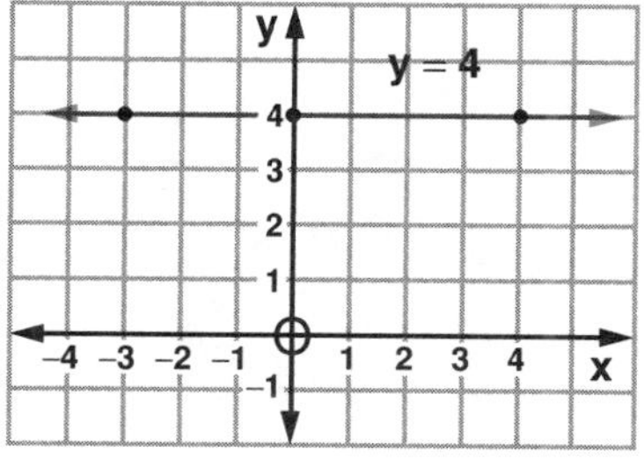

EXAMPLE 5 Make a table of three solutions of $x = -3$ and graph the equation.

The only value that can be selected for x is –3, but –3 can be paired with any value of y. The graph is the vertical line 3 units to the left of the y-axis.

Answer:

x	y	(x, y)
–3	–5	(–3, –5)
–3	0	(–3, 0)
–3	2	(–3, 2)

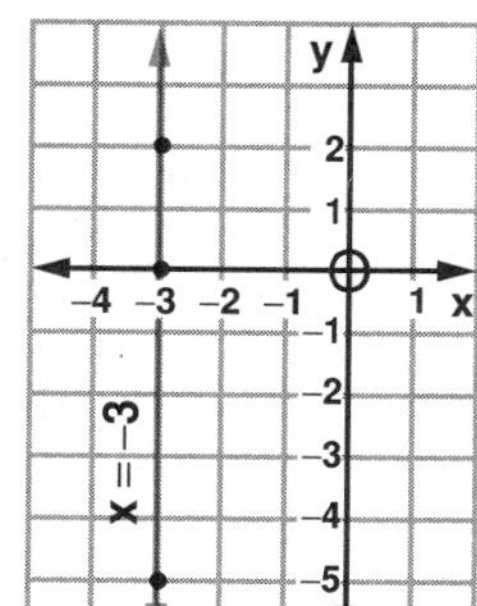

CLASS EXERCISES

1. Tell whether each ordered pair is a solution of $y = 3x - 9$.

a. $(1, -6)$ **b.** $(-3, -12)$ **c.** $(4, 2)$ **d.** $(0, -9)$

2. Find the ordered pair with the given x-coordinate that is a solution of $y = 5x + 2$.

a. $x = -1$ **b.** $x = 0$ **c.** $x = 4$ **d.** $x = -3$

3. For each equation, find the value of y that corresponds to the given value of x.

	Equation	*Value of x*		*Equation*	*Value of x*
a.	$y = 4x - 9$	$x = -3$	**b.**	$y = -2x + 7$	$x = 4$
c.	$y = \frac{1}{2}x + 1$	$x = 6$	**d.**	$y = -2$	$x = 10$
e.	$y = 2.1x - 3$	$x = 10$	**f.**	$y = \frac{1}{4}x - 2$	$x = -8$

4. Make a table of three solutions and graph each equation.

a. $y = 4x + 1$ **b.** $y = -x + 5$ **c.** $y = 4$ **d.** $y = 3x$

HOMEWORK EXERCISES

1. Tell whether each ordered pair is a solution of $y = -2x + 6$.

a. $(0, 5)$ **b.** $(2, 2)$ **c.** $(3, 0)$ **d.** $(4, -1)$

2. Tell whether each ordered pair is a solution of $x + 2y = 4$.

a. $(0, 2)$ **b.** $(2, 1)$ **c.** $(4, 1)$ **d.** $(3, \frac{1}{2})$

3. Find the ordered pair with the given x-coordinate that is a solution of $y = 3x - 9$.

a. $x = -4$ **b.** $x = -2$ **c.** $x = 2$ **d.** $x = 5$

4. For each equation, find the value of y that corresponds to the given value of x.

	Equation	*Value of x*		*Equation*	*Value of x*
a.	$y = x - 7$	$x = 2$	**b.**	$y = 2x - 10$	$x = -1$
c.	$y = -3x + 4$	$x = 7$	**d.**	$2x + y = 6$	$x = 2$
e.	$y = 5x + 6$	$x = -4$	**f.**	$y - 3x = 9$	$x = -2$

5. Complete the table for $y = 2x - 3$ and graph the equation.

x	y	(x, y)
-2		
2		
4		

6. Make a table of three solutions and graph each equation.

a. $y = x + 4$ **b.** $y = -2x + 5$ **c.** $y = -3x - 2$ **d.** $y = 3x - 6$ **e.** $y = 4$

f. $y = x - 9$ **g.** $y = -x + 7$ **h.** $x = -6$ **i.** $y = -3x$ **j.** $x = 0$

SPIRAL REVIEW EXERCISES

1. Use an ordered pair to tell the coordinates of each of the given points.

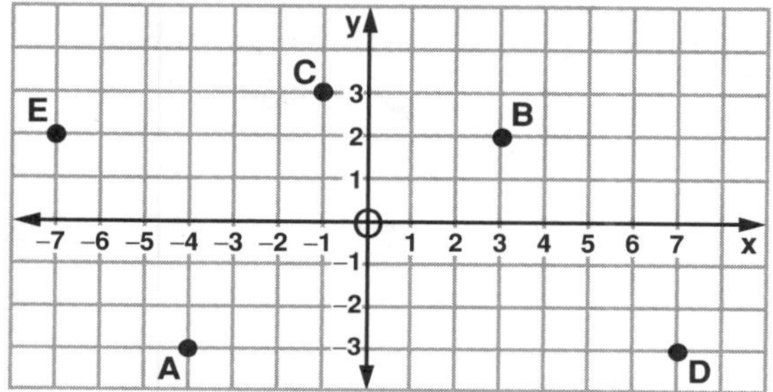

2. On the graph below, what are the coordinates of point A?
   (a) (4, 1)
   (b) (–1, 4)
   (c) (–4, 1)
   (d) (–4, –1)

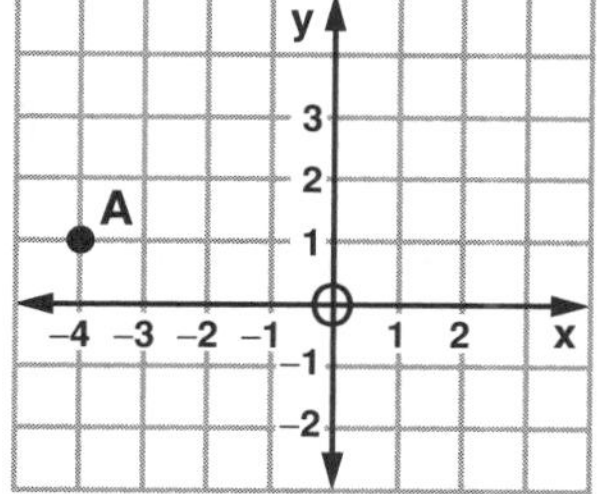

3. On the graph below, which point has coordinates (2, –3)?
   (a) D
   (b) E
   (c) F
   (d) G

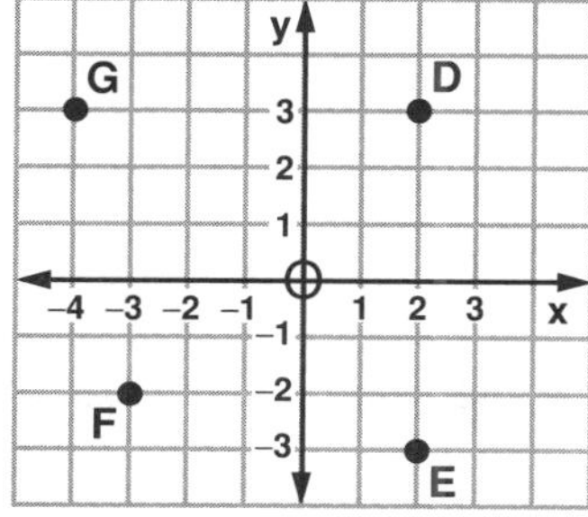

4. Find the measure of $\angle A$.

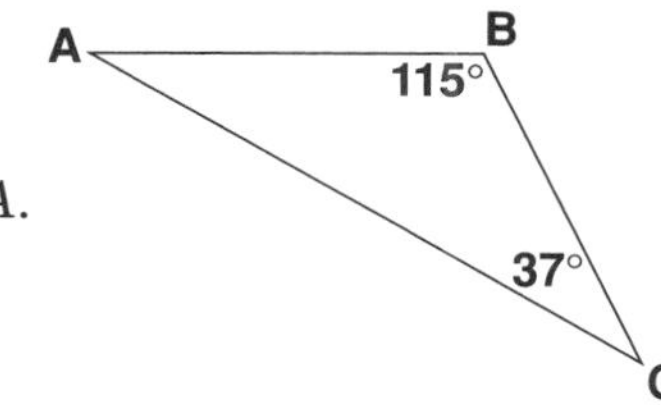

5. Tell whether a triangle whose sides measure 6 cm, 8 cm, and 12 cm is a right triangle.

6. 25 cm is equivalent to
   (a) 25 m (b) 2.5 m
   (c) 0.25 m (d) 0.025 m

7. If a coin is tossed 150 times, what is the number of times that you would expect it to come up "heads"?

8. 120% is equivalent to
   (a) $\frac{1}{5}$ (b) $\frac{3}{5}$ (c) $\frac{4}{5}$ (d) $\frac{6}{5}$

9. When Rachel used her calculator to divide the cost of her party by the difference between the number of people she invited and the number of people who actually came to the party, her calculator displayed an error message. This means that:
   (a) The cost of the party per person cannot be determined.
   (b) Too many people were invited to the party.
   (c) No one came to the party.
   (d) Everyone who was invited came to the party.

10. A poll revealed that students in 29% of 900 households used the kitchen for doing homework. In how many of those households was the kitchen used for homework?

11. The Tse family is planning a trip that takes $4\frac{1}{2}$ hours. They want to take two rest stops of one-half hour each. At what time should they leave if they want to arrive at 4:30 P.M.?

UNIT 18–4 The Slope of a Line; Scatter Plots

THE SLOPE OF A LINE

THE MAIN IDEA

1. The ***slope*** of a line is a number that tells how much and in what direction a line slants. The slope of a line may be a positive number, a negative number, or zero.

Positive Slope

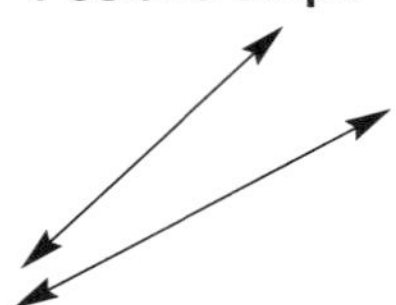

Lines with positive slope slant up from left to right.

Negative Slope

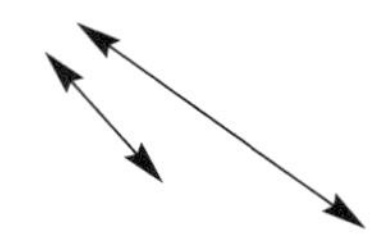

Lines with negative slope slant down from left to right.

Zero Slope

Lines with a slope of zero are horizontal.

2. A vertical line has no slope.
3. Any two points on a line can be used to find its slope.
4. To find the slope of a line between two points:
   a. Imagine moving an object along the line from *left* to *right* from the first point to the second point.
   b. Write a signed number to represent the distance the object would move up (+) or down (–). This number is the ***rise***.
   c. Write a signed number to represent the distance the object would move right (+). This number is the ***run***.
   d. Write and simplify the following ratio: $\text{slope} = \dfrac{\text{rise}}{\text{run}}$

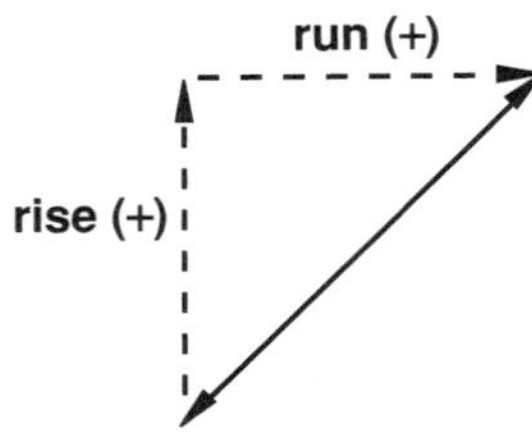

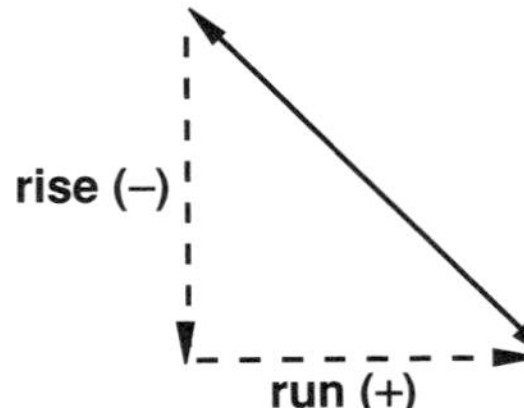

5. If you know the coordinates of two points on a line, you can subtract to find the rise and the run:

$$\text{rise} = \text{second } y\text{-coordinate} - \text{first } y\text{-coordinate}$$
$$\text{run} = \text{second } x\text{-coordinate} - \text{first } x\text{-coordinate}$$

EXAMPLE 1 For each line, tell if the slope is positive, negative, zero, or does not exist.

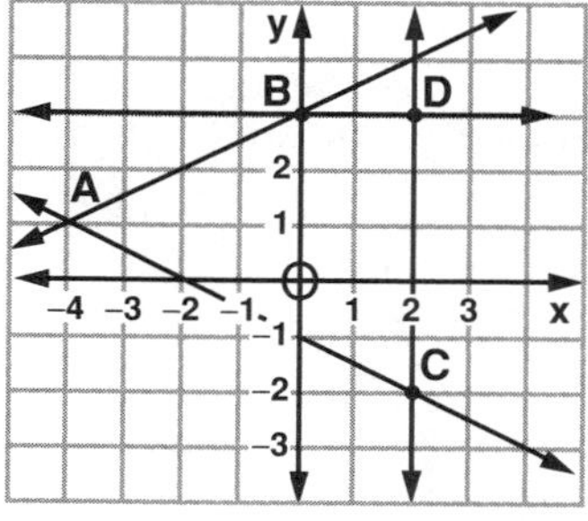

a. $\overleftrightarrow{AB}$ $\overleftrightarrow{AB}$ slants up from left to right.

Answer: The slope of $\overleftrightarrow{AB}$ is positive.

b. $\overleftrightarrow{AC}$ $\overleftrightarrow{AC}$ slants down from left to right.

Answer: The slope of $\overleftrightarrow{AC}$ is negative.

c. $\overleftrightarrow{BD}$ $\overleftrightarrow{BD}$ is a horizontal line.

Answer: The slope of $\overleftrightarrow{BD}$ is 0.

d. $\overleftrightarrow{CD}$ $\overleftrightarrow{CD}$ is a vertical line.
A vertical line does not have a slope.

Answer: $\overleftrightarrow{CD}$ has no slope.

EXAMPLE 2 Write signed numbers to represent the rise, the run, and the slope of each line.

a. 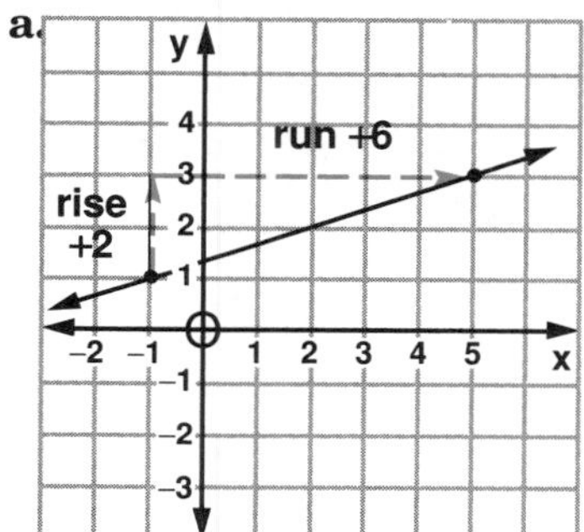

b. 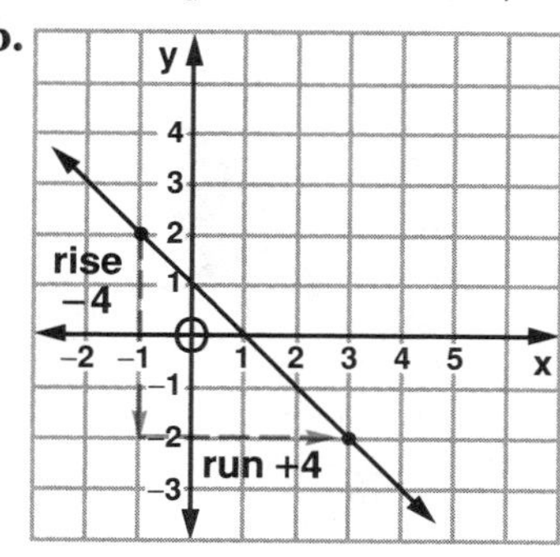

c. 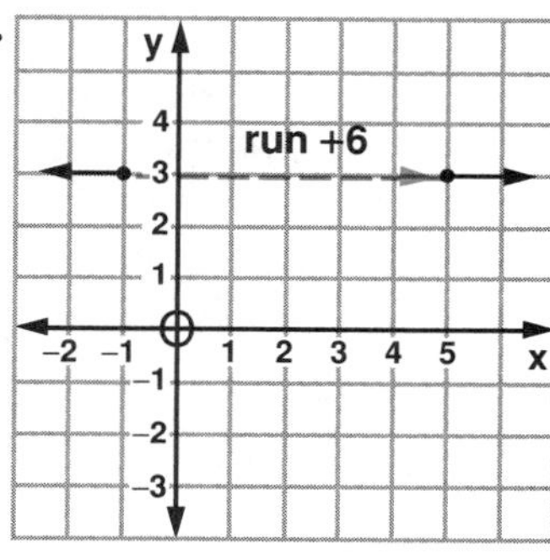

Answer:

a. rise = 2
run = 6
slope = $\frac{2}{6} = \frac{1}{3}$

b. rise = –4
run = 4
slope = $\frac{-4}{4} = -1$

c. rise = 0
run = 6
slope = $\frac{0}{6} = 0$

EXAMPLE 3 Find the slope of the line that contains the points $A(2, 1)$ and $B(4, 7)$, and describe the direction of the line.

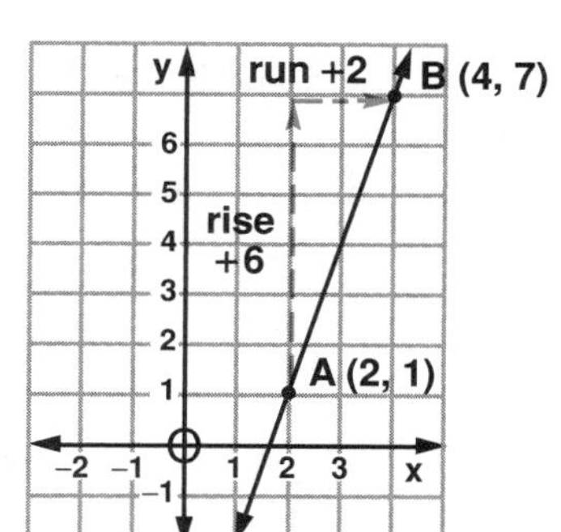

rise = 7 – 1 = 6 run = 4 – 2 = 2 slope of $\overleftrightarrow{AB} = \frac{6}{2} = 3$

To find *both* the rise and the run, the coordinate for *A* was subtracted from the coordinate for *B*.

Answer: The slope of $\overleftrightarrow{AB}$ is 3. The line slants upward from left to right and goes up 3 units for every 1 unit to the right.

CLASS EXERCISES

1. For each line, tell if the slope is positive, negative, zero, or does not exist.

   a. $\overleftrightarrow{AB}$ **b.** $\overleftrightarrow{AC}$ **c.** $\overleftrightarrow{AD}$
   d. $\overleftrightarrow{BD}$ **e.** $\overleftrightarrow{BE}$ **f.** $\overleftrightarrow{CD}$

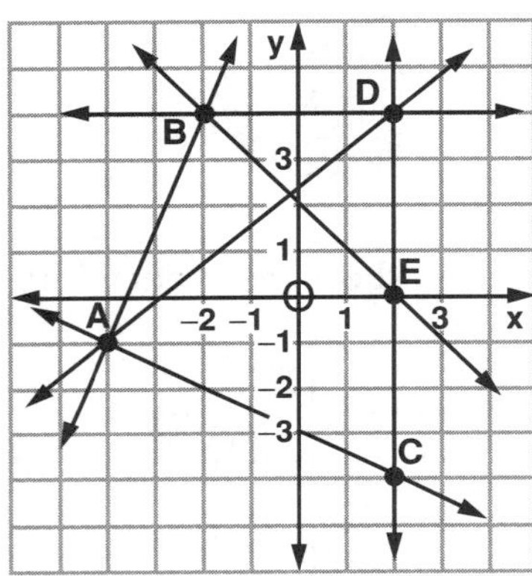

2. Find the slope of the ramp to the truck.

3. Write a signed number to represent the rise, the run, and the slope of the line through A and B.

a.

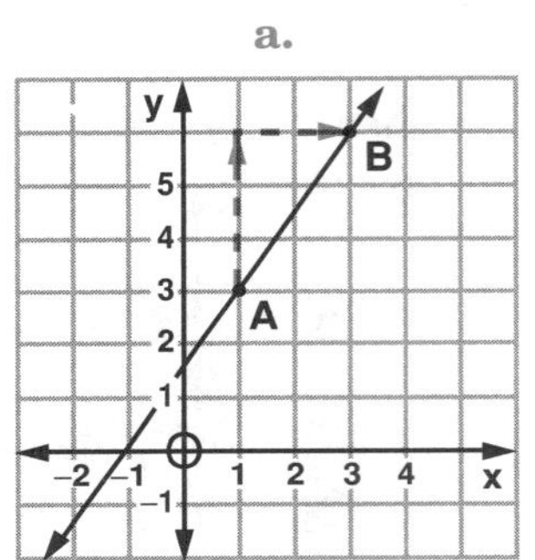

b.

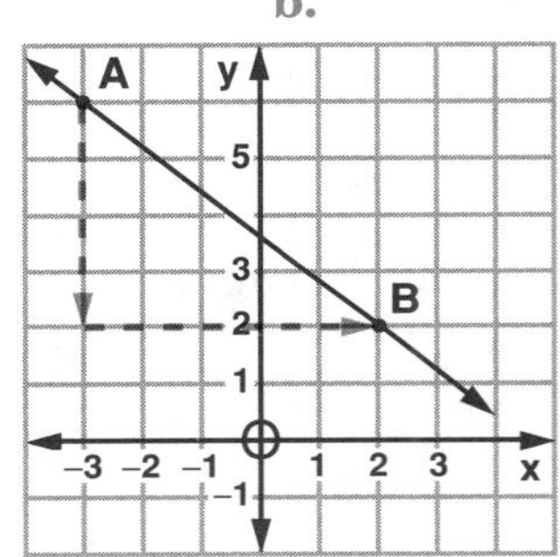

4. A line contains the points $E(5, -3)$ and $F(2, -9)$. The direction of the line is
   (a) up from left to right (b) down from left to right (c) horizontal (d) vertical

5. Through each pair of points, graph the line, find its slope, and describe its direction.
   a. $A(8, 7), B(5, 7)$ **b.** $C(3, 4), D(-2, 6)$ **c.** $E(-4, 8), F(-4, 10)$ **d.** $G(-4, -3), H(-6, -7)$

6. Use graph paper and a protractor to do the following:
   a. Draw the x- and y-axes and label the origin as point O.
      (1) Graph point $P(6, 3)$ and draw line OP.
      (2) Find the slope of line OP.
      (3) Measure the acute angle formed by line OP and the x-axis.
   b. Repeat Exercise 6a, using: (1) (5, 5) for P_1 (2) (3, 6) for P_2 (3) (1, 8) for P_3.
   c. As the slope of line OP increases, how does the measure of the angle formed by line OP and the x-axis change? Does it increase or does it decrease?

7. Joel was asked to find the slope of the line through $A(3, 5)$ and $B(5, 4)$. He said that, since the rise was $5 - 4$ and the run was $5 - 3$, the slope must be $\frac{5-4}{5-3}$, or $\frac{1}{2}$. Write a note to Joel, telling him whether you agree with him or not, and why.

SCATTER PLOTS

THE MAIN IDEA

1. A ***scatter plot*** is a kind of graph that is used to tell if one set of data is related to another set of data. Points on a scatter plot have x-coordinates from one set of data and y-coordinates from the other set of data.
2. If the points on a scatter plot are scattered randomly, there is no relationship between the two sets of data. If most of the points are near a straight line, then the two sets of data are related.
3. The slope of the line tells how the two sets of data are related. A line with a positive slope indicates that one quantity increases as the other increases. A line with a negative slope indicates that one quantity increases as the other decreases.

EXAMPLE 4 An aerobics instructor recorded the heights and weights of students in her classes. Point A, for example, represents a student whose height is 59 inches and whose weight is 125 pounds. The scatter plot shows the data together with a line that fits the data.

Heights and Weights For Aerobics Classes

a. How many points are more than two squares away from the line?

Answer: Only 3 points (A, B, and C).

b. Describe the general relationship between the heights and weights of members of this class. Are there any exceptions?

Answer: Since the slope of the line is positive, in general, as height increases, weight increases also. There are a few exceptions. For example, the person represented by C is taller than A but weighs less.

EXAMPLE 5 A neighborhood was surveyed to determine how the cost of admission would affect the frequency of movie attendance. Only those answers given by 4 or more people were recorded. The scatter plot shows the results. For example, the dot at (3, 5) means that, at a cost of $3 a ticket, moviegoers would attend 5 times a month.

a. Describe a line that fits the data, if any.

Position a ruler to connect two points such as $A(1, 8)$ and $B(9, 1)$. *Do not draw in this book.* Most of the points are on or near line AB.

Answer: A line that fits the data is the line through $A(1, 8)$ and $B(9, 1)$.

b. Describe how the cost of a ticket is related to the number of times people would attend the movies each month.

Answer: The line AB has a negative slope, indicating that as the cost increases, attendance decreases.

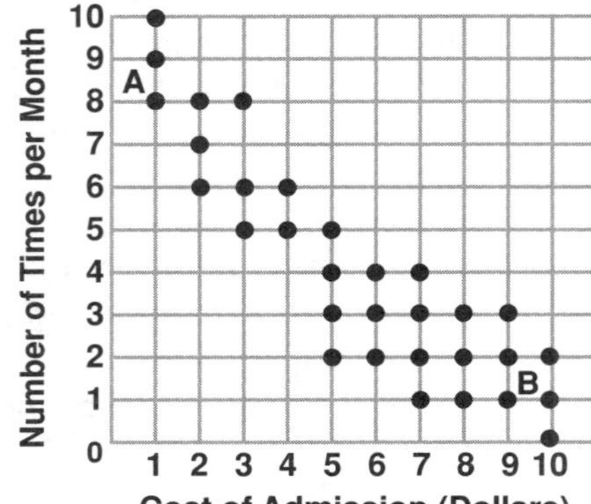

EXAMPLE 6 The scatter plot compares scores on a Math exam and an English exam for a class of students. What relationship, if any, exists between the two sets of scores?

Answer: The points are scattered and do not cluster around a line. This shows that there is no clear relationship between the Math scores and English scores for these students.

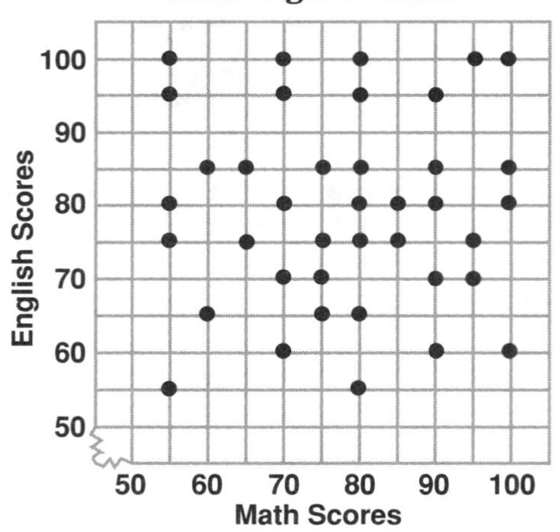

CLASS EXERCISES

1. A psychology experiment tested volunteers to see what percent of a vocabulary list they could remember after several days had elapsed. Results are shown in the scatter plot.
   a. Position the edge of a ruler on a line that fits the data, if there is one. State the coordinates of two points on the line.
   b. If there is a line in part **a**, describe the slope and the relationship that it indicates.

Percent of Vocabulary List Remembered After Elapsed Time

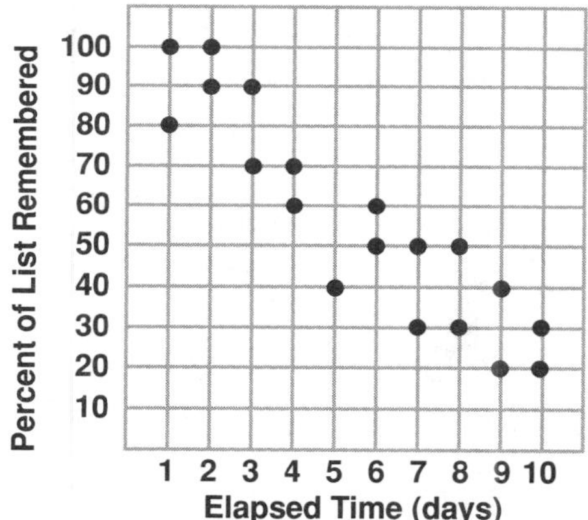

2. In a biology experiment, fruit trees were given a new fertilizer to increase their production. The scatter plot shows the amount of fertilizer and the average number of blossoms per branch.
   a. Position the edge of a ruler on a line that fits the data, it there is one. State the coordinates of two points on the line.
   b. If there is a line in part **a**, describe the slope and the relationship that it indicates.

Results of Fertilizer on Fruit Trees

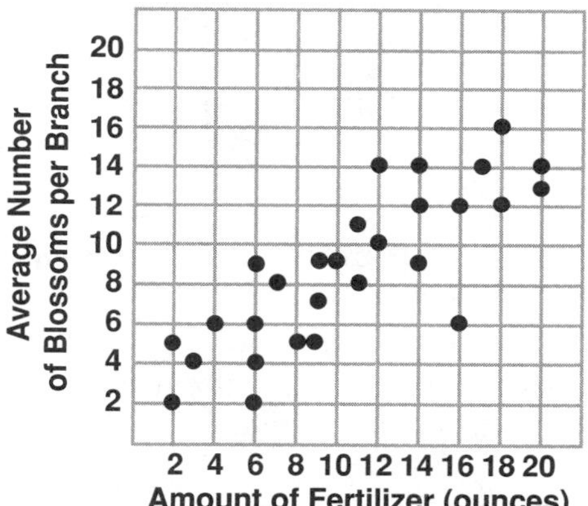

3. The scatter plot shows the U.S. motor-vehicle production, both as millions of vehicles and as a percent of worldwide production, every 5 years from 1950 to 1990. Each dot represents the data for a 5-year interval.
   a. Position the edge of a ruler on a line that fits the data, if there is one. State the coordinates of two points on the line.
   b. If there is a line in part **a**, describe the slope and the relationship that it indicates.

Motor Vehicle Production in 5-year Intervals, 1950-1990

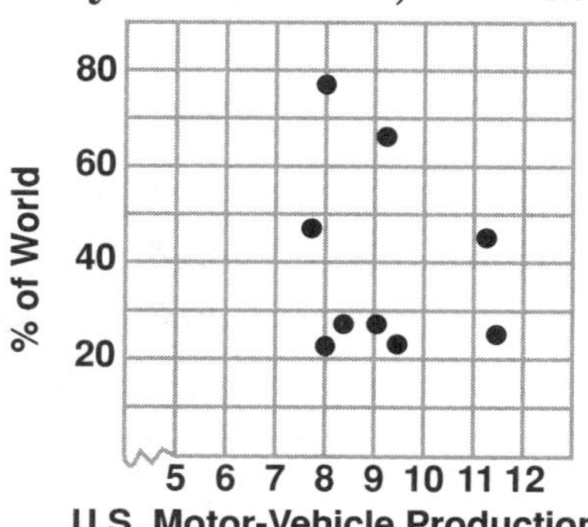

HOMEWORK EXERCISES

1. For each line, tell if the slope is positive, negative, zero, or does not exist.

   a. $\overleftrightarrow{AD}$ **b.** $\overleftrightarrow{AE}$ **c.** $\overleftrightarrow{BD}$
   d. $\overleftrightarrow{BF}$ **e.** $\overleftrightarrow{CD}$ **f.** $\overleftrightarrow{DG}$

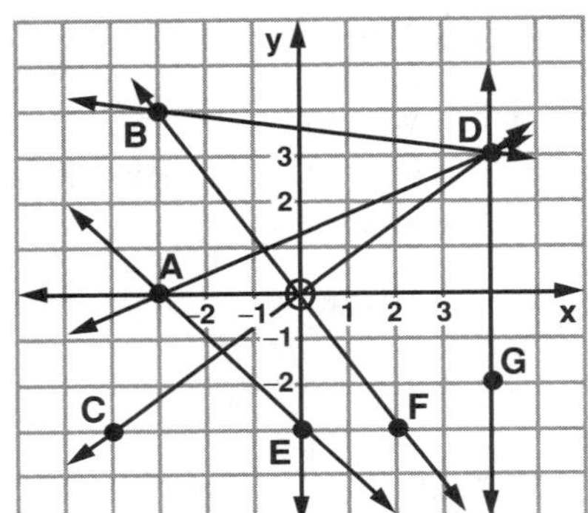

2. For each graph, write signed numbers to represent the rise, the run, and the slope of the line through A and B.

a.

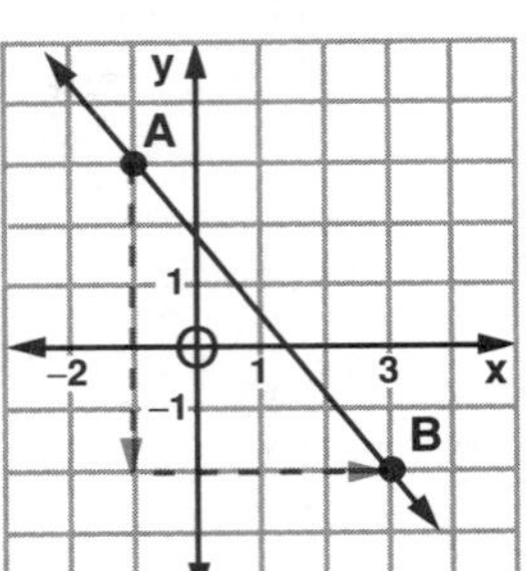

b.

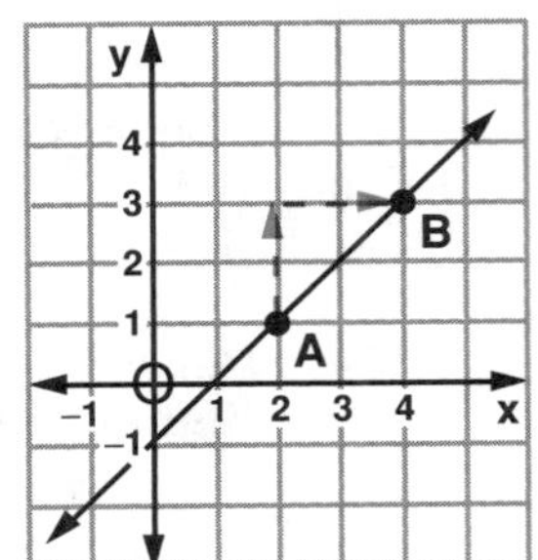

3. A line contains the points $P(-4, 8)$ and $Q(-2, 8)$. The direction of the line can be described as
   (a) horizontal (b) vertical
   (c) slanting up from left to right
   (d) slanting down from left to right

4. Find the slope of the ramp.

5. Graph the line through each pair of points. Then, find the slope of the line and describe its direction.

   a. $A(4, 9)$, $B(6, 17)$ **b.** $C(9, 12)$, $D(-5, 12)$ **c.** $E(-2, 4)$, $F(2, 0)$
   d. $G(-3, -6)$, $H(-1, -8)$ **e.** $K(2, -5)$, $L(3, -3)$ **f.** $M(4, 6)$, $N(4, 9)$

6. Graph each linear equation and find the slope of the line, using any two points on the line.

   a. $y = x$ **b.** $y = x - 3$ **c.** $y = 2x + 1$

7. **a.** On one set of axes, graph the following lines:
   (1) the line through $A(-2, 3)$ and $B(0, 5)$ **(2)** the line through $C(0, -4)$ and $D(2, -2)$
   (3) the line through $E(3, 1)$ and $F(5, 3)$

   b. Find the slope of: **(1)** line AB **(2)** line CD **(3)** line EF

   c. Does it appear as though any of these three lines will intersect? Explain.

8. For each part, **a-d**, graph the three lines on one set of axes, find the slope of each line, and select the statement (I, II, III, or IV) that describes the lines and their slopes.

I. The lines are parallel and the slopes are equal.
II. The lines intersect and the slopes are equal.
III. The lines are parallel and the slopes are unequal.
IV. The lines intersect and the slopes are unequal.

a. the line through: **(1)** (–3, 2) and (0, 1) **(2)** (1, 4) and (4, 3) **(3)** (0, 5) and (3, 4)
b. the line through: **(1)** (0, –4) and (4, 1) **(2)** (–3, 2) and (2, –3) **(3)** (0, 6) and (4, –1)
c. the line whose equation is: **(1)** $y = 2x$ **(2)** $y = 2x - 3$ **(3)** $y = 2x + 4$
d. the line whose equation is: **(1)** $y = 4x$ **(2)** $y = x - 4$ **(3)** $y = -x + 1$

9. A botanist recorded the heights of seedlings every day for 2 weeks after the seeds first sprouted. Results are shown on the scatter plot.

a. Position the edge of a ruler on a line that fits the data, if there is one. State the coordinates of 2 points on the line.

b. If there is a line in part **a**, describe the slope, and the relationship that it indicates.

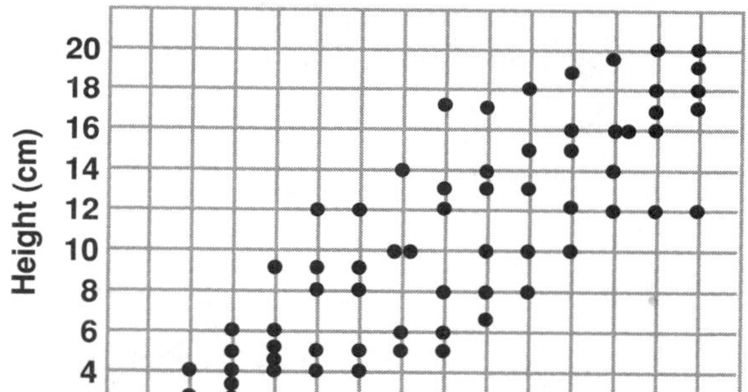

10. When students registered for library cards last fall at Phillips Public Library, they were asked how many weeks had passed since their last birthday and how many books they had read on their summer vacation. Results are shown on the scatter plot.

a. Position the edge of a ruler on a line that fits the data, if there is one. State the coordinates of two points on the line.

b. If there is a line in part **a**, describe the slope, and the relationship that it indicates.

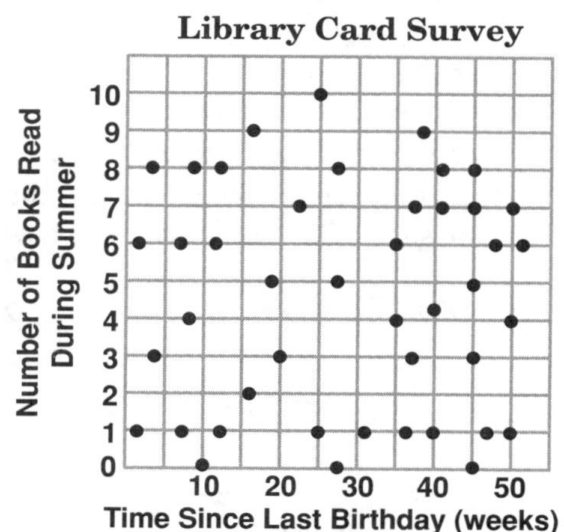

11. The table gives data for the average speed of the most commonly used types of airplanes and the number of gallons of fuel that are used per hour.

a. Use the data in the table to make a scatter plot. Round speeds and fuel to the nearest hundred.

b. Use the scatter plot to describe the relationship between the speed of these planes and their fuel usage.

Model	***Speed*** (mph)	***Fuel*** (gal./hr.)	***Model***	***Speed*** (mph)	***Fuel*** (gal./hr.)
B747-100	519	3,529	MD-80	416	882
L-1011	498	2,215	B737-300	413	732
DC-10-10	484	2,174	DC-9-50	378	848
A300B4	460	1,482	B727-100	422	1,104
A310-300	473	1,574	B737-100	388	806
B767-300	478	1,503	F-100	360	631
B767-200	475	1,377	DC-9-30	377	804
B757-200	449	985	DC-9-10	376	764
B727-200	427	1,249			

12. A cross section of part of a marathon race course is shown in the graph below.

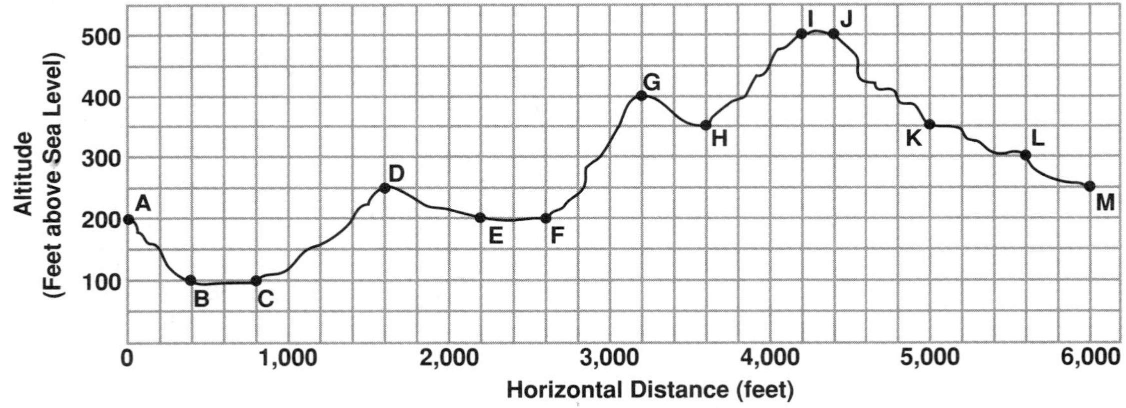

a. Which section of the race course would you expect to be hardest to climb? Use the idea of slope to explain your answer.

b. Which downhill section of the course is steepest? Use slopes to explain your answer.

c. Marty wrote to his friend saying that he expects the part of the course from A to M to be rather easy. His explanation is shown below. Write a reply to Marty about his reasoning.

Coordinates for A and M are A (0, 200) and M (6,000, 250).

rise = 250 − 200 = 50 run = 6,000 − 0 = 6,000

slope = $\frac{\text{rise}}{\text{run}} = \frac{50}{6{,}000} = \frac{1}{120}$ It should be easy to climb just 1 foot for every 120 ft. of horizontal distance.

SPIRAL REVIEW EXERCISES

1. Which of the ordered pairs is not a solution of $y = -3x + 1$?

(a) (0, 1) (b) (–1, 4) (c) (1, –2) (d) (3, –7)

2. If $\triangle ABC \cong \triangle DEF$, find $m\angle C$.

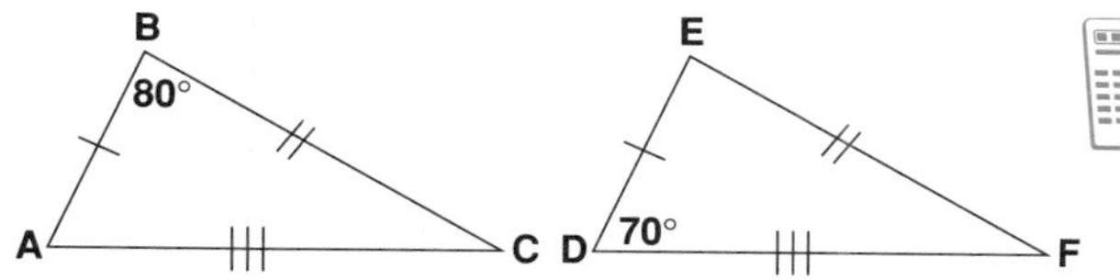

3. The angle whose measure is closest to 30° is

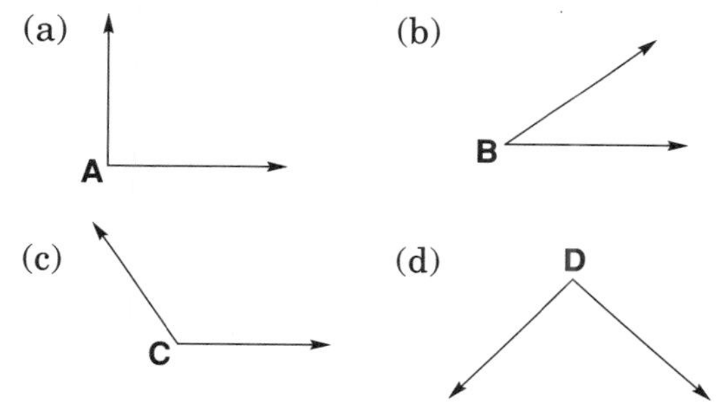

4. The solution set of the inequality $-5 - 2x < -11$ is

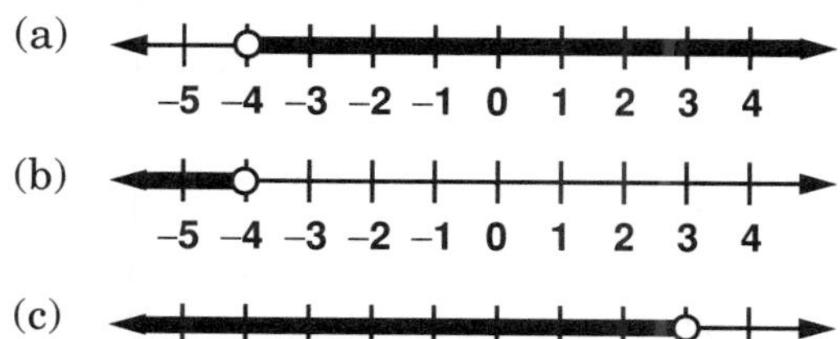

5. Change each of the given probabilities, which are expressed as decimals, into fractions in simplest form.

a. 0.17 **b.** 0.05 **c.** 0.8 **d.** 0.019

6. Find the largest perfect square less than 1,000.

7. A 20-foot ladder leans against a wall. If the top of the ladder is 16 feet from the ground, how far is the foot of the ladder from the wall?

8. Erin is saving \$24 each week from her paycheck. If she has \$59 in her account, in how many weeks will her balance be \$275?

9. The next number in the pattern 0.7, 1.5, 2.4, 3.4, 4.5, . . . is

(a) 5.0 (b) 5.2 (c) 5.7 (d) 5.9

10. Name the figure:

(a) $\overrightarrow{ED}$ (b) $\overrightarrow{DE}$ (c) $\overline{DE}$ (d) $\overleftrightarrow{DE}$

11. Find the average (mean) of the following test scores: 77, 83, 94, 86, 85

12. A car travels 316 miles using 14 gallons of gas. At this rate, how many gallons of gas will be used to travel 948 miles?

13. John answered 18 out of 20 questions correctly on a quiz. What percent did he answer correctly?

UNIT 18–5 Geometric Transformations

THE MAIN IDEA

1. There are three ways to move a geometric figure without changing its size or shape.

A ***translation*** slides a figure to a new position.

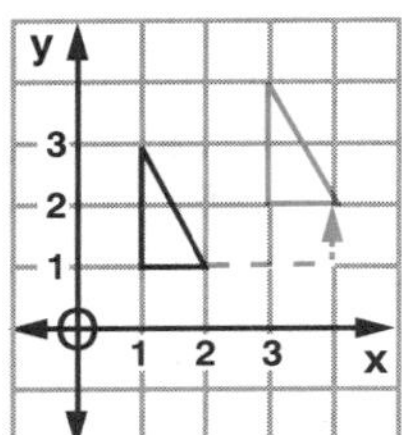

Translation 2 units right and 1 unit up

A ***rotation*** turns a figure about a point.

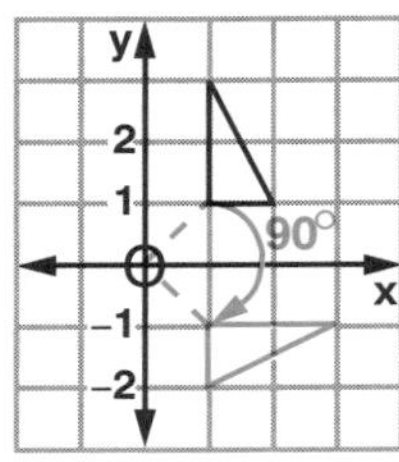

Rotation 90° clockwise about the origin

A ***reflection*** flips a figure to produce a mirror image.

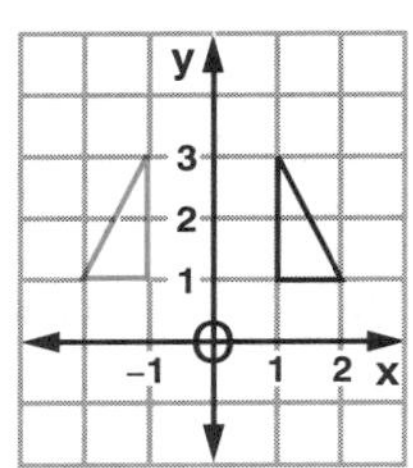

Reflection through the y-axis

2. The figure that results after one of these movements, or ***transformations***, is called the ***image*** of the original figure.
3. A figure is congruent to its image after a translation, a rotation, or a reflection.
4. A figure has ***line symmetry*** if it is possible to draw a line that cuts the figure into two parts such that one part is a reflection of the other. A figure may have more than one line of symmetry.

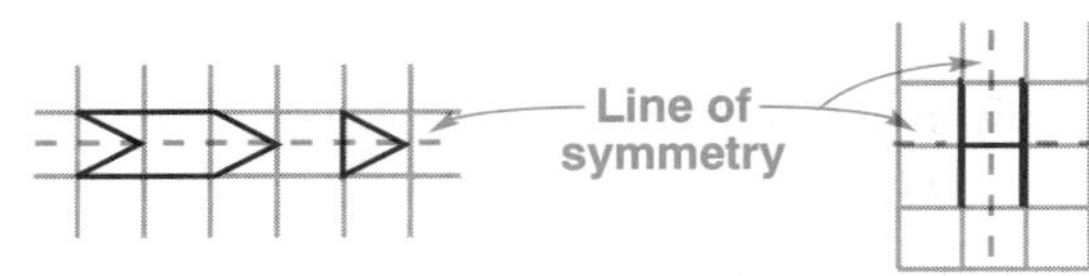

5. A figure has ***point symmetry*** if there is a point that is the midpoint of the line segment between any point and its image.

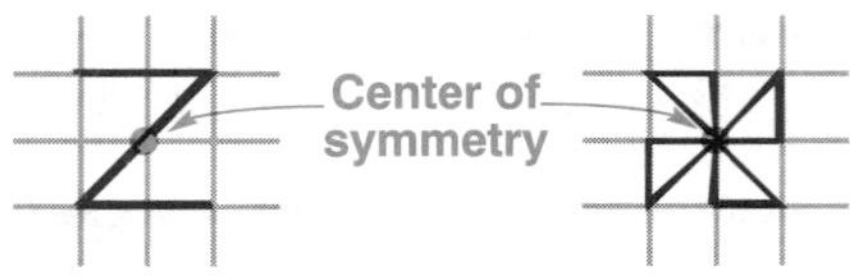

EXAMPLE 1 For the figure shown, state the coordinates of each vertex after the transformation described.

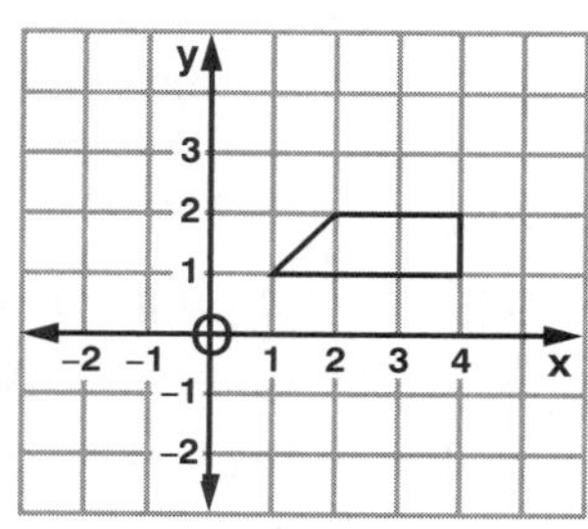

a. Translation left 5 and up 1

Copy the figure and draw the image of the translation. Count spaces to move each vertex left 5 and up 1.

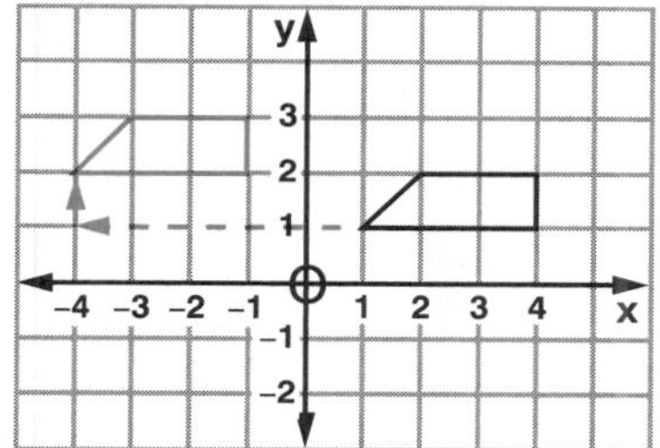

Answer: (–4, 2), (–3, 3), (–1, 3), (–1, 2)

b. Rotation 90° counterclockwise about (1, 1)

The vertex at (1, 1) is the only point that does not move. Think of turning each vertex through 90°.

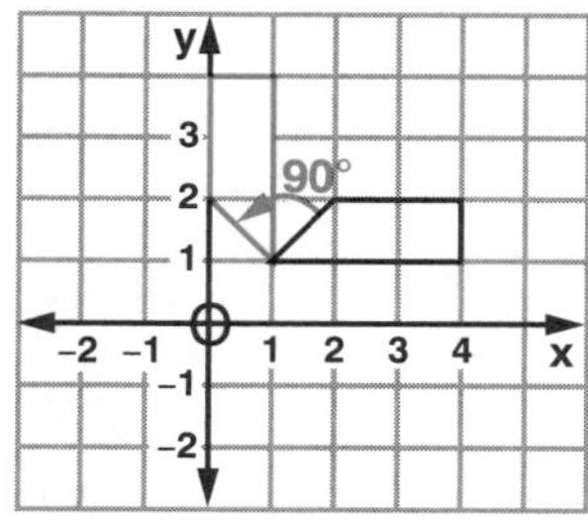

Answer: (1, 1), (0, 2), (0, 4), (1, 4)

c. Reflection through the x-axis

Think of folding the paper on the x-axis as a mirror and draw the image of the trapezoid where it would fall below the x-axis.

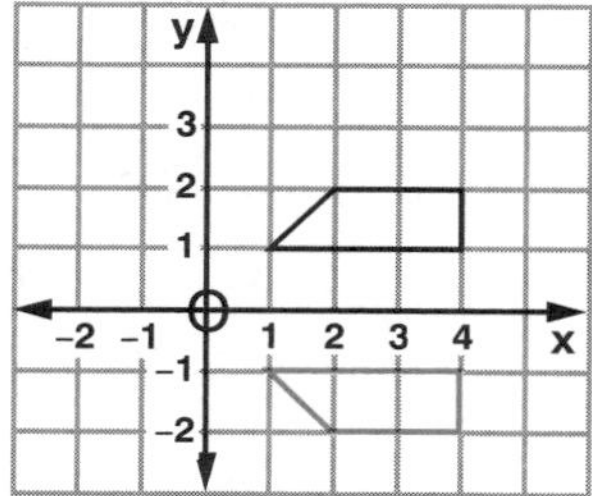

Answer: (1, –1), (2, –2), (4, –2), (4, –1)

d. Rotation 90° clockwise about the origin

To rotate each vertex through 90°, imagine a line segment from the vertex to the origin and rotate that segment to form a right angle. Do this for each vertex. Then, connect the images of the vertices.

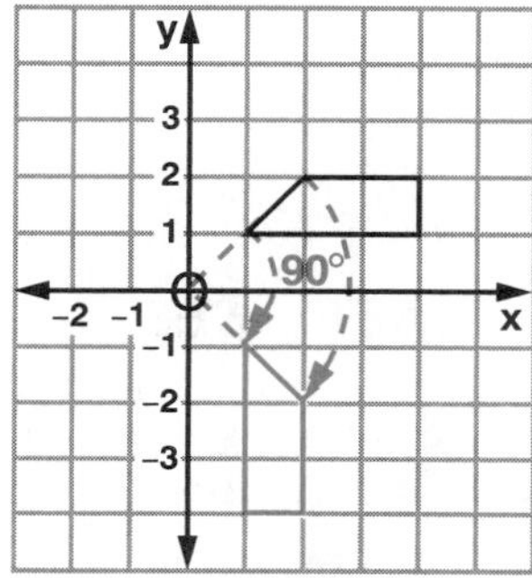

Answer: (1, –1), (2, –2), (2, –4), (1, –4)

EXAMPLE 2 Describe the transformation of the black letter "J" that is needed to produce each image.

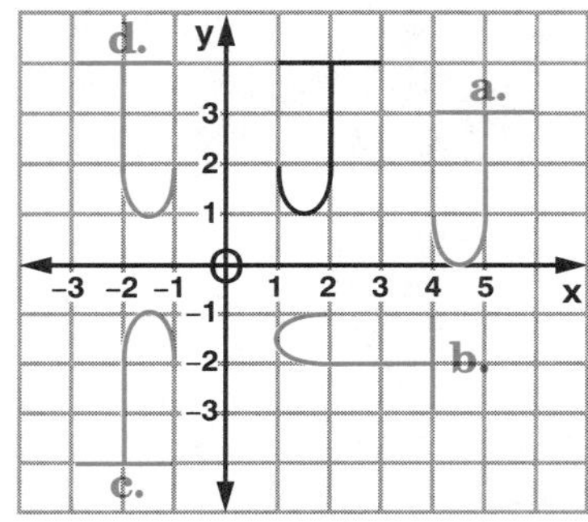

Answers

a. Translation down 1 and right 3

b. Rotation 90° clockwise about the origin

c. Rotation of 180° either clockwise or counterclockwise about the origin

d. Reflection through the y-axis

EXAMPLE 3 How many lines of symmetry does each figure have?

	Figure	*Lines of Symmetry*
a.		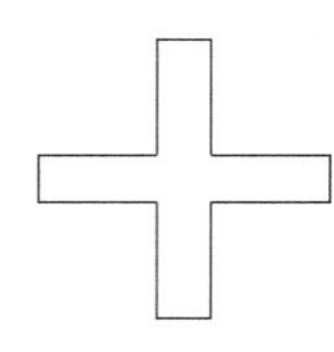
b.	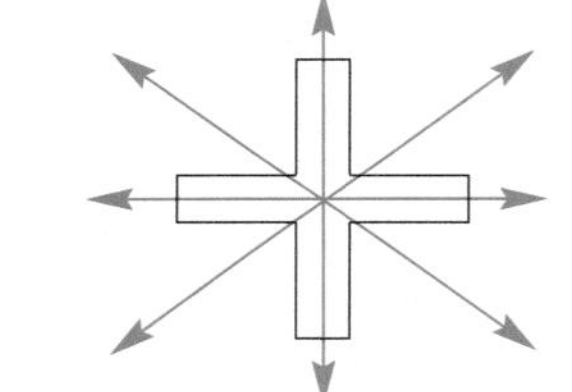	

Answer: 3 lines of symmetry

Answer: 4 lines of symmetry

EXAMPLE 4 Which of the following figures have point symmetry?

(a) 

(b) 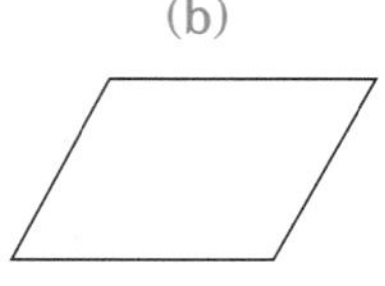

(c) 

(d) 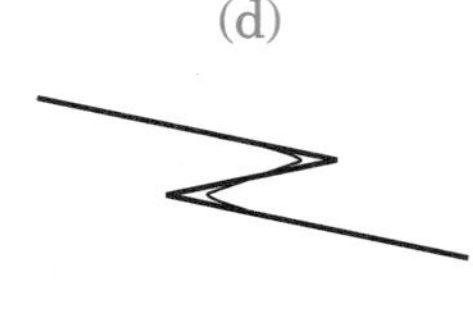

Answer: (a), (b), and (d)

CLASS EXERCISES

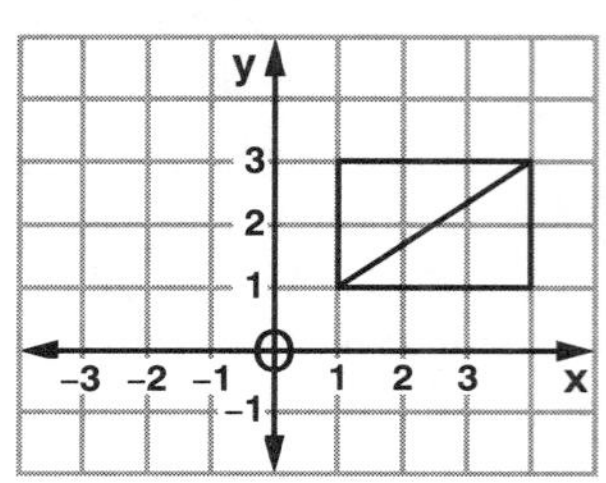

1. For the figure shown, graph the image of the figure after each transformation.

a. Translation down 1 and left 3 **b.** Reflection through the y-axis

c. Rotation 180° about the origin **d.** Reflection through the x-axis

e. Rotation 90° clockwise about (5, 1)

2. Describe the transformation of the arrow in black that is needed to produce each image.

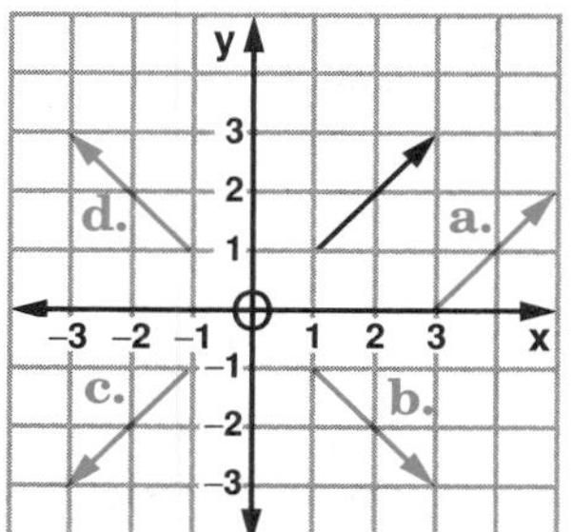

3. Triangle *ADC* is the image of isosceles triangle *ABC* after it is reflected through the x-axis.

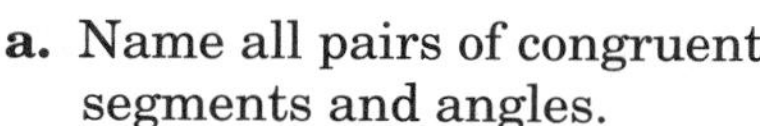

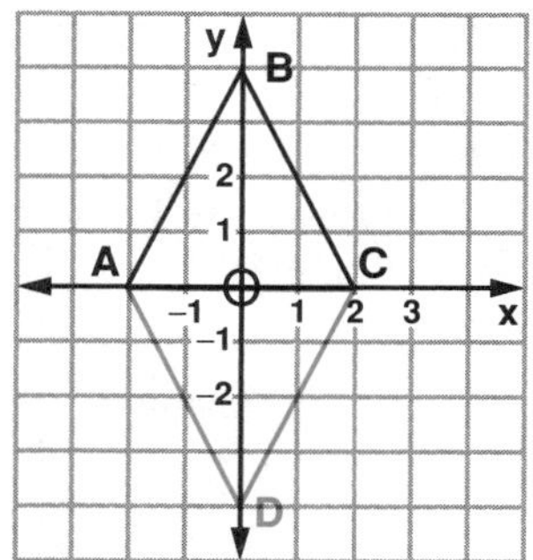

a. Name all pairs of congruent segments and angles.

b. What kind of quadrilateral is *ABCD*?

c. Describe two special properties of the diagonals of quadrilateral *ABCD*.

4. Copy each figure and draw all the lines of symmetry. Then tell how many lines of symmetry each figure has.

a. 

b. 

c. 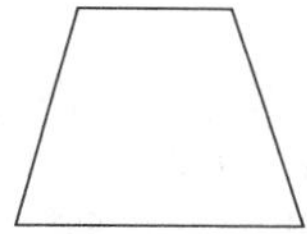

d. 

5. Which of the following figures have point symmetry?

(a)

(b) 

(c) 

(d) 

HOMEWORK EXERCISES

1. For the figure shown, graph the image of the figure after each transformation.

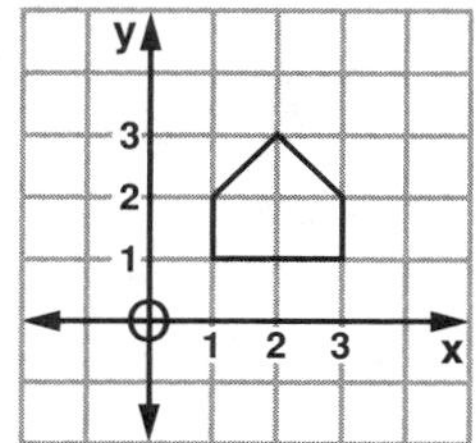

a. Translation down 2 and right 4

b. Reflection through the y-axis

c. Rotation 180° about the origin

d. Reflection through the x-axis

e. Rotation 90° clockwise about (3, 1)

2. Quadrilateral *PQRS* with vertices $P(-2, 2)$, $Q(1, 4)$, $R(2, 1)$, and $S(-1, 1)$ is translated so that the image of P is at the origin. Write the coordinates of the images of vertices Q, R, and S.

3. $\triangle WXY$ has vertices $W(-4, 4)$, $X(-1, 3)$, and $Y(-2, 1)$. Write the coordinates of each vertex of the image if $\triangle WXY$ is reflected through: **a.** the y-axis **b.** the x-axis

4. Describe the transformation of the figure in black that is needed to produce each image.

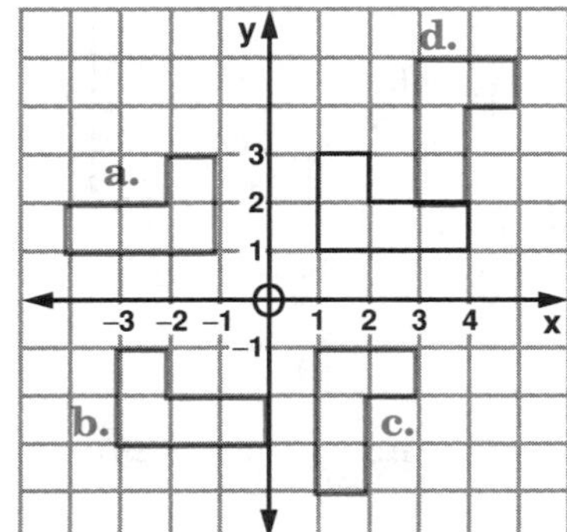

5. Line segments AB and CP are rotated 180° about point P. The image of $\overline{AB}$ is $\overline{YW}$ and the image of $\overline{CP}$ is $\overline{ZP}$.

a. Will line AB ever intersect line YW? Explain.

b. What is the image of $\angle CRB$ after a reflection through point P?

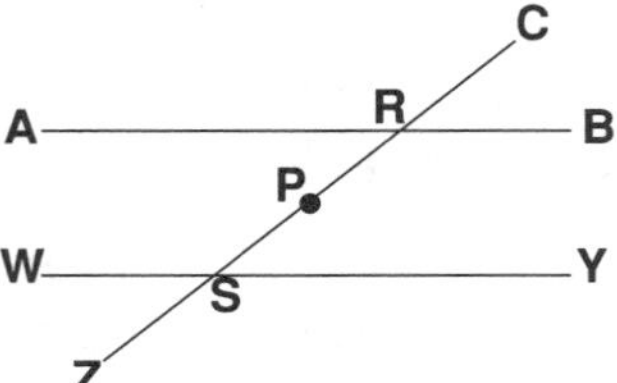

6. A figure is reflected through the y-axis. Then, its image is reflected through the x-axis. Finally, the second image is reflected through the y-axis. Explain how the final image could have been produced with only one transformation.

7. Which of the following transformations brings a figure back to its original position?

(a) a translation of 2 units up (b) a rotation of 360°

(c) a reflection about the y-axis (d) a reflection about the x-axis

8. The figure at the right is rotated clockwise 90° about point P. The image is:

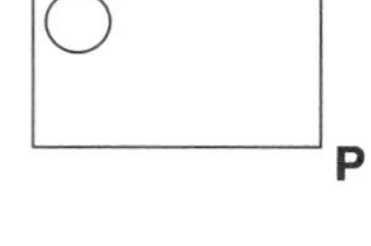

(a) 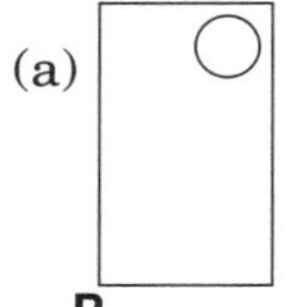

(b) 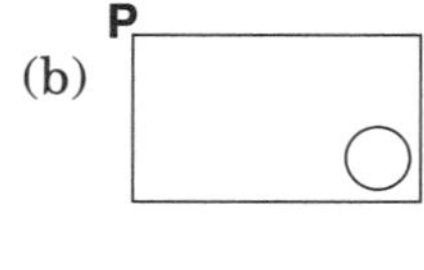

(c) 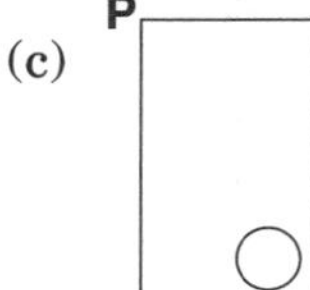

(d) 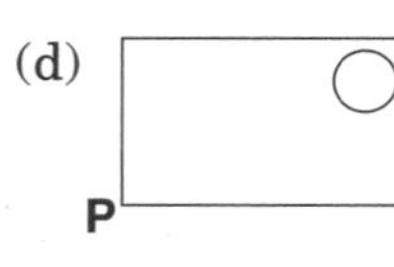

9. a. Which of the following letters have both line symmetry and point symmetry?

ABCDEFGHIJKLMNOPQRSTUVWXYZ

b. Copy the consonants that have both line and point symmetry, and draw all lines of symmetry and centers of symmetry.

10. After a figure was reflected once through the x-axis, its image was identical to the original figure. Explain how this could have happened.

SPIRAL REVIEW EXERCISES

1. If $(3, y)$ is a solution of $y = 2x - 7$, then the value of y is
(a) 5 (b) 3 (c) 2 (d) –1

2. The length of a line segment having endpoints (–9, 8) and (–15, 8) is
(a) 24 (b) 16 (c) 6 (d) 2

3. The hypotenuse of a 45° –45° –90° triangle has a length of 12 inches. The length of each leg is best approximated by
(a) 18 in. (b) 10 in. (c) 8 in. (d) 6 in.

4. A sports jacket that usually costs \$90 is on sale for 20% off the regular price. If the rate of sales tax is 5%, what is the total cost of the jacket?

5. $1.45 \times 60{,}000$ written in scientific notation is
(a) 8.7×10^{-2} (b) 8.7×10^{3}
(c) 8.7×10^{4} (d) 8.7×10^{5}

6. The mode of the data 51, 49, 40, 55, 51, 57, 60, 58, 56 is
(a) 20 (b) 51 (c) 53 (d) 55

7. Find the probability that the arrow on the spinner shown will stop on:
a. the number "5"
b. an even number
c. an odd number
d. a prime number
e. a number less than 3

8. The price of a stock increased from $10\frac{5}{8}$ to $12\frac{1}{4}$ per share. How much was the increase?

9. Solve and check: $18 - 2x = 12$

10. Write a key sequence that can be used to find the remainder when 537 is divided by 29.

11. The ratio of expenses to income for the Sharpe Printing Company is 5 to 9. If the income last week was \$2,700, what were the expenses for that week?

12. The cost of a coin-box telephone call from New York City to Miami is \$1.80 for the first minute and \$0.20 for each additional minute. What is the cost of a 15-minute call?

13. Mr. Loman had a balance of \$1,247.58 in his checking account. He wrote checks for \$198.50, \$250.00, and \$37.95. What was his new balance?

14. At least 152 students from Westwood High School attended a recent concert. If n represents the number of students from Westwood at the concert, then the open sentence that best describes this situation is
(a) $n = 152$ (b) $n \geq 152$
(c) $n > 152$ (d) $n \leq 152$

UNIT 19–1 Perimeter

THE MAIN IDEA

1. The *perimeter* of a polygon is the distance around the polygon.
2. To find the perimeter of a polygon, find the sum of the lengths of all its sides.

Figure	*Find the Sum of*	*Formula*
Triangle (sides a, b, c)	3 sides	$P = a + b + c$
Rectangle (sides ℓ, w, w, ℓ)	4 sides, two equal pairs	$P = 2\ell + 2w$
Square (sides s, s, s, s)	4 sides, all of which are equal in measure	$P = 4s$

EXAMPLE 1 Find the perimeter of the polygon.

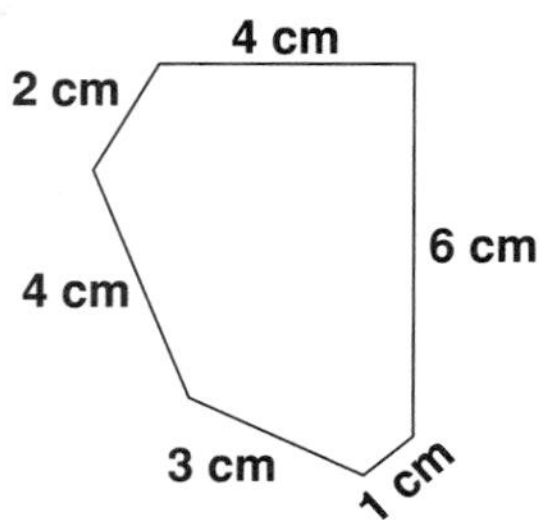

Find the sum of the lengths of all the sides.

$P = 2 + 4 + 6 + 1 + 3 + 4$
$P = 20$

Answer: The perimeter is 20 cm.

EXAMPLE 2 Find the perimeter of each figure.

Figure	*Diagram*	*By Formula*
a. a triangle whose sides measure 10 cm, 12 cm, and 20 cm	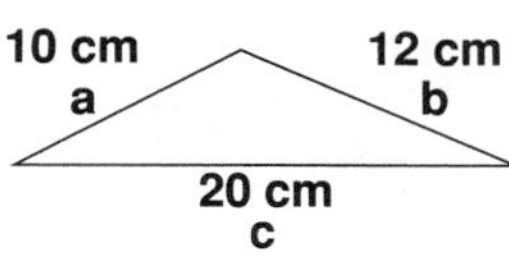	$P = a + b + c$ $= 10 + 12 + 20$ $= 42$ cm *Ans.*
b. a rectangle whose length measures 9 in. and whose width measures 5 in.	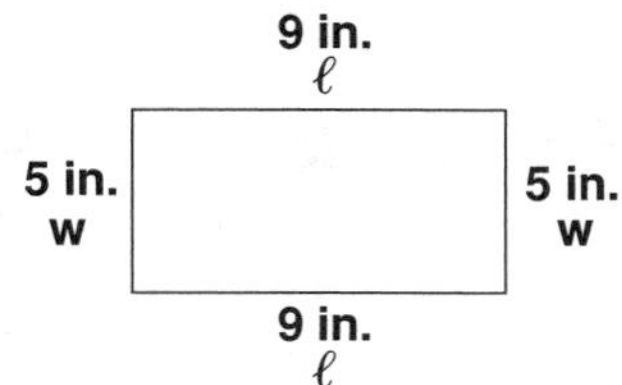	$P = 2\ell + 2w$ $= 2(9) + 2(5)$ $= 18 + 10$ $= 28$ in. *Ans.*
c. a square, each of whose sides measures 22 m	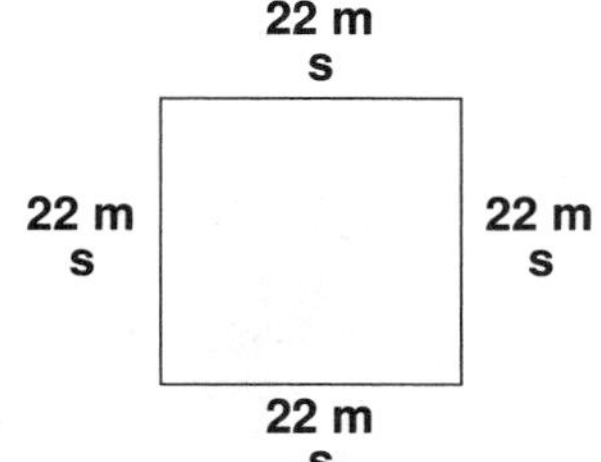	$P = 4s$ $= 4(22)$ $= 88$ m *Ans.*

EXAMPLE 3 Find the measure of each side of a square whose perimeter is 100 cm.

THINKING ABOUT THE PROBLEM

Draw a diagram and let x equal the length of each side of the square.

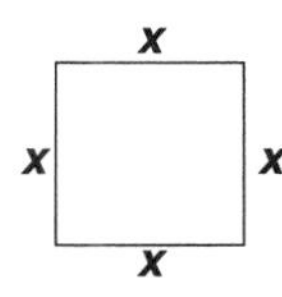

Write an equation that represents the perimeter. $4x = 100$

Solve for x.

$$\frac{4x}{4} = \frac{100}{4}$$

$$x = 25$$

Answer: The length of a side of the square is 25 cm.

EXAMPLE 4 If the perimeter of the given rectangle is 140 cm, find the dimensions of the rectangle.

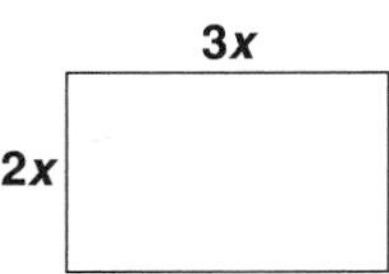

Opposite sides of a rectangle are equal in length. The sum of the four sides is 140 cm.

$$2x + 3x + 2x + 3x = 140$$

$$10x = 140$$

$$\frac{1}{10}(10x) = \frac{1}{10}(140)$$

$$x = 14$$

To find the dimensions of the rectangle:

width $= 2x$
$= 2(14) = 28$ cm

length $= 3x$
$= 3(14) = 42$ cm

CLASS EXERCISES

1. Find the perimeter of each of the following polygons.

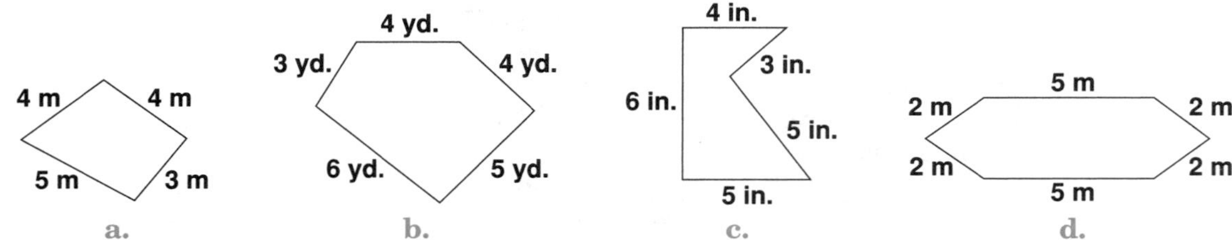

2. Find the perimeter of a triangle whose sides measure:
   a. 5 in., 9 in., 7 in. **b.** 5 cm, 12 cm, 13 cm **c.** 3.1 m, 5.2 m, 4.5 m **d.** $1\frac{1}{2}$ ft., $3\frac{1}{3}$ ft., $2\frac{1}{6}$ ft.

3. Find the perimeter of a rectangle whose measurements are:
   a. $\ell = 5$ cm, $w = 2$ cm **b.** $\ell = 10.5$ in., $w = 4.6$ in. **c.** $\ell = 5\frac{2}{3}$ yd., $w = 3\frac{1}{2}$ yd.

4. Find the perimeter of a square whose side measures: **a.** 17 ft. **b.** 5.9 cm **c.** $8\frac{2}{3}$ m

5. What is the perimeter of an isosceles triangle whose base measures 10 ft. and each of whose legs measures 14.1 ft.?

6. Find the length of a side of a square whose perimeter is 200 m.

7. Find the length of a side of an equilateral triangle whose perimeter is 300 yd.

8. Find the perimeter.
   Note: All angles are right angles.

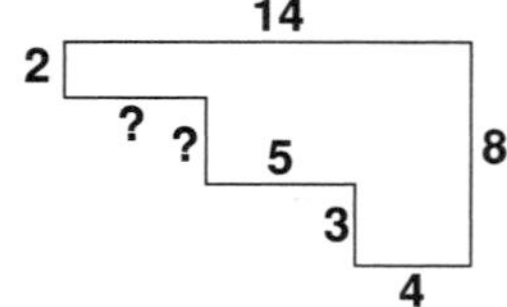

9. The perimeter of the figure shown is 94 cm. Find the length of the unknown side.

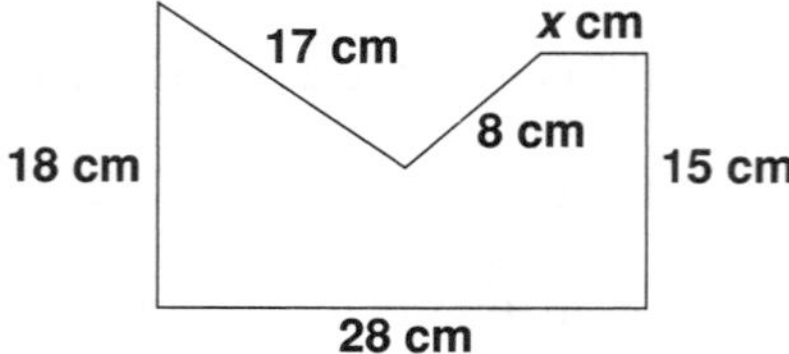

10. The perimeter of the isosceles triangle shown is 108 cm. Find the length of each side.

11. How many feet of baseboard are needed to go around a rectangular room that measures 14 feet by 10 feet?

HOMEWORK EXERCISES

1. Find the perimeter of each of the following polygons.

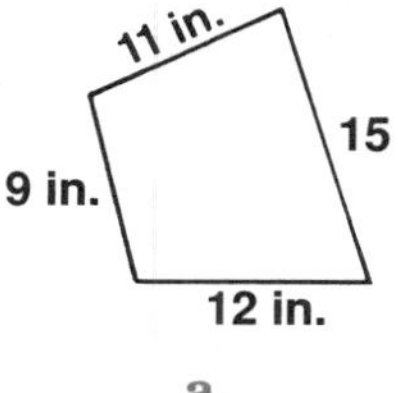

a.

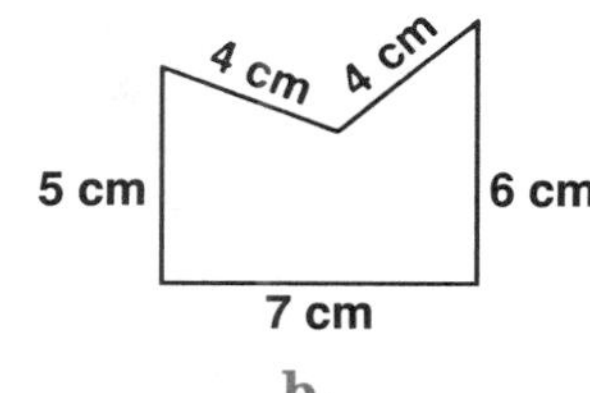

b.

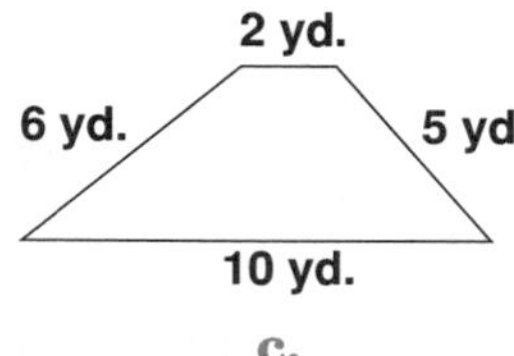

c.

2. Find the perimeter of a triangle whose sides measure:
a. 15 cm, 20 cm, 25 cm **b.** 6.3 m, 9.2 m, 14.9 m **c.** $3\frac{2}{3}$ yd., $4\frac{1}{8}$ yd., $6\frac{1}{2}$ yd.

3. Find the perimeter of a rectangle whose measurements are:
a. $\ell = 12$ m, $w = 10$ m **b.** $\ell = 9.6$ in., $w = 8.2$ in. **c.** $\ell = 2\frac{1}{8}$ ft., $w = 1\frac{3}{4}$ ft.

4. Find the perimeter of a square whose side measures: **a.** 11 cm **b.** 9.9 ft. **c.** $7\frac{3}{8}$ yd.

5. Find the perimeter of an equilateral triangle whose side measures: **a.** $3\frac{1}{3}$ ft. **b.** 5.9 cm

6. What is the perimeter of an isosceles triangle whose legs each measure 7 inches and whose base measures $5\frac{1}{2}$ inches?

7. Find the length of a side of a square whose perimeter is: **a.** 48 yd. **b.** 14.4 m **c.** 4.48 ft.

8. Find the length of a side of an equilateral triangle whose perimeter is: **a.** 150 ft. **b.** 180 m

9. Find the missing lengths and calculate the perimeter. Note: All angles are right angles.

a.

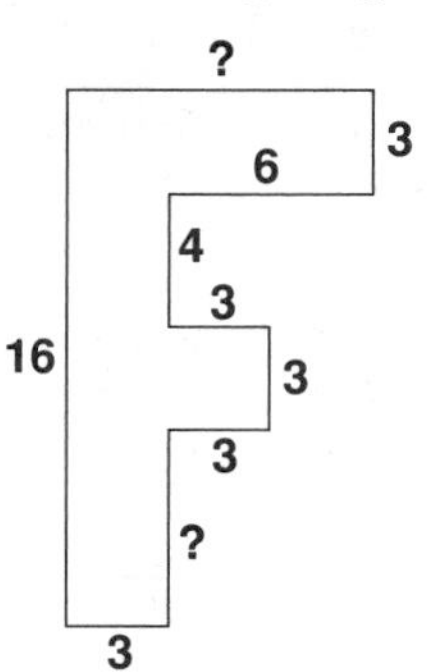

b.

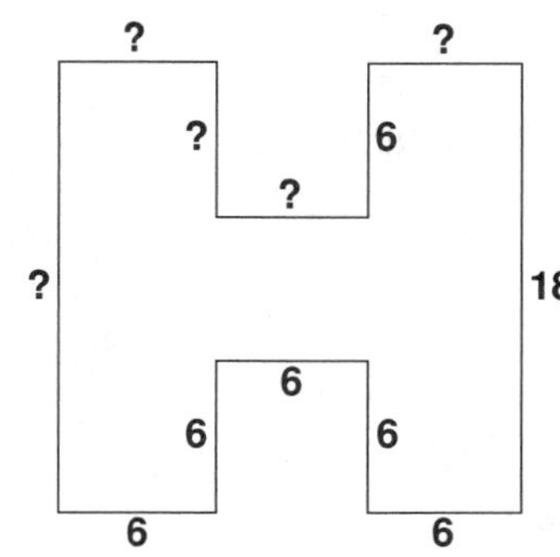

10. The perimeter of the swimming pool shown is 59 feet. Write and solve an equation to find the value of y.

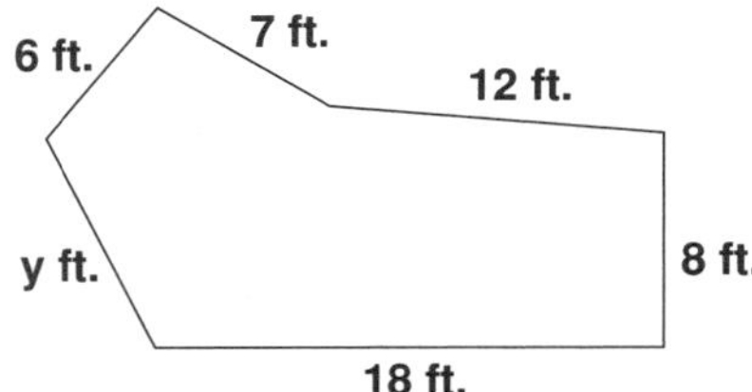

11. A play area is to be shaped like the trapezoid shown. The area will be enclosed with 190 feet of fencing. Find the length of each side.

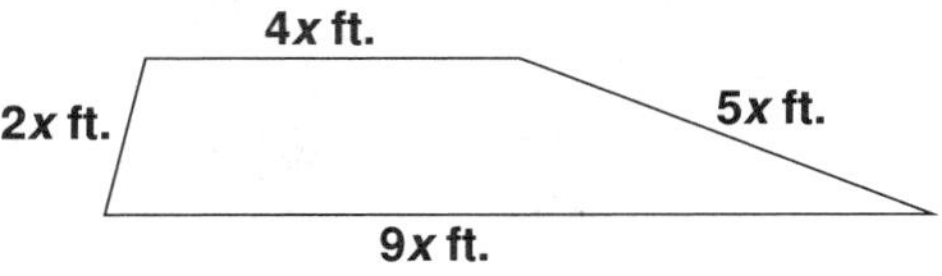

12. Find the perimeter of a right triangle whose legs measure 15 cm and 20 cm.

13. The perimeter of an isosceles triangle is 66 inches. If the length of each leg is 5 times the length of the base, write and solve an equation to find the length of each side of the triangle.

14. How many inches of fringe are needed to go around a rectangular rug that is 9 feet by 6 feet?

1. Find the measure of ∠B.

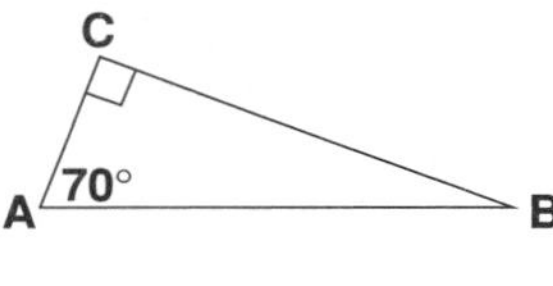

2. Find the missing length.

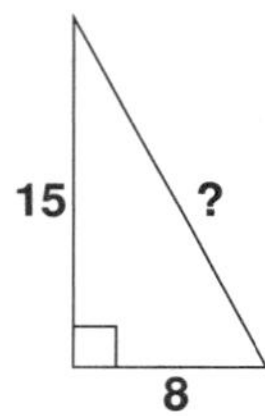

3. An angle that measures 98° is
(a) acute (b) right
(c) obtuse (d) straight

4. If 44 students out of 200 are girls, what percent are girls?

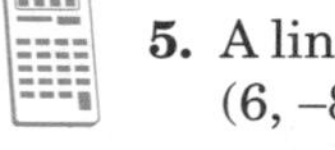

5. A line segment has endpoints at (–2, 5) and (6, –8). Approximate the length of the segment to the nearest tenth.

6. If the rates for a 900 telephone call are:
$3.75 for the first minute
$2.50 for each additional minute
what is the cost of an 11-minute call?

7. Jeff worked from 10:45 A.M. to 1:15 P.M. How long did he work?

8. 5 liters is equivalent to
(a) 5,000 mL (b) 500 mL
(c) 0.5 kL (d) 0.05 kL

9. 20 is 40% of
(a) 40 (b) 50 (c) 80 (d) 100

10. On the graph shown, which point has the coordinates (3, –2)?

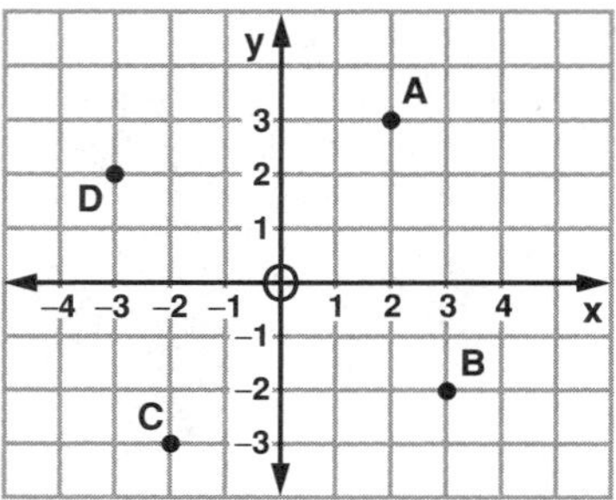

11. Write 0.00000076 in scientific notation.

UNIT 19-2 Area: Rectangle; Parallelogram

THE AREA OF A RECTANGLE

THE MAIN IDEA

1. ***Area*** means the measurement of the space contained within a flat, closed surface.
2. Area is measured in ***square units***. A square centimeter (written 1 cm^2) is an example of a unit of area. It is the space contained in a square that measures 1 centimeter on each side.

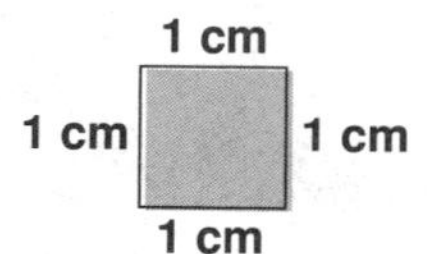

3. Units of area include the square of any unit used to measure a line. The product of 2 linear units is a square unit. For example, $2 \text{ m} \times 3 \text{ m} = 6 \text{ m}^2$.
4. To find the area of a rectangle, multiply the measure of its length (base) by the measure of its width (height or altitude).

$A_{\text{rectangle}} = \ell w$

5. A square is a rectangle with four sides that are equal in measure. To find the area of a square, multiply the measure of its side by itself.

$A_{\text{square}} = s^2$

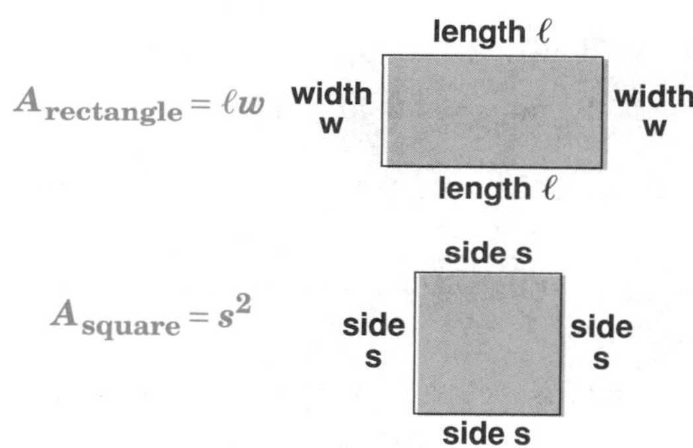

EXAMPLE 1 Find the area of the rectangle.

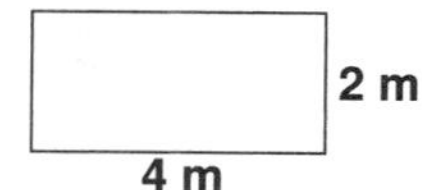

In the formula for the area of a rectangle, substitute 4 for ℓ and 2 for w.

$$A_{\text{rectangle}} = \ell w$$
$$A_{\text{rectangle}} = 4 \times 2$$
$$= 8 \text{ m}^2 \quad \textit{Ans.}$$

Observe that there are 8 squares contained in the rectangle, with each side of each square measuring 1 meter. Thus, there are 8 square meters of area in this rectangle.

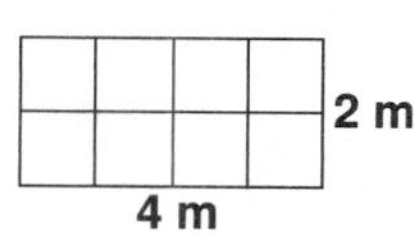

EXAMPLE 2 How much will it cost to cover a rectangular room that is 3 yards wide and 5 yards long with carpet at \$18.50 per square yard?

$$A_{\text{rectangle}} = \ell w$$
$$= 5 \times 3$$
$$= 15 \text{ sq. yd. of carpet are needed}$$

$\$18.50 \times 15 = \$277.50$ *Ans.*

EXAMPLE 3 The area of a rectangle whose dimensions are 3 feet by 4 inches is
(a) 12 sq. in. (b) 144 sq. in. (c) 12 sq. ft. (d) 2 sq. ft.

THINKING ABOUT THE PROBLEM

One dimension is given in feet and the other is in inches. To find the area, you must first write both dimensions in the same unit.

$$\begin{aligned} A_{\text{rectangle}} &= \ell w \\ &= 3 \text{ ft.} \times 4 \text{ in.} \\ &= 36 \text{ in.} \times 4 \text{ in.} \\ &= 144 \text{ in.}^2 \end{aligned}$$

Answer: (b)

CLASS EXERCISES

1. Find the area of a rectangle in which:

a. $\ell = 9$ cm, $w = 4$ cm **b.** $\ell = 15$ in., $w = 8$ in. **c.** $\ell = 22.4$ m, $w = 8.7$ m

d. $\ell = 15.9$ cm, $w = 4.2$ cm **e.** $\ell = 4\frac{3}{8}$ ft., $w = 2$ ft. **f.** $\ell = 10\frac{3}{4}$ in., $w = 5\frac{1}{2}$ in.

2. Find the number of square yards of linoleum needed to cover a kitchen floor that is 4 yards wide and 6 yards long.

3. How many square tiles, each with area of 1 sq. ft., are needed to cover a floor 9 ft. wide and 16 ft. long?

4. The area of a rectangle whose dimensions are 10 feet by 3 yards is
(a) 30 square feet (b) 30 square yards (c) 9 square yards (d) 90 square feet

5. The area of a rectangle whose dimensions are 25 cm by 1 m is
(a) 2,500 cm^2 (b) 250 cm^2 (c) 25 m^2 (d) 250 m^2

6. The cost of carpet is $22 a square yard. How much will it cost to cover a rectangular floor that is 10 yards wide and 15 yards long?

7. The cost to clean a carpet is $1.15 a square foot. How much will it cost to clean a rectangular carpet 9 ft. by 12 ft.?

8. Find the area of a square each of whose sides measures:

a. 5 in. **b.** 10 cm **c.** $\frac{3}{4}$ ft. **d.** 9.7 m

9. Find the measure of a side of a square whose area is:

a. 9 cm^2 **b.** 144 ft.^{2} **c.** 36 m^2 **d.** 400 cm^2

10. Approximate, to the nearest tenth, the length of a side of a square with the given area.

a. 12 in.^{2} **b.** 20 cm^2 **c.** 180 m^2 **d.** 250 ft.^{2}

THE AREA OF A PARALLELOGRAM

THE MAIN IDEA

1. By moving $\triangle$I, a parallelogram can be changed into a rectangle.

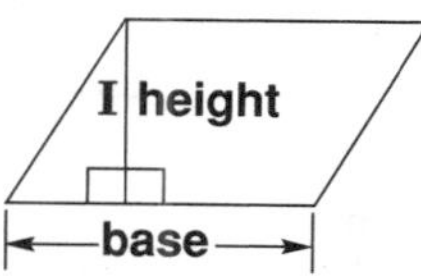

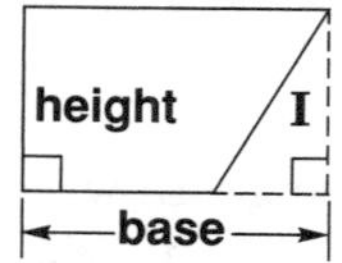

Thus, the area of a parallelogram is equal to the area of the rectangle that has the same length and width (or base and height).

2. $A_{\text{parallelogram}} = bh$

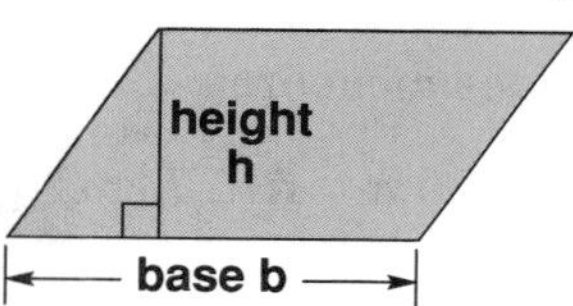

EXAMPLE 4 Find the area of the parallelogram.

In the formula for the area of a parallelogram, substitute 10 for b and 5 for h.

$A_{\text{parallelogram}} = bh$
$A_{\text{parallelogram}} = 10 \times 5$
$= 50 \text{ m}^2$ *Ans.*

CLASS EXERCISES

1. Find the area of each parallelogram.

a. 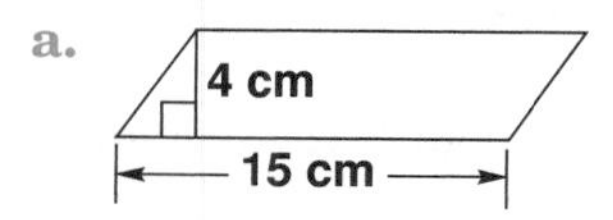

b. 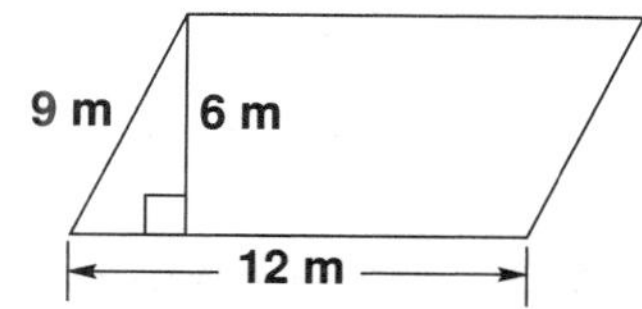

c. 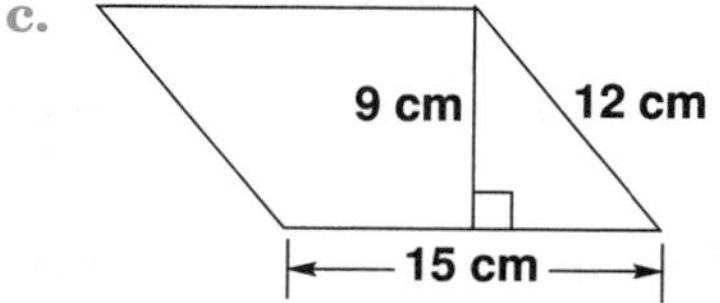

2. Find the area of a parallelogram in which:

a. $b = 9$ in., $h = 11$ in. **b.** $b = 11.3$ m, $h = 6.4$ m **c.** $b = 12$ ft., $h = 2\frac{1}{3}$ ft.

d. $b = 14$ cm, $h = 7$ cm **e.** $b = 20.7$ m, $h = 8.6$ m **f.** $b = 7\frac{5}{8}$ in., $h = 3\frac{1}{4}$ in.

HOMEWORK EXERCISES

1. Find the area of a rectangle in which:

a. $\ell = 5$ in., $w = 3$ in. **b.** $\ell = 8$ cm, $w = 5$ cm **c.** $\ell = 14$ m, $w = 10$ m

d. $\ell = 6.7$ cm, $w = 3.4$ cm **e.** $\ell = 5\frac{1}{2}$ in., $w = 2\frac{1}{4}$ in. **f.** $\ell = 10$ ft., $w = 6\frac{3}{4}$ ft.

2. Find the area of a square each of whose sides measures:

a. 6 cm **b.** 9 in. **c.** 11 m **d.** 15 yd. **e.** 8.9 m **f.** 2.7 cm **g.** $\frac{1}{2}$ in. **h.** $3\frac{1}{4}$ ft.

3. Find the measure of a side of a square whose area is:

a. 16 m^2 **b.** 121 sq. in. **c.** 900 sq. ft. **d.** 64 cm^2

4. Approximate to the nearest tenth, the length of a side of a square with the given area.

a. 40 in.^{2} **b.** 500 cm^2 **c.** 1,200 mm^2 **d.** 16,000 ft.^{2}

5. Find the area of each parallelogram.

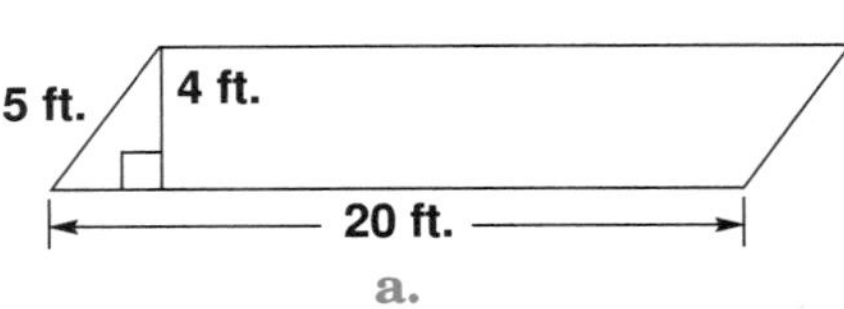

a.

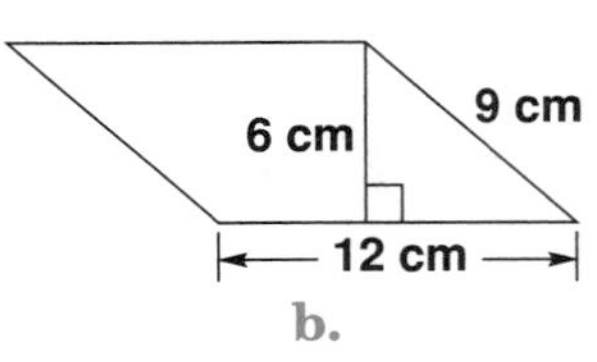

b.

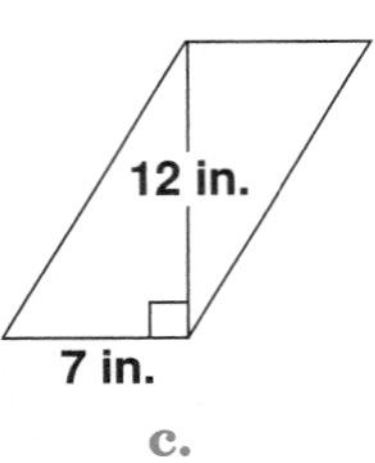

c.

6. Find the area of a parallelogram in which:

a. $b = 20$ in., $h = 11$ in. **b.** b = 45 cm, $h = 12.9$ cm **c.** $b = 4.8$ m, $h = 2.6$ m

d. $b = 24$ yd., $h = 10\frac{1}{2}$ yd. **e.** $b = 5\frac{1}{2}$ in., $h = 2\frac{3}{8}$ in. **f.** $b = 17.5$ cm, $h = 9.2$ cm

7. Find the number of square yards of carpet needed to cover a living room floor 3 yards wide and 7 yards long.

8. How many square tiles, each with area 1 sq. in., are needed to cover a countertop 14 inches wide and 30 inches long?

9. The area of a rectangle whose dimensions are 5 feet by 10 inches is
(a) 50 sq. in. (b) 600 sq. in. (c) 50 sq. ft. (d) 60 sq. in.

10. The area of a rectangle whose dimensions are 2 m by 50 cm is
(a) 100 cm^2 (b) 1,000 cm^2 (c) 10,000 cm^2 (d) 10,000 m^2

11. The cost of carpet is $27.80 a square yard. How much will it cost to cover a rectangular floor that is 12 yards long and 8 yards wide?

12. A gardener charges $0.60 per square foot to seed a lawn. How much will he charge to seed a rectangular lawn 20 feet by 12 feet?

13. The sides of a square are doubled in length. What effect does this have on the area of the square?

14. The area of a square is 64 cm^2. What is its perimeter?

15. A square and a rectangle each have an area of 100 cm^2. The length of the rectangle is 25 cm. What is the ratio of the perimeter of the square to the perimeter of the rectangle?

16. If a rectangular room is 14 feet long by 10 feet wide, how many square yards of carpet are needed for the room?

SPIRAL REVIEW EXERCISES

1. Find the perimeter of a square each of whose sides measures 5.8 cm.

2. The measure of a side of an equilateral triangle that has a perimeter of 18 m is
(a) 9 m (b) 6 m (c) 54 m (d) 72 m

3. If the rate of sales tax is 7%, how much tax will be charged on a $20 purchase?

4. 17 meters is equal to
(a) 170 cm (b) 1,700 cm
(c) 17,000 cm (d) 0.17 cm

5. The value of $3x - 9$ when x is 10 is
(a) 3 (b) 39 (c) 21 (d) 18

6. When written as a fraction, 55% is
(a) $\frac{11}{20}$ (b) $\frac{1}{2}$ (c) $\frac{2}{3}$ (d) $\frac{2}{5}$

7. $\frac{1}{2} + \frac{3}{8} - \frac{1}{8}$ is equal to
(a) $\frac{3}{8}$ (b) $\frac{1}{4}$ (c) $\frac{5}{8}$ (d) $\frac{3}{4}$

8. Tell the coordinates of each vertex of parallelogram *ABCD*.

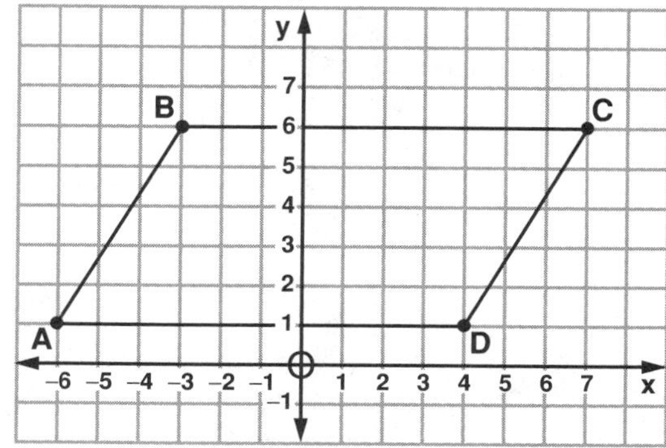

9. The equation of a line is $y = 0.687x - 13.58$. Find the y-coordinate of the point on the line whose x-coordinate is 25.2.

10. Find the complement of an angle whose measure is 37°.

11. Solve for x: $3x + 12 \geq 9$

12. Find the mean, median, mode, and range of the following numbers:
22, 18, 38, 25, 15, 40, 38

UNIT 19-3 Area: Triangle; Trapezoid; Composite Figure

THE AREA OF A TRIANGLE

THE MAIN IDEA

1. A *diagonal* of a parallelogram divides the parallelogram into two triangles that are equal in area. Thus, the area of one triangle is one-half the area of the parallelogram.

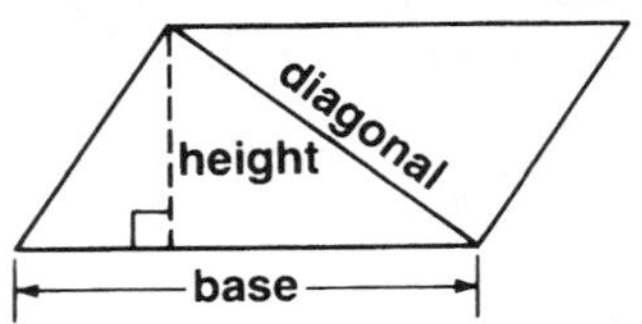

2. $A_{triangle} = \frac{1}{2}bh$

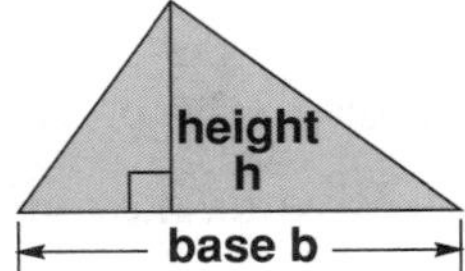

EXAMPLE 1 Find the area of each triangle.

a.

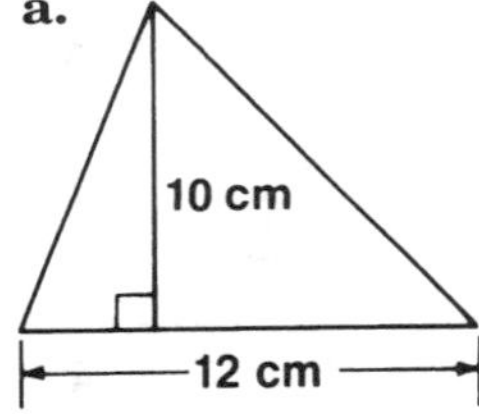

$A_{triangle} = \frac{1}{2}bh$

$A_{triangle} = \frac{1}{2} \times 12 \times 10$

$= \frac{1}{2} \times 120$

$= 60 \text{ cm}^2$

b.

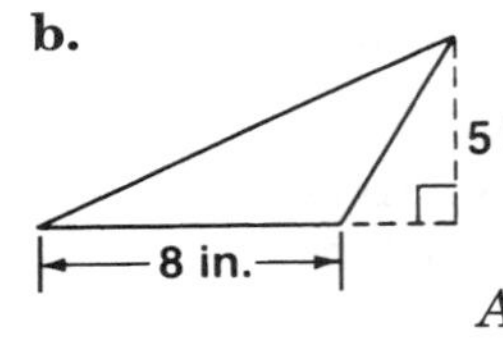

In an obtuse triangle, the height is measured outside the triangle.

$A_{triangle} = \frac{1}{2}bh$

$A_{triangle} = \frac{1}{2} \times 8 \times 5$

$= \frac{1}{2} \times 40 = 20 \text{ in.}^2$

c.

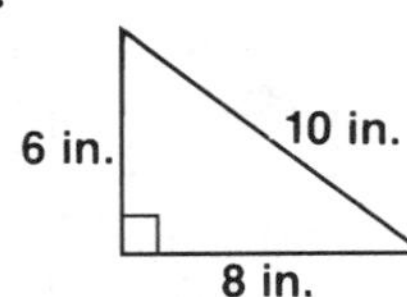

In a right triangle, one leg is the base and the other leg is the height.

$A_{triangle} = \frac{1}{2}bh$

$A_{triangle} = \frac{1}{2} \times 8 \times 6 = \frac{1}{2} \times 48 = 24 \text{ in.}^2$

CLASS EXERCISES

1. Find the area of each triangle.

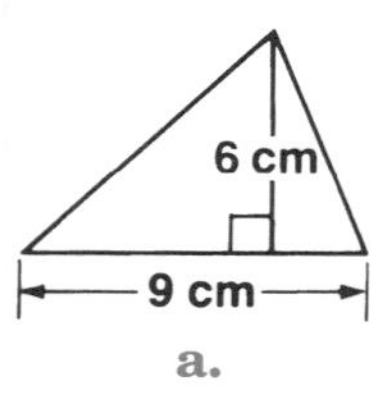

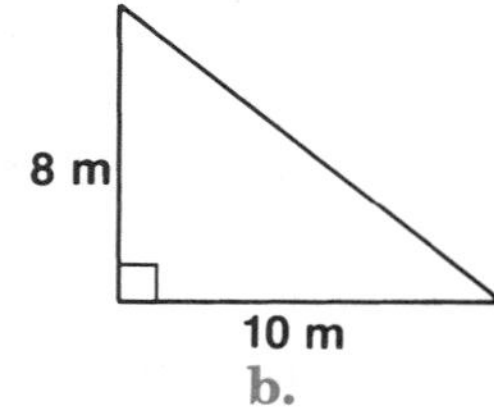

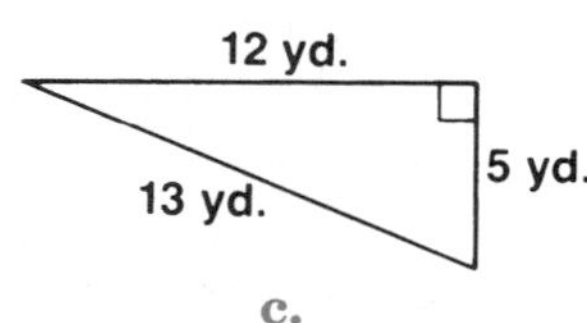

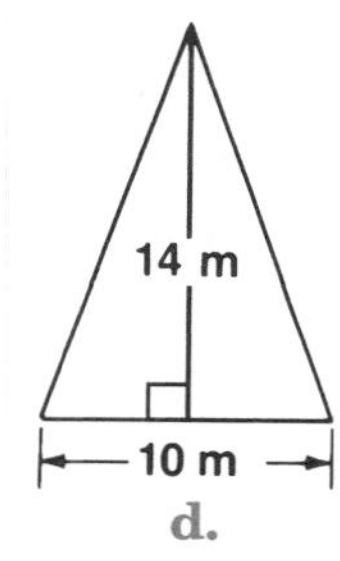

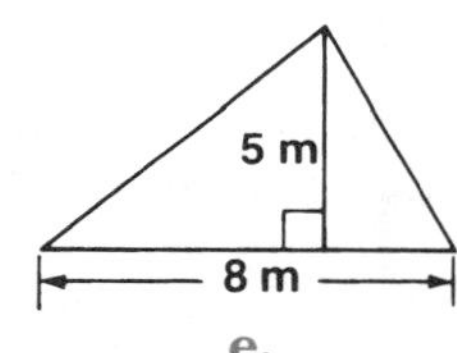

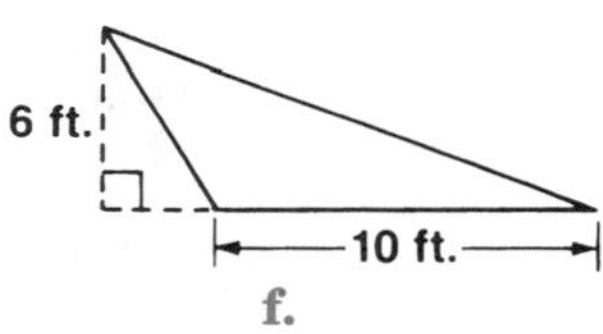

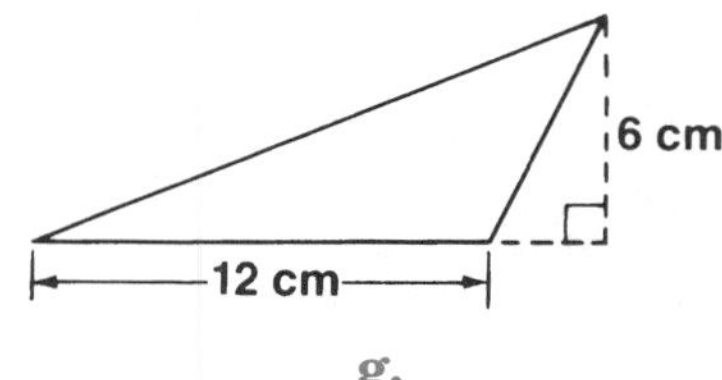

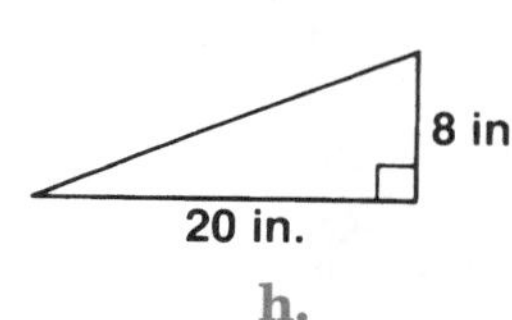

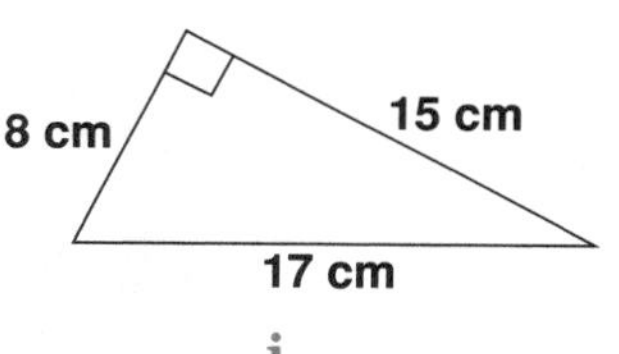

2. Find the area of a triangle in which:

a. $b = 12$ cm, $h = 10$ cm **b.** $b = 5$ ft., $h = 4$ ft. **c.** $b = 2.5$ m, $h = 1.6$ m

d. $b = 5\frac{1}{2}$ ft., $h = 4$ ft. **e.** $b = 4.8$ m, $h = 2.2$ m **f.** $b = 8\frac{1}{4}$ in., $h = 5\frac{1}{2}$ in.

3. Find the area of each right triangle, using the given lengths of the legs.

a. 8 cm, 5 cm **b.** 8.4 m, 10 m **c.** 6.8 cm, 4.6 cm

4. **a.** Part of a rectangular lobby is to be tiled at \$5.75 per square yard. Use the diagram to determine the cost of the area to be tiled.

b. The remainder of the lobby is to be carpeted at \$25 per square yard. Find the cost of the carpeted area.

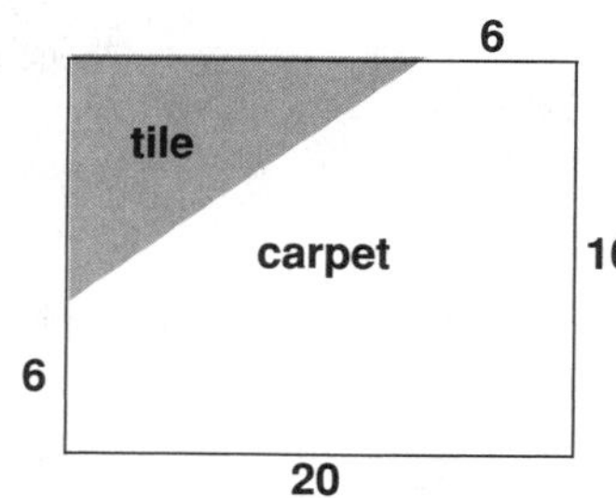

THE AREA OF A TRAPEZOID

THE MAIN IDEA

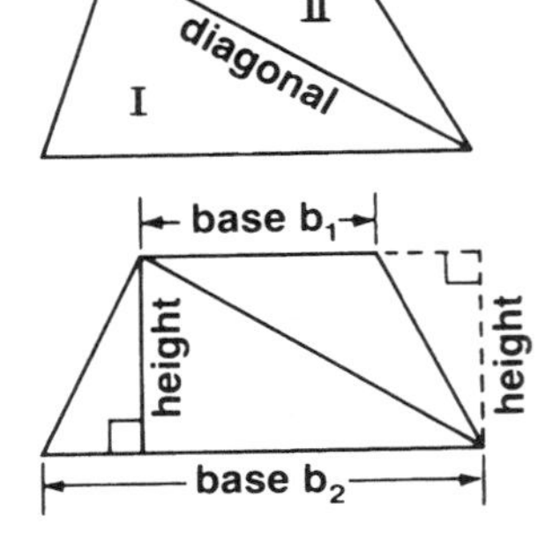

1. A diagonal of a trapezoid divides the trapezoid into two triangles. Thus, the area of the trapezoid is the sum of the areas of the triangles.

   Each triangle contains one base of the trapezoid (the parallel sides). The height of each triangle is the same as the height of the trapezoid.

2. $A_{trapezoid} = \frac{1}{2}h(b_1 + b_2)$

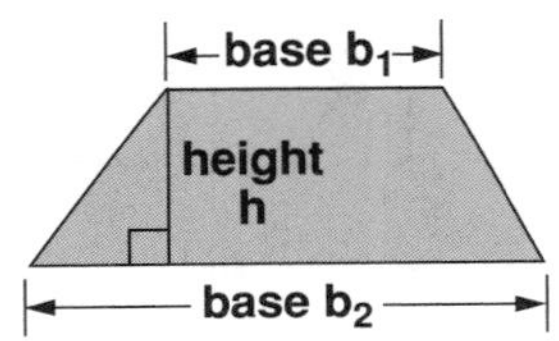

3. When applying this formula, be sure to follow the order of operations.

EXAMPLE 2 Find the area of each trapezoid.

a.

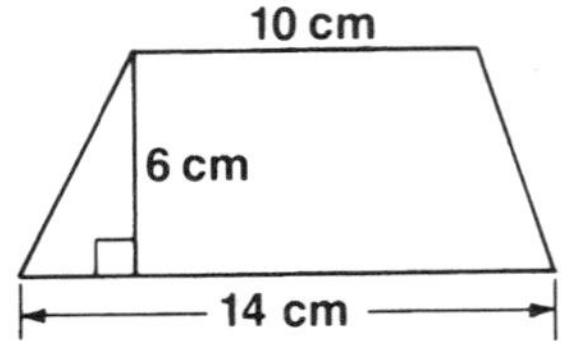

$$A_{trapezoid} = \frac{1}{2}h(b_1 + b_2)$$

$$A_{trapezoid} = \frac{1}{2} \times 6 \times (10 + 14)$$

$$= \frac{1}{2} \times 6 \times (24)$$

$$= 3(24) = 72 \text{ cm}^2$$

b.

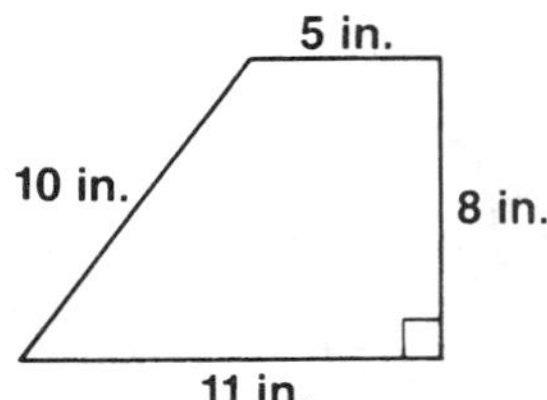

The right angle in the diagram tells that the 8-inch side is the height of the trapezoid.

$$A_{trapezoid} = \frac{1}{2}h(b_1 + b_2)$$

$$A_{trapezoid} = \frac{1}{2} \times 8 \times (5 + 11)$$

$$= \frac{1}{2} \times 8 \times (16) = 4(16) = 64 \text{ in.}^2$$

CLASS EXERCISES

1. Find the area of each trapezoid.

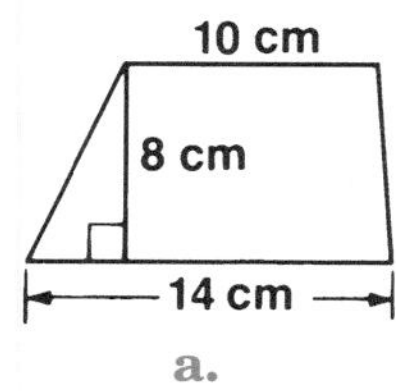

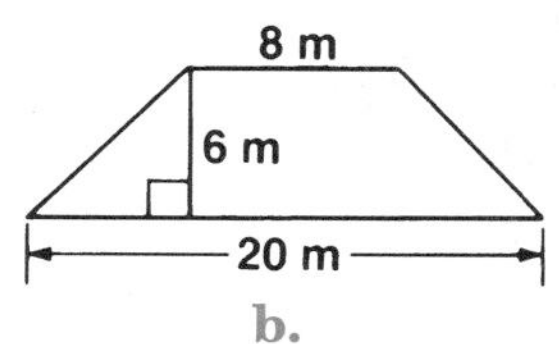

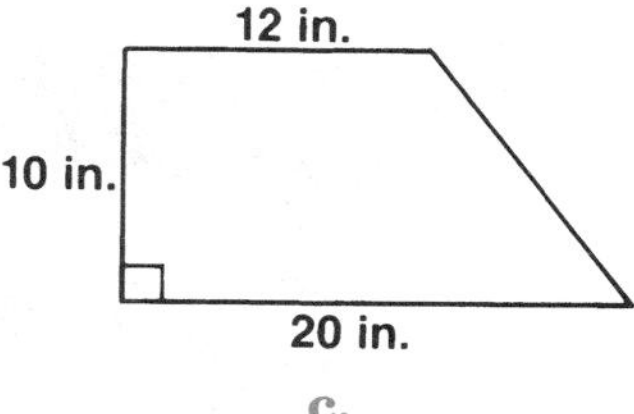

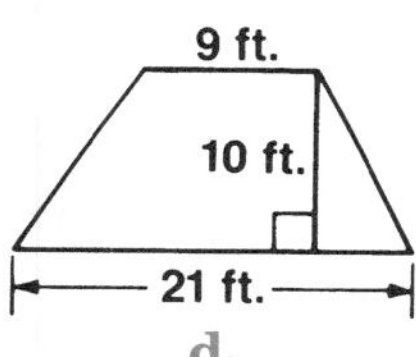

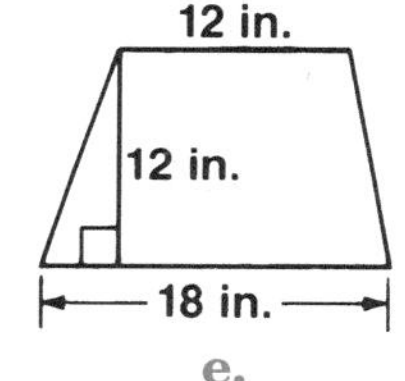

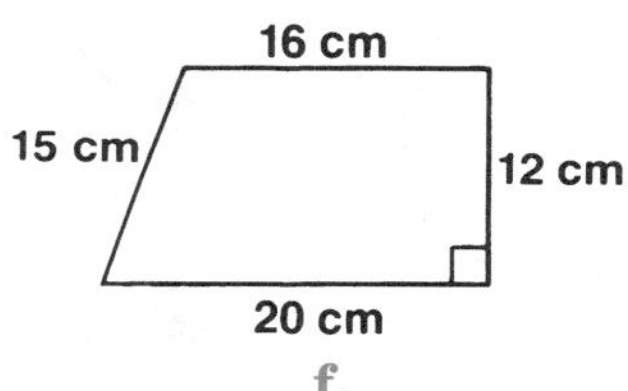

2. Find the area of a trapezoid in which:

a. $h = 6$ m, $b_1 = 8$ m, $b_2 = 14$ m
b. $h = 5$ in., $b_1 = 11$ in., $b_2 = 19$ in.
c. $h = 10$ ft., $b_1 = 9$ ft., $b_2 = 11$ ft.
d. $h = 9$ m, $b_1 = 13$m, $b_2 = 17$ m
e. $h = 10$ cm, $b_1 = 5.2$ cm, $b_2 = 12.8$ cm
f. $h = 16$ in., $b_1 = 20$ in., $b_2 = 40$ in.
g. $h = 5.6$ m, $b_1 = 10.4$ m, $b_2 = 21.6$ m
h. $h = 2\frac{1}{2}$ ft., $b_1 = 8$ ft., $b_2 = 12$ ft.

THE AREA OF A COMPOSITE FIGURE

THE MAIN IDEA

1. A ***composite figure*** is a figure that is made up of two or more of the basic figures for which there are area formulas.
2. To find the area of a composite figure:
   a. Divide the figure into smaller, familiar figures for which area formulas are known. Sometimes this can be done in more than one way.
   b. Find the area of each of the smaller figures.
   c. Add the areas of the smaller figures.
3. Some geometric figures enclose an empty space. To find such an area, subtract the area of the empty space from the area of the surrounding figure.

EXAMPLE 3 Find the area of the figure. All angles are right angles.

3 m
7 m
12 m
6 m

There is more than one way to break the figure into rectangles that do not overlap.

Method 1:

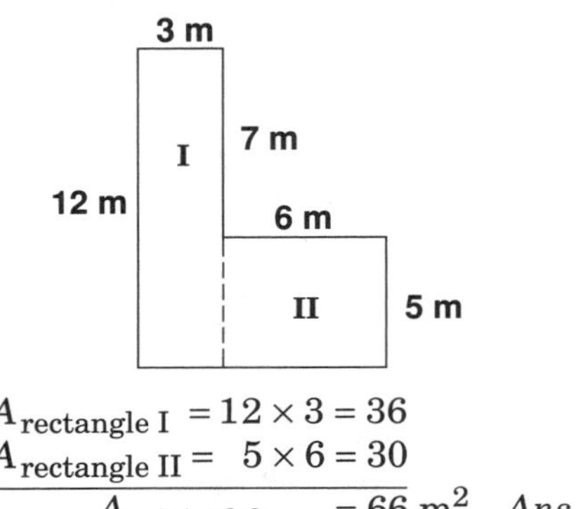

$$\begin{aligned} A_{\text{rectangle I}} &= 12 \times 3 = 36 \\ A_{\text{rectangle II}} &= \ \ 5 \times 6 = 30 \\ \hline A_{\text{original figure}} &= 66 \text{ m}^2 \quad \textit{Ans.} \end{aligned}$$

Method 2:

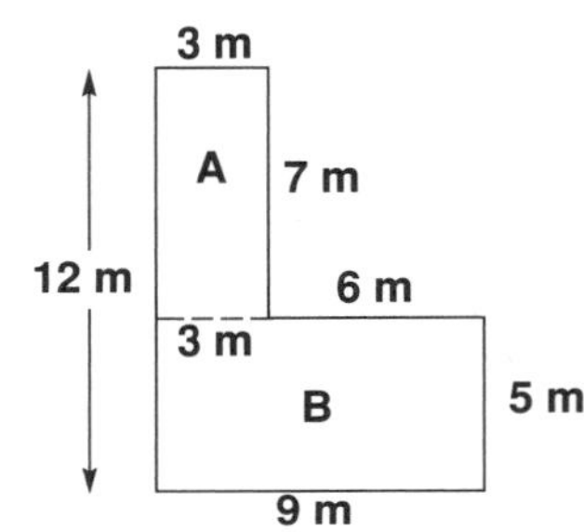

$$\begin{aligned} A_{\text{rectangle } A} &= 7 \times 3 = 21 \\ A_{\text{rectangle } B} &= 5 \times 9 = 45 \\ \hline A_{\text{original figure}} &= 66 \text{ m}^2 \quad \textit{Ans.} \end{aligned}$$

EXAMPLE 4 A square hole with sides 5 centimeters long is cut out of a rectangular plate that is 12 centimeters long and 9 centimeters wide. What is the area of the remaining part of the plate?

Draw a diagram, find the area of each known figure, and subtract to find the required area.

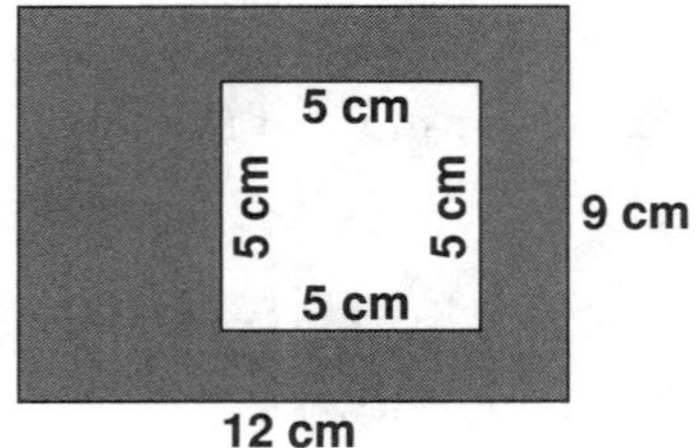

$$\begin{aligned} A_{\text{rectangle}} &= 12 \times 9 = 108 \\ A_{\text{square}} &= 5^2 \quad = \ \ 25 \\ \hline A_{\text{shaded figure}} &= \ \ 83 \text{ cm}^2 \quad \textit{Ans.} \end{aligned}$$

CLASS EXERCISES

1. Find the area of each figure. (You may assume right angles as shown.)

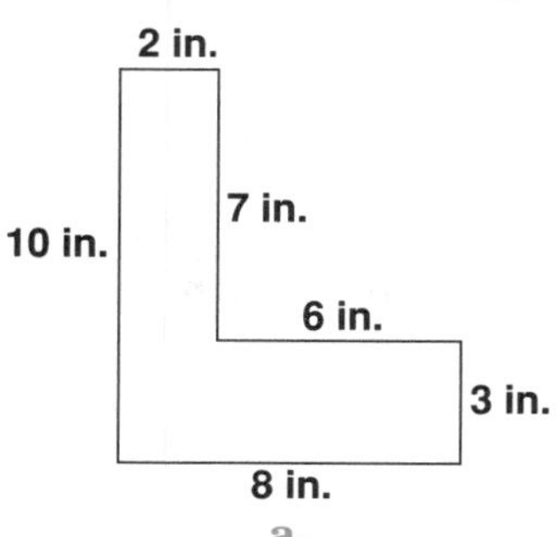

a.

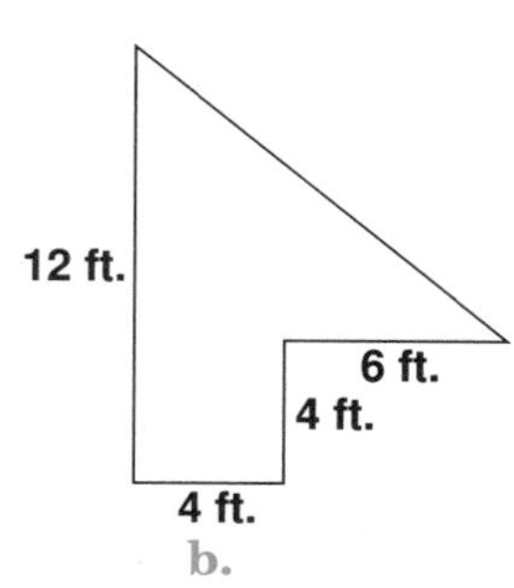

b.

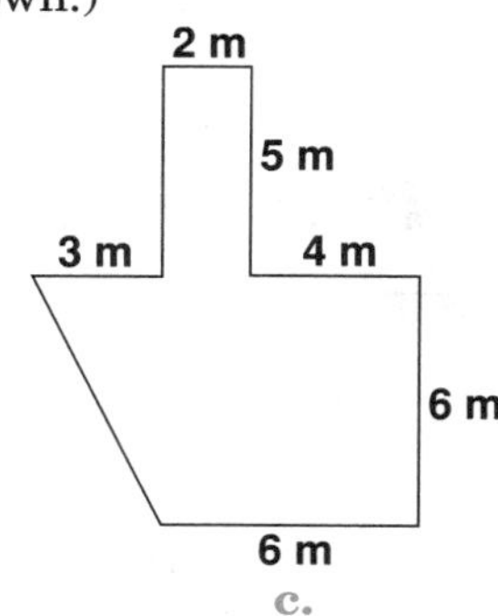

c.

2. Find the area of the shaded part of each figure.

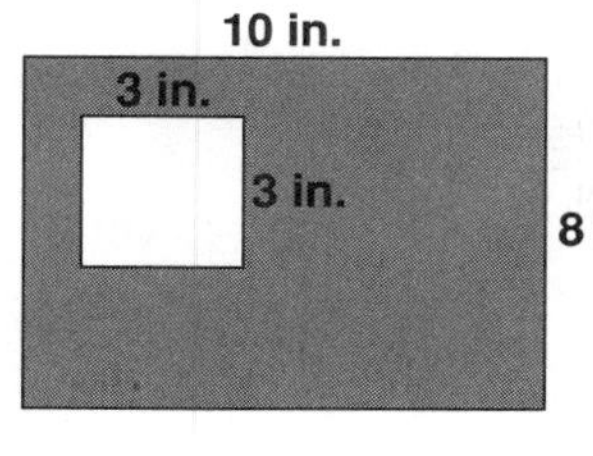

a.

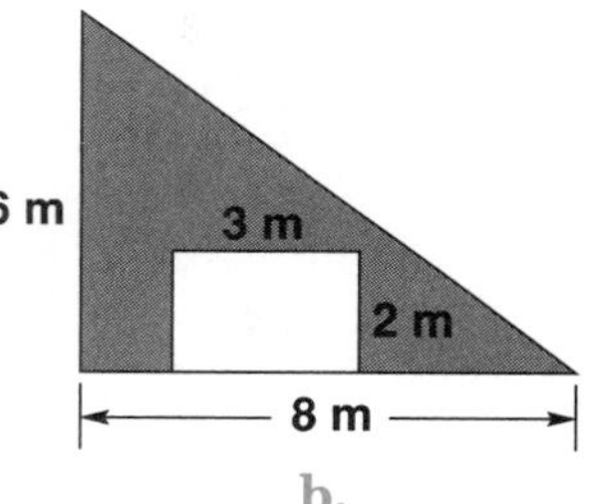

b.

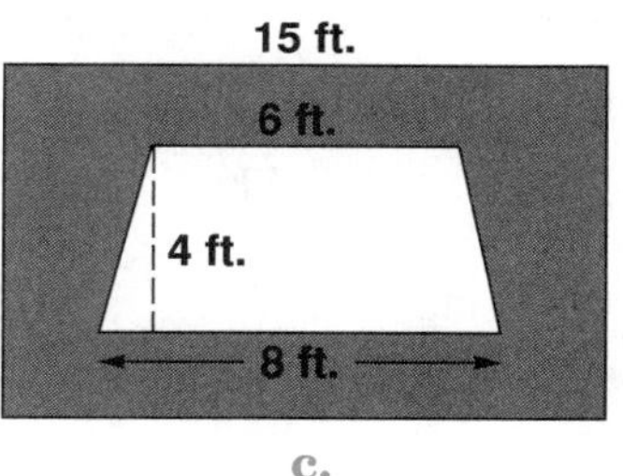

c.

HOMEWORK EXERCISES

1. Find the area of each triangle.

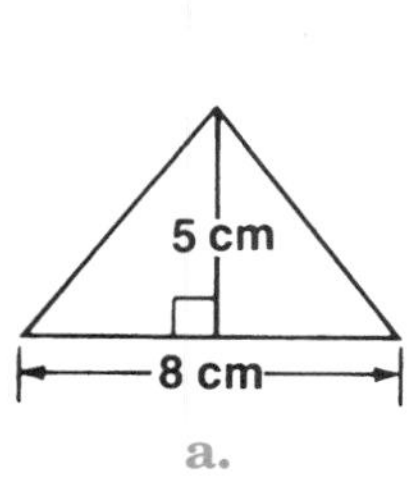

a.

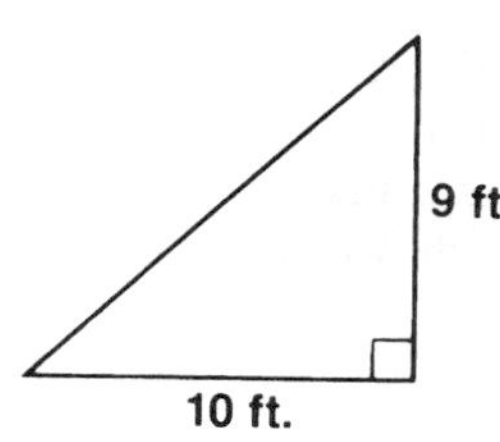

b.

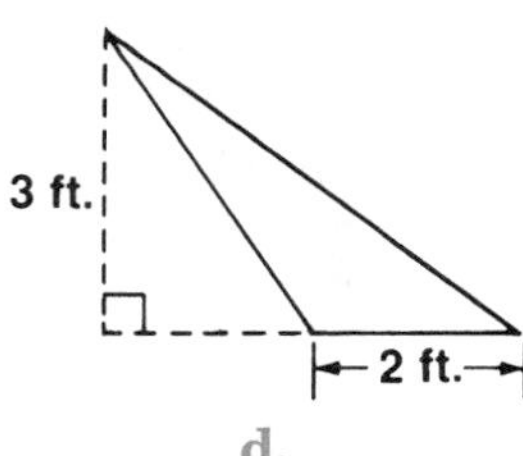

c.

d.

2. Find the area of a triangle in which:

a. $b = 8$ cm, $h = 6$ cm **b.** $b = 8.7$ cm, $h = 6.5$ cm **c.** $b = 14.2$ m, $h = 10.7$ m

d. $b = 18$ in., $h = 6\frac{1}{2}$ in. **e.** $b = 4\frac{1}{2}$ ft., $h = 2\frac{3}{4}$ ft. **f.** $b = 20$ in., $h = 10\frac{1}{4}$ in.

3. Find the area of each right triangle, using the given lengths of the legs.

a. 5 in., 6 in. **b.** 14 cm, 12 cm **c.** 9.2 m, 8.6 m

d. 10 in., $4\frac{1}{2}$ in. **e.** 15.8 cm, 12 cm **f.** $12\frac{1}{2}$ ft., 10 ft.

4. Find the area of each trapezoid.

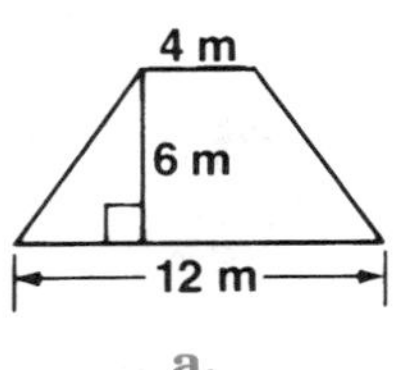

a.

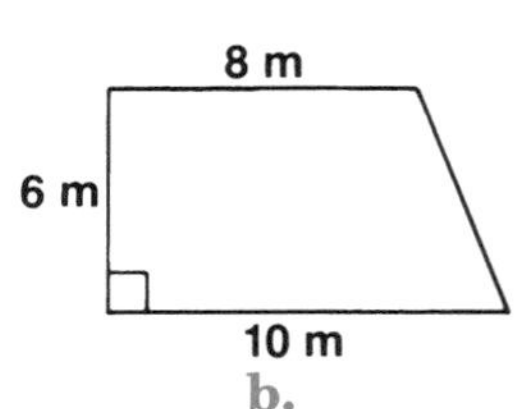

b.

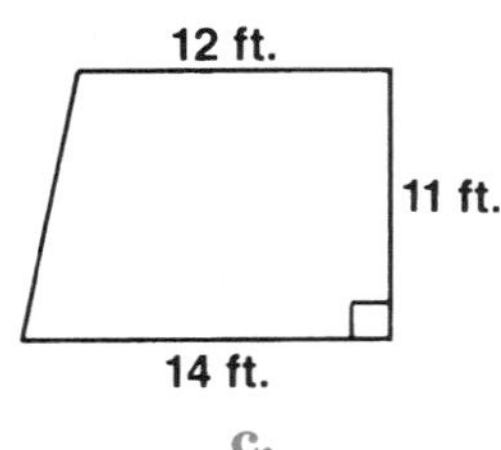

c.

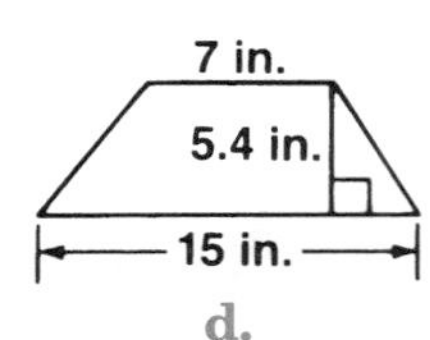

d.

5. Find the area of a trapezoid in which:

a. $h = 4$ cm, $b_1 = 6$ cm, $b_2 = 12$ cm **b.** $h = 6$ ft., $b_1 = 9$ ft., $b_2 = 13$ ft.

c. $h = 12$ cm, $b_1 = 18.6$ cm, $b_2 = 21.4$ cm **d.** $h = 12$ in., $b_1 = 9\frac{1}{4}$ in., $b_2 = 20\frac{3}{4}$ in.

6. Find the area of each figure. All angles are right angles.

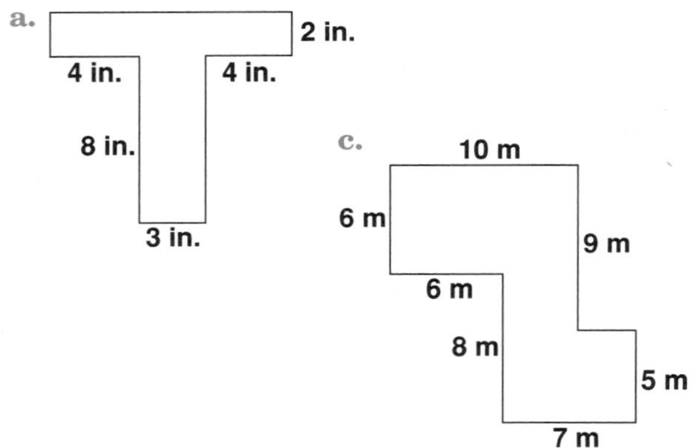

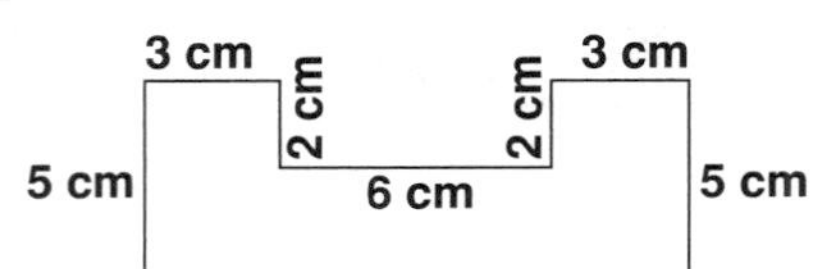

d.

7. The Howard Heights Health Club is making felt emblems to sew on the club sweaters. The emblem, in the shape of an H, has the dimensions shown. How many square centimeters of felt are needed for each emblem?

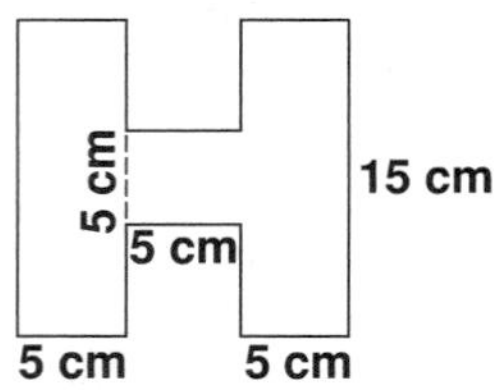

8. Find the area of the shaded part of each figure. (You may assume right angles as shown.)

a.

7 m
4 m
4 m
4 m
4 m
15 m

b.

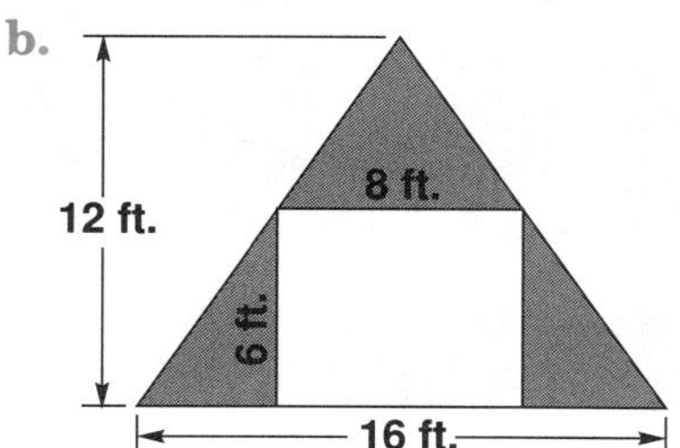

c.

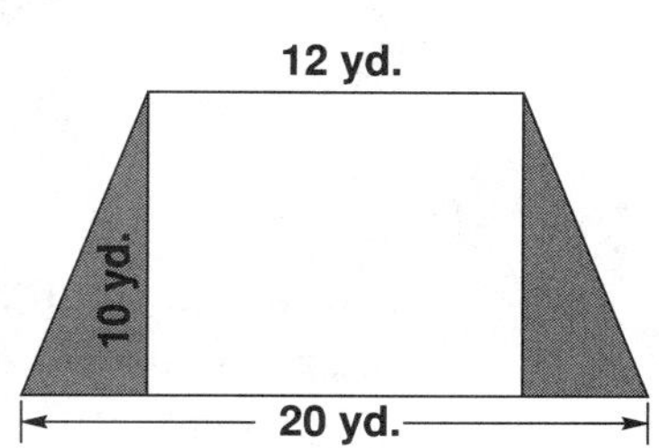

9. A photograph 4 inches wide and 6 inches high is bordered by a frame 7 inches wide and 9 inches high. What is the area of the frame, without the picture?

10. **a.** If two polygons are congruent, must their areas be equal? Explain.
    b. If two polygons have equal areas, must the polygons be congruent? Explain.

11. Graph the points, draw the polygon, and find the area.
    a. Square *GHJK*: $G(-3, 2)$, $H(1, 2)$, $J(1, -2)$, $K(-3, -2)$
    b. Rectangle *LMNP*: $L(-1, -3)$, $M(-1, -1)$, $N(5, -1)$, $P(5, -3)$
    c. Triangle *DEF*: $D(2, 1)$, $E(4, 5)$, $F(8,1)$
    d. Parallelogram *QRST*: $Q(-2, 3)$, $R(3,3)$, $S(4, 0)$, $T(-1, 0)$
    e. Trapezoid *UVWX*: $U(-2, -4)$, $V(-1, -1)$, $W(3, -1)$, $X(6, -4)$
    f. Triangle *ABC*: $A(-4, -3)$, $B(-4, 5)$, $C(7, -3)$

12. Max wants to fence a rectangular yard for his dog. He has 24 feet of fencing and he wants the length of each side to be a whole number.
    a. Make a list of all the possible combinations of length and width that satisfy these conditions.
    b. What are the dimensions of the rectangle that will give the greatest possible area for the dog?

1. Find the area of a rectangle whose dimensions are 25 cm by 10.6 cm.

2. The area of a square is 64 cm^2.
   a. What is the measure of a side?
   b. Find the perimeter of the square.

3. Mr. Flint charges $3.50 per square foot to tile a floor. How much will he charge to tile a rectangular floor 15 feet by 11 feet?

4. If 3 quarts of milk cost $2.88, what is the cost of 7 quarts of milk?

5. $\frac{15}{4}$ is equivalent to
   (a) $2\frac{3}{4}$ (b) $3\frac{3}{4}$ (c) $4\frac{1}{4}$ (d) $4\frac{1}{2}$

6. The length of the hypotenuse of a right triangle is 12.4 cm and the measure of one leg is 4.9 cm. Find the measure of the other leg, correct to the nearest tenth of a centimeter.

7. Divide: $\frac{1}{6} \div 12$

8. Find the mode of the following group of numbers: 1, 3, 6, 8, 9, 9, 9, 12

THE CIRCLE AND ITS LINE SEGMENTS

THE MAIN IDEA

1. A *circle* is a flat, closed curve that has all of its points the same distance from an inside point called the *center*.

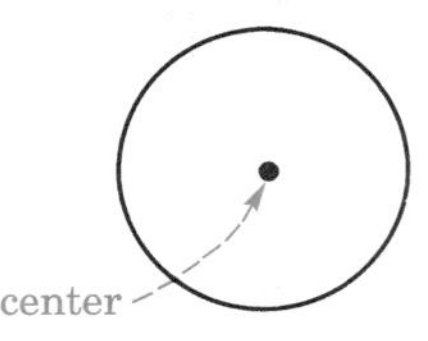

2. A *radius* of a circle is a line segment that has one endpoint at the center of the circle and the other endpoint on the circle. The plural of radius is *radii*.

   In a circle, the measures of all the radii are equal.

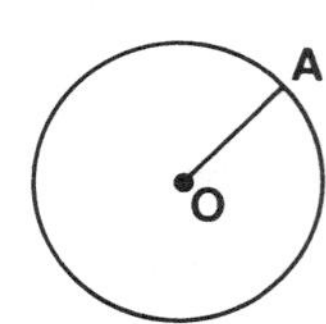

Line segment OA is a radius.

3. A *chord* of a circle is a line segment that has both of its endpoints on the circle. A *diameter* of a circle is a chord that passes through the center of the circle.

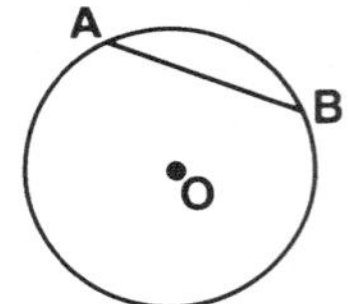

Line segment AB is a chord.

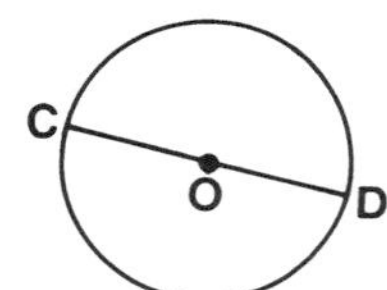

Chord COD is a diameter.

4. A diameter of a circle is twice as long as a radius of that circle:

$$\text{diameter} = 2 \times \text{radius} \quad \text{or} \quad d = 2r$$

   A radius of a circle is one-half as long as a diameter of that circle:

$$\text{radius} = \frac{1}{2} \times \text{diameter} \quad \text{or} \quad r = \frac{1}{2}d$$

EXAMPLE 1 In circle O, tell whether each line segment is a radius, a chord, a diameter, or no special line segment. Explain.

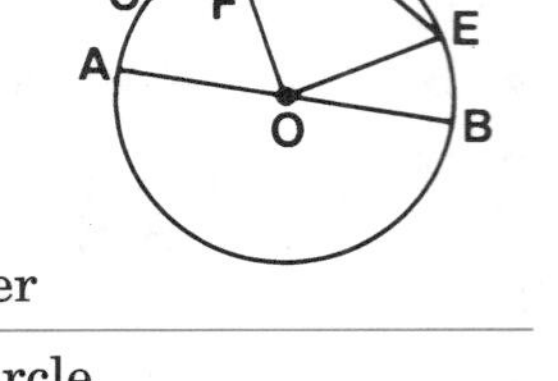

	Line Segment	*Part of Circle*	*Explanation*
a.	$\overline{AB}$	diameter	connects two points on the circle and goes through center
b.	$\overline{DE}$	chord	connects two points on the circle

	Line Segment	*Part of Circle*	*Explanation*
c.	$\overline{CD}$	chord	connects two points on the circle
d.	$\overline{OE}$	radius	connects center to a point on the circle
e.	$\overline{OA}$	radius	connects center to a point on the circle
f.	$\overline{OF}$	no special line segment	F is not a point on the circle.

EXAMPLE 2 Find the measure of:

a. the radius of a circle whose diameter measures 10 cm.

$$\text{radius} = \frac{1}{2} \times \text{diameter}$$
$$= \frac{1}{2} \times 10$$
$$= 5 \text{ cm} \quad \textit{Ans.}$$

b. the diameter of a circle whose radius measures 4 in.

$$\text{diameter} = 2 \times \text{radius}$$
$$= 2 \times 4$$
$$= 8 \text{ in.} \quad \textit{Ans.}$$

CLASS EXERCISES

1. Tell whether each of the line segments is a radius, a chord, a diameter, or no special line segment.

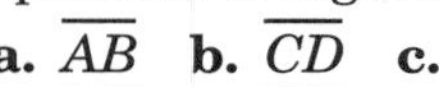

a. $\overline{AB}$ **b.** $\overline{CD}$ **c.** $\overline{EF}$
d. $\overline{OF}$ **e.** $\overline{OG}$ **f.** $\overline{CH}$

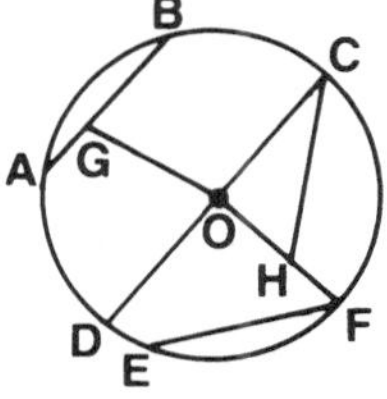

2. **a.** Explain why chord CD is not a diameter.

b. Explain why line segment OA is not a radius.

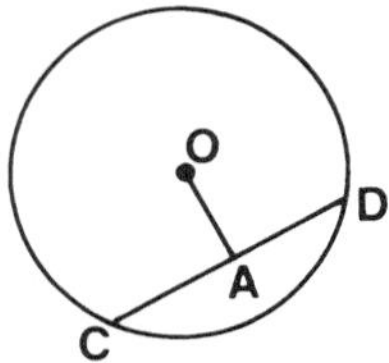

3. Find the measure of the radius of a circle whose diameter measures:

a. 6 cm **b.** 18 in. **c.** 12 m **d.** $\frac{3}{4}$ in. **e.** 5 yd. **f.** 3.7 in.

4. Find the measure of the diameter of a circle whose radius measures:

a. 3 in. **b.** 9 cm **c.** 11 m **d.** $\frac{5}{8}$ in. **e.** 2.8 yd. **f.** $3\frac{1}{2}$ cm

5. What are the measures of the radius and the diameter of each circle?

a.
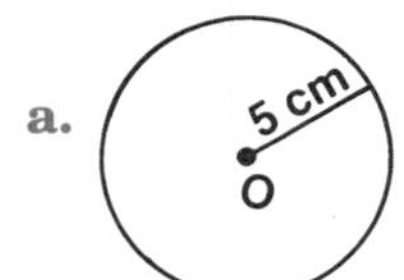

b.

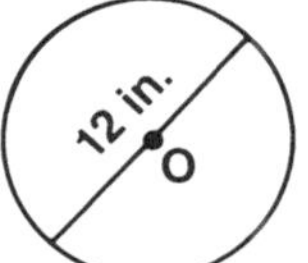

c.

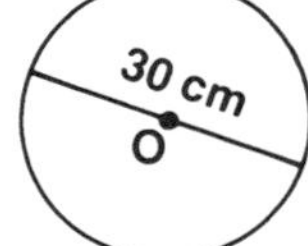

d.
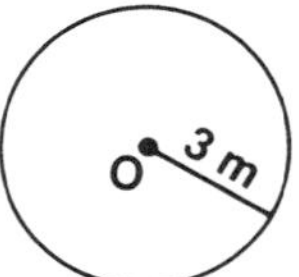

THE CIRCUMFERENCE OF A CIRCLE

THE MAIN IDEA

1. The ***circumference*** of a circle is the distance around the circle.
2. The ratio of the circumference of a circle to the diameter of that circle is the same for all circles. The Greek letter π(pi) is used to represent this number:

$$\frac{C \leftarrow \text{circumference}}{d \leftarrow \text{diameter}} = \pi \quad \text{or} \quad C = \pi d$$

3. The number π is approximately equal to 3.14 or $\frac{22}{7}$.

   $\pi \approx 3.14$ or $\pi \approx \frac{22}{7}$ The symbol $\approx$ means "is about equal to."
4. To find the circumference of a circle:
   a. Use the formula $C = \pi d$ when you are given the measure of the diameter.
   b. Use the formula $C = 2\pi r$ when you are given the measure of the radius.

EXAMPLE 3 Using $\pi \approx 3.14$, find the circumference of a circle if:

a. the diameter measures 40 in.

$C = \pi d$

$C \approx 3.14 \times 40$

≈ 125.6 in. *Ans.*

b. the radius measures 10 cm

$C = 2\pi r$

$C \approx 2 \times 3.14 \times 10$

≈ 62.8 cm. *Ans.*

EXAMPLE 4 Using $\pi \approx \frac{22}{7}$, find the circumference of each circle.

a.

$C = 2\pi r$

$C \approx 2 \times \frac{22}{7} \times 14$

$\approx 2 \times \frac{22}{\cancel{7}_1} \times \cancel{14}^2$

≈ 88 cm *Ans.*

b.

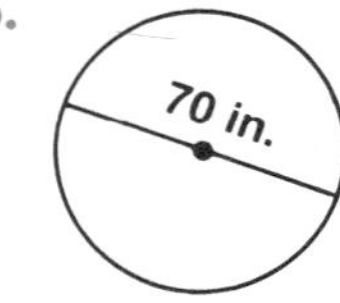

$C = \pi d$

$C \approx \frac{22}{7} \times 70$

$\approx \frac{22}{\cancel{7}_1} \times \cancel{70}^{10}$

≈ 220 in. *Ans.*

EXAMPLE 5 Use a calculator to find the length of the diameter of a circle with a circumference of 20.41 inches. [Use $\pi \approx 3.14$]

Rewrite the formula $C = \pi d$ so that the subject is d.

Divide both sides by π. $\quad \dfrac{C}{\pi} = \dfrac{\pi d}{\pi}$

$$\frac{C}{\pi} = d$$

Replace C with 20.41 and π with 3.14. $\quad d \approx \dfrac{20.41}{3.14}$

Key Sequence	***Display***
20.41 [÷] 3.14 [=]	6.5

Answer: The diameter is about equal to 6.5 inches.

CLASS EXERCISES

1. Using $\pi \approx 3.14$, find the circumference of a circle if the length of the radius is:

a. 20 cm **b.** 15 m **c.** 8.6 in. **d.** 30.5 ft.

2. Using $\pi \approx \frac{22}{7}$, find the circumference of a circle if the length of the radius is:

a. 14 in. **b.** 35 m **c.** $3\frac{1}{2}$ ft. **d.** $1\frac{3}{11}$ yd.

3. Using $\pi \approx 3.14$, find the circumference of a circle if the length of the diameter is:

a. 10 cm **b.** 22 in. **c.** 9.8 m **d.** 24.5 yd.

4. Using $\pi \approx \frac{22}{7}$, find the circumference of a circle if the length of the diameter is:

a. 21 in. **b.** 49 cm **c.** $1\frac{3}{4}$ ft. **d.** $2\frac{1}{3}$ yd.

5. Using $\pi \approx 3.14$, find the circumference of each circle.

a. 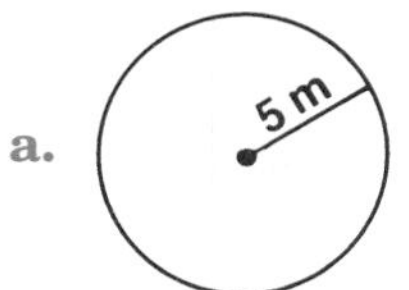

b. 8.8 cm

c. 

d. 11.2 ft.

6. Using $\pi \approx 3.14$, find the measure of the diameter of a circle whose circumference is 31.4 m.

7. Using $\pi \approx \frac{22}{7}$, find the measure of the diameter of a circle whose circumference is 66 cm.

8. A circular swimming pool has a circumference of 32.656 m. Use a calculator and $\pi \approx 3.14$ to find:
   a. the diameter of the pool b. the radius of the pool

9. The diameter of one circle is 2 times the radius of another circle. How are their circumferences related?

10. On the number line below, d represents the diameter of a quarter. Follow the directions below to find the circumference of a quarter. Each unit on the number line is divided into tenths.

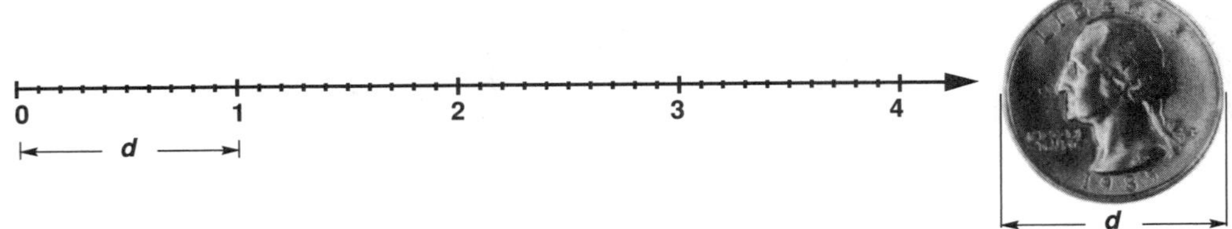

   a. Position the edge of the quarter on the zero of the number line, using a "reminder point" on the quarter such as the L in LIBERTY. Then, carefully roll the quarter along the number line for one complete turn of the quarter. Try not to let the coin slide.
   b. Approximate the number on the number line that represents the end of one complete turn of the quarter.
   c. What number times the diameter of the quarter equals the circumference?
   d. If d equals about $\frac{15}{16}$ of 1 inch, what is the approximate circumference of the quarter?

THE AREA OF A CIRCLE

THE MAIN IDEA

1. As with any flat, closed surface, the area of a circle is the measure of the space contained inside.
2. $A_{\text{circle}} = \pi r^2$

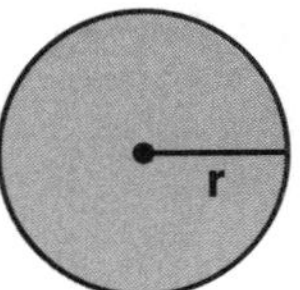

EXAMPLE 6 Using $\pi \approx 3.14$, find the area of a circle whose radius measures 5 cm.

In the formula for the area of a circle, substitute 3.14 for π and 5 for r.

$A_{\text{circle}} = \pi r^2$
$A_{\text{circle}} \approx 3.14 \times 5^2$

Evaluate, following the order of operations.

$\approx 3.14 \times 5 \times 5$
$\approx 3.14 \times 25 \approx 78.5 \text{ cm}^2$

CLASS EXERCISES

1. Using $\pi \approx 3.14$, find the area of a circle whose radius is:

a. 10 in. **b.** 4 cm **c.** 8 yd. **d.** 2.3 m

2. Using $\pi \approx \frac{22}{7}$, find the area of a circle whose radius is:

a. 14 cm **b.** 28 in. **c.** 2.1 m **d.** $3\frac{1}{2}$ ft.

3. Using $\pi \approx 3.14$, find the area of a circle whose diameter is:

a. 2 cm **b.** 8 in. **c.** 20 m **d.** 9.6 ft.

4. Using $\pi \approx \frac{22}{7}$, find the area of a circle whose diameter is:

a. 14 in. **b.** 42 cm **c.** 2.8 m **d.** 7 yd.

5. Using $\pi \approx 3.14$, find the area of each circle.

a.

b.
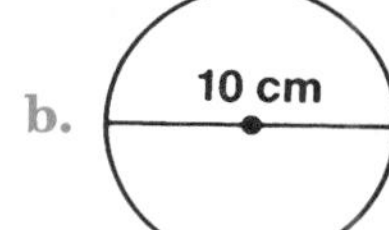

c.

d.
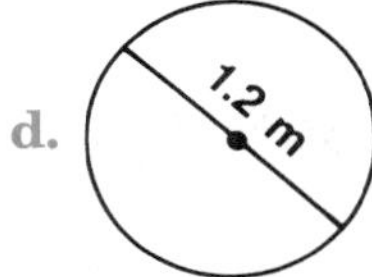

6. Use $\pi \approx 3.14$ to find the area of a circle whose radius equals 2.54 cm.

7. Find the area of the shaded portion of the figure. Use $\pi \approx \frac{22}{7}$.

a.

7 in.

b.
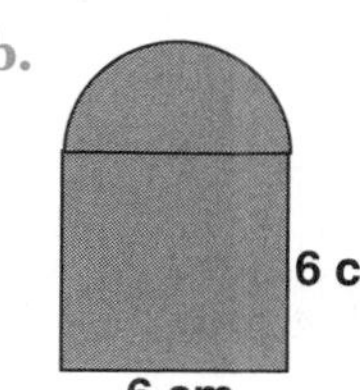

c.
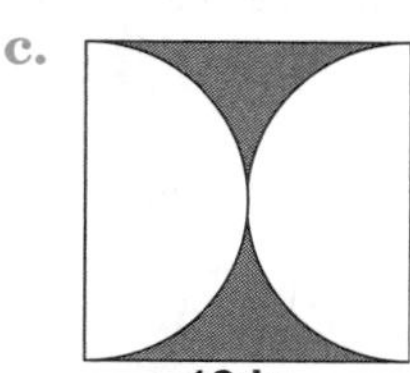

8. Two circles are congruent if their radii are congruent. If two circles are congruent, explain the relation in their: **a.** areas **b.** circumferences **c.** diameters

HOMEWORK EXERCISES

1. Tell whether each of the line segments is a radius, a chord, a diameter, or no special line segment.

a. $\overline{AD}$ **b.** $\overline{BF}$ **c.** $\overline{OB}$
d. $\overline{OC}$ **e.** $\overline{CE}$ **f.** $\overline{CQ}$

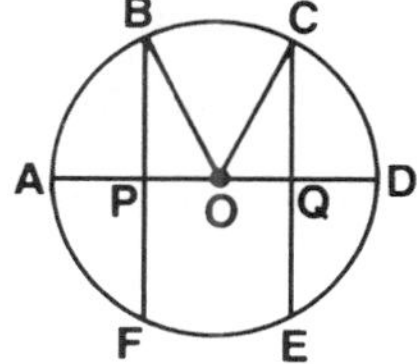

2. a. Explain why chord AC is not a diameter.

b. Explain why line segment OP is not a radius.

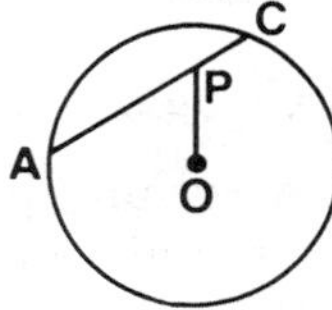

3. Find the measure of the radius of a circle whose diameter measures:

a. 28 cm **b.** 8.2 in. **c.** $\frac{5}{16}$ in. **d.** 9.6 yd. **e.** $4\frac{1}{3}$ in.

4. Find the measure of the diameter of a circle whose radius measures:

a. 11 in. **b.** 14.7 m **c.** $\frac{3}{4}$ ft. **d.** 2.95 m **e.** $4\frac{5}{8}$ in.

5. What are the measures of the radius and the diameter of each circle?

a. 3 in.

b. 2.4 m

c.

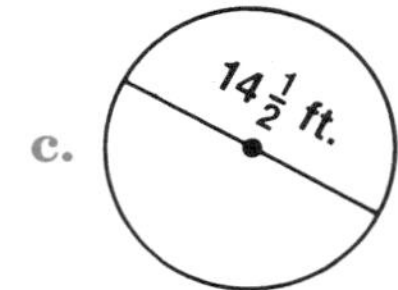

d. 29.6 cm

6. Using $\pi \approx 3.14$, find the circumference of a circle if the length of the radius is:

a. 30 cm **b.** 55 in. **c.** 9.4 m **d.** 28.3 ft. **e.** $10\frac{1}{3}$ yd.

7. Using $\pi \approx \frac{22}{7}$, find the circumference of a circle if the length of the radius is:

a. 28 cm **b.** 42 in. **c.** $2\frac{1}{3}$ ft. **d.** $4\frac{3}{8}$ yd. **e.** 19.6 m

8. Using $\pi \approx 3.14$, find the circumference of a circle if the length of the diameter is:

a. 50 in. **b.** 100 m **c.** 10.2 cm **d.** 9.8 yd. **e.** 7.25 in.

9. Using $\pi \approx \frac{22}{7}$, find the circumference of a circle if the length of the diameter is:

a. 7 cm **b.** 56 in. **c.** $5\frac{1}{4}$ yd. **d.** $8\frac{3}{4}$ ft. **e.** 62.4 mm

10. Using $\pi \approx 3.14$, find the circumference of each circle.

a.

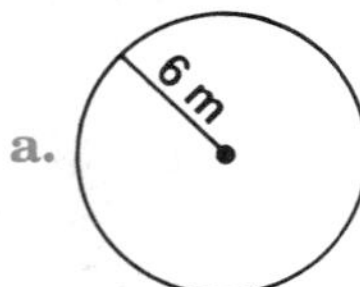

b. 36 m

c.

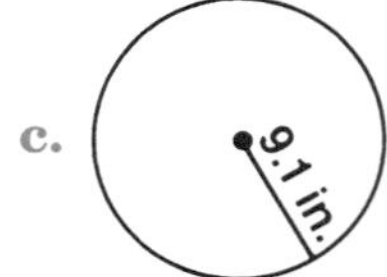

d.

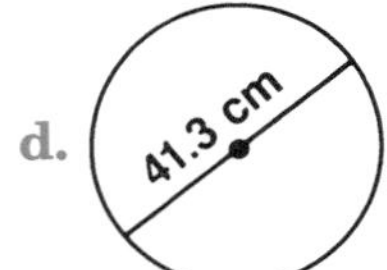

11. Using $\pi \approx \frac{22}{7}$, find the measure of the diameter of a circle whose circumference is 110 m.

12. Using $\pi \approx 3.14$, find the measure of the diameter of a circle whose circumference is 628 cm.

13. A circular driveway has a diameter of length 50.6 m. Use a calculator and $\pi \approx 3.14$ to find the circumference of the driveway.

14. Using $\pi \approx 3.14$, find the area of a circle whose radius is: **a.** 3 cm **b.** 9 in. **c.** 8.2 cm

15. Using $\pi \approx \frac{22}{7}$, find the area of a circle whose radius is: **a.** 14 m **b.** 2.8 cm **c.** $2\frac{1}{3}$ ft.

16. Using $\pi \approx 3.14$, find the area of a circle whose diameter is: **a.** 40 in. **b.** 1.8 cm **c.** 5.8 m

17. Using $\pi \approx \frac{22}{7}$, find the area of a circle whose diameter is: **a.** 70 in. **b.** 4.2 m **c.** 9.8 m

18. Using $\pi \approx 3.14$, find the area of each circle.

a. 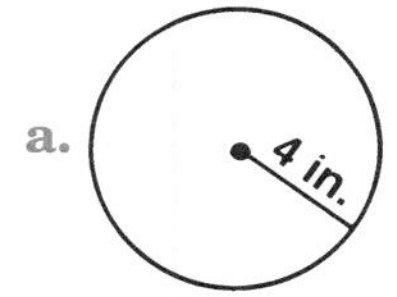

b. 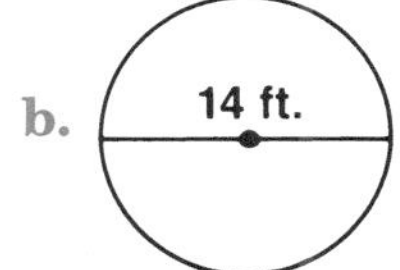

c. 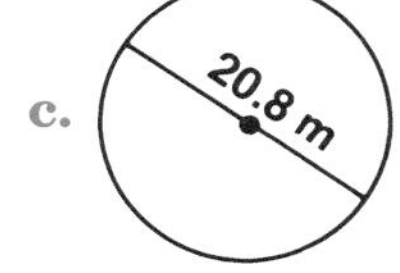

d. 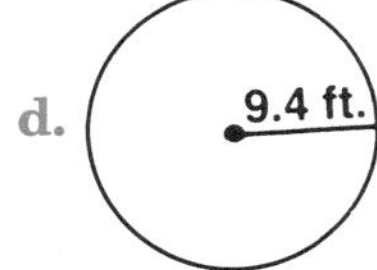

19. A circular flower bed has a diameter that measures 8.4 feet.

a. Use $\pi \approx 3.14$ to find the area of the flower bed correct to the nearest hundredth of a square foot.

b. If topsoil costs $3.50 per square foot, find the total cost of topsoil needed to cover the flower bed.

20. In the diagram, the perimeter of the square is 56 cm. Find the area of the circle. Use $\pi \approx \frac{22}{7}$.

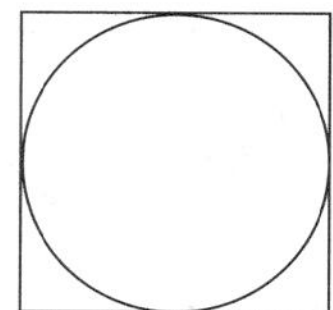

21. Find the area of the shaded portion of each figure. Use $\pi \approx 3.14$.

a.

7 m
7 m
7 m
20 m
50 m

b. 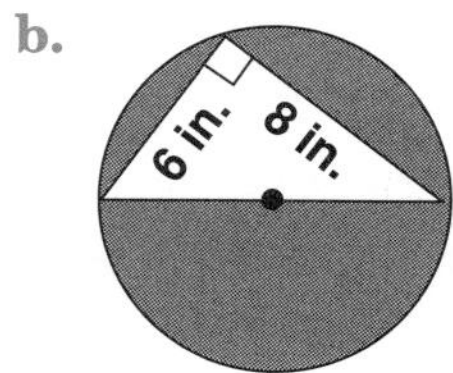

c. 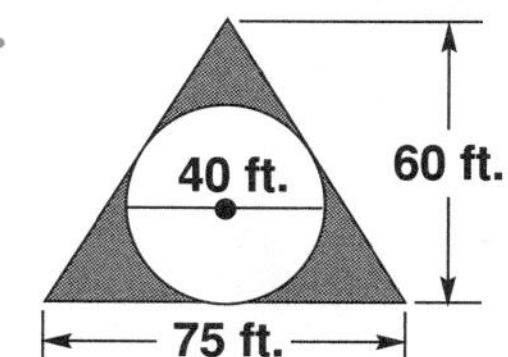

22. The value for π, correct to the nearest ten-thousandth, is $\pi \approx 3.1416$.

a. Using $\pi \approx 3.1416$, find the circumference of a circle with a diameter of 100 cm.

b. Find the circumference of the same circle using $\pi \approx 3.14$.

c. Find the circumference of the same circle using $\pi \approx \frac{22}{7}$.

d. Assuming that 3.1416 is the closest of the three approximations for π, which of the other two approximations for π gives the "better" approximation for the circumference of a circle, 3.14 or $\frac{22}{7}$?

1. a. Find the perimeter of the rectangle.

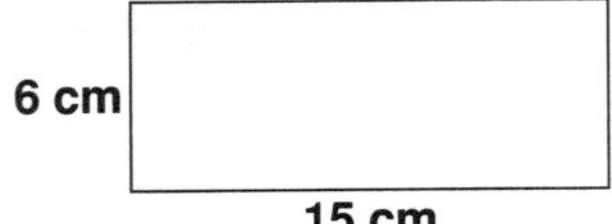

b. Find the area of the rectangle.

2. What is the length of a side of a square whose perimeter is 50.8 meters?

3. A square has an area of 144 square inches. What is the measure of each side?

(a) 18 inches (b) 3 feet
(c) 24 inches (d) 1 foot

4. Find the area of this parallelogram:

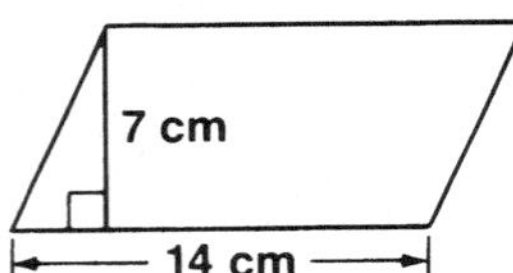

5. On the graph, which point has the coordinates (–2, –4)?

(a) Q (b) R (c) S (d) T

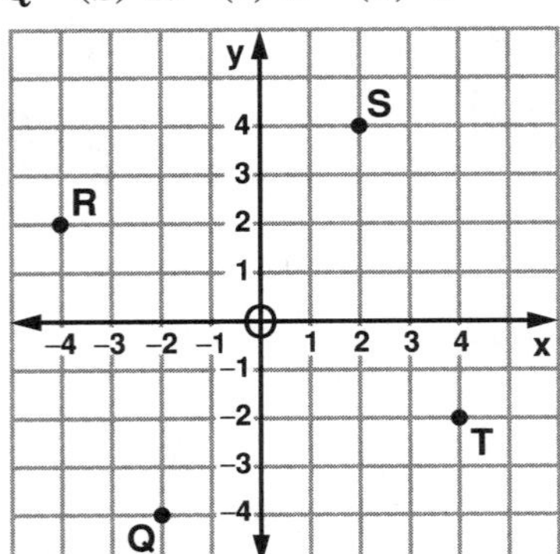

6. Which open sentence is represented by the graph?

(a) $x > -2$

(b) $x \leq 5$

(c) $-2 \leq x \leq 5$

(d) $-2 < x \leq 5$

7. A rectangular room measures 15 feet by 12 feet. Which key sequence can be used to calculate the price of carpeting the room at a cost of $17 per square yard?

(a) 15 [×] 12 [×] 17 [=]

(b) 15 [×] 12 [÷] 9 [×] 17 [=]

(c) 15 [×] 12 [÷] 3 [×] 17 [=]

(d) 15 [×] 12 [×] 3 [×] 17 [=]

8. Solve: $3x - 14 = 19$

9. 255% is equivalent to

(a) 0.255 (b) 25.5 (c) $2\frac{11}{20}$ (d) $2\frac{3}{5}$

10. 4.89×10^{-6} written as a decimal is

(a) 0.000000489 (b) 0.00000489
(c) 0.00489 (d) 0.489

11. Find the value of $-5x^2$ when $x = -3$.

UNIT 19-5 Volume

THE VOLUME OF A RECTANGULAR SOLID AND OF A CUBE

THE MAIN IDEA

1. ***Volume*** means the measurement of the amount of space contained within a solid figure.

2. Volume is measured in ***cubic units***. A cubic centimeter (written 1 cm^3) is an example of a unit of volume. It is the space contained in a cube that measures 1 centimeter on each side.

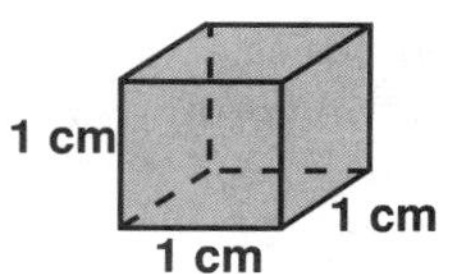

3. Units of volume include the cube of any unit used to measure a line. The product of 3 linear units is a cubic unit. For example: $2 \text{ m} \times 3 \text{ m} \times 4 \text{ m} = 24 \text{ m}^3$

4. To find the volume of a ***rectangular solid***, multiply the measure of its length by the measure of its width and then multiply by the measure of its height.

$$V_{\text{rectangular solid}} = \ell wh$$

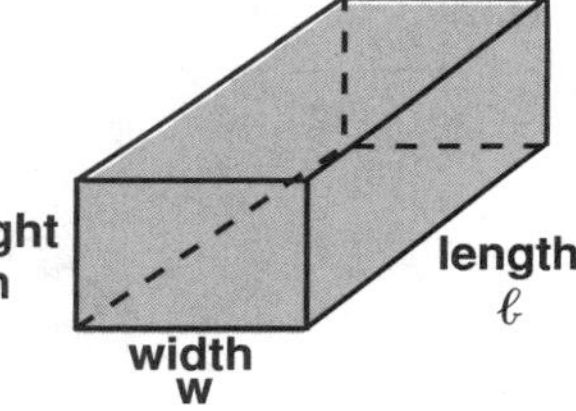

5. A ***cube*** is a rectangular solid, all 12 edges of which are congruent. To find the volume of a cube, use the measure of an edge as a factor three times.

$$V_{\text{cube}} = e^3$$

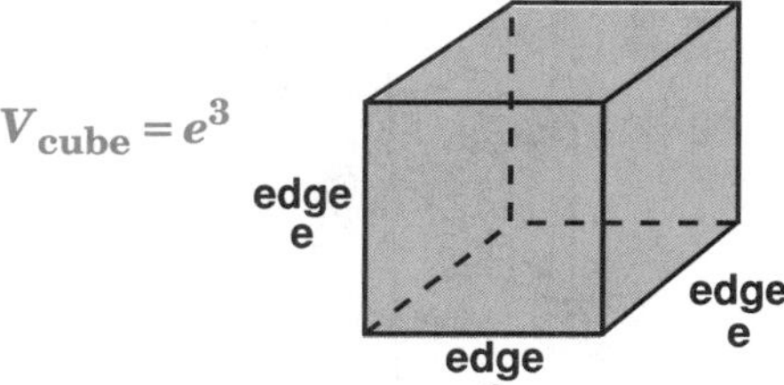

EXAMPLE 1 Find the volume of the rectangular solid.

In the formula for the volume of a rectangular solid, substitute 8 for ℓ, 5 for w, and 3 for h.

$$V_{\text{rectangular solid}} = \ell wh$$
$$V_{\text{rectangular solid}} = 8 \times 5 \times 3$$
$$= 120 \text{ cm}^3 \quad \textit{Ans.}$$

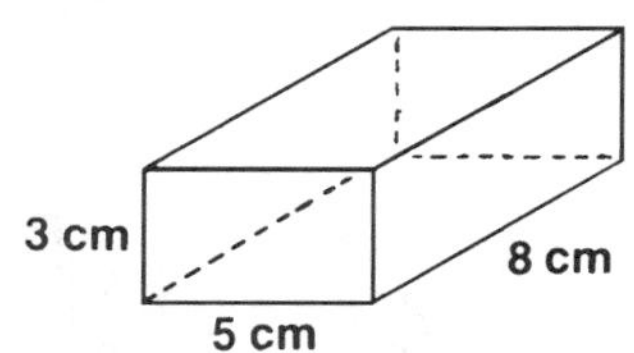

EXAMPLE 2 Find the volume of the cube.

In the formula for the volume of a cube, substitute 2 for e.

$V_{cube} = e^3$
$V_{cube} = 2^3$
$= 8 \text{ m}^3$ *Ans.*

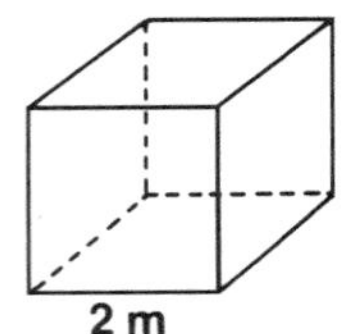

CLASS EXERCISES

1. Find the volume of each rectangular solid.

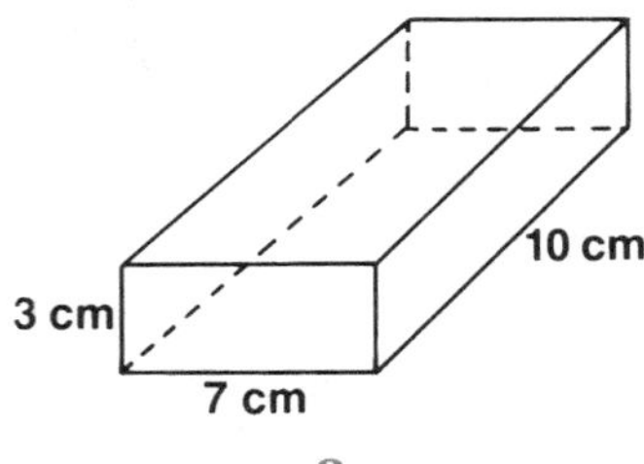

a.

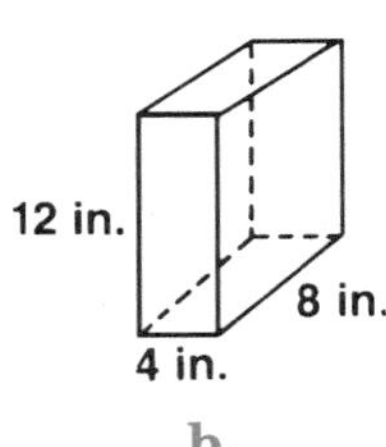

b.

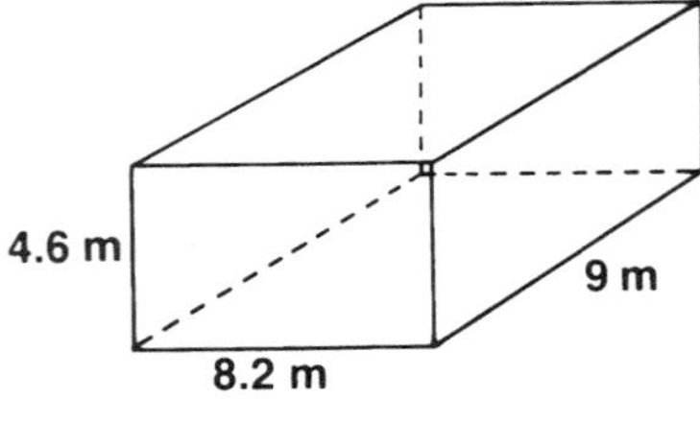

c.

11 in.
14 in.
$\frac{1}{2}$ in.

d.

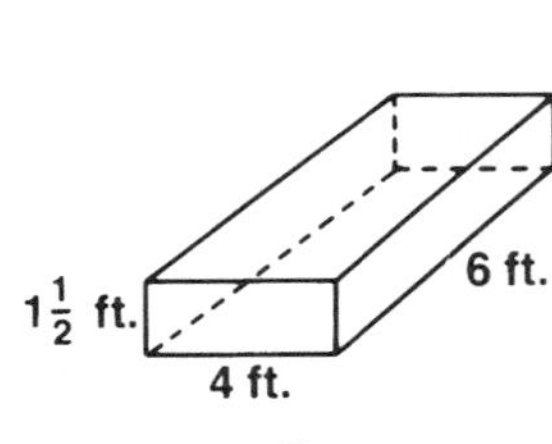

e.

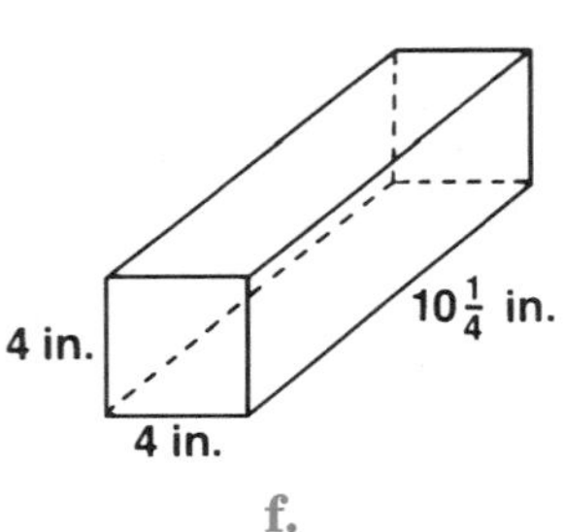

f.

2. Find the volume of the rectangular solids having dimensions:

a. $\ell = 2$ cm, $w = 1$ cm, $h = 6$ cm **b.** $\ell = 8$ in., $w = 5$ in., $h = 10$ in.

c. $\ell = 14$ m, $w = 10$ m, $h = 1.5$ m **d.** $\ell = 8$ ft., $w = 6$ ft., $h = 3\frac{1}{2}$ ft.

e. $\ell = 10$ cm, $w = 4.5$ cm, $h = 2.8$ cm **f.** $\ell = 6$ yd., $w = 2$ yd., $h = 9\frac{1}{4}$ yd.

3. What is the volume of a rectangular box that is 12 inches long, 6 inches wide, and 3 inches high?

4. How many cubic feet are there in a rectangular shipping case 4 feet long, 2 feet wide, and $5\frac{1}{2}$ feet high?

5. Find the volume of a cube each of whose edges measures:

a. 2 in. **b.** 4 cm **c.** 10 m **d.** 8 ft. **e.** $\frac{3}{4}$ in. **f.** 1.5 cm

THE VOLUME OF A CYLINDER AND OF A SPHERE

THE MAIN IDEA

1. A ***right circular cylinder*** is a solid figure that has two parallel circular bases, and a height that is perpendicular to the bases.

$V_{\text{cylinder}} = \pi r^2 h$

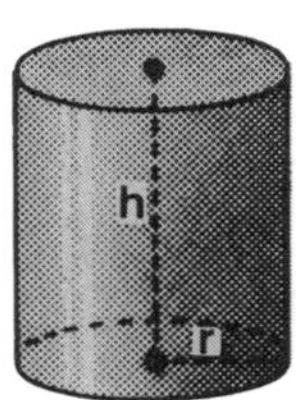

2. A ***sphere*** is a solid figure in the shape of a globe.

$V_{\text{sphere}} = \frac{4}{3}\pi r^3$

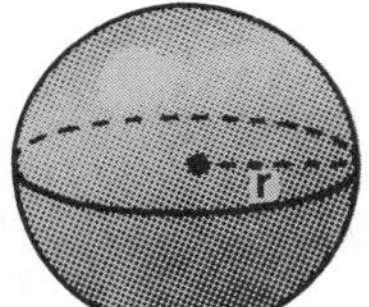

EXAMPLE 3 Using $\pi \approx 3.14$, find the volume of the cylinder.

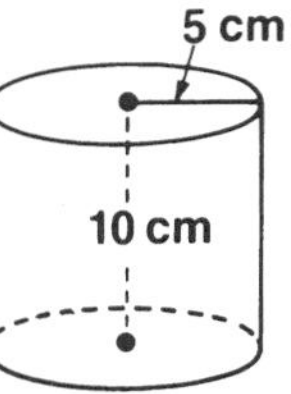

In the formula for the volume of a cylinder, substitute 3.14 for π, 5 for r, and 10 for h.

$$V_{\text{cylinder}} = \pi r^2 h$$
$$V_{\text{cylinder}} \approx 3.14 \times 5^2 \times 10$$
$$\approx 3.14 \times 25 \times 10$$
$$\approx 785 \text{ cm}^3 \quad \textit{Ans.}$$

EXAMPLE 4 Using $\pi \approx \frac{22}{7}$, find the volume of a sphere that has a diameter of 42 m.

Divide the diameter by 2 to find the measure of the radius.

In the formula for the volume of a sphere, substitute $\frac{22}{7}$ for π and 21 for r.

$$V_{\text{sphere}} = \frac{4}{3}\pi r^3$$
$$V_{\text{sphere}} \approx \frac{4}{3} \times \frac{22}{7} \times 21^3$$
$$\approx \frac{4}{\cancelto{1}{3}} \times \frac{22}{\cancelto{1}{7}} \times \cancelto{7}{21} \times \cancelto{3}{21} \times 21$$
$$\approx 38{,}808 \text{ m}^3 \quad \textit{Ans.}$$

CLASS EXERCISES

1. Using $\pi \approx 3.14$, find the volume:

a. 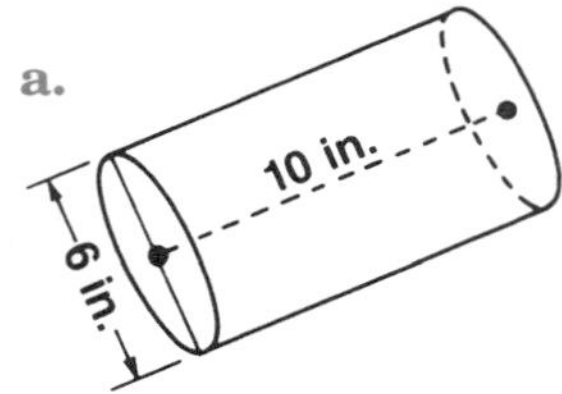

b. 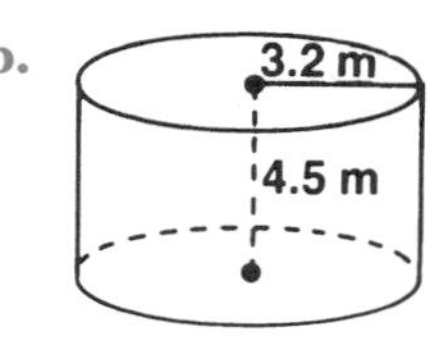

2. Using $\pi \approx \frac{22}{7}$, find the volume of each cylinder, given the measures:

   a. $r = 7$ cm, $h = 8$ cm **b.** $r = 10$ in., $h = 7$ in. **c.** $r = 28$ cm, $h = 1\frac{1}{2}$ cm

3. Using $\pi \approx 3.14$, find the volume of a cylindrical can with $r = 2$ in. and $h = 5$ in.

4. Using $\pi \approx 3.14$, find the volume of a sphere that has a diameter of:

   a. 4 cm **b.** 6.3 in. **c.** $7\frac{1}{2}$ ft. **d.** 70.1 mm

5. Using $\pi \approx \frac{22}{7}$, find the volume of a sphere that has a diameter of 84 centimeters.

THE VOLUME OF A PYRAMID AND OF A CONE

THE MAIN IDEA

1. A ***rectangular pyramid*** is a solid figure whose base is a rectangle and whose sides are triangles.

   The volume of such a pyramid is one-third the volume of a rectangular solid with the same length, width, and height.

$V_{\text{pyramid}} = \frac{1}{3} \times V_{\text{rectangular solid}}$

$V_{\text{pyramid}} = \frac{1}{3}\ \ell w h$

height h
width w
length ℓ

2. A ***right circular cone*** has a base that is a circle, and a height perpendicular to the base.

   The volume of such a cone is one-third the volume of a right circular cylinder with the same height and radius.

$V_{\text{cone}} = \frac{1}{3} \times V_{\text{cylinder}}$

$V_{\text{cone}} = \frac{1}{3}\ \pi r^2 h$

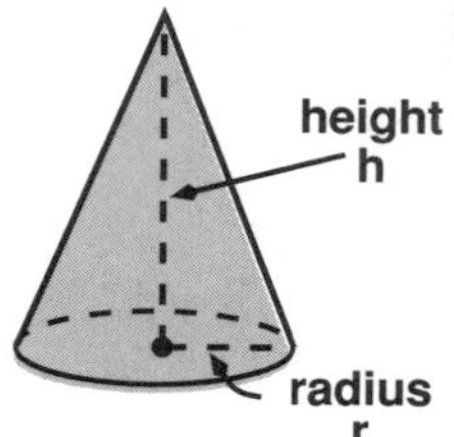

EXAMPLE 5 Find the volume of the rectangular pyramid.

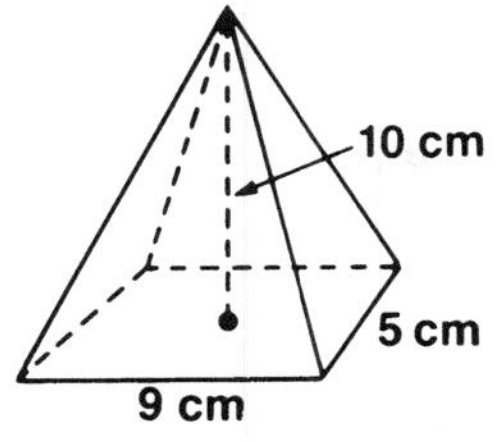

$V_{\text{pyramid}} = \frac{1}{3}\ \ell wh$

$V_{\text{pyramid}} \approx \frac{1}{3} \times 9 \times 5 \times 10$

$\approx \frac{1}{\cancel{3}_1} \times \overset{3}{\cancel{9}} \times 5 \times 10$

$\approx 150\ \text{cm}^3$ *Ans.*

EXAMPLE 6 Using $\pi \approx 3.14$, find the volume of the cone.

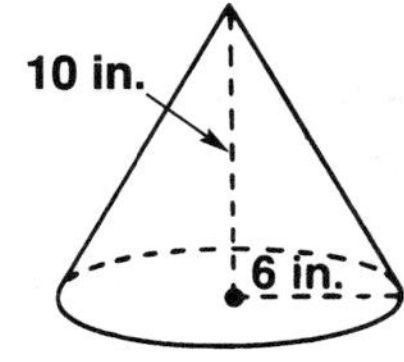

$V_{\text{cone}} = \frac{1}{3}\ \pi r^2 h$

$V_{\text{cone}} \approx \frac{1}{3} \times 3.14 \times 6^2 \times 10$

$\approx \frac{1}{3} \times 3.14 \times 36 \times 10$

$\approx \frac{1}{\cancel{3}_1} \times 3.14 \times \overset{12}{\cancel{36}} \times 10$

$\approx 376.8\ \text{in.}^3$ *Ans.*

CLASS EXERCISES

1. Find the volume of each rectangular pyramid.

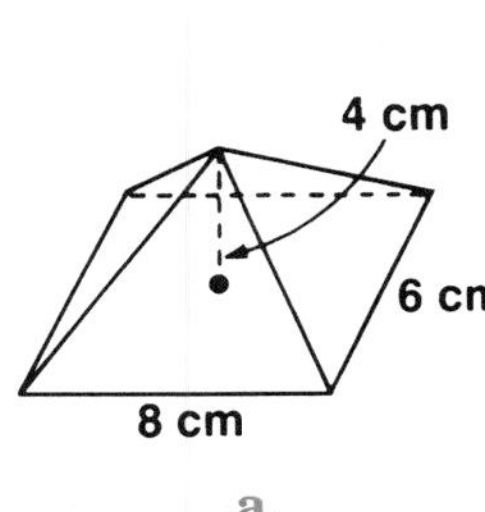

a.

b.

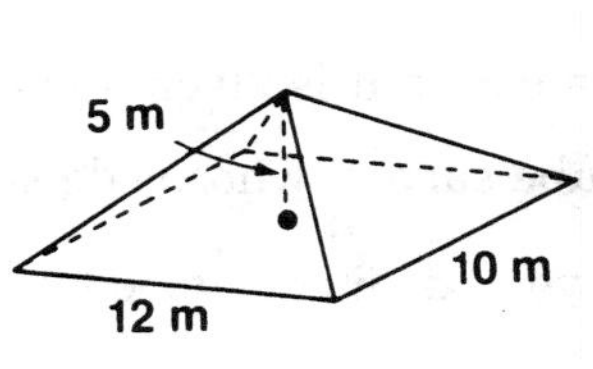

c.

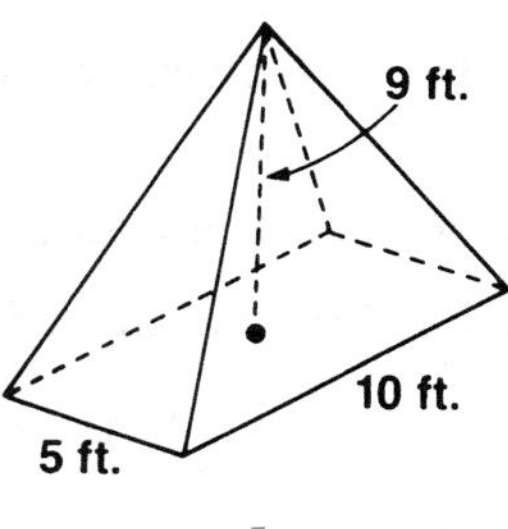

d.

2. Find the volume of each rectangular pyramid, given the dimensions:

a. $\ell = 20$ in., $w = 10$ in., $h = 6$ in. **b.** $\ell = 8$ cm, $w = 4$ cm, $h = 15$ cm

c. $\ell = 9$ m, $w = 5\frac{1}{2}$ m, $h = 2$ m **d.** $\ell = 1.2$ ft., $w = 8$ ft., $h = 10$ ft.

3. Using $\pi \approx 3.14$, find the volume:

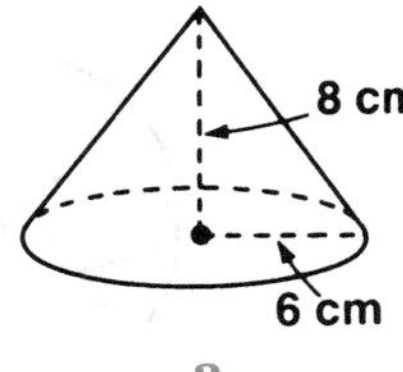

a.

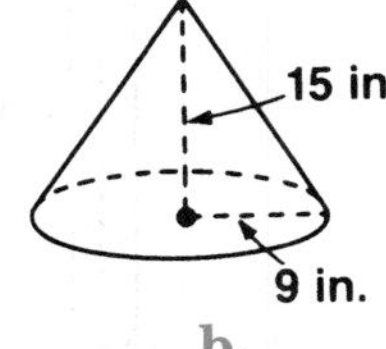

b.

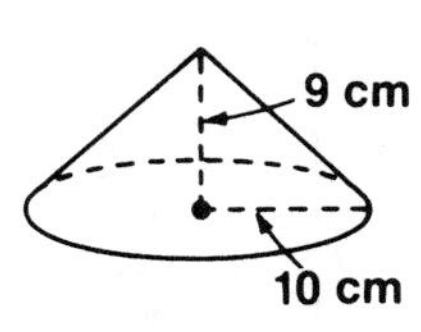

c.

4. Using $\pi \approx 3.14$, find the volume of each cone, given the dimensions:

a. $r = 5$ cm, $h = 3$ cm **b.** $r = 0.6$ in., $h = 20$ in. **c.** $r = 1\frac{1}{2}$ m, $h = 6$ m

HOMEWORK EXERCISES

1. Find the volume of each rectangular solid.

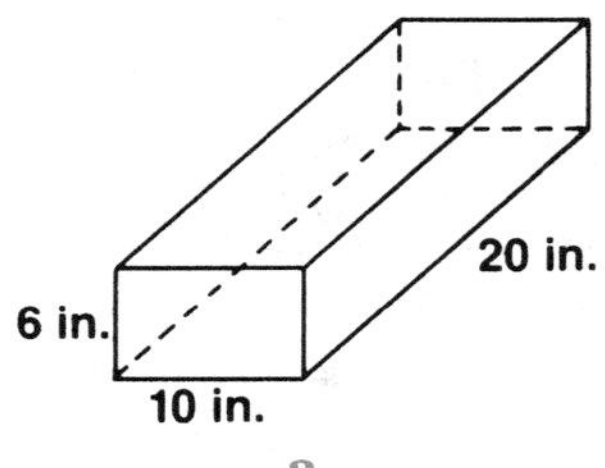

a.

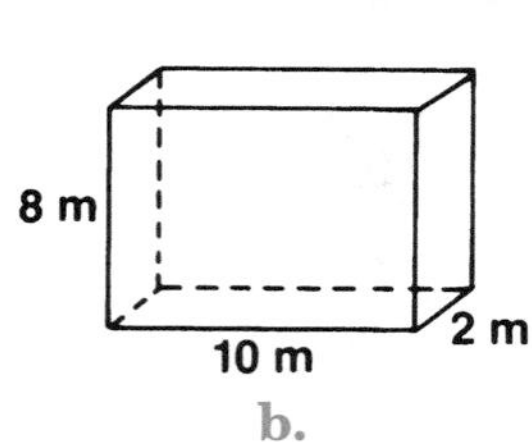

b.

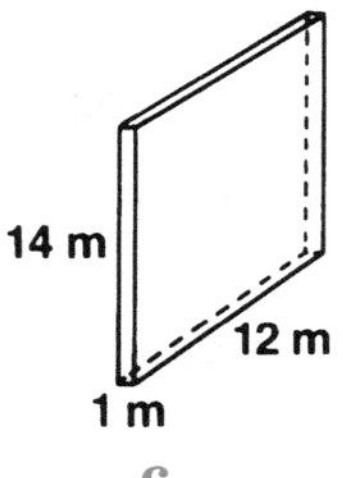

c.

2. Find the volume of each rectangular solid, given the dimensions:

a. $\ell = 9$ m, $w = 8$ m, $h = 7$ m **b.** $\ell = 10$ ft., $w = 3\frac{1}{2}$ ft., $h = 8$ ft.

c. $\ell = 14$ cm, $w = 8.2$ cm, $h = 4.1$ cm **d.** $\ell = 12$ m, $w = 10.6$ m, $h = 0.8$ m

3. What is the volume of a rectangular fish tank that is 20 inches long, 12 inches high, and 10 inches deep?

4. How many cubic feet are there in a rectangular room that is 9 feet wide, 12 feet long, and 8 feet high?

5. What is the volume of a box that is 20 cm long, 10 cm wide, and 6.5 cm high?

6. Find the volume of a cube each of whose edges measures:

a. 1 m **b.** 3 ft. **c.** 7 yd. **d.** 8 cm **e.** 0.6 m **f.** $\frac{3}{4}$ in. **g.** 1.2 cm **h.** $5\frac{2}{3}$ in.

7. Using the constant feature of multiplication, write a key sequence that can be used to find the volume of a cube each of whose edges measures 1.25 cm.

8. Using $\pi \approx 3.14$, find the volume of each cylinder or sphere.

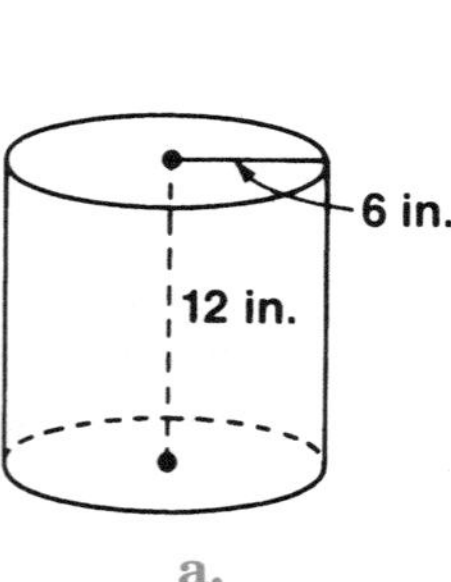

a.

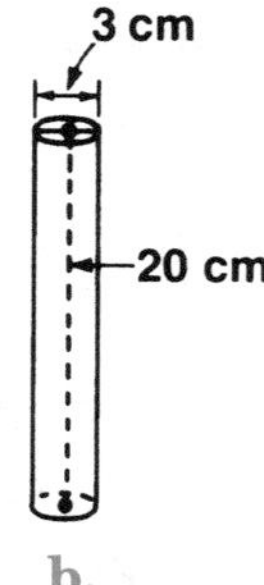

b.

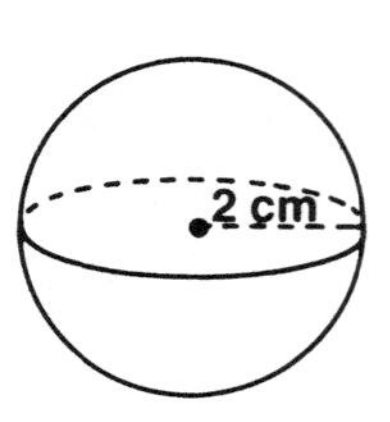

c.

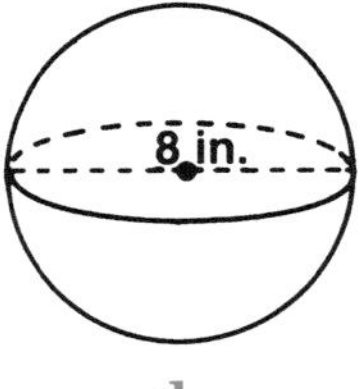

d.

9. Using $\pi \approx \frac{22}{7}$, find the volume of each cylinder, given the measures:

a. $r = 14$ cm, $h = 10$ cm **b.** $r = 28$ cm, $h = 6$ cm **c.** $d = 14$ in., $h = 5.2$ in.

d. $d = 42$ ft., $h = 3\frac{1}{2}$ ft. **e.** $r = 49$ cm, $h = 10$ cm **f.** $d = 1.4$ m, $h = 2$ m

10. Using $\pi \approx 3.14$, find the volume of a cylindrical can whose radius measures 3 inches and whose height measures 8 inches.

11. Using $\pi \approx \frac{22}{7}$, find the volume of a globe that has a radius of 21 cm.

12. Find the volume of each rectangular pyramid.

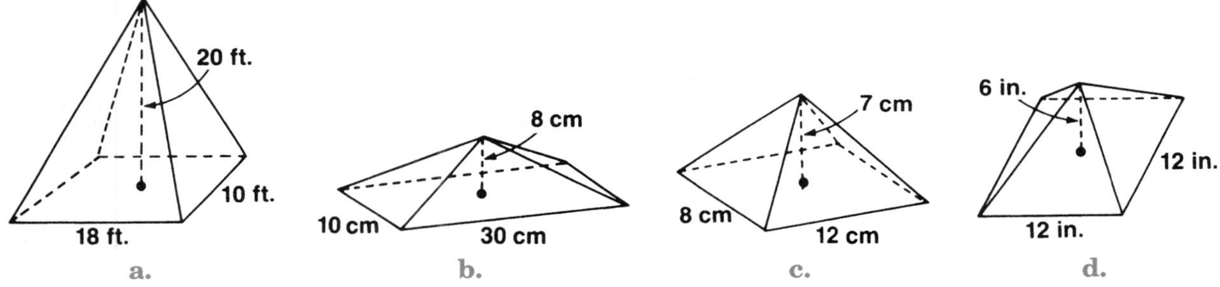

13. Find the volume of each rectangular pyramid, given the dimensions:

a. $\ell = 6$ cm, $w = 4$ cm, $h = 2$ cm **b.** $\ell = 9$ in., $w = 6$ in., $h = 20$ in.

c. $\ell = 15$ ft., $w = 8$ ft., $h = 10$ ft. **d.** $\ell = 24$ m, $w = 10$ m, $h = 30$ m

14. Using $\pi \approx 3.14$, find the volume of each cone.

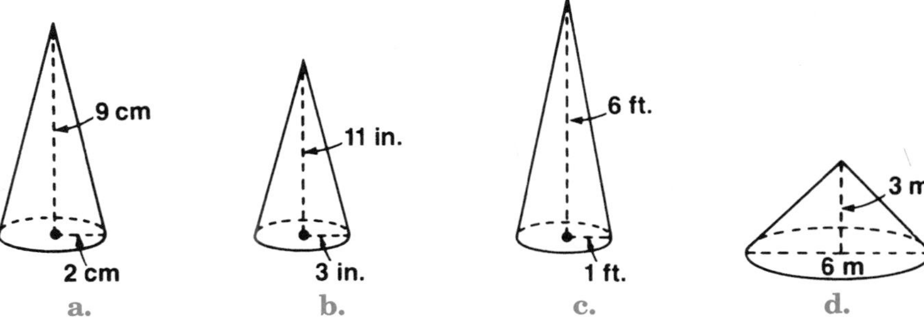

15. Using $\pi \approx 3.14$, find the volume of each cone, given the dimensions:

a. $r = 2$ cm, $h = 6$ cm **b.** $r = 6$ cm, $h = 2$ cm **c.** $r = 8$ in., $h = 3$ in.

16. A ball placed in a cubical box just touches each side of the box. If the volume of the box is 343 cm^3, find the volume of the ball. Use $\pi \approx \frac{22}{7}$.

17. How many cubic yards of concrete are needed for a rectangular patio that is to be 15 feet long and 10 feet wide if the concrete slab is to be 6 inches deep?

SPIRAL REVIEW EXERCISES

1. a. Find the area of the trapezoid.

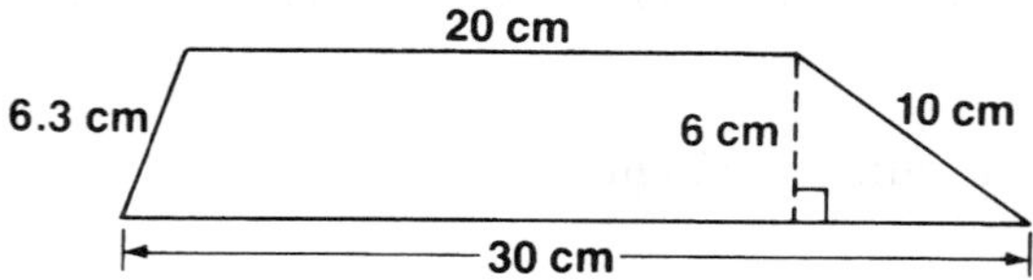

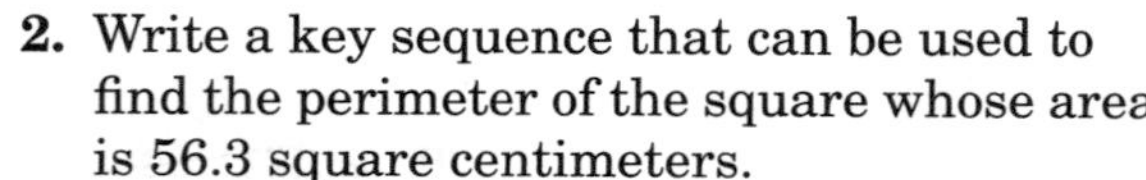

b. Find the perimeter of the trapezoid.

2. Write a key sequence that can be used to find the perimeter of the square whose area is 56.3 square centimeters.

3. A $120 jacket is on sale at 20% off. What is the sale price of the jacket?

4. 0.57 is equivalent to

(a) 5.7% (b) 57% (c) $\frac{57}{1000}$ (d) $\frac{1}{2}$

5. $\frac{2}{3} - \frac{1}{4}$ is (a) $\frac{5}{12}$ (b) 1 (c) $\frac{1}{6}$ (d) $\frac{3}{8}$

6. 9 is what percent of 36?

7. If $a = 5$ and $b = 2$, the value of $3a - 5b$ is

(a) 10 (b) 5 (c) 20 (d) 22

8. This circle graph shows the kinds of vehicles in a high school parking lot. What kind of vehicle is about 25% of the total?

Vehicles in School Parking Lot

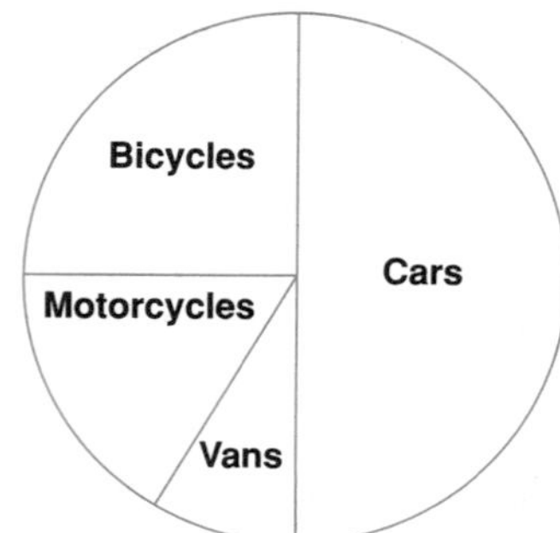

9. Which pair of lines is drawn parallel?

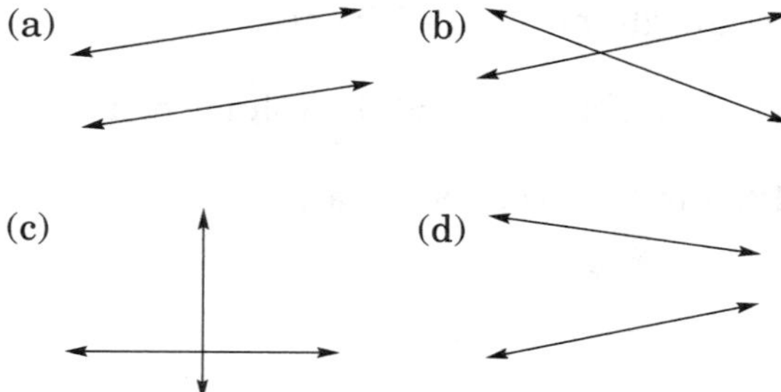

10. Find the greatest common factor (GCF) of 36, 42, and 90.

11. Six students ran in a race. If there are no ties, in how many different orders can the first-, second-, and third-place winners finish?

12. On a map, $\frac{1}{2}$ inch represents 20 miles. If the cities of Tomasville and Fulton are 2.5 inches apart on the map, find the actual distance between these cities.

13. In a student-council election, there are five candidates running for president, four for vice-president, three for secretary, and three for treasurer. In how many different ways can the four offices be filled?

14. Between which two consecutive integers does $\sqrt{87}$ lie?

(a) 6 and 7 (b) 7 and 8

(c) 8 and 9 (d) 9 and 10

15. An investor bought 3,600 shares of a stock at $29\frac{1}{2}$ (dollars per share). If the current price is $33\frac{1}{4}$, how much more is the total value of the shares?

UNIT 19–6 Surface Area

THE MAIN IDEA

1. To find the ***surface area*** of a rectangular solid or pyramid:
   (1) Use area formulas to find the area of each ***face***.
   (2) Then, add the areas of all the faces.

Face

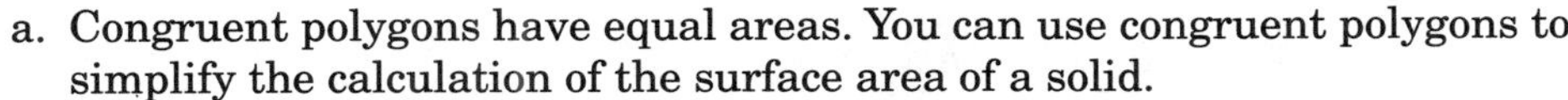

a. Congruent polygons have equal areas. You can use congruent polygons to simplify the calculation of the surface area of a solid.

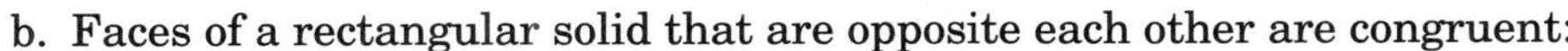

b. Faces of a rectangular solid that are opposite each other are congruent:

Top Face ≅ Bottom Face

Front Face ≅ Back Face

Left Side Face ≅ Right Side Face

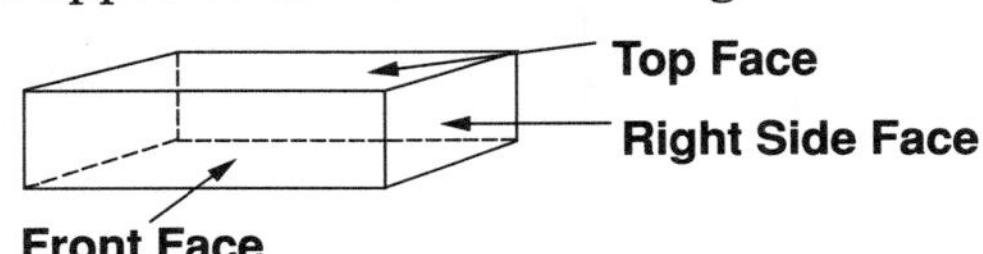

2. The surface of a right circular cylinder consists of a rectangle and two congruent circular bases.

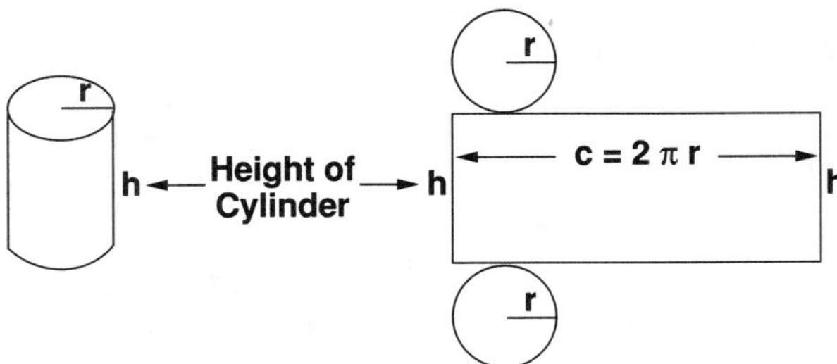

The height of the rectangle is the height of the cylinder. The width of the rectangle is the circumference of one of the bases.

To find the surface area of a cylinder:

(1) Find the area of one of the circular bases.

(2) Multiply the circumference of a base by the height of the cylinder to find the area of the rectangle.

(3) Add 2 times the area of one of the bases and the area of the rectangle.

EXAMPLE 1 Find the surface area of the rectangular solid.

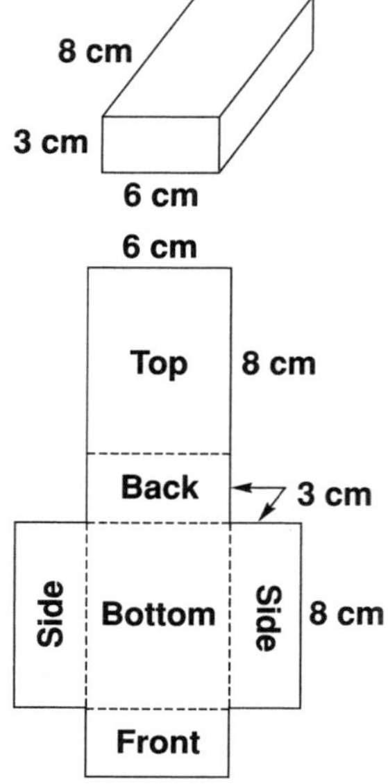

Draw a diagram of the rectangular solid as it would appear if it were a cardboard box that has been unfolded. Write the dimensions for each rectangle on the diagram. Then, find the area of each face.

Congruent Faces	*Area of 1 Face*	*Area of Both Faces*
Top and Bottom	$A = lw = 8(6) = 48 \text{ cm}^2$	$2(48) = 96 \text{ cm}^2$
Back and Front	$A = lw = 6(3) = 18 \text{ cm}^2$	$2(18) = 36 \text{ cm}^2$
Left and Right Sides	$A = lw = 8(3) = 24 \text{ cm}^2$	$2(24) = 48 \text{ cm}^2$

Surface Area $= 96 + 36 + 48 = 180 \text{ cm}^2$ *Ans.*

EXAMPLE 2 Find the surface area of a cube if each edge is 3 cm long.

A cube has 6 congruent square faces.

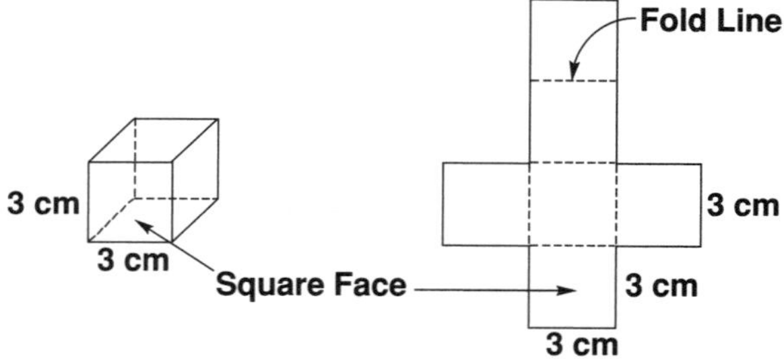

Find the area of one of the square faces.

$$A_{\text{face}} = s^2 = 3^2 = 9 \text{ cm}^2$$

Multiply the area of one of the faces by 6.

Surface Area $= 6(9 \text{ cm}^2)$
$= 54 \text{ cm}^2$ *Ans.*

EXAMPLE 3 Find the surface area of the cylinder, using $\pi \approx 3.14$.

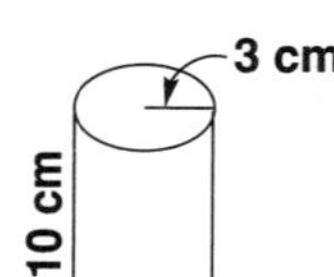

First, find the area of one circular base.

$$A_{\text{base}} = \pi r^2$$
$$A_{\text{base}} \approx (3.14)(3 \text{ cm})^2 \approx (3.14)(9 \text{ cm}^2) \approx 28.26 \text{ cm}^2$$

Then, find the area of the rectangle that would result if the cylinder were opened flat.

$$A_{\text{rectangle}} = \ell \cdot w$$
$$A_{\text{rectangle}} = (2\pi r)(w) \approx (2 \times 3.14 \times 3 \text{ cm})(10 \text{ cm}) \approx 188.4 \text{ cm}^2$$

Add 2 times the area of a circular base to the area of the rectangle.

Surface Area $= 2\,A_{\text{base}} + A_{\text{rectangle}}$

Surface Area $\approx 2\,(28.26 \text{ cm}^2) + 188.4 \text{ cm}^2$
$\approx 56.52 \text{ cm}^2 + 188.4 \text{ cm}^2$
$\approx 244.92 \text{ cm}^2$ *Ans.*

EXAMPLE 4 Find the surface area of the square pyramid.

The square pyramid has four congruent triangular faces.

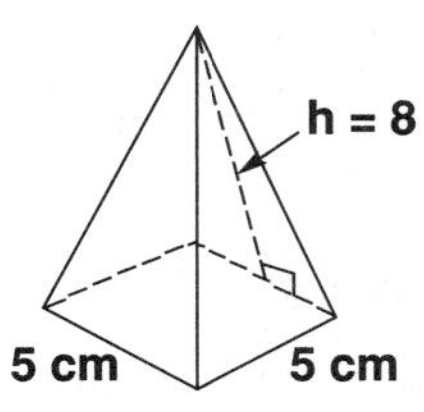

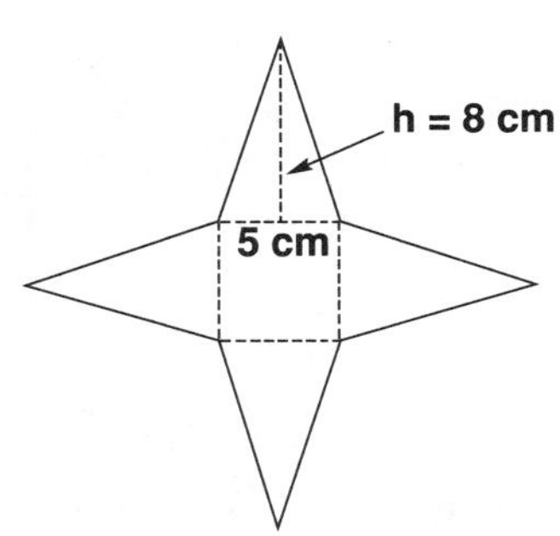

Find the area of the square base.

$$A_{\text{square}} = s^2 = 5^2 = 25 \text{ cm}^2$$

Find the area of one triangular face.

$$A_{\text{face}} = \frac{1}{2} bh = \frac{1}{2}(5)(8) = 20 \text{ cm}^2$$

To find the surface area of the square pyramid, add the area of the base to 4 times the area of one triangular face.

$$\text{Surface Area} = A_{\text{base}} + 4A_{\text{face}}$$
$$= 25 \text{ cm}^2 + 4(20 \text{ cm}^2)$$
$$= 25 \text{ cm}^2 + 80 \text{ cm}^2 = 105 \text{ cm}^2 \quad \textit{Ans.}$$

CLASS EXERCISES

1. Find the surface area of each rectangular solid.

a.

10 cm
3 cm
4 cm

b.

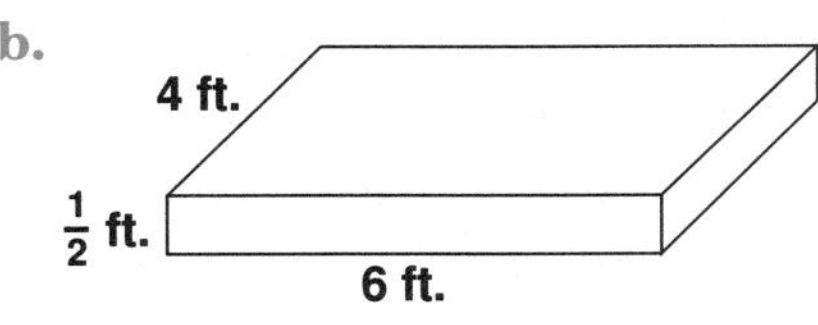

2. Find the surface area of a cube if each edge measures: **a.** 5 cm **b.** 9 m **c.** 1.5 ft.

3. Which of the following patterns can be used to form a cube by folding along the dotted lines?

Pattern I

Pattern II

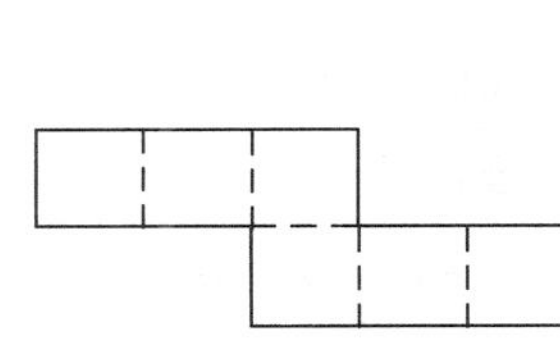

Pattern III

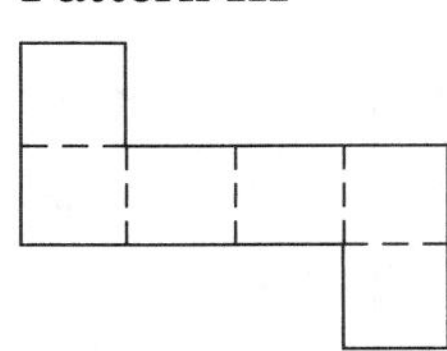

(a) I only (b) II only (c) III only (d) I and II only
(e) I and III only (f) II and III only (g) I, II, and III

4. If the height of a box is doubled and the other dimensions remain unchanged, which of the following areas are doubled?

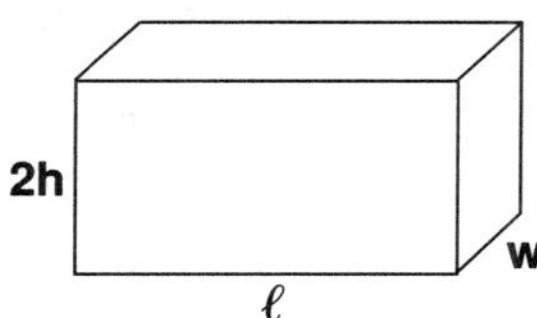

I. Top and Bottom II. Front and Back III. Left and Right Sides IV. Total Surface Area

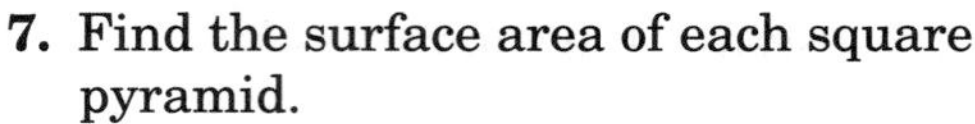

(a) I only (b) II only (c) III only (d) II and III only (e) II, III, and IV only

5. If the length of an edge of a cube is doubled, how does this change the surface area of the cube?

6. Using $\pi \approx 3.14$, find the surface area of each of the following cylinders.

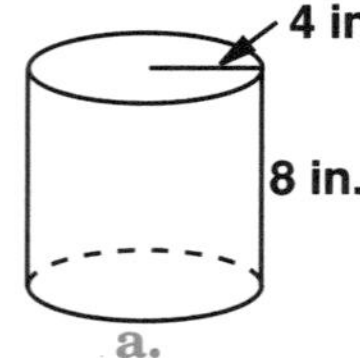

a.

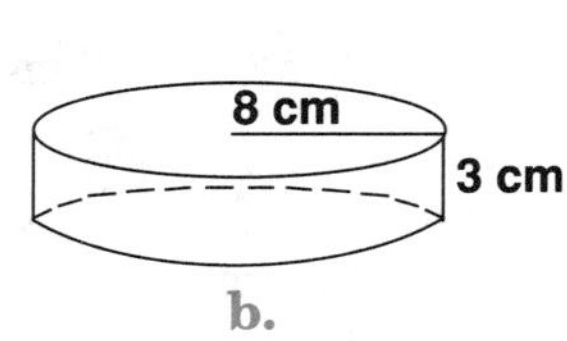

b.

7. Find the surface area of each square pyramid.

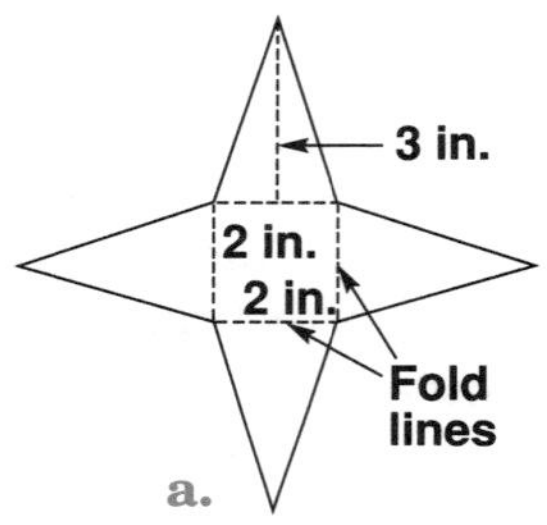

a.

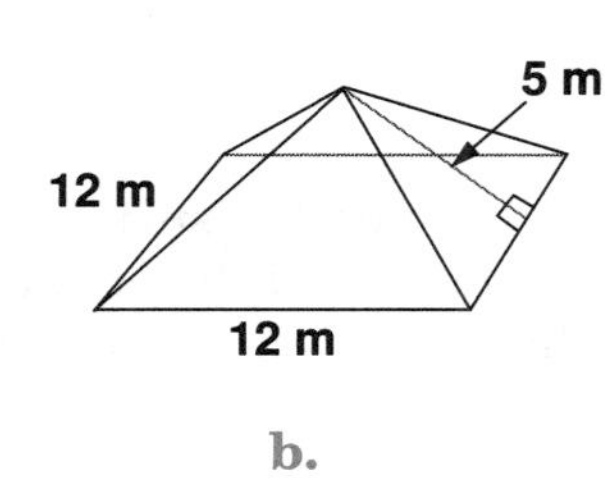

b.

8. Using $\pi \approx \frac{22}{7}$, find the surface area of the cylinder with $r = 7$ ft. and $h = 10$ ft.

HOMEWORK EXERCISES

1. Find the surface area of each rectangular solid.

a.

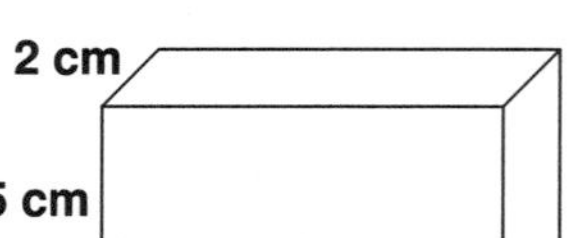

b.

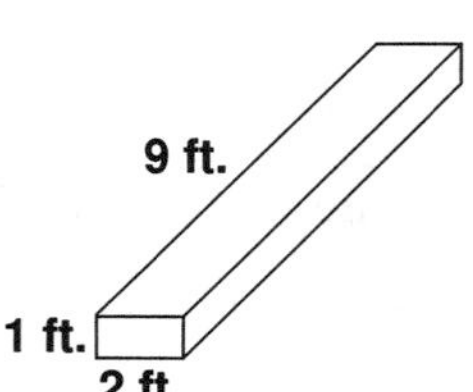

c.

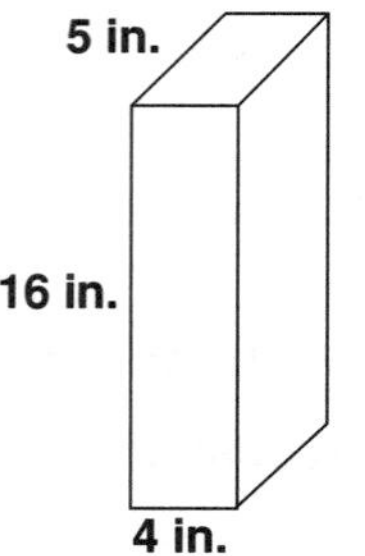

d.

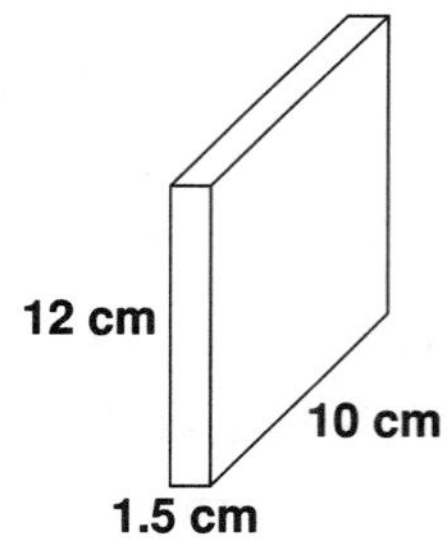

2. Find the surface area of a cube each of whose edges measures: **a.** 9 m **b.** $\frac{3}{4}$ in. **c.** 2.5 ft.

3. If the edge of a cube is tripled in length, what change occurs in its surface area?

4. Mr. Orlando wants to reinforce his rectangular swimming pool that is 20 feet wide, 32 feet long, and 5 feet deep.

   a. What is the total number of square feet to be reinforced?

   b. If it costs $2.80 per square foot to reinforce the pool, how much will Mr. Orlando spend?

5. Find the surface area of each cylinder, using $\pi \approx 3.14$.

a.

10 cm

20 cm

b.

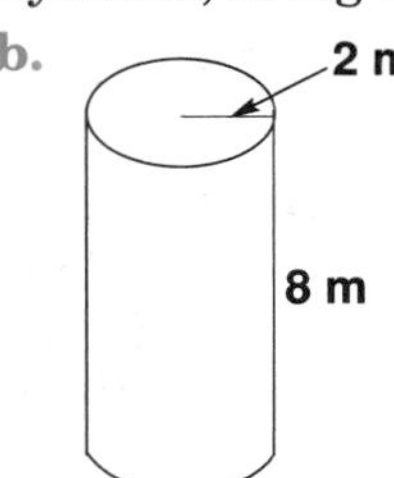

c.

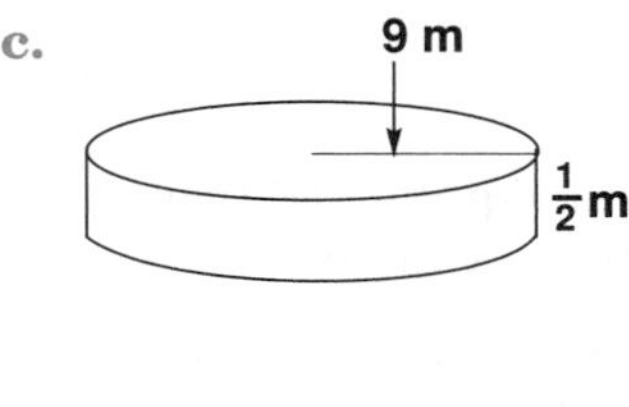

6. Using $\pi \approx \frac{22}{7}$, find the surface area of the cylinders with the given dimensions.

a. $r = 14$ cm, $h = 5$ cm **b.** $r = 4$ in., $h = 14$ in.

7. A roll of paper towels is 12 inches wide and 6 inches in diameter. The area of plastic needed to wrap the roll is closest to (a) 72 in.^{2} (b) 141 in.^{2} (c) 274 in.^{2} (d) 283 in.^{2}

8. Find the surface area of the square pyramid.

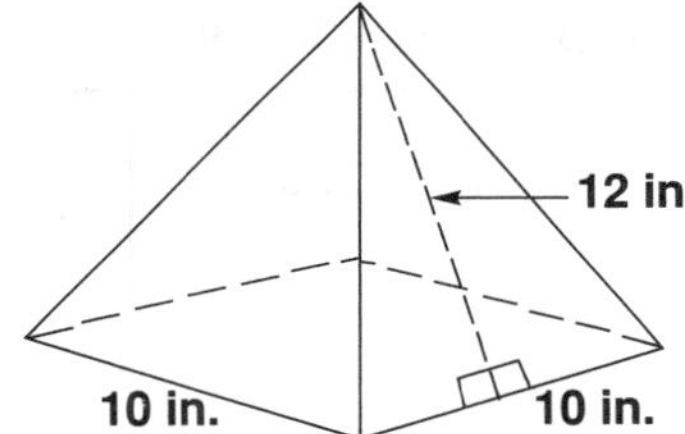

9. As part of a theatrical set for an opera, a square pyramid is to be built of plywood. If one side of the square base is to be 8 feet long and the height of each triangular face of the pyramid is to be 6 feet, what is the area of plywood needed to build the floor and sides of the pyramid?

10. A tent maker wants to make a nylon tent in the shape of a square pyramid. The floor of the tent will measure 6 feet on each side, and the height of each triangular side will be 6 feet. How much nylon fabric is needed to make the four sides of this tent?

11. Manufacturers usually use the surface area of a package for advertising and to provide consumers with information about the product. Which of the containers below has:
a. the greatest surface area that can be used for advertising or other information?
b. the greatest volume for the contents?

Duz-All Detergent

Kleen Detergent

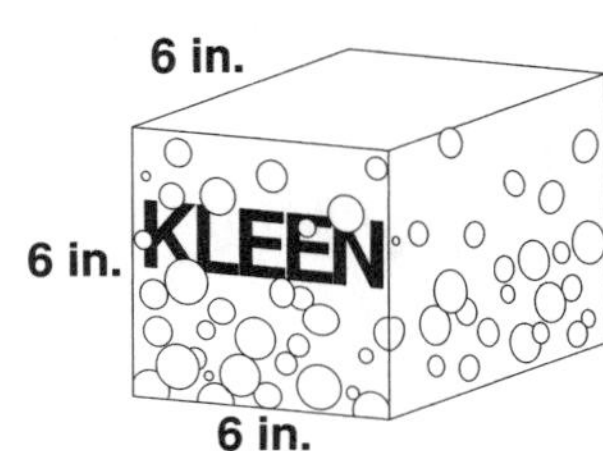

Brite Detergent

SPIRAL REVIEW EXERCISES

1. One side of an equilateral triangle is 15 cm long. Find the perimeter.

2. Find the least common multiple (LCM) of 24, 40, and 90.

3. Solve and graph the solution set on a number line: $16 - 2x > 8$

4. If $\triangle DEF \sim \triangle XYZ$, find YZ.

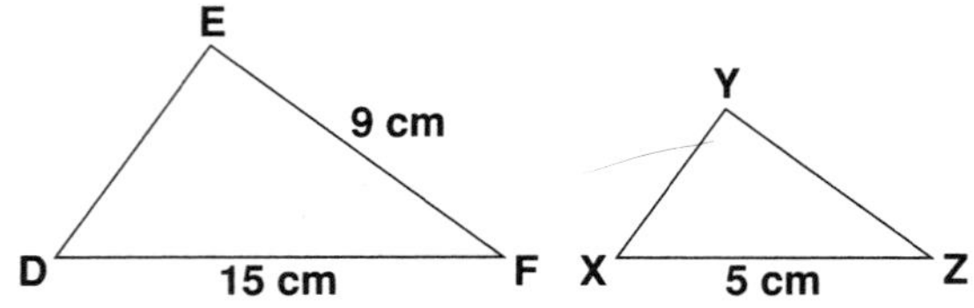

5. If carpet costs $21 per square yard, how much will it cost to carpet a rectangular room that measures 12 feet by 18 feet?

6. A $200 jacket is on sale for $140. What is the percent of the discount?

7. Find the area of the shaded region.

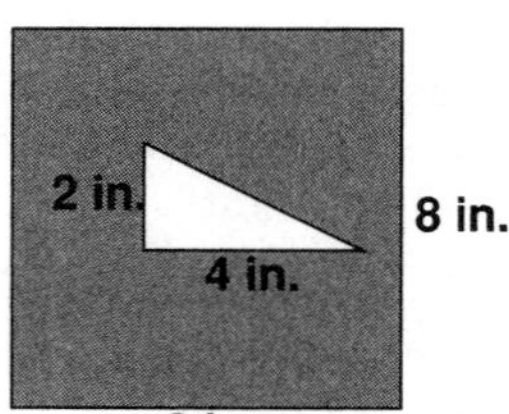

8. At Happy Days Diner, the price of a sandwich special is $2.79. Which of the following values is not a possible total price for a number of sandwich specials?
   (a) $5.68 (b) $8.37
   (c) $13.95 (d) $16.74

9. The ordered pair (2, –3) is a solution of
   (a) $x + y = 5$ (b) $x - y = 5$
   (c) $y = x + 1$ (d) $y = x - 1$

10. A digit is selected at random from among the digits of the number 862,428. The probability that the digit selected is a multiple of 2 is

    (a) 0 (b) $\frac{2}{3}$ (c) $\frac{5}{6}$ (d) 1

11. Ribbon costs $3 per yard. The cost of 4 feet of ribbon is
    (a) $2 (b) $2.33 (c) $3 (d) $4

12. This pictograph shows the number of records sold at A & B Music Shop in one week.

CDs Sold in One Week

Each ◎ = 24 CDs

Monday

Tuesday

Wednesday

Thursday

Friday

Saturday

How many CDs were sold on Saturday?
(a) 5 (b) 6 (c) 132 (d) 55

GLOSSARY

These informal definitions are intended to be brief descriptions of the terms listed.

absolute value the distance of a signed number from 0 on a number line: $|+6| = |-6| = 6$

acute angle an angle that measures more than 0° and less than 90°

addend a number to be added. In 2 + 3, 2 and 3 are the addends.

angle a figure formed by two rays with a common endpoint

annual yearly

area the number of square units contained within a closed, two-dimensional figure

associative property the way numbers are grouped in a series of additions or of multiplications does not change the result

average the value found by adding a set of values and dividing by the number of values; the mean

balance an amount of money kept in a financial account

bar graph a graph in which numerical facts are shown as bars of different lengths

base the number that is raised to a power in an expression with an exponent. In 3^5, 3 is the base

base the side of a polygon from which the height is measured

box-and-whisker plot a graph that uses quartiles to show distribution of data

budget a plan for organizing expenses

carrying charge the increase in cost of an item purchased on an installment plan

Celsius a scale for measuring temperature

center of a circle the point inside a circle that is the same distance from any point on the circle

central tendency, measures of statistical measures—mean, median, and mode

check a written order authorizing a bank to pay out money from an account

checking account a financial account for drawing money against deposits

chord a line segment that has both endpoints on a circle

circle a flat, closed curve that has all its points the same distance from an inside point called the center

circle graph a graph in which numerical facts are represented by sectors of a circle

circumference the distance around a circle

coefficient a number multiplying a variable

combination a selection of items that need not be in any particular order

commission a percentage of the money received for a sale, paid to the salesperson who made the sale

common factor a number that divides evenly into two or more given numbers. 3 is a common factor of 9 and 12.

commutative property the order in which numbers are added, or multiplied, does not affect the result

complementary angles angle pairs whose measures have a sum of 90°

composite number a whole number that has factors other than 1 and itself. Since $6 = 3 \times 2$, the number 6 is a composite.

compound interest calculated on the original principal plus interest previously earned

congruent figures figures that have the same size and the same shape

constant feature of a calculator provides a shortcut for repeated operations with the same number

coordinates the numbers describing the location of a point on a graph

counting principle used to find the number of possible selections if one item is chosen from each of two or more sets

cross-multiply to multiply the diagonally opposite numerators and denominators (means and extremes) in a proportion

cross products the results of cross-multiplying

cube a three-dimensional figure whose sides are all squares

cube to raise a number to the third power (use a number three times as a factor)

cubic unit a unit used to measure volume

customary measures a system of measures, commonly used in the U.S., that includes units such as the foot, the pound, and the gallon

data numerical facts

decagon a ten-sided figure

decimal a way of writing a fraction whose denominator is a power of 10. The fraction $\frac{3}{100}$ can be written as the decimal .03.

degree a unit of measure for angles

degree a unit of measure for temperatures

denominator the nonzero numeral below the division line in a fraction, showing into how many parts the whole is divided. In the fraction $\frac{3}{5}$, the denominator is 5.

deposit an amount of money added to the balance in a bank account

diagonal a line segment connecting two nonconsecutive vertices of a polygon

diameter a chord that passes through the center of a circle

dice cubes with six sides that represent the numbers 1 through 6 (singular: die)

difference the result of a subtraction. In $7 - 2 = 5$, the difference is 5.

dimension a measure of length, width, or height

directed number a signed number (positive, negative, or zero)

discount the amount by which the price of an item is lowered

distributive property multiplication distributes over addition or subtraction: $6 \times (5 + 7) = 6 \times 5 + 6 \times 7$

dividend a number that is divided. In $16 \div 2$, 16 is the dividend.

divisor a number by which another number is divided. In $16 \div 2$, the divisor is 2.

down payment an original payment at the time of purchase on an installment plan

equation a mathematical sentence stating that two quantities are equal. $5 - 2 = 3$ and $x + 1 = 7$ are equations.

equilateral a word used to describe a polygon whose sides are all equal in measure

equivalent equal in value

estimate to find an approximate value

evaluate to find the value of a mathematical expression by carrying out the indicated operations

event a specified result in a probability experiment

exact divisor a number that divides into a quantity evenly, leaving a remainder of 0; a factor. 4 is an exact divisor of 12.

exponent a number that shows how many times a base is used as a factor. In 3^5, the exponent is 5.

extremes the first and fourth numbers in a proportion. In the proportion $\frac{2}{3} = \frac{8}{12}$, 2 and 12 are the extremes.

factor an exact divisor. 4 is a factor of 12.

factorial notation representing the product of successive natural numbers down to 1. 4 factorial is $4! = 4 \times 3 \times 2 \times 1$

Fahrenheit a scale for measuring temperature

finance charge a fee charged on the unpaid balance in credit-card purchases

formula a mathematical sentence that shows how variables are related to each other. $A = bh$ is a formula.

fraction a mathematical expression, such as $\frac{5}{6}$, that shows the quotient of two numbers

frequency the number of times a value appears when a tally is made

frequency table a list of the frequencies of a group of values

graph a visual way to organize and present data

greatest common factor the largest exact divisor of two or more numbers. 6 is the greatest common factor of 12 and 18.

height the perpendicular distance between a point and a line segment

hexagon a six-sided polygon

histogram a bar graph with no separation between intervals

horizontal in a direction of side to side; across

hypotenuse the side opposite the right angle in a right triangle

identity 0 or 1; adding 0 or multiplying by 1 leaves a number unchanged

improper fraction a fraction in which the numerator is greater than or equal to the denominator. $\frac{7}{5}$ and $\frac{5}{5}$ are improper fractions

inequality a mathematical sentence stating that two quantities are not equal. $2 < 3$, $3 > 2$, and $2 \neq 3$ are inequalities

installment plan the purchase of an item by making regular equal payments over a period of time

integers the set of numbers consisting of the whole numbers and their opposites. $\{\ldots, -2, -1, 0, 1, 2, \ldots\}$

interest a charge for money that is borrowed or a return on an investment

inverse operations two operations, such as addition or subtraction, that "undo" each other

investment the use of money to earn money

irrational numbers numbers that cannot be expressed as a quotient of integers. π and $\sqrt{2}$ are irrational numbers.

isosceles triangle a triangle that contains two sides equal in measure

least common denominator the least common multiple of the denominators of two or more fractions. 24 is the least common denominator of the fractions $\frac{1}{8}$ and $\frac{5}{12}$.

least common multiple the smallest whole number that is a multiple of two or more whole numbers. 24 is the least common multiple of 8 and 12.

like fractions fractions that have the same denominator. $\frac{2}{5}$ and $\frac{3}{5}$ are like fractions.

linear unit a unit used to measure distance

line graph a graph in which numerical facts are represented by points with connecting line segments

line plot a display of numerical data on a number line

line segment a part of a line with two endpoints

list price the original price of an item

mean the value found by adding a set of values and dividing by the number of values; the average

means the second and third numbers in a proportion. In the proportion $\frac{2}{3} = \frac{8}{12}$, 3 and 8 are the means.

median the middle value in a set of values that are arranged in order of size

metric measures a system of measures that uses powers-of-ten multiples of the meter, the gram, and the liter

minuend a number from which another number is subtracted. In $7 - 2$, the minuend is 7.

mixed number a numeral consisting of a whole number and a proper fraction. $7\frac{1}{3}$ is a mixed number.

mode the value that has the greatest frequency in a set of data

multiplicand a number to be multiplied. In $1{,}436 \times 5$, the number 1,436 is the multiplicand.

negative number a number less than 0

number line a graph that represents numbers by marked points on a line

numeral the written symbol for a number

numerator the numeral above the division line in a fraction. In the fraction $\frac{3}{5}$, 3 is the numerator.

obtuse angle an angle that measures more than 90° and less than 180°

obtuse triangle a triangle that contains an obtuse angle

octagon an eight-sided polygon

open sentence a sentence that contains a variable. $x + 2 = 10$ and $x > 4$ are open sentences.

opposites two signed numbers that are the same distance from 0 on a number line. +5 and –5 are opposites.

ordered pair the coordinates that represent the horizontal and vertical locations of a point on a graph. (1, –4) is an ordered pair.

order of operations rules that decide how a numerical expression is evaluated

origin the point where the x-axis and y-axis of a graph intersect

overtime the money earned for working extra hours, generally at the rate of $1\frac{1}{2}$ or 2 times the regular rate of pay

parallel lines two lines that extend in the same direction and never meet

parallelogram a quadrilateral that has both pairs of opposite sides parallel

pentagon a five-sided polygon

percent hundredths

percentiles measures used to compare values in a ranked set of data. The 60th percentile is greater than or equal to 60% of the data.

perfect square a rational number whose square root is also a rational number. 9 is a perfect square since $9 = 3^2$ and $\sqrt{9} = 3$.

perimeter the distance around a polygon; the sum of the measures of its sides

perpendicular lines two lines that meet at right angles

pictograph a graph in which numerical facts are represented by pictures

pie graph a graph in which numerical facts are represented by sectors of a circle (also called a circle graph)

place value the number by which a digit is multiplied to find its value in a numeral. The place value of 3 in 35 is 10.

polygon a flat, closed figure whose sides are line segments

polynomial an algebraic term, or a sum of algebraic terms, with no variable in the denominator. $\frac{1}{2}x^2 - 5x + 3$ is a polynomial.

positive number a number greater than 0

power the product obtained when a number is multiplied by itself a given number of times. 8, or 2^3, is the third power of 2.

prime number a whole number that has 1 and itself as its only factors. 2, 3, 5, and 7 are prime numbers.

principal the amount of money that is invested or borrowed

probability a number describing the likelihood that an event will occur

product the result of a multiplication. In $5 \times 2 = 10$, the product is 10.

proper fraction a fraction in which the numerator is less than the denominator. $\frac{3}{5}$ is a proper fraction.

proportion a statement that two ratios are equal. $\frac{2}{3} = \frac{8}{12}$ is a proportion.

protractor an instrument that is used to measure angles

pyramid a solid figure whose base is a polygon and whose sides are triangles

Pythagorean relationship the relation between the lengths of the sides of a right triangle. In a right triangle, the square of the hypotenuse equals the sum of the squares of the other two sides.

quadrilateral a four-sided polygon

quartiles numbers that divide ranked data into four sets of values with equal frequencies

quotient the result of a division. In $16 \div 2 = 8$, the quotient is 8.

radius a line segment that has one endpoint at the center of a circle and the other endpoint on the circle (plural: radii)

random chosen in no particular order

range the difference between the greatest and least values in a set of data

rate of discount the percent by which the price of an item is lowered

rate of interest the percent of principal paid on a loan or an investment

ratio a comparison between two numbers. 2 to 3, also written as 2 : 3 or $\frac{2}{3}$, is a ratio.

rational number a number that can be written as a fraction whose numerator and denominator are integers and whose denominator is not 0. Some examples are 5 or $\frac{5}{1}$, $\frac{-2}{3}$, 397 and 0.

ray a part of a line consisting of a fixed point and all points to one side of it (a ray has a beginning, but no end)

real numbers all numbers that can be associated with points on the number line; includes all rational and irrational numbers

reciprocal the result of inverting a fraction, or exchanging the numerator and denominator of the fraction. $\frac{7}{4}$ is the reciprocal of $\frac{4}{7}$.

rectangle a parallelogram that has four right angles

rectangular solid a three-dimensional figure whose sides are all rectangles; a box

reflection gives a mirror image of a geometric figure

remainder the quantity left after division. 2 is the remainder when 17 is divided by 3.

repeating decimal a decimal that results when a division never ends. $\frac{1}{3} = .3333\ldots$ and $\frac{5}{11} = .454545\ldots$ are examples.

rhombus a parallelogram that has all four sides equal in measure

right angle an angle that measures 90°

right triangle a triangle that contains one right angle

rotation turning a geometric figure about a given point

rounded number an approximation for a number, to a given place value. To the nearest hundred, 324 rounds to 300.

salary the amount of money earned by an employee

sales tax a state or local tax on purchases

sample space the set of all possible outcomes in a probability experiment

scale drawing a drawing in which the actual dimensions of an object are enlarged or reduced proportionally

scalene triangle a triangle that contains no two sides equal in measure

scatter plot a graph that indicates whether two sets of data are related

scientific notation a form of writing very large or very small numbers as a product in which the first factor is a number between 1 and 10 and the second factor is a power of 10. 9.4×10^5 is the scientific notation for 940,000.

sector a section of a circle between two radii

signed numbers numbers that are positive, negative, or zero

similar figures figures that have the same shape

simple interest a percent paid on a loan or investment

simplify to carry out indicated operations as far as possible. In simplest form, $6x + y + 2x = 8x + y$, and $\frac{12}{15} = \frac{3}{5}$.
slope of a line a number indicating steepness and direction
solution the answer to a problem
solving an equation to find the values of the variable for which an equation is true
sphere a three-dimensional figure that has all of its points the same distance from its center; a ball
square a rectangle that has all four sides equal in measure, or a rhombus that has four right angles
square to multiply a number by itself; to raise to the second power (use a number two times as a factor). 5 squared, or 5^2, is 25.
square root one of two equal factors of a number. 5 is a square root of 25.
square unit a unit used to measure area
statistics the study of numerical facts, or data
stem-and-leaf display a way to organize and display numerical data
stepped rates rates that change according to the quantity purchased. A long-distance telephone cost is a stepped rate. There is an initial charge followed by a decrease in charge for additional time.
straight angle an angle that measures 180°
subtrahend a number to be subtracted. In $7 - 2$, the subtrahend is 2.
sum the result of an addition. In $2 + 3 = 5$, 5 is the sum.
supplementary angles angle pairs whose measures have a sum of 180°
surface area sum of areas of the surfaces of a solid
symmetry the part of a figure on one side of a line or a point is an exact reflection of the part on the other side

take-home pay net earnings remaining after taxes and other deductions
tally to count, using tally marks to keep score
tax money collected by governments to pay for public services
term an algebraic expression using no operations other than multiplication or division. $5x^2$ and $\frac{1}{x}$ are terms.
terminating decimal a decimal that results when a division has a zero remainder; for example, $\frac{1}{4} = .25$
translation slides a geometric figure to a new position
trapezoid a quadrilateral that has only one pair of opposite sides parallel
tree diagram its branches show all possible outcomes of successive activities
triangle a three-sided polygon

unit a quantity chosen as a standard by which other quantities are to be expressed
unit pricing cost per unit of measure
unlike fractions fractions that do not have the same denominator. $\frac{1}{2}$ and $\frac{1}{3}$ are unlike fractions.

variable a symbol, usually a letter, used to represent an unknown number
vertex the point at which the two rays of an angle meet (plural: vertices)
vertical in a direction of up and down; upright
vertical angles opposite angle pairs formed by intersecting lines
volume the number of cubic units contained within a closed, three-dimensional figure

wages an amount of money earned, usually by the hour
whole number numbers that are greater than or equal to zero and that contain no fractions: {0, 1, 2, 3, . . .}

***x*-axis** the horizontal number line on a coordinate graph

***y*-axis** the vertical number line on a coordinate axis

INDEX